List of Elements with Their Symbols and Atomic Masses[a]

Element	Symbol	Atomic number	Atomic mass	Element	Symbol	Atomic number	Atomic mass
Actinium	Ac	89	(227)[b]	Meitnerium	Mt	109	(268)
Aluminum	Al	13	26.98	Mendelevium	Md	101	(256)
Americium	Am	95	(243)	Mercury	Hg	80	200.6
Antimony	Sb	51	121.8	Molybdenum	Mo	42	95.94
Argon	Ar	18	39.95	Neodymium	Nd	60	144.2
Arsenic	As	33	74.92	Neon	Ne	10	20.18
Astatine	At	85	(210)	Neptunium	Np	93	(237)
Barium	Ba	56	137.3	Nickel	Ni	28	58.69
Berkelium	Bk	97	(247)	Niobium	Nb	41	92.91
Beryllium	Be	4	9.012	Nitrogen	N	7	14.01
Bismuth	Bi	83	209.0	Nobelium	No	102	(253)
Bohrium	Bh	107	(262)	Osmium	Os	76	190.2
Boron	B	5	10.81	Oxygen	O	8	16.00
Bromine	Br	35	79.90	Palladium	Pd	46	106.4
Cadmium	Cd	48	112.4	Phosphorus	P	15	30.97
Calcium	Ca	20	40.08	Platinum	Pt	78	195.1
Californium	Cf	98	(249)	Plutonium	Pu	94	(242)
Carbon	C	6	12.01	Polonium	Po	84	(210)
Cerium	Ce	58	140.1	Potassium	K	19	39.10
Cesium	Cs	55	132.9	Praseodymium	Pr	59	140.9
Chlorine	Cl	17	35.45	Promethium	Pm	61	(147)
Chromium	Cr	24	52.00	Protactinium	Pa	91	(231)
Cobalt	Co	27	58.93	Radium	Ra	88	(226)
Copper	Cu	29	63.55	Radon	Rn	86	(222)
Curium	Cm	96	(247)	Rhenium	Re	75	186.2
Dubnium	Db	104	(261)	Rhodium	Rh	45	102.9
Dysprosium	Dy	66	162.5	Rubidium	Rb	37	85.47
Einsteinium	Es	99	(254)	Ruthenium	Ru	44	101.1
Erbium	Er	68	167.3	Rutherfordium	Rf	106	(263)
Europium	Eu	63	152.0	Samarium	Sm	62	150.4
Fermium	Fm	100	(253)	Scandium	Sc	21	44.96
Fluorine	F	9	19.00	Selenium	Se	34	78.96
Francium	Fr	87	(223)	Silicon	Si	14	28.09
Gadolinium	Gd	64	157.3	Silver	Ag	47	107.9
Gallium	Ga	31	69.72	Sodium	Na	11	22.99
Germanium	Ge	32	72.59	Strontium	Sr	38	87.62
Gold	Au	79	197.0	Sulfur	S	16	32.07
Hafnium	Hf	72	178.5	Tantalum	Ta	73	180.9
Hahnium	Hn	108	(265)	Technetium	Tc	43	(99)
Helium	He	2	4.003	Tellurium	Te	52	127.6
Holmium	Ho	67	164.9	Terbium	Tb	65	158.9
Hydrogen	H	1	1.008	Thallium	Tl	81	204.4
Indium	In	49	114.8	Thorium	Th	90	232.0
Iodine	I	53	126.9	Thulium	Tm	69	168.9
Iridium	Ir	77	192.2	Tin	Sn	50	118.7
Iron	Fe	26	55.85	Titanium	Ti	22	47.88
Joliotium	Jl	105	(262)	Tungsten	W	74	183.9
Krypton	Kr	36	83.80	Uranium	U	92	238.0
Lanthanum	La	57	138.9	Vanadium	V	23	50.94
Lawrencium	Lr	103	(257)	Xenon	Xe	54	131.3
Lead	Pb	82	207.2	Ytterbium	Yb	70	173.0
Lithium	Li	3	6.941	Yttrium	Y	39	88.91
Lutetium	Lu	71	175.0	Zinc	Zn	30	65.39
Magnesium	Mg	12	24.31	Zirconium	Zr	40	91.22
Manganese	Mn	25	54.94				

[a] All atomic masses have four significant figures. These values are recommended by the Committee on Teaching of Chemistry, International Union of Pure and Applied Chemistry.

[b] Approximate values of atomic masses for radioactive elements are given in parentheses.

Geochemistry

Second Edition

Arthur H. Brownlow

Boston University

PRENTICE HALL
Upper Saddle River, New Jersey 07458

Library of Congress Cataloging in Publication Data
Brownlow, Arthur H.
Geochemistry / Arthur H. Brownlow. — 2nd ed.
p. cm.
Includes bibliographical references and indexes.
ISBN 0-13-398272-6 (hardcover)
1. Geochemistry. I. Title.
QE515.B77 1996
551.9—dc20 95-37089
CIP

Acquisitions editor: Robert McConnin
Editor-in-chief: Paul Corey
Editorial director: Tim Bozick
Editorial/production supervision: ETP/Harrison
Project liaison: Barbara DeVries
Director of marketing: Gary June
Manufacturing manager: Trudy Pisciotti
Copy editor: Corleigh Stixrud
Cover design director: Jayne Conte
Cover designer: Bruce Kenselaar
Editorial assistant: Grace Anspake

A Pearson Education Company
Upper Saddle River, NJ 07458

Printed in the United States of America

10 9 8 7 6 5 4 3 2 1

ISBN 0-13-398272-6

Prentice-Hall International (UK) Limited,London
Prentice-Hall of Australia Pty. Limited, Sydney
Prentice-Hall Canada Inc., Toronto
Prentice-Hall Hispanoamericana, S.A., Mexico
Prentice-Hall of India Private Limited, New Delhi
Prentice-Hall of Japan, Inc., Tokyo
Pearson Education Asia Pte. Ltd., Singapore
Editora Prentice-Hall do Brasil, Ltda., Rio de Janeiro

Contents

Preface

The purpose of the second edition of this book is the same as that of the first edition: to serve as a textbook for undergraduate and graduate students taking an introductory course in geochemistry. At the same time, it can be used as a reference text by professionals in geology and related fields. Although knowledge of general geology, particularly mineralogy, is needed to fully appreciate some of the material in this book, scientists and engineers in fields such as biology, chemistry, and civil and environmental engineering will find useful explanations and data in every chapter.

I have again attempted the impossible—a textbook that covers the *entire* field of geochemistry. In the years since publication of the first edition, geochemistry has grown rapidly into an ever more active and ever larger field of endeavor. Geologists and other scientists are using geochemical principles and techniques (old and new) in their work, both academic and applied, to a greater and greater extent. This made it very difficult to decide what new topics and applications to add and what old material to delete. Because of the enormous expansion in geochemistry, this edition required a complete, major rewrite of the book rather than a revision. I have added many sections of new material while retaining or rewriting the best of the first edition. At the same time I have tried to keep the length of the book at a reasonable level.

New or expanded topics in this edition include geochemical modeling, reaction kinetics, crystal bonding forces, nutrient cycling, chemical weathering, mineral distribution coefficients, geothermometry-geobarometry, and environmental geochemistry, to name a few. The chapter on isotope geology has been expanded and the many uses of isotopic data are emphasized throughout the book. As in the first edition, the important role of thermodynamics is pointed out in many of the chapters. Organic geochemistry continues to grow in importance, and in addition to having its own chapter, is integrated into the chapters on water chemistry and on sedimentary rocks.

This edition has twice as many figures and tables as the first edition. Thus, although the format is the same as that of the first edition, this is essentially a new book.

Each of the first six chapters is devoted to a specific area of geochemistry. Pertinent chemical principles and concepts are reviewed, the basic data are summarized, and examples (recent and classic) of applications to geological problems are given. I assume the reader has very little chemical and mathematical background. The first chapter deals with the basic units of geochemistry, the elements. Their origin, properties, abundances, and classification are discussed, along with some aspects of cosmochemistry. The compositions of the various parts of the Earth are discussed in some detail. The second chapter reviews the extremely active field of isotope geology. All of the major isotope systems are covered. The myriad uses of both radioactive and stable isotopes are emphasized. This was the toughest chapter to keep at a reasonable length. The third chapter explores concepts and applications of thermodynamics in geology. Geologic uses are described for the Gibbs function, fugacities, activities, distribution coefficients, and various types of phase diagrams. The uses and the limitations of thermodynamic calculations are discussed. These first three chapters provide the foundation for the rest of the book.

Chapter Four summarizes our knowledge of water chemistry. Emphasis is placed on the properties and chemical role of water at the surface of the Earth. Major topics include oxidation-reduction reactions, kinetics, and modeling of water chemistry. Practical uses of water chemistry are also emphasized. The chemical properties of minerals are covered in the next chapter on crystal chemistry. Mineral structures, along with bonding forces, solid solution, order-disorder, and trace element geochemistry are discussed in some detail. Uses of thermodynamics in the study of mineral chemistry are pointed out. Chapter Six is a summary of the organic geochemistry of soils, sediments, and rocks. Carbon, nutrient, and trace element cycles are reviewed. Brief discussions are provided on coal, petroleum, and the origin of life.

The final three chapters use the material in the preceding chapters to describe the results of geochemical research on sedimentary, igneous, and metamorphic rocks. In Chapter Seven (sedimentary rocks) emphasis is placed on chemical weathering, clay mineralogy, carbonate chemistry, and diagenesis. The role of organisms and organic geochemistry is emphasized throughout. The nature of brines and evaporite deposits is discussed, along with their relationship to metal-rich sediments. The next chapter, on igneous rocks, discusses occurrence and composition, the chemistry of the crust and mantle, the origin and history of magmas, and details of the crystallization of specific types of silicate melts. Recent research on the role of volatiles in magmas is summarized. Throughout the chapter the use of isotope and trace element data is highlighted. The final chapter (metamorphic rocks) reviews the variables of metamorphism and the nature of metamorphic reactions. Methods of estimating and depicting temperatures and pressures of metamorphism are discussed in detail. Other topics include uses of isotopic data, metasomatism, metamorphic facies, and the relation of composition and fluid-solid interaction to mineralogy. In both of the last two chapters, relationships to plate tectonics are pointed out.

A two-semester sequence is required to cover adequately all of the material presented. However, this text also can be used in a one-semester course on "soft-rock" geochemistry (covering Chapters One, Two, Three, Four, Six, and Seven) or in a one-semester course on "hard-rock" geochemistry (covering Chapters One, Two, Three, Five, Eight, and Nine). I emphasize the importance of the questions at the end of each chapter. It has been my experience that students do not really understand geochemical principles until they are required to apply the principles to specific problems. Each chapter has about twenty questions based on chapter material.

Answers to all of the questions are given at the end of the book. Also, a large number of references (classic and recent) is provided at the end of each chapter for further study of particular topics. Throughout the text, I have mainly used cgs units rather than the mks units of the International System of Units (SI) because most tables of data consulted by geologists are still given in cgs units. Table 3-1 summarizes and relates the cgs units to the SI system and SI base and derived units are listed inside the back cover.

In developing this book, I was greatly influenced by the work of V. M. Goldschmidt. He is often called, and rightly so, the "Father of Modern Geochemistry." Goldschmidt's publications have inspired me ever since I discovered them as an undergraduate student. As a young geologist, I visited several of the locations of Goldschmidt's field work in Norway and was privileged to discuss his work with scientists who knew him personally. In 1992 the Geochemical Society published an excellent biography of Goldschmidt by Brian Mason (who has also made many important contributions to our field).* All science is affected by the human element and Mason's book brings out the very human side of a true scientific genius. I strongly recommend this book to all who have any interest in geochemistry or in the history of science. In particular, all students of geochemistry need to be aware of the enormous contributions to the field by Goldschmidt.

The following friends and colleagues reviewed drafts of individual chapters: Mary Jo Baedecker, U.S. Geological Survey; Sherm Bloomer, Oregon State University; Tim Drever, University of Wyoming; Jerry Gibbs, Virginia Polytechnic Institute; Chris Hepburn, Boston College; Carla Montgomery, Northern Illinois University; Rick Murray, Boston University; Charlie Prewitt, Geophysical Laboratory; and John Wood, Harvard University. I am very grateful to all of them. Their comments led to significant improvement of the chapters.

The entire manuscript was evaluated by the following reviewers obtained by Prentice Hall: G. N. Eby, University of Massachusetts, Lowell; M. D. Feigenson, Rutgers the State University of New Jersey; A. Kilinc, University of Cincinnati; R. Kretz, University of Ottawa; and S. C. Slaymaker, California State University. Further improvements came from their thoughtful reviews. My thanks to them as well.

I am grateful to the following people at Prentice Hall for their fine work in getting this book from a working manuscript to a professional final product: Robert McConnin, Acquisitions Editor; Paul Corey, Editor-in-Chief; Grace Anspake, Editorial Assistant; and Barbara DeVries, Project Liaison. Corleigh Stixrud, Project Manager for ETP/Harrison, did an excellent job handling all the details of book production.

Linda Coates and Elaine Chaison provided important help in the final stages of manuscript preparation. Edward LeBlanc constructed a wonderful working environment for me. I am thankful to them for their contributions.

A major source of encouragement and help throughout the preparation of this book has been my friend and colleague Dan Hawkins. Dan is now Professor Emeritus at the University of Alaska, Fairbanks. Besides being a fellow Montanan, Dan is an excellent geochemist who kept me going with his evaluations, suggestions, and supportive observations. He carefully read the entire manuscript when it was at its roughest, continuing even after he realized I wasn't going to mention his favorite subject (zeolites). Dan provided continuous enthusiasm, support, and encouragement and deserves more credit than this brief mention.

*Mason, B., 1992, *Victor Moritz Goldschmidt: Father of Modern Geochemistry*, Special Publication No. 4, The Geochemical Society, San Antonio, TX.

This book has been written over a long period of time. My wife, Jean Rekemeyer, has been a patient, caring partner and friend throughout the writing. Even though she had to give up a lot of quality time with me, she still provided the continuous and extensive support I needed for my writing. She challenged me with thought-provoking questions and paid special attention to the nonscientific details of manuscript preparation. Thus, she added an extra dimension of richness to the overall quality of my book. Jean, more than anyone, knows how much time and energy went into this publication. My dedication of this book to her does not adequately indicate the role she has played. I am truly grateful to her.

Both Jean and I are deeply grateful to our entire family for their unqualified support, intense interest, and boundless concern and love throughout the writing of this book.

Arthur H. Brownlow
Boston, Massachusetts

The Elements

Geochemistry deals with the abundance and distribution of the elements. An *element* is matter represented by a particular kind of atom (an atom whose nucleus has a specific number of protons). The abundance and distribution of the elements depend on their physical and chemical properties. In turn, these properties depend upon the electronic structure of the atoms and vary systematically with the nuclear charge (atomic number). There is a periodic recurrence of characteristic properties as the atomic number increases. This periodicity is represented by the *periodic table* (Figure 1-1). A thorough understanding of the periodic table is basic to all aspects of geochemistry.

ELECTRONIC STRUCTURE OF ATOMS

Atoms consist of a nucleus of protons and neutrons surrounded by electrons. The modern concept of the atom, based on extensive experimental evidence, views electrons as being able to occur at any distance from the nucleus, but as most likely occurring concentrated at various specific distances from the nucleus in the form of *orbitals.* An orbital can be thought of as a volume of space in which an electron is most likely to occur (Figure 1-2). An electron in a specific orbital has a specific energy state associated with it. Orbitals are clustered in various energy regions or *shells* around the nucleus. In these shells the orbitals occur in different *subshells.*

This model of atomic structure uses the principles of quantum mechanics, which, among other things, provide explanations for the way that atoms absorb and emit radiant energy. Energy is not absorbed or emitted in a continuous pattern; instead energy changes occur in certain discrete amounts. When an atom absorbs or emits radiation, the energy absorbed or emitted is the amount involved in the movement of an electron to a different energy level. A basic equation of

1A																	8A
1 H	2A											3A	4A	5A	6A	7A	2 He
3 Li	4 Be											5 B	6 C	7 N	8 O	9 F	10 Ne
11 Na	12 Mg	3B	4B	5B	6B	7B		8		1B	2B	13 Al	14 Si	15 P	16 S	17 Cl	18 Ar
19 K	20 Ca	21 Sc	22 Ti	23 V	24 Cr	25 Mn	26 Fe	27 Co	28 Ni	29 Cu	30 Zn	31 Ga	32 Ge	33 As	34 Se	35 Br	36 Kr
37 Rb	38 Sr	39 Y	40 Zr	41 Nb	42 Mo	43 (Tc)	44 Ru	45 Rh	46 Pd	47 Ag	48 Cd	49 In	50 Sn	51 Sb	52 Te	53 I	54 Xe
55 Cs	56 Ba	57 La*	72 Hf	73 Ta	74 W	75 Re	76 Os	77 Ir	78 Pt	79 Au	80 Hg	81 Tl	82 Pb	83 Bi	84 Po	85 (At)	86 Rn
87 (Fr)	88 Ra	89 Ac**															

*	58 Ce	59 Pr	60 Nd	61 (Pm)	62 Sm	63 Eu	64 Gd	65 Tb	66 Dy	67 Ho	68 Er	69 Tm	70 Yb	71 Lu
**	90 Th	91 Pa	92 U	93 (Np)	94 (Pu)	95 (Am)	96 (Cm)	97 (Bk)	98 (Cf)	99 (Es)	100 (Fm)	101 (Md)	102 (No)	103 (Lw)

Figure 1-1 Periodic table of the elements. Elements whose symbols are enclosed in parentheses are extremely rare or do not occur in nature. All the atoms of elements 43 (technetium) and 61 (promethium) are radioactive with short half-lives and thus are not found on the Earth today (technetium has been detected in some stars). Similarly, elements of atomic number 93 and higher have short half-lives and do not occur naturally. They have been produced artificially in the laboratory. Atoms of elements 84 to 92 are also radioactive, but some of them have long half-lives and thus they are found in nature since they have not completely broken down in the time since the formation of the elements. Elements marked ◪ are metalloids and elements marked ◥ are nonmetals. All the other elements are metals.

quantum mechanics (Schrödinger's wave equation) relates the energy of an atom to the location and energy levels of its electrons. The most stable energy states (possible energy levels) for these electrons can be calculated from the wave equation. These solutions are essentially a statement of the distribution and spatial location of the electrons in an atom under steady-state conditions (no radiation given off or absorbed). When an electron jumps from one subshell to another, energy in the form of electromagnetic radiation is given off if the electron goes to a lower energy subshell or is absorbed if the electron goes to a higher energy subshell. When this radiation is dispersed by a grating or a prism, a spectrum consisting of lines is formed (Figure 1-3). The quality and location of these lines can be related to the different subshells found in atoms.

The state of an electron is described in quantum theory by four quantum numbers: (1) the *principal quantum number, n,* which determines the shell in which orbitals will occur; (2) the *subsidiary quantum number, l,* which determines the shape and subshell of an orbital; (3) the *magnetic quantum number, m,* which determines the orientation of an orbital within the appropriate subshell; and (4) the *spin quantum number, s,* specifying which of two possible ways an electron is spinning (Table 1-1). The shells, from the nucleus outward, are numbered 1, 2, 3, 4, . . . or lettered K, L, M, N, ... The four main types of subshells are labeled *s, p, d,* and *f.*

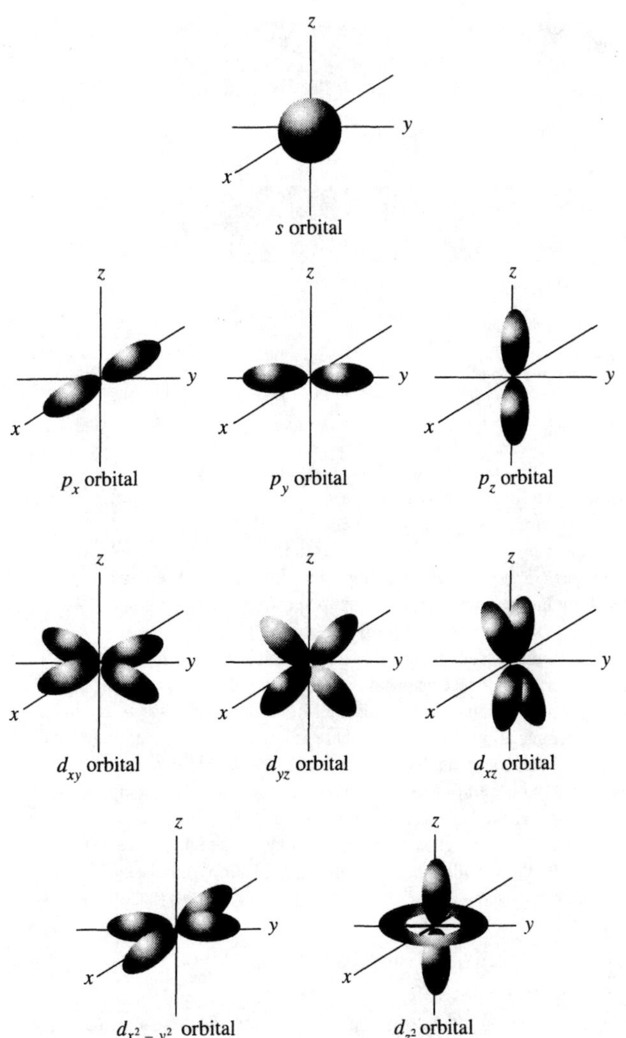

Figure 1-2 "Shapes" of electron clouds for various types of atomic orbitals. Because electrons exhibit wavelike properties we cannot define their precise locations. However, we can calculate the probability that an electron will occur in a particular region. The electron clouds shown here represent the regions most likely to contain electrons. The nucleus of the atom is at the center of each orbital. The electron clouds of orbitals retain the same shape but are larger when they occur in outer shells as compared to inner shells. (After W. S. Fyfe, *Geochemistry of Solids*, p. 19. Copyright 1964 by McGraw-Hill Book Company. Used with permission of McGraw-Hill Book Company.)

Niels Bohr's 1913 hypothesis for the hydrogen atom assumed circular electron orbits in shells, and his original quantum number n (the principal quantum number) reflected the radius of a circular orbit. As more spectroscopic measurements were made, it became clear that elliptical orbits with the nucleus at a focus were also possible. A secondary or subsidiary quantum number (l) was then introduced to describe the shape of a given ellipse. Thus the subshells associated with a particular principal quantum number (shell) each have orbitals of unique shapes and energies.

Under the effect of an external magnetic field, the plane of the circular or elliptical orbits can have only certain orientations in space (because of the magnetic field established by each moving electron). Each differently oriented orbital plane represents a slightly different energy state. The magnetic quantum number m identifies these different possible orientations of the

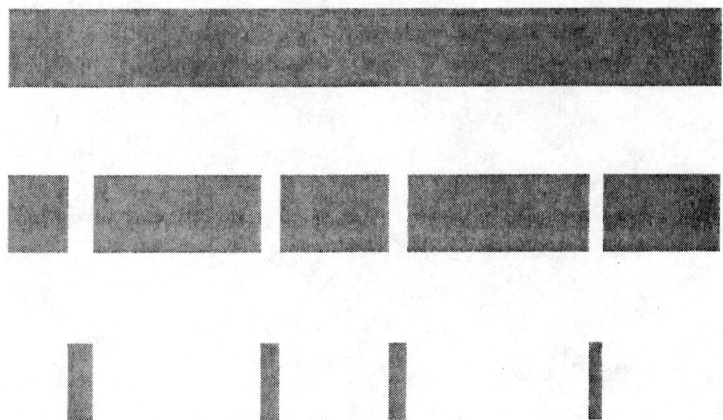

Figure 1-3 Different types of spectra. A continuous spectrum (*top*) is an uninterrupted series of images of the spectroscope slit through which light is passed. Such a spectrum comes from a hot solid or liquid body. In a solid or liquid the atoms are practically in contact with one another and cannot act independently. As a result, a radiating solid or liquid emits electromagnetic waves of all wavelengths. In contrast, the atoms in a gas are at relatively large distances from one another and thus are capable of radiating and absorbing independently, resulting in a spectrum containing relatively few wavelengths. Therefore, the line spectrum produced by the atoms of a gas is characteristic of these atoms. Each line is an image of the spectroscope slit. Absorption lines (*middle*) are formed when white light shines through a cool gas. These lines have less intensity than the continuous spectrum on either side of them. Light from the main body of the sun emits a continuous spectrum, which passes through a layer of cooler gas above the surface where atoms absorb their characteristic spectrum lines. This allows determination of the composition of the solar atmosphere, which is believed to be representative of the outer part of the Sun's mass. Emission lines (*bottom*) are produced by energizing the atoms of a gas. These have a greater intensity than the background. The position of the lines depend on the number and arrangement of the outer electrons of the atoms. The spectra are shown as negatives. (After P. W. Merrill, *Space Chemistry*, p. 24. Copyright © by the University of Michigan, 1963. All rights reserved.)

orbitals. Finally, as an electron rotates about a nucleus, it also spins on its own axis. The spin quantum number s identifies the direction of spin rotation.

All of the above types of electron motion are used to explain the different energy states originally identified experimentally by high-resolution spectrometers. In other words, the spectra of the elements are explained by allowing (quantizing) only certain types of motion.

Certain principles of quantum theory are used in describing the electronic structure of the elements. The *aufbau,* or *building-up, principle* states that electrons are put into orbitals in order of increasing subshell energy. A representation of relative energies for the four main types of subshells is given in Figure 1-4. These are labeled *s, p, d,* and *f.* The relationships of their levels for each shell (1, 2, 3, etc.) have been determined by use of the wave equation in combination with laboratory study of atomic spectra. The *Pauli exclusion principle* states that a maximum of two electrons can occupy an orbital and that the two electrons must spin in opposite directions. Another way of stating this principle is that two or more electrons cannot exist in the same state at the same time.

If, as in minerals and other solids, there is an external magnetic field acting on the electrons of individual atoms, orbitals exist with slightly different energy levels and with different spatial locations (Figure 1-2). Thus there are three different p orbitals, each with two electrons, and each

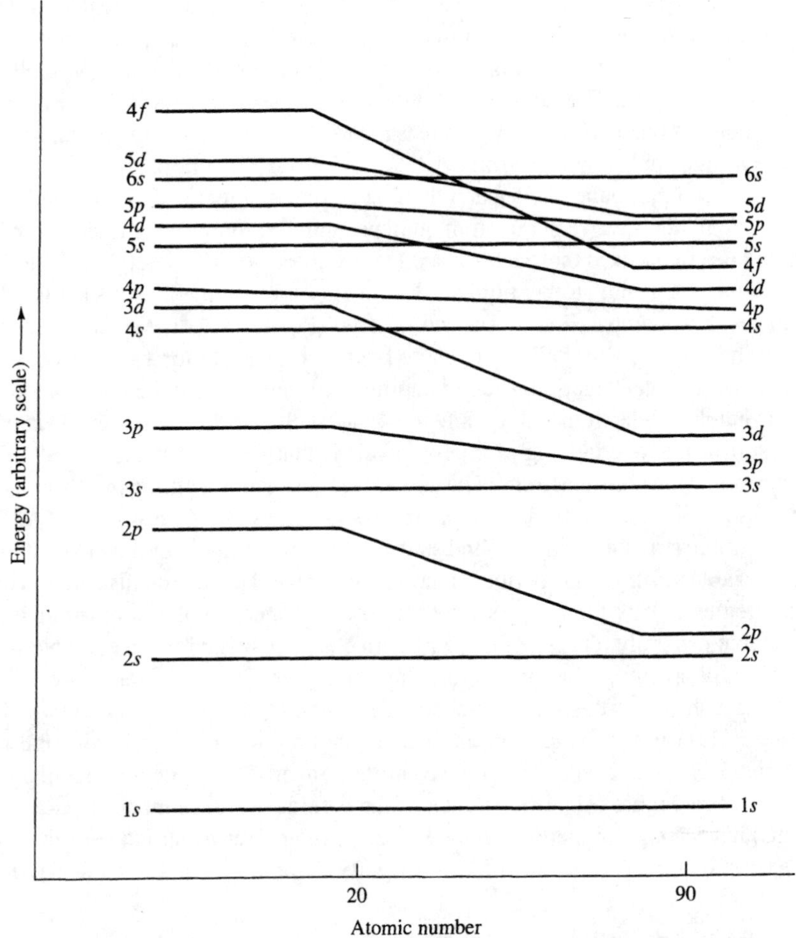

Figure 1-4 Variation of energy levels of atomic subshells as a function of atomic number. The stability of an electron is determined by how strongly it is attracted by the nucleus of an atom. A 1*s* electron is closer to the nucleus than a 2*s* electron and is thus more strongly attracted. A 2*s* electron is more strongly attracted than a 2*p* electron because a 2*s* electron spends more time near the nucleus. A more strongly attracted electron is more stable, that is, has a lower energy level. We can also say that more energy is required to remove a 2*s* electron from an atom than is required to remove a 2*p* electron. (After R. C. Evans, 1964, *Crystal Chemistry*, 2nd Edition, Cambridge University Press, p. 19.)

with a different general location. (The *s* subshell has only one orbital, whereas the *d* subshell has five and the *f* subshell has seven.) *Hund's rule* states that one electron goes into each of these orbitals before two electrons can be placed in any one of them. This rule also requires that the electrons in singly occupied orbitals have parallel spins.

Energy relationships are such that the maximum possible number of electrons in any shell is equal to $2n^2$. Thus the innermost, or first, shell can hold 2 electrons, the second 8, the third 18, the fourth 32, the fifth 50, etc. Electrons of the different subshells are designated, in order of increasing subshell energy, as *s*, *p*, *d*, and *f* electrons.

A particular electron is identified by a number representing the major shell in which it occurs, by a letter representing its type of subshell, and by a subscript identifying its orbital orientation. For example, a $3p_x$ electron occurs in the third shell out from the nucleus (M shell) and in the p_x orbital of the *p* subshell. The number of orbitals in each shell increases outward from the nucleus. The K shell has one *s* orbital only, whereas the M shell has one *s*, three *p*, and five *d* orbitals. Each of these orbitals holds two electrons. All the orbitals of a given subshell (*s, p, d,* or *f*) have the same energy in a particular shell when there is no external magnetic field affecting an individual atom. The relation between quantum numbers and electron orbitals is shown in Table 1-1.

It should be emphasized that each subshell represents a different energy level and that these energy levels may vary with atomic number. The energy level of the 2*p* subshell is different for atomic number 15 as compared to atomic number 90 (Figure 1-4). The energy differences between the subshells of a given shell can be quite large. It is possible for a subshell of one shell to have a higher energy level than a subshell of another shell farther from the nucleus. For example, at low atomic numbers, electrons will go into the 4*s* subshell before entering the 3*d* subshell. One important result of this overlapping of energy levels is that the outermost shell of an atom can never contain more than eight electrons. The general sequence for the filling of subshells as atomic number increases is 1*s*, 2*s*, 2*p*, 3*s*, 3*p*, 4*s*, 3*d*, 4*p*, 5*s*, 4*d*, 5*p*, 6*s*, 5*d*, 4*f*, 6*p*, 7*s*, 6*d*, 5*f*.

The electronic structure of the ground state (lowest energy state) of the elements is given in Table 1-2. This table illustrates the periodic recurrence of a similar electron distribution in the outer shells of the elements. This similarity is what makes the elements in the same group (column) of the periodic table generally resemble one another in chemical behavior. An example is the series lithium, sodium, potassium, rubidium, cesium, and francium. These elements have one electron in their outermost shell, and the outer two shells are similar for all except lithium (which has only three electrons). In a few groups, such as those containing metals, metalloids, and nonmetals (groups 4A and 5A), the elements do not have similar properties, despite their similar outer electron distribution (see Figure 1-1). In the center of the periodic table, elements in each row (transition elements) have similar properties because of the similar electron structure of their outermost shells. An example of the periodic nature of a property (atomic size) is shown in Figure 1-5.

TABLE 1-1 Relation Between Quantum Numbers and Electron Orbitals

n	l	m	Number of subshells	Number of orbitals	Designations of electron orbitals
1	0	0	1	1	1*s*
2	0	0	2	1	2*s*
	1	−1, 0, 1		3	$2p_x, 2p_y, 2p_z$
3	0	0	3	1	3s
	1	−1, 0, 1		3	$3p_x, 3p_y, 3p_z$
	2	−2, −1, 0, 1, 2		5	$3d_{xy}, 3d_{yz}, 3d_{xz}, 3d_{z^2}, 3d_{x^2-y^2}$
⋮	⋮	⋮	⋮	⋮	⋮

Source: R. Chang, *General Chemistry*, p. 228. Copyright 1986 by McGraw-Hill, Inc. Used with permission of McGraw Hill, Inc.

TABLE 1-2 Electronic Configurations of the Ground States of the Elements

Period, element, and atomic number			K	L		M			N				O				P			Q
			1*s*	2*s*	2*p*	3*s*	3*p*	3*d*	4*s*	4*p*	4*d*	4*f*	5*s*	5*p*	5*d*	5*f*	6*s*	6*p*	6*d*	7*s*
1	H	1	1																	
	He	2	2																	
2	Li	3	2	1																
	Be	4	2	2																
	B	5	2	2	1															
	C	6	2	2	2															
	N	7	2	2	3															
	O	8	2	2	4															
	F	9	2	2	5															
	Ne	10	2	2	6															
3	Na	11	2	2	6	1														
	Mg	12	2	2	6	2														
	Al	13	2	2	6	2	1													
	Si	14	2	2	6	2	2													
	P	15	2	2	6	2	3													
	S	16	2	2	6	2	4													
	Cl	17	2	2	6	2	5													
	Ar	18	2	2	6	2	6													
4	K	19	2	2	6	2	6		1											
	Ca	20	2	2	6	2	6		2											
	Sc	21	2	2	6	2	6	1	2											
	Ti	22	2	2	6	2	6	2	2											
	V	23	2	2	6	2	6	3	2											
	Cr	24	2	2	6	2	6	5	1											
	Mn	25	2	2	6	2	6	5	2											
	Fe	26	2	2	6	2	6	6	2											
	Co	27	2	2	6	2	6	7	2											
	Ni	28	2	2	6	2	6	8	2											
	Cu	29	2	2	6	2	6	10	1											
	Zn	30	2	2	6	2	6	10	2											
	Ga	31	2	2	6	2	6	10	2	1										
	Ge	32	2	2	6	2	6	10	2	2										
	As	33	2	2	6	2	6	10	2	3										
	Se	34	2	2	6	2	6	10	2	4										
	Br	35	2	2	6	2	6	10	2	5										
	Kr	36	2	2	6	2	6	10	2	6										
5	Rb	37	2	2	6	2	6	10	2	6			1							
	Sr	38	2	2	6	2	6	10	2	6			2							
	Y	39	2	2	6	2	6	10	2	6	1		2							
	Zr	40	2	2	6	2	6	10	2	6	2		2							
	Nb	41	2	2	6	2	6	10	2	6	4		1							
	Mo	42	2	2	6	2	6	10	2	6	5		1							
	Tc	43	2	2	6	2	6	10	2	6	5		2							
	Ru	44	2	2	6	2	6	10	2	6	7		1							

TABLE 1-2 *(cont.)*

Period, element, and atomic number			K	L		M			N				O				P			Q
			1s	2s	2p	3s	3p	3d	4s	4p	4d	4f	5s	5p	5d	5f	6s	6p	6d	7s
	Rh	45	2	2	6	2	6	10	2	6	8		1							
	Pd	46	2	2	6	2	6	10	2	6	10									
	Ag	47	2	2	6	2	6	10	2	6	10		1							
	Cd	48	2	2	6	2	6	10	2	6	10		2							
	In	49	2	2	6	2	6	10	2	6	10		2	1						
	Sn	50	2	2	6	2	6	10	2	6	10		2	2						
	Sb	51	2	2	6	2	6	10	2	6	10		2	3						
	Te	52	2	2	6	2	6	10	2	6	10		2	4						
	I	53	2	2	6	2	6	10	2	6	10		2	5						
	Xe	54	2	2	6	2	6	10	2	6	10		2	6						
6	Cs	55	2	2	6	2	6	10	2	6	10		2	6			1			
	Ba	56	2	2	6	2	6	10	2	6	10		2	6			2			
	La	57	2	2	6	2	6	10	2	6	10		2	6	1		2			
	Ce	58	2	2	6	2	6	10	2	6	10	1	2	6	1		2			
	Pr	59	2	2	6	2	6	10	2	6	10	3	2	6			2			
	Nd	60	2	2	6	2	6	10	2	6	10	4	2	6			2			
	Pm	61	2	2	6	2	6	10	2	6	10	5	2	6			2			
	Sm	62	2	2	6	2	6	10	2	6	10	6	2	6			2			
	Eu	63	2	2	6	2	6	10	2	6	10	7	2	6			2			
	Gd	64	2	2	6	2	6	10	2	6	10	7	2	6	1		2			
	Tb	65	2	2	6	2	6	10	2	6	10	9	2	6			2			
	Dy	66	2	2	6	2	6	10	2	6	10	10	2	6			2			
	Ho	67	2	2	6	2	6	10	2	6	10	11	2	6			2			
	Er	68	2	2	6	2	6	10	2	6	10	12	2	6			2			
	Tm	69	2	2	6	2	6	10	2	6	10	13	2	6			2			
	Yb	70	2	2	6	2	6	10	2	6	10	14	2	6			2			
	Lu	71	2	2	6	2	6	10	2	6	10	14	2	6	1		2			
	Hf	72	2	2	6	2	6	10	2	6	10	14	2	6	2		2			
	Ta	73	2	2	6	2	6	10	2	6	10	14	2	6	3		2			
	W	74	2	2	6	2	6	10	2	6	10	14	2	6	4		2			
	Re	75	2	2	6	2	6	10	2	6	10	14	2	6	5		2			
	Os	76	2	2	6	2	6	10	2	6	10	14	2	6	6		2			
	Ir	77	2	2	6	2	6	10	2	6	10	14	2	6	7		2			
	Pt	78	2	2	6	2	6	10	2	6	10	14	2	6	9		1			
	Au	79	2	2	6	2	6	10	2	6	10	14	2	6	10		1			
	Hg	80	2	2	6	2	6	10	2	6	10	14	2	6	10		2			
	Tl	81	2	2	6	2	6	10	2	6	10	14	2	6	10		2	1		
	Pb	82	2	2	6	2	6	10	2	6	10	14	2	6	10		2	2		
	Bi	83	2	2	6	2	6	10	2	6	10	14	2	6	10		2	3		
	Po	84	2	2	6	2	6	10	2	6	10	14	2	6	10		2	4		
	At	85	2	2	6	2	6	10	2	6	10	14	2	6	10		2	5		
	Rn	86	2	2	6	2	6	10	2	6	10	14	2	6	10		2	6		
7	Fr	87	2	2	6	2	6	10	2	6	10	14	2	6	10		2	6		1
	Ra	88	2	2	6	2	6	10	2	6	10	14	2	6	10		2	6		2
	Ac	89	2	2	6	2	6	10	2	6	10	14	2	6	10		2	6	1	2
	Th	90	2	2	6	2	6	10	2	6	10	14	2	6	10		2	6	2	2

TABLE 1-2 *(cont.)*

Period, element, and atomic number		K	L		M			N				O				P			Q
		1*s*	2*s*	2*p*	3*s*	3*p*	3*d*	4*s*	4*p*	4*d*	4*f*	5*s*	5*p*	5*d*	5*f*	6*s*	6*p*	6*d*	7*s*
Pa	91	2	2	6	2	6	10	2	6	10	14	2	6	10	2	2	6	1	2
U	92	2	2	6	2	6	10	2	6	10	14	2	6	10	3	2	6	1	2
Np	93	2	2	6	2	6	10	2	6	10	14	2	6	10	4	2	6	1	2
Pu	94	2	2	6	2	6	10	2	6	10	14	2	6	10	6	2	6		2
Am	95	2	2	6	2	6	10	2	6	10	14	2	6	10	7	2	6		2
Cm	96	2	2	6	2	6	10	2	6	10	14	2	6	10	7	2	6	1	2
Bk	97	2	2	6	2	6	10	2	6	10	14	2	6	10	9	2	6		2
Cf	98	2	2	6	2	6	10	2	6	10	14	2	6	10	10	2	6		2
E	99	2	2	6	2	6	10	2	6	10	14	2	6	10	11	2	6		2
Fm	100	2	2	6	2	6	10	2	6	10	14	2	6	10	12	2	6		2
Mv	101	2	2	6	2	6	10	2	6	10	14	2	6	10	13	2	6		2
No	102	2	2	6	2	6	10	2	6	10	14	2	6	10	14	2	6		2
Lr	103	2	2	6	2	6	10	2	6	10	14	2	6	10	14	2	6	1	2

Note: Slight irregularities in the generally regular distribution of the electrons (for example, compare the configuration of chromium with its neighbors vanadium and manganese) are due to small differences in stability between two different possible distributions of electrons in outer shells.

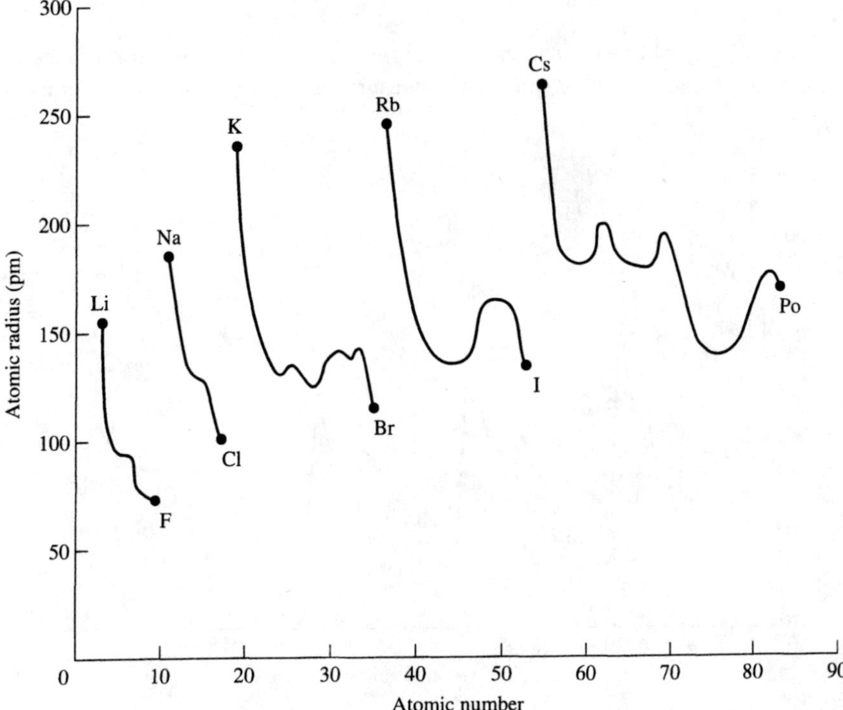

Figure 1-5 Plot of atomic radii (in picometers) of elements against their atomic numbers. (After R. Chang, *General Chemistry*, p. 257. Copyright 1986 by McGraw-Hill, Inc. Used with permission of McGraw-Hill, Inc.)

Many properties of the elements depend only on the electronic structure of the individual atom. An important example is bonding character, which generally depends on the outermost electrons. Elements with similar electronic structures form compounds that also show similarities in their physical-chemical properties. (See Chapter Five for further discussion of the role of bonding in determining the properties of solids.)

EXCITATION OF ATOMS

As indicated in the previous section, atoms can exist in a ground state of lowest energy and in a series of excited states produced by increasing the energies of their electrons. One way to produce excited states is to cause atoms to collide with electrons, molecules, ions, or other particles. By using this technique, it is possible to measure the amount of energy that must be applied to an element in the gaseous state to drive off an electron from one of its atoms. The atom is said to be *ionized,* and the energy required is known as the *ionization potential.* (First ionization potentials refer to the removal of one electron, and second ionization potentials to the removal of a second electron.) There is a relationship between the location of elements in the periodic table and their ionization potentials (Figure 1-6). The inert gases have very high ionization potentials and show little tendency to form ions. Elements in groups 1A and 2A have low ionization potentials and lose one or two electrons relatively easily, forming positive ions. In contrast, elements of groups 6A and 7A do not lose electrons easily, but they do have a high electron affinity, tending to form negative ions.

Electron affinity can be defined as the energy change when an electron is accepted by an atom in the gaseous state. Ionization potentials measure the attraction of an atom for its own outer

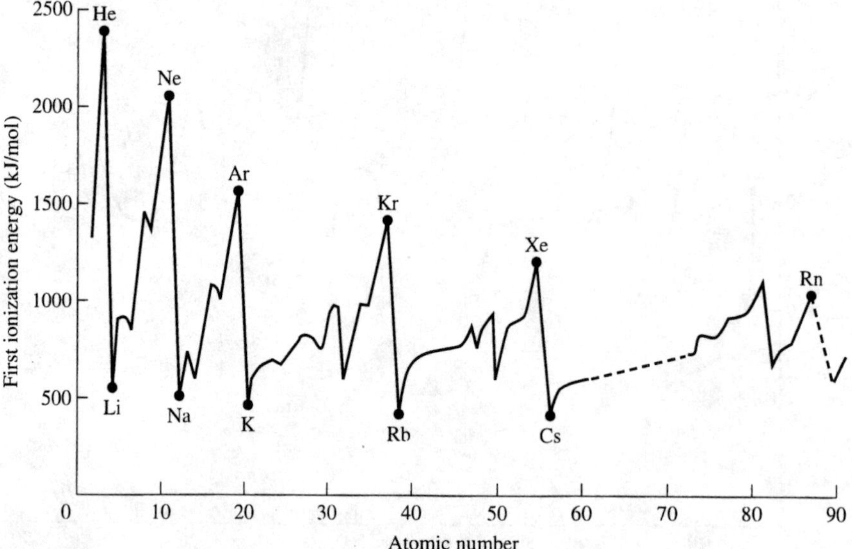

Figure 1-6 Variation of the first ionization potential with atomic number. Note that the noble gases have high ionization energies, whereas the alkali metals and alkaline earth metals have low ionization energies. (After R. Chang, *General Chemistry*, p. 262. Copyright 1986 by McGraw-Hill, Inc. Used with permission of McGraw-Hill, Inc.)

electrons while electron affinities are a measure of the attraction of an atom for electrons from some other source. Ionization potentials and electron affinities thus explain why a given element forms ions with specific valences (charges). Study of ionization potentials also provides strong evidence for the existence of electron shells. The potentials found for the removal of various electrons from an element fall into groups. Each group of similar potentials represents the energies required to remove electrons from a specific shell. For instance, the measured ionization potentials for $2s$ and $2p$ electrons would be similar and significantly higher than the ionization potentials for $3s$ and $3p$ electrons.

An atom can acquire lesser amounts of energy than that needed to ionize it. Such energies are known as *excitation potentials.* The occurrence of such an excited state can cause the atom to give off radiation of a definite frequency, as follows. Energy by particle collision or other means is imparted to the atom. This causes one of the atom's electrons to jump to a higher shell or subshell. The electron is in an unstable (excited) state and soon drops to a lower level, releasing all or part of its extra energy in the form of radiation.

It is possible to determine experimentally a series of excitation potentials and, at a higher energy input, the ionization of potentials of elements in the gaseous state. For instance, hydrogen atoms have been shown to have five particular excitation potentials; these potentials correspond to the excess energy that the hydrogen electron of the K shell would have if it were forced into one of the L through P shells and had one of the quantum numbers $n = 2, 3, 4, 5$, or 6 instead of the normal or ground-state of $n = 1$ (Figure 1-7). The experimentally determined excitation potentials thus agree very well with those predicted by quantum theory. Likewise, the first ionization potential found by experiment agrees with the value predicted by theory. Such experiments are carried out by impacting hydrogen atoms with high-velocity electrons and measuring the radiation emitted by the atoms.

Excitation potentials can be specifically related to spectroscopic data. The spectral lines emitted by a given element are caused by transitions of atoms between different excitation states. Since the atoms of each element have unique spacings of energy levels (and thus unique excitation states), spectral lines can be assigned to each element. The relationship of the frequency of emitted radiation to the difference between initial and final energy levels is $E_i - E_f = h\nu$, where h is a constant (Planck's constant) and ν is the frequency of radiation in seconds^{-1}. (Light energy occurs in discrete amounts known as *photons*, with each photon having energy equal to $h\nu$.) A correspondence is thus found between the energy difference for any two shells or subshells of an atom and the frequency of emitted radiation. For example, the series of hydrogen spectroscopic lines in the ultraviolet portion of the electromagnetic spectrum known as the Lyman series represents energy differences between quantum number 1 and higher-energy levels (Figure 1-7).

One way to produce spectra is by introducing an element into a flame. Some compounds under such conditions are vaporized and dissociated into atoms. Collisions may impart enough energy to the atoms of a given element to elevate them to an excited state. When the excited atoms return to the ground state, characteristic radiation is emitted. If the energy difference between the two states is of the right magnitude, visible light will be emitted. Examples of such characteristic visible light are the red color seen for lithium compounds and the yellow color seen for sodium compounds. When higher temperatures are used, such as in the carbon arc of a spectrograph, many more excited states of a given element can be attained and many more spectral lines are produced. The radiations emitted by elements of low atomic number have frequencies representative of the visible, infrared, and ultraviolet portions of the electromagnetic spectrum. These types of

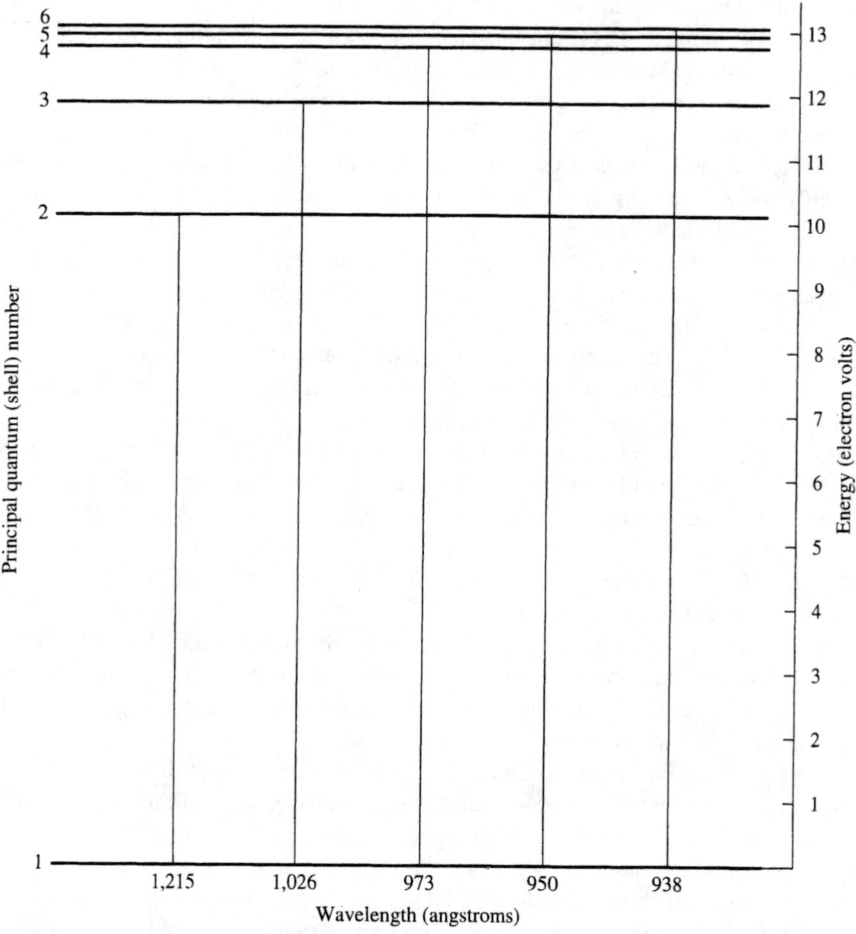

Figure 1-7 Lyman series of the hydrogen spectrum. The spectral line in the emission spectrum of hydrogen with a wavelength of 1,215 angstroms represents the energy emitted when an electron falls from shell two to shell one. The other lines represent transitions from shells further out from the innermost shell (shell one). The brightness of a particular line depends on how many photons of the same wavelength are emitted. The Balmer series of lines (not shown) represent energies emitted when electrons fall from various outer shells to shell two. The lines of these two series and of other known series are all produced by transitions between energy states that have different principal quantum numbers. The energy values for lines of the Lyman series cause them to occur in the ultraviolet portion of the electromagnetic spectrum, while lines of the Balmer series occur in the visible and ultraviolet portions. Hydrogen lines were first found in the spectra of stars and later measured in the laboratory (see Figure 1-3).

radiation often involve only outer electrons and can be produced with low amounts of activation energy. In addition to these types of radiation, the heavier elements also emit radiations having the frequencies of X rays. High-energy inputs can cause electrons in inner shells to move to shells farther from the nucleus or to leave an atom entirely. Other electrons fall to these lower levels and energy is given off in the form of X rays. X-ray spectra are grouped into families to which the names K series, L series, M series, etc., have been given. Spectra of the K series are the result of outer electrons filling vacancies in the K shell of an atom, those of the L series are the result of vacancies in the L shell, and so on.

When the study of spectra was a relatively new field, the various series of spectral lines produced by elements were referred to in a descriptive manner. The four series found for many elements came to be known as *sharp, principal, diffuse*, and *fundamental.* The first letters of these words are now used to describe the four different subshells (*s, p, d,* and *f*) that are responsible for the four different series of spectral lines.

PERIODIC TABLE

In the periodic table (Figure 1-1) the elements are arranged in rows (periods) such that elements in a given column (group) display similar properties. The elements in a group have similar properties because they have similar electronic structures. Refer to Figures 1-1 and 1-4 and Table 1-2 as you read the following. A period begins with the filling of a new outer shell and ends when this shell has eight electrons. The first period has only two elements, since the innermost shell can hold only two electrons. The next period goes from lithium (3) to neon (10) as electrons are added up to a total of eight in the L shell. A similar sequence occurs for the elements sodium (11) to argon (18). The fourth period starts off as do the previous periods, but a significant change occurs with scandium (21). The filling of the outermost shell is not continued, because the energy levels of the subshells are such that an electron is added to the unfilled M shell rather than to the outer N shell (see Figure 1-4). As the atomic number increases, the inner M shell is filled, and then, beginning with gallium (31), electrons are again added to the outer N shell. The group of elements from scandium (21) to copper (29) are known as *transition elements*. A similar sequence is found in the fifth period, resulting in the transition elements yttrium (39) to silver (47). The transition elements have variable ionic charges because some of the electrons in the two outer shells have similar energy levels. An iron ion with a charge of +2 is formed when the two electrons of the outermost N shell are removed. Little additional energy is needed to remove an electron from the 3*d* subshell of the M shell, giving Fe^{3+}.

A new complication develops in the sixth period. Lanthanum (57) starts a third transition series. However, the energy levels of the subshells are such that electrons now begin to fill up the 4*f*-subshell of the unfilled N shell. The series of elements formed by the filling of this shell (58–71) is known as the lanthanide or *rare-earth group.* The properties of these 14 elements are very similar because their outer two shells have similar electron distributions. The transition series of the sixth period is continued with hafnium (72) and ends with mercury (80). The end of the period is represented by radon (86), with eight electrons in its outer shell. Period 7 starts off as did period 6, with actinium (89) starting a new series of transition elements, and then thorium (90) beginning a new subsidiary series of elements in which the third shell in from the outside is filled, as occurs with the rare-earth elements. This series (90–103) is known as the actinide series.*

Thus the elements can be divided into categories according to their electron structure. All the elements in groups 1A through 7A have incompletely filled *s* or *p* subshells of highest principal quantum number. With the exception of helium, the inert gases all have a completely filled *p*

*Elements 104 to 111 have been discovered (produced artificially) and belong to the series of transition elements that starts with actinium.

subshell. The transition elements (groups 1B and 3B through 8) all have an incompletely filled *d* subshell or form cations with an incompletely filled *d* subshell. The lanthanides and actinides have an incompletely filled *f* subshell.

The three group 2B elements resemble the alkaline earth group (group 2A) because each has two electrons in its outer shell. However, the chemical properties of the two groups are different because of differences in the next-to-outermost shell; the alkaline earth elements have eight electrons there while the group 2B elements have eighteen electrons there. For example, group 2B elements lose electrons less easily and are thus less reactive than group 2A elements.

A broader classification of the elements can be made into the three categories: metals, nonmetals, and metalloids (see Figure 1-1). A metalloid has properties between those of metals and nonmetals. Most of the elements are metals (good conductors of heat and electricity). Boron, silicon, germanium, arsenic, antimony, tellurium, polonium, and astatine are metalloids; the elements to the right of them in the periodic table are nonmetals.

The periodic table is useful to geologists because it helps them to predict and better understand the behavior of the elements in nature. The behavior of an element that forms ions and ionic compounds depends essentially on two properties: (1) the size of the ions, and (2) the valence (charge) of the ions. These are both controlled by electronic structure. The outer electrons of an atom, which are those involved in the formation of ions and in chemical bonding, are referred to as *valence electrons*. Ions tend to form that have a stable arrangement of the outer electrons. The most stable arrangements are those exhibited by the inert gases (eight electrons in the outermost shell, with the exception of helium, which has only two electrons). Because of their stable electronic structure, the inert gases are not very active chemically.

To obtain the inert-gas structure, the elements of group 1A must lose one electron (giving ions with a $+1$ charge); elements of group 2A must lose two electrons (giving ions with a $+2$ charge). Elements of group 7A, in contrast, need to gain an electron (forming ions with a -1 charge). Using the periodic table, we would predict that rubidium occurs as a trace element in potassium minerals and that strontium occurs as a trace element in calcium minerals. Similarly, we would expect to find bromine in chlorine minerals. The transition elements, because of their more complicated electronic relationships, can form ions that do not have the inert-gas structure. However, similar chemical behavior is still found for transition elements with similar electron structure.

Some elements, such as carbon, are nonionic in their behavior. These elements can most easily attain the inert-gas structure by sharing electrons between atoms (covalent behavior). Such sharing can occur extensively only if each of the atoms involved has a single electron in one or more orbitals. Thus only certain elements can show strongly covalent behavior.

The group 2B elements show a greater tendency to form covalent compounds as compared to group 2A elements (elements of both groups have two electrons in their outer shell) because their nuclear charge makes it more difficult for them to lose the outer electrons. We shall discuss ionic and covalent behavior further in Chapter Five.

In summary, the periodic table can be considered as a representation of the electronic structure of the elements and, by extension, as a representation of the predictable behavior of the elements. Geologists must consider, in addition to the properties of the element itself, factors such as pressure, temperature, and chemical environment (see Chapter Three). An ion in a silicate magma behaves differently from the same ion in river water. The great variety of natural conditions makes the prediction and understanding of elemental behavior no easy task.

ABUNDANCE OF THE ELEMENTS

One goal of geochemistry is to determine the abundance of the elements in nature. This information is needed to develop hypotheses for the origin of the elements and of the structure of the universe. These hypotheses are only as good as the chemical data and assumptions on which they are based. These data are obtained in many different ways, and the quality and quantity of the results vary for different materials and from element to element.

Meteorites

The universe is made up of galaxies of stars, gaseous intergalactic nebulae, and interstellar gas and dust (Harrison 1981). Our knowledge of the abundance of the elements in the universe comes from two major sources: (1) spectroscopic analyses of the light from our sun, other stars, and nebulae (particularly for volatile elements); and (2) meteorites (for nonvolatile elements). The spectroscopic analyses involve comparisons of the intensities of elemental lines in stellar and nebulae spectra. Meteorite analyses (by various chemical methods) have long been an important source of information on the abundance of the elements. In recent years these samples of the solar system have become available in larger numbers due to the discovery and collection of a wide variety of meteorites in Antarctica.

Most meteorites are probably fragments from the asteroid belt of the solar system.* Their chemical variety indicates that they were once parts of several larger planetary bodies, each with its own unique chemical properties. Because they represent material created at the beginning of the solar system, they are an important source of our knowledge about the formation, early history, and composition of planetary bodies (Wasson 1985).

Meteorites are classified into three major groups: stones (chondrites and achondrites), stony irons, and irons (Table 1-3). Stones are mainly made up of silicate minerals, while irons consist largely of an iron-nickel alloy. Stony irons have an abundance of both silicate minerals and iron-nickel alloy. Stony meteorites make up over ninety percent of all meteorites (Table 1-4) and can be subdivided into three groups: (1) chondrites, characterized by abundant, millimeter-sized, rounded masses known as chondrules (King 1983; Wasson 1993); (2) achondrites, which do not have chondrules; and (3) carbonaceous chondrites, characterized by their content of hydrocarbons, water, and volatile elements. Chondrites other than carbonaceous chondrites are subdivided into enstatite (E), high iron (H), low iron (L), and very low iron (LL) chondrites (Van Schmus and Wood 1967). The H, L, and LL types are also referred to as ordinary chondrites. Carbonaceous chondrites are divided into three types on the basis of chemical and mineralogical criteria (Types I, II, and III or C1, C2, and C3).

Stony meteorites show significant differences in texture. Some chondrites have chondrules consisting of glass and various minerals distributed through a matrix of similar mineralogy. Others consist of angular fragments welded together (breccias), with each fragment containing chondrules, irregular inclusions, and matrix. The various materials making up chondrites probably formed under a variety of conditions and were subsequently accreted. For example, the spheroidal shapes and igneous nature of chondrules indicates that they were molten droplets at some stage in their history.

*In 1982 a meteorite from the Moon was discovered in Antarctica. Since then several other meteorites have been identified as coming from the Moon. Some unusual achondrites (SNC meteorites) with very young ages may have come from Mars (see discussion in text).

TABLE 1-3 Classification of Meteorites

Class	Symbol	Principal minerals
		Chondrites
Enstatite	E	Enstatite, nickel-iron
Bronzite	H	Olivine, bronzite, nickel-iron
Hypersthene	L	Olivine, hypersthene, nickel-iron
Amphoterite	LL	Olivine, hypersthene, nickel-iron
Carbonaceous	C	Serpentine, olivine
		Achondrites
Aubrites	Ae	Enstatite
Diogenites	Ah	Hypersthene
Chassignite	Ac	Olivine
Ureilites	Au	Olivine, clinobronzite, nickel-iron
Angrite	Aa	Augite
Nakhlite	An	Diopside, olivine
Howardites	Aho	Hypersthene, plagioclase
Eucrites	Aeu	Pigeonite, plagioclase
		Stony irons
Pallasites	P	Olivine, nickel-iron
Siderophyre	S	Orthopyroxene, nickel-iron
Lodranite	Lo	Orthopyroxene, olivine, nickel-iron
Mesosiderites	M	Pyroxene, plagioclase, nickel-iron
		Irons
Hexahedrites	Hx	Kamacite
Octahedrites	O	Kamacite, taenite
Ataxites	D	Taenite

Source: Mason (1979).

Carbonaceous chondrites have chondrules similar to those of other chondrites (Type 1 carbonaceous chondrites do not have chondrules but are still considered chondrites due to their mineralogy and chemistry). In addition they may have irregular inclusions made up of minerals such as spinel and perovskite that are not generally found in the chondrules of other chondrites. Also the mineralogy of their matrix is different from that of other chondrites. The matrix consists of hydrous and other minerals stable only at relatively low temperatures, while the minerals found in the matrix, chondrules, and inclusions of other chondrites are all anhydrous, high-temperature minerals. Small amounts of various organic compounds also occur in the matrix of carbonaceous chondrites. Thus all the material making up other chondrites formed at high temperatures, while only the chondrules and inclusions of carbonaceous chondrites formed at high temperatures, and the matrix in which they are embedded must have formed at much lower temperatures.

As mentioned earlier, achondrites do not have chondrules; they are generally similar in texture and mineralogy to mafic and ultramafic igneous rocks (including lunar basalts) and to breccias made up of these rocks. In detail, these relatively rare meteorites are very complex. Some of the more common ones are closely related in chemistry and mineralogy to chondrites, while others are represented by only one or two samples and have very unusual properties. Some unusual achondrites (SNC meteorites) are discussed later.

TABLE 1-4 Abundances of Meteorites

	Falls	Finds	Total
Stones			
Chondrites			
Enstatite (E)	13	11	24
High Iron (H)	276	405	681
Low iron (L)	319	350	669
Very low iron (LL)	66	30	96
Carbonaceous (C)	35	32	67
Unclassified	75	69	144
Total chondrites	784	897	1681
Achondrites	69	63	132
Total stones	853	960	1813
Stony irons	10	63	73
Irons	42	683	725
Total meteorites	905	1706	2611

Note: *Falls* refers to meteorites actually seen to fall. These numbers are a better indication of the relative abundance of the various types than percentages obtained from the total number of meteorites found. Most *finds* are iron meteorites. The table includes all well-authenticated meteorites identified up to January, 1984.

Source: Graham et al. (1985).

Another way to classify meteorites is as undifferentiated (chondrites) and differentiated (irons, stony irons, and achondrites). Their chemical compositions indicate that the parent bodies of chondrites have never been melted and fractionated, while the other meteorites reflect the result of melting prior to their formation. In other words, the parent bodies of differentiated meteorites separated into masses of varying composition and, as a result, these meteorites have compositions that are different from that of the solar atmosphere (which is believed to be representative of the nebula from which the solar system formed). Thus chondrites, with element ratios similar to those in the solar atmosphere, are believed to be representative of the primitive material from which the bodies of the solar system formed. Estimated compositions for Type I carbonaceous chondrites and for the solar atmosphere are given in Table 1-5. A detailed review of meteorite compositions can be found in Mason (1979).

In terms of composition, the most primitive (from an undifferentiated parent and least altered since formation) meteorites are the carbonaceous chondrites (Figure 1-8). Most chondrites, although formed from undifferentiated parent bodies, show signs of having undergone slight to extensive thermal metamorphism. In contrast, carbonaceous chondrites have not been significantly metamorphosed. If they had, they would have lost their water, volatiles, and hydrocarbons. As shown in Figure 1-8 and Table 1-5, some carbonaceous chondrites (Type I) are very similar in composition to the Sun. It is believed that these meteorites formed from the same material and at the same time as the Sun and have not been significantly changed in composition since their formation. They have been subjected only to aqueous alteration, resulting in the formation of hydrous minerals.

The parent bodies of chondritic meteorites probably formed by accretion of solid material with an overall composition similar to that of our Sun. The compositions of meteorites indicate that elements and compounds separated from a gaseous phase over a range of temperatures, with

TABLE 1-5 Atomic Abundances of the Elements (Relative to Si = 10^6 Atoms)

Elements	Type I carbonaceous chondrites	Solar atmosphere	Solar system	Earth	Primitive mantle	Earth oceanic crust	Earth continental crust
H	—	2.20×10^{10}	2.79×10^{10}	1.52×10^{4}	—	—	—
He	—	2.30×10^{9}	2.72×10^{9}	—	—	—	—
Li	6.05×10^{1}	2.20×10^{-1}	5.71×10^{1}	7.61×10^{1}	1.50×10^{1}	—	1.49×10^{3}
Be	8.01×10^{-1}	3.20×10^{-1}	7.30×10^{-1}	1.20×10^{0}	1.48×10^{0}	—	1.71×10^{1}
B	2.47×10^{1}	9.10×10^{0}	2.12×10^{1}	8.30×10^{0}	9.90×10^{0}	—	—
C	—	1.00×10^{7}	1.01×10^{7}	5.71×10^{3}	—	—	—
N	—	2.20×10^{6}	3.13×10^{6}	1.27×10^{2}	—	—	—
O	—	1.90×10^{7}	2.38×10^{7}	3.49×10^{6}	—	—	—
F	9.01×10^{2}	8.10×10^{2}	8.43×10^{2}	5.46×10^{2}	—	—	—
Ne	—	2.60×10^{6}	3.44×10^{6}	—	—	—	—
Na	5.59×10^{4}	4.70×10^{4}	5.74×10^{4}	1.35×10^{4}	1.45×10^{4}	1.10×10^{5}	1.17×10^{5}
Mg	1.07×10^{6}	9.60×10^{5}	1.07×10^{6}	1.06×10^{6}	1.32×10^{6}	2.32×10^{5}	8.99×10^{4}
Al	8.53×10^{4}	7.40×10^{4}	8.49×10^{4}	1.28×10^{5}	8.67×10^{4}	3.68×10^{5}	3.65×10^{5}
Si	1.00×10^{6}	1.00×10^{6}	1.00×10^{6}	1.00×10^{6}	1.00×10^{6}	1.00×10^{6}	1.00×10^{6}
P	8.81×10^{3}	6.90×10^{3}	1.04×10^{4}	1.36×10^{4}	—	—	—
S	4.90×10^{5}	3.90×10^{5}	5.15×10^{5}	1.12×10^{5}	—	—	—
Cl	5.13×10^{3}	7.10×10^{3}	5.24×10^{3}	1.38×10^{2}	—	—	—
Ar	—	8.50×10^{4}	1.01×10^{5}	—	—	—	—
K	3.83×10^{3}	3.20×10^{3}	3.77×10^{3}	8.51×10^{2}	6.16×10^{2}	3.89×10^{3}	3.31×10^{4}
Ca	6.14×10^{4}	5.00×10^{4}	6.11×10^{4}	9.43×10^{4}	6.31×10^{4}	2.45×10^{5}	1.39×10^{5}
Sc	3.45×10^{1}	2.80×10^{1}	3.42×10^{1}	5.26×10^{1}	3.14×10^{1}	1.03×10^{2}	6.91×10^{1}
Ti	2.35×10^{3}	2.60×10^{3}	2.40×10^{3}	4.21×10^{3}	2.51×10^{3}	2.28×10^{4}	1.04×10^{4}
V	2.94×10^{2}	3.00×10^{2}	2.93×10^{2}	3.96×10^{2}	2.21×10^{2}	5.97×10^{2}	3.56×10^{2}
Cr	1.36×10^{4}	1.30×10^{4}	1.35×10^{4}	1.80×10^{4}	7.72×10^{3}	6.31×10^{2}	1.10×10^{2}
Mn	9.25×10^{3}	6.00×10^{3}	9.55×10^{3}	2.10×10^{3}	2.43×10^{3}	2.21×10^{3}	2.08×10^{3}
Fe	8.72×10^{5}	8.50×10^{5}	9.00×10^{5}	1.26×10^{6}	1.48×10^{5}	1.78×10^{5}	1.08×10^{5}
Co	2.31×10^{3}	2.00×10^{3}	2.25×10^{3}	3.12×10^{3}	2.27×10^{2}	9.69×10^{1}	4.39×10^{1}
Ni	4.88×10^{4}	4.30×10^{4}	4.93×10^{4}	6.81×10^{4}	4.56×10^{3}	2.79×10^{2}	5.29×10^{1}
Cu	5.09×10^{2}	2.60×10^{2}	5.22×10^{2}	1.76×10^{2}	5.89×10^{1}	1.65×10^{2}	9.78×10^{1}
Zn	1.28×10^{3}	8.70×10^{2}	1.26×10^{3}	2.79×10^{2}	1.02×10^{2}	1.58×10^{2}	—
Ga	3.76×10^{1}	1.40×10^{1}	3.78×10^{1}	1.54×10^{1}	5.70×10^{0}	2.96×10^{1}	2.67×10^{1}
Ge	1.22×10^{2}	7.10×10^{1}	1.19×10^{2}	3.72×10^{1}	—	—	—
As	6.57×10^{0}	8.90×10^{0}	6.56×10^{0}	9.40×10^{0}	—	—	—
Se	6.64×10^{1}	1.80×10^{1}	6.21×10^{1}	1.51×10^{1}	—	—	—
Br	1.21×10^{1}	—	1.18×10^{1}	3.30×10^{-1}	—	—	—
Kr	—	4.70×10^{1}	4.50×10^{1}	—	—	—	—
Rb	6.95×10^{0}	8.90×10^{0}	7.09×10^{0}	1.30×10^{0}	7.00×10^{-1}	2.10×10^{0}	5.09×10^{1}
Sr	2.41×10^{1}	1.80×10^{1}	2.35×10^{1}	4.06×10^{1}	2.36×10^{1}	1.80×10^{2}	4.73×10^{2}
Y	4.33×10^{0}	4.00×10^{0}	4.64×10^{0}	7.20×10^{0}	4.30×10^{0}	4.37×10^{1}	2.56×10^{1}
Zr	1.11×10^{1}	1.00×10^{1}	1.14×10^{1}	4.22×10^{1}	1.14×10^{1}	1.07×10^{2}	1.14×10^{2}
Nb	7.77×10^{-1}	1.80×10^{0}	6.98×10^{-1}	2.10×10^{0}	8.00×10^{-1}	2.80×10^{0}	1.22×10^{1}
Mo	2.57×10^{0}	1.90×10^{0}	2.55×10^{0}	6.00×10^{0}	—	—	—
Ru	1.88×10^{0}	1.50×10^{0}	1.86×10^{0}	2.80×10^{0}	—	—	—
Rh	3.48×10^{-1}	5.60×10^{-1}	3.44×10^{-1}	6.00×10^{-1}	—	—	—
Pd	1.41×10^{0}	7.10×10^{-1}	1.39×10^{0}	1.80×10^{0}	—	—	—
Ag	5.16×10^{-1}	1.60×10^{-1}	4.86×10^{-1}	1.50×10^{-1}	—	—	—
Cd	1.55×10^{0}	1.60×10^{0}	1.61×10^{0}	3.70×10^{-2}	—	—	—

TABLE 1-5 (*cont.*)

Elements	Type I carbonaceous chondrites	Solar atmosphere	Solar system	Earth	Primitive mantle	Earth oceanic crust	Earth continental crust
In	1.86×10^{-1}	1.00×10^{0}	1.84×10^{-1}	4.60×10^{-3}	—	—	—
Sn	3.88×10^{0}	2.00×10^{0}	3.82×10^{0}	1.16×10^{0}	—	—	—
Sb	3.36×10^{-1}	2.20×10^{-1}	3.09×10^{-1}	1.03×10^{-1}	—	—	—
Te	5.03×10^{0}	—	4.81×10^{0}	1.43×10^{0}	—	—	—
I	1.05×10^{0}	—	9.00×10^{-1}	2.60×10^{-2}	—	—	—
Xe	—	5.40×10^{0}	4.70×10^{0}	—	—	—	—
Cs	3.68×10^{-1}	$<1.80 \times 10^{0}$	3.72×10^{-1}	8.70×10^{-2}	$<2.00 \times 10^{-2}$	1.37×10^{-2}	1.30×10^{0}
Ba	4.48×10^{0}	2.80×10^{0}	4.49×10^{0}	7.20×10^{0}	4.70×10^{0}	1.32×10^{1}	2.64×10^{2}
La	4.55×10^{-1}	3.00×10^{-1}	4.46×10^{-1}	6.75×10^{-1}	4.00×10^{-1}	3.20×10^{0}	1.41×10^{1}
Ce	1.18×10^{0}	7.90×10^{-1}	1.14×10^{0}	1.78×10^{0}	1.20×10^{0}	9.90×10^{0}	2.81×10^{1}
Pr	1.76×10^{-1}	1.20×10^{-1}	1.67×10^{-1}	2.26×10^{-1}	1.00×10^{-1}	1.50×10^{0}	3.10×10^{0}
Nd	8.48×10^{-1}	3.80×10^{-1}	8.28×10^{-1}	1.18×10^{0}	8.00×10^{-1}	8.40×10^{0}	1.14×10^{1}
Sm	2.65×10^{-1}	1.40×10^{-1}	2.58×10^{-1}	3.38×10^{-1}	2.00×10^{-1}	2.60×10^{0}	2.50×10^{0}
Eu	9.86×10^{-2}	1.10×10^{-1}	9.73×10^{-2}	1.29×10^{-1}	9.00×10^{-2}	1.00×10^{0}	7.00×10^{-1}
Gd	3.35×10^{-1}	3.00×10^{-1}	3.30×10^{-1}	4.61×10^{-1}	3.00×10^{-1}	3.50×10^{0}	2.30×10^{0}
Tb	5.98×10^{-2}	—	6.03×10^{-2}	8.30×10^{-2}	5.00×10^{-2}	6.50×10^{-1}	4.00×10^{-1}
Dy	4.03×10^{-1}	2.60×10^{-1}	3.94×10^{-1}	5.42×10^{-1}	4.00×10^{-1}	4.20×10^{0}	2.30×10^{0}
Ho	8.87×10^{-2}	—	8.89×10^{-2}	1.20×10^{-1}	8.50×10^{-2}	9.50×10^{-1}	5.00×10^{-1}
Er	2.56×10^{-1}	1.30×10^{-1}	2.51×10^{-1}	3.39×10^{-1}	2.00×10^{-1}	2.60×10^{0}	1.40×10^{0}
Tm	3.91×10^{-2}	4.10×10^{-2}	3.78×10^{-2}	5.10×10^{-2}	3.30×10^{-2}	3.80×10^{-1}	2.00×10^{-1}
Yb	2.46×10^{-1}	1.80×10^{-1}	2.48×10^{-1}	3.28×10^{-1}	2.00×10^{-1}	3.50×10^{0}	1.30×10^{0}
Lu	3.75×10^{-2}	1.30×10^{-1}	3.67×10^{-2}	5.50×10^{-2}	3.20×10^{-2}	3.90×10^{-1}	2.00×10^{-1}
Hf	1.80×10^{-1}	1.70×10^{-1}	1.54×10^{-1}	3.18×10^{-1}	1.00×10^{-1}	1.60×10^{0}	1.70×10^{0}
Ta	2.51×10^{-2}	—	2.07×10^{-2}	3.10×10^{-2}	—	—	—
W	1.46×10^{-1}	1.10×10^{0}	1.33×10^{-1}	2.66×10^{-1}	—	—	—
Re	5.32×10^{-2}	$<1.10 \times 10^{-2}$	5.17×10^{-2}	8.00×10^{-2}	—	—	—
Os	6.89×10^{-1}	1.10×10^{-1}	6.75×10^{-1}	1.13×10^{0}	—	—	—
Ir	6.40×10^{-1}	6.50×10^{-1}	6.61×10^{-1}	1.08×10^{0}	—	—	—
Pt	1.36×10^{0}	1.30×10^{0}	1.34×10^{0}	2.11×10^{0}	—	—	—
Au	1.96×10^{-1}	1.30×10^{-1}	1.87×10^{-1}	2.88×10^{-1}	—	—	—
Hg	5.20×10^{-1}	$<2.80 \times 10^{0}$	3.40×10^{-1}	9.70×10^{-3}	—	—	—
Tl	1.86×10^{-1}	1.80×10^{-1}	1.84×10^{-1}	4.70×10^{-3}	—	—	—
Pb	3.10×10^{0}	1.90×10^{0}	3.15×10^{0}	1.23×10^{-1}	—	—	5.00×10^{0}
Bi	1.41×10^{-1}	$<1.80 \times 10^{0}$	1.44×10^{-1}	3.47×10^{-3}	—	—	—
Th	4.34×10^{-2}	3.50×10^{-2}	3.35×10^{-2}	5.48×10^{-2}	3.60×10^{-2}	1.50×10^{-1}	2.10×10^{0}
U	9.22×10^{0}	$<8.90 \times 10^{-2}$	9.00×10^{-3}	1.48×10^{-2}	1.00×10^{-2}	2.90×10^{-1}	5.00×10^{-1}

Columns one and two (Type I carbonaceous chondrites and Solar atmosphere: Wasson (1985, 236–239).

Column three (Solar system): Anders and Grevesse (1989, 198).

Column four (Earth): Ganapathy and Anders (1974, 1181).

Columns five, six, and seven (primitive mantle, Earth oceanic crust, Earth continental crust): Taylor (1982, 382–384).

refractory (high boiling point) elements and compounds separating first and the most volatile (low boiling point) elements and compounds forming last and at the lowest temperatures (Table 1-6). However, the process was probably very complex, with repeated melting, evaporation, and recondensation of nebula dust grains. Wood (1988) points out that the original idea of simple cooling and condensation of a homogeneous nebula is no longer accepted. It is still possible that

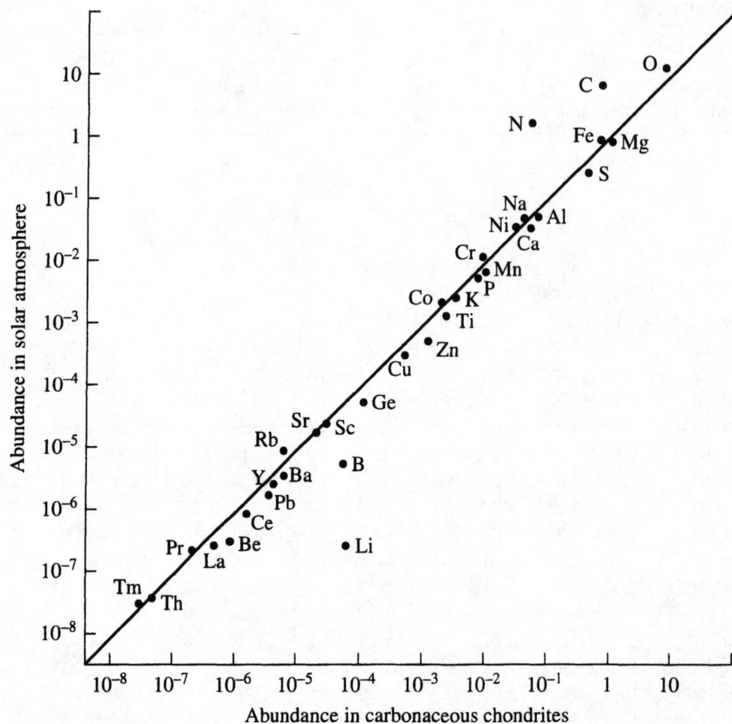

Figure 1-8 Comparison of abundances of elements in Type 1 carbonaceous chondrites with abundances in the solar atmosphere. In both cases, abundances are in terms of atoms per silicon atom. Type 1 carbonaceous chondrites are chondrites with volatile elements in approximately the same proportions as in the Sun. They are representative of primitive material from which the solar system formed. (After Wood, J. A., *The Solar System*. Copyright © 1979. p. 103. Adapted by permission of Prentice Hall, Inc., Upper Saddle River, NJ.)

inclusions in carbonaceous chondrites, which have a high content of refractory elements such as aluminum and titanium, represent the initial material formed from the solar system nebula. As mentioned earlier, carbonaceous chondrites in their final form represent a low-temperature product of the condensation process. Thus these meteorites probably provide evidence on both the early and late stages of the condensation process.

TABLE 1-6 Element Classification Based on Relative Volatility

Refractory	Al	As	Au	Ba	Be	Ca	Cr	Co	Fe
(>1300 K)	Hf	Ir	Li	Mg	Mo	Nb	Ni	Os	P
	Pt	REE	Re	Rh	Ru	Sc	Si	Sr	Ta
	Ti	U	V	W	Y	Zr			
Volatile	Ag	Cs	Cu	F	Ga	Ge	K	Mn	Na
(1300–600 K)	Rb	S	Sb	Se	Sn	Te	Zn		
Very volatile	B	Bi	Br	C	Cd	Cl	Hg	I	In
(<600 K)	Pb	Tl	Inert gases						

Source: Taylor (1982).

There are two important time periods related to meteorite formation. One is the time between the formation of elements by nucleosynthesis and the formation by accretion of the meteorites themselves (formation interval). The other is the time from meteorite formation to the present (meteorite age). The evidence indicates that there were several different episodes of presolar nucleosynthesis. The formation intervals associated with some of these events were short compared to meteorite age. Most measured meteorite ages (using radioactive isotopes—see Chapter Two) range from 4,300 to 4,700 million years.* Isotopic studies have identified formation intervals ranging from a few million years to several hundred million years (Figure 1-9).

Eight achondrite meteorites have ages (about 1,300 million years) that are approximately 3,000 million years younger than those of all other meteorites. These stones are known as SNC meteorites.† Recent research has shown that all the SNC meteorites have chemical and isotopic similarities (suggesting a common parent) and some of their chemical properties are similar to those found in studies of the atmosphere and soil of Mars. It is now generally accepted that these meteorites escaped from Mars as a result of impact on that planet of an asteroid or comet. If the SNC meteorites are from Mars, then they are a significant new source of data that can be used to study planetary formation, evolution, and composition of our solar system. A collection of twenty papers on the SNC meteorites can be found in *Geochimica et Cosmochimica Acta* (no. 6, vol. 50, 1986). A review of more recent work on the SNC meteorites, including a report on the newly discovered ninth member of the group, is given by Harvey et al. (1993).

Stellar Evolution

An estimation of the relative abundance of the elements in the solar system is given in the third column of Table 1-5. The figures for the nonvolatile elements are mainly from analyses of carbonaceous chondrites; it is assumed that they represent the material that aggregated to form the planets of our solar system. Data for the volatile gases is taken from analyses of the solar atmosphere. The third column of Table 1-5 is not based directly on the first two columns, but is an independent estimate of solar-system abundances using data similar to that in the first two columns.

The values in Table 1-5 are estimates that are constantly changing and improving as we obtain new data. Anders and Grevesse (1989) state that their values in column three of Table 1-5 seem to be accurate to ± 10 percent or better. In any event, the relative abundances indicated are probably correct for most of the elements. These abundances can be considered to represent the primitive solar nebula, that is, this was the composition of the solar system at the time it formed. (Some abundances have to be modified to take into account radioactive decay.)

Spectroscopic study of stars other than our Sun indicate a range of compositions representing various stages of stellar evolution. Studies have also been made of the composition of interstellar gas. The most abundant element by far is hydrogen; the abundance of the other elements is not well known. Interstellar dust grains appear to be made up of hydrogen, oxygen, carbon, nitrogen, magnesium, silicon, and iron.

*The range in ages is due to errors in dating methods. It is believed that almost all meteorites actually formed in a very short time—about 4,550 million years ago.

†The name is from the names of the three meteorite classes to which they belong: shergottites, nakhlites, and chassignites. The three class names are from the localities where three of the achondrites fell: Shergotty, India; Nakhla, Egypt; and Chassigny, France.

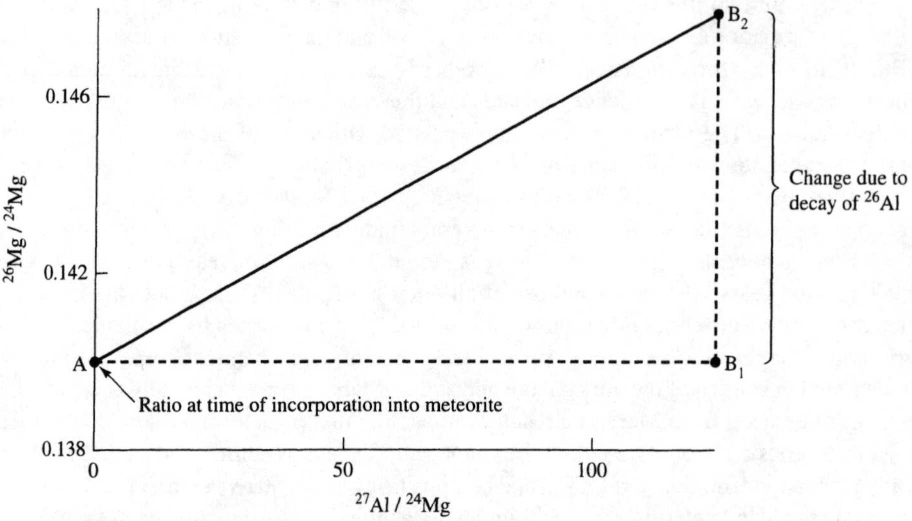

Figure 1-9 Method of determining meteorite formation interval. Assume the radioactive isotope ^{26}Al formed along with other nuclides by stellar evolution and then became part of the solar nebula. The half-life (see Chapter Two) of ^{26}Al is 7×10^5 years and it decays to ^{26}Mg. An aluminum-free mineral that became part of a meteorite would plot at A at the time of meteorite formation. An aluminum-bearing mineral would plot at B_1 initially but, if it contained ^{26}Al in addition to the stable isotope ^{27}Al, it would plot at B_2 at the present time because ^{26}Al would decay over time to ^{26}Mg. If a large amount of time had elapsed from the time of formation of ^{26}Al (time of nucleosynthesis) to the time of meteorites formation (long formation interval), then all of the minerals of meteorites would plot on the horizontal line A–B_1, since all ^{26}Al would have decayed to ^{26}Mg before meteorite formation. Anorthite from an inclusion in the Allende carbonaceous chondrite plots at B_2, indicating a short formation interval. The slope of the line A–B_2 can be used to calculate formation interval. A steep line indicates a short formation interval. (After G. C. Brown and A. E. Mussett, *The Inaccessible Earth*. Copyright © 1981 by Chapman & Hall. Used by permission of Chapman & Hall, London.)

When the relative solar-system abundances of Table 1-5 are plotted against atomic number (Figure 1-10), certain regularities and differences are observed. Any attempt to explain the origin of the elements must explain the following:

1. The extreme abundance of hydrogen and helium (more than 99 percent of all atoms).
2. The general decrease in abundance with increasing atomic number.
3. The relatively low abundance of some elements (such as lithium, beryllium, boron, and scandium) and the relatively high abundance of other elements (such as iron, nickel, and lead).
4. The greater abundance of even atomic number elements as compared to odd atomic number elements (shown by the zigzag pattern of Figure 1-10).

Formation of the elements, as part of stellar evolution, started well before the Earth solidified about 4.6 billion years ago and continues to the present time. Trimble (1977) gives the following sequence for stellar evolution, starting about 15 billion years ago. A hot, dense early universe expanded (the big bang theory of cosmology) and cooled so that it consisted of local

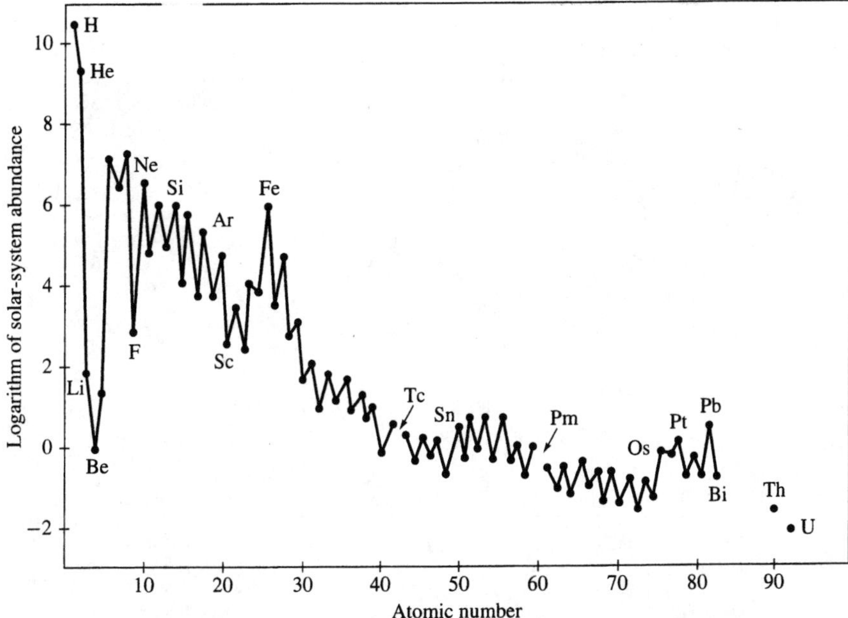

Figure 1-10 Plot of the abundance of the elements in the solar system versus their atomic number. The abundances are expressed as the logarithm of the numbers of atoms of each element relative to 10^6 atoms of silicon. (After G. Faure, *Principles of Isotope Geology,* 2d ed. Copyright © 1986 by John Wiley & Sons. Reprinted by permission of John Wiley & Sons, Inc.) Data from Anders and Ebihara (1982).

concentrations of hydrogen and (to a lesser extent) helium (Silk 1980).* These concentrations grew and developed into galaxies, as stars began to form by gravitational collapse from clouds of gas and dust (nebulas). The center of each stellar mass eventually became hot enough for hydrogen to fuse ("burn") to form helium (Figure 1-11). Hydrogen burning involves a number of different sequences of thermonuclear reactions. An example would be the following:

1. Two hydrogen nuclei (protons) interact to form a deuterium nucleus with one proton and one neutron (^2_1D).
2. The deuterium captures a proton to form helium with two protons and one neutron (^3_2He).
3. Two ^3_2He nuclei react to form ^4_2He (which has two protons and two neutrons) with the release of two protons.

Energy is released by each of these reactions. Element building by such sequences (known as nucleosynthesis) continues to occur, billions of years after the "big bang."

The hydrogen-burning stage of stellar evolution is called the *main sequence stage* (Figures 1-12 and 1-13). This stage ends when the inner core of hydrogen has been used up, resulting in expansion of the outer part and contraction of the inner part of the star. The product is a red-giant star (Figure 1-13), and there is now further fusing of hydrogen, which is newly added to the core by

*A recent modification of the big bang model is known as the inflationary universe model. It postulates enormous, extremely rapid expansion of the early universe in the first fraction of a second of the big bang. All matter and energy were created during this initial, rapid inflation of a very dense, primeval fireball between 10 and 20 billion years ago. The most startling aspect of the inflationary model is the indication that all matter and energy in the observable universe (and in the initial fireball) may have formed from virtually nothing (empty space)!

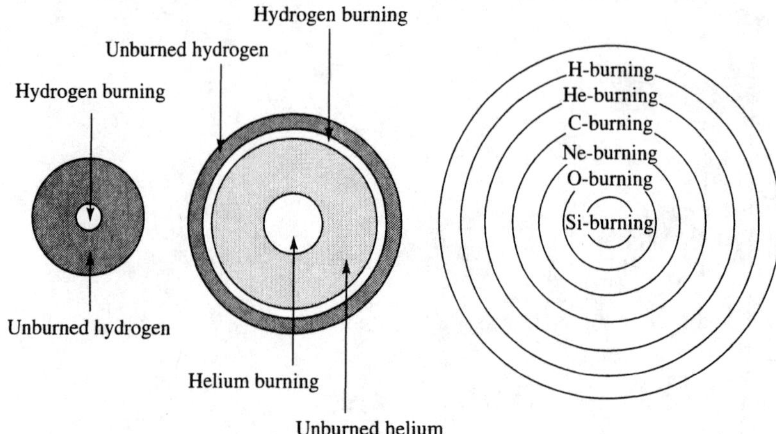

Name of process	Fuel	Products	Temperature
Hydrogen-burning	H	He	60×10^6 K
Helium-burning	He	C, O	200×10^6 K
Carbon-burning	C	O, Ne, Na, Mg	800×10^6 K
Neon-burning	Ne	O, Mg	1500×10^6 K
Oxygen-burning	O	Mg to S	2000×10^6 K
Silicon-burning	Mg to S	Elements near Fe	3000×10^6 K

Figure 1-11 Three stars with progressively hotter nuclear fires. Like our Sun, the star at the left burns hydrogen to form helium in its core; this core is surrounded by unburned fuel. The middle star is burning helium to form carbon and oxygen in its core. This core is surrounded by a layer of unburned helium. Outside of this is a layer in which hydrogen burns to produce helium. Finally there is an outer layer of unburned hydrogen. The star on the right has a multilayered fire. The successive nuclear fires are separated by layers in which no reaction occurs. These layers contain the same fuel as is being consumed in the underlying fire. These layers are depleted in the ingredient being consumed in the overlying fire. The approximate temperatures required to ignite the successive fuels are also given. After Broecker (1985).

contraction. Eventually, the core becomes hot enough to start the burning of helium to produce carbon and oxygen (Figures 1-11 and 1-12). Huge stars remain in the red-giant region (Figure 1-12), while smaller stars now move into the horizontal branch of Figure 1-13. Once the hydrogen and helium in the core are used up, the outer part of a star again expands, along with further collapse of the inner part into the core area. Low-mass stars return to the red-giant region (asymptotic giants of Figure 1-13). No other elements are produced in the lower-mass stars, and they eventually shed their outer layers, moving to the region of Figure 1-13 labeled "nuclei of nebulae" and then to the white-dwarf region. In the larger stars the carbon-oxygen core gets hot enough to burn, producing elements such as magnesium, silicon, and sulfur (Figures 1-11 and 1-12). Further burning of these elements produces a core of iron and nickel. The evolution of a star beyond this point is usually not shown on Hertzsprung-Russell diagrams, such as those in Figures 1-12 and 1-13. Continued collapse in the core area of the star produces a neutron star. In this type of star, addition of neutrons to iron in the core forms heavier elements up to at least plutonium. The reactions go through a series of steps in which elements of increasingly higher atomic number are produced.

The above description is for a first-generation star, one that forms from only hydrogen and helium. However, we know that a neutron star enters a supernova stage in which masses of material,

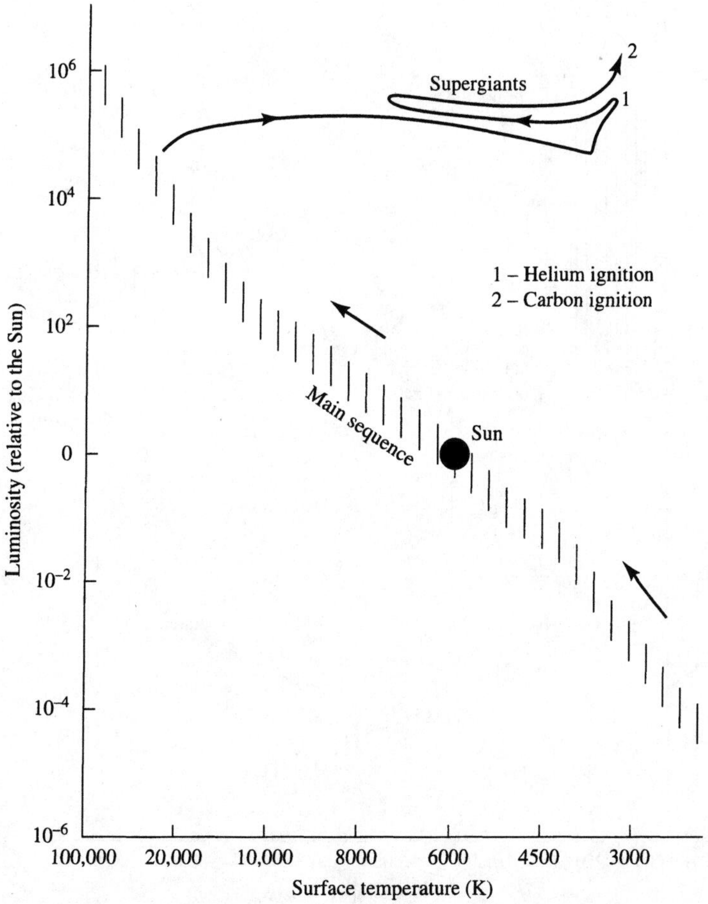

Figure 1-12 Hertzsprung-Russell diagram for a representative group of stars young enough that the entire main sequence is still populated. Vertical scale is luminosity relative to that of our Sun (which plots in the center of the diagram). Horizontal scale is surface temperature of the stars. Evolution of a typical massive star through the supergiant region shown, along with points at which helium and carbon burning start. Most of the time the stellar evolution is spent close to the main sequence and in the supergiant region. Large stars consume their hydrogen more rapidly and therefore spend less time on the main sequence. After Trimble (1977, 78).

including heavy elements, are expelled back into the interstellar medium. Thus second and later generations of stars form from material containing a small percentage of other elements, in addition to the main components hydrogen and helium. Such a starting material allows a greater variety of nuclear reactions during stellar evolution, and thus other elements and isotopes (different forms of an element, such as $^{3}_{2}He$ and $^{4}_{2}He$; see Chapter Two) can be produced that were not produced during the evolution of a first-generation star. Some of the general relationships, but not the details, represented by Figure 1-10 can be explained by the above outline of stellar evolution. For instance, oxygen and carbon are more abundant than all other elements except hydrogen and helium because they are the final products of helium burning. Lithium, boron, and beryllium have a low abundance because their isotopes either do not form in the normal chain of thermonuclear reactions (because they

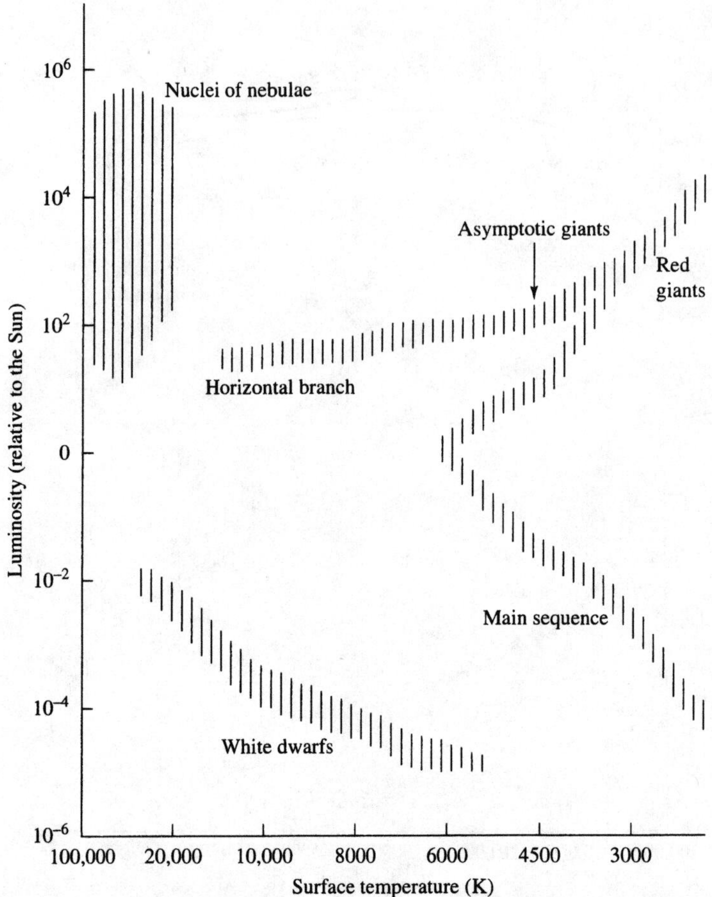

Figure 1-13 Hertzsprung-Russell diagram for a cluster of stars old enough that the stars (like the Sun) are leaving the main sequence. Scales are the same as in Figure 1-12. The evolutionary track of a single star would most likely go from the main sequence up into the red-giant region, down to the horizontal branch (where He is ignited), back up to the red-giant region along the asymptotic branch, then horizontally (and very rapidly) from right to left as a nebula is shed, and finally down into the white-dwarf region. Other paths are possible. White dwarfs eventually become black dwarfs, which do not radiate visible light. After Trimble (1977, 79).

are unstable) or those that do form are used up as part of the hydrogen-burning process. (It is now thought that stable forms of these three elements formed later than those of most other elements by spallation (fragmentation) of still heavier elements as a result of cosmic ray effects.)

It is clear that the abundance of the elements is related to their *nuclear* properties rather than their *chemical* properties. Thus, it is also useful, in studying the origin of the elements, to plot abundances as a function of mass number (protons plus neutrons) for the various nuclides (nuclear species characterized by their number of protons and number of neutrons). This type of plot, although similar in general outline to Figure 1-10, brings out the greater abundance of certain nuclides (such as $^{56}_{26}Fe$), which are the most stable products of nucleosynthesis in stars. A table of the abundance of the nuclides is given by Anders and Grevesse (1989).

TABLE 1-7 Distribution of Stable Nuclides Depending on "Evenness" or "Oddness" of A, Z, and N

A	Z	N	Number of stable nuclides
Even	Even	Even	157
Odd	Even	Odd	53
Odd	Odd	Even	50
Even	Odd	Odd	4
Total number of stable nuclides			264

Source: After Holden and Walker (1972). From G. Faure, *Principles of Isotope Geology*, 2d ed. Copyright © 1986 by John Wiley & Sons. Reprinted by permission of John Wiley & Sons, Inc.

Table 1-7 illustrates the relationships between the stability of nuclides and their mass numbers (A), number of protons (Z), and number of neutrons (N). The most stable (and thus most abundant) nuclides have nearly equal and even (rather than odd) values of N and Z. Thus the iron nuclide $^{56}_{26}Fe$ is more stable and more abundant than $^{55}_{25}Mn$. Also the nuclide $^{55}_{26}Fe$ is unstable and does not occur in nature. The stability of a nuclide is related to the binding energy that holds its nucleus together. Calculations of the binding energies of the nuclides show an increase, as mass number increases to $^{56}_{26}Fe$, and then a steady decrease for higher mass numbers.

Calculations of abundances from equilibrium processes of nuclear burning show fairly good agreement with the measured abundances of the elements and of their nuclides. A review of research on the origin of the elements is given by Fowler (1984).

The Solar System

Our Sun is a typical star that is currently in the main sequence stage of evolution. It is not, however, a first-generation star, but instead is made up of matter produced by earlier stars. The composition of the Sun depends on the material from which it formed and on its present stage of evolution. Column two of Table 1-5 gives an estimate of the composition of the outer part of the Sun from analyses of spectra of the solar atmosphere. This estimate is probably representative of the composition of the entire Sun, since it is believed that the Sun is compositionally homogeneous (with the exception of variations in the relative abundances of hydrogen and helium).

Almost all of the naturally occurring elements have been detected in the Sun's spectrum, with hydrogen and helium by far the most abundant. The properties of the Sun indicate that its outer layers consist of about 73 percent hydrogen, 25 percent helium, 0.5 percent carbon, 0.5 percent oxygen, and 1 percent other elements. Many of the numbers in the first three columns of Table 1-5 are similar for individual elements, and the third column is mainly based on data from the Sun's spectrum (column two) and from meteorite analyses (column one). This third column represents an estimate of the overall composition of our solar system, but actually is also an estimate of the Sun's composition, since the Sun makes up 99.87 percent of the mass of the solar system.

Table 1-8 compares various features of the planets of the solar system. The nine planets can be subdivided into the outer Jovian planets (Jupiter, Saturn, Uranus, and Neptune), the inner terrestrial planets (Mercury, Venus, Earth, and Mars), and the least known and outermost planet, Pluto. The Jovian planets have large masses and low densities, while the terrestrial planets show the reverse relationships. The major chemical difference among the planets lies in the abundance

TABLE 1-8 The Solar System

	Mercury	Venus	Earth	Mars	Jupiter	Saturn	Uranus	Neptune	Pluto
Mean solar distance $\times 10^6$ km	57.9	108.2	149.6	227.9	778.3	1427	2869.6	4496.6	5900
Mass (relative to Earth)	0.055	0.815	1	0.108	317.9	95.2	14.6	17.2	0.0024
Volume (relative to Earth)	0.06	0.88	1	0.15	1316	755	67	57	0.012
Density (g/cm^3)	5.4	5.2	5.5	3.9	1.3	0.7	1.2	1.7	1.14
Atmosphere (main components)	None	CO_2, Ar, N_2	N_2, O_2	CO_2, N_2 Ar	H, He	H, He	H, He, CH_4	H, He, CH_4	None (?)
Known satellites	0	0	1	2	16	21	5	3	1

Source: From Henderson (1982). Reprinted by permission of Butterworth-Heinemann Ltd.

or scarcity of hydrogen and helium. The inner planets are small and do not have the gravitational fields to retain these gases. Also it seems likely that the inner planets were much hotter during their early stages of formation; as a result, hydrogen and helium were lost. The Jovian planets, on the other hand, are mainly composed of gaseous hydrogen and helium and of "ices" (frozen volatiles of H_2O, CH_4, and NH_3). In addition to a scarcity of hydrogen and helium, the terrestrial planets (and meteorites) appear to have low ratios of volatile elements to refractory elements compared to these ratios in the primordial solar nebula (Taylor 1987). (An example is the ratio K/U.)

It is difficult to determine differences that are thought to exist in the overall chemical composition of the terrestrial planets. Based on density, Mercury, Venus, and Mars all evidently have a crust, mantle, and core similar in composition and structure to those of Earth. Several types of data put constraints on the composition of each planet. Some examples are size, density, moment of inertia, heat flow, seismic data, remote-sensing of surfaces, atmosphere composition, magnetic-field data, and so forth. For obvious reasons the greatest amount of data is available for the Earth, and an estimate of its overall composition is included in Table 1-5. Tentative values, based on very limited data, are given in Table 1-9 for the other three terrestrial planets.

As pointed out above, the chemistry of the Jovian planets is very different (Table 1-10). Jupiter and Saturn are composed mostly of hydrogen and helium, while Uranus and Neptune are rich in "ices" of H_2O, CH_4, and NH_3. Note that all of the Jovian planets are mainly made up of the very lightest and most abundant elements in the solar system (H, He, O, C, and N). Models of the interiors of the Jovian planets suggest a rock-ice core, a liquid middle layer of protons and electrons (metallic hydrogen liquid), and an outer liquid layer of mixed molecular hydrogen and helium.

As indicated in Table 1-8, the atmospheres of the planets vary considerably (Wayne 1991). Some of the differences can be explained. For example, Mercury has no atmosphere because of its small size (weak gravity field) and high surface temperature. Chemical differences, such as in the relative abundances of nitrogen and carbon dioxide, are not easily explained. It does appear that planetary atmospheres do not represent nebular gases captured at the time of planet formation, but instead represent the results of subsequent processes, including surface-atmosphere interaction and volcanic outgassing.

The asteroids are located in a belt between the terrestrial and Jovian planets. These small bodies are believed to have undergone very little change since their formation at the same time

TABLE 1-9 Chemical Compositions of Mercury, Venus, Earth, and Mars

Mantle and crust (%)	Mercury	Venus	Earth	Mars
SiO_2	43.6–47.1	40.4–49.8	45	36.8–49.6
TiO_2	0.33	0.2–0.3	0.15	0.2–0.3
Al_2O_3	4.7–6.4	3.4–4.1	3.3	3.1–6.4
MgO	33.7–54.6	33.3–38.0	40	30
FeO	3.7	5.4–18.7	8.0	15.8–26.8
CaO	1.8–5.2	3.2–3.4	2.65	2.4–5.2
Na_2O	0.08	0.1–0.28	0.34	0.1–0.2
Core (%)				
Fe	93.5–94.5	79—89	85.5–86.2	64–88
Ni	5.5	4.8–5.5	4.8–5.5	8.0–8.2
S	0–0.35	1.0–5.1	1.0–9.0	3.5–9.3
O	—	8.0–9.8	0–8.0	0–18.7
Relative masses				
Mantle and crust	32.0–35.2	68.0–76.4	67.6	81–82
Core	64.8–68.0	23.6–32.0	32.4	18–19

Data from Taylor (1982, 385–386 and 403).

as the planets. The compositions of the asteroids can be estimated from observing properties such as spectral reflectivities, and results indicate a regular change in composition with distance from the Sun (Gradie and Tedesco 1982). This result implies accretion of the asteroids from the solar nebula at their present location and it is consistent with models developed for formation of the planets over a period of time by accretion from a primordial solar nebula.

The Moon is significantly different in composition from most other bodies of the solar system (Figure 1-14). It is strongly depleted (relative to Type 1 carbonaceous chondrites) in volatile elements such as potassium, more so than the terrestrial planets. And it seems to have an enrichment of refractory elements, such as uranium, compared to the rest of the solar system. Also the

TABLE 1-10 Composition of Jovian Planets

		Inferred overall composition (%)		
	Compounds observed in atmospheres	H_2, He (gas)	H_2O, CH_4, NH_3 ("ice")	SiO_2, MgO, Fe ("earth")
Jupiter	H_2, He, CH_4, NH_3	82	5	13
Saturn	H_2, CH_4, NH_3	67	12	21
Uranus	H_2, CH_4	15	60	25
Neptune	H_2, CH_4	10	70	20

Source: Wood, J. A., *The Solar System.* Copyright © 1979. p. 88. Adapted by permission of Prentice-Hall, Inc., Upper Saddle River, NJ.

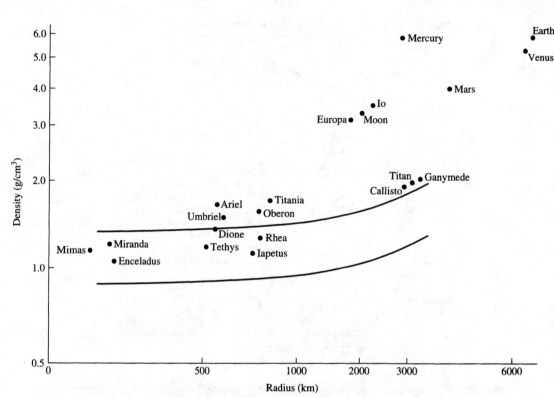

Figure 1-14 The Moon's composition is unusual in comparison to that of most other satellites. Most satellites are varying mixtures of ice (composed of compounds such as H_2O, CH_4, and NH_3) and rock, which give them low densities. Io and the Moon are different. Io is only six Jupiter radii distant from that giant planet, a fact to which it owes both its violent volcanic activity and its high density. Europa, farther out, is about 10 percent ice, while Ganymede and Callisto formed sufficiently far out in the proto-Jovian nebula for water and ice to condense readily. The bone-dry Moon remains the exception among satellites. This plot of density versus radius includes the inner planets and those satellites for which density is known. The lower curve is for the bodies made of 100 percent water ice and the upper curve is for bodies made of 60 percent ice and 40 percent silicate rock. In general, the further planets and satellites are from the Sun, the higher their content of volatile materials (ices). For further information on satellites, see Burns and Matthews (1986). Shown here, in addition to the Jupiter satellites (Io, Europa, Ganymede, and Callisto), are satellites of Saturn (Titan, Rhea, Iapetus, Tethys, Dione, Enceladus, and Mimas) and of Uranus (Titania, Oberon, Umbriel, Ariel, and Miranda). After Taylor (1987).

abundance values of other elements, such as iron, are unique to the Moon. Finally, the Moon's composition clearly does not match that of the Earth or of the primordial solar nebula (as represented by Type 1 carbonaceous chondrites). This geochemical evidence, along with other evidence, suggests a unique origin for the Moon.

The general geology of the Moon is simple: the relatively smooth, dark maria and the more rugged, lighter highlands. The maria consist of a wide variety of basaltic rocks, while the highlands are composed of anorthosite and related rocks. (The lunar surface is covered with a blanket of debris formed by meteorite impacts, and outcrops of bedrock are rare.) The highland rocks are as old as 4.5 billion years, while the maria basalts are as young as 3.0 billion years.

Overall, the Moon is thus not a primitive, relatively unchanged mass similar in composition to carbonaceous chondrites. The lunar crust represents strong chemical fractionation, with a chemical composition distinctly different from the bulk lunar composition.* We know less about the Moon's mantle, but several types of data suggest the mantle is heterogeneous and surrounds a small core.

Wood (1986) reviews the various hypotheses for formation of the Moon. These include: capture of an originally independent body, formation from material of the Earth's mantle by fission, formation from a ring of accreting material surrounding the newly formed Earth, and impact with the Earth by a Mars-sized planetesimal. The impact hypothesis, with the material making up the Moon coming mainly from the impacting body, is one of the best of the current explanations for the chemical composition and physical properties of the Moon (Taylor 1987). According to this hypothesis, the impact, which produced a mass of partly vaporized material in orbit around the Earth, occurred shortly after the Earth accreted from a population of planetesimals.

Taylor (1987) gives the following sequence for the evolution of the Moon. After the impact described above, the material of the Moon accreted and became partially or completely molten (forming a magma ocean). Differentiation of the molten mass resulted in a feldspathic crust and a fractionated lunar mantle. The mantle probably surrounded a small metallic core. Crystallization of the magma ocean produced residual melts of various types, some of which later invaded the crust.† This stage concluded at about 4.4 billion years and was followed by partial melting (due to radioactive heating) of the mantle and ejection of the basaltic lavas of the maria from 4.2 to 3.0 billion years ago. No further igneous activity occurred after this time.

We can develop the following general model for the origin of the solar system (Cameron 1988; Runcorn et al. 1988; Taylor 1992). Approximately 4.6 billion years ago condensation began from a solar nebula made up mainly of gas. Solid material condensed from the gaseous mass as it cooled. This separation would tend to occur in a certain chemical order if simple condensation from a homogenous, totally gaseous nebula occurred. In a classic study, Grossman (1972) determined the order in which various minerals would condense from such a vapor. The composition of the vapor was assumed to be similar to the cosmic abundances of the elements as then known. Wood and Hashimoto (1993) point out that more recent studies of the solar nebula do not assume a high-temperature, homogeneous, wholly vaporized initial state. It is now believed that the vapor had some solid material containing presolar stable isotope anomalies (Clayton 1978). Current models suggest a complex multistage history involving both partial evaporation and condensation; this seems to be required to explain the properties of meteorite chondrules and refractory inclusions. In addition, some mineral assemblages probably formed from gaseous material that differed from the initial material as a result of fractionation of portions of the original nebula. Some of the possible mineral assemblages have been calculated by Wood and Hashimoto (1993); one of these assemblages for a fractionated system is shown in Figure 1-15.

*Taylor (1992, 216) suggests that planetary crusts form in three ways. Primary crusts form due to differentiation resulting from planetwide melting during, or soon after, accretion. An example is the lunar highland crust. Secondary crusts form later in planetary history as a result of partial melting within the mantle of a planetary body. Examples are the lunar maria and the Earth's oceanic crust. Tertiary crusts develop as a result of further melting and differentiation of material composing the secondary crusts. An example is the continental crust of the Earth. Taylor (1992, 268) also points out that the Moon has a thick crust (represented by the highland rocks) that makes up about 12 percent of its volume and that formed shortly after accretion from a deep magma ocean. In contrast, the Earth's continental crust is less than 0.5 percent of its volume and has formed and grown throughout the Earth's history.

†Some residual liquids produced by fractional crystallization of a magma tend to have high concentrations of potassium (K), rare-earth elements (REE), and phosphorus (P). KREEP-rich rock fragments have been found on the Moon, and these are believed to have formed as a result of intrusion of the lunar crust by residual melts.

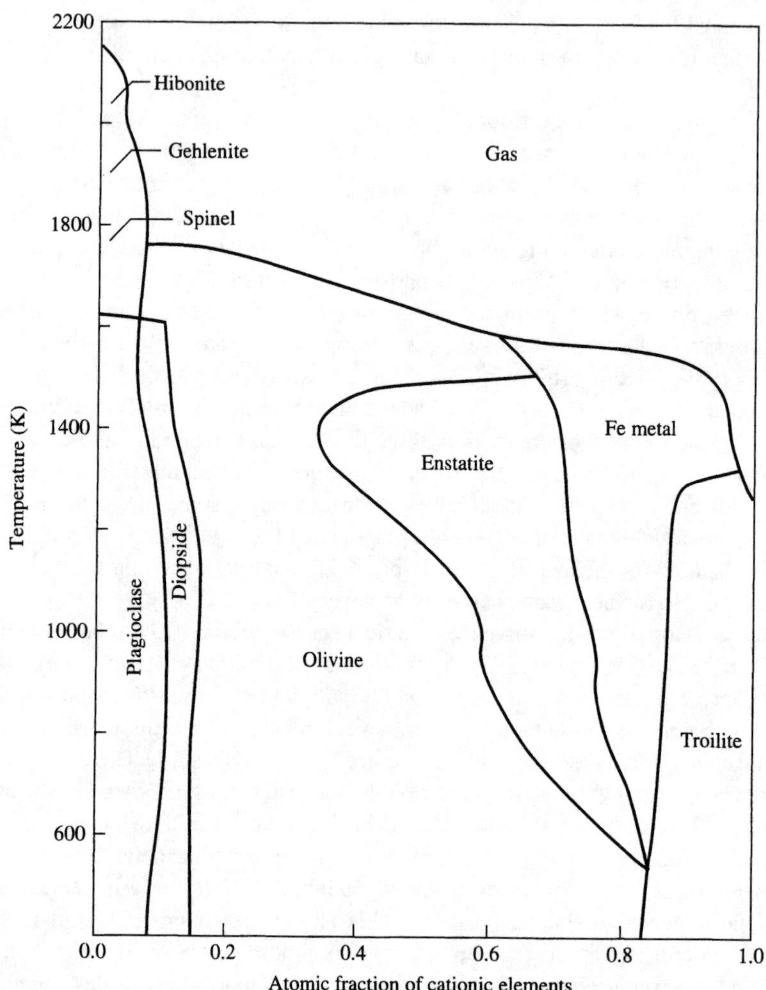

Figure 1-15 Equilibrium mineral sequence for a chemically fractionated nebular system, at a pressure of 10^{-5} bar. The composition of the initial gas is enriched in the refractory dust component (metal, oxides, FeS) relative to other components (carbonaceous matter, ices, H_2 gas) by a factor of 1000 as compared to a system of cosmic composition. If chemical fractionation occurred in the nebula, it would likely involve separation of refractory solids from the gas phase. Widths of the various phase fields, at each temperature, are proportional to the relative numbers of atoms of cationic elements (Si, Mg, Fe, Ca, Al, Na) incorporated in each. The calculations for this diagram determine the set of phases and gas species that minimize the net Gibbs free energy of the system. (Reprinted from *Geochimica et Cosmochimica Acta*, v. 57, J. A. Wood and A. Hashimoto, Mineral equilibrium in fractionated nebular systems, pp. 2377–2388, Copyright 1993, with kind permission from Elsevier Science Ltd., The Boulevard, Langford Lane, Kidlington OX5 1GB, UK.)

Thus it is currently accepted that there was not one unique, simple condensation sequence of minerals as cooling proceeded. This implies that different parts of the solar system have different histories of mineral formation and thus somewhat different final compositions. The mineral assemblages that formed in the various areas of the nebula established the general distribution of the elements in the Earth and in the other terrestrial planets. If, as seems likely, aggregation of solid particles and masses began before the overall condensation process had run its full course, the initial accreting objects would receive an abundance of the early, high-temperature reaction products. All of the planets probably formed over a period of time by accretion of small masses into larger, kilometer-sized masses (planetesimals) and finally to their present size.

As the planets formed, there were probably chemical differences among the materials from which they developed (heterogeneous accretion). Also final compositions were strongly affected by the different locations of the accreting masses with respect to the Sun. The overall process of solar system formation was thus very complex. Whatever actually occurred, it seems clear that the compositions of the planets and their satellites are very different from the composition of the original solar nebula as represented by Type 1 carbonaceous chondrites.

As indicated earlier, much of our knowledge of the early history of the solar system comes from the study of meteorites. A particularly interesting meteorite is Allende, which fell in Mexico in February, 1969. Allende samples are carbonaceous chondrites containing inclusions rich in the refractory elements calcium, aluminum, and titanium. Overall, Allende's composition is very similar to that of the Sun's atmosphere (Table 1-5). Its age, 4.57 billion years, is exactly the same as that generally accepted for the age of the solar system. The minerals listed at the top of Figure 1-15 are similar to those found in some of the Allende inclusions.Thus these inclusions appear to represent the first material to condense from the solar nebula. Other inclusions have minerals representing later stages of condensation. Isotopic study of the inclusions from Allende and other carbonaceous chondrites suggest that they formed from material of two different compositions (Figure 1-16). These two components of the early solar system probably formed prior to the formation of the solar nebula. Thus carbonaceous chondrites not only tell us about compositions and processes in the early solar system, they also provide information about pre-solar system events (Clayton 1993).

The Earth

Seismic evidence has shown that the Earth is divided into three zones—crust, mantle, and core—differing in physical properties and chemical composition. We have direct evidence of the composition of the crust, but must rely mostly on indirect evidence for the other two. Since the mantle and core represent over 99 percent of the Earth's mass, our knowledge of the overall composition of the Earth is limited. In particular, it would be useful to know in greater detail the composition and mineralogy of the upper mantle, where most magmas and earthquakes originate.

Knowledge of the density and magnetic field of the Earth, of the physical properties of the core (from seismic studies), and of iron meteorites has led to general agreement that the core of the Earth is composed predominantly of iron. Nickel is probably also present in about the proportion found in the iron-nickel alloys of meteorites. Geophysical data suggest that a small amount of silicon or sulfur is also likely to be present. The absence of shear waves in the core has led geophysicists to conclude that at least the outer part of the core is liquid. Limitations on the composition of the mantle are not as stringent. Seismic evidence indicates that the mantle is heterogeneous both vertically and laterally. Of rock types known at the surface, only three—dunite, peridotite, and eclogite—have the elastic properties that could produce the observed seismic velocities in the upper mantle. These rocks all have compositions roughly similar to that of chondritic meteorites. The essential elements are magnesium, iron, oxygen, and silicon, and these probably occur in the upper mantle in the form of dense magnesium-iron silicates. Various oxides

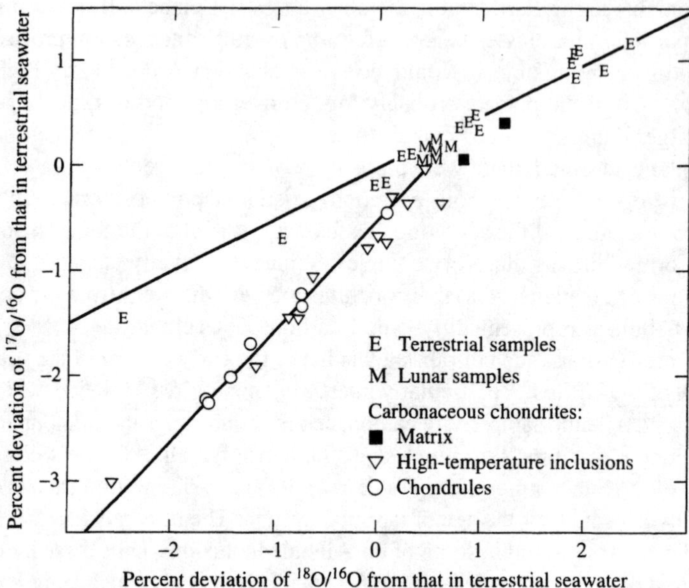

Figure 1-16 Isotopic compositions of oxygen in terrestrial, lunar, and meteoritic materials. Some fractionation of oxygen isotopes occurs during any geochemical process (e.g., separation of crystals from a melt); the nature of these fractionations is understood, and is represented by the upper line in this plot. Differences in the composition of oxygen among terrestrial and lunar samples can be accounted for in this way. However, the components of carbonaceous chondrites display variability in oxygen composition (lower curve) that cannot be explained by geochemical fractionation processes. Instead, these samples are understood to represent mixtures in various proportions of two fundamentally different types of oxygen that entered the early solar system: a component similar to that in Earth and the Moon (upper end of the chondrite curve); and a component very much enriched in ^{16}O, which would plot somewhere far to the lower left of the chondrite curve. For further discussion see Clayton (1993). After Clayton et al. (1973, 485–488). (From Wood, J. A. *The Solar System*. Copyright © 1979. Adapted by permission of Prentice Hall, Inc., Upper Saddle River, NJ.)

(for example, perovskite) may become abundant in the deeper mantle. Seismic discontinuities in the mantle are most likely due to mineral transitions caused by increasing pressure with depth. Examples of pertinent changes from low- to high-pressure forms are enstatite to olivine (spinel structure) plus stishovite (high-density SiO_2), and albite to jadeite plus quartz.

Assuming that the core and mantle have the average composition of iron meteorites and chondritic meteorites, respectively, the overall composition of the Earth can be calculated, since the sizes of these zones are known from seismic data. (The crust does not have to be considered in this calculation because of its small contribution to the total mass.) A more sophisticated approach involves estimation of the types and amounts of condensation products from the solar nebula. Ganapathy and Anders (1974) present a model of the Earth's composition using seven different condensation products similar in composition to various types of chondritic meteorites. Particularly important components are (1) early, refractory-rich condensate; (2) nickel-iron; and (3) magnesium silicates. The compositions of these three components are given by (1) inclusions in carbonaceous chondrites; (2) iron meteorites; and (3) ordinary chondrites. Proportions of the various components can be estimated using geochemical data, such as the probable uranium content of the Earth at the time of its formation (determined from heat-flow and other data). Because element concentration ratios appear to have remained constant during various nucleosynthetic processes (as shown by analyses of primitive solar system samples—meteorites and lunar samples—with a range of concentrations), these ratios (from estimates of solar system composition) can be used to determine the proportions in the primitive Earth of the various condensates from the solar nebula.

An example of the above approach is the use of the volatile/refractory ratio Tl/U to get the proportion of late-forming, volatile-rich condensate containing volatile elements such as thallium.* The results of this approach to determining the Earth's composition are given in column four of Table 1-5. No matter what approach is taken to estimate the Earth's bulk composition, there is general agreement that the Earth is made up of oxygen, iron, silicon, and magnesium, with lesser amounts of nickel, sulfur, and aluminum. All the other elements are insignificant in abundance.

Also given in Table 1-5 is an estimate of the Earth's primitive mantle (present mantle plus the crust and not including the core). Estimates of the composition of the mantle can be obtained as discussed above for the whole Earth or they can be modified (or independently determined) by using data such as that from ultramafic nodules found in volcanic rocks. These nodules appear to be mantle fragments brought to the surface by intruding magmas, and thus they represent direct samples of the mantle. Note, however, that the nodules represent a very limited sample from only the upper region of the mantle. Also it should be pointed out that the composition of the present mantle is different from that of the primitive mantle due to formation of the Earth's crust (see next section). The lower mantle may have a similar composition to that of the upper mantle, or it may be very different in composition. As mentioned above, the evidence so far indicates that the mantle is heterogeneous both vertically and horizontally. Further discussion of the mantle's composition can be found in Chapter Eight.

The Earth's Crust

When we compare the estimated abundance of the elements in the Earth with their abundance in the Earth's continental and oceanic crusts, a completely different situation is found (Table 1-5). The Earth is a highly fractionated body, with elements such as potassium strongly concentrated in the crust. This is particularly true for the very old continental crust, but also holds true to a lesser extent for the young (less than 250 million years) oceanic crust. This fractionation may represent the result of heterogeneous accretion from a condensing solar nebula, as described earlier, or it may represent segregation within an initially homogeneous body. In the latter model, the initially solid mass became molten owing to heat from radioactive decay. Molten drops of iron and other heavy metals sank to the core, and light elements moved upward to form the crust. The body then cooled and became solid, with the exception of part of the core. Thus the formation of the

*The reasoning is as follows. We select a Tl/U ratio for the solar nebula using estimates of solar system composition. The uranium content of the primitive Earth is estimated from heat-flow and other data. Knowing the absolute amount of uranium in the primitive Earth and assuming the Tl/U ratio has not changed with time, we can determine the absolute amount of thallium in the primitive Earth. This in turn tells us what proportion of the late-forming, thallium-containing condensate (whose composition has been calculated using vapor-solid equilibria and an estimate of original nebula composition) would be needed to form the primitive Earth. Knowing the proportion of the late-forming, volatile-rich condensate tells us the proportions of the other elements occurring with thallium in the condensate. Finally, this information can be used with other information obtained in the same way to estimate the overall composition of the Earth.

core, mantle, and crust may have been the result of accretion, without a liquid stage, or it may have occurred as part of a molten period after accretion.

Other models of the Earth's early history postulate formation of the continental crust significantly later than the initial development of a core and primitive mantle (Taylor 1992). In any case, further segregation of elements in the upper mantle and in the crust occurred owing to magma formation (as a result of the convective movement of mantle material to lower pressure and of radiogenic heating), movement, and fractional crystallization. The continents may represent the altered remnants of an originally global continental crust, or they may have formed by tectonic additions to initial continental nuclei (Taylor and McLennan 1985).

Our knowledge of the chemical composition of the Earth's continental and oceanic crusts comes from rock analyses and from geophysical evidence on the structure of the crust. The crust averages 35 kilometers in thickness under the continents and 10 kilometers beneath the oceans, and is not homogeneous either vertically or laterally. Thus assumptions must be made about the size and compositions of various subdivisions of the crust. The gross subdivisions usually considered are (1) sediments and sedimentary rocks at the surface of the crust, (2) granitic material in the upper part of the continental crust, and (3) basaltic material in the lower part of the continental crust and making up the oceanic crust. A more detailed approach is to divide the crust into geological divisions—shield areas, geosynclinal areas, continental shelves, etc.—and estimate the composition and volume of each. A major problem is uncertainty concerning the material making up the lower part of the continental crust. It is interesting to note that, despite the increased knowledge of rock compositions and crustal structure, current estimates of the composition of the crust are very similar to the first well-documented estimate made by F. W. Clarke and H. S. Washington in 1924. This pioneering effort assumed that the crust has a composition equal to the average composition of igneous rocks. Table 1-5 contains a "modern" estimate of the composition of the continental and oceanic crusts.

Oxygen is the dominant element in the crust (about 47 percent by weight and 94 percent by volume). The other major element is silicon, which is about 28 percent by weight (but less than 1 percent by volume because of the small size of the silicon atom). The processes involved in the formation of the present crust separated certain elements from the main body of the Earth and concentrated them in the crust. Even though many of these elements are not abundant in the crust, they have a higher concentration there than in the rest of the Earth. If we compare abundances of various elements in the primitive mantle and in the continental crust (Table 1-5), we find the following elements strongly enriched in the crust: Li, Be, Na, Al, K, Ti, Rb, Sr, Y, Zr, Nb, Cs, Ba, La, the rare earths, Hf, Pb, Th, and U. These elements, known as incompatible elements, are preferentially concentrated in a magma formed by partial fusion of a source rock (see Chapter Eight).* The separation of crustal material from the mantle was (and is) primarily an igneous process, and these elements were apparently brought to the crust as it formed. Many of these same elements are found in concentrations greater than their crustal average in certain types of igneous rock (such as pegmatite). Sedimentary processes produce a further fractionation of the elements at the surface of the crust. The chemical compositions and general chemistry of igneous, sedimentary, and

*Many of the incompatible elements also belong to the group of elements known as the LIL (large ion lithophile) elements. These elements tend to occur in silicate minerals (as opposed to sulfide and native iron phases) and have ionic radii larger than those of the common rock-forming elements. Barium and uranium are examples of LIL elements. See the discussion of lithophile elements later in this chapter. Appropriate data are not shown in Table 1-5 for other incompatible elements: B, F, In, I, Ta, W, Tl, and Bi.

metamorphic rocks are discussed in Chapters Seven through Nine. Also discussed in these chapters are surface variations in the composition of the Earth's crust.

The Oceans and the Atmosphere

The compositions of seawater and the Earth's atmosphere are very different from that of the Earth's solid material (Tables 1-11 and 1-12; note that these two tables have abundances expressed

TABLE 1-11 Composition of Seawater

Major constituent	Concentration (parts per million by weight)	Log residence time (years)
Sodium	10,760	7.7
Magnesium	1,294	7.0
Calcium	412	5.9
Potassium	399	6.8
Strontium	7.9	6.6
Chloride	19,350	7.9
Sulfate	2,712	6.9
Bicarbonate	145	4.9
Bromide	67	8.0
Boron	4.6	7.0
Fluoride	1.3	5.7

Minor constituent	Concentration (parts per million by weight)	Log residence time (years)
AG	8×10^{-6}	5.0
JAl	2×10^{-3}	2.0
Ar	4×10^{-3}	—
As	3×10^{-3}	5.0
Au	5×10^{-6}	5.0
Ba	20×10^{-3}	4.5
Be	6×10^{-6}	2.0
Bi	20×10^{-6}	—
Cd	1×10^{-4}	4.7
Ce	1×10^{-6}	4.7
Co	50×10^{-6}	4.5
Cr	3×10^{-4}	3.0
Cs	4×10^{-4}	5.8
Cu	3×10^{-4}	4.0
Dy	9×10^{-7}	—
Er	9×10^{-7}	—
Eu	1×10^{-7}	—
Fe	2×10^{-3}	2.0
Ga	30×10^{-6}	4.0
Gd	7×10^{-7}	—
Ge	6×10^{-5}	—
He	72×10^{-7}	—
Hf	$<8\times10^{-6}$	—
Hg	2×10^{-5}	5.0
Ho	2×10^{-7}	—
I	6×10^{-2}	6.0

TABLE 1-11 *(cont.)*

Minor constituent	Concentration (parts per million by weight)	Log residence time (years)
In	1×10^{-7}	—
Kr	21×10^{-5}	—
La	34×10^{-7}	—
Li	18×10^{-2}	6.3
Lu	1×10^{-7}	—
Mn	2×10^{-4}	4.0
Mo	1×10^{-2}	5.0
N	15	6.3
Nb	15×10^{-6}	—
Nd	28×10^{-7}	—
Ne	12×10^{-5}	—
Ni	6×10^{-4}	4.0
P	6×10^{-2}	4.0
Pa	2×10^{-13}	—
Pb	3×10^{-5}	2.6
Pr	6×10^{-7}	—
Ra	1×10^{-6}	6.6
Rb	12×10^{-2}	6.4
Sb	33×10^{-5}	4.0
Sc	6×10^{-7}	4.6
Se	9×10^{-5}	4.0
Si	2.9	3.8
Sm	4×10^{-7}	—
Sn	1×10^{-5}	—
Ta	$<25\times10^{-7}$	—
Tb	1×10^{-7}	—
Th	15×10^{-7}	2.0
Ti	1×10^{-3}	4.0
Tl	1×10^{-5}	—
Tm	2×10^{-7}	—
U	33×10^{-4}	6.4
V	25×10^{-4}	5.0
W	4×10^{-6}	—
Xe	47×10^{-6}	—
Y	1×10^{-6}	—
Yb	8×10^{-7}	—
Zn	3×10^{-3}	4.0
Zr	3×10^{-5}	—

Source: H. D. Holland, *The Chemistry of the Atmosphere and Oceans.* Copyright ©1978 by John Wiley & Sons. Reprinted by permission of John Wiley & Sons, Inc.

in units of ppm by weight and ppm by volume, respectively, whereas those in Table 1-5 are in terms of atoms). The ocean and atmosphere originally consisted of liquid and gaseous material separated from the main mass of the Earth during its early history (Holland 1984). Since that time, their composition has been altered as a result of chemical weathering, biological activity, and igneous eruptions. In historical time, humans have begun to alter their composition. Both seawater and the atmosphere are quite homogeneous compared to the Earth and the Earth's crust.

TABLE 1-12 Composition of the Atmosphere at Ground Level

Component	Concentration (ppm by volume)	Residence time
Nitrogen and its compounds		
Nitrogen, N_2	78.084×10^4	4×10^8 years for cycling through sediments
Nitrous oxide, N_2O	0.33	5 to 50 years
Nitric oxide, NO; Nitrogen dioxide, NO_2	0.001	Less than one month
Ammonia, NH_3	0.006–0.020	Approximately one day
Oxygen, hydrogen, and their compounds		
Oxygen	20.946×10^4	6000 years for cycling through the biosphere
Hydrogen	0.55	4 to 7 years
Water	40–40,000	
Ozone	0.01–0.03	
Carbon compounds		
Carbon dioxide, CO_2	320	About ten years for cycling through the biosphere
Carbon monoxide, CO	0.06–0.2	0.5 year
Methane, CH_4	1.4	2.6 to 8 years
Sulfur compounds		
Sulfur dioxide, SO_2	0.001–0.004	Hours to weeks
Hydrogen sulfide, H_2S	<0.0002	Less than one day
Noble gases		
Helium	5.24	2×10^6 years for escape
Neon	18.18	Largely accumulating
Argon	9340	Largely accumulating
Krypton	1.14	Largely accumulating
Xenon	0.087	Largely accumulating

Note: Values given are for atmospheric background, unaffected by local pollution.

Source: H. D. Holland, *The Chemistry of the Atmosphere and Oceans.* Copyright © 1978 by John Wiley & Sons. Reprinted by permission of John Wiley & Sons, Inc.

In seawater, the relative proportions of the major elements (Cl, Na, Mg, S, Ca, and K) are essentially constant (calcium and magnesium vary slightly from place to place in their ratios with other elements). This is true even though the total amount of dissolved salts (salinity) is very low and shows some variation from place to place. Those elements that are involved in biological activity (such as C, O, N, and P) vary in concentration with depth, mainly because photosynthesis can occur near the surface but not in deeper water. Other factors are the high concentrations of organisms in surface layers and chemical exchange between surface water and the atmosphere. The major dissolved gases in seawater (N_2, O_2, and CO_2) all show large variations in concentration. Only 15 elements occur in seawater with a concentration over 1 part per million (ppm) by weight (note that hydrogen and oxygen are not listed in Table 1-11). Some of the elements that occur in trace amounts show large regional variations; others appear to have a constant value.

Also listed in Table 1-11 are estimates of the residence time of individual elements. The residence time is the average time an atom of a particular element spends in the ocean. Various amounts of the elements are continually entering and leaving the ocean. Knowledge of element residence times and of the incoming and outgoing fluxes (masses transported per unit time) for the ocean allows us to identify the major factors controlling ocean chemistry. Note that the residence times in Table 1-11 vary from 10^2 to 10^8 years. Holland (1978) has pointed out that the residence times of the dissolved constituents are a small fraction of the age of the Earth. Thus most elements have moved through the ocean many times over since it first formed. Further discussion of the chemistry of the ocean can be found in Chapters Three, Four, Six, and Seven.

As with the ocean, the atmosphere can be viewed as a reservoir with incoming and outgoing fluxes. The residence times listed in Table 1-12 vary from less than one day to over one million years. Some elements, such as oxygen and nitrogen, have had constant concentrations during the recent history of the Earth. Other elements and compounds, with very short residence times, show highly variable concentrations. Examples are H_2O, NH_3, and SO_2.

The composition of the Earth's early atmosphere was determined by the processes that formed the Earth from planetary material (Holland 1984). The gases originally exhaled from the evolving Earth went through a series of reactions to reach some type of equilibrium state. Gases that were probably abundant at that time include NH_3, H_2, H_2O, CH_4, and CO. With the passage of time, additional material was added by volcanic activity. This early atmosphere was anaerobic (no free oxygen) and the most critical change with time was the development of free oxygen as a result of biological photosynthesis. This change took place in late Precambrian time, and study of the rocks formed since then indicates that the composition of the atmosphere has remained relatively constant for about the last one billion years. The composition of the atmosphere has regulated chemical weathering and the formation of sedimentary rocks through its control of the oxygen and carbon dioxide contents of surface waters. The history of seawater probably parallels that of the atmosphere; it formed from water vapor escaping from the solid earth and changed in composition with time until stabilizing in the late Precambrian.

GEOCHEMICAL MODELING

In the previous section the concept of residence time was introduced. This can be mathematically defined as

$$T = \frac{A}{(dA/dt)} \tag{1-1}$$

where A is the total amount of an element (or species) in a reservoir and dA/dt is the rate of influx or efflux of that element. This concept is very useful in modeling the natural chemical cycles of the elements. The general procedure involves setting up a model consisting of arbitrarily defined reservoirs (storage areas) and estimation of the fluxes into and out of the reservoirs. One of several different possible models might be appropriate in a given case. For example, a steady state model would apply when there is one reservoir of constant size and the input and output for the reservoir are constant and equal. The residence time for each element for a steady state model is constant over time. For other models the residence times may continuously increase or decrease and they may be different with respect to input and output (because the fluxes are different for input and output).

For non-steady-state models, the concept of replacement (renewal) time is useful. This is the time needed, at a given rate of influx, to fill a reservoir to its present volume. Completely replacing the material in a reservoir, without any mixing of old and new masses, is unlikely in most natural systems. Thus we also have to consider the mixing rate, that is, how fast the old and new masses are uniformly intermingled. Fast mixing and a long residence time results in a homogeneous reservoir. Slow mixing and a short residence time results in a reservoir with different compositions in different parts of the reservoir.

An example of a simple model has been given by Berner and Berner (1987, 247–249) for the study of a lake. The lake is represented by a single reservoir, with input of a dissolved substance (Q) from streams and one surface outlet. The following abbreviations are used:

F_i = rate of water inflow from streams (volume per unit time)
F_o = rate of water outflow through outlet (volume per unit time)
M = total mass of the dissolved substance Q in the lake
R_p = rate of removal of Q via adsorption and precipitation with sedimentation to the bottom (mass per unit time)
R_d = rate of addition of Q via desorption and/or dissolution of solids (mass per unit time)
Ci = concentration of Q in stream water (mass per unit volume)
C = concentration of Q in lake water
t = time

If the lake has a constant volume of water (steady state), the rate of change of mass of Q with time in the lake is

$$\frac{\Delta M}{\Delta t} = C_i F_i - C F_o + R_d - R_p \tag{1-2}$$

If there is a steady state also with respect to Q, $\Delta M/\Delta t = 0$, then

$$C_i F_i - C F_o + R_d - R_p = 0 \tag{1-3}$$

Now let us apply this model to a specific problem. Assume that a small lake, noted for good fishing, starts receiving pesticide-containing stream water as a result of new pesticide use at local farms. It is known that a concentration of 0.02 g/m^3 of the pesticide is toxic to fish. Pesticide concentration in the stream water draining the farms that surround the lake is found to be 0.1 g/m^3. Study of the lake sediment shows removal of the pesticide by adsorption on clay particles could be as much as 0.05 g/day. No additions of pesticide to the lake water by desorption or other process is found to occur. Given this data, what will the concentration of the pesticide in the lake be if a steady state situation for the pesticide is reached (input balances output) as pesticide use continues?

We need two other numbers to use our model: stream inflow rate is found to be 1000 m^3/day and lakewater (streamwater plus rainfall) outflow rate is found to be 1200 m^3/day. Using equation 1-3, and solving for C,

$$C = \frac{C_i F_i - R_p}{F_o} = \frac{0.1\ g/m^3 \times 1000\ m^3/day - 0.05\ g/day}{1200\ m^3 day}$$

$$= \frac{(0.1 - 0.05)\ g/day}{1.2\ m^3/day} = 0.042\ g/m^3$$

This tells us that the water will become toxic to the fish in the lake, since our result is slightly more than two times the toxic value of 0.02 g/m^3.

In many natural systems it is useful and/or necessary to deal with more than one reservoir and to have a large number of fluxes. Figure 1-17 illustrates a model of the atmospheric sulfur cycle. This cycle is probably close to a steady state system. (Most of the element cycling at the Earth's surface over recent geologic time has, to a first approximation, been under steady state conditions.) In Figure 1-17 the atmosphere is divided into two reservoirs, air over the continents and air over the oceans. For each reservoir the total influx and total outflux are equal. This model can be used to describe and quantify atmospheric pollution. Berner and Berner (1987) have concluded that, for the world as a whole, about forty percent of the sulfur input to the atmosphere is anthropogenic (mainly due to the burning of fossil fuels). As shown in Figure 1-17, about seventy-five percent of the sulfur input to the air over the continents is due to pollution.

Walker (1991) provides many examples of the application of computational methods to geochemical systems. A set of examples in his book involves the exchange of carbon dioxide between the oceans and the atmosphere. For example, a question that could be asked about this system is: what would be the response of the oceans and atmosphere to the gradual input over time of fossil fuel carbon dioxide? This is a concern because of the possibility of global warming due to the increase of carbon dioxide in the atmosphere. Using a computer program relating dissolved and solid carbon species in the oceans to the partial pressure of carbon dioxide in the atmosphere, Walker obtained the results shown in Figure 1-18. The partial pressure of carbon dioxide increases from an assigned value of one for the current value to about four times the current value in A.D. 2350 and then slowly decreases to a value about 2.5 times the current value in a few thousand years. This mathematical simulation is based on a number of assumptions and may have serious errors. However, it illustrates that quantitatively simulating part or all of a given geochemical system or cycle allows us to predict, at least roughly, future changes in our environment.

GEOCHEMICAL CLASSIFICATION OF THE ELEMENTS

The first suggestion of a similarity between the bulk composition of the Earth and average meteorite composition was made in 1850. In 1922, V. M. Goldschmidt suggested details of this analogy, many of which are still accepted today. Goldschmidt postulated that the three main phases of meteorites—native iron, sulfide, and silicate—also represented the major zones of the Earth, with the Earth having an outer silicate layer (and crust), a sulfide-oxide layer, and an iron core. (The existence of a sulfide-oxide layer as such in the mantle has been ruled out by later geophysical studies. Material of this composition may be dispersed in the lower mantle.) Goldschmidt also suggested that the distribution of the elements in the Earth is similar to that among the meteorite phases. It was known in 1922 that precious metals such as gold and platinum occurred in the iron phase of at least some meteorites. Goldschmidt felt that *all* such material would contain these metals, and he suggested that the scarcity of such metals in the Earth's crust was due to their concentration in the iron core of the Earth at the time of the primary differentiation of the Earth into crust, mantle, and core.

Chemical analyses of the three phases of meteorites by Goldschmidt and other investigators confirmed his prediction of the association of the precious metals with the iron phase. In addition, data were obtained for elements concentrated in the sulfide and silicate portions of meteorites. A similar development of three phases was known in the smelting of ores, the products

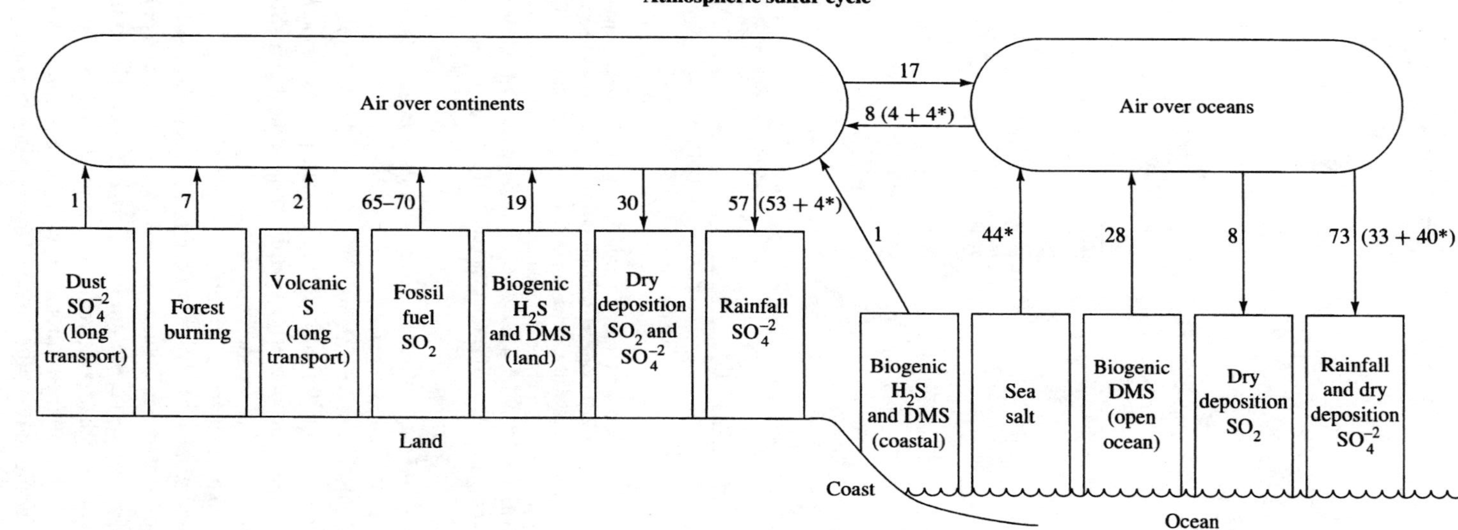

Figure 1-17 The atmospheric sulfur cycle. Values are fluxes in Tg S/yr (Tg = 10^{12} g). Values denoted by an asterisk refer to sea salt. DMS = dimethyl sulfide. (From Berner, E. K., and Berner, R. A. *The Global Water Cycle*. Copyright © 1987. Adapted by permission of Prentice Hall, Inc., Upper Saddle River, NJ.)

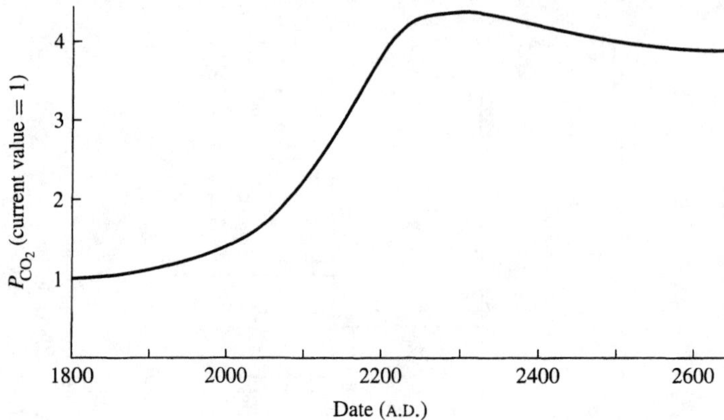

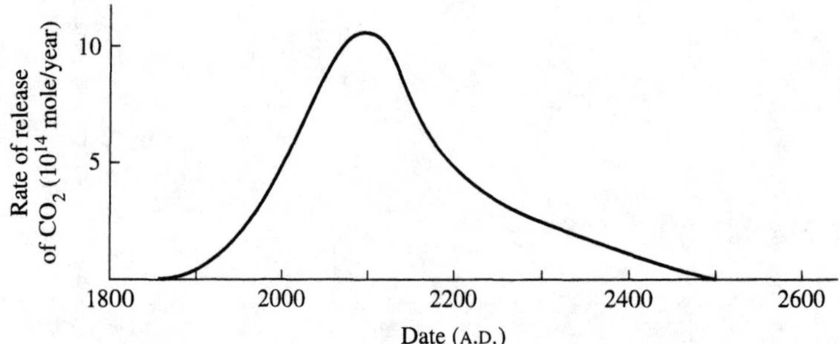

Figure 1-18 Predicted increase in atmospheric carbon dioxide due to fossil burning. The model assumes a rate of release of carbon dioxide to the atmosphere that increases to the year 2100 and then decreases (bottom graph). This mathematical simulation of the response of the Earth's oceans and atmosphere to the assumed input of carbon dioxide predicts that the pressure of carbon dioxide will increase until the year 2350 and then slowly decrease (top graph). Data for bottom graph from Broecker and Peng (1982). Simulation from Walker (1991).

being metal, matte (sulfide), and slag (silicate). Studies on these materials tended to confirm the distribution of the elements found in meteorites.

In his 1922 paper, Goldschmidt used the terms *siderophile*, *chalcophile*, and *lithophile* to describe elements concentrated in the iron, sulfide, and silicate phases, respectively. He also used the term *atmophile* for those elements occurring mainly in the atmosphere. Studies on meteorites and smelter products allowed the elements to be assigned to these groups (Table 1-13). Some elements appeared to belong to more than one group, and Goldschmidt suggested that their behavior in a given situation would depend on the relative abundances of iron, sulfur, and oxygen. For example, when an excess of iron is available, as in meteorites, nickel and cobalt are siderophile (occur with the iron phase), whereas chromium is chalcophile and goes to the sulfide phase. In the Earth's crust, iron is not so abundant relative to total oxygen and sulfur. In this case, the nickel and cobalt are chalcophile and combine with sulfur, whereas chromium is lithophile and combines with oxygen. The distribution of a given metal *M* would depend on the relative stability of iron silicate, iron sulfide, *M* silicate, and *M* sulfide in a given environment.

TABLE 1-13 Geochemical Classification of the Elements

Siderophile	Chalcophile: Meteorites	Chalcophile: Crust	Lithophile	Atmophile
C P Fe	P S V	S Fe Co	Li Be B O F	H C N
Co Ni Ge	Cr Mn Cu	Ni Cu Zn	Na Mg Al Si Cl	O Cl Br
Mo Ru Rh	Zn As Se	Ga As Se	K Ca Sc Ti V	I Inert gases
Pd Sn Ta	Ag Cd Te	Mo Rh Pd	Cr Mn Br Rb Sr	
Re Os Ir		Ag Cd In	Y Zr Nb I Cs	
Pt Au		Sb Te Hg	Ba La Hf Ta W	
		Tl Pb Bi	Th U Rare earths	

Note: Some elements are listed in more than one group.

Goldschmidt's classification is qualitative only and cannot be used to explain many of the details of element occurrence and distribution in the crust of the Earth. It does provide a useful hypothesis for the primary fractionation of the Earth and serves as a useful general geochemical classification of the elements. The general validity of the classification is due to the similarity in electron configuration of the various groups of elements. Siderophile elements are those whose outer (valence) electrons under certain chemical conditions are not readily available for combination with other elements. Thus they tend to occur in the native state. Their electrons are not readily available because their electronic structure is such that the positive charge on the nucleus can exert a strong pull on the outer electrons. The electrons of lithophile and chalcophile elements are more available; thus these elements tend to form ions. However, the ions of these two groups have different properties (again, because of electronic structure), with chalcophile elements tending to form covalent bonds with sulfur, and lithophile elements tending to form more ionic bonds with oxygen, as in the silicate minerals. Lithophile elements, in general, have their valence electrons outside a shell of eight electrons, whereas chalcophile elements have valence electrons that are outside a shell of 18 electrons. There are gradational changes in these electronic properties from element to element; thus some elements show a stronger tendency to belong to a given group than do other elements. Gold, for example, is more siderophile than molybdenum. Similarly, one element may show a tendency to belong to two groups. An example is manganese, which is both lithophile and chalcophile.

SUMMARY

Knowledge of the periodic table and of the electronic structure of the elements is the first step in studying the abundance and distribution of the elements. The most general characteristic of the elements is a periodic recurrence of similar properties with increasing atomic number. The chemical behavior of a particular element depends mainly on the number and location of its outer electrons. Geologists combine knowledge of the chemical behavior of the elements with chemical analyses of natural materials and with knowledge of natural conditions to explain the abundance and distribution of the elements.

Meteorites are an important source of information about the formation, early history, and composition of planetary bodies. Most meteorites are made up of silicate minerals (stony meteorites) and most of these (chondrites) contain rounded masses known as chondrules. The most primitive (least altered since formation) meteorites are the carbonaceous chondrites. Their composition is representative of the material from which the bodies of the solar system formed.

The most abundant elements in the universe are hydrogen and helium. All the other elements probably formed by nuclear reactions (in stars) that started with these two simple elements. Our Sun is composed mostly of hydrogen and helium, and the composition of other bodies in the solar system has been governed by their mode of origin, particularly the extent of their loss of volatile elements. In addition, the composition of the different parts of the Moon, meteorites, and the planets has been determined by the extent of chemical fractionation of nonvolatile elements.

In the case of the Moon, chemical evidence suggests that its unique composition is due to formation from material resulting from the impact of a planetesimal with the Earth. Fractionation of the initial impact-produced mass resulted in the present lunar crust and mantle. Prior to the Moon's formation most other planetary bodies, including the Earth, formed from a mixture of diverse materials that resulted from reactions involving solids and vapor in a cooling solar nebula. Later these bodies underwent various degrees of chemical fractionation.

The materials of greatest interest in geochemistry—the Earth's crust, oceans, and atmosphere—represent extreme fractionation of the primordial material from which the Earth was formed. This fractionation has been, and is, governed by chemical environment and by the chemical properties of the elements, which can be used to explain Goldschmidt's empirical geochemical classification of the elements. Variations in the distribution and abundance of the elements on the Earth can be understood in terms of their chemical cycles. Various models can be used to describe both natural and anthropogenic changes and to determine the residence times of the elements in specific reservoirs.

QUESTIONS

1. Group the following electron configurations in pairs that would represent similar chemical properties of their atoms.
   (a) $1s^2\,2s^2\,2p^6\,3s^2\,3p^5$
   (b) $1s^2\,2s^2\,2p^6\,3s^2$
   (c) $1s^2\,2s^2\,2p^3$
   (d) $1s^2\,2s^2\,2p^6\,3s^2\,3p^6\,4s^2\,3d^{10}\,4p^6$
   (e) $1s^2\,2s^2$
   (f) $1s^2\,2s^2\,2p^6$
   (g) $1s^2\,2s^2\,2p^6\,3s^2\,3p^3$
   (h) $1s^2\,2s^2\,2p^5$
2. The element manganese can have valences of +2, +3, +4, +6, and +7. In terms of electron structure, explain why these different valences are possible.
3. Some chemists feel that zinc, cadmium, and mercury should not be considered transition elements. Suggest a reason for this.
4. In 1964 the Russians announced that they had made (discovered) element 104 by nuclear bombardment techniques. Scientists at the University of California (Berkeley) claimed that they were the first ones to produce element 104 and that the Russian tests on their material did not prove it to be element 104. One way of checking this material is to compare its compounds with those of another element that should form similar compounds. Name an element whose chemical properties should be similar to those of element 104.

5. Why does group 8 of the periodic table have three elements in the same period (example: iron, cobalt, and nickel), whereas no other group has more than one element in the same period (excluding the rare earths and actinides)?
6. Why are the inert gases the most stable chemical elements? Why are they sometimes referred to as the group 0 (zero) elements?
7. The elements lanthanum, cerium, yttrium, and thorium all substitute for each other in the mineral monazite. Use the periodic table to explain why this happens.
8. Given that the abundance of platinum in the Earth's crust is 0.0001268 percent by weight, what is the atomic abundance of platinum relative to silicon $= 10^6$ atoms? Assume that the SiO_2 content of the crust is 60 percent by weight.
9. Which one of the meteorite classes listed in Table 1-3 may represent material from the core-mantle interface of a parent body of differentiated meteorites?
10. As pointed out in the section on the periodic table, the rare-earth elements (REE) are all very similar in their behavior. However, the natural abundances of the REE vary by a factor of 30. In order to study overall REE geochemical behavior, the element-to-element variations can be removed ("normalized") by dividing REE concentrations in a sample by REE concentrations in ordinary chondrites. When the REE values for some differentiated meteorites are normalized, the values obtained are similar except for europium (see diagram below). Suggest a reason for the unique behavior of europium as compared to the other REE.

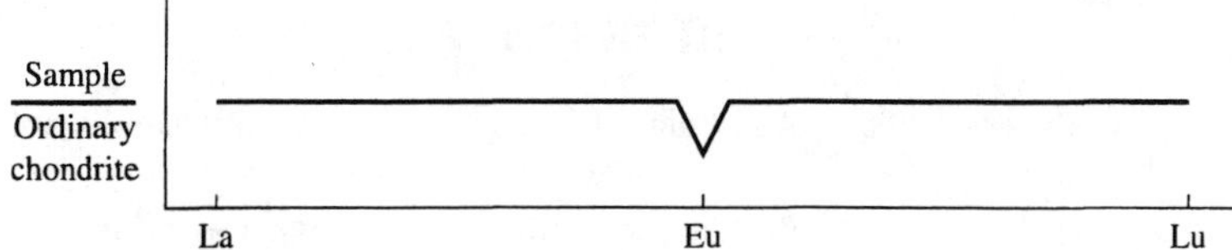

11. Since the separation (partitioning) of oxygen isotopes is related to the mass differences of the isotopes (see Chapter Two), suggest a reason why the slope of the upper line in Figure 1-15 is 1:2. What does the different slope of the lower line (it is 1:1) tell us about the early history of the solar system?
12. **(a)** What element is formed during the main sequence stage of stellar evolution?
    (b) Which one of the following elements would you expect not to occur in a first-generation white-dwarf star: carbon, sodium, boron?
    (c) List the following elements in the order of abundance you would expect to find in a massive star from the core to the outside: helium, silicon, hydrogen, iron, oxygen.
13. Barium and rubidium exhibit similar geochemical behavior due to similar ionic radii. It has been found that lunar anorthosite (highland) samples and mare basalt (lowland) samples all have similar barium-rubidium ratios (Ba/Rb equals approximately 60). In contrast, differentiated meteorites have very different barium-rubidium ratios (about 130) from the lunar samples and from the ratios of undifferentiated meteorites (about one). Using these results, what conclusions could be reached about the origin and history of the Moon?
14. In carbonaceous chondrites the matrix contains iron that is oxidized and present in oxides and silicates. The chondrules of these meteorites have iron mainly in the form of metallic iron. What does this tell us about the way in which these chondrites formed?
15. The overall compositions of the inner planets may be very similar. If so, then the variations in density (Table 1-8) are probably due to the oxidation state of iron at the time of planet formation. How could oxidation-reduction conditions, in addition to condensation temperatures (Figure 1-15), affect the final density of the planets?
16. An assumption can be made that no relative fractionation of refractory elements (listed in Table 1-6) occurred during accretion of the Earth. The abundance of major refractory elements such as calcium and

aluminum in the Earth's mantle can be estimated from various sources, such as analyses of mantle-derived ultramafic nodules in volcanic rocks. How could the ratio of calcium or aluminum in the primitive mantle to calcium or aluminum in Type I carbonaceous chondrites (Table 1-5) be used to estimate the *primitive* mantle abundances of refractory trace elements such as barium and the rare-earth elements? (The abundances of calcium and aluminum in the present mantle can be assumed to be nearly equal to their abundances in the primitive mantle, that is, present mantle plus crust.) Even if the above assumptions are correct, what would be a major source of error in using this method to estimate the abundance of refractory trace elements in the *present* mantle?

17. The mixing time of deep ocean water is about 1000 years (Holland 1978). Therefore elements with residence times significantly longer than 1000 years should be homogeneously distributed in the oceans. This is true for many elements such as sodium and magnesium. However, other elements with long residence times, such as phosphorus and silicon, are not homogeneously distributed. Suggest a reason for this.

18. Using data from Figure 1-17 calculate the residence time of fossil fuel SO_2 in the atmosphere over the continents. Assume that the total amount of SO_2 in the atmosphere over the continents is 0.2×10^{11} moles. Compare your answer with the value given in Table 1-12.

19. On a periodic table of the elements mark four regions for the following four groups of elements: lithophile, chalcophile, siderophile, and atmophile.

REFERENCES

ANDERS, E., and M. EBIHARA. 1982. Solar-system abundances of the elements. *Geochim. Cosmochim. Acta,* v. 46:2363–2380.

ANDERS, E., and N. GREVESSE. 1989. Abundances of the elements: meteoritic and solar. *Geochim. Cosmochim. Acta,* v. 53:197–214.

BERNER, E. K., and R. A. BERNER. 1987. *The Global Water Cycle.* Prentice-Hall, Englewood Cliffs, NJ.

BROECKER, W. S. 1985. *How to Build a Habitable Planet.* Eldigio Press, Palisades, NY.

BROECKER, W. S., and T.-H. PENG. 1982. *Tracers in the Sea.* Eldigio Press, Palisades, NY.

BROWN, G. C., and A. E. MUSSETT. 1981. *The Inaccessible Earth.* George Allen & Unwin, London.

BURNS, J. A., and M. S. MATTHEWS, eds. 1986. *Satellites.* University of Arizona Press, Tucson, AZ.

CAMERON, A. G. W. 1988. Origin of the solar system. In *Ann. Rev. of Astronomy and Astrophysics,* G. Burbridge, D. Layzer, and J. G. Phillips, eds., v. 26:441–472.

CHANG, R. 1986. *General Chemistry.* Random House, New York.

CLAYTON, R. N. 1978. Isotopic anomalies in the early solar system. *Ann. Rev. Nuclear and Particle Sciences,* J. D. Jackson, ed., v. 28:501–522.

CLAYTON, R. N. 1993. Oxygen isotopes in meteorites. In *Ann. Rev. of Earth and Planetary Sciences,* v. 21, G. W. Wetherill, ed., Annual Reviews Inc., Palo Alto, CA. 115–149.

CLAYTON, R. N., L. GROSSMAN, and T. K. MAYEDA. 1973. A component of primitive nuclear composition in carbonaceous meteorites. *Science,* v. 182:485–488.

EVANS, R. C. 1964. *Crystal Chemistry.* 2d ed. Cambridge University Press, New York.

FAIRBRIDGE, R. W., ed. 1972. *The Encyclopedia of Geochemistry and Environmental Sciences.* Dowden, Hutchinson & Ross, Stroudsburg, PA.

FAURE, G. 1986. *Principles of Isotope Geology.* 2d ed. John Wiley & Sons, New York.

FLEISCHER, M., ed. 1962 (on), The Data of Geochemistry, 6th ed. U.S. Geol. Survey Profess. Paper 440.

FOWLER, W. A. 1984. The quest for the origin of the elements. *Science,* v. 226:922–935.

FYFE, W. S. 1964. *Geochemistry of Solids.* McGraw-Hill Book Co., New York.

GANAPATHY, R., and E. ANDERS. 1974. Bulk compositions of the moon and earth, estimated from meteorites. *Proc. 5th Lunar Sci. Conf.*, Lunar and Planetary Institute, Houston, TX.

GLASS, B. P. 1982. *Introduction to Planetary Geology.* Cambridge University Press, Cambridge.

GOLDSCHMIDT, V. M. 1954. *Geochemistry.* Oxford University Press, London.

GRADIE, J., and E. TEDESCO. 1982. Compositional structure of the asteroid belt. *Science,* v. 216:1405–1407.

GRAHAM, A. L., A. W. R. BEVAN, and R. HUTCHISON. 1985. *Catalog of Meteorites,* 4th ed. University of Arizona Press, Tucson, AZ.

GREENWOOD, N. N., and A. EARNSHAW. 1984. *Chemistry of the Elements.* Pergamon Press, Oxford.

GROSSMAN, L. 1972. Condensation in the primitive solar nebula. *Geochim. Cosmochim. Acta,* v. 36:597–619.

GROSSMAN, L. and J. W. LARIMER. 1974. Early chemical history of the solar system. *Rev. Geophys. Space Phys.,* v. 12:71–101.

HARRISON, E. R. 1981. *Cosmology.* Cambridge University Press, Cambridge.

HARTMAN, W. K., R. J. PHILLIPS, and G. J. TAYLOR, eds. 1986. *Origin of the Moon.* Lunar and Planetary Institute, Houston, TX.

HARVEY, R. P., M. WADHWA, H. W. MCSWEEN, JR., and G. CROZAZ. 1993. Petrography, mineral chemistry, and petrogenesis of Antarctic Shergottite LEW88516. *Geochim. Cosmochim. Acta,* v. 57:4769–4783.

HEIKEN, G. H., D. T. VANIMAN, and B. M. FRENCH. 1991. *Lunar Sourcebook: A User's Guide to the Moon.* Cambridge University Press, Cambridge.

HENDERSON, P. 1982. *Inorganic Geochemistry.* Pergamon Press, Oxford.

HOLDEN, N. E., and F. W. WALKER. 1972. *Chart of the Nuclides,* 11th ed. General Electric Co., Schenectady, NY.

HOLLAND, H. D. 1978. *The Chemistry of the Atmosphere and Oceans.* John Wiley & Sons, New York.

HOLLAND, H. D. 1984. *The Chemical Evolution of the Atmosphere and Oceans,* Princeton University Press, Princeton, NJ.

KERRIDGE, J. F., and M. S. MATTHEWS, eds. 1988. *Meteorites and the Early Solar System.* University of Arizona Press, Tucson, AZ.

KING, E. A., ed. 1983. *Chondrules and their Origins.* Lunar and Planetary Institute, Houston, TX.

MASON, B. 1979. Cosmochemistry Part 1. Meteorites. U.S. Geol. Survey Profess. Paper 440-B-1.

MCSWEEN, H. Y., JR. 1988. Chondritic meteorites and the formation of the planets. *Am. Scientist,* v. 77:146–153.

MERRILL, P. W. 1963. *Space Chemistry.* University of Michigan Press, Ann Arbor, MI.

RUNCORN, S. K., G. TURNER, and M. M. WOOLFSON, eds. 1988. The solar system: chemistry as a key to its origin. *Proc. Trans. Royal Soc. London A.,* v. 325:391–641.

SILK, J. 1980. *The Big Bang.* W. H. Freeman and Company, New York.

TAYLOR, S. R. 1982. *Planetary Science: A Lunar Perspective.* Lunar and Planetary Institute, Houston, TX.

TAYLOR, S. R. 1987. The origin of the moon. *Am. Scientist,* v. 75:469–477.

TAYLOR, S. R. 1992. *Solar System Evolution.* Cambridge University Press, Cambridge.

TAYLOR, S. R., and S. M. MCLENNAN. 1985. *The Continental Crust: Its Composition and Evolution.* Blackwell Scientific, Palo Alto, CA.

TRIMBLE, V. 1977. Cosmology: man's place in the universe. *Am. Scientist,* v. 65:76–86.

VAN SCHMUS, W. R., and J. A. WOOD. 1967. A chemical-petrologic classification for the chondritic meteorites. *Geochim. Cosmochim. Acta,* v. 31:747–765.

WALKER, J. C. G. 1991. *Numerical Adventures with Geochemical Cycles*. Oxford University Press, New York.

WASSON, J. T. 1985. *Meteorites*. W. H. Freeman and Company, New York.

WASSON, J. T. 1993. Constraints on chondrule origins. *Meteoritics*, v. 28: 14–28.

WAYNE, R. P. 1992. *Chemistry of Atmospheres*, 2d ed. Oxford University Press, New York.

WEDEPOHL, K. H., ed. 1969 (on). *Handbook of Geochemistry*, vols. 1 and 2 (parts 1–4). Springer-Verlag, Berlin.

WILHELMS, D. E. 1987. The Geologic History of the Moon. U.S. Geol. Survey Profess. Paper 1348.

WOOD, J. A. 1979. *The Solar System*. Prentice-Hall, Inc., Englewood Cliffs, NJ.

WOOD, J. A. 1986. Moon over Mauna Loa: a review of hypotheses of formation of Earth's moon. In *Origin of the Moon*, W. K. Hartman, R. J. Phillips, and G. J. Taylor, eds. Lunar and Planetary Institute, Houston, TX. 17–56.

WOOD, J. A. 1988. Chondritic meteorites and the solar nebula. In *Ann. Rev. Earth and Planetary Sciences*, v. 16, G. W. Wetherill, ed., Annual Reviews Inc., Palo Alto, CA. 53–72.

WOOD, J. A., and A. HASHIMOTO. 1993. Mineral equilibrium in fractionated nebular systems. *Geochim. Cosmochim. Acta*, v. 57: 2377–2388.

Isotope Geology

Studies of the abundance and distribution of the isotopes in nature have become an integral part of geology. Applications of these studies have been made in practically every subfield of geology, and all geologists should be familiar with the basic concepts of isotope geology. We shall begin with a review of the chemistry of isotopes. This will be followed by a discussion of the use of radioactive (unstable) isotopes in age dating of natural materials and in studying the Earth's history. Finally, the role of stable isotopes in geologic studies will be examined.

ISOTOPES AND THE PERIODIC TABLE

The *periodic law* (represented by the periodic table), as originally conceived by Mendeleev in 1869, involved a correlation between the atomic weights and the chemical properties of the elements. It was eventually found that such an arrangement resulted in anomalies for three pairs of elements, argon and potassium, cobalt and nickel, and tellurium and iodine. When placed in the periodic table according to increasing atomic weight, these elements appeared to be reversed with respect to the positions suggested by their chemical properties.

This problem was solved as a result of the proposal by F. Soddy in 1913 of the existence of isotopes. He used the term *isotopes* for two or more substances of different mass but occupying the same place in the periodic table. The existence of isotopes of the various elements was soon confirmed, and this, along with other theoretical and experimental work, led to the understanding that the fundamental numbers of the periodic table are the atomic numbers and not the atomic weights. It was also shown that the atomic number of an element was equal to its nuclear charge and was characteristic of that element. A modern definition is that the *isotopes of an element are atoms of the same atomic number but with different atomic masses.*

It is now known that there are three isotopes of potassium, ^{39}K, ^{40}K, and ^{41}K. Their percentages of abundance in nature are 93.3, 0.01, and 6.7 percent, respectively. The numbers 39, 40, and 41 are known as *atomic mass numbers* and are equal to the total number of protons and neutrons in each type of isotope. All three isotopes have 19 protons (and all three therefore have the chemical properties of potassium), but the number of neutrons is different (20, 21, and 22, respectively). The *atomic weights* of the elements are defined as the average relative weights (masses) of their atoms for the natural isotopic composition of each element. (The mass of an atom is mainly determined by the number of protons and neutrons).* Potassium has an atomic weight of 39.098, which is close to the mass (39) of the isotope making up most of the potassium in nature. Argon has three isotopes, ^{36}Ar (0.34 percent), ^{38}Ar (0.06 percent), and ^{40}Ar (99.6 percent). Most argon consists of ^{40}Ar; thus its atomic weight, 39.948, is close to 40. As a result of the relative abundances of the potassium and argon isotopes, argon has a lower atomic number but a higher atomic weight than potassium. The other two anomalies of the early periodic table, cobalt-nickel and tellurium-iodine, can be explained in a similar manner.

Soddy's isotope hypothesis was largely a result of research by many people in the new field of radioactivity. Radioactivity was discovered in 1896, one year after the discovery of X rays. Within six years the general nature of radioactivity was understood, even though the concept of the atomic nucleus had not yet been developed. (We now know that radioactivity is a result of changes in the nucleus of an atom, whereas X rays can be caused by a change in either the nucleus or the electron shells.) By 1913 a large number of *radioelements* (substances produced as a result of radioactivity) had been discovered. In fact, too many had been discovered to fit into the periodic table. Furthermore, many of these radioelements seemed to be chemically similar to each other or to other elements not associated with radioactivity. Soddy's concept of isotopes solved these bothersome relationships. It was soon shown, for example, that the stable product resulting from uranium decay and ordinary lead represent different combinations of isotopes of the same element. In a short time all the radioelements were found to fit nicely into the periodic table.

RADIOACTIVITY AND GEOCHRONOLOGY

Radioactivity can be defined as the spontaneous adjustment of the nuclei of unstable atoms to a more stable state. Radiation is given off as a result of changes in the nuclei of such atoms. Alpha and beta particles are high-speed particles corresponding to helium nuclei and electrons, respectively. The third type of radiation (gamma rays) consists of electromagnetic waves that are similar in character to X rays. (Gamma rays contain more energy and have a shorter wave length than X rays.) In addition to these three types of nuclear radiation, X rays may also be produced as a result of radioactive decay. The radioactive properties of an isotope depend upon its nucleus and not upon its electron structure. In the simplest terms, the reason for the occurrence of radioactivity is the existence of nuclei in a state of high energy that is not stable.

Since alpha particles have two protons, the effect of the removal of an alpha particle from a nucleus is to change the atom to two places lower in the periodic table. The mass number is decreased by four units, because alpha particles have two neutrons in addition to two protons.

*By international agreement, one atomic mass unit (also known as a dalton) is defined as a mass exactly equal to one-twelfth the mass of one carbon-12 atom. A hydrogen atom (^{1}H) has a mass that is 8.4 percent that of ^{12}C and thus the atomic mass of hydrogen (which is almost 100 percent ^{1}H; see Table 2-13) is 0.084×12 or 1.008. The atomic mass of carbon is 12.01 because natural carbon is a mixture of the isotopes ^{12}C, ^{13}C, and ^{14}C.

Beta decay may be defined as any radioactive process in which the mass number remains unchanged while the atomic number is changed. Two types are important in determining geologic ages. *Ordinary beta decay* involves a neutron changing to a proton with the emission of an electron. The atomic number of the atom is increased by one unit. In the other type of beta decay, *electron capture*, a proton is changed to a neutron as a result of the movement of an electron from the K shell of an atom to its nucleus. The atomic number is decreased by one unit. No beta radiation results from this process, but X rays and gamma rays are produced as a result of filling the vacancy created in the K shell.

Gamma rays are produced as a part of any change in which the resultant nucleus is left in an excited state. When the nucleus returns to its ground state, energy in the form of gamma radiation is given off. There is no change in atomic number or mass number as a result of gamma radiation. When radioactive decay does cause a change in atomic number, the original (unstable) isotope is known as the *parent*, and any new isotope formed is called a *daughter*.

The experimentally measured rates of decay of radioactive isotopes indicate that such decay is a first-order reaction; that is, the number of atoms that decompose in unit time is proportional to the number present:

$$\frac{dN}{dt} = -\lambda N \tag{2-1}$$

where N is the number of parent atoms at a time t, and λ is a constant characteristic of the decay of a given radioactive isotope. The constant λ can be expressed as number of decayed atoms per atom per unit of time. For example, if the rate constant is 0.01 seconds^{-1}, the equation says that during each second, $1/100$ of the atoms present would decompose. The minus sign indicates that N is decreasing.

If we integrate the decay equation from $t = 0$ to t and from N_0 to N (where N_0 equals the number of atoms present when $t = 0$), and take the antilog, we get

$$N = N_0 e^{-\lambda t} \tag{2-2}$$

A particularly useful term in discussing rate of decay is *half-life*, the time required for half of an initial number of atoms to break down. We can get a relationship between the decay constant λ and the half-life $t_{1/2}$ by substituting in the previous equation $t = t_{1/2}$ and $N = \frac{N_0}{2}$:

$$\frac{1}{2} = e^{-\lambda t_{1/2}} \quad \text{or} \quad t_{1/2} = \frac{\ln 2}{\lambda} \tag{2-3}$$

The possible usefulness of radioactivity to geologists was pointed out in 1904 by E. Rutherford. At that time B. B. Boltwood was studying the ratio of lead to uranium in various minerals. He concluded that the lead was a product of the radioactive breakdown of uranium, and suggested in 1905 that this ratio could be used to determine the absolute age of rocks and minerals since it would vary with age. Boltwood's basic premise was correct, although we now know that it is isotopes and not elements that must be measured. Ideally, all that is necessary is to measure the amount of parent (for instance, ^{238}U), and daughter (^{206}Pb, which is the product of the decay of ^{238}U), assuming that the rate of breakdown λ is constant and is known experimentally. The rate of decay for almost all radioactive transitions has been found to be constant and unaffected by changes in temperature, pressure, chemical surroundings, or physical surroundings. Very small changes have been produced experimentally in the rate of electron capture, which is a rather special type of decay, since it includes changes in electron distribution as well as nuclear composition.

Returning to the equation

$$N = N_0 e^{-\lambda t} \tag{2-2}$$

we can substitute for N the number of parent atoms currently present in a mineral (P) and for N_0 the number of parent atoms originally present when the mineral formed (P_0). The number of daughter atoms now present (D) is equal to P_0 minus P, since the breakdown of one atom of P produces one atom of D. Thus $P_0 = P + D$, and the decay equation becomes

$$P = (P + D)e^{-\lambda t} \tag{2-4}$$

or

$$D = P(e^{\lambda t} - 1) \tag{2-5}$$

Rearranging, and solving for t,

$$\frac{P}{P + D} = e^{-\lambda t} \qquad e^{\lambda t} = \frac{P + D}{P}$$

$$\lambda t = \ln\left(1 + \frac{D}{P}\right) \qquad t = \frac{1}{\lambda}\ln\left(1 + \frac{D}{P}\right) \tag{2-6}$$

Substituting values for D/P and λ, we can solve for t, the time of formation (age) of the sample. The ratio D/P is a ratio of numbers of atoms.

Geochronology is a general term for the subfield of geology that is concerned with determination of the isotopic age and isotopic history of geologic materials. There are a number of naturally occurring radioactive isotopes, several of which have been found useful in geochronology (Faure 1986; Geyh and Schleicher 1990). Because of their abundance, the uranium isotopes were the first to be studied in detail. The two most abundant uranium isotopes, ^{235}U and ^{238}U, were found to be the long-lived parents of separate decay series consisting of a number of short-lived radioactive daughters and ending in the stable isotopes ^{207}Pb and ^{206}Pb, respectively (Tables 2-1 and 2-2). (One important result of this work was discovery of the fact that a nonradioactive element such as lead can exist in isotopic form. This led to the successful search for isotopes of other stable elements.) Note that the half-life (4.47×10^9 years) of ^{238}U is, strictly speaking, the amount of time it takes for half of an initial number of ^{238}U atoms to break down to ^{234}Th. However, it is also essentially the time it takes for half of the ^{238}U atoms to change to ^{206}Pb atoms, since the other half-lives in Table 2-1 all add up to about 0.00033×10^9 years.

A third radioactive series was found, beginning with ^{232}Th and ending with stable ^{208}Pb (Table 2-3). Work on these series revealed that all elements with atomic number greater than 83 (bismuth) occur in nature only as radioactive isotopes.* Most of the other naturally occurring radioactive isotopes involve only a parent isotope and a stable daughter isotope. The isotopic pairs that are most used in geochronology are listed in Table 2-4 and reviewed in the following sections. With the exception of potassium, the pertinent elements are not abundant in common rocks

*One of the elements formed in each of the three series is radon. Radon gas can accumulate in buildings, a situation that can cause lung cancer in residents. The gas is produced by decay of uranium and thorium in rocks underneath buildings and enters through cracks in basement floors and walls. The actual hazard is from the daughter polonium isotopes of radon parent isotopes. Radioactive polonium isotopes stick to the lining of the lungs and irritate the surrounding tissue. Radon's properties are such that it does not stay in contact with lung tissues.

TABLE 2-1 Decay Series of Uranium 238

Isotope	Particle emitted	Half-life
$^{238}_{92}U$	α	4.468×10^9 yr
$^{234}_{90}Th$	β	24.101 days
$^{234}_{91}Pa$	β	1.175 min
$^{234}_{92}U$	α	2.475×10^5 yr
$^{230}_{90}Th$	α	8.0×10^4 yr
$^{226}_{88}Ra$	α	1,622.0 yr
$^{222}_{86}Rn$	α	3.825 days
$^{218}_{84}Po$	α	3.05 min
$^{214}_{82}Pb$	β	26.8 min
$^{214}_{83}Bi$	α (0.04%), β (99.96%)	19.72 min
$^{214}_{84}Po$	α	163.7 μsec
$^{210}_{81}Tl$	β	1.32 min
$^{210}_{82}Pb$	β	22.5 yr
$^{210}_{83}Bi$	β	4.989 days
$^{210}_{84}Po$	α	138.374 days
$^{206}_{82}Pb$		Stable

Source: Data from Russell and Farquhar (1960, 3).

TABLE 2-2 Decay Series of Uranium 235

Isotope	Particle emitted	Half-life
$^{235}_{92}U$	α	0.7038×10^9 yr
$^{231}_{90}Th$	β	25.6 hr
$^{231}_{91}Pa$	α	3.43×10^4 yr
$^{227}_{89}Ac$	β	22.0 yr
$^{227}_{90}Th$	α	18.6 hr
$^{223}_{88}Ra$	α	11.2 days
$^{219}_{86}Rn$	α	3.917 sec
$^{215}_{84}Po$	α	1.83×10^{-3} sec
$^{211}_{82}Pb$	β	36.1 min
$^{211}_{83}Bi$	α (99.7%), β (0.3%)	2.16 min
$^{211}_{84}Po$	α	0.52 sec
$^{207}_{81}Tl$	β	4.79 min
$^{207}_{82}Pb$		Stable

Source: Data from Russell and Farquhar (1960, 4).

TABLE 2-3 Decay Series of Thorium 232

Isotope	Particle emitted	Half-life
$^{232}_{90}Th$	α	1.401×10^{10} yr
$^{228}_{88}Ra$	β	6.7 yr
$^{228}_{89}Ac$	β	6.13 hr
$^{228}_{90}Th$	α	1.90 yr
$^{224}_{88}Ra$	α	3.64 days
$^{220}_{86}Rn$	α	54.53 sec
$^{216}_{84}Po$	α	0.158 sec
$^{212}_{82}Pb$	β	10.67 hr
$^{212}_{83}Bi$	α (33.7%), β (66.3%)	60.48 min
$^{212}_{84}Po$	α	0.29 μsec
$^{208}_{81}Tl$	β	3.1 min
$^{208}_{82}Pb$		Stable

Source: Data from Russell and Farquhar (1960, 5).

and minerals. Typical concentrations of these elements in various rocks and minerals are given in Table 2-5. Note that, with the exception of carbon, nitrogen, and calcium, all the parent and daughter elements in Table 2-4 are large-ion lithophile (LIL) elements (see Chapters One and Eight). These elements have, over time, been preferentially concentrated in the Earth's crust and, as will be discussed later in this chapter, can be used to reconstruct the chemical evolution of the Earth's mantle and crust.*

RUBIDIUM-STRONTIUM SYSTEM

The rubidium-strontium system involves the simple decay of radioactive ^{87}Rb to ^{87}Sr. Only two isotopes of rubidium occur in nature (85 and 87), with ^{87}Rb making up about 28 percent of the total rubidium. Natural strontium consists of four stable isotopes, ^{84}Sr, ^{86}Sr, ^{87}Sr, and ^{88}Sr. The expression for the decay of ^{87}Rb to ^{87}Sr comes by substitution into equation (2-5):

$$^{87}\mathrm{Sr} = {}^{87}\mathrm{Rb}(e^{\lambda t} - 1) \quad (2\text{-}7)$$

The total number of ^{87}Sr atoms in a mineral or rock sample will be

$$^{87}\mathrm{Sr}_{\mathrm{measured}} = {}^{87}\mathrm{Sr}_{\mathrm{initial}} + {}^{87}\mathrm{Rb}(e^{\lambda t} - 1) \quad (2\text{-}8)$$

where $^{87}Sr_{initial}$ represents atoms originally present and taken into the sample at the time of its formation. Since ratios of atoms are more easily and accurately measured than absolute abundances, we can write equation (2-8) in the form

$$\left(\frac{^{87}\mathrm{Sr}}{^{86}\mathrm{Sr}}\right)_{\mathrm{measured}} = \left(\frac{^{87}\mathrm{Sr}}{^{86}\mathrm{Sr}}\right)_{\mathrm{initial}} + \frac{^{87}\mathrm{Rb}}{^{86}\mathrm{Sr}}(e^{\lambda t} - 1) \quad (2\text{-}9)$$

Note that a nonradiogenic isotope has been used in the denominator.

*Large ions that occur as trace elements do not fit well in the dense crystal structures of mantle minerals and are partitioned to magmas formed by partial melting. Thus magmas from the mantle have, throughout the Earth's history, brought LIL elements to the crust where they occur in less dense minerals.

TABLE 2-4 Radioactive Systems Used in Geochronology

Parent/daughter	Type of decay	λ (yr^{-1})	Half-life (yr)	Effective range (yr) (T_0 = age of earth)	Isotopic abundance of parent and daughter	Typical materials dated
$^{238}U/^{206}Pb$	8 Alpha + 6 beta	1.55125×10^{-10}	4.468×10^9	$10^7 - T_0$	0.9928 g/g U 0.252 g/g Pb	Zircon, uraninite, monazite, lead-bearing minerals
$^{235}U/^{207}Pb$	7 Alpha + 4 beta	9.8485×10^{-10}	0.7038×10^9	$10^7 - T_0$	0.0072 g/g U 0.215 g/g Pb	Zircon, uraninite, monazite, lead-bearing minerals
$^{232}Th/^{208}Pb$	6 Alpha + 4 beta	4.9475×10^{-11}	14.010×10^9	$10^7 - T_0$	1.00 g/g Th 0.520 g/g Pb	Zircon, uraninite, monazite, lead-bearing minerals
$^{87}Rb/^{87}Sr$	Beta	1.42×10^{-11}	48.8×10^9	$10^7 - T_0$	0.278 g/g Rb 0.07 g/g Sr	Biotite, muscovite, microcline, whole rocks
$^{40}K/^{40}Ar$	Electron capture	0.581×10^{-10} 4.962×10^{-10}	1.250×10^9 (total)			
$^{40}K/^{40}Ca$	Beta			$5{,}000 - T_0$	0.0001 g/g K 0.996 g/g Ar	Biotite, muscovite, hornblende, whole rocks
$^{147}Sm/^{143}Nd$	Alpha	0.654×10^{-11}	106×10^9	$0 - T_0$	0.150 g/g Sm 0.122 g/g Nd	Feldspars, pyroxenes, amphiboles, whole rocks
$^{14}C/^{14}N$	Beta	1.209×10^{-4}	5,730	0–70,000	10^{-12} g/g C 0.996 g/g N	Charcoal, wood, peat

Note: Ages of rocks and other materials obtained by use of radioactive systems are expressed in three different forms: (1) descriptive (millions of years, etc.); (2) numerical notation (10^6 years, etc.); and (3) by use of Standard International (SI) units (Ma and Ga, which equal 10^6 and 10^9 years respectively).

TABLE 2-5 Typical Concentrations (in ppm) of Radioactive and Daughter Elements in Selected Rocks and Minerals

Rock or mineral	U	Th	Pb	Rb	Sr	K	Sm	Nd
Chondrites	0.014	0.04	1.2	2.8	11	860	0.2	0.6
Basalt	0.8	2.7	4.5	30.0	470	8,500	7.0	37.0
Granite	4.4	16.0	20.0	170.0	290	33,000	8.0	43.0
Shale	3.8	12.0	20.0	140.0	300	28,000	10.0	50.0
Biotite	20.0	25.0	25.0	1,000.0	20	70,000	37.0	172.0
Muscovite	20.0	25.0	25.0	700.0	80	70,000	—	—
K-feldspar	1.5	5.0	25.0	500.0	100	90,000	4.0	26.0
Plagioclase	2.5	1.5	0.1	30.0	500	6,000	0.5	1.9
Amphibole	15.0	25.0	7.0	10.0	150	3,000	6.0	17.0
Pyroxene	20.0	13.0	7.0	10.0	150	5,000	3.0	9.0
Zircon	2,500.0	2,000.0	100.0	—	—	—	—	—

To obtain the age of a mineral sample, we have to determine by chemical analysis the present $^{87}Sr/^{86}Sr$ and $^{87}Rb/^{86}Sr$ ratios of the sample. Furthermore, the initial $^{87}Sr/^{86}Sr$ has to be estimated and a value selected for the decay constant. With this information, equation (2-9) can be solved for the age t. One way of getting the initial $^{87}Sr/^{86}Sr$ is by measuring this ratio in coexisting undisturbed minerals that contain little or no rubidium (examples are apatite and calcic plagioclase). Ages calculated by making certain assumptions about the history of a sample (closed system, initial ratios, etc.) are sometimes referred to as "model ages."

In the case of whole-rock dating of igneous rocks, it is possible to determine both time of crystallization and initial $^{87}Sr/^{86}Sr$ ratio. This is done by use of an *isochron diagram* (Figure 2-1). The diagram is obtained by plotting $^{87}Sr/^{86}Sr$ versus $^{87}Rb/^{86}Sr$ for the various minerals of a rock or for different rock samples from a particular body of rock. If the individual samples have differing Rb/Sr ratios, a straight line (isochron) is obtained. The equation for this line is given by (2-9). The intercept of the lines is the $^{87}Sr/^{86}Sr$ ratio when $t = 0$ and the slope of the line is $e^{\lambda t} - 1$. Such a diagram thus allows determination of both the age of the material being studied (from the slope) and the composition of initial strontium at the time of formation (from the intercept on the $^{87}Sr/^{86}Sr$ axis).

In the past there has been some uncertainty in the value of the rubidium decay constant. Laboratory measurement of the constant is difficult because of the nature of the rubidium decay (normal beta decay with a high percentage of low-energy particles). The value of the constant has also been determined by analyzing samples of known age and substituting values for t, $^{87}Sr/^{86}Sr$, and $^{87}Rb/^{86}Sr$ into the basic age equation. The λ values obtained in the past by these two methods have differed by as much as 5 percent. As a result of this problem, age values reported in the literature have been based on a variety of values for λ. The currently accepted value for λ is $1.42 \times 10^{-11}\text{year}^{-1}$.

Rubidium commonly substitutes for potassium in minerals; thus, most dating has been of potassium-bearing minerals and of whole rocks containing these minerals. The analytical uncertainty for $^{87}Sr/^{86}Sr$ ratios is ± 0.1 percent, and for $^{87}Rb/^{86}Sr$ ratios is ± 2 percent. The largest errors in age determinations occur for samples with a high value of initial $^{87}Sr/^{86}Sr$ and/or a low value of $^{87}Rb/^{86}Sr$. Ages for unaltered samples can usually be determined with a precision of ± 2 percent. In addition to errors due to analytical measurements and to uncertainty in the decay

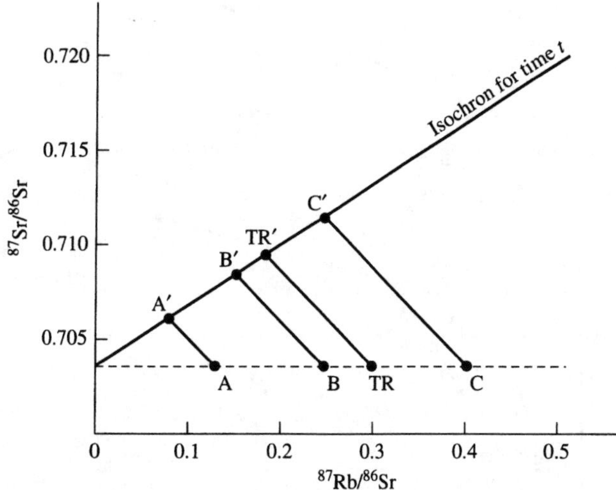

Figure 2-1 A simple isochron for an igneous rock that has not been subsequently altered. The inital composition of three different minerals of the rock is given by A, B, and C; their composition after time t is given by A′, B′, and C′. The zero intercept of the isochron gives the initial $^{87}Sr/^{86}Sr$ for the rock, and the age of the rock (t) can be calculated from the slope of the isochron, which equals $e^{\lambda t} - 1$. TR represents the bulk rock. The $^{87}Sr/^{86}Sr$ value for the bulk rock is determined by the $^{87}Sr/^{86}Sr$ values and relative proportions of the minerals in the rock. (Modified from Faul, *Ages of Rocks, Planets and Stars*, p. 34. Copyright 1966 by McGraw-Hill Book Company. Used with permission of McGraw-Hill Book Company.) After Lanphere et al. (1964).

constant, the radiometric age found for a sample may also be in error owing to gain or loss of rubidium and strontium after mineral or rock crystallization. Two types of alteration are known to affect the rubidium-strontium content of minerals: (1) a temperature effect due to metamorphism (which causes diffusion of strontium), and (2) chemical exchange with circulating water. Metamorphism can cause extensive and unequal strontium redistribution among the minerals of a rock or between different rock types. The result will be, for age measurements, a variety of "ages," none of which is a measure of a particular event (*discordant ages*).

In some cases it is possible to determine for a metamorphosed igneous rock both the time of metamorphism and the time of original crystallization. Consider an igneous rock mass that has undergone metamorphism, resulting in a redistribution of strontium among its minerals (Figure 2-2). If individual portions of the rock mass remain a closed system during metamorphism, the isochrons found for *rock* samples will still give the primary age and initial strontium composition. However, if the minerals of each rock sample have a new, homogeneous distribution of ^{87}Sr and ^{86}Sr as a result of metamorphism, their "clocks are reset" and mineral analysis will give, for a particular rock, an isochron with different slope and larger intercept than that found for rock samples. The mineral isochrons represent the time of metamorphism and the rock isochron the time of initial rock formation. This approach is not very useful for rocks that have undergone partial homogenization of isotopic ratios or that have been metamorphosed several times (see Table 2-6). Scattering of plotted points for rock samples on an isochron diagram indicates that the samples have not been a closed system.

The initial $^{87}Sr/^{86}Sr$ ratios found for oceanic and continental rocks have been used to interpret the history of the Earth's crust and to define the source region of igneous rocks. To do this,

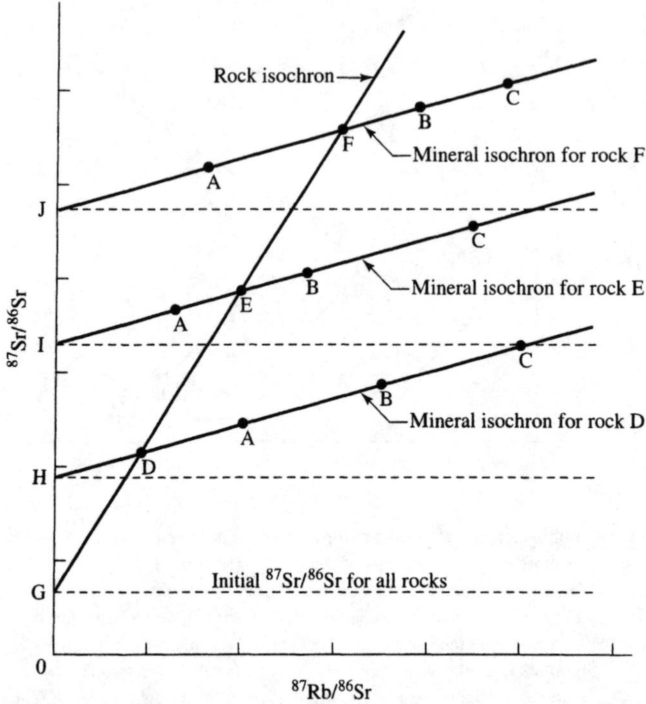

Figure 2-2 Isochron diagram for three related igneous rocks of similar age that have undergone subsequent metamorphism. The rocks D, E, and F define an isochron representing the age of the rocks. The minerals A, B, and C in each rock define individual isochrons with parallel slopes determined by the time of a later metamorphism. At the time of metamorphism the rocks have differing ratios of $^{87}Sr/^{86}Sr$ (because of differing original Rb/Sr ratios). G is the initial $^{87}Sr/^{86}Sr$ for all rocks; H, I, and J are the $^{87}Sr/^{86}Sr$ ratios for rocks D, E, and F at the end of the metamorphism. (Modified from H. Faul, 1966, *Ages of Rocks, Planets and Stars*, McGraw-Hill, New York, p. 37. Copyright 1966 by McGraw-Hill Book Company. Used with permission of McGraw-Hill Book Company.) After Lanphere et al. (1964).

assumptions have to be made about the primeval $^{87}Sr/^{86}Sr$ ratio and about the extent of fractionation of ^{87}Rb between crust and mantle (Figure 2-3). Because of the geochemical character of rubidium (see Chapter One), the crust has a higher Rb/Sr value than the mantle. Thus a magma from the mantle that forms new igneous rock in the crust will have a different initial $^{87}Sr/^{86}Sr$ ratio than will a magma formed in the crust. This ratio will depend on the original ratio in the source area and on the time of melting. Although igneous rocks have a restricted range of variation for their initial $^{87}Sr/^{86}Sr$ ratio, it is possible to characterize individual bodies by their ratio. For example, oceanic basalts have an initial $^{87}Sr/^{86}Sr$ ratio of 0.702 to 0.707. This has been interpreted as indicating a source in the mantle. The wider variation in ratios found for continental basalts is believed to be due to contamination of their mantle-derived magmas by crustal material. Most magmas formed entirely in the crust should produce rocks with a higher (greater than about 0.707) initial $^{87}Sr/^{86}Sr$ ratio. This type of approach has been used to study the question of whether the various continents have grown continuously throughout geologic time or if, instead, an initial crust has been periodically regenerated. Either process would explain the decrease in ages found toward continental margins.

TABLE 2-6 Discordant Mineral Ages from Granites of Central Transvaal, South Africa

Rock	Mineral	Radiogenic ^{87}Sr (ppm)	^{87}Rb (ppm)	Apparent age (m. y.)
Halfway-House Granite	Feldspar	2.58	59.10	3,060
	Biotite	10.75	327.5	2,310
	Chlorite	1.09	26.50	2,890
	Muscovite A	10.36	159.4	4,540
	Muscovite B	9.95	182.9	3,820
Witkoppen Granite	Feldspar	1.74	47.25	2,580
	Biotite	6.08	202.4	2,120
Halfway-House Pegmatite	Feldspar	7.54	176.4	3,000
	Biotite	22.13	514.4	3,010
Witkoppen Pegmatite	Feldspar	3.32	88.86	2,620
Corlett Drive Pegmatite	Feldspar	16.48	413.8	2,800

Note: Isotopic analysis of whole-rock samples gives an isochron age of 3,200 million years for the five granites. However, a plot of these data for the minerals of the granites does not give results similar to those shown in Figure 2-2. The discordant mineral ages suggest that the rocks have undergone partial homogenization or have been affected by several metamorphic events sometime after initial formation.

Source: Data from E. I. Hamilton, *Applied Geochronology*, p. 101. Copyright © 1965 by Academic Press, Inc. Reprinted by permission of Academic Press, Inc., London.

It is sometimes possible to date sedimentary rocks using the Rb-Sr method. Authigenic, Rb-bearing minerals such as glauconite can give the time elapsed since sedimentation. Fine-grained clastic sedimentary rocks can give whole-rock isochrons if the rocks contain mainly clay minerals formed authigenically or altered during diagenesis. Separate dating of the authigenic and detrital fractions of a rock may yield an age of deposition and a "provenance age," which helps to identify the source areas of the rock (Morton 1985). A detailed review of the strontium isotope geology of various rock types is given by Faure and Powell (1972) and Faure (1986).

SAMARIUM-NEODYMIUM SYSTEM

The study of natural materials using this system is based on the same principles as those described above for the rubidium-strontium system. However, there are significant differences between the two systems. The decay of ^{147}Sm to ^{143}Nd involves an alpha particle (compared to a beta particle for rubidium decay) and the two elements in the Sm-Nd system both belong to the rare-earth group and thus are geochemically similar (rubidium and strontium are not geochemically similar and therefore are affected differently by natural processes). Because of analytical challenges involved in separating one rare-earth element from another, and because of the need for very high precision in neodymium analyses, use of the samarium-neodymium system in geochemical applications occurred much later than the use of the rubidium-strontium system.

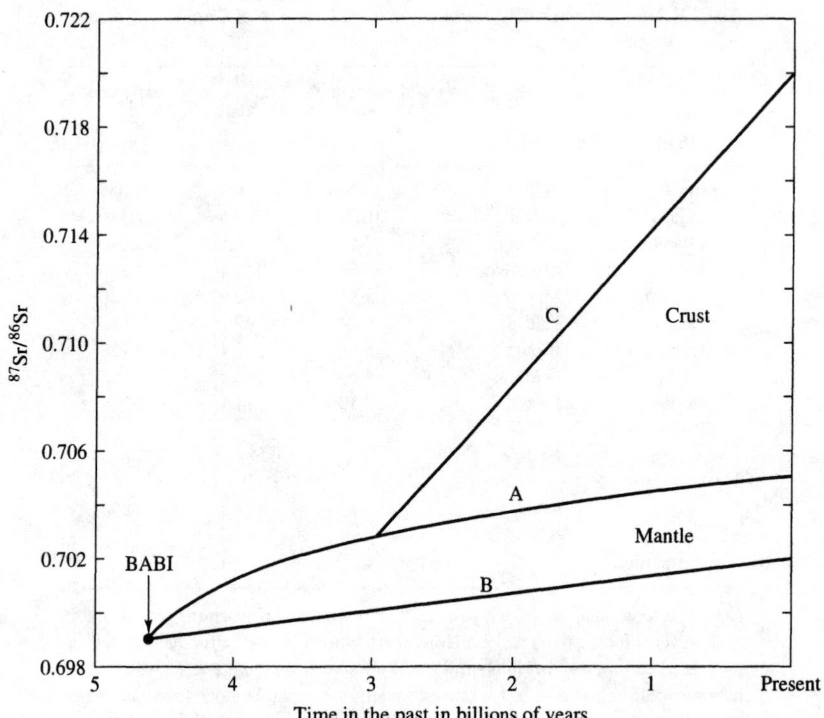

Figure 2-3 Isotopic evolution of terrestrial strontium. The point labeled BABI is representative of the $^{87}Sr/^{86}Sr$ ratio at the time of formation of the Earth and other planetary objects. The value of BABI is 0.69897 and it comes from measurements of $^{87}Sr/^{86}Sr$ in basaltic achondrites (which resemble terrestrial igneous rocks and have very low Rb/Sr ratios). BABI stands for "basaltic achondrite best initial." Line A represents a hypothetical evolutionary path of strontium in the unaltered primary mantle under the continents. The curvature of the line implies a time-dependent decrease in the Rb/Sr ratio of the upper mantle (as a result of magma formation and consequent growth of the crust). The straight line B represents strontium evolution in mantle regions initially depleted in rubidium. Line C traces the change in $^{87}Sr/^{86}Sr$ for strontium that was withdrawn from the mantle about 2.9 billion years ago and subsequently resided in a part of the crust having a Rb/Sr ratio of 0.15. The average Rb/Sr ratio of the present mantle is believed to be about 0.027 and the present $^{87}Sr/^{86}Sr$ ratios of the mantle lie within the interval 0.704 ± 0.002. The present $^{87}Sr/^{86}Sr$ ratios of the crust are much higher (because of higher Rb/Sr ratios) and have a greater range of values than those of the mantle. (After G. Faure, *Principles of Isotope Geology.* Copyright © 1986 by John Wiley & Sons. Reprinted by permission of John Wiley & Sons, Inc.)

As with the rubidium-strontium system, we can write an equation for the total number of ^{143}Nd atoms in a sample as

$$^{143}\mathrm{Nd}_{\mathrm{measured}} = {}^{143}\mathrm{Nd}_{\mathrm{initial}} + {}^{147}\mathrm{Sm}\,(e^{\lambda t} - 1) \qquad \text{(2-10)}$$

To express this relationship in the form of isotopic abundance ratios we use a neodymium isotope, ^{144}Nd, that is not part of any radioactive system,

$$\left(\frac{^{143}\mathrm{Nd}}{^{144}\mathrm{Nd}}\right)_{\mathrm{measured}} = \left(\frac{^{143}\mathrm{Nd}}{^{144}\mathrm{Nd}}\right)_{\mathrm{initial}} + \left(\frac{^{147}\mathrm{Sm}}{^{144}\mathrm{Nd}}\right)(e^{\lambda t} - 1) \qquad \text{(2-11)}$$

The usual method of dating with this system involves measurement of mineral samples separated from a single rock type or of cogenetic rock samples and the construction of an isochron diagram (Figure 2-4). Sm-Nd isochrons are based on the same principles as Rb-Sr isochrons. A plot of measured $^{143}Nd/^{144}Nd$ versus measured $^{147}Sm/^{144}Nd$ gives a straight line with the intercept on the $^{143}Nd/^{144}Nd$ axis representing initial $^{143}Nd/^{144}Nd$ and the slope of the line given by $(e^{\lambda t} - 1)$.

The Sm-Nd system is used extensively to date igneous and metamorphic rocks, particularly those that cannot be dated by other methods because of low abundances of the appropriate elements or, in the case of metamorphic or badly weathered rocks, because these elements have been gained or lost during metamorphism or weathering. Minerals commonly analyzed include feldspars, pyroxenes, amphiboles, garnets, titanite, and zircon. DePaolo (1988) states that the age

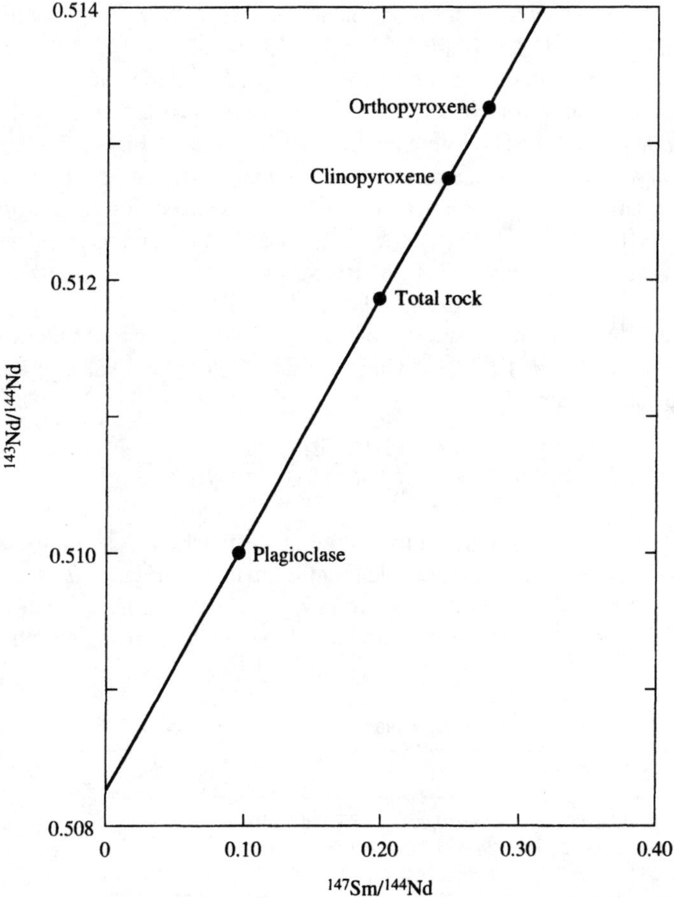

Figure 2-4 Mineral isochron for a sample of gabbro from the Stillwater intrusion, Montana. The $^{143}Nd/^{144}Nd$ value for "total rock" is determined by the $^{143}Nd/^{144}Nd$ values and relative proportions of the minerals in the rock. The age given by the isochron is 2701 ± 8 million years. (After D. J. DePaolo, *Neodymium Isotope Geochemistry*. Copyright 1988. Reprinted by permission of Springer-Verlag, New York.)

uncertainty (due to analytical uncertainty) at the 95 percent confidence level is approximately ±20 million years. This value does not take into account a ±1 percent uncertainty in the value of the decay constant of ^{147}Sm. Rocks of all ages can be dated. Table 2-7 shows that Sm-Nd ages can agree very well with ages obtained by other methods on the same samples. Because small but significant isotopic fractionation (separation) can occur during chemical analysis, it is necessary to correct (normalize) measured isotopic ratios to an assigned value for two stable nonradiogenic isotopes. Different laboratories have used different assigned values and this must be kept in mind when comparing published Sm-Nd measurements from different sources. For further details see DePaolo (1988).

Because of their similar geochemical properties, it is likely that no fractionation of Sm from Nd occurred during the early stages of the Earth's formation by nebular condensation (see discussion in Chapter One). However, the later formation and separation of magmas from the Earth's mantle has caused significant fractionation (see later discussion). Thus the Sm-Nd system is very useful in studying the history of crust formation on the Earth. The process involves measurement of the age and initial $^{143}Nd/^{144}Nd$ ratio of a rock. The initial $^{143}Nd/^{144}Nd$ ratio of a rock is the ratio the rock material had at the time its magma separated from the mantle or crust (Figure 2-5). In fact, the ratio can often be used to tell whether the parent magma formed in the mantle or crust (Farmer and DePaolo 1983). The reference line (CHUR line) in Figure 2-5 represents material in the mantle that has not undergone chemical differentiation over time. In other words, this material has had the same Sm/Nd ratio since the Earth formed except for the uniform change due to ^{147}Sm decay. The CHUR line is derived from measurements on chondrites, and CHUR stands for "*ch*ondritic *u*niform *r*eservoir." Chondrites also provide an estimate of present-day $^{143}Nd/^{144}Nd$ and $^{147}Sm/^{144}Nd$ for the total Earth.

The $^{143}Nd/^{144}Nd$ ratio for samples is often expressed relative to the CHUR line. Thus ϵ_{Nd} represents, in units of parts in 10^4, the deviation of the $^{143}Nd/^{144}Nd$ value of a sample from that of the CHUR line at the same time t:

$$\epsilon_{Nd} = \left[\frac{(^{143}Nd/^{144}Nd)^t_{sample}}{(^{143}Nd/^{144}Nd)^t_{CHUR}} - 1 \right] \times 10^4 \qquad (2\text{-}12)$$

Significant fractionation during magma formation in the mantle results in a scattering of initial values for ϵ_{Nd} of samples derived from fractionation magmas (Figures 2-5a and 2-5b). When fractionation due to partial melting occurs, the newly formed magma will have a lower Sm/Nd ratio than CHUR and the residual solid material will have a higher Sm/Nd ratio. (The element

TABLE 2-7 Comparison of Ages Obtained on the Same Rock by Different Methods

Rock	Sm/Nd	Rb/Sr	U/Pb	K/Ar
Lunar basalt 10062	3.88 ± 0.06	3.93 ± 0.11	—	3.82 ± 0.06
Lunar basalt 10072	3.57 ± 0.03	3.56 ± 0.05	—	3.57 ± 0.04
Stillwater intrusion	2.701 ± 0.008	—	2.713 ± 0.004	—
Isua metavolcanics	3.75 ± 0.04	—	3.77 ± 0.01	—

Ages given in units of 10^9 years.

Source: D. J. DePaolo, *Neodymium Isotope Geochemistry*. Copyright 1988. Reprinted by permission of Springer-Verlag, New York.

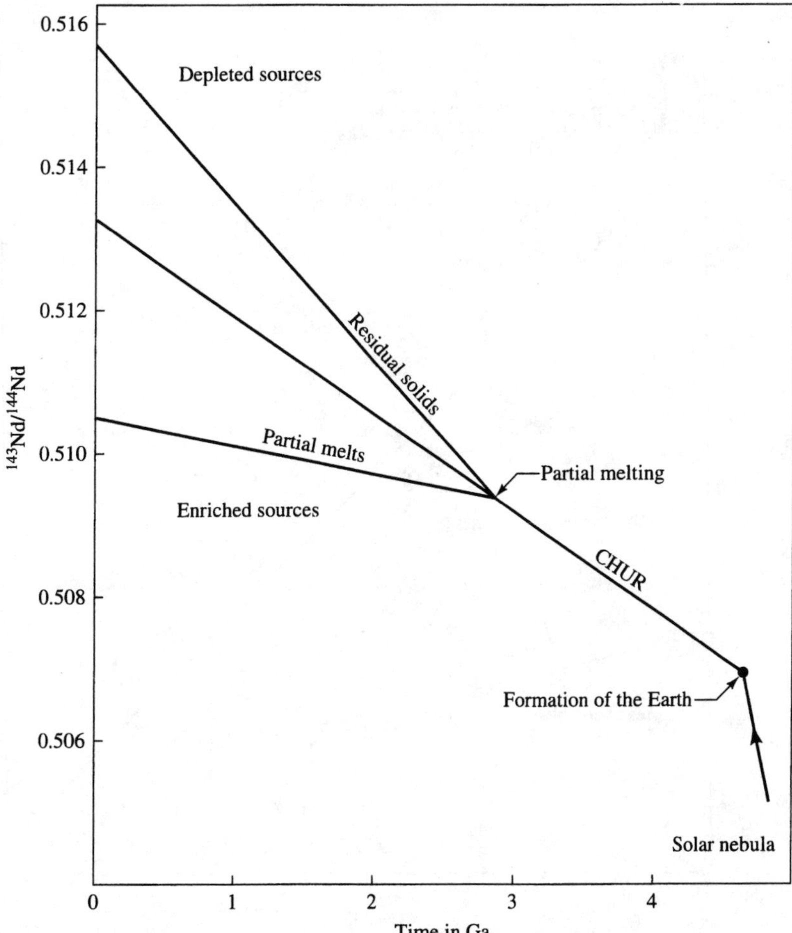

Figure 2-5a Isotopic evolution of Nd in a chondritic uniform reservoir (CHUR). Magma formed by partial melting has a lower Sm/Nd ratio than CHUR, whereas the residual solids have a higher Sm/Nd ratio. As a result, the present-day $^{143}Nd/^{144}Nd$ ratio of the rocks formed from the silicate liquid is less than that of CHUR, whereas the $^{143}Nd/^{144}Nd$ ratio of the residual solids is higher. If magma forms by a second episode of melting of such "depleted" mantle, the resulting igneous rocks have initial $^{143}Nd/^{144}Nd$ ratios that are greater than CHUR (positive ϵ_{Nd}). (After G. Faure, *Principles of Isotope Geology.* Copyright © 1986 by John Wiley & Sons. Reprinted by permission of John Wiley & Sons, Inc.)

with the larger ionic radius, neodymium, is concentrated in the magma; however, the $^{143}Nd/^{144}Nd$ ratio would not change as a result of partial melting.)

A positive initial ϵ_{Nd} value for a rock indicates that its source has undergone periods of partial melting (differentiation). At least two periods of partial melting are indicated by a positive initial ϵ_{Nd} value. The first episode produces residual rock with a high Sm/Nd ratio and the second episode affects this residual rock and produces magmas and rocks with initial $^{143}Nd/^{144}Nd$ ratios that are greater than that of CHUR material (Figure 2-5a). A negative initial ϵ_{Nd} value

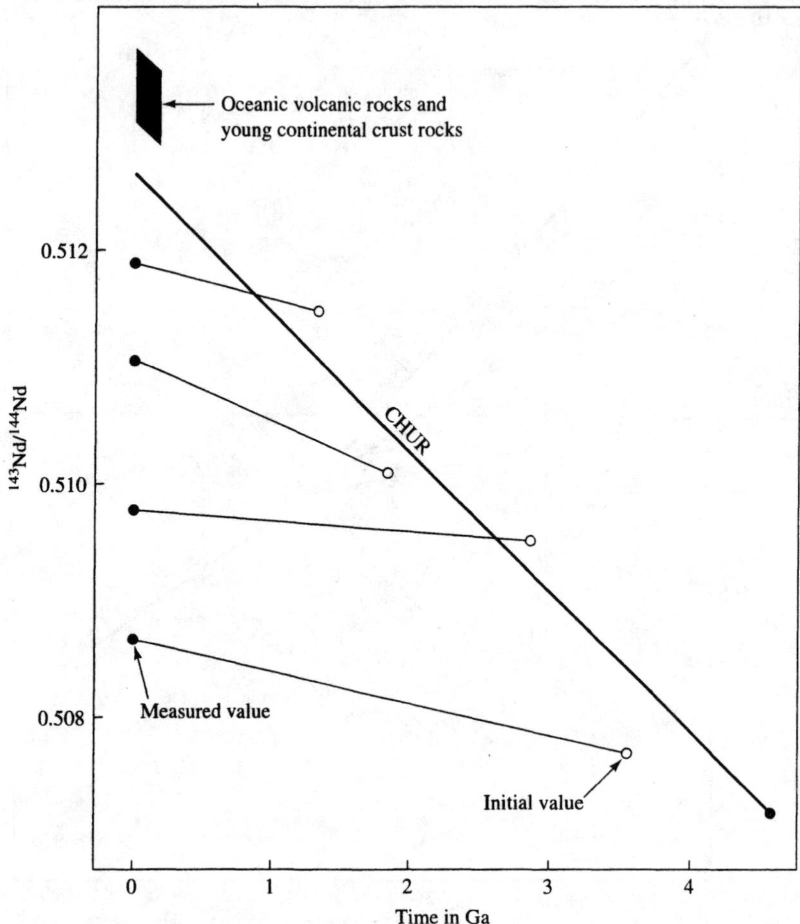

Figure 2-5b Application of the Sm-Nd isotope system to the determination of the evolution of the Earth's mantle and crust through geologic time. Measured and initial (calculated) values of $^{143}Nd/^{144}Nd$ are shown for four igneous rocks from the continental crust. The initial values represent the $^{143}Nd/^{144}Nd$ ratios the rocks had at the time they came from the mantle. ϵ_{Nd} (see text) at the time of the initial value represents the difference between the calculated initial $^{143}Nd/^{144}Nd$ of rocks when they came from the mantle and the calculated value of $^{143}Nd/^{144}Nd$ of CHUR material at that time. ϵ_{Nd} at the present time (measured value at $t = 0$) represents the difference between the measured value of $^{143}Nd/^{144}Nd$ in the rock and the present value of $^{143}Nd/^{144}Nd$ for CHUR material. Deviation of initial values from the CHUR line indicates chemical differentiation in the mantle and can be interpreted in terms of models of Earth evolution. Initial values can be obtained from an isochron diagram (not easy to do without large uncertainty), or by using a measured age for the rock to correct the measured $^{143}Nd/^{144}Nd$ value back to its initial value. Similar measured and initial values are found for the $^{143}Nd/^{144}Nd$ ratio in oceanic volcanic rocks (which are all relatively young) and in young continental igneous rocks. These numbers give an indication of the value and variability of the $^{143}Nd/^{144}Nd$ ratio in the upper mantle today. (After D. J. DePaolo, *Neodymium Isotope Geochemistry*. Copyright 1988. Reprinted by permission of Springer-Verlag, New York.)

suggests that the magma source represents material that earlier separated as a liquid from the primary chondritic uniform reservoir (Figure 2-5a).

If initial ϵ_{Nd} is assumed to equal zero for a sample, then the calculated interception of a sample line with the CHUR line in Figure 2-5b gives us a "model age" for the sample. The intercept represents the time in the past when the $^{143}Nd/^{144}Nd$ ratio of the sample equaled the $^{143}Nd/^{144}Nd$ of CHUR. In other words, this is when separation from CHUR *could* have occurred. A model age has significance only if the Sm/Nd ratio of the sample has not changed since the time of separation from CHUR.

Thus isotopic systems such as Rb-Sr and Sm-Nd can be used in two ways. First, they can give us numerical dates for rock samples. Second, rock ages plus isotopic abundance measurements (initial $^{143}Nd/^{144}Nd$ values, etc.) can be used to study the history of the Earth's evolution. Changes in isotopic composition of daughter elements can be used to identify chemical fractionations that occurred at various times in the past. The CHUR line discussed above can be considered a reference evolution curve for undifferentiated solar system material. Samples that do not fall on this line tell us something about planetary evolution processes. Such samples have ϵ_{Nd} values either greater than or less than zero. In other words the fractionation history of parent and daughter isotopes can be deduced from abundance variations found for various times and locations.

Another system that has been found useful is the Lu-Hf isotopic system (Patchett 1983). For this system, and for the Rb-Sr system, ϵ_{Hf} and ϵ_{Sr} can be defined in the same manner as ϵ_{Nd}. Figure 2-6 summarizes data for mid-ocean ridge basalts (MORB) for these three systems. These data indicate that the source regions of MORB in the mantle were "depleted" (affected by earlier melting episodes). Other data confirm that portions of the mantle are depleted in large-ion lithophile (LIL) elements such as neodymium, rubidium, potassium, thorium, and uranium. Depleted mantle, which produces magmas with unusual isotopic composition such as high initial $^{143}Nd/^{144}Nd$ ratios, is said to have a distinctive "isotopic signature." Figure 2-7 is a plot of ϵ_{Nd} evolution in time for depleted mantle and average upper continental crust.

Patchett and Arndt (1986) have used the Sm-Nd system to study the formation of early Proterozoic crust. They found that about 50 percent of the Earth's crust formed from 1.9 to 1.7 Ga is juvenile; that is, it represents magmas derived from the mantle. The other half of the crust formed at this time represents recycled Archean crust. Using this type of information Patchett and Arndt (1986) calculated a crustal production rate of about 1.2 km^3/yr, a value similar to production rates for crustal growth that have been calculated by other investigators for island-arc accretion. A review of the various isotopic and chemical constraints on mantle-crust evolution has been presented by Jacobsen (1988).

Arndt and Goldstein (1987) have pointed out that Nd model ages for crustal rocks do *not* necessarily correspond to a specific "crust-formation" age, but may instead reflect mixing of material derived from the mantle at different times. Thus one of the requirements for a sample to provide a true crust-formation age is that the sample not contain a mixture of younger and older crustal material. All the material in the sample must have come from the mantle as the result of a single event. Single-event samples can be identified by using independent evidence, such as ages derived from the uranium-thorium-lead system (see next section). If the Sm-Nd and U-Th-Pb ages agree, a single crust-forming event is indicated.

Because they are derived from depleted mantle or recycled oceanic crust (Sm/Nd greater than CHUR), oceanic crustal rocks (mid-ocean basalts, island arc rocks, etc.) are characterized

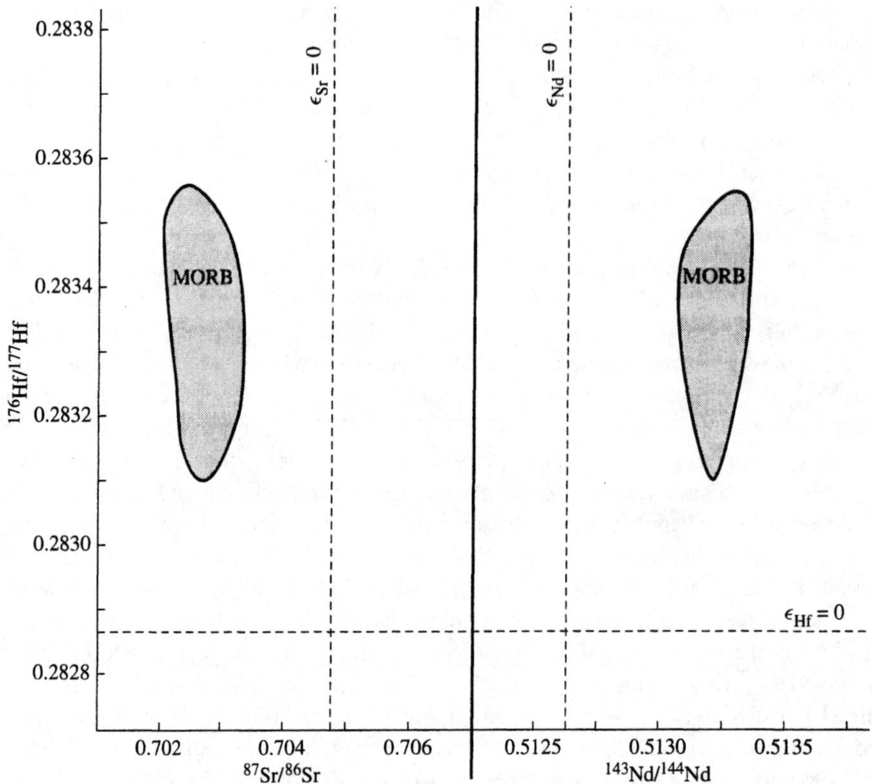

Figure 2-6 Hf-Nd-Sr isotopic relationships for mid-ocean ridge basalts (MORB). The lines $\epsilon_{Hf} = 0$, $\epsilon_{Nd} = 0$, and $\epsilon_{Sr} = 0$ represent the present-day isotopic composition of undifferentiated chondritic material, material that has undergone no change with respect to the indicated isotopic system (other than radioactive decay of parent isotopes) since the formation of the Earth. MORB samples have positive ϵ_{Hf} and ϵ_{Nd} and negative ϵ_{Sr}, indicating a source in the sub-ocean mantle that had *higher* Sm/Nd and Lu/Hf ratios and *lower* Rb/Sr ratios than undifferentiated chondritic material. Such regions have undergone one or more differentiation processes and are said to be "depleted." The plotted ratios are "initial" ratios, since MORB are very young rocks. (Reprinted from *Geochimica et Cosmochimica Acta*, v. 47, P. J. Patchett, Importance of the Lu-Hf isotopic system in studies of planetary chronology and chemical evolution. p. 84, Copyright 1983, with kind permission from Elsevier Science Ltd, The Boulevard, Langford Lane, Kidlington OX5 1GB, UK.)

by positive present-day ϵ_{Nd} values. Continental crustal rocks generally have negative values for present-day ϵ_{Nd} as a result of derivation from a parent with Sm/Nd less than CHUR.

Thus these two different sources of rare-earth elements to the oceans can be identified by their isotopic composition. The large variations in ϵ_{Nd} in ocean water have also been used to study the transport behavior of the rare-earth elements in the oceans (Piepgras and Jacobsen 1988). In addition, the Sm-Nd system can be applied to the study of the provenance of detrital sedimentary rocks (Nelson and DePaolo 1988). It appears that Sm/Nd ratios of clastic materials are not altered during the formation of nonchemical sedimentary rocks and thus the model ages of these sedimentary rocks are indicative of the sediment source area.

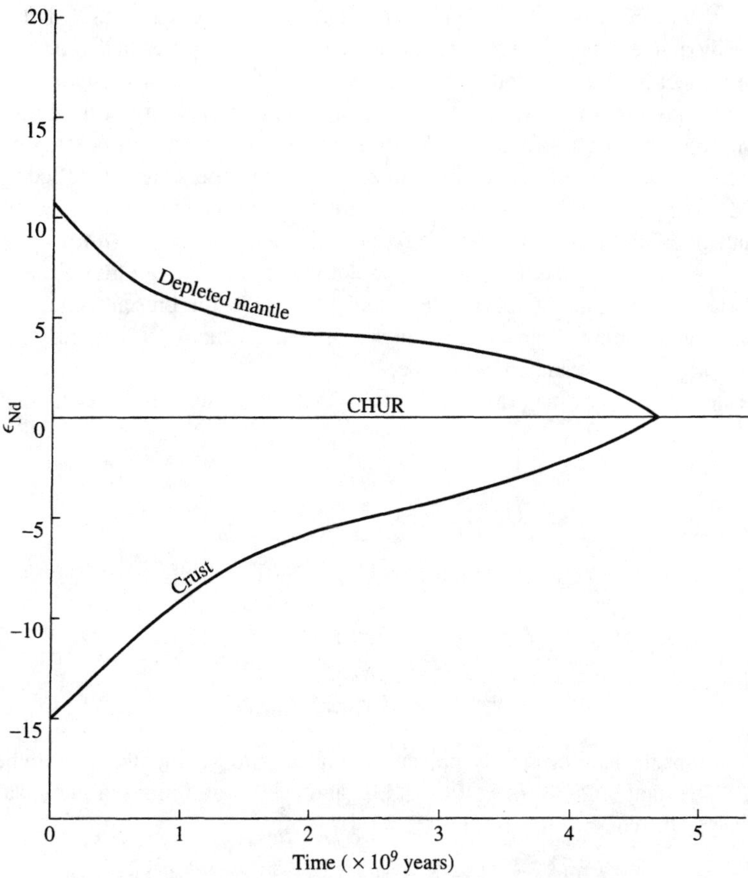

Figure 2-7 Evolution of ϵ_{Nd} over the Earth's history for depleted mantle and average upper continental crust. CHUR stands for "chondritic uniform reservoir." The mantle trend is defined by various terrestrial igneous rocks with high ϵ_{Nd} values, terminating at ϵ_{Nd} equal to about eleven for mid-ocean ridge basalts (MORB). The crustal trend has been obtained from shales. Analyses of rocks from the Sudbury Igneous Complex gives an initial ϵ_{Nd} value of -7.54 ± 1.1 and an age of $1840 \pm 21 \times 10^6$ years. Thus the Sudbury Complex plots near the crustal trend, evidence for its formation by meteoritic impact rather than by intrusion of mantle-derived magma. (Reprinted with permission from Faggart et al., 1985, Origin of the Sudbury Complex by meteoric impact: neodymium isotopic evidence. *Science*, v. 230, pp. 436–439. Copyright 1985 by the American Association for the Advancement of Science.)

URANIUM-THORIUM-LEAD SYSTEM

The uranium-thorium-lead system is the most complicated of the major methods used in geochronology for two reasons: (1) three different parent isotopes are involved, and (2) each isotope goes through an extensive decay series before a stable lead isotope is formed. On the other hand, it has been found that these complications can be used to gain additional information about the history of rocks and minerals. As mentioned earlier, the three decay series of the system are ${}^{238}U \rightarrow {}^{206}Pb$, ${}^{235}U \rightarrow {}^{207}Pb$, and ${}^{232}Th \rightarrow {}^{208}Pb$ (Tables 2-1, 2-2, and 2-3). The occurrence of radioactive daughter products in each decay series requires that age measurements be made only on material that is in equilibrium with its daughter products; this is so when the rate of generation of any intermediate member is equal to its rate of decay. Equilibrium is reached about one million years after formation of a uranium-bearing or thorium-bearing mineral. After this period of time, the rates of decay of the intermediate members are controlled by the rates of the slowest decaying isotopes in the series, the parent uranium and thorium isotopes. Thus, because the half-lives of ${}^{238}U$, ${}^{235}U$, and ${}^{232}Th$ are all much longer than those of their various daughters, we can deal with the decay of the parent isotopes as though they decayed directly to the ultimate daughter isotopes (${}^{206}Pb$, ${}^{207}Pb$, and ${}^{208}Pb$). At the present time, ${}^{238}U$ makes up 99.27 percent of all uranium. Since ${}^{235}U$ decays more rapidly than ${}^{238}U$, the relative abundances of these two isotopes have changed with time.

The basic equations of the uranium-thorium-lead system used for age calculations are obtained from

$$D = P(e^{\lambda t} - 1) \tag{2-5}$$

Use of this equation gives

$${}^{206}Pb = {}^{238}U(e^{\lambda_{238}t} - 1) \tag{2-13}$$

$${}^{207}Pb = {}^{235}U(e^{\lambda_{235}t} - 1) \tag{2-14}$$

$${}^{208}Pb = {}^{232}Th(e^{\lambda_{232}t} - 1) \tag{2-15}$$

We have to account for lead originally present and taken into a mineral sample when it formed (*initial lead*). The total amounts of ${}^{208}Pb$, ${}^{207}Pb$, and${}^{206}Pb$ found in a sample equal initial lead plus radiogenic lead. Thus

$${}^{206}Pb_{measured} = {}^{206}Pb_{initial} + {}^{238}U(e^{\lambda_{238}t} - 1) \tag{2-16}$$

$${}^{207}Pb_{measured} = {}^{207}Pb_{initial} + {}^{235}U(e^{\lambda_{235}t} - 1) \tag{2-17}$$

$${}^{208}Pb_{measured} = {}^{208}Pb_{initial} + {}^{232}U(e^{\lambda_{232}t} - 1) \tag{2-18}$$

Finally, more accurate values can be obtained by measuring ratios of isotopes than by measuring their absolute abundances. Since a nonradiogenic isotope of lead (${}^{204}Pb$) occurs with the other isotopes, we divide the above equations by the abundance of this isotope to get the basic age equations of the uranium-thorium-lead system:

$$\left(\frac{{}^{206}Pb}{{}^{204}Pb}\right)_{measured} = \left(\frac{{}^{206}Pb}{{}^{204}Pb}\right)_{initial} + \left(\frac{{}^{238}U}{{}^{204}Pb}\right)(e^{\lambda_{238}t} - 1) \tag{2-19}$$

$$\left(\frac{{}^{207}Pb}{{}^{204}Pb}\right)_{measured} = \left(\frac{{}^{207}Pb}{{}^{204}Pb}\right)_{initial} + \left(\frac{{}^{235}U}{{}^{204}Pb}\right)(e^{\lambda_{235}t} - 1) \tag{2-20}$$

$$\left(\frac{{}^{208}Pb}{{}^{204}Pb}\right)_{measured} = \left(\frac{{}^{208}Pb}{{}^{204}Pb}\right)_{initial} + \left(\frac{{}^{232}Th}{{}^{204}Pb}\right)(e^{\lambda_{232}t} - 1) \tag{2-21}$$

The values for initial lead are obtained by determining the isotopic composition of lead in the environment at the time of sample formation. One way of making the correction is to analyze a lead mineral (such as galena) in the same rock sample or geologic setting. If no lead-bearing mineral is present with the sample, the correction is often made by estimating the average oceanic lead composition that existed at the approximate time that the sample formed. This procedure clearly introduces an uncertainty in the calculated age of the sample. The first method, analyzing a coexisting lead mineral, can also produce an error in the estimated age, since the isotopic composition of lead has been found to vary somewhat, even in different samples of the same mineral from the same locality. Another way to solve this problem is to use a mineral like zircon, which often contains so little initial lead that a measurement or estimate of initial lead is not needed.

The major source of error in using the age equations is uncertainty (limited precision) in the measurement of the lead isotope ratios. To obtain the age of a sample, the appropriate isotope ratios are measured and these, along with the known values of λ, allow solving of equations (2-19) to (2-21) for time of formation t. For example,

$$t_{206} = \frac{1}{\lambda_{238}} \ln \left[\frac{\left(\frac{^{206}\mathrm{Pb}}{^{204}\mathrm{Pb}}\right)_{\text{measured}} - \left(\frac{^{206}\mathrm{Pb}}{^{204}\mathrm{Pb}}\right)_{\text{initial}}}{\frac{^{238}\mathrm{U}}{^{204}\mathrm{Pb}}} + 1 \right] \tag{2-22}$$

Because ^{238}U and ^{235}U decay at different rates, a fourth age equation can be developed from the ratio of equations (2-20) and (2-19):

$$\frac{(^{207}\mathrm{Pb}/^{204}\mathrm{Pb})_{\text{measured}} - (^{207}\mathrm{Pb}/^{204}\mathrm{Pb})_{\text{initial}}}{(^{206}\mathrm{Pb}/^{204}\mathrm{Pb})_{\text{measured}} - (^{206}\mathrm{Pb}/^{204}\mathrm{Pb})_{\text{initial}}} = \left(\frac{^{235}\mathrm{U}}{^{238}\mathrm{U}}\right)_{\text{measured}} \frac{e^{\lambda_{235} t} - 1}{e^{\lambda_{238} t} - 1} \tag{2-23}$$

Use of this equation is referred to as the *lead-lead*, or *207-206, method*. The ratio $^{235}U/^{238}U$ has a value at the present time of 1/137.88 and is independent of age and history of the source material from which the uranium came.* Thus the only measurements required to get an age using equation (2-23) are those for the lead isotopes. This method also has the advantage that it will be least affected by any loss of lead from a sample during geologic time, since the lead would probably be lost in the isotopic proportion in which it is present. If lead loss did occur, ages obtained using equations (2-19) through (2-21) would be less than the true age. On the other hand, these three methods can be used as a check on each other and on the lead-lead method. Equation (2-23) is not directly soluble for t, and the age for various values of the isotope ratios must be obtained by successive approximations or from published graphs or tables [for example, Stacey and Stern (1973)].

Analytical techniques allow measurement of the various isotope ratios to a precision of about $\pm 0.1-0.2$ percent. If no gain or loss of material has occurred, the 238/206, 235/207, 232/208, and 207/206 ages should agree (*concordant ages*). Gain or loss of material can be caused by metamorphism or weathering (only fresh samples should be analyzed). Such concordant ages allow the age of a sample to be determined with a precision of about ± 2 percent.

When two or more age determinations do not agree, the results are said to be *discordant*. This frequently occurs in use of the uranium-thorium-lead system (Table 2-8). Assuming careful

*An exception to this statement is the uranium ore found at the Oklo mine, Gabon. This ore is depleted in ^{235}U, apparently as a result of a fission chain reaction that occurred during Precambrian time. For further information see Kuroda (1982).

TABLE 2-8 Concordant and Discordant Uranium-Thorium-Lead Isotope Ages

Mineral analyzed	Sample locality	Isotope ages (m. y.)			
		$^{238}U/^{206}Pb$	$^{235}U/^{207}Pb$	$^{232}Th/^{208}Pb$	$^{207}Pb/^{206}Pb$
Concordant ages					
Samarskite	Spruce Pine, N.C., U.S.A.	314	316	302	342
Zircon	Witchita Mts., Okla., U.S.A.	520	527	506	550
Zircon	Ceylon	540	544	538	555
Pitchblende	Katanga, Belgian Congo	575	595	—	630
Uraninite	Romteland, Norway	890	892	900	920
Thucolite	Witwatersrand, South Africa	2,110	2,080	—	2,070
Discordant ages					
Zircon	Capetown, South Africa	330	356	238	530
Zircon	Beartooth Mts., Mont., U.S.A.	770	1,400	—	2,580
Zircon	Quartz Creek, Colo., U.S.A.	930	1,130	515	1,540
Xenotime	Uncompahgre, Colo, U.S.A.	3,180	2,065	1,100	1,640
Euxenite	Wakefield, Quebec, Canada	620	710	550	1,000
Zircon	Beartooth Mts., Mont, U.S.A.	1,660	2,380	870	3,080

Source: Data from E. I, Hamilton, *Applied Geochronology*, p. 101. Copyright © 1965 by Academic Press, Inc. Reprinted by permission of Academic Press, Inc., London.

sample collection, purification, chemical analysis, and initial lead correction, this discordance is probably due to gain or loss of parent or daughter isotopes. In terms of general geochemical behavior, uranium and thorium (lithophile) are quite different from chalcophile lead. (The radii of the ions of uranium and thorium differ significantly from that of lead, and this is a major cause of their dissimilar behavior.)

As can be seen in Table 2-8, when ages are discordant for a sample, the lead-lead (207-206) method usually gives the oldest (and probably the most accurate) age.

If lead has been lost (the most likely cause of discordance), it is possible to use the data to obtain not only the age of the sample, but also the time when the lead was lost, assuming that this was a rapid process (Wetherill 1956). This is done by plotting a *concordia diagram* (Figure 2-8). The curve on this diagram represents the ratios $^{206}Pb/^{238}U$ and $^{207}Pb/^{235}U$ for all concordant ages [see equations (2-13) and (2-14)]. Thus the curve is the locus of points having equal ages for the two age methods. When a geologic event (usually metamorphism) alters the lead content of the minerals of a rock in a nonuniform way, a plot of the ratios for each sample of a particular mineral (zircon, for instance) will define a straight line below the curve (because $^{206}Pb/^{207}Pb$ is constant for all samples). This chord can be extrapolated back to the time when the rock formed

(T_1) and forward to the time when the lead was lost (T_2). Because of the nature of the curve, this method is most useful for T_1 ages greater than 500 million years.

The time T_2 of Figure 2-8 may not have much meaning, depending upon the manner in which the lead was lost. The straight line of Figure 2-8 represents *episodic lead loss*; that is, lead is lost at a single instant, T_2. When plotted on concordia diagrams, the results for some rocks suggested that the hypothetical times of lead loss appear to be related to the true ages of the samples. In other words, the younger the true age, the younger the time of lead loss. This type of relationship could not be explained by one or more times of episodic lead loss. Thus a second hypothesis, *continuous-diffusion lead loss*, was developed. Continuous diffusion of lead out of a mineral over a long period of time would favor the removal of ^{207}Pb relative to ^{206}Pb (because of the difference in half-life of the two isotopes; this preferential removal would not occur with episodic loss). Instead of the straight line of Figure 2-8, calculation of the expected results for continuous diffusion for a rock suite of age T_1 gives the dashed line of Figure 2-8. It can be seen that the results for episodic and diffusion loss differ mainly in the region near the origin. Diffusion loss could thus explain the apparent relationship of true ages and times of lead loss mentioned

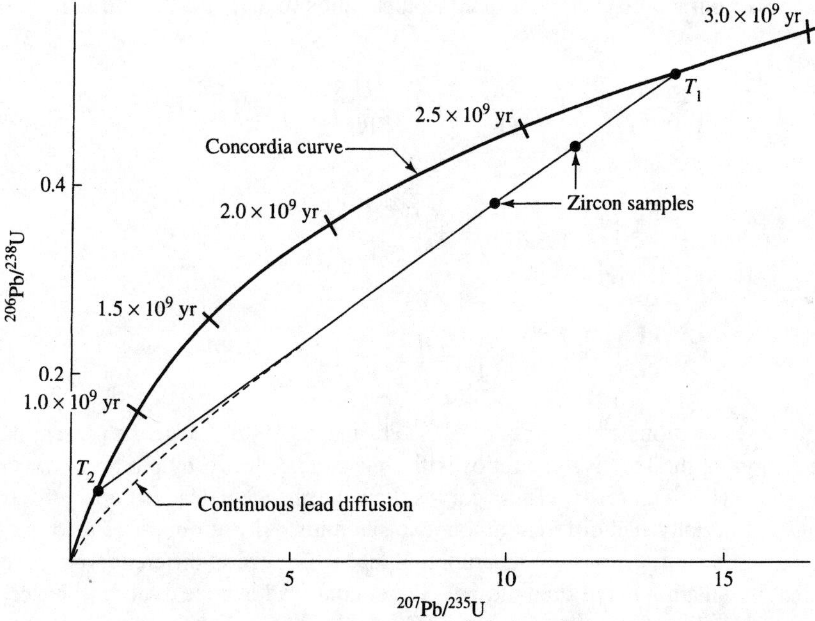

Figure 2-8 Concordia diagram. The rock mass from which the zircon samples were taken crystallized at time T_1 and suffered a nonuniform loss of lead at time T_2; that is, different zircon samples lost different amounts of lead. The individual samples would give discordant ages. No matter how much lead was lost, the ratio of ^{206}Pb to ^{207}Pb in the samples would remain the same, even though the total amounts of ^{206}Pb and ^{207}Pb would be different for each sample. Because $^{206}Pb/^{207}Pb$ is constant for all samples, these values plot on a straight line. If varying amounts of lead were taken from zircons at the present time, they would plot on a straight line running from T_1 to the origin. See text for significance of the dashed line. The concordia curve represents a plot of all concordant ages. Points on the curve (calculated using equations 2-19 and 2-20) have compatible ratios of ^{206}Pb to ^{238}U and ^{207}Pb to ^{235}U for each selected age.

above. Therefore, a time T_2 obtained by plotting a concordia diagram for a group of samples has no significance if diffusion loss of lead has occurred. However, when a group of samples plot on a straight line on a concordia diagram, the true age of the suite (T_1) can be estimated regardless of the cause of the discordance. Readers interested in further discussion of the interpretation of concordia diagrams, including a number of specific examples, are referred to Cantanzaro (1968) and Ludwig (1980).

We now turn to a discussion of lead-isotope abundance variations. The term *common lead* refers to lead in a mineral in which no significant radiogenic lead has been produced since the mineral formed. Examples of such minerals are galena, K-feldspar, and the micas. Common leads can also be described as leads that have an insignificant amount of uranium and thorium associated with them. Certain rocks, particularly very young ones, can be said to have only common lead. Common lead is made up of lead existing at the time the Earth formed (*primeval lead*), plus additions of lead formed since then by radioactive decay of uranium and thorium (*radiogenic lead*) and separated from the uranium and thorium in some manner.

Let us consider a large mass of igneous rock that crystallized at the time the Earth formed (T). It initially had primeval lead and a given amount of uranium and thorium. The decay of the uranium and thorium would change the lead isotope ratios so that, at a later time t, the ratios are given by

$$\left(\frac{^{206}\text{Pb}}{^{204}\text{Pb}}\right)_t = \left(\frac{^{206}\text{Pb}}{^{204}\text{Pb}}\right)_T + \left(\frac{^{238}\text{U}}{^{204}\text{Pb}}\right)_{\text{present}} (e^{\lambda_{238}T} - e^{\lambda_{238}t}) \tag{2-24}$$

$$\left(\frac{^{207}\text{Pb}}{^{204}\text{Pb}}\right)_t = \left(\frac{^{207}\text{Pb}}{^{204}\text{Pb}}\right)_T + \left(\frac{^{235}\text{U}}{^{204}\text{Pb}}\right)_{\text{present}} (e^{\lambda_{235}T} - e^{\lambda_{235}t}) \tag{2-25}$$

$$\left(\frac{^{208}\text{Pb}}{^{204}\text{Pb}}\right)_t = \left(\frac{^{208}\text{Pb}}{^{204}\text{Pb}}\right)_T + \left(\frac{^{232}\text{Th}}{^{204}\text{Pb}}\right)_{\text{present}} (e^{\lambda_{232}T} - e^{\lambda_{232}t}) \tag{2-26}$$

Note that in these equations $^{238}\text{U}/^{204}\text{Pb}$, $^{235}\text{U}/^{204}\text{Pb}$, and $^{232}\text{Th}/^{204}\text{Pb}$ are the present-day ratios.

The change of the lead isotope ratios with time can be shown by plotting *growth curves*, such as those shown in Figure 2-9. These curves show the change in lead isotope ratios over time for systems that initially had different amounts of uranium. If a group of related ore deposits formed at a given time in the past by separation of lead from several different source areas, each with different initial amounts of uranium but with similar lead isotope evolution histories, plotting lead isotope values for these samples would define a straight line known as an *isochron* (Figure 2-9). The isochron line represents the ratio of equation (2-25) to equation (2-24):

$$\frac{(^{207}\text{Pb}/^{204}\text{Pb})_t - (^{207}\text{Pb}/^{204}\text{Pb})_T}{(^{206}\text{Pb}/^{204}\text{Pb})_t - (^{206}\text{Pb}/^{204}\text{Pb})_T} = \left(\frac{^{235}\text{U}}{^{238}\text{U}}\right)_{\text{present}} \frac{e^{\lambda_{235}T} - e^{\lambda_{235}t}}{e^{\lambda_{238}T} - e^{\lambda_{238}t}} \tag{2-27}$$

and can be expressed (Kanasewich 1968) as

$$y = b_0 + A(x - a_0) \tag{2-28}$$

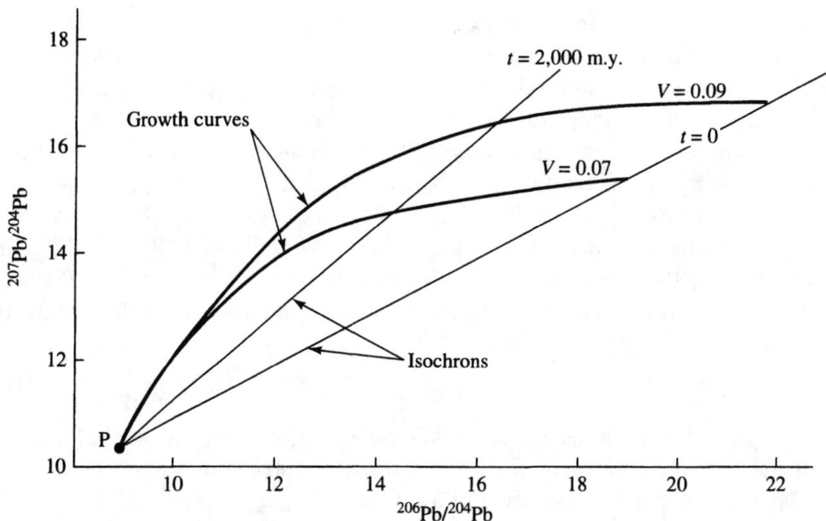

Figure 2-9 Theoretical growth curves for lead isotope ratios. Each growth curve represents the change in lead isotope ratios for a system presently containing particular $^{238}U/^{204}Pb$ and $^{235}U/^{204}Pb$ ratios. $V = {}^{235}U/^{204}Pb$, $^{238}U/^{204}Pb = 137.88\ {}^{235}U/^{204}Pb$, and m.y. = million years. Time t is the time when the lead was isolated from uranium, where $t = 0$ is the present. The isochron lines are a plot of equation (2-27) and pass through a point P, which represents the ratios of $^{207}Pb/^{204}Pb$ and $^{206}Pb/^{204}Pb$ at the time the Earth formed (primeval ratios). The slope of a given isochron depends only on times t and T of equation (2-27). Samples containing lead separated 2,000 million years ago from several different locally closed systems, each with differing initial amounts of uranium, would plot on the 2,000-m.y. isochron.

where

$$a_0 = \left(\frac{^{206}\text{Pb}}{^{204}\text{Pb}}\right)_T \qquad b_0 = \left(\frac{^{207}\text{Pb}}{^{204}\text{Pb}}\right)_T$$

$$A = \frac{e^{\lambda_{235}T} - e^{\lambda_{235}t}}{137.88(e^{\lambda_{238}T} - e^{\lambda_{238}t})} \quad \text{and} \quad \left(\frac{^{238}\text{U}}{^{238}\text{U}}\right)_{\text{present}} = 137.88$$

The isochron passes through the primeval abundances a_0 and b_0 and has a slope A determined only by T and t.

Equation (2-27) can be used to estimate the age of the Earth by putting it in the form

$$\frac{(^{207}\text{Pb}/^{204}\text{Pb})_{\text{present}} - (^{207}\text{Pb}/^{204}\text{Pb})_T}{(^{206}\text{Pb}/^{204}\text{Pb})_{\text{present}} - (^{206}\text{Pb}/^{204}\text{Pb})_T} = \left(\frac{^{235}\text{U}}{^{238}\text{U}}\right)_{\text{present}} \frac{e^{\lambda_{235}T} - 1}{e^{\lambda_{238}T} - 1} \tag{2-29}$$

where T = age of the Earth. Present-day lead isotopic composition for the Earth is estimated from analyses of various rocks and minerals (or of sediments formed from them), and the initial (primeval) composition is assumed to be that of troilite (FeS) in iron meteorites (this troilite has negligible uranium and thorium). The ratios in these meteorites are found to be $(^{207}Pb/^{204}Pb)_T = 10.29$ and $(^{206}Pb/^{204}Pb)_T = 9.31$ (Chen and Wasserburg 1983). Because modern lead (lead found in recently formed rocks and minerals) has a range of isotopic composition, calculations of the age of the Earth range from 4,550 to 4,750 million years. Another method to estimate the Earth's age involves plotting lead analyses of meteorites to obtain an isochron. Assuming that meteorites formed at the same time as the Earth, the age of the Earth is calculated from the slope of the isochron (given in the next paragraph). Direct dating of meteorites using the uranium-thorium-lead system and the rubidium-strontium system generally gives ages of 4,000 to 4,750 million years.

Equation (2-29) can be used to determine rock ages by substituting A for T, where A = the time of formation of a group of cogenetic igneous rocks. In this case the slope of the isochron is

$$\frac{e^{\lambda_{235}A} - 1}{137.88\,(e^{\lambda_{238}A} - 1)} \tag{2-30}$$

Equation (2-27) and other forms of it such as equation (2-29) cannot be solved for an age by conventional algebraic methods. They can be solved by a graphical method or by using a table giving the slopes of isochrons as a function of age (Table 2-9).

An age can be obtained for the ore district described above by plotting an isochron using measured lead ratios or by using these plus the calculated age of the Earth for T in equation (2-27). It should be noted that the "age" represented by the lead isotopic composition is the time at which the lead separated from its sources and not the time of crystallization of the ore minerals. In many cases, lead composition in ore districts is *anomalous* and does not fit such a simple model (*single-stage model*) of evolution (see Figure 2-10). Districts with anomalous lead must have had a more complicated history (such as contamination by lead in the upper crust or remobilization after initial deposition), and the resultant isotope ratios provide clues (but not definitive answers) to this multistage history. Similarly, the lead isotopic composition of large masses of the

TABLE 2-9 Numerical Values of $e^{\lambda_1 t} - 1$, $e^{\lambda_2 t} - 1$, and of the $^{207}Pb/^{206}Pb$ Isochron Slope Ratio as a Function of Age (t)

$\times 10^9$ years	$e^{\lambda_1 t} - 1$	$e^{\lambda_2 t} - 1$	$^{207}Pb/^{206}Pb$ (Slope)
0.0	0.0000	0.0000	0.04604
0.4	0.0640	0.4828	0.05471
0.8	0.1321	1.1987	0.06581
1.2	0.2046	2.2603	0.08012
1.6	0.2817	3.8344	0.09872
2.0	0.3638	6.1685	0.12298
2.4	0.4511	9.6296	0.15482
2.8	0.5440	14.7617	0.19680
3.2	0.6428	22.3716	0.25241
3.6	0.7480	33.6556	0.32634
4.0	0.8599	50.3878	0.42498
4.4	0.9789	75.1984	0.55714

$\lambda_1(^{238}U) = 1.55125 \times 10^{-10}\ y^{-1}$
$\lambda_2(^{235}U) = 9.8485 \times 10^{-10}\ y^{-1}$
$(^{207}Pb/^{206}Pb) = 1/137.88(e^{\lambda_2 t} - 1)/(e^{\lambda_1 t} - 1)$

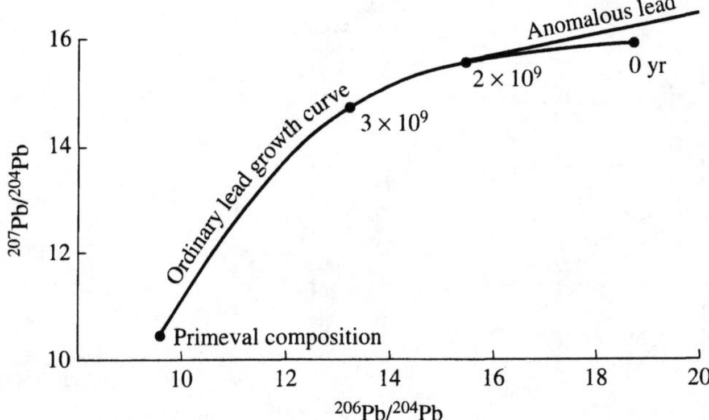

Figure 2-10 Diagram of $^{207}Pb/^{204}Pb$ versus $^{206}Pb/^{204}Pb$, indicating the primeval isotopic composition and a growth curve for ordinary lead corresponding to a source region in which $^{238}U/^{204}Pb$ (today) $= 9.08$. The isotopic composition of lead varies as shown according to the time of isolation from the source region. The straight line is made up of an array of points derived from anomalous lead from the Sudbury, Ontario mining district. Most ore deposits have anomalous lead. It may have formed by the addition of crustal radiogenic lead to ordinary lead (lead with isotope ratios that can be explained by a simple model of lead evolution). For a simple, single-stage model, all lead analyses from a single locality would plot on the ordinary lead growth curve at a point representing the time of separation from the source region. The estimated time of separation from the source is known as a "model age" and may not represent the actual time of formation of the analyzed lead-bearing minerals (which may have occurred much later). [After L. E. Long, 1972, *The Encyclopedia of Geochemistry and Environmental Sciences*, Rhodes W. Fairbridge, ed., Dowden, Hutchinson & Ross (Stroudsburg, Pa.) p. 647.]

Earth's crust can be studied for evidence of prior history and composition. Various mathematical and empirical models have been proposed as representative of lead development in the Earth. Other geologic evidence must be combined with the isotopic data in applying these models. Further information can be found in Kanasewich (1968), Doe (1970), Russell and Farquhar (1960), Albarede and Juteau (1984), and Faure (1986).

An example of the combined use of the Rb-Sr, Sm-Nd, and U-Th-Pb systems is given by Neumann et al. (1988). Isotopic analyses of a group of igneous rocks from the Oslo rift province of Norway were used to develop a mantle-crust mixing model for the origin of the rocks (Figure 2-11). The data suggest a complex origin for these rocks, involving mixing of materials from isotopically different source regions. The mildly depleted mantle proposed as the original source region for the Oslo rocks must have had periods of relatively high U/Pb and Th/Pb ratios (possibly caused by metasomatic enrichment in lithophile elements), that is, a multistage history. Isotopic studies such as this suggest that the source regions for continental basalts are isotopically distinct from mantle source areas of mid-ocean ridge basalts and of ocean island basalts.

Uranium-series isotopes have been used to investigate numerous problems in surface and groundwater hydrology, oceanic and estuarine chemistry, paleoclimatology, and archaeology (Ivanovich and Harmon 1992). For example, the uranium-thorium-lead system has been used to date young (less than one million years) sediments in the oceans (Krishnaswami and Lal 1982). In this case, abundances of the short-lived intermediate (radioactive) daughter isotopes of the uranium and thorium series are measured. The two methods that have been found must useful

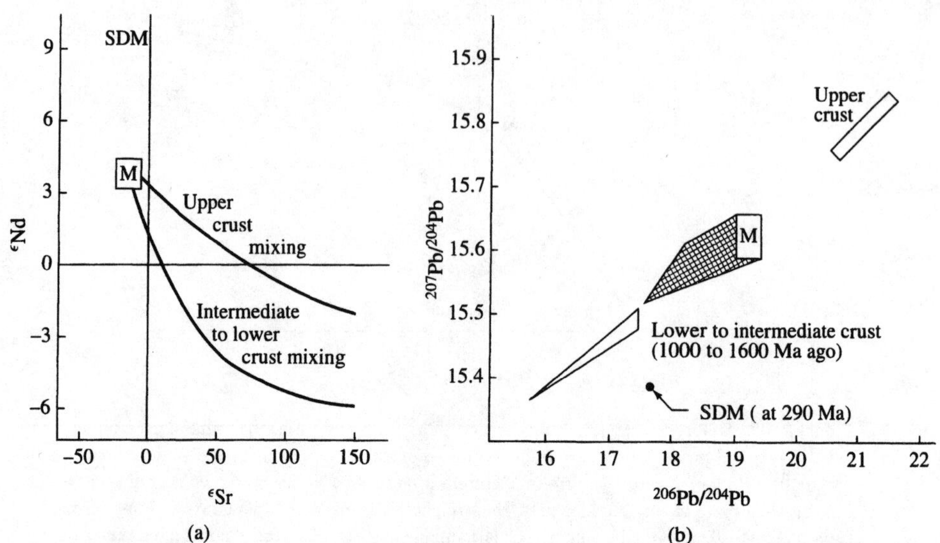

Figure 2-11 (a) Initial ϵ_{Nd}-ϵ_{Sr} relations among Oslo, Norway rift rocks. Most of the rocks analyzed fall in the box marked M or between the two curved lines, which represent various degrees of crystal contamination of material from a mantle source with the composition of M (a mildly depleted, somewhat heterogeneous mantle source region). Mixing occurred by assimilation of crustal rocks in basaltic magmas derived from the mantle. The majority of the samples plot near the lower curved line, indicating mixing with intermediate to lower crustal material. Samples plotting furthest from M are the most contaminated, with as much as 50 percent contamination indicated for some samples. SDM: strongly depleted mantle (source of mid-ocean ridge basalts). (b) Mixing model for Oslo, Norway rift rocks in terms of initial ^{207}Pb-^{206}Pb relations. Samples that plot in the box marked M are believed to be uncontaminated in the ϵ_{Nd}-ϵ_{Sr} system and plot at M in part (a). Samples believed to be contaminated by crustal material fall in the hatched area, indicating contamination by intermediate to lower crustal material. Samples plotting furthest from M are the most contaminated. M: a mildly depleted, somewhat heterogeneous mantle source region. SDM: strongly depleted mantle (source of mid-ocean ridge basalts). (Reprinted from *Geochimica et Cosmochimica Acta*, v. 52, E. R. Newmann, G. R. Tilton, and E. Tuen, Sr, Nd, and Pb isotope geochemistry of the Oslo rift igneous province, southeast Norway, pp. 1997–2007, Copyright 1988, with kind permission from Elsevier Science Ltd, The Boulevard, Langford Lane, Kidlington OX5 1GB, UK.)

involve the ratios ^{230}Th/^{232}Th (ionium-thorium method) and ^{231}Pa/^{230}Th (ionium-protactinium method). The radioactive lead isotope ^{210}Pb has been used to date recent (less than 200 years) events in lacustrine, estuarine, and marine sediment (Schell 1982). There are a number of problems in using these methods. Examples are (1) the possibility that migration of the pertinent elements occurs after sediment deposition, and (2) occurrence in the sediments of detrital material containing these elements.

In addition to alpha decay, ^{238}U also breaks down by spontaneous fission, producing various isotopes as fragments. The rate of breakdown is extremely slow compared to normal radioactive decay, but a decay constant can be determined for this process. The time of formation of a mineral containing ^{238}U can be found if the amount of ^{238}U and the amount of daughter (fission) products are measured. The daughter products have a great deal of energy, and they act as projectiles, which form tubular damage tracks in a mineral. The tracks are about 10 micrometers

(μm) long and can be revealed clearly by etching a mineral sample in an appropriate solution, such as hydrofluoric acid. Thus the amount of daughter products produced in a sample can be approximated by counting the number of *fission tracks.* The amount of ^{238}U is estimated by bombarding the sample with neutrons in a nuclear reactor. This causes fission of ^{235}U, and the number of new tracks produced in the sample is a measure of the sample's uranium content and thus of its ^{238}U content. A standard of known uranium content may be irradiated and analyzed at the same time to calibrate the neutron flux. An alternative method involves use of an external detector to measure the induced fission.

A general formula for obtaining an age using the fission track method is

$$F = {}^{238}\mathrm{U}\frac{\lambda_f}{\lambda}k(e^{\lambda t} - 1) \qquad (2\text{-}31)$$

where F is the number of fission tracks produced in time t per unit volume, ^{238}U is the number of ^{238}U atoms per unit volume, λ_f is the decay constant for spontaneous fission (8.46×10^{-17} y^{-1}), λ is the total decay constant (which is much larger than λ_f and essentially equal to the alpha-decay constant), and k is a constant determined by the range of tracks and their probability of observation on a two-dimensional surface.

Fission track ages can be obtained from a large variety of minerals and from glassy material such as obsidian. The method can be used to date very old (millions of years) to very young (thousands of years) samples. The precision of the method depends on analytical precision in measuring parent and daughter, on some uncertainty in the value of the spontaneous-fission decay constant (values used vary by as much as 15 percent), and on the careful selection of a value for k in the above equation. The method does not work well for rocks that have been heated as a result of deep burial or metamorphism, since the fission tracks are destroyed by annealing. Further information can be found in Fleischer et al. (1975).

POTASSIUM-ARGON SYSTEM

The radioactive decay of ^{40}K produces two daughter products ^{40}Ca and ^{40}Ar, formed as a result of normal beta decay and electron capture, respectively. Each daughter has a related decay constant (λ_β and λ_K), and the total decay constant for ^{40}K is the sum of these. Because argon is an inert element it is less likely to be taken into newly formed minerals, and most, if not all, argon in a mineral is a result of radioactive decay. The most abundant calcium isotope is ^{40}Ca, and calcium is more abundant than potassium in the Earth's crust. For these reasons, age dating is more reliable using the daughter argon isotope than the daughter calcium isotope, even though 89 percent of all ^{40}K decay is to ^{40}Ca.

Equation (2-5), $D = P(e^{\lambda t} - 1)$, must be modified for potassium decay as follows:

$$^{40}\mathrm{Ar} = {}^{40}\mathrm{K}\frac{\lambda_K}{\lambda_K + \lambda_\beta}[e^{(\lambda_K+\lambda_\beta)t} - 1] \qquad (2\text{-}32)$$

since only the ^{40}K that decays to ^{40}Ar is to be considered. The ratio λ_K/λ_β is known as the *branching ratio.* Solving the previous equation for t in years gives the age equation for the potassium-argon system:

$$t = \frac{1}{\lambda_K + \lambda_\beta}\ln\left[1 + \left(\frac{\lambda_K + \lambda_\beta}{\lambda_K}\right)\frac{^{40}\mathrm{Ar}}{^{40}\mathrm{K}}\right] \qquad (2\text{-}33)$$

The accepted values (by laboratory measurement) for λ_K and λ_β are 0.581×10^{-10} and 4.962×10^{-10}, respectively, and the final equation is

$$t = 1.804 \times 10^9 \ln\left(1 + 9.540\frac{^{40}\text{Ar}}{^{40}\text{K}}\right) \tag{2-34}$$

The amount of ^{40}K in a sample is determined directly with a mass spectrometer or by measurement of the total potassium content. In the latter case, the ^{40}K can be determined by using the ratio ^{40}K/total K, which is essentially constant in natural samples ($^{40}\text{K} = 0.01167$ percent of total K). The precision of these measurements is ± 2 percent. Although the analytical technique is more complicated, similar precision can be obtained in argon measurement. Taking into account uncertainties in the values of the decay constants, an age can be obtained (for samples that have not gained or lost argon) with a precision of ± 3 percent.

The main source of error in potassium-argon age determinations is argon leakage from a sample. This can occur fairly easily, because argon is a gas and is not chemically bound in a mineral. Heating due to metamorphism appears to be the main cause of argon diffusion. If all the argon is driven out of a mineral or rock by a metamorphic event and no further loss occurs, the time of metamorphism can be determined by dating the material. Laboratory and field studies indicate that hornblende is very resistant to argon loss; biotite is less resistant, and K-feldspars lose argon quite easily. Many K-Ar whole-rock ages are too low because of argon leakage from some minerals. Thus most dating is carried out on separates of minerals that tend to retain their argon.

Certain rocks and minerals, because of unusual conditions existing during their formation, contain *excess* ^{40}Ar. Examples include pillow basalts and minerals occurring in pegmatites and kimberlite pipes. In this case measured ages are too high.

A dating technique (the ^{40}Ar-^{39}Ar method) has been developed that can sometimes be used to distinguish samples that have experienced argon loss or gain (*disturbed samples*) from those that have remained closed systems since their original formation. In this method, a known fraction of the ^{39}K in a sample is converted to ^{39}Ar by irradiation with fast neutrons in a nuclear reactor. The argon is then released from the sample fractionally by stepwise heating. As fractions of the total ^{39}Ar and radiogenic ^{40}Ar are released, a series of apparent ages (known as an *age spectrum* or *release curve*) is obtained. The age calculations are based on the ratios ^{40}Ar/^{39}Ar and ^{40}K/^{39}K. [For further details see Mitchell (1968), Faure (1986), and McDougall and Harrison (1988).] If the sample has had an undisturbed history, the ratio of ^{40}Ar to ^{39}Ar will be the same for each increment of gas released, and each increment will therefore have the same ^{40}Ar-^{39}Ar age. If partial argon loss or gain has occurred during postcrystallization metamorphism, the apparent age of one or more of the gas increments will be different, and the age spectrum will not show a series of similar ages. Argon released at low laboratory temperatures during stepwise heating comes from sample sites that lose argon readily and represents argon accumulated since metamorphism. Argon at more retentive sites inside a sample may not have been affected by metamorphism and, when released at higher laboratory temperatures, indicates the age of original sample formation. Thus, in some cases, both the age of initial rock formation and the age of later metamorphism can be estimated (Figure 2-12). Therefore this technique is a valuable tool for studying samples from geologically complex areas.

The potassium-argon method is particularly useful for several reasons. Potassium is abundant and found in many minerals. Sedimentary rocks, which usually cannot be dated by other radioactive methods, can be dated if they contain the authigenic mineral glauconite. Volcanic rocks interlayered with sedimentary rocks can also be dated. In addition, direct ages can be obtained for

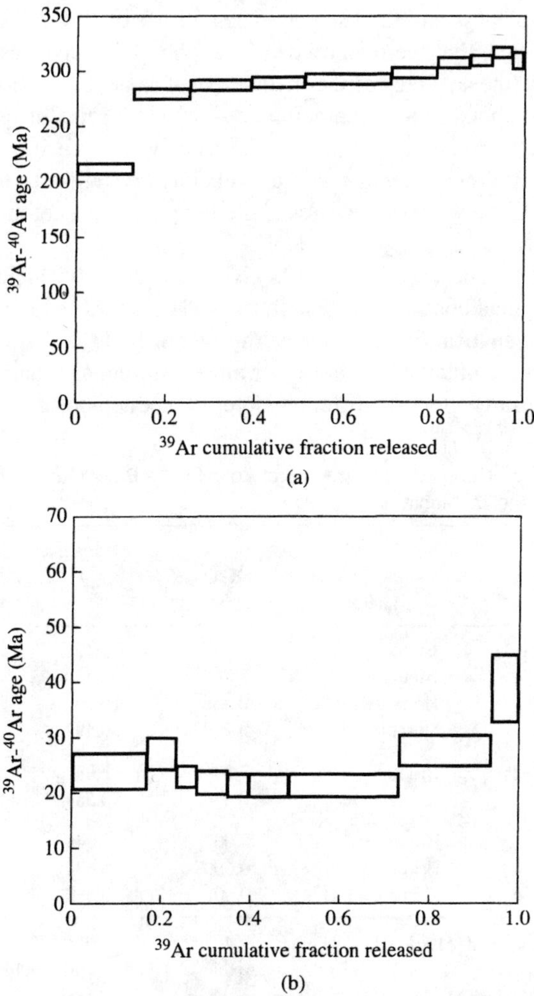

Figure 2-12 (a) Plot of determined ^{39}Ar-^{40}Ar ages as a function of cumulative fraction of ^{39}Ar released from a granitic clast from the Ries, Germany impact crater. This sample of shocked granite exhibits solid state deformation features and has a ^{39}Ar-^{40}Ar age of about 300–320 Ma. If the sample had been unaffected by the impact event at 15 Ma, the data would plot along a horizontal line at 310–320 Ma (the previously established crystallization age of the granite). The slightly lower apparent ages for small fraction releases suggests that the sample lost about 10 percent of its radiogenic argon as a result of the impact. The analytical uncertainty is given by the width of individual data "boxes." (b) Plot of determined ^{39}Ar-^{40}Ar ages as a function of cumulative fraction of ^{39}Ar released from a glassy "bomb" from the Ries, Germany impact crater. This strongly shocked (and partially melted) sample lost almost all of its radiogenic argon as a result of the impact event. Thus the sample's K-Ar "clock" was reset by the impact. The ^{39}Ar-^{40}Ar age is about 20–25 Ma; independent evidence dates the impact at 15 Ma. The analytical uncertainty is given by the width of individual data "boxes." (Reprinted from *Geochimica et Cosmochimica Acta*, v. 52, D. Bogard, F. Hörz, and D. Stöffler, Loss of radiogenic argon from shocked granitic clasts in suevite deposits from the Ries Crater, pp. 2639–2649, Copyright 1988, with kind permission from Elsevier Science Ltd, The Boulevard, Langford Lane, Kidlington OX5 1GB, UK.)

metamorphic rocks such as slates and schists. The half-life of ^{40}K and the ready detectability of radiogenic ^{40}Ar are such that the method can be applied to samples as old as the Earth and as young as 5,000 years (the accuracy of the method is not as good for very young samples). Since rubidium is found with potassium in many minerals, two independent ages can often be obtained for the same sample. Correction for initial argon usually is not necessary, since only minerals formed under unusual circumstances (such as minerals crystallized under very high pressure) have original ^{40}Ar. All these characteristics make the potassium-argon method one of the most versatile of the radioactive age methods.

An example of concordant K-Ar ages is given in Table 2-10. Included with the age values are estimates of analytical uncertainty. Also in the table are three examples of a second analysis of a mineral from a given locality. Even though the ages in Table 2-10 show very good agreement, the most reliable ages are obtained when two or more different age methods give similar results. Table 2-11 presents results when six different isotopic age methods are applied to the same rock.

TABLE 2-10 Concordant K-Ar Mineral Ages for the Butte Quartz Monzonite, Boulder Batholith, Montana

Sample locality	Mineral	K_2O (weight %)	Radiogenic ^{40}Ar (10^{-10} mol/g)	Age (m. y.)
Kain Quarry, Clancy	Biotite A	8.74	9.42	71.7 ± 2.3
	Biotite B	8.74	9.45	71.9 ± 2.2
	Hornblende A	0.456	0.503	73.3 ± 2.5
	Hornblende B	0.459	0.492	71.4 ± 2.5
Homestake Pass	Biotite	7.33	8.08	72.9 ± 3.6
	Hornblende	0.550	0.589	71.2 ± 3.6
Nine-Mile Station	Biotite	9.00	9.84	72.7 ± 3.0
	Hornblende A	0.600	0.691	76.5 ± 2.4
	Hornblende B	0.602	0.680	75.0 ± 3.4

Source: Tilling et al. (1968, 671–689).

TABLE 2-11 Comparison of Ages Obtained Using Six Major Isotope Age Methods

		Age (m. y.)					
Sample locality	Rock type	$^{238}U/^{206}Pb$	$^{235}U/^{207}Pb$	$^{232}Th/^{208}Pb$	$^{207}Pb/^{206}Pb$	$^{87}Rb/^{87}Sr$	$^{40}K/^{40}Ar$
Wichita Mts., Okla.	Granite	520	527	506	550	500	480
Llano, Tex.	Granite	950	990	890	1,070	1,100	1,090
Capetown, South Africa	Granite	330	356	238	530	600	530
Pikes Peak, Colo.	Granite	624	707	313	980	1,020	1,030
Bagdad, Ariz.	Granite	630	770	270	1,210	1,390	1,410
Bear Mt., N.Y.	Gneiss	1,140	1,150	1,030	1,170	930	900

Note: The U-Th-Pb ages were obtained from zircons and the Rb and K ages from micas.

Source: Tilton and Davis (1959, 190–216).

The granite from Oklahoma gives concordant results for all six methods. An average of the six age values is probably very close to the true age of the rock. The other rocks show varying degrees of discordancy. Note that even when the U-Th-Pb ages are discordant, the K-Ar and Rb-Sr ages tend to be concordant. The $^{207}Pb/^{206}Pb$ ages show closes agreement with the K-Ar and Rb-Sr ages. The lead-lead method often gives higher ages than the U-Pb and Th-Pb methods, since it is most sensitive to slight differences in the $^{238}U/^{206}Pb$ and $^{235}U/^{207}Pb$ ages and is least affected by any loss of lead from a rock.

Detailed discussions of the K-Ar method of dating are given by Hunziker (1979) and Faure (1986).

RADIOCARBON (^{14}C) SYSTEM

The ^{14}C age method differs from the previously discussed methods in several important ways. Radiocarbon, in contrast to the other radioactive isotopes, is continuously being produced in nature. The half-life of ^{14}C (5,730 years) is much shorter than that of the other radioactive isotopes, and this limits the application of the method to very young samples (less than 100,000 years). The daughter product (^{14}N) is not measured in applying the method. Finally, the material dated is usually organic in origin rather than inorganic. Greatest use of the method has been in the fields of Pleistocene geology and archaeology.

Atoms of ^{14}C are produced in the upper atmosphere as a result of reaction between cosmic-ray-produced neutrons and ^{14}N:

$$^{14}_{7}N + ^{1}_{0}n = ^{14}_{6}C + ^{1}_{1}H \tag{2-35}$$

The ^{14}C atoms are unstable and eventually decay by beta emission to ^{14}N. Radiocarbon combines with oxygen to form carbon dioxide, which mixes with carbon dioxide containing the other carbon isotopes (^{12}C and ^{13}C). Natural carbon is composed of about 98.9 percent ^{12}C and 1.1 percent ^{13}C, with a very minor amount of ^{14}C. Atmospheric mixing brings the $^{14}CO_2$ to the surface of the Earth where, along with other carbon dioxide, it is taken up by plants through photosynthesis. As long as a plant is alive it maintains a $^{14}C/^{12}C$ ratio representing the ratio in the atmosphere. It is assumed that equilibrium between ^{14}C production and decay has existed in the atmosphere during the past and thus that the concentration of ^{14}C has remained constant with time (actually small variations have occurred). Animals acquire ^{14}C from plants (or from carbon dioxide dissolved in water); thus, all living organisms have a $^{14}C/^{12}C$ ratio that is related to the ratio in the atmosphere. When an organism dies, it ceases to take up ^{14}C, so it can no longer maintain a constant $^{14}C/^{12}C$ ratio, and the ^{14}C content decreases with time owing to radioactive decay. The time of death of a plant or animal can be calculated by determining the present ^{14}C content of a portion of the remains. This is done by measuring the beta activity of the sample in terms of the number of disintegrations (decaying atoms) per minute per gram of sample. This number is proportional to the amount of ^{14}C in the sample.

The equation for radioactive decay, $N = N_0 e^{-\lambda t}$, can be written

$$\ln \frac{N}{N_0} = -\lambda t \tag{2-36}$$

Since the decay constant is related to half-life by $t_{1/2} = \ln 2/\lambda$, we can write

$$t = -\frac{t_{1/2}}{\ln 2}\ln\frac{N}{N_0} \tag{2-37}$$

$$t = -8{,}266 \ln\frac{N}{N_0} \tag{2-38}$$

where t is the age in years of a sample (time of death) and N/N_0 is the ratio of ^{14}C concentration in the sample (in terms of atoms per gram of carbon) to ^{14}C concentration in the atmosphere (initial ^{14}C concentration of the sample). Since N/N_0 is a fraction, the natural log will be a negative number, canceling the negative sign in the equation. What is actually measured in a sample is its activity due to ^{14}C. The ratio used in place of N/N_0 is A/A_0, where A is the activity of a sample and A_0 is the activity of ^{14}C in living plants and animals. Published ^{14}C dates are based on the equation

$$t = -8{,}035 \ln\frac{A}{A_0} \tag{2-39}$$

because the half-life of ^{14}C was originally believed to be 5,568 years; this number is still used to facilitate comparison of older with more recently published ages. The age of samples calculated with the lower half-life value should be increased by 3 percent to agree with the more accurate half-life value of 5,730 years.

Radiocarbon dating can be done on a wide range of materials. The most commonly dated substances are charcoal, wood, and peat. Also usually suitable for dating are shells, bones, paper, cloth, parchment, hair, hydrocarbons, and soils containing organic material. Practically any carbon-bearing object can be dated if it has enough organic content. Inorganic carbonate minerals and time of recharge of groundwater can sometimes be dated. For all materials the carbon present must have been derived directly or indirectly from the atmosphere with no subsequent alteration or recrystallization.

Radiocarbon analysis involves purification of a sample, conversion of carbon to carbon dioxide or other suitable gas, and measurement of radioactivitiy in a beta counter. The main analytical limitation is the statistical error in counting ^{14}C disintegrations. Younger samples have less error because of greater activity. The precision for good sample ranges from about ± 1 percent for very young samples to ± 3 percent for older samples.

A precise and sensitive instrument, the tandem accelerator mass spectrometer, has recently made it possible to directly count ^{14}C ions and thus to date carbon samples weighing less than one milligram (the beta counting method requires at least one gram of carbon).

Radiocarbon ages are usually reported as years *before present* (B.P.), with 1950 used as the reference year. Because of the short half-life of ^{14}C, samples older than about 70,000 years cannot be dated by use of the beta counting method. With an ultrasensitive mass spectrometer, direct counting of ^{14}C ions allows dating of samples as old as 100,000 years.

The basic assumptions of the method are that the production of ^{14}C in the atmosphere is (and has been) constant, and that a uniform distribution of ^{14}C exists (and has existed) between the atmosphere, surface and groundwater, and organic life. Small variations in the ^{14}C content of seawater and freshwater have been found, but these are not important in ^{14}C dating. A more serious problem is past variation in atmospheric content of radiocarbon. It is now known that the ^{14}C content of the atmosphere increased by about 10 percent during the period 6,000 to 2,000 B.P. This has been found by radiocarbon dating of material of known age, such as very old trees and material of Egyptian dynasties. Particularly useful has been a tree-ring chronology covering the last 7,100 years, which has been developed by study of the bristlecone pine and other plant species (Suess 1980). The cause of the change in ^{14}C content of the atmosphere is not known, but it is probably due either to variation in cosmic-ray intensity or to disequilibrium in ^{14}C distribution in natural materials. Variations of the order of ± 2 percent in atmospheric ^{14}C have been found for the past 2,000 years. Finally, variations in ^{14}C concentration have been produced since the nineteenth century by the burning of fossil fuel (which lowers the $^{14}C/^{12}C$ ratio) and by explosion of nuclear bombs (which increases the ratio). These known variations over the last 6,000 years result in an additional uncertainty (up to about 10 percent) in age measurements for this period. Similar uncertainties should probably be assumed for older samples.

A possible major source of error in radiocarbon dating is contamination of samples by younger carbon (an example would be a plant root growing into older peat, which is to be dated). Addition of a small amount of young carbon to a very old sample can cause a large error in age determination. Careful selection and treatment of samples can usually eliminate this problem. A number of articles dealing with ^{14}C dating can be found in Currie (1982) and Taylor et al. (1992).

Recent improvements in analytical techniques have made it possible to now use other cosmogenic radionuclides for dating of various natural materials and to study many different geologic processes. Table 2-12 lists these nuclides and their principal users. Applications are discussed by Faure (1986).

STABLE ISOTOPES AND GEOLOGY

We have already discussed the use in geology of stable isotopes formed as a result of radioactive decay. Other stable isotopes are also used in geologic research, and we now turn our attention to these isotopes. The total number of natural stable isotopes of a particular element varies from a high of ten for tin to a low of zero for 21 elements.* The average is about three stable isotopes per element. The total number of naturally occurring isotopes is over 300 (including both stable and unstable isotopes); these vary in abundance from large to extremely small amounts (Table 2-13). The most abundant isotopes are those that have a relatively stable type of nucleus (even number of protons and even number of neutrons).

Although a large number of stable isotopes occur naturally, only those of a few elements (H, O, S, C, and N) have the necessary properties (discussed later) to be of extensive use in geochemical studies. The primary requirement is that the ratios of the isotopes of an element vary among natural substances. This variation is due to completely different reasons from those that cause variations in the parent and daughter isotopes, discussed earlier in this chapter. Time as such is not a factor for abundance variations of isotopes like hydrogen and oxygen, which are not radioactive and which do not form from a radioactive parent isotope. Radioactive isotopes provide

*These 21 elements consist of one kind of atom only (a nuclide) and cannot be said to consist of one isotope according to the accepted definition of isotope (see discussion at beginning of this chapter). The term *nuclide* is a broader term than isotope and is used to refer to any particular nuclear species characterized by its number of protons and its number of neutrons.

TABLE 2-12 Principal Long-Lived Cosmogenic Radionuclides and Their Uses in Isotope Geoscience

Nuclide	Half-life (years)	Decay constant (years^{-1})	Principal uses
^{10}Be	1.5×10^6	0.462×10^{-6}	Dating marine sediment, Mn-nodules, glacial ice, quartz in rock exposures, terrestrial age of meteorites, and petrogenesis of island-arc volcanics
^{14}C	5730	0.1209×10^{-3}	Dating of biogenic carbon, calcium carbonate, terrestrial age of meteorites
^{26}Al	0.716×10^6	0.968×10^{-6}	Dating marine sediment, Mn-nodules, glacial ice, quartz in rock exposures, terrestrial age of meteorites
^{32}Si	276	0.251×10^{-2}	Dating biogenic silica, glacial ice
^{36}Cl	0.308×10^6	2.25×10^{-6}	Dating glacial ice, exposures of volcanic rocks, groundwater, terrestrial age of meteorites
^{39}Ar	269	0.257×10^{-2}	Dating glacial ice, groundwater
^{53}Mn	3.7×10^6	0.187×10^{-6}	Terrestrial age of meteorites, abundance of extraterrestrial dust in ice and sediment
^{59}Ni	8×10^4	0.086×10^{-4}	Terrestrial age of meteorites, abundance of extraterrestrial dust in ice and sediment
^{81}Kr	0.213×10^6	3.25×10^{-6}	Dating glacial ice, cosmic-ray exposure age of meteorites

Source: From G. Faure, *Principles of Isotope Geology.* Copyright ©1986 by John Wiley & Sons. Reprinted by permission of John Wiley & Sons, Inc.

us with age measurements and information on the history and source of rocks. As we will see, stable isotopes allow us to determine temperatures of formation for rocks and to identify sources of fluids involved in geologic processes.

The definition of atomic weight on the basis of the usual isotopic composition of an element is complicated by the fact that the relative proportions of the isotopes of some elements in nature are not constant. Thus, even though many atomic weights can be measured with a precision of about 1 in 10^6, atomic weight values are limited in precision by the variable isotopic composition in nature of many elements. For example, the natural variation in the relative abundance of ^{10}B and ^{11}B results in a variation of boron's atomic weight from 10.807 to 10.819. Original determinations of atomic weights in the nineteenth century were based on an arbitrary weight of 16.000 assigned to the oxygen atom. After the discovery, in 1929, that oxygen consists of three isotopes (^{16}O, ^{17}O, and ^{18}O), atomic weights were based on either natural oxygen (chemical atomic weights) or on the basis of ^{16}O set equal to 16.000 (physical atomic weights). In 1961 an international agreement led to the adoption of ^{12}C with a mass of exactly 12.000 as the reference standard for atomic mass values.

It is not surprising that lead in nature has been found to show a variation in atomic weight. Obviously, lead from a sample of galena will have a different atomic weight from that of lead in a zircon that has formed by radioactive decay of uranium and thorium (variations found span the range from 204 to 208). Less expected was the relatively large variation (± 0.01) found in the atomic weight of natural sulfur. Other elements that show natural isotopic variations large enough

TABLE 2-13 Selected Isotopic Abundances

Atomic number	Symbol	Mass number	Abundance (%) (by weight)	Atomic number	Symbol	Mass number	Abundance (%) (by weight)
1	H	1	99.9844	50	Sn	112	1.0
		2	0.0156			114	0.66
2	He	3	10^{-4}–10^{-5}			115	0.35
		4	100			116	14.4
6	C	12	98.89			117	7.6
		13	1.11			118	24.1
		14	10^{-10}			119	8.6
7	N	14	99.64			120	32.8
		15	0.36			122	4.7
8	O	16	99.763			124	5.8
		17	0.0375	60	Nd	142	27.1
		18	0.1995			143	12.2
16	S	32	95.02			144	23.9
		33	0.75			145	8.3
		34	4.21			146	17.2
		36	0.02			148	5.7
18	Ar	36	0.337			150	5.6
		38	0.063	62	Sm	144	3.1
		40	99.60			147	15.0
19	K	39	93.2581			148	11.2
		40	0.01167			149	13.8
		41	6.7302			150	7.4
20	Ca	40	96.9821			152	26.7
		42	0.6421			154	22.8
		43	0.1334	82	Pb	204	1.4
		44	2.0567			206	25.2
		46	0.0031			207	21.5
		48	0.1824			208	52.0
37	Rb	85	72.1654	90	Th	232	100
		87	27.8346	92	U	234	0.0057
38	Sr	84	0.56			235	0.7200
		86	9.87			238	99.2743
		87	7.04				
		88	82.53				

Sources: Faure (1986) and Kyser (1987). A listing of abundances (atoms/10^6 Si and abundance percent) of all the naturally occurring nuclides is given by Anders and Grevesse (1989).

to be measured are hydrogen, helium, lithium, boron, carbon, nitrogen, oxygen, silicon, copper, and molybdenum. There are a number of causes for these variations. We shall discuss them for the elements hydrogen, oxygen, sulfur, carbon, and nitrogen and show how study of variations in the abundance of the stable isotopes of these elements can be used to solve geologic problems. Before doing this, we need to review the theoretical basis of isotopic fractionation in natural processes.

Fractionation (change in relative abundance) of the isotopes of an element can occur in nature as a result of chemical, physical, or biological processes. This results in small differences in the abundance of the isotopes in various compounds. Complete separation of the isotopes of an element produces compounds with measurable differences in their properties (Table 2-14). Physical processes such as diffusion or evaporation are important in certain situations, but the

TABLE 2-14 Characteristic Constants of $H_2{}^{16}O$, D_2O, and $H_2{}^{18}O$

Property	$H_2{}^{16}O$	D_2O	$H_2{}^{18}O$
Density (20°C, in g cm^{-3})	0.9979	1.1051	1.1106
Temperature of greatest density (°C)	3.98	11.24	4.30
Melting point (760 Torr, in °C)	0.00	3.81	0.28
Boiling point (760 Torr, in °C)	100.00	101.42	100.14
Vapor pressure (at 100°C, in Torr)	760.00	721.60	—
Viscosity (at 20°C, in centipoise)	1.002	1.247	1.056

Source: J. Hoefs, *Stable Isotope Geochemistry*, 3rd ed. Copyright 1987. Reprinted by permission of Springer-Verlag, New York.

most significant fractionations are mainly due to chemical and biological reactions. The biological processes are complex and not well understood. On the other hand, chemical isotope effects have been studied in detail and can be explained on the basis of the differences in the masses and bonding properties of isotopes.

The energy of a molecule can be described in terms of its electronic energy (energy of electrons in relation to one another and to the nucleus), translational energy (that due to linear motion), rotational energy, and vibrational energy (except for the hydrogen isotopes). The vibrational frequency of a molecule depends on the forces holding its atoms together and on the masses of the atoms in the molecule. Vibrational motion occurs as a result of the stretching and compressing of the chemical bonds between atoms. Because isotopes differ in mass, two molecules with the same chemical formula (such as H_2O) but with different isotopes ($H_2{}^{16}O$ and $H_2{}^{18}O$ will have different vibrational frequencies. The molecule with the heavier isotope will be more stable (the forces holding it together will be stronger). At higher temperatures the energy differences between the two molecules decrease and thus less fractionation will occur in chemical processes.

The chemical properties of ^{16}O and ^{18}O are very similar because of similar electronic structure, but the existence of the different vibrational frequencies causes them to show slightly different behavior in chemical reactions. The free energy differences are not large enough to cause chemical reactions to occur, but they are large enough to produce fractionation when these reactions are produced by much larger free energy differences due to other causes (see Chapter Three). The degree of fractionation for a particular element depends on the relative difference in masses of the element's isotopes. Hydrogen (mass 1) and deuterium (mass 2) show a much greater difference in isotopic behavior than do ^{16}O and ^{18}O. Hydrogen isotope fractionations are about ten times greater than those shown by oxygen and other elements. Relative isotope mass differences decrease with atomic number and thus the extent of chemical isotopic fractionation shown by the elements decreases with increasing atomic number. Fortunately for radiometric dating, isotopic fractionation effects are insignificant in the relatively heavy elements used in these methods, except for carbon and, to a small extent, potassium.

Consider the following isotopic exchange reaction involving carbonate ion in water:

$$\frac{1}{3}C^{16}O_3^{2-} + H_2{}^{18}O = \frac{1}{3}C^{18}O_3^{2-} + H_2{}^{16}O \tag{2-40}$$

$$K = \frac{(C^{18}O_3^{2-}/C^{16}O_3^{2-})^{1/3}}{(H_2{}^{18}O/H_2{}^{16}O)} \tag{2-41}$$

The equilibrium constant is written in the form of ratios because ratios are used in analyzing actual samples. Samples and standards are compared by measuring differences in absolute isotopic ratios. Differences can be measured more precisely than the absolute ratios themselves.

It is possible to calculate the value of the above equilibrium constant by using partition functions, which are mathematical terms describing the energy relationships of molecules. Thus, using spectroscopic data, theoretical equilibrium constants for the reaction can be determined for different temperatures. These equilibrium constants can also be obtained by laboratory experiment (Figure 2-13) and by empirical methods using natural samples. For many isotopic reactions, use of the three different methods gives three different results that are not in close agreement. The theoretical value for the carbonate-water reaction above at 273 K is 1.033 and the experimental value is 1.036, representing relatively good agreement. Since the number is greater than one, this means that the $^{18}O/^{16}O$ ratio in carbonate ion is greater than the ratio in liquid water.

Because of problems in calculating equilibrium constants for complex molecules, Urey (1947) proposed the use of fractionation factors in place of equilibrium constants. The fractionation factor α is defined as the overall ratio of two isotopes of an element in one compound as compared to this ratio in a second compound:

$$\alpha_{A-B} = \frac{R_A}{R_B} \tag{2-42}$$

where A and B represent the two compounds (Friedman and O'Neil 1977). For the carbonate-water reaction above,

$$\alpha_{CO_3^{2-}-H_2O} = \frac{(^{18}O/^{16}O)_{CO_3^{2-}}}{(^{18}O/^{16}O)_{H_2O}} \tag{2-43}$$

By assuming the random distribution of the two different isotopes in the carbonate and water

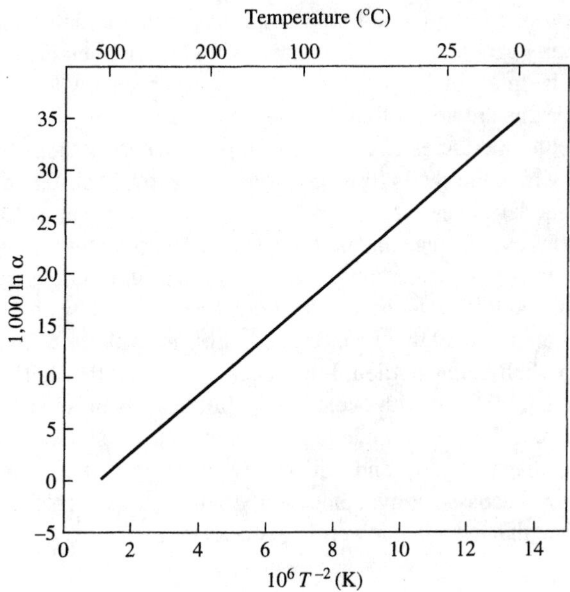

Figure 2-13 Oxygen isotopic fractionation between calcium carbonate and water. The low-temperature part of the line is based on slow precipitation of calcium carbonate under laboratory control; the high-temperature portion comes from experiments in which calcium carbonate was recrystallized in water in high-pressure hydrothermal apparatus. The results can be represented by the linear equation $1{,}000 \ln \alpha = 2.78(10^6 T^{-2}) - 3.39$, where α equals the overall ratio of ^{18}O to ^{16}O in calcium carbonate as compared to this ratio in water. The apparent straight-line behavior of the experimental systems is an approximation to a sigmoid curve, which is concave upward at high temperatures and concave downward at low temperatures, with an inflection point in the region of 200 to 300°C. (After J. R. O'Neil, R. N. Clayton, and T. K. Mayeda, 1969, Oxygen isotope fractionation in divalent metal carbonates, *J. Chem. Phys.*, v. 51, p. 5552.)

molecules, it can be shown that the fractionation factor α is related to the equilibrium constant as follows:

$$\alpha = K^{1/n} \qquad (2\text{-}44)$$

where n is the number of atoms exchanged. In the carbonate-water reaction above, only one atom is exchanged and $\alpha = K$.

Urey (1947) was the first to relate isotopic equilibrium constants to geologic problems. He suggested that the temperature of formation of calcium carbonate precipitated in the ocean could be determined by measuring the oxygen isotope composition of the carbonate. As noted above, the value of the calculated oxygen isotope equilibrium constant (and thus α) for the carbonate-water reaction at 273 K is 1.033. The calculated value at 298 K is 1.026 and there is a regular change in the equilibrium constant (and α) with temperature (Figure 2-13). If the $^{18}O/^{16}O$ ratio in ocean water is constant regionally and through time, then a specific change in temperature will cause a specific change in the $^{18}O/^{16}O$ ratio in carbonate ion precipitated from the ocean as part of carbonate minerals.

Urey's carbonate paleotemperature technique consists of the following steps: (1) measurement of the $^{18}O/^{16}O$ ratio in carbonate material formed by inorganic or biological precipitation in ancient seawater (some biological precipitation of chemical compounds seems to follow the rules of inorganic chemistry and does not produce complex "biological fractionation"); (2) assumption that the ancient seawater had an $^{18}O/^{16}O$ value similar to that of present-day seawater; and (3) comparison of the value of α found using the two isotopic ratios with theoretically calculated values at different temperatures or with a plot of α versus temperature found for the reaction by experiment (Figure 2-13). The change of the carbonate $^{18}O/^{16}O$ ratio with temperature is very small, but analytical techniques are such that variations in isotopic ratios as small as ± 0.0001 can be measured. This means that ideally temperature differences of less than 1°C can be determined. Obviously, samples to be analyzed must have formed under equilibrium conditions and must not have undergone alteration after original formation. It has been found that samples from Paleozoic and older rocks usually cannot be used; they have probably had their isotopic compositions altered by circulating fluids. In general, limestones become progressively lower in ^{18}O with increasing age. The reasons for this are not well understood.

In the first determinations of paleotemperatures by oxygen isotope analysis of shells, the isotopic composition of the seawater in which the shells formed was assumed to be constant and equal to the mean value found for the modern ocean. It is now known that the oxygen isotope composition of present-day seawater varies over a range that is equivalent to a temperature change of 9°C. Various methods of compensating for this uncertainty have been devised. Relative temperature changes can still be determined, and this problem can theoretically be solved by analyzing another compound (such as a phosphate mineral) that formed in equilibrium with the original seawater at the same time as the rock or shell being studied. It has been found that the shells of some organisms do not form in isotopic equilibrium with ocean water. Studies have also shown that aragonite is the best type of carbonate mineral to sample (calcite as the more stable form of $CaCO_3$ may be formed from aragonite diagenetically and with a new isotopic composition). Despite the problems and sources of error discussed above, paleotemperatures have been extensively measured and it has been established that a gradual lowering of ocean temperature occurred during the Tertiary (Figure 2-14).

Most of the discussion so far has dealt with equilibrium isotopic exchange. Nonequilibrium processes in nature, particularly evaporation, diffusion, and unidirectional chemical and biologi-

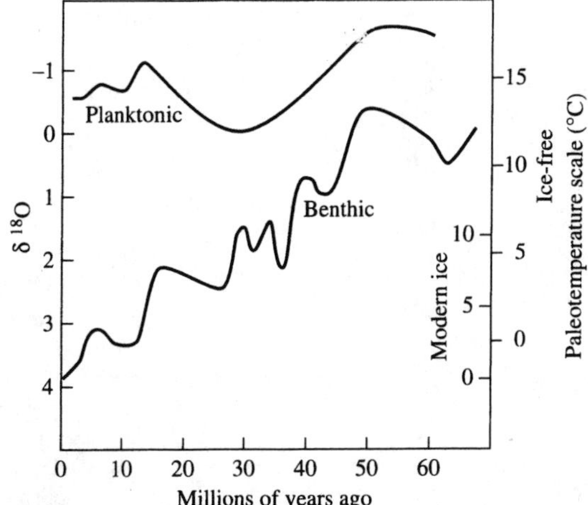

Figure 2-14 Oxygen isotopic paleotemperature curves for planktonic and benthic foraminifera from the North Pacific Ocean. The change in isotopic composition of the foraminifera (and the implied change in ocean isotopic composition) is due to two separate effects, namely, temperature change and ice volume change. As long as polar ice caps exist, ocean bottom water (which is produced at high latitudes) should have a relatively constant temperature. Thus the change in bottom water represented by the benthic line in this figure may be mainly due to changes in the isotopic composition of seawater as a result of changes in the amounts of polar ice (ice-volume effect). However, studies of Tertiary sediments from throughout the Pacific Ocean indicate that the temperature of bottom waters (and thus high-latitude surface waters) has irregularly decreased throughout the Tertiary for a total change of about 10°C (temperature effect). The planktonic line represents surface waters and the effect of changes in both temperatures and ice volume. It is not possible to quantitatively evaluate the relative effects of temperature and ice volume. Two benthic temperature scales are shown, one for times when polar ice caps are present (modern ice) and one for times with an absence of polar ice. To estimate isotopic temperatures for the planktonic foraminifera add approximately 2.5°C. (After J. R. O'Neil, Stable isotope geochemistry of rocks and minerals, In *Lectures in Isotope Geology*, E. Jäger and J. C. Hunziker, eds. Copyright 1979. Reprinted by permission of Springer-Verlag, Berlin.)

cal reactions, can also produce significant isotopic fractionations. When water evaporates, the greater average translational velocities of isotopically lighter $H_2{}^{16}O$ molecules allows them to escape preferentially.* As a result of continual evaporation, the water vapor above the oceans is depleted in ^{18}O compared to seawater (Figure 2-15). Condensation, in contrast to evaporation, is mainly an equilibrium process. Fractionation of isotopes as a result of diffusion in the solid state is significant in certain geologic processes, such as hydrothermal alteration of rocks. Light isotopes tend to be more mobile than heavy isotopes. Finally, many isotopic fractionations in nature result from unidirectional reactions. For example, bacterially-mediated chemical reactions are of this type. The significance of these nonequilibrium fractionations will be discussed later in the chapter.

Before discussing further uses of stable isotope measurements in studying geologic problems, we need to define the terms used in reporting these data. For most elements isotopic abundance measurements are carried out on gaseous species prepared from geologic samples. (Sample

*If the vapor phase reaches saturation, then equilibrium fractionation, which produces less isotopic separation than nonequilibrium fractionation, is the result. The oxygen isotopic fractionation factor between liquid water and water vapor at equilibrium is 1.0098 at 0°C.

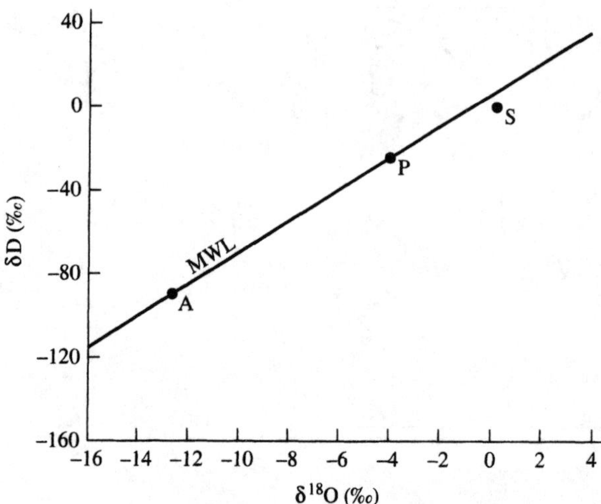

Figure 2-15 Variation in δ D and $\delta\ ^{18}O$ in the hydrologic cycle. MWL = meteoric water line; S = surface ocean water; P = mean precipitation over the oceans; A = mean atmospheric water vapor over the oceans. Evaporation of seawater produces vapor depleted in D and ^{18}O (point A). Condensation of the vapor produces precipitation enriched in ^{18}O and D (point P). After Welhan (1987).

collection and preparation, which is determined by the nature of the material being studied, must be done very carefully to obtain representative, homogeneous, and uncontaminated samples.) For example, oxygen isotope ratios are measured by chemically extracting oxygen from a solid sample and converting it to carbon dioxide gas. For water samples CO_2 gas is isotopically equilibrated with the water at a known temperature. The gas is introduced into a mass spectrometer, which separates and measures molecules on the basis of their masses.* The isotopic ratio found for a sample is given as the deviation from a standard by *delta values*, δ, in per mil (‰, or parts per thousand) units:

$$\delta\ ‰ = \left(\frac{\text{isotopic ratio for sample}}{\text{isotopic ratio for standard}} - 1\right) 1{,}000 \tag{2-45}$$

Ratios are of heavier to lighter isotopes ($^{18}O/^{16}O$, $^{34}S/^{32}S$), etc.). A higher ratio in the sample compared to the standard yields a positive number; a lower ratio a negative number. The delta value is the *relative difference* between a sample and a standard. For many elements the precision of a measured delta value is better than ± 0.05 per mil. For oxygen isotopes precision is on the order of ± 0.2 per mil.

The delta value represents isotopic measurement of a single sample. We also need a way to express the fractionation factor α for two different materials in terms of their measured delta values. We do this with the Δ value:

$$\Delta_{A-B} = \delta_A - \delta_B \approx 1000 \ln \alpha_{A-B} \approx \alpha_{A-B} \text{ on a per mil basis} \tag{2-46}$$

The Δ_{A-B} value is a close approximation to the per mil fractionation factor α for two substances A and B. For example, if $\delta_A = 4.2$ and $\delta_B = 0.6$, then $\alpha_{A-B} = 1.0036$ and $10^3 \ln \alpha_{A-B} = 3.6$ $= \Delta_{A-B} = \alpha$ on a per mil basis. Another way of expressing the relationship between α_{A-B} and δ_A and δ_B is

$$\alpha_{A-B} = \frac{R_A}{R_B} = \frac{1000 + \delta_A}{1000 + \delta_B} \tag{2-47}$$

*Isotopic ratios can be measured on very small samples using laser microprobe and ion microprobe analysis techniques. For more information see Shanks and Criss (1989).

In order to have uniformity in the reporting of delta values, internationally accepted reference standards are used (O'Neil 1986). The original standard for oxygen and hydrogen isotopes was a hypothetical water sample known as SMOW (Standard Mean Ocean Water); this was defined by Craig (1961). More recently, a new water standard (V-SMOW) has been prepared that has the same oxygen and hydrogen isotopic compositions as the original hypothetical SMOW. V-SMOW and a second standard (Standard Light Antarctic Precipitation, or SLAP) are available from the International Atomic Energy Agency in Vienna, Austria. For carbonate analyses a third standard, PDB (Peedee Belemnite) has been used. This is a solid belemnite sample from the Peedee formation in South Carolina. The original material is no longer available; however, secondary standards developed from it are still used. It has been redefined as V-PDB (see below). A newer carbonate standard, NBS-19, is used for both oxygen and carbon isotopes.

The $\delta\,^{18}O$ values of the various reference standards are *defined* as

$$\delta\,^{18}O_{SMOW} = 0.0$$

$$\delta\,^{18}O_{SLAP/V\text{-}SMOW} = -55.50$$

$$\delta\,^{18}O_{NBS\text{-}19/V\text{-}PDB} = -2.20$$

where $\delta\,^{18}O_{SLAP/V\text{-}SMOW}$ means the $\delta\,^{18}O$ value of the SLAP standard relative to the value of the V-SMOW standard. Thus the $\delta\,^{18}O$ value of SLAP is -55.50 per mil relative to V-SMOW. Tables 2-15 and 2-16 give compositions for various hydrogen and oxygen reference standards. Table 2-15 also lists the isotopic reference standard for sulfur, troilite from the Cañon Diablo meteorite, and the reference standard for nitrogen, atmospheric air (which has a constant isotopic composition everywhere).

TABLE 2-15 Primary Stable Isotope Reference Standards

Element	Standard	Comments
Hydrogen	V-SMOW	Vienna Standard Mean Ocean Water; identical to SMOW; D/H = 155.76×10^{-6}
Carbon	PDB	Peedee Belemnite; $^{13}C/^{12}C = 1123.75 \times 10^{-5}$
Oxygen	V-SMOW	Vienna Standard Mean Ocean Water; $^{18}O/^{16}O = 2005.2 \times 10^{-6}$
	PDB	Peedee Belemnite; $^{18}O/^{16}O = 2067.2 \times 10^{-6}$
Nitrogen	Air	NBS-14; $^{15}N/^{14}N = 367.6 \times 10^{-5}$
Sulfur	CDT	Cañon Diablo Troilite; $^{34}S/^{32}S = 449.94 \times 10^{-4}$

NBS = National Bureau of Standards.

Source: Kyser (1987, 448).

OXYGEN AND HYDROGEN ISOTOPES

Natural Waters

Water has two independent isotopic ratios, D/H and $^{18}O/^{16}O$, that exhibit measurable variations and that can be used to trace its source and history. There is a linear relationship between δ D and

TABLE 2-16 Isotopic Composition of Stable Isotope Standards

Standard	Description	δD	$\delta^{18}O$	$\delta^{13}C$
V-SMOW	Water	0.00	0.00	—
V-SLAP	Standard light antarctic precipitation	−428.0	−55.5	—
PDB	Calcite	—	+30.91	0.00
NBS-19	Toilet seat limestone	—	+28.65 (−2.20 PDB)	+1.95
NBS-18	Carbonatite	—	+7.20 (−23.00 PDB)	−5.00
NBS-30	Biotite	−65	+5.10	—
NBS-22	Oil	—	—	−29.63

δD and $\delta^{18}O$ values relative to SMOW (or PDB). $\delta^{13}C$ values relative to PDB.
Source: Kyser (1987, 449). NBS = National Bureau of Standards.

$\delta^{18}O$ for present-day meteoric waters (Figure 2-15). The diagonal line in Figure 2-15 is known as the meteoric water line (MWL) and is represented by the equation [from Craig (1961)]

$$\delta D = 8\delta^{18}O + 10 \tag{2-48}$$

The δD and $\delta^{18}O$ values of present-day meteoric waters vary because of fractionations due to evaporation of seawater and because of the nature of condensation processes in air masses (see below). Evaporation produces vapor depleted in the heavy isotopes D and ^{18}O. The liquid phase is enriched in ^{18}O and D during condensation. Seawater itself shows isotopic variations that correlate with salinity. Extensive evaporation leads to high salinity and enrichment in the heavy isotopes.

Welhan (1987) describes a model for condensation processes that defines the MWL in $\delta D - \delta^{18}O$ space as a function of the relative amount and rate of condensate removal from air masses relative to the isotopic evolution of the air masses. The model (a combined Rayleigh-batch separation model) predicts what is actually found:

1. atmospheric vapor has an isotopic composition with δ values that are more negative (the water contains less of the heavy isotopes) than atmospheric precipitation (Figure 2-15);
2. the isotopic compositions (δ values) of vapor and precipitation correlate with latitude, becoming more negative at higher latitudes;
3. in addition to the variation in latitude, there is a similar variation from coastal regions toward inland areas.

Thus the delta values of precipitation are much lower in polar regions as compared to equatorial regions and higher along the coast of a continent as compared to the interior of the continent (Figure 2-16). Sheppard (1986) points out that the mean isotopic compositions of meteoric waters at a specific locality are determined by many factors: temperature, latitude, altitude, distance from coasts, intensity of precipitation, local climate, and local topography. Each local area has its own meteoric water line that differs somewhat from the global average line illustrated in Figure 2-15.

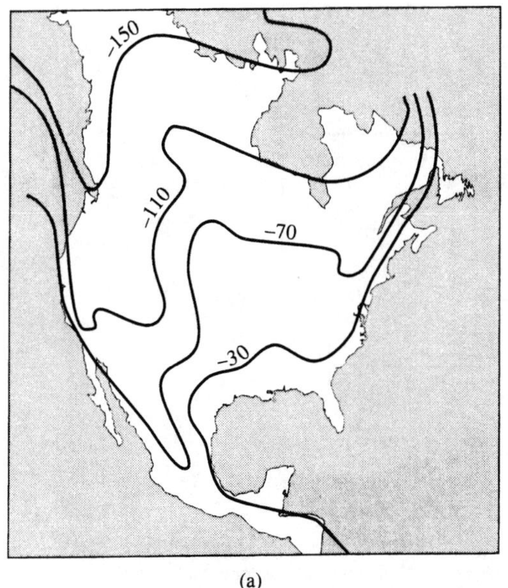

(a)

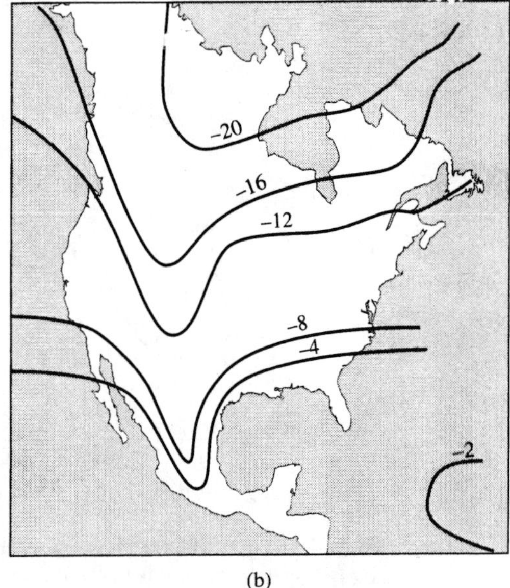

(b)

Figure 2-16 (a) Map of North America showing generalized contours of average δD values of present-day meteoric waters. Local climatic and topographic effects can distort the regional pattern. (After H. P. Taylor, Jr. Reproduced from *Economic Geology*, 1974, vol. 69, p. 850.) (b) Map of North America showing generalized contours of average $\delta^{18}O$ value of present-day meteoric waters. Local climatic and topographic effects can distort the regional pattern. After Sheppard (1986).

While we know the mean and range of isotopic compositions of present-day seawater and of meteoric waters, we can only approximate these values for other natural waters. Modern groundwaters are generally assumed to have the same isotopic compositions as those of the local precipitation (unless modified by surface evaporation). Ancient seawater, meteoric waters, and groundwaters probably had similar values to their modern counterparts. The isotopic compositions of other natural waters, ancient and modern, can be estimated, but are generally not well known (Figure 2-17). Often they are determined indirectly. For example, the isotopic composition of magmatic waters in Figure 2-17 has been calculated by using mineral analyses and mineral-water isotopic fractionation factors at appropriate temperatures. The general range of $^{18}O/^{16}O$ ratios in various minerals, rocks, and waters are listed in Table 2-17. Some examples of the use of this type of information to solve geologic problems are given in the next section.

Applications

One use of oxygen and hydrogen isotopes is in the study of surface water–groundwater interactions. For example, isotopic measurements have been used to determine the ground water contribution to streamflow during storm events (Figure 2-18a). It has been found that direct runoff on the surface does not contribute as much to total stream discharge as does groundwater discharge to the stream. Another example of this type of application is the determination of the amount of river water contribution to groundwater in the vicinity of a river (Figure 2-18b). In this case, as in the study shown in Figure 2-18a, the mixing of waters of two different isotopic compositions allows quantitative measurement of the relative amounts of the two components.

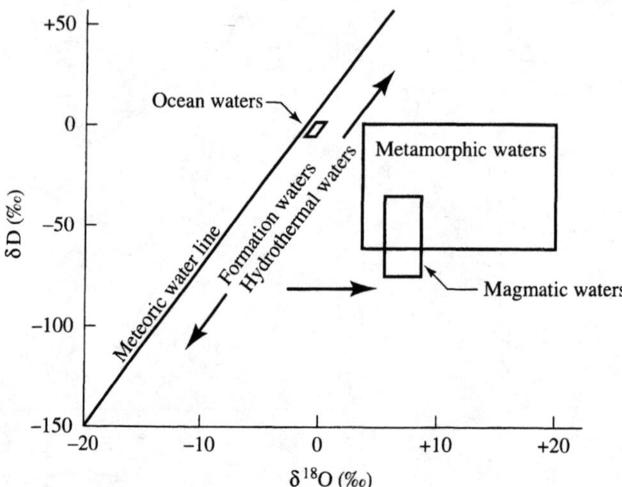

Figure 2-17 Isotopic fields for ocean waters, magmatic waters, metamorphic waters, sedimentary formation waters, and hydrothermal waters. The isotopic compositions of formation waters and hydrothermal waters scatter over a large area, reflecting the variety of ways in which they are formed and altered. SMOW (Standard Mean Ocean Water) plots at the upper right edge of the field for ocean waters. Most meteoric waters vary in a systematic way and plot along the meteoric water line (see discussion in text of Figure 2-15). After Sheppard (1986).

Isotopic fractionations between minerals can be used to determine equilibrium temperatures of formation. Theoretical studies have shown that the fractionation factor, α, between two phases can be described, at high temperatures, by

$$\ln \alpha = A + B/T^2 \tag{2-49}$$

where A and B are constants (O'Neil 1986). A similar relationship is found experimentally for mineral-water fractionations (Figure 2-13) and two different mineral-water fractionation factors can be used to get a mineral-mineral fractionation factor. Figure 2-19 shows calculated fractionation factors between quartz and several minerals.

Fractionation factors determined experimentally do not always agree with those calculated theoretically. This introduces uncertainty in the use of oxygen isotopes for geothermometry. Other possible sources of error are nonattainment of equilibrium and alteration of isotopic compositions after rock formation.

An example of the use of quartz-mineral pairs to obtain equilibration temperatures for metamorphic rocks (metapelites) from Zimbabwe is shown in Table 2-18. For these rocks $\delta^{18}O$ values of coexisting quartz and garnet yield temperatures of 621–846°C. Quartz-biotite and quartz-feldspar pairs give nearly concordant, but lower, temperatures of 384–643°C.

The source of an ore-forming fluid is indicated by the δD and $\delta^{18}O$ composition of the fluid. The isotopic composition can be directly estimated by measurements on fluids from fluid inclusions or indirectly estimated by using the delta values of minerals, mineral-water fractiona-

TABLE 2-17 General Range of $^{18}O/^{16}O$ Ratios (relative to standard mean ocean water)

Material	δ value range (‰)
Meteorites	
Achondrites	4–5
Chondrites	5–6
Carbonaceous chondrites	−1 to +12
Igneous rocks	
Granite pegmatites	7–14
Granites and quartz monzonites	7–9
Basalts and gabbros	6–7
Ultramafic rocks	5–6
Igneous minerals	
Quartz	8.9–10.3
K-feldspar	7.0–9.1
Plagioclase	6.5–9.1
Hornblende	5.9–6.9
Pyroxene	5.5–6.3
Biotite	4.4–6.6
Magnetite	1.0–3.0
Metamorphic rocks	
Marbles	15–27
Pelitic schists	12–18
Quartzites	10–15
Amphibolites	7–13
Metamorphic minerals	
Quartz	8–19
Plagioclase	7–14
Muscovite	6–20
Biotite	4–11
Garnet	4–12
Ilmenite	3–7
Magnetite	6–7
Chlorite	3–9
Sedimentary rocks	
Marine limestones	22–30
Freshwater limestones	18–25
Arkosic sandstones	12–16
Shales	14–19
Cherts and diatomites	28–36
Waters	
Ocean water	−0.5 to +0.5
Temperate fresh waters	−10 to −4
Snow and ice	−60 to −20
Meteoric and geothermal waters	−24 to +7

Source: Modified from Taylor (1967, 109–142).

tion factors, and an estimated temperature of formation. Figure 2-20 gives isotopic data for ore-forming fluids of some major porphyry-type ore deposits. It is not clear from these data whether the fluids that formed these deposits were mainly magmatic ($\delta\,^{18}O$ about +7 per mil to +13 per mil, δD about −9 per mil to −30 per mil) or mainly meteoric ($\delta\,^{18}O$ about −24 per mil to +7 per mil, δD about −160 per mil to 0 per mil). The earlier stages of ore formation may have involved magmatic water (isotopic values furthest from the meteoric water line) with the later stages dominated by meteoric water (values closer to the MWL). More detailed data for the Butte deposit are given in Figures 2-21 and 2-22.

Oxygen isotopes have been used to study the interaction of seawater with oceanic crust. Submarine weathering of ocean-floor basalts results in the formation of hydrated secondary minerals with higher $\delta\,^{18}O$ values than the primary igneous minerals. In addition to this low-temperature alteration, higher temperature hydrothermal alteration of igneous crust by hot seawater produces metamorphic rocks in areas of high heat flow, such as spreading ridges. Oxygen isotope data for a hole drilled to a depth of over 1000 meters is shown in Figure 2-23a. The first 600 meters consists of basalt that has undergone low-temperature alteration and that has $\delta\,^{18}O$ values averaging about 7 per mil (Alt et al. 1986; Muehlenbachs 1986). At 624 meters below the ocean

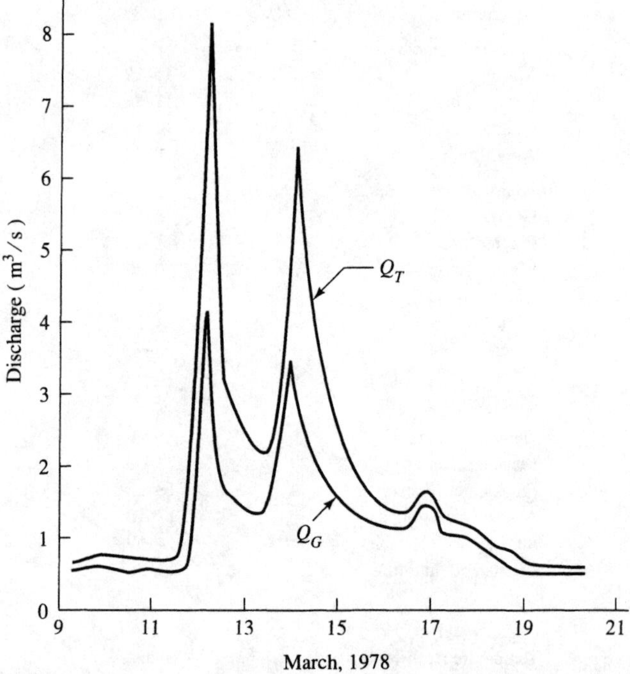

Figure 2-18a Discharge hydrograph of Lainbach Creek, Bavarian Alps, during three storm events. From isotopic measurements of precipitation and groundwater, the variation of δD in stream discharge can be used to separate the contribution of groundwater discharge to the stream, Q_G, from the total stream discharge, Q_T. These and similar studies demonstrate the importance of groundwater in basin storm runoff. The difference in isotopic composition between surface water and groundwater is mainly due to evaporative enrichment of heavy isotopes in surface water. After Welhan (1987).

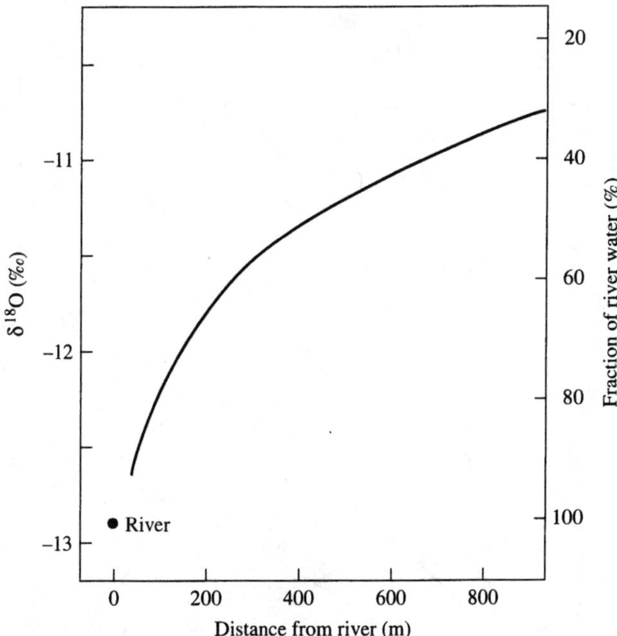

Figure 2-18b $\delta^{18}O$ of the Aare River and of groundwater in the Aare valley between Thun and Bern, Switzerland, as a function of distance from the river. Water from the river infiltrates into groundwater, and by measuring $\delta^{18}O$ the fraction of river water in the groundwater can be determined as a function of distance from the channel (right-hand scale). The Aare, with its catchment basin in the Alps at relatively high altitudes, has a lower $\delta^{18}O$ value (about −12.9 per mil) than groundwater formed locally (about −9.8 per mil). Thus isotopic measurements can be used to determine the relative amount of river water in the groundwater. (After U. Siegenthaler, Stable hydrogen and oxygen isotopes in the water cycle. In *Lectures in Isotope Geology*, E. Jäger and J. C. Hunziker, eds. Copyright 1979. Reprinted by permission of Springer-Verlag, Berlin.)

floor greenschist facies minerals are found and rocks from this depth to the bottom of the hole are depleted in ^{18}O, with $\delta^{18}O$ values ranging from 4.7 to 7.0 per mil. The oxygen isotope data reflect differences in the two types of alteration, one at low temperatures and with a high water/rock ratio (10/1 to 100/1) and one at higher temperatures and with water/rock ratios of about 1/1 (Muehlenbachs 1986).

Ophiolites represent obducted oceanic crust and provide the opportunity to study seawater-crust interaction in more detail than is possible with present-day oceanic crust. Figure 2-23b gives $\delta^{18}O$ values for a section through the Macquarie Island ophiolite (Cocker et al. 1986). The upper 500 meters represents low-temperature submarine weathering, as shown in Figure 2-23a for present-day crust. At greater depths the various rock types of the ophiolite exhibit generally lower $\delta^{18}O$ values due to high-temperature hydrothermal alteration. Thus we find that two types of isotopic change occur, one that increases the ^{18}O of the crust and one that decreases it. Using this type of data, Holland (1984) estimated the ^{18}O budget of the world's oceans and concluded that no net transfer occurs between seawater and crust. Despite large isotopic exchanges occurring at the present time, the $\delta^{18}O$ value of modern oceans appears to be fixed at about 0 per mil (Muehlenbachs 1986). It is not clear that this value has been constant throughout geologic time.

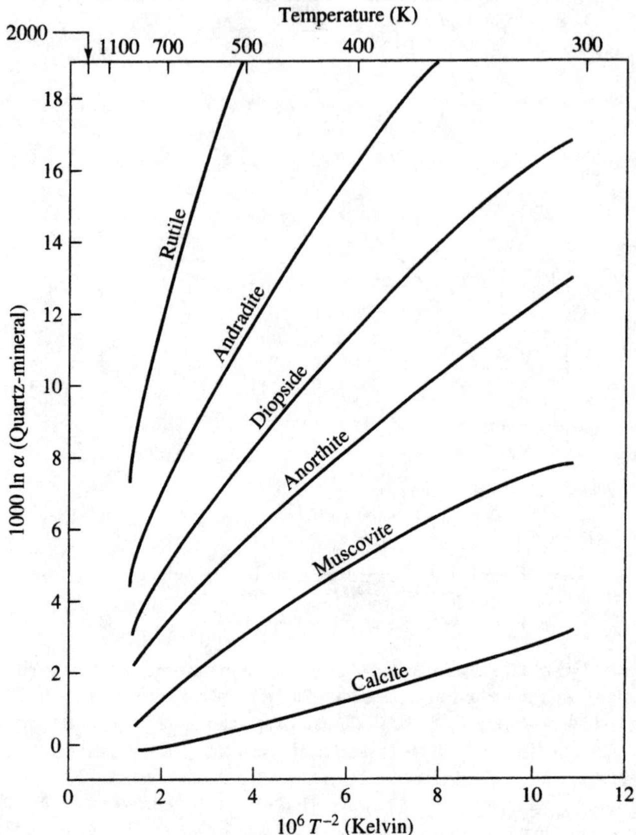

Figure 2-19 Calculated oxygen-isotope fractionation factors between quartz and the minerals indicated. After Kieffer (1982) and O'Neil (1986).

TABLE 2-18 Apparent Equilibration Temperatures (°C) of Quartz-Mineral Pairs from Zimbabwe Metapelites

Sample	Quartz-feldspar	Quartz-garnet	Quartz-biotite
Bb 11b	562	683	532
Bb 11d	463	661	511
Bb 12a	460	750	539
Bb 13b	464	707	505
Bb 13e	384	621	—
Bb 16c	538	846	561
Bb 17a	437	779	518
Bb 20	541	846	528
Bb 25c	—	707	643

Fractionation factors used from Bottinga and Javoy (1975, 401–408).
Source: Huebner et al. (1986, 1343–1353).

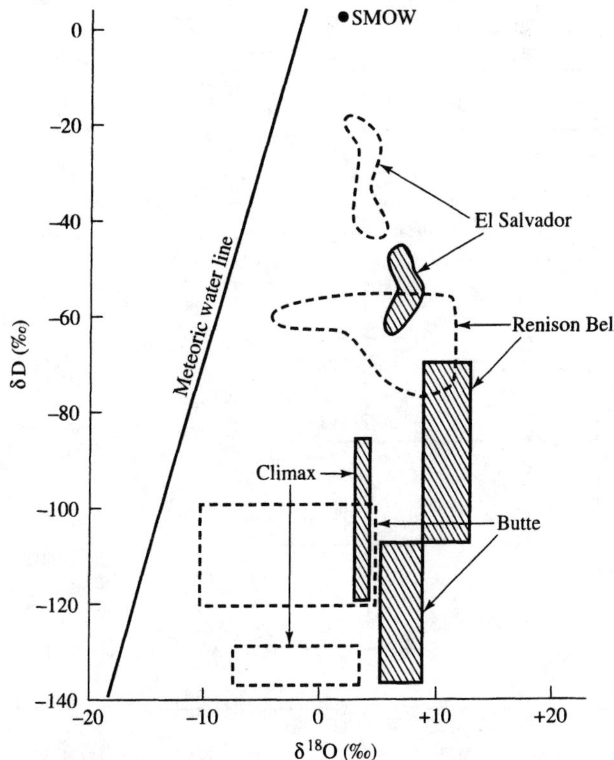

Figure 2-20 δD–$\delta^{18}O$ characteristics of ore-forming fluids for selected deposits associated with porphyries. Hatched areas = early fluids and open areas = later fluids. SMOW = Standard Mean Ocean Water. After Ohmoto (1986).

SULFUR ISOTOPES

Sulfur has four stable isotopes (^{32}S, ^{33}S, ^{34}S, and ^{36}S); ^{32}S (95.02 percent) and ^{34}S (4.21 percent) are the most abundant. Although more variation probably occurs in the $^{32}S/^{36}S$ ratio because of the greater difference in mass, the $^{32}S/^{34}S$ ratio is usually measured, since the far greater abundance of ^{34}S compared to ^{36}S allows the latter ratio to be measured with much better precision. The isotopic composition of sulfur is commonly expressed as $\delta^{34}S$ in a manner similar to that used for δD and $\delta^{18}O$.

Measurement of sulfur isotope ratios is carried out by separating the sulfur-bearing material from a sample, converting the material to sulfur dioxide, and analyzing the gas in a mass spectrometer. Values of $\delta^{34}S/^{32}S$ can be determined to ± 0.02 per mil. The $^{32}S/^{34}S$ ratios found for the troilite phase of meteorites show only slight variation from the standard value of 22.22 adopted for the Cañon Diablo troilite. It is generally assumed that primordial earth sulfur and the Earth's mantle also have a $^{32}S/^{34}S$ value of 22.22 ($^{34}S/^{32}S = 0.0450045$ and $\delta^{34}S/^{32}S = 0$).

The relatively large variation in atomic weight of sulfur (due to variations in the relative abundances of its isotopes) compared to other elements is due to the existence in nature of several valences for sulfur (-2, 0, $+4$, and $+6$). Isotopic fractionation not only occurs in a manner similar to that discussed for oxygen, but such fractionation is enhanced as a consequence of oxidation-reduction reactions. These reactions involving isotopes often have, in addition to a change in mass, a significant change in bonding, and this increases the resulting difference in molecular vibrational energies. Oxidation-reduction reactions can produce large fractionations and can occur in nature inorganically or through biological processes. It is generally true that the heavier

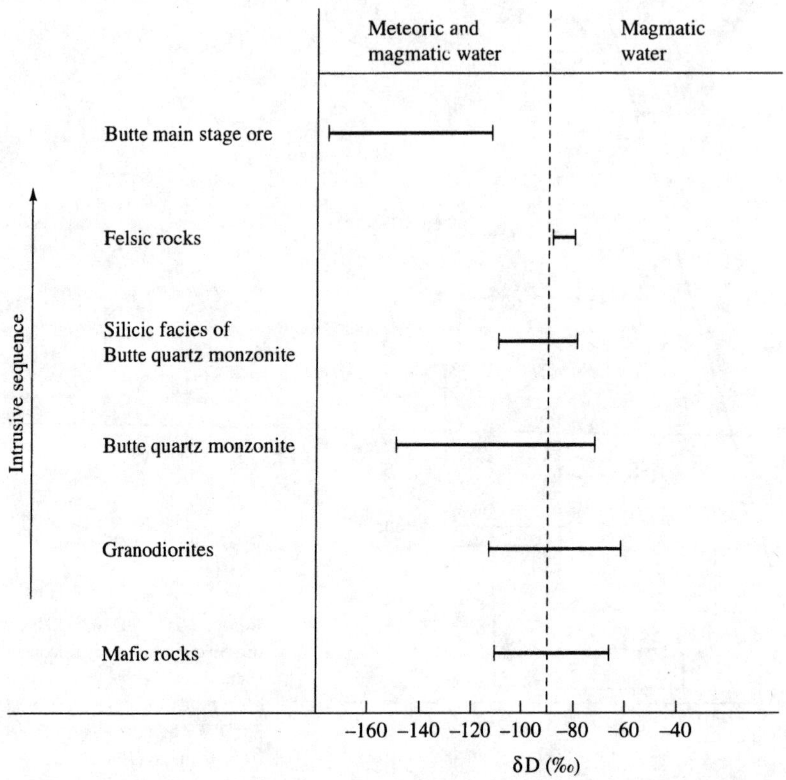

Figure 2-21 Range of δD values for mineral and whole-rock samples from the Boulder batholith and the associated Butte ore deposit. Biotite and hornblende were analyzed for the batholith rocks. Hydrothermal alteration minerals (biotite, sericite, dickite, kaolinite, pyrophyllite, and montmorillonite) were analyzed for the ore deposit. There is no correlation between position within the intrusive sequence and δD values. All the batholith minerals and rocks probably initially had δD values controlled by magmatic waters and therefore values more positive than −90. Local interaction of some of the rocks with low-δD groundwater at the end of batholith formation produced δD values more negative than −90. The δD values for the Butte ore deposit are interpreted as indicating formation from hydrothermal fluids that were dominantly meteoric in origin. Data from Sheppard and Taylor (1974).

of two isotopes goes preferentially to the compound in which the element in question is most strongly bonded. In the case of sulfur, the bond strengths of oxidation products are usually much greater than those of reduction products. Therefore, the partial oxidation of a sulfide mineral to a sulfate mineral results in an enrichment of ^{34}S over ^{32}S in the sulfate.

In addition to classifying sulfur isotope fractionations as inorganic versus organic (biologic), we can also divide the various fractionations that occur into equilibrium and kinetic fractionations. Equilibrium fractionations are generally inorganic and can be approximated, for two coexisting materials A and B, by

$$\delta_A - \delta_B \approx 1000 \ln \alpha_{A-B} \tag{2-50}$$

as was discussed earlier for oxygen isotopes. If equilibrium has been reached by the coexisting materials, then it may be possible to use a sulfur isotope thermometer (Figure 2-24).

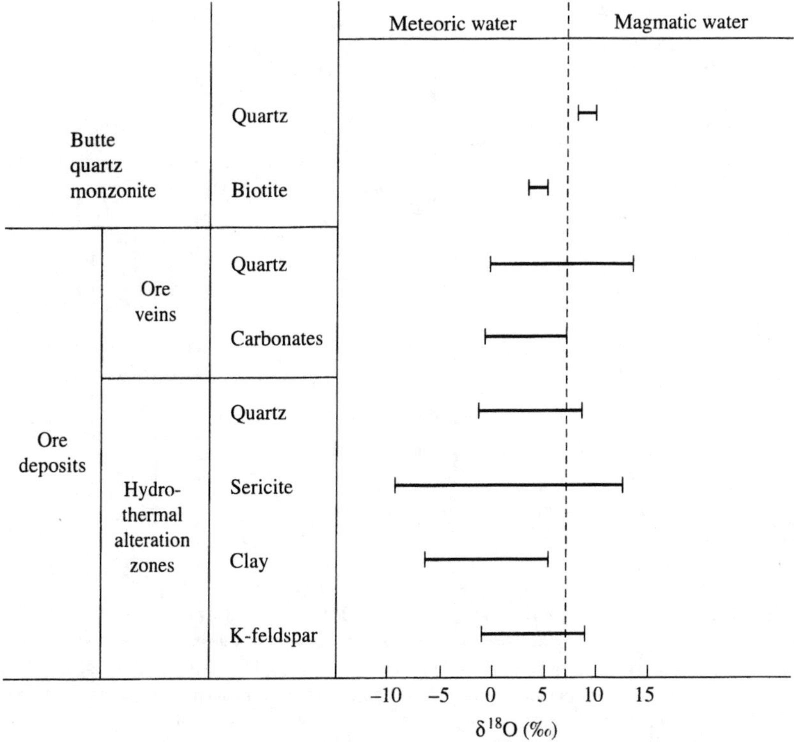

Figure 2-22 Range of $\delta\,^{18}O$ for minerals of the Butte ore deposit and of the host Butte Quartz Monzonite. Note the narrow range of values for the host rock minerals and the wide range of values for the minerals formed as a result of hydrothermal ore deposition. In contrast to this, δ D values show a large range for biotites of the Butte Quartz Monzonite (−155 to −78) and a large range for alteration minerals of the ore deposit (−179 to −116). The biotites of the Butte Quartz Monzonite that give δ D values similar to those of the alteration zones come from rocks that show signs of deuteric or hydrothermal alteration (Figure 2-21). The rocks showing this alteration have $\delta\,^{18}O$ values characteristic of meteoric water. Data from Sheppard and Taylor (1974).

An example of kinetic fractionation, and a major cause of variations in sulfur isotopic composition in nature, is the process of bacterial sulfate reduction. Certain anaerobic bacteria use the reduction of sulfate ion to hydrogen sulfide for their metabolism. Fractionation occurs because the rate at which $^{34}SO_4^{2-}$ goes to $H_2{}^{34}S$ is significantly slower than the rate at which $^{32}SO_4^{2-}$ goes to $H_2{}^{32}S$. Thus the result is light sulfide and heavy sulfate. Large bacterial fractionations from this process have been found in laboratory experiments. In nature the extent of bacterial fractionation is quite variable, depending on environmental conditions (Trudinger and Chambers 1973; McCready et al. 1974). The overall fractionation patterns in the sulfur cycle due to biological processes are shown in Figure 2-25. Because of complications due to biological and kinetic effects in most natural environments, use of sulfur equilibrium fractionations to determine paleotemperatures has been limited mainly to the study of hydrothermal ore deposits (Figure 2-24).

The general range of $\delta\,^{34}S$ values found in natural materials is given in Table 2-19. Sulfides from igneous rocks show a narrow range of values, with most slightly enriched in ^{34}S. Other types of material, particularly sedimentary sulfides and sulfides from hydrothermal ore deposits, show

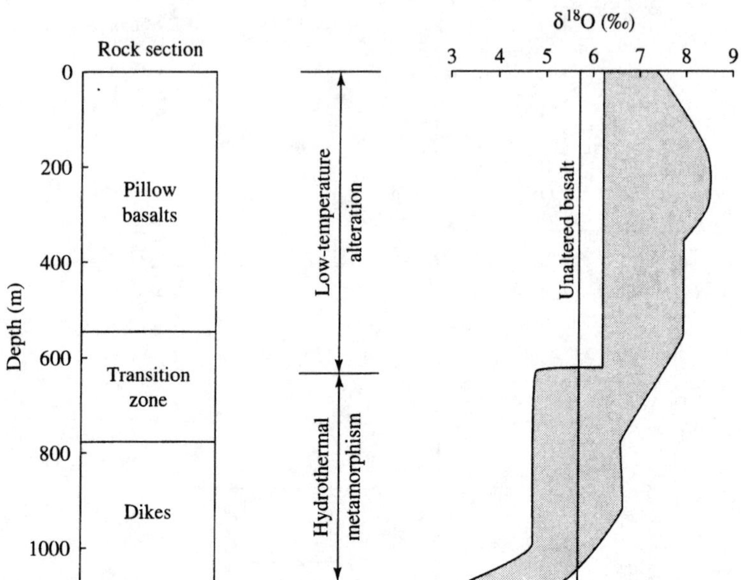

Figure 2-23a Whole-rock δ^{18}O data versus depth in Deep Sea Drilling Project Hole 504B. The vertical line at 5.7 per mil indicates the value for primary, unaltered basalt. Shallow basalt has undergone submarine weathering and low-temperature (less than 150°C) alteration as indicated by δ^{18}O values from about 6 per mil to over 8 per mil. Deeper rocks have been subjected to hydrothermal metamorphism (temperatures from 100 to 400°C), resulting in δ^{18}O values as low as 3.6 per mil. After Alt et al. (1986) and Meuhlenbachs (1986).

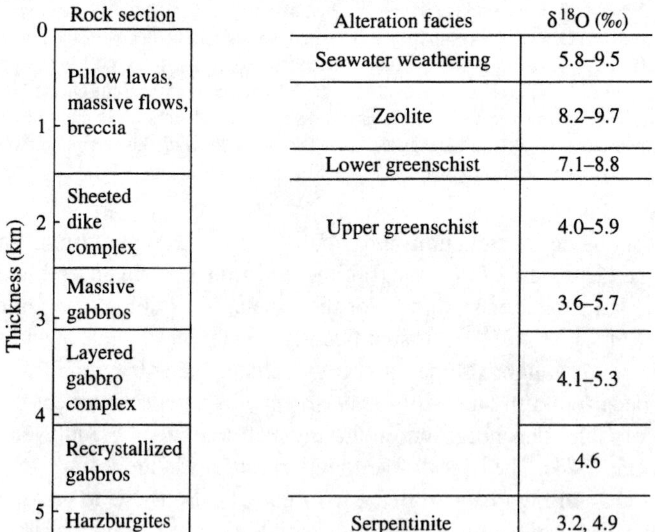

Thickness (km)	Rock section	Alteration facies	δ^{18}O (‰)
0–1	Pillow lavas, massive flows, breccia	Seawater weathering	5.8–9.5
		Zeolite	8.2–9.7
		Lower greenschist	7.1–8.8
2	Sheeted dike complex	Upper greenschist	4.0–5.9
3	Massive gabbros		3.6–5.7
4	Layered gabbro complex		4.1–5.3
	Recrystallized gabbros		4.6
5	Harzburgites	Serpentinite	3.2, 4.9

Figure 2-23b Whole-rock δ^{18}O data and alteration facies for a section of the Macquarie Island ophiolite. The rocks represent on-land exposure of old ocean crust. The upper part of this section is similar to that shown for present-day ocean crust in Figure 2-23a. After Cocker et al. (1982) and Meuhlenbachs (1986).

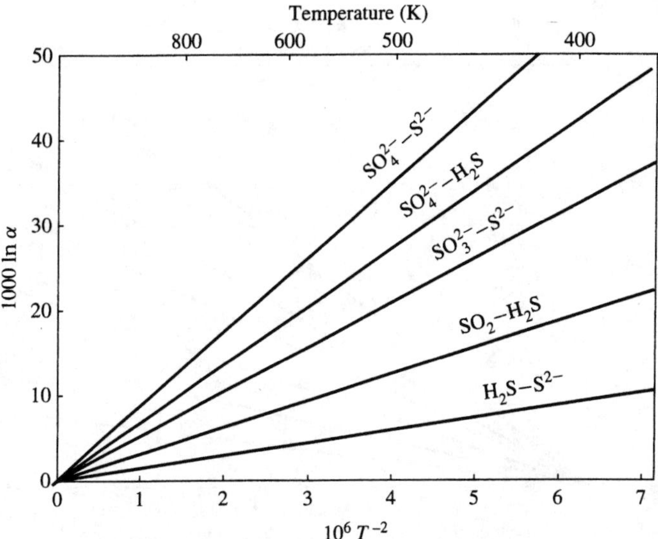

Figure 2-24a Calculated temperature curves of 1000 ln α for some major pairs of sulfur compounds under isotope exchange equilibrium. Experimental and theoretical studies generally agree on the order of ^{34}S enrichment among various sulfur compounds. However, the actual values of the fractionation factors obtained by different researchers do not show good agreement. (After H. Nielsen, Sulfur isotopes. In *Lectures in Isotope Geology,* E. Jäger and J. C. Hunziker, eds. Copyright 1979. Reprinted by permission of Springer-Verlag, Berlin.)

wide variations in isotopic composition. Such variations probably reflect the variety of processes and geologic conditions involved in forming these sulfides. The lightest sulfur values have been found in sulfide minerals, hydrogen sulfide gas, and native sulfur. Heaviest sulfur occurs in sulfate minerals; the most extreme values have been found in anhydrite from the calcite cap rock of salt domes. Carbon and sulfur isotopic studies of native sulfur deposits associated with salt domes have established that the sulfur deposits formed as a result of bacterial reduction of anhydrite and gypsum to H_2S, which was subsequently oxidized to native sulfur.

We now review briefly some examples of the use of sulfur isotopes to study geologic problems. The nature of sulfur isotopic variation in recent marine environments has been summarized by Kaplan (1983). He reports that sulfur isotope studies on recent marine sediments show:

1. dissolved hydrogen sulfide in the sediment pore water is isotopically enriched in ^{32}S by -40 per mil compared to dissolved seawater sulfate;
2. dissolved sulfate in the pore waters decreases with depth due to bacterial reduction of sulfate ion to dissolved hydrogen sulfide (which increases with depth);
3. associated with the depletion of sulfate ion with depth is an increase with depth in $\delta^{34}S$ of sulfate ion;
4. dissolved hydrogen sulfide also becomes increasingly enriched in $\delta^{34}S$ with depth; and
5. total sulfur decreases with depth due to precipitation of sulfur as sulfide minerals (mainly pyrite).

All of the above results are consistent with bacterial reduction of seawater sulfate as the dominant process controlling sulfur distribution in marine sediments.

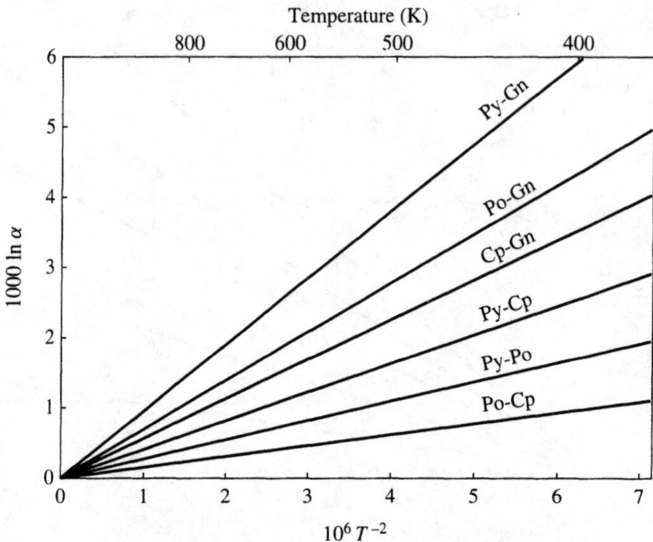

Figure 2-24b Calibration curves for sulfide pair thermometers. The sulfides are each assumed to be in isotopic exchange equilibrium with the dissolved sulfide in a common fluid. Data from calculations and laboratory experiments. Ohmoto (1972) has pointed out that the $\delta^{34}S$ values of a set of hydrothermal minerals are controlled by, in addition to temperature, the ionic strength, pH value, isotopic composition, oxidation state, and total sulfur content of the parent hydrothermal fluid. Py = pyrite, Gn = galena, Cp = chalcopyrite, Po = pyrrhotite. (After H. Nielsen, Sulfur isotopes. In *Lectures in Isotope Geology*, E. Jäger and J. C. Hunziker, eds. Copyright 1979. Reprinted by permission of Springer-Verlag, Berlin.)

Cameron (1982) has used plots of mean $\delta^{34}S$ composition versus standard deviation $\delta^{34}S$ to distinguish different types of Precambrian sedimentary sulfide. He proposes three types: biogenic, hydrothermal, and closed system. Biogenic sulfide is formed by the process discussed above for modern marine sediments. The sulfide is depleted in ^{34}S (δ values of about -20 per mil) and shows a relatively large variation in $\delta^{34}S$ values due to the inherent variation in environmental conditions (sedimentation rate, amount of organic matter present, etc.) of marine sediments. Hydrothermal sulfide forms in sediment as a result of hot springs and submarine volcanic eruptions on the sea floor. The sulfur for sulfides in this case may come from magmas ($\delta^{34}S \approx 0$) or from the high-temperature, inorganic reduction of seawater sulfate (giving less loss of ^{32}S than in biogenic reduction). Hydrothermal sulfide is less strongly depleted in ^{34}S (values near $+5$ per mil) and tends to show less variation in $\delta^{34}S$ among samples. Closed system sulfides form in environments where free exchange with a large amount of seawater is not possible (for example, a basin closed off from the main ocean). Sulfides formed under these conditions have high values of $\delta^{34}S$ (near $+17$ per mil, greater than those of hydrothermal sulfides, which in turn are greater than those of biogenic sulfides) and show large variations in $\delta^{34}S$ from sample to sample. Cameron (1983) analyzed sulfides from seven different Proterozoic iron formations. He found that the mean $\delta^{34}S$ composition of the seven units varied between -4.9 per mil and $+6.6$ per mil and that the sample variance is low within each unit. These data thus support a hydrothermal origin for sulfide-facies iron formation and, by extension, for the other facies of iron formation (carbonate, silicate, and oxide).

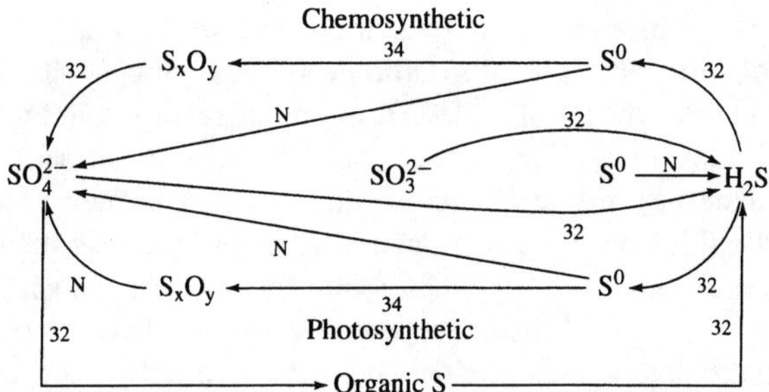

Figure 2-25 Fractionation patterns of ^{32}S and ^{34}S by biological processes in the sulfur cycle. The isotope enrichment is indicated in the final and intermediate products. For example, ^{32}S is enriched in S° formed by chemosynthesis oxidation of H_2S. No fractionation is designated by N. Photosynthetic processes are carried out by organisms that get their energy from sunlight in photosynthesis. Chemosynthesis processes are carried out or caused by organisms that utilize energy from chemical reactions. S_xO_y refers to compounds such as hyposulfite, S_2O_4. The most important organisms involved in the sulfur cycle are microbes (microorganisms) such as bacteria and algae. (After I. R. Kaplan, 1983, Stable isotopes of sulfur, nitrogen and deuterium in Recent marine environments, in M. A. Arthur, T. F. Anderson, I. R. Kaplan, J. Veizer, and L. S. Land, eds., *Stable Isotopes in Sedimentary Geology*, p. 2-1 to 2-108, SEPM Short Course No. 10, P. O. Box 4756, Tulsa, OK.)

TABLE 2-19 General Range of $^{32}S/^{34}S$ Ratios and of $\delta\ ^{34}S$

Material	$^{32}S/^{34}S$	$\delta\ ^{34}S$ (‰)
Troilite from Cañon Diablo meteorite (standard)	22.22	0
Sulfide minerals		
Meteorites	22.18–22.24	+1.3 to −1.2
Igneous rocks	21.99–22.26	+10 to −2
Sedimentary rocks	21.28–23.21	+42 to −45
Ore deposits		
In igneous rocks	22.09–22.35	+7 to −7
In sedimentary rocks	21.18–23.28	+50 to −45
Hydrogen sulfide		
Volcanic	22.27–22.43	−3 to −10
Biogenic	22.08–22.71	+5 to −24
Native sulfur		
Volcanic	22.10–22.57	+5 to −16
Biogenic	21.73–22.57	+21 to −16
Sulfur dioxide		
Volcanic	21.82–22.39	+17 to −9
Rainwater	22.04–22.15	+9 to +3
Sulfate		
Seawater	21.75–21.79	+21 to +19
Evaporite	21.56–22.13	+30 to +4
Salt dome cap rock	20.84–21.95	+61 to +12
Hydrothermal	21.44–22.34	+35 to −6

Source: Data mainly from Ault (1959, 241–259).

Sulfur isotopic measurements have been used extensively as an additional source of information in the study of sulfide ore deposits (Ohmoto 1986; Taylor 1987). Those deposits that have similar values throughout probably formed from homogeneous solutions with the source suggested by the $\delta^{34}S$ values found. For instance, sulfur brought to the crust from the mantle would probably be lighter than biogenic sulfur mobilized from a sedimentary rock. Sulfur deposited at high temperatures should show less fractionation between minerals than that deposited at low temperatures. In general, the $^{32}S/^{34}S$ values found for deposits associated with igneous rocks have a narrow range, and the $\delta^{34}S$ numbers are close to zero. Other deposits formed as a result of metamorphism, sedimentation, or groundwater activity usually show a wider range of values for a particular deposit. For some deposits it has been possible to conclude that the composition of the mineralizing solutions changed with time, as indicated by regularly changing $^{32}S/^{34}S$ ratios found for samples with known age relationships.

The work of Shelton and Rye (1982) provides an example of a detailed sulfur isotopic study of an ore deposit. Sulfur isotopes were measured on samples from a porphyry-type ore deposit in Quebec, Canada. Samples consisted of sulfides and sulfates in the ore and sulfides in the sedimentary rocks that were invaded by a porphyry intrusive. These data were used to obtain apparent temperatures for ore formation and to obtain information on the nature and source of the ore-forming fluids. The temperature values were obtained by using the sulfur thermometers (Table 2-20) published by Ohmoto and Rye (1979). Table 2-21 gives some of the results, including an independent determination of temperatures using fluid inclusion measurements. Because of large differences in the temperature values obtained from sulfate-sulfide pairs, Shelton and Rye (1982) conclude that there was consistent isotopic disequilibria between sulfate and H_2S in the ore-forming fluid. However, pyrite-chalcopyrite pairs from anhydrite-bearing assemblages yield isotopic temperatures in general agreement with the fluid inclusion data (Table 2-21), suggesting that these two

TABLE 2-20 Sulfur Isotopic Thermometers

Mineral pair	Equation (T in Kelvin)
Sulfates-chalcopyrite	$T = \dfrac{2.85 \times 10^3}{(\Delta \pm 1)^{1/2}}$
Sulfates-pyrite	$T = \dfrac{2.76 \times 10^3}{(\Delta \pm 1)^{1/2}}$
Pyrite-galena	$T = \dfrac{(1.01 \pm 0.04) \times 10^3}{\Delta^{1/2}}$
Sphalerite (pyrrhotite)-galena	$T = \dfrac{(0.85 \pm 0.03) \times 10^3}{\Delta^{1/2}}$
Pyrite-chalcopyrite	$T = \dfrac{(0.67 \pm 0.04) \times 10^3}{\Delta^{1/2}}$
Pyrite-pyrrhotite (sphalerite)	$T = \dfrac{(0.55 \pm 0.04) \times 10^3}{\Delta^{1/2}}$

$\Delta = \delta^{34}S_a - \delta^{34}S_b$

Uncertainties for the calculated temperatures range from ±5 to ±55 K.

Source: After H. Ohmoto and R. Rye, Isotopes of sulfur and carbon. In *Geochemistry of Hydrothermal Ore Deposits*, 2d ed. Copyright © 1979 by John Wiley & Sons. Reprinted by permission of John Wiley & Sons, Inc.

TABLE 2-21 Sulfur Isotopic Data from Anhydrite-bearing Late Stage Veins from Mines Gaspé, Quebec

	$\delta^{34}S$ (‰)			Sulfur thermometer temperatures						
Sample	anh	cp	py	$\Delta^{34}S_{py-cp}$	T°C	$\Delta^{34}S_{anh-cp}$	T°C	$\Delta^{34}S_{anh-py}$	T°C	Fluid inclusion temperatures (°C)
S258–1804	8.1	−2.2	−1.1	1.1	366	10.3	615	9.2	637	350–400
S258–2004	7.3	−2.1	−1.0	1.1	366	9.4	656	8.3	685	350–400
S675–2641	7.7	−1.0	−0.4	0.6	552	8.7	693	8.1	696	350–400
U3587–1262	7.0	−1.1	0.0	1.1	366	8.1	728	7.0	770	350–400
U3587–1715	6.7	−1.5	−0.6	0.9	433	8.2	722	7.3	748	350–400
U3587–1269	8.1	−3.9	−2.7	1.2	388	12.0	550	10.8	567	200–300

Abbreviations: anh = anhydrite, cp = chalcopyrite, py = pyrite.

Source: K. L. Shelton and D. M. Rye. Reproduced from *Economic Geology*, 1982, vol. 7, p. 1693.

sulfides formed in isotopic equilibrium with the ore fluid (a necessary condition for an isotope thermometer to give correct results). Thus the pyrite-chalcopyrite thermometer, along with the fluid inclusion results, appears to give correct temperatures for ore formation.

Figure 2-26 illustrates another use of sulfur isotopes by Shelton and Rye (1982). In this case information is obtained on the source of sulfur in sulfide veins that cut through the porphyry intrusive and the surrounding sedimentary rocks. The veins represented in Figure 2-26 may have formed from fluid with a meteoric component as indicated by their large range of $\delta^{34}S$ values, which also suggests that the vein sulfur came from the surrounding sedimentary rocks. Slightly older veins, which occur in the porphyry intrusive, have a narrow range of $\delta^{34}S$, indicating a magmatic source for the vein sulfur and the ore fluid.

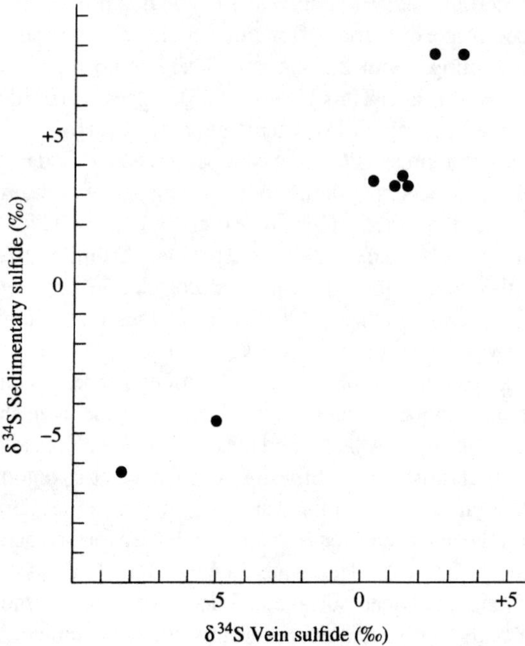

Figure 2-26 Sulfur isotopic data for vein and associated sedimentary sulfides at Mines Gaspé, Quebec, Canada. The porphyry-type veins extend beyond a quartz-porphyry intrusion into sedimentary rocks containing sulfides. The linear trend near $x = y$ suggests that the veins obtained a significant portion of their sulfur from the sedimentary sulfides. This conclusion is also suggested by the large range of $\delta^{34}S$ values found for the veins. (After K. L. Shelton and D. M. Rye. Reproduced from *Economic Geology*, 1982, vol. 7, p. 1695.)

CARBON ISOTOPES

Most carbon in nature is made up of ^{12}C (98.9 percent), with the rest mainly consisting of ^{13}C (1.1 percent). Although it is very useful in radioactive age measurements, radiocarbon (^{14}C) is negligible in terms of abundance. As with sulfur, the two stable carbon isotopes are fractionated by both organic and inorganic reactions. Many of these are not equilibrium processes. Values found for $\delta\,^{13}C$ range from +6 per mil to −38 per mil, where the standard, a belemnite from South Carolina (PDB), has been assigned a zero per mil value and a $^{12}C/^{13}C$ ratio of 88.99 (Figure 2-27). In general, organic carbon is lighter than inorganic carbon, apparently owing to fractionation during photosynthesis. To obtain the greatest precision in radiocarbon dating, it is necessary to correct for ^{14}C fractionation during formation of organic material. This can be done by measuring $^{12}C/^{13}C$ in the sample, since it has been found that $^{12}C-^{14}C$ fractionation is twice that of $^{12}C-^{13}C$.

Carbon isotopic analyses have been used to investigate organic matter found in Precambrian rocks. In particular, attempts have been made to determine the timing of two important events on the Earth: (1) the beginning of biological activity and (2) the advent of oxygen production by photosynthesis. Study of the present biological carbon cycle indicates that the change of inorganic carbon to organic matter by green plants and other organisms is accompanied by a strong fractionation of the carbon isotopes. Organic carbon is enriched, on the average, by about 25 per mil in ^{12}C as compared to inorganic carbon as represented by the carbon of modern sedimentary carbonate rocks (Figure 2-27). We can compare these results with results on organic matter and carbonates in Phanerozoic and Precambrian rocks. [It has been found that changes of isotopic composition during burial and diagenesis of carbonate muds and of organic matter are relatively small; larger changes occur during metamorphism. See the article by Schidlowski et al. (1983).] There is an approximately constant average fractionation of about 25 per mil between organic carbon and carbonate carbon back to at least 3.5 Ga ago (Schidlowski et al. 1983). Thus it appears that carbon fixation by autotrophic life forms (plants and certain bacteria) began as early as that time.

Hayes (1983) has pointed out the occurrence of marked variations in the relative abundance of ^{13}C in sedimentary organic matter beginning about 2.8 Ga ago. There is no recognizable change in the isotopic record for carbonates of this age. Thus Hayes (1983) suggests that this perturbation of the isotopic record represents the beginning of oxygenic photosynthesis.

We now move from Precambrian time to the present. Carbon isotopes are being used to study modern oceanic chemistry. A major process that occurs at depth in the oceans is the oxidation of organic matter settling from the near-surface environment. This process causes the $\delta\,^{13}C$ value of total dissolved inorganic carbon to decrease from surface water to deep water. A similar effect is also observed in lakes (Figure 2-28). In the oceans there is a general correlation between the amount of dissolved oxygen in deep-water masses and their $\delta\,^{13}C$ values for total dissolved inorganic carbon. Lower values of dissolved oxygen accompany lower values of $\delta\,^{13}C$. There is also a relationship between the ages of water masses and their $\delta\,^{13}C$ values, with older masses having lower values due to more extensive oxidation of organic matter. $\delta\,^{13}C$ measurements have shown that the youngest deep-water masses occur in the North Atlantic and the oldest in the North Pacific.

Anderson and Arthur (1983) point out that most marine limestones and calcareous organisms tend to reflect the $\delta\,^{13}C$ of total dissolved inorganic carbon of the water mass in which they formed. Thus pelagic limestones provide evidence of the surface water $\delta\,^{13}C$ in which they formed and fossils of certain benthic organisms indicate the $\delta\,^{13}C$ of the deep water in which they lived. As mentioned above, a significant $\delta\,^{13}C$ gradient exists between surface and deep water of the modern oceans. Variations also exist regionally for both surface and deeper waters due to oceanic circula-

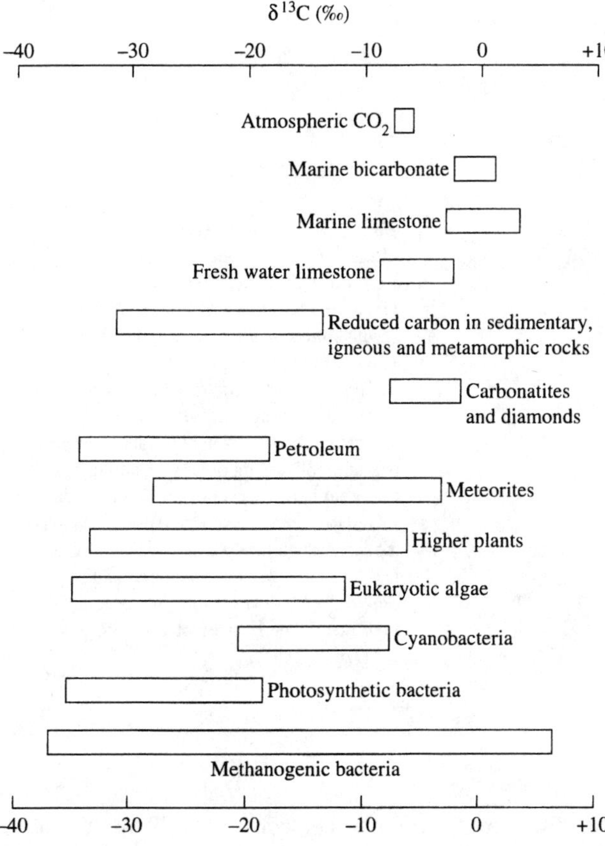

Figure 2-27 Carbon isotopic variation in nature. (After H. Ohmoto and R. O. Rye, Isotopes of sulfur and carbon. In *Geochemistry of Hydrothermal Ore Deposits,* 2d ed. Copyright © 1979 by John Wiley & Sons. Reprinted by permission of John Wiley & Sons, Inc. Also after Schidlowski, M., Hayes, J. M., and Kaplan, I. R., Isotopic inferences of ancient biochemistries: carbon, sulfur, hydrogen, and nitrogen. In *Earth's Earliest Biosphere—Its Origin and Evolution.* Copyright © 1983. Reprinted by permission of Princeton University Press.)

tion patterns. Thus $\delta\,^{13}C$ measurements of limestones and of the remains of organisms such as foraminifers provide significant information in the field of paleoceanography (Broecker 1981).

Stable carbon isotopes are being used to study organic carbon in various biogeochemical systems. An example is the work of Gearing et al. (1984) on a temperate estuary (Narragansett Bay, Rhode Island). They measured $\delta\,^{13}C$ in phytoplankton (the single major carbon source), sediments, zooplankton, larval fish, and benthic fauna to quantify the variability of $\delta\,^{13}C$ and determine its causes. Some of their results are as follows: (1) carbon from primary producers (phytoplankton) varied with taxon and size, ranging from -20.3 ± 0.6 per mil for diatoms to -22.2 ± 0.6 per mil for nanoplankton; (2) there is little isotopic change in planktonic carbon during aquatic decomposition to detritus; and (3) there is an overall pattern of increasing isotope ratios with trophic level, progressing from diatoms (-20.3 per mil) to zooplankton (-19.8 per mil), meiofauna (-19.5 per mil), noncarnivorous macrofauna (-18.6 per mil), and benthic predators (-16.6 per mil). This type of data is helpful in evaluating attempts to use carbon isotopes to trace organic carbon in systems having two or more isotopically distinguishable sources. It also supports the use of isotopic composition as an indicator of trophic position.

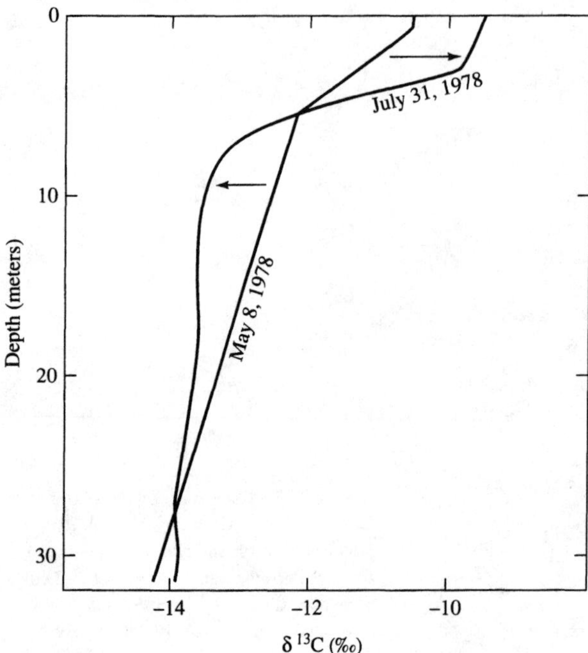

Figure 2-28 Spring-summer depth profiles of the ^{13}C content of total dissolved CO_2 from Lake Greifensee, Switzerland. The arrows indicate the direction of shift from spring values of $\delta\,^{13}C$ to summer values. The shift toward more positive $\delta\,^{13}C$ in the surface waters results from biological activity and the shift toward more negative $\delta\,^{13}C$ values in the lower waters results from the oxidation of sinking organic material. The lower shift corresponds to depletion of dissolved oxygen in Lake Greifensee during the summer months. (After W. Stumm and J. J. Morgan, *Aquatic Chemistry*, 2nd ed. Copyright © 1981 by John Wiley & Sons. Reprinted by permission of John Wiley & Sons, Inc.)

NITROGEN ISOTOPES

Nitrogen has two stable isotopes, ^{14}N and ^{15}N. Atmospheric nitrogen consists of 99.64 percent ^{14}N and 0.36 percent ^{15}N. The N_2 of the atmosphere is used as a standard (0 per mil) when the isotopic composition of samples is expressed as $\delta\,^{15}N$. Figure 2-29 shows the general distribution of $\delta\,^{15}N$ for various natural substances.

Nitrogen isotopes can be used to study the chemical behavior of nitrate ion in groundwater. It is often found that the nitrate content of groundwater decreases either with depth or in the direction of flow. The cause of such a decrease could be denitrification (nitrate reduction to N_2O or N_2), assimilation through root systems, or dilution through mixing with nitrate-free waters. The first process brings about a fractionation of the nitrogen isotopes, while the other two processes generally do not cause fractionation. In the denitrification process remaining nitrate becomes enriched in ^{15}N ($\delta\,^{15}N$ increases) and N_2 and N_2O are depleted in ^{15}N. Mariotti et al. (1988) measured $\delta\,^{15}N$ in an aquifer in northern France to determine the cause of a decrease in nitrate concentration in the direction of groundwater flow. The isotopic data indicated that major denitrification occurs in the groundwater and this was confirmed by water chemistry and bacteriological observations (Figure 2-30). Thus a low concentration of nitrate ion in groundwater should not be considered a result of denitrification unless the nitrate is strongly enriched in ^{15}N.

Kaplan (1983) has shown that $\delta\,^{15}N$ can be used to determine the relative contributions of terrestrial and marine organic nitrogen to the marine sediment of the Santa Barbara Basin (Figure 2-31). Since nitrogen plays an important role in the biological world, nitrogen isotopes, as well as carbon isotopes, can be employed to study trophic dynamics in water masses (Estep

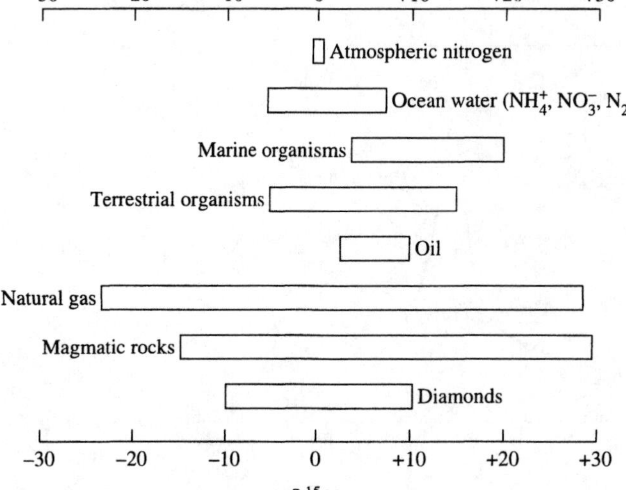

Figure 2-29 Range of δ^{15}N for selected natural materials. Values are given relative to atmospheric nitrogen. (After J. Hoefs, *Stable Isotope Geochemistry*, 3rd ed. Copyright 1987. Reprinted by permission of Springer-Verlag, New York.)

and Vigg 1985). The isotopic composition of nitrogen in soils has been found to vary widely due to variations in the type and isotopic composition of nitrogen absorbed by plants. In addition, variable amounts of fractionation occur during plant decay. As can be seen from this brief review, there are many potential uses for nitrogen isotopes in carrying out biogeochemical research.

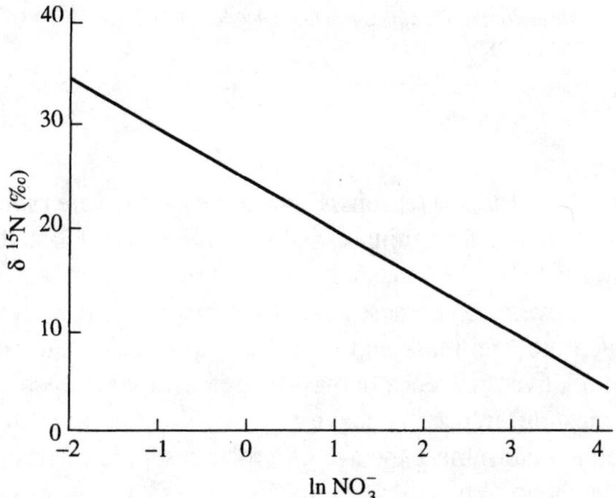

Figure 2-30 Relation between concentration and isotopic composition of nitrate in the chalk aquifer of Northern France. Nitrate concentrations range from 0.23 to 27.7 mg/l with a steady decrease in the direction of groundwater flow. The change of the δ^{15}N value with the NO_3^- concentration is characteristic of isotopic fractionation and not of dilution. The process that explains the disappearance of nitrate, accompanied by isotopic enrichment of ^{15}N, is denitrification. (Reprinted from *Geochimica et Cosmochimica Acta*, v. 52, A. Mariotti, A. Landreau, and B. Simon, ^{15}N isotope biochemistry and natural denitrification process in groundwater: application to the chalk aquifer of northern France. pp. 1869–1878, Copyright 1988, with kind permission from Elsevier Science Ltd, The Boulevard, Langford Lane, Kidlington OX5 1GB, UK.)

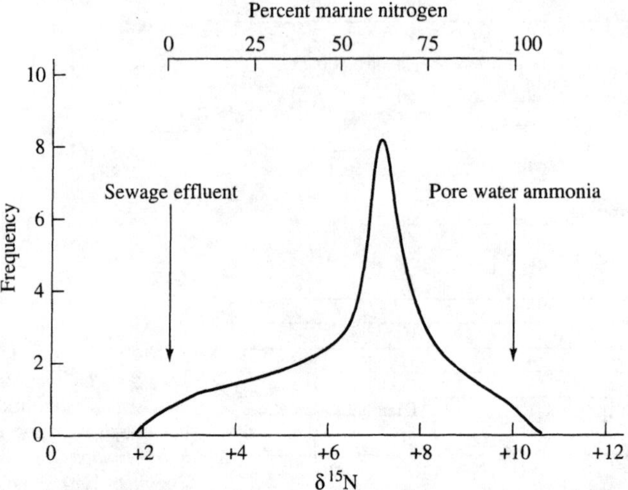

Figure 2-31 Frequency distribution of $\delta\,^{15}N$ for total nitrogen measurements of Santa Barbara Basin sediment. Analyses of forty-two samples produced a range of $\delta\,^{15}N$ values from +2.8 per mil to +9.4 per mil with a mean of 6.8 per mil. The upper scale represents the corresponding percent marine nitrogen for each $\delta\,^{15}N$ value. Terrestrial organic nitrogen, as represented by sewage effluent, has a $\delta\,^{15}N$ value of +2.5 per mil. The marine nitrogen, represented by pore water ammonia resulting from degradation of planktonic organic matter by bacteria, has a $\delta\,^{15}N$ value of about +10.0 per mil. The isotopic measurements indicate that nitrogen in the sediment is generally 50 to 75 percent marine in origin. (After I. R. Kaplan, 1983, Stable isotopes of sulfur, nitrogen, and deuterium in Recent marine environments, in M. A. Arthur, T. F. Anderson, I. R. Kaplan, J. Veizer, and L. S. Land, eds., *Stable Isotopes in Sedimentary Geology*, p. 2-1 to 2-108, SEPM Short Course No. 10, P. O. Box 4756, Tulsa, OK.)

SUMMARY

Isotopes are useful in geology for two reasons. First, some of them are radioactive and this property, when carefully used, allows determination of the time of crystallization, formation, or last alteration of natural materials. In addition, study of the changing abundance of daughter products with time can be used to develop hypotheses about the history of natural materials. Second, all isotopes differ from each other in mass and thus, for a given element, show slightly different chemical behavior. The relative difference in mass is greatest for isotopes of low atomic number, and these show the greatest difference in behavior.

The best approach to determining the age of a rock is separate measurements using two or more independent age methods. When these are in agreement, the ages are said to be concordant. Ages that do not agree (discordant ages) can sometimes be interpreted to gain information about the history of a rock in addition to its time of original crystallization. The cause of discordance is usually loss of the daughter isotope. The precision and accuracy of age measurements are also affected by uncertainty in the values of decay constants, analytical errors, and sample alteration.

Each commonly used age method has strong points and weak points. Because of its complexity, a great deal of information can be obtained using the uranium-lead system. Its major limitations are the lack of uranium- and thorium-bearing minerals in many rocks and the complications involved in interpreting lead isotope abundances. The rubidium-strontium system is widely used because most rocks have potassium minerals, and these usually have measurable amounts of rubidium. In addition, information on initial $^{87}Sr/^{86}Sr$ ratios has been found to be useful in interpreting the history of the Earth's crust and mantle. Sources of error in the rubidium-strontium method are the correction for initial strontium and a relatively large uncertainty in the value for the decay constant. The potassium-argon method may be used in conjunction with the rubidium-strontium method, since the same minerals can be dated by both methods. Rocks that cannot be dated by other methods can sometimes be dated by whole-rock, potassium-argon analysis. The main source of error for the potassium-argon system is argon leakage. Since a correction for initial argon is usually not necessary, potassium-argon ages are generally considered to be reliable minimum ages.

The samarium-neodymium system is unique, since both elements belong to the rare-earth group and thus are geochemically similar. This makes the system very resistant to geological disturbances, such as metamorphism or weathering. The Sm-Nd system is particularly useful in studying crust-mantle evolution and the origin and history of igneous rocks. The low abundances of samarium and neodymium and very long half-life of ^{147}Sm means that use of the system requires extremely careful analytical work. In contrast to other methods, the radiocarbon age method can only be used for dating very young (usually organic) material. It has proved to be of great utility in studying the geologic and human history of the last 100,000 years. There is a relatively large uncertainty in the ages found because of past variation in the atmospheric content of radiocarbon.

Because of the difference in mass, two different isotopes of an element will be fractionated during an exchange reaction between two phases. This results in differing ratios for the isotopes in the two phases. The extent of equilibrium isotopic fractionation is related to temperature, and the first geologic use of stable isotopes not related to radioactivity was in determining ocean paleotemperatures by oxygen isotope analysis of carbonate material. Oxygen isotope measurements have since been extended to coexisting mineral phases. Ideally, information can be obtained on such matters as temperature of formation, degree of equilibrium attained, and source of waters active in mineral formation. The "temperatures" and other information found in oxygen isotope studies need to be confirmed by other types of geologic information, since there are a number of assumptions (such as equilibrium exchange of isotopes) and sources of error involved in applying the method.

Oxygen isotope measurements are often combined with hydrogen isotope measurements to derive information about natural waters. A wide variety of applications is possible, involving both low-temperature processes such as surface water–groundwater interaction, and high-temperature processes such as hydrothermal alteration of oceanic crust. Sulfur isotopic fractionation occurs mainly in oxidation-reduction reactions and is largely due to biological processes. Because most sulfur fractionation processes appear to involve nonequilibrium reactions, sulfur isotopes have limited use in the determination of paleotemperatures. They have been used to resolve questions on the source of sulfur in variouis natural materials (such as marine sediments and ore deposits) and to specify the conditions under which they formed. The major problem in sulfur isotope studies is a lack of detailed knowledge of fractionation processes. Again, any conclusions reached should be checked with independent evidence.

Carbon isotopes are used to study both modern and ancient organic matter. Because of the abundance of carbon in nature, in both inorganic and organic forms, carbon isotopes are particularly useful in studying biogeochemical processes. Nitrogen isotopes can also be applied in this type of research.

QUESTIONS

1. Why is the atomic weight of uranium given as 238.03 in tables of atomic weight when the heaviest naturally occurring isotope of uranium is ^{238}U? Since the three naturally occurring isotopes of uranium are ^{238}U, ^{235}U, and ^{234}U, shouldn't the atomic weight of uranium be less than 238?
2. Using the following information on the chemical composition of a mineral, calculate its age.
$$^{87}Rb/^{86}Sr = 17.59 \qquad ^{87}Sr/^{86}Sr = 0.705$$
Assume that all the ^{87}Sr is radiogenic.
3. Minerals from an igneous rock give the following values for $^{87}Rb/^{86}Sr$ and $^{87}Sr/^{86}Sr$:

	$^{87}Rb/^{86}Sr$	$^{87}Sr/^{86}Sr$
Biotite	4.5	0.92
Muscovite	3.4	0.86
Microcline	2.1	0.79

What is the age of the rock and what was its $^{87}Sr/^{86}Sr$ ratio at the time of crystallization?
4. When dating a rock by the Sm-Nd method, the best results are obtained when the minerals of the rock have large differences in their values of $^{147}Sm/^{144}Nd$. Why?
5. Given the following diagram from the analysis of an igneous rock. What do points A, B, C, and D on the figure represent?

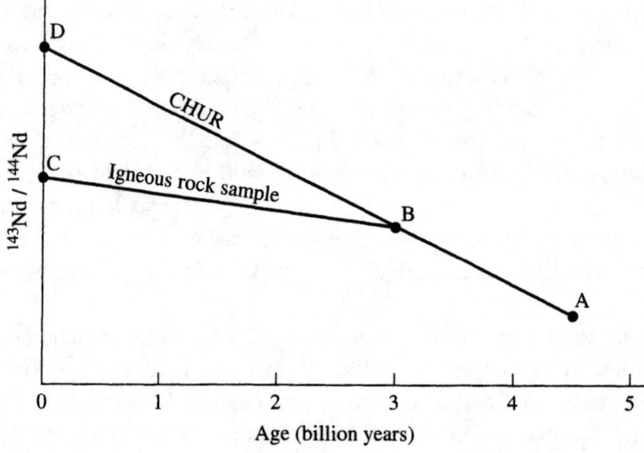

6. Given are the following analytical results for minerals separated from three samples of an igneous rock (Faggart et al. 1985).

Mineral	$^{147}Sm/^{144}Nd$	$^{143}Nd/^{144}Nd$
Plagioclase 1	0.0909	0.5109
Plagioclase 2	0.0868	0.5109
Plagioclase 3	0.0780	0.5108
Pyroxene 1	0.1460	0.5116
Pyroxene 2	0.1524	0.5118
Pyroxene 3	0.1716	0.5120

(a) Calculate the time of crystallization and initial $^{143}Nd/^{144}Nd$ ratio for the rock.
(b) Calculate an ϵ value relative to CHUR for the time of crystallization. Assume the CHUR value at the time of crystallization was 0.5110 for $^{143}Nd/^{144}Nd$.
(c) What does the ϵ value tell you about the source of the magma responsible for the rock?

7. The age equation for ^{238}U decay (2-19) contains the ratio $^{238}U/^{204}Pb$ in terms of atoms. If the analysis of a sample gives the amounts of ^{238}U and ^{204}Pb in parts per million (ppm) by weight, how can these values be used in the age equation?
8. For the $^{207}Pb/^{206}Pb$ age method, the older the sample, the more reliable the measured age. Suggest a reason for this.
9. Given are the following data for a group of associated igneous rocks. Plot a lead isochron and determine the age of the rocks.

$^{206}Pb/^{204}Pb$	$^{207}Pb/^{204}Pb$
19.2	14.8
30.0	16.9
40.1	18.8

10. In extending the $^{40}K/^{40}Ar$ method to very young samples, a major limitation is how well a correction for contamination by atmospheric ^{40}Ar can be made. How could the presence of atmospheric ^{40}Ar in a sample be confirmed and how could the correction be made?
11. The analysis of a biotite yields the following results: K = 7.34 percent (by weight) and $^{40}Ar = 24.5 \times 10^{-7}$ cubic centimeters per gram of sample (25°C and 1 atmosphere). Assuming that all the ^{40}Ar is radiogenic, what is the age of the sample?
12. Given is a piece of wood with a measured ^{14}C activity of 0.03 disintegrations per minute per gram (dpm/g). Assume that the concentration of ^{14}C in the atmosphere at the time that the wood grew was similar to the value found today of about 16 dpm/g. How old is the sample? Use 5,730 years for the half-life of ^{14}C.
13. It is found that carbonate material from animals grown in isotopic equilibrium with seawater has the following $\delta\ ^{18}O/^{16}O$ values: +25 per mil at 25°C, +29 per mil at 15°C, and +35 per mil at 0°C. Assume V-SMOW was used as a standard in calculating the δ values.
(a) Derive an equation relating temperature and shell isotopic composition (δ).
(b) What is the $^{18}O/^{16}O$ ratio in the material that has a δ value of 25 per mil?
14. The data in Table 2-17 indicate that different minerals in igneous and metamorphic rocks tend to have different $^{18}O/^{16}O$ ratios. For example, quartz in igneous rocks usually has a higher $\delta\ ^{18}O/^{16}O$ value than K-feldspar. What property of minerals would be important in controlling $^{18}O/^{16}O$ ratios?
15. Several samples from a limestone formation are analyzed for $^{18}O/^{16}O$ ratios to determine a temperature of formation for the rock. The results show two subgroups, one with much higher $^{18}O/^{16}O$ ratios than the other. Rocks in which subgroup, ^{18}O-rich or ^{18}O-poor, are most likely to have retained their original isotopic composition?
16. In Figure 2-17, the meteoric water line (MWL) does not pass through the SMOW point (which represents average seawater composition). Why doesn't it pass through this point even though the MWL represents water vapor and precipitation formed from seawater?
17. If the $^{32}S/^{34}S$ ratio in a sulfide mineral is 23.18, what would be the value of $\delta\ ^{34}S/^{32}S$ for the mineral? What type of rock, igneous or sedimentary, would you guess the mineral came from?
18. Explain why, in marine sediments, both sulfate ion and dissolved hydrogen sulfide in pore water show increases in $\delta\ ^{34}S$ with depth (see discussion in chapter on sulfur isotopes in marine sediments).

19. **(a)** Calculate the fractionation factor α for the pyrite-chalcopyrite pair listed in Table 2-21 for sample S258-1804.

(b) Assume that the value of $\Delta\,^{34}S$ for the sample in part (a) is wrong because of analytical error or because of sample impurity. If the correct value is 1.3 per mil rather than the 1.1 per mil value listed in Table 2-21, how much in error is the temperature value of 366°C in Table 2-21?

20. **(a)** Figure 2-28 illustrates spring and summer depth profiles of $\delta\,^{13}C$ for Lake Greifensee, Switzerland. What does the figure tell us about the fractionation of carbon isotopes by photosynthesis in surface waters?

(b) This lake has a vertical $\delta\,^{13}C$ profile (no $\delta\,^{13}C$ gradient) in the winter. Why?

21. A $\delta\,^{15}N$ measurement of the organic matter in a marine sediment gives a result of +6.5 per mil. Using information from Figure 2-31, derive an equation that will give the exact percent of marine nitrogen and calculate this value for +6.5 per mil.

REFERENCES

ALBAREDE, F., and M. JUTEAU. 1984. Unscrambling the lead model ages. *Geochim. Cosmochim. Acta*, v. 48:207–212.

ALT, J. C., K. MUEHLENBACHS, and J. HONNOREZ. 1986. An oxygen isotopic profile through the upper kilometer of the oceanic crust, DSDP Hole 504B. *Earth Planet. Sci. Lett.*, v. 80:217–229.

ANDERS, E., and N. GREVESSE. 1989. Abundances of the elements: Meteoritic and solar. *Geochim. Cosmochim. Acta*, v. 53:197–214.

ANDERSON, T. F., and M. A. ARTHUR. 1983. Stable isotopes of oxygen and carbon and their application to sedimentologic and paleoenvironmental problems. In *Stable Isotopes in Sedimentary Geology*. Society of Economic Paleontologists and Mineralogists, Tulsa, OK. 1–1 to 1–151.

ARNDT, N. T., and S. L. GOLDSTEIN. 1987. Use and abuse of crust-formation ages. *Geology*, v. 15:893–895.

AULT, W. U. 1959. Isotopic fractionation of sulfur in geochemical processes. In *Researches in Geochemistry*, v. 1, P. H. Abelson, ed. John Wiley and Sons, Inc., New York. 241–259.

BOGARD, D. , F. HÖRZ, and D. STÖFFLER. 1988. Loss of radiogenic argon from shocked granitic clasts in suevite deposits from the Ries Crater. *Geochim. Cosmochim. Acta*, v. 52:2639–2649.

BOTTINGA, Y., and M. JAVOY. 1975. Oxygen isotope partitioning among the minerals in igneous and metamorphic rocks. *Rev. Geophys. and Space Geophys.*, v. 13:401–408.

BROECKER, W. S. 1981. Glacial to interglacial changes in ocean and atmospheric chemistry. In *Climatic Variations and Variability, Facts and Theories*, A. Berger, ed. D. Riedel Pub. Co., Dordrecht, The Netherlands. 109–20.

CAMERON, E. M. 1982. Sulphate and sulphate reduction in early Precambrian oceans. *Nature*, v. 296:145–148.

CAMERON, E. M. 1983. Genesis of Proterozoic iron-formation: sulfur isotope evidence. *Geochim. Cosmochim. Acta*, v. 47:1069–1074.

CATANZARO, E. J. 1968. The interpretation of zircon ages. In *Radiometric Dating for Geologists*, E. I. Hamilton and R. M. Farquhar, eds. John Wiley & Sons, Inc., New York. 225–258.

CHEN, J. H., and G. J. WASSERBURG. 1983. The least radiogenic Pb in iron meteorites, Fourteenth Lunar and Planetary Science Conference, Abstracts, Part 1. Lunar and Planetary Institute, Houston, TX. 103–104.

COCKER, J. D., B. J. GRIFFIN, and K. MUEHLENBACHS. 1982. Oxygen and carbon isotope evidence for seawater-hydrothermal alteration of the Macquarie Island ophiolite. *Earth Planet. Sci. Lett.*, v. 61:112–122.

CRAIG, H. 1961. Standard for reporting concentrations of deuterium and oxygen-18 in natural waters. *Science*, v. 133:1833–1834.

CURRIE, L. A., ed. 1982. *Nuclear and Chemical Dating Techniques*, American Chemical Society, Washington, D.C.

DEPAOLO, D. J. 1988. *Neodymium Isotope Geochemistry*. Springer-Verlag, New York.

Doe, B. R. 1970. *Lead Isotopes*. Springer-Verlag, Berlin.

Estep, M. L. F., and S. Vigg, 1985. Stable carbon and nitrogen isotope tracers of trophic dynamics in natural populations and fisheries of the Lahontan Lake System, Nevada. *Can. J. Fisheries Aquat. Sci.,* v. 42:1712–1719.

Faggart, B. E., Jr., A. R. Basu, and M. Tatsumoto. 1985. Origin of the Sudbury Complex by meteoritic impact: neodymium isotopic evidence. *Science*, v. 230:436–439.

Farmer, G. L., and D. J. DePaolo. 1983. Origin of Mesozoic and Tertiary granite in the western U.S. and implications for pre-Mesozoic crustal structure. I. Nd and Sr isotopic studies in the geocline of the northern Great Basin. *J. Geophys. Res.*, v. 88:3379–3401.

Faul, H. 1966. *Ages of Rocks, Planets and Stars*. McGraw-Hill, New York. 37.

Faure, G. 1986. *Principles of Isotope Geology.* John Wiley & Sons, Inc., New York.

Faure, G. and J. L. Powell. 1972. *Strontium Isotope Geology.* Springer-Verlag, New York.

Fleischer, R. L., P. B. Price, and R. M. Walker. 1975. *Nuclear Tracks in Solids*. University of California Press, Berkeley, CA.

Friedman, I., and J. R. O'Neill. 1977. Compilation of stable isotope fractionation factors of geochemical interest. In *Data of Geochemistry*, 6th ed. M. Fleischer, ed. U.S. Geol. Survey Prof. Paper.

Gearing, J. N., P. J. Gearing, D. T. Rudnick, A. G. Requejo, and M. J. Hutchings. 1984. Isotopic variability of organic carbon in a phytoplankton-based, temperate estuary. *Geochim. Cosmochim. Acta*, v. 48:1089–1098.

Geyh, M. A., and H. Schleicher. 1990. *Absolute Age Determination.* Springer-Verlag, New York.

Hamilton, E. I. 1965. *Applied Geochronology.* Academic Press, Inc., London.

Hayes, J. M. 1983. Geochemical evidence bearing on the origin of aerobiosis, a speculative hypothesis. In J. W. Schopf, ed. *Earth's Earliest Biosphere.* Princeton University Press, Princeton, NJ.

Hoefs, J. 1987. *Stable Isotope Geochemistry,* 3d ed. Springer-Verlag, New York.

Holland, H. D. 1984. *The Chemical Evolution of the Atmosphere and Oceans.* Princeton University Press, Princeton, NJ.

Huebner, M., T. K. Kyser, and E. G. Nisbet. 1986. Stable-isotope geochemistry of high-grade metapelites from the Central zone of the Limpopo belt. *Am. Min.,* v. 71:1343–1353.

Hunziker, J. C., 1979. Potassium argon dating. In *Lectures in Isotope Geology,* E. Jäger and J. C. Hunziker, eds. Springer-Verlag, Berlin. 52–76.

Ivanovich, M., and R. S. Harmon, eds. 1992. *Uranium-Series Disequilibrium*, 2d ed. Oxford University Press, New York.

Jacobsen, S. B. 1988. Isotopic and chemical constraints on mantle-crust evolution. *Geochim. Cosmochim. Acta*, v. 52:1341–1350.

Jäger, E., and J. C. Hunziker. 1979. *Lectures in Isotope Geology.* Springer-Verlag, Berlin.

Kanasewich, E. R. 1968. The interpretation of lead isotopes and their geological significance. In *Radiometric Dating for Geologists*, E. I. Hamilton and R. M. Farquhar, eds. John Wiley & Sons, Inc., New York. 147–223.

Kaplan, I. R. 1983. Stable isotopes of sulfur, nitrogen, and deuterium in Recent marine environments. In *Stable Isotopes in Sedimentary Geology*, M. A. Arthur, T. F. Anderson, I. R. Kaplan, J. Veizer, and L. S. Land, eds. Society of Economic Paleontologists and Mineralogists, Tulsa, OK. 2-1 to 2-108.

Kieffer, S. W. 1982. Thermodynamics and lattice vibrations of minerals: 5. Applications to phase equilibria, isotopic fractionation, and high pressure thermodynamic properties. *Rev. Geophys. Space Phys.,* v. 20:827–849.

Krishnaswami, S., and D. Lal, 1982. Deep-sea sedimentation processes and chronology. In *Nuclear and Chemical Dating Techniques,* L. A. Currie, ed. American Chemical Society, Washington, D.C., 363–388.

Kuroda, P. K. 1982. *The Origin of the Chemical Elements and the Oklo Phenomenon.* Springer-Verlag, Berlin.

Kyser, T. K., ed. 1987. *Stable Isotope Geochemistry of Low Temperature Processes.* Mineralogical Association of Canada, Toronto.

Lanphere, M. A., G. J. Wasserburg, and A. L. Albee. 1964. Redistribution of strontium and rubidium isotopes during metamorphism, World Beater Complex, Panamint Range, California. In *Isotopic and Cosmic Chemistry*, H. Craig, S. L. Miller, and G. J. Wasserburg, eds. North-Holland Publishing Company, Amsterdam. 269–320.

Long, L. E. 1972. Lead: interpretation of stable isotope abundances. In *Encyclopedia of Geochemistry and Environmental Sciences*, W. Fairbridge, ed. Dowden, Hutchinson & Ross, Stroudsburg, PA. 647.

Ludwig, K. R. 1980. Calculation of uncertainties of U-Pb isotopic data. *Earth Planet. Sci. Lett.*, v. 46:212–220.

Mariotti, A., A. Landreau, and B. Simon. 1988. ^{15}N isotope biogeochemistry and natural denitrification process in groundwater: application to the chalk aquifer of northern France. *Geochim. Cosmochim. Acta*, v. 52:1869–1878.

McCready, R. G. L., I. R. Kaplan, and G. A. Din. 1974. Fractionation of sulfur isotopes by the yeast *Saccharomyces cerevisiae*. *Geochim. Cosmochim. Acta*, v. 38:1239–1253.

McDougall, I., and T. M. Harrison. 1988. *Geochronology and Thermochronology by the* $^{40}Ar/^{39}Ar$ *Method.* Oxford University Press, New York.

Mitchell, J. G. 1968. The argon-40/argon-39 method for potassium-argon age determination. *Geochim. Cosmochim. Acta*, v. 32:781–790.

Morton, J. P. 1985. Rb-Sr dating of diagenesis and source age of clays in Upper Devonian black shale of Texas. *Geol. Soc. Am. Bull.*, v. 96:1043–1049.

Muehlenbachs, K. 1986. Alteration of the oceanic crust and the ^{18}O history of seawater. In *Stable Isotopes in High Temperature Geological Processes,* J. W. Valley, H. P. Taylor, Jr., and J. R. O'Neil, eds. Mineralogical Society of America, Washington, D.C.

Nelson, B. K., and D. J. DePaolo. 1988. Application of Sm-Nd and Rb-Sr isotope systematics to studies of provenance and basin analysis. *J. Sediment. Petrol.*, v. 58:348–357.

Neumann, E-R., G. R. Tilton, and E. Tuen. 1988. Sr, Nd, and Pb isotope geochemistry of the Oslo rift igneous province, southeast Norway. *Geochim. Cosmochim. Acta,* v. 52:1997–2007.

Nielsen, H. 1979. Sulfur isotopes. In *Lectures in Isotope Geology*, E. Jäger and J. C. Hunziker, eds. Springer-Verlag, Berlin. 283–312.

Ohmoto, H., 1972. Systematics of sulfur and carbon isotopes in hydrothermal ore deposits. *Econ. Geol.*, v. 67:551–578.

Ohmoto, H. 1986. Stable isotope geochemistry of ore deposits. In *Stable Isotopes in High Temperature Geological Processes,* J. W. Valley, H. P. Taylor, Jr., and J. R. O'Neil, eds. Mineralogical Society of America, Washington, D.C. 491–559.

Ohmoto, H., and R. O. Rye. 1979. Isotopes of sulfur and carbon. In *Geochemistry of Hydrothermal Ore Deposits*, 2nd ed., H. L. Barnes, ed. John Wiley & Sons, Inc., New York. 509–567.

O'Neil, J. R. 1979. Stable isotope geochemistry of rocks and minerals. In *Lectures in Isotope Geology,* E. Jäger and J. C. Hunziker, eds. Springer-Verlag, Berlin. 235–263.

O'Neil, J. R. 1986. Theoretical and experimental aspects of isotope fractionation. In *Stable Isotopes in High Temperature Geological Processes,* J. W. Valley, H. P. Taylor, Jr., and J. R. O'Neil, eds. Mineralogical Society of America, Washington, D.C. 1–40.

O'Neil, J. R., R. N. Clayton, and T. N. Mayeda. 1969. Oxygen isotope fractionation in divalent metal carbonates. *J. Chem. Phys.*, v. 51:5547–5558.

Patchett, P. J. 1983. Importance of the Lu-Hf isotopic system in studies of planetary chronology and chemical evolution. *Geochim. Cosmochim. Acta,* v. 47:81–91.

Patchett, P. J., and N. T. Arndt. 1986. Nd isotopes and tectonics of 1.9–1.7 Ga crustal genesis. *Earth Planet. Sci. Lett.*, v. 78:329–338.

Piepgras, D. J., and S. B. Jacobsen. 1988. The isotopic composition of neodymium in the North Pacific. *Geochim. Cosmochim. Acta,* v. 52: 1373–1381.

RUSSELL, R. D., and R. M. FARQUHAR. 1960. *Lead Isotopes in Geology.* John Wiley & Sons, Inc., New York.

SCHELL, W. R. 1982. Dating recent (200 years) events in sediments from lakes, estuaries, and deep ocean environments using lead-210. In *Nuclear and Chemical Dating Techniques,* L. A. Currie, ed. American Chemical Society, Washington, D.C. 331–361.

SCHIDLOWSKI, M., J. M. HAYES, and I. R. KAPLAN. 1983. Isotopic inferences of ancient biochemistries: carbon, sulfur, hydrogen, and nitrogen. In *Earth's Earliest Biosphere—Its Origin and Evolution.* J. W. Schopf, ed. Princeton University Press, Princeton, NJ.

SHANKS, W. C., III, and R. E. CRISS, eds. 1989. New frontiers in stable isotope research: laser probes, ion probes, and small-sample analysis. U.S. Geol. Survey Bull. 1890.

SHELTON, K. L., and D. M. RYE. 1982. Sulfur isotopic compositions of ores from Mines Gaspé, Quebec: An example of sulfate-sulfide isotopic disequilibria in ore-forming fluids with applications to other porphyry-type deposits. *Econ. Geol.,* v. 77:1688–1709.

SHEPPARD, S. M. F. 1986. Characterization and isotopic variations in natural waters. In *Stable Isotopes in High Temperature Geological Processes,* J. W. Valley, H. P. Taylor, Jr., and J. R. O'Neil, eds. Mineralogical Society of America, Washington, D.C. 165–183.

SHEPPARD, S. M. F., and H. P. TAYLOR, JR. 1974. Hydrogen and oxygen isotope evidence for the origins of water in the Boulder batholith and the Butte ore deposits, Montana. *Econ. Geol.,* v. 69:926–946.

SIEGENTHALER, U. 1979. Stable hydrogen and oxygen isotopes in the water cycle. In *Lectures in Isotope Geology,* E. Jäger, and J. C. Hunziker, eds. Springer-Verlag, Berlin. 264–273.

Society of Economic Paleontologists and Mineralogists. 1983. *Stable Isotopes in Sedimentary Geology.* Tulsa, OK.

STACEY, J. S., and T. W. STERN. 1973. Revised tables for the calculation of lead isotope ages. National Technical Information Service No. PB2-20919, U.S. Dept. of Commerce, Springfield, VA.

STUMM, W., and J. J. MORGAN. 1981. *Aquatic Chemistry,* 2d ed. John Wiley & Sons, New York. 220.

SUESS, H. 1980. The radiocarbon record in tree rings of the last 8000 years. *Radiocarbon,* v. 22:200–209.

TAYLOR, B. E. 1987. Stable isotope geochemistry of ore-forming fluids. In *Stable Isotope Geochemistry of Low Temperature Fluids,* T. K. Kyser, ed. Mineralogical Association of Canada, Toronto. 337–445.

TAYLOR, H. P., JR. 1967. Oxygen isotope studies of hydrothermal mineral deposits. In *Geochemistry of Hydrothermal Ore Deposits,* H. L. Barnes, ed. Holt, Rinehart and Winston, Inc., New York. 109–142.

TAYLOR, H. P., JR. 1974. The application of oxygen and hydrogen isotope studies to problems of hydrothermal alteration and ore deposition. *Econ. Geol.,* v. 69:843–883.

TAYLOR, R. E., A. LONG, and R. S. KRA, eds. 1992. *Radiocarbon After Four Decades.* Springer-Verlag, New York.

TILLING, R. I., M. R. Klepper, and J. D. OBRADOVICH. 1968. K-Ar ages and time span of emplacement of the Boulder batholith, Montana. *Am. J. Sci.,* v. 266:671–689.

TILTON, G. R., and G. L. DAVIS. 1959. Geochronology. In *Researches in Geochemistry,* P. H. Abelson, ed. John Wiley & Sons, Inc., New York. 190–216.

TRUDINGER, P.A., and L. A. CHAMBERS. 1973. Reversibility of bacterial sulfate reduction and its relevance to isotopic fractionation. *Geochim. Cosmochim. Acta,* v. 37:1775–1778.

UREY, H. C. 1947. The thermodynamic properties of isotopic substances. *J. Chem. Soc.,* v. 1947:562–581.

VALLEY, J. W., H. P. TAYLOR, JR., and J. R. O'NEIL, eds. 1986. *Stable Isotopes in High Temperature Geological Processes.* Mineralogical Society of America, Washington, D.C.

WELHAN, J. A. 1987. Stable isotope hydrology. In *Stable Isotope Geochemisty of Low Temperature Fluids,* T. K. Kyser, ed. Mineralogical Association of Canada, Toronto. 129–161.

WETHERILL, G. W. 1956. Discordant uranium-lead ages. *Trans. Am. Geophys. Union,* v. 37:320–326.

Thermodynamics

The universe as we know it consists of matter, which can be defined as the material of which any physical object is composed. Matter occurs in the form of solids, liquids, and gases. These forms of matter are interchangeable. In addition, when we have two separate pieces of matter of two different chemical compositions, such as liquid water and solid anhydrite, a chemical reaction may occur when they are brought together. These changes, from one state of matter to another, or from one set of compositions to another, are brought about by the performance of work or by the flow of heat. The science that deals with the physical and chemical changes of matter due to work and heat flow is known as *thermodynamics*.

The rocks of the Earth are the end product of complex processes that cannot, in many cases, be duplicated in a laboratory. However, laboratory measurements of the thermodynamic properties of rocks and minerals can be combined with thermodynamic reasoning to allow geologists to make reasonable guesses as to the causes and nature of the processes that resulted in these products. In the twentieth century there has been increasing use of thermodynamics by geologists; this has brought about important advances in our knowledge and understanding of the natural world. There is no doubt that the future will see even greater use of thermodynamics by geologists.

In the following pages a brief introduction to thermodynamics will be given. Some applications to geologic problems will be included. This introduction will certainly not provide a full understanding of the significance of thermodynamic principles. However, it will introduce the reader to the potential that thermodynamics has for solving geologic problems and will thus hopefully develop a desire for further study. A good source for further study is Fletcher (1993).

HISTORICAL DEVELOPMENT

One concern of physicists in the eighteenth century was the flow of heat in various physical processes. This study came to be called thermodynamics (heat movement). Among other things, chemists were concerned with the determination of the direction in which a chemical reaction would go. These separate concerns of physicists and chemists were found to be related as a result of experimentation and theoretical work by various nineteenth-century scientists. It was discovered that the law of conservation of energy, used by physicists studying the relation of work done to heat flow, also applied to chemical processes. At the same time, studies of physical changes associated with chemical reactions were being carried out. These studies, together with thermodynamics, were grouped into the scientific field called *physical chemistry*.

From the study of the relationship between heat flow and chemical reactions came the idea that these reactions would always go in the direction that resulted in heat being given off. This was soon found to be in error. But out of the study of the relationship of heat flow and work done came a quantity, named *entropy* in 1854 by R. J. E. Clausius, that could be used to determine the direction in which all natural processes will go. One important result of these studies was the development of the first and second law of thermodynamics. The first law deals with conservation of energy; the second is concerned with the direction in which a reaction will proceed.

The detailed application of these laws of thermodynamics to changes of state of matter and to chemical reactions was outlined by physicist J. W. Gibbs in papers published in the period 1875–1878. In these papers he derived the phase rule and developed the concepts of the various thermodynamic potentials. Gibbs is generally considered to be the founder of the field of modern thermodynamics. Very few significant ideas were added to the theory of thermodynamics after he laid the foundation. Only one other development need be mentioned here. The related concepts of *fugacity* and *activity*, which are of great use in thermodynamic studies, were introduced by G. N. Lewis in 1901 and 1907, respectively.

The first application of thermodynamics to a geologic problem was a series of experimental studies, directed by J. H. Van't Hoff, which were intended to explain the origin of the Permian salt deposits of Germany. This work covered the period 1896–1909 and was successful in qualitatively explaining the mineral sequence and assemblages found in the evaporite deposits. Six years after the evaporite experiments were started, a committee of the Carnegie Institution of Washington suggested the establishment of a Laboratory of Geophysics in Washington, D.C. The purpose of the laboratory would be to apply the principles of physics and chemistry to geologic problems in a quantitative manner. Van't Hoff's work was cited as part of the justification for the proposed laboratory.

In 1907 the Geophysical Laboratory became a reality and immediately began producing significant results by applying experimental petrology to the origin of igneous rocks. The leader in this field for the next 40 years was N. L. Bowen, whose Ph.D. thesis was completed at the Geophysical Laboratory in 1912. A Ph.D. thesis by V. M. Goldschmidt, completed in Norway one year earlier, marked the first application of thermodynamics to the study of metamorphic rocks. Goldschmidt's use of the phase rule to explain contact metamorphism in the Kristiania (Oslo) region of Norway led directly to the concept of metamorphic facies put forth by P. Eskola a few years later. Theoretical and experimental applications of thermodynamics to igneous and metamorphic rocks became increasingly important in the years following the initial work by Bowen, Goldschmidt, and others. In contrast, such applications were not continued on sedimentary rocks after the work of Van't Hoff and his colleagues was completed. Thermodynamic study of sedimentary rocks and surficial processes began again after World War II and is now a very active field.

BASIC CONCEPTS AND TERMS

When applying thermodynamics to geologic or other problems, the material of interest is referred to as the *system* and the material around the system as the *surroundings.* A system may contain gases, liquids, solids, or any combination of these. In studying a system it is necessary to know the nature and number of phases present. A *phase* is defined as a physically homogeneous portion of a system with a definite boundary.* It is also desirable to specify the chemical composition of the system of interest. This is done by listing the amounts of the *components* (chemical constituents) of the system. The number of constituents should be the smallest number needed to describe all possible variations in the composition of every phase. The number of components for the system gypsum-anhydrite-water is two, $CaSO_4$ and H_2O, since gypsum can be described by a combination of these two components.

Each system under given conditions has a certain set of physical and chemical properties that can be used to describe it completely. When there is a change in one or more of these properties, it is said that a *change of state* has occurred. Note that this use of the term *state* is not the same as the more general use when talking about changes from the solid to the liquid state or from the liquid to the gaseous state. In thermodynamics these represent a change of phase. An equation that relates certain properties of the system is called an *equation of state.* For example, the properties of an ideal gas can be related by the equation $PV = nRT$. The properties that describe the gas (system) are pressure (P), volume (V), temperature (T), and the mass of the gas, expressed as n, the number of moles. R represents a universal proportionality constant, which does not depend on the kind of ideal gas being considered (Table 3-1). Note that only three of the four variables need to be known to describe the gas completely, since the value of the fourth variable can be found by using the equation of state.

A general definition of *energy* from the field of mechanics is that energy is the capacity to do work. In mechanics, *work* is defined as the result of a force acting through a displacement. Work done on a body that displaces the entire body, such as lifting it or accelerating it, produces an increase in the energy of the body. If the body is lifted, there is an increase in its potential energy. If it is accelerated, the kinetic energy of the body is increased. These are examples of energy that are determined by the relationship of the entire body to some reference point. In both examples there is no change in the internal conditions of the body.

Thermodynamics is concerned with the *internal energy* of a body or system. Thus changes that take place within a system, owing to interaction of the system and its surroundings, are said to cause a change in its internal energy. In thermodynamics, work can be done by a system on its surroundings or by the surroundings on the system. When this work is done, the internal energy of the system is changed. In many cases work that is done causes either an increase or decrease in the volume of the system. If a gas confined in a container with movable walls expands and increases the size of the container, the gas has done work on its surroundings. An increase in the

*In actuality some physical properties (e.g., optical parameters, density) may not be homogeneous throughout what is considered normally to be a phase (e.g., a solid solution such as plagioclase feldspar). A phase also in theory should be capable of being mechanically separable from the system.

TABLE 3-1 Units Commonly Used in Thermodynamics

Units of the International System of Units (SI):

Newton: the force that will accelerate a mass of 1 kilogram by 1 meter per second squared
Pascal: the pressure resulting when a force of 1 newton acts uniformly over an area of 1 square meter
Joule: the work done when 1 newton of force produces a displacement of 1 meter in the direction of the force
Kelvin: the fraction 1/273.16 of the thermodynamic temperature of the triple point of water

Comparison of SI units and older units:

Quantity	SI unit	Old unit	Value of old unit in SI units
Force	Newton (N)	Dyne	10^{-5} N
Pressure	Pascal (Pa)	Atmosphere	1.013×10^5 Pa
Energy	Joule (J)	Calorie	4.184 J
Temperature	Kelvin (K)	°C and °K	1 K

Conversion factors (including non-SI units still commonly used):

Energy:

1 calorie (gram-calorie)	= 4.184 joules
	= 4.184×10^7 ergs
1 joule	= 0.239 calorie
	= 1×10^{-7} ergs
	= 1 volt-coulomb
1 erg	= 1 dyne-centimeter
	= 2.39×10^{-8} calorie
	= 1×10^{-7} joule
1 volt-faraday	= 96,500 volt-coulombs
	= 23,060 calories

Pressure:

1 atmosphere	= 1.013 bars
	= 1.013×10^6 dynes/square centimeter
1 bar	= 0.987 atmosphere
	= 1×10^6 dynes/square centimeter
1 dyne per square centimeter	= 9.869×10^7 atmospheres
	= 1×10^{-6} bar

Gas constant values:

Pressure-volume product in:	Gas constant (R)
Milliliters-atmospheres	82.06 cm^3-atm/deg-mol
Ergs	8.315×10^7 ergs/deg-mol
Joules	8.315 joules/deg-mol
Calories	1.987 cal/deg-mol

volume of the system does not necessarily change the location of the entire system with respect to some external reference point. But there is a change in the properties of the system, and this is called a change in internal energy.

Of all the concepts of thermodynamics, that of *temperature* is the most basic. Temperature can be considered a measure of the intensity of motion of the atoms and molecules in a system. If two systems are both in thermal equilibrium with a third system (no heat flows between either of them and the third system), then they are in thermal equilibrium with each other. This is known as the *zeroth principle* or *law of thermodynamics.* Systems in thermal equilibrium are said to have the same temperature. Temperature can be measured by using a number of different scales. The best temperature scale is a scale, originally called the absolute scale, that allows measurements in a manner independent of the properties of any particular substance. This scale, now known as the *Kelvin scale* after its proposer Lord Kelvin, has its numerical values fixed by using an arbitrary temperature value for the triple point of water (the temperature and pressure at which ice, liquid water, and water vapor coexist in equilibrium). The value chosen was 273.16 K (the pressure for the triple point is 0.006 atmosphere). The thermodynamic unit of temperature, the Kelvin (K), is defined as 1/273.16 of the temperature of the triple point of water.

On the Kelvin scale the freezing point of water is 273.15 K and the boiling point of water is 373.15 K. The difference of 0.01 K between the freezing point of water and the triple point of water is due to a difference in the conditions of the two types of equilibrium. The freezing point involves equilibrium between air-saturated water and pure ice at 1 atmosphere pressure; the triple point involves equilibrium between three phases, ice, liquid water, and water vapor, in the absence of air. The other temperature scale commonly used by scientists is the Celsius scale. It employs a degree of the same magnitude as that of the Kelvin scale, but the triple point of water is fixed at 0.01 degree Celsius; thus, the freezing point of water is 0°C. Thermodynamic temperatures are expressed on the Kelvin scale by adding 273.15 to the Celsius temperature.

The concept of *equilibrium* can be applied to different phases that do not react chemically (physical equilibrium) or to different substances that are capable of reacting with each other (chemical equilibrium). Simply stated, a system in equilibrium does not change with time. Various reactions may be occurring on a minute scale, but, if so, they cancel each other out and the net effect is no change in the system. Thus in the chemical equilibrium

$$A + B \leftrightharpoons C + D \tag{3-1}$$

A and B may be continuously reacting to form C and D, but C and D are also continuously reacting to form A and B. At equilibrium there is a constant concentration of each substance.

It should be pointed out that equilibrium is a relative matter. A piece of calcite in a laboratory is in equilibrium with respect to the pressure and temperature of the laboratory, but it is not in equilibrium with respect to hydrochloric acid. Thus a system can be in equilibrium with respect to one set of conditions in its surroundings but not with respect to another set. When a substance occurs in a form that is not the most stable form under the existing conditions, but does not spontaneously change to a more stable form, it is said to be in a *metastable* state. Often this condition is due to a very slow rate of reaction. Given enough time, the most stable form will develop. Thermodynamics, however, tells us nothing about the rate of reaction; it tells us only about which form is the most stable under the given conditions. Many minerals occurring at room temperature and pressure are metastable. Diamond is a good example. Most metamorphic rocks, which have formed under conditions of high temperature and/or pressure, are metastable under conditions existing at the surface of the Earth.

As mentioned above, thermodynamics tells us nothing about how fast a reaction occurs. It also does not tell us how a reaction occurs. Often there is more than one mechanism or path by which a system can get from one state to another. Further, the path that is followed may have several different intermediate steps. Thermodynamics tells us only what the final state should be for a given change in the variables affecting the system. Studies of how and when, if ever, that final state is reached, belong to the field of *kinetics.*

A common approach to understanding the kinetics of a chemical reaction starts with an experimental determination of the rate of reaction. Changes in concentration are measured as a function of time. These rate data can then be used to deduce information about the mechanism. We use macroscopically observable changes to try to explain changes that must be occurring on the microscopic molecular and atomic scale.

We find experimentally that the rates of reactions depend on temperature and are proportional to the concentrations of the reacting molecules. The mechanistic steps that occur appear to be dependent on the numbers of reacting molecules and on the energies of the collisions between molecules. A minimum amount of collision energy is needed to activate a molecule to a state where it will react. In general small activation energies result in very fast reactions. Even though we can determine rates of reaction, we can only propose possible, simple mechanistic steps that, taken together, will explain the measured rates.

Nothing further will be said about kinetic theory in this chapter. For more information on kinetics and a review of the use of kinetics in studying geochemical processes, see Lasaga and Kirkpatrick (1981).

FIRST LAW OF THERMODYNAMICS

In mechanical systems characterized by kinetic and potential energy, it can be shown experimentally that the total energy of a system is constant. This relationship is called the *principle* or *law of conservation of energy.* The experimental work of physicist J. P. Joule during the period 1840–1849 on the relation between heat flow and work done led scientists to apply the principle of conservation of energy to thermodynamic processes. The important result of this experimental work was to show that the amount of work required to raise the temperature of a system through a given range is independent of the type of work done. This result was interpreted in the following manner. The change in temperature of the system due to the work performed represents an internal change in the system. A change in the state of the system occurs. This change can be considered a change in a property of the system that will be called *internal energy.* Thus a system in a given state can be said to possess a certain value of internal energy, just as physical bodies can be said to have a certain value of kinetic or potential energy. We can only measure changes or differences in internal energy since, by analogy with mechanics, this energy is measured relative to some arbitrary zero point.

For an *adiabatic* system (no heat flow in or out of the system), the amount of work done on the system is equal to the change in internal energy. When work is done on a system, the internal energy change is by convention said to be positive, and thus the internal energy of the system increases. Now what happens if the system is not enclosed by an adiabatic wall? In this case, heat can flow between the system and the surroundings. Experiments of this type show that a given amount of work does not produce the same change in temperature for a nonadiabatic system as it does for an adiabatic system. Thus, the change in internal energy is different for the two systems. The difference between the actual work done on a nonadiabatic system and the change in internal energy of the system is explained as being due to a transfer of energy in the form of heat. Thus the total change in internal energy is equal to the work done plus the heat transferred. This interpretation is consistent with the principle of conservation of energy, which, as previously mentioned, was found to hold true in the field of mechanics.

The statement of equivalence of internal energy change with work done plus heat transferred is known as the *first law of thermodynamics*. The relationship holds true in all cases where it has been tested experimentally or theoretically. Mathematically, we can state the first law as

$$dU = dQ - dW \quad \text{(infinitesimal change)}$$

and

$$\Delta U = Q - W \quad \text{(finite change)} \tag{3-2}$$

where ΔU is the change in internal energy, Q is heat transferred, and W is work done. Internal energy can be expressed in the units of work or those of heat. Prior to the recommendation in 1960 of the adoption of the joule as the standard unit of energy, work was usually given in joules and heat in calories. The calorie is still used as a heat-energy unit. One calorie equals 4.184 joules (Table 3-1). When work is done by a system, the internal energy of the system is decreased. This is the reason for the negative sign in the first law equation. When heat is absorbed by the system, Q is said to be positive and the internal energy is increased.

Let us summarize the significance of the first law. It postulates the existence of a quantity called the internal energy. This quantity depends only on the state of the system and not on how this state was attained.* Each state has a characteristic internal energy. The first law also represents the principle of conservation of energy as applied to systems involving work and heat flow. Finally, the first law includes in its statement the concept of heat as a mode of transport of energy. Thus heat can be defined as the difference between the work done and the internal energy change in a nonadiabatic process.

At this point we shall return to geology and apply the first law to some geologic materials. Suppose we can measure the heat change that occurs when aragonite changes to calcite at 25°C and 1 atmosphere pressure. We find that heat is absorbed by the system and amounts to 59 calories per mole of $CaCO_3$. The volume of the calcite is 36.94 cubic centimeters per mole (cm^3/mol) and that of the aragonite is 34.16 cm^3/mol. With this information we can calculate the change in internal energy for the reaction aragonite to calcite. We use the equation

$$\Delta U = Q - P\,\Delta V \tag{3-3}$$

where the work is represented by the change in volume times the pressure (force times distance equals force/area times distance change times area equals pressure times volume change). See Table 3-1 for conversion factors.

$$\begin{aligned}
\Delta U &= 59 \text{ cal/mol} - 1 \text{ atm} \times (36.94 - 34.16) \text{ cm}^3/\text{mol} \\
&= 59 \text{ cal/mol} - 1.013 \times 10^6 \text{ dynes/cm}^2 \times 2.78 \text{ cm}^3/\text{mol} \\
&= 59 \text{ cal/mol} - 2.82 \times 10^6 \text{ dyne-cm/mol} \\
&= 59 \text{ cal/mol} - 2.82 \times 10^6 \times 2.39 \times 10^{-8} \text{ cal/mol} \\
&= 59 \text{ cal/mol} - 0.067 \text{ cal/mol} = 58.933 \text{ cal/mol}
\end{aligned}$$

Thus there is an increase in internal energy in changing aragonite to calcite. Practically all the change is due to absorption of heat by the system. A slight amount of work is done by the system,

*Although the value of ΔU depends only on the initial and final state of a system, this is *not* true for Q and W. The magnitudes of Q and W depend on the particular way in which the system changes. However, the algebraic sum of Q and W is a constant for a given change, no matter how that change occurs. The dependence of Q and W on the manner of change can be expressed mathematically by stating that dQ and dW are inexact differentials. In contrast, dU is an exact differential, which means that the integral of dU is independent of the way in which the system changes.

resulting in an increase in volume. The absorption of heat causes an increase in internal energy, while the work done decreases the internal energy. Keep in mind that the internal energy change occurs at a fixed temperature (i.e., there is no change in the temperature of the system).

If we make a similar study of the polymorphs graphite and diamond, we find that for the change of graphite to diamond the internal energy change is 453.046 calories per mole at 25°C and 1 atmosphere pressure. The figures used to obtain this value are 453 calories per mole absorbed by the system and a volume decrease of 1.882 cubic centimeters per mole. Both the heat and work changes cause an increase in the internal energy, since the heat is absorbed and work is done on the system to decrease its volume. Our results show that almost 400 calories per mole more energy is required to change graphite to diamond than is required to change aragonite to calcite. Furthermore, the amount of energy needed to change aragonite to calcite is relatively small. We might guess from this that, at 25°C and 1 atmosphere pressure, the change from aragonite to calcite can occur much more easily than the change from graphite to diamond. Studies of these materials in nature confirm this conclusion. Calcite is often found as a pseudomorph after aragonite, whereas such a relationship is not found for graphite and diamond.

It has been found that the internal energy is most useful in studying changes that take place at constant volume. In these cases, the internal energy change is equal to the heat flow.* However, chemical reactions and phase changes more often occur under constant pressure conditions. Thus it would be useful to have another characteristic of a system, in addition to U, that is equal to the heat flow when pressure is constant. We define this new function as $H = U + PV$. The function H is called the *enthalpy.* As in the case of the internal energy, only changes in enthalpy can be measured. These changes can be expressed as

$$dH = dU + P\,dV + V\,dP \tag{3-4}$$

and, using equation (3-2), when PV work is performed

$$\begin{aligned} dH &= dQ - P\,dV + P\,dV + V\,dP \\ &= dQ + V\,dP \quad \text{(infinitesimal change)} \end{aligned} \tag{3-5}$$

or

$$\Delta H = Q + V\,\Delta P \quad \text{(finite change)}$$

Note that $\Delta H = Q$ when the pressure is constant.

The values of ΔH for certain processes, such as a phase change from solid to liquid, are given general names. For example, the *heat of fusion* is the heat absorbed (which is equal to ΔH) by a solid when it melts under constant pressure. The *heat of vaporization* refers to the heat absorbed when a liquid changes to a gas. In chemical reactions at constant pressure, the heat absorbed or given off is called the *heat of reaction.*

A particularly useful type of enthalpy change is the *heat of formation* (Table 3-2). This is also known as the enthalpy of formation or the standard heat of formation. The heat of formation of a compound is the heat change produced in the reaction of the necessary elements under standard conditions to form that compound. These values for various compounds are determined experimentally and can then be used to determine heats of reaction for other chemical changes. The heat of formation of the elements is arbitrarily set at zero for that form of the element which is stable at 1 atmosphere pressure and 25°C (298.15 K).

*At this point we are considering only pressure-volume work. Other kinds of work may be important in specific cases. Electrochemical work is discussed in Chapter Four.

TABLE 3-2 Standard Heats of Formation, Entropies, and Free Energies of Formation at 298.15 K (25°C)

Formula	State	ΔH_f^0	S^0	ΔG_f^0	Source
Al_2O_3	Corundum	−400,400	12.18	−378,082	R & W
Al_2SiO_5	Andalusite	−619,390	22.28	−584,134	R & W
Al_2SiO_5	Kyanite	−619,930	20.02	−584,000	R & W
Al_2SiO_5	Sillimanite	−618,650	22.97	−583,600	R & W
C	Graphite	0	1.372	0	R & W
C	Diamond	453	0.568	693	R & W
CO_2	Gas	−94,051	51.06	−94,257	R & W
CO_3^{2-}	Aqueous	−161,630	−12.7	−126,220	K
Ca^{2+}	Aqueous	−129,800	13.2	−132,200	K
$CaCO_3$	Calcite	−288,592	22.15	−269,908	R & W
$CaCO_3$	Aragonite	−288,651	21.18	−269,678	R & W
$CaMg(CO_3)_2$	Dolomite	−557,613	37.09	−518,734	R & W
$CaSiO_3$	Wollastonite	−390,640	19.60	−370,313	R & W
$CaSO_4$	Anhydrite	−343,321	25.5	−316,475	R & W
$CaSO_4 \cdot 2H_2O$	Gypsum	−483,981	46.36	−430,137	R & W
Cu^{2+}	Aqueous	14,500	−26.5	15,500	K
$Cu_2(OH)_2CO_3$	Malachite	—	—	−216,440	R & W
Fe^{2+}	Aqueous	−21,000	−27.1	−20,300	K
Fe_2O_3	Hematite	−197,300	20.89	−177,728	R & W
H_2O	Aqueous	−68,315	16.71	−56,688	R & W
H^+	Aqueous	0	0	0	By convention
Mg^{2+}	Aqueous	−111,520	−32.7	−108,800	K
OH^-	Aqueous	−54,960	−2.52	−37,600	K
O_2	Gas	0	48.996	0	R & W
Pb	Solid	0	15.55	0	R & W
PbS	Galena	−23,353	21.84	−22,962	R & W
S	Orthorhombic solid	0	7.60	0	R & W
S_2	Gas	30,840	54.51	19,120	R & W
SO_4^{2-}	Aqueous	−216,900	4.4	−177,340	K
SiO_2	Cristobalite	−216,930	10.38	−204,075	R & W
SiO_2	Quartz	−217,650	9.88	−204,646	R & W
SiO_2	Tridymite	−216,895	10.50	−204,076	R & W

Note: Values are given in calories per mole for ΔH_f^0 and ΔG_f^0 and in calories per mole per degree for S^0. Not all zeroes shown are significant. ΔH_f^0 and ΔG_f^0 refer to formation from the elements in their standard states. The standard states are as follows:

Solids: pure material at 1 atmosphere pressure, most stable form

Gases: pure gas at 1 atmosphere pressure and assuming ideal gas behavior

Ions: ideal 1-molal solution in water under 1 atmosphere pressure

Some data sources report these values in joules/mole rather than calories/mole. Also the standard states are often given as having one bar pressure rather than one atmosphere pressure. Since one bar equals 0.987 atmosphere, the differences are generally not significant. A number of thermodynamic data compilations have been published since the 1967 and 1968 references used here. These are listed in Appendix E of the book by Nordstrom and Munoz (1994), who also provide an extensive discussion (Chapters 11 and 12) of measurement techniques, sources of error, and data evaluation methods. A short list of the most widely used sources can be found in Appendix B of Richardson and McSween (1989). Some of these references are listed at the end of this chapter (Robie et al. 1978; Helgeson et al. 1978; Helgeson 1982, 1985; Helgeson and Flowers 1982; Wagman et al. 1982; Woods and Garrels 1986; Berman 1988; and Cox et al. 1989).

Sources: Krauskopf (1967, App. VIII) and Robie and Waldbaum (1968, 11–25).

We learn in elementary chemistry courses that two equations for two chemical reactions can be added or subtracted to get a third equation representing a third reaction. When we are interested in heat changes associated with chemical reactions, we can also add and subtract equations. For example, if we wish to obtain the heat of reaction for the change of graphite to diamond at 25°C and 1 atmosphere, we find that it cannot be measured directly because of experimental difficulties. So we measure the heat changes in the following two reactions and then combine them to get the desired heat of reaction:

$$\text{C (graphite)} + \text{O}_2 \text{ (gas)} \rightleftharpoons \text{CO}_2 \text{ (gas)} + 94{,}051 \text{ cal}$$

$$\text{C (diamond)} + \text{O}_2 \text{ (gas)} \rightleftharpoons \text{CO}_2 \text{ (gas)} + 94{,}504 \text{ cal}$$

Subtracting the second equation from the first,

$$\text{C (graphite)} \rightleftharpoons \text{C (diamond)} - 453 \text{ cal}$$

The heat values given are for 1 gram-atom (mole) of carbon taking part in the reactions.

We find that 453 calories of heat must be added to 1 mole of graphite to change it to 1 mole of diamond. The heat change of the first reaction represents the standard heat of formation of carbon dioxide from carbon and oxygen at 1 atmosphere and 25°C. In a table giving heats of formation of substances, the value for CO_2 (gas) would be given as −94,051 calories per mole, since the heat is given off in the reaction. The equation relating graphite and diamond shows that the heat of formation of diamond is +453 calories per mole. The heat of formation of graphite is zero, since it is the stable form of carbon at 1 atmosphere and 25°C.

ENTROPY AND THE SECOND LAW OF THERMODYNAMICS

In our discussion of the historical development of thermodynamics, it was indicated that the second law deals with the direction in which a process will proceed. Our approach will be based on answering the question: What makes a chemical reaction occur? We have already discussed the heat change involved in a reaction. It was once thought that the difference in internal energy represented by this flow is the cause of chemical reactions, for most chemical reactions result in the release of heat (exothermic reactions). Thus we could reason that reactions occur when a decrease in the internal energy of the system being considered is possible, this decrease occurring by the release of heat from the system. Similarly, a chemical reaction would not occur if it would result in an absorption of heat (endothermic reaction) unless energy were added from an outside source. But this reasoning cannot explain reactions that do occur in nature and are accompanied by an absorption of heat. For example, if a piece of the mineral halite is put in water, a chemical reaction occurs and the halite dissolves. Measurement of the heat change would show that approximately 1,200 calories per mole of halite dissolved is absorbed. What makes this reaction occur? There must be some other factor that affects chemical reactions beside heat changes.

If we review a large number of natural processes, we find that in general spontaneous processes in nature tend toward the formation of less ordered (or less organized) structures. Rocks tend to fall apart and minerals tend to dissolve at the surface of the Earth. Vegetation decays when it dies and can no longer make use of energy from the Sun. The degree of ordering is thus the other factor, in addition to heat flow, that helps to determine whether a given process occurs or not. We need a thermodynamic function that takes into account both heat flow and degree of ordering. Such a function was introduced by J. W. Gibbs in 1875 and is now known as the *Gibbs function.* We shall soon define it more precisely.

We can now explain why halite dissolves in water. Although the system halite-water absorbs heat as it reacts, the total change is such that the system undergoes a decrease in the value of the Gibbs function. After the reaction is finished, there is less order in the system. Instead of having two separate phases with the sodium and chlorine atoms ordered and separate from the water molecules, we now have a random distribution of the sodium and chlorine as ions throughout the system. The decrease in the Gibbs function due to the change in order is greater than the increase in the Gibbs function due to the absorption of heat. Thus the value of the Gibbs function of the system decreases and the reaction occurs. If the reaction were such that it required the Gibbs function of the system to increase, the reaction would not occur (unless energy from an outside source were available).

The fact that those processes that occur in nature are in general the ones that result in less order can be related to the concept of probability. The processes that do occur are the processes that are the most probable. Disorder is more probable than order, or, in other words, randomness is a more probable state than one of structure or organization. When a landslide occurs, it is not very likely that the rock masses taking part will end up in the same arrangement (or a more ordered arrangement) as before the slide. Thus the degree of disorder will increase. In another sense, the more probable a chemical reaction is, the more it goes to completion (i.e., there are large amounts of the products as compared to the reactants).

The rate of a reaction between substances A and B depends on how often a molecule of substance A comes into contact with a molecule of substance B. The more often this contact occurs, the more probable it is that they will react. Thus the halite will dissolve faster in the water if we put in several small pieces rather than one large piece. And the reaction will be even faster if we stir the mixture with a stirring rod. In both cases we have increased the probability of a contact and thus of reaction. This can be compared with driving a car on a highway. If the highway is crowded, there is a greater probability of a collision occurring.

The change in the Gibbs function of a system due to a change in order is related to the change of a quantity known as *entropy.* The entropy of a substance in a given state can be considered a measure of the probability that the substance will occur in that state. Whether or not a reaction among substances will occur depends on the balance between entropy and heat flow. If the Gibbs function decreases, the reaction will occur. Loss of heat decreases the Gibbs function, and increase of entropy (change to a more probable state) decreases the Gibbs function. At equilibrium these two quantities are in balance, and the value of the Gibbs function of the reactants is equal to that of the products. We can summarize the effect of changes in entropy and heat content in the following manner:

1. Reactions that give off heat and produce less order (an increase in entropy) will always occur. An example is the burning of coal. Heat is given off and the complicated, highly ordered molecules making up coal are reduced to simpler molecules, such as those of carbon dioxide.
2. Reactions that give off heat and result in more order will occur only if the heat energy lost is greater than the increase in the Gibbs function due to the decrease in entropy. The change of liquid water to ice is an example.

3. Reactions that absorb heat and produce less order will occur only if the decrease in the Gibbs function due to the increase in entropy is greater than the increase in the Gibbs function due to heat absorption. An example is the change of liquid water to water vapor.
4. Reactions that absorb heat and produce more order will not occur. Heating liquid water does not produce ice.

Theoretical study of the relationship between heat and temperature has shown that it is possible to develop an absolute temperature scale that holds for all substances. This is the Kelvin scale mentioned earlier. Study of the relationship between heat and absolute temperature gives the following equality:

$$dS = \frac{dQ_r}{T} \quad \text{(infinitesimal change)}$$

and

$$\Delta S = \frac{Q_r}{T} \quad \text{(finite change)} \tag{3-6}$$

where Q_r is the heat absorbed from the surroundings for an ideal type of process, called a *reversible process;* T is absolute temperature; and ΔS is the change in entropy when a system is changed from one state to another state by a reversible process at a constant temperature.* A reversible process is a process that is conducted in infinitesimal steps so that the system being dealt with is always in equilibrium. Such a process can be reversed at any time by making correspondingly small, but opposite, changes in the surroundings. All changes in nature do not satisfy the strict requirements of a reversible process (a few closely approximate a reversible process). It can be shown experimentally that for all natural (irreversible) processes

$$dS > \frac{dQ}{T} \quad \text{(infinitesimal change)}$$

and

$$\Delta S > \frac{Q}{T} \quad \text{(finite change)} \tag{3-7}$$

Equations (3-6) and (3-7) define the fundamental quantity known as entropy and represent the *second law of thermodynamics.* A nonmathematical way of stating the second law is that heat cannot be converted into work with 100 percent efficiency.† In a reversible (ideal) process the heat absorbed is equal to the work done and to $T\,\Delta S$. (Note that ΔS is *not* an energy term, but $T\,\Delta S$ is.) In natural processes the same amount of heat will, because of inefficiency, do less work and a greater entropy change is produced. In other words, an irreversible (spontaneous) change requires a larger entropy change, and thus less work can be done for a given value of Q.

The practical significance of entropy can be summarized as follows. It is possible to calculate the values of S for phases (such as minerals) and for systems (Table 3-2). Furthermore, it can be shown that, for an adiabatic system (no heat flow in or out), a change is possible if it results in an increase in entropy (due to internal changes) and impossible if it results in a decrease in en-

*The magnitude of Q_r depends on the way a system changes from one state to another, whereas the magnitude of ΔS depends only on the initial and final states of the system. Stated mathematically, dQ_r is an inexact differential and dS is an exact differential. In all the expressions in this chapter, dQ is an inexact differential. Equation (3-6) holds true only for a closed system, which is defined as a system that allows energy but not matter to move across its boundaries. For a geologic example of the difference between an open and a closed system, see Figure 3-1.

†Nordstrom and Munoz (1985) provide this earthy statement of the second law: You can't shovel manure into the rear end of a horse and expect to get hay out of its mouth.

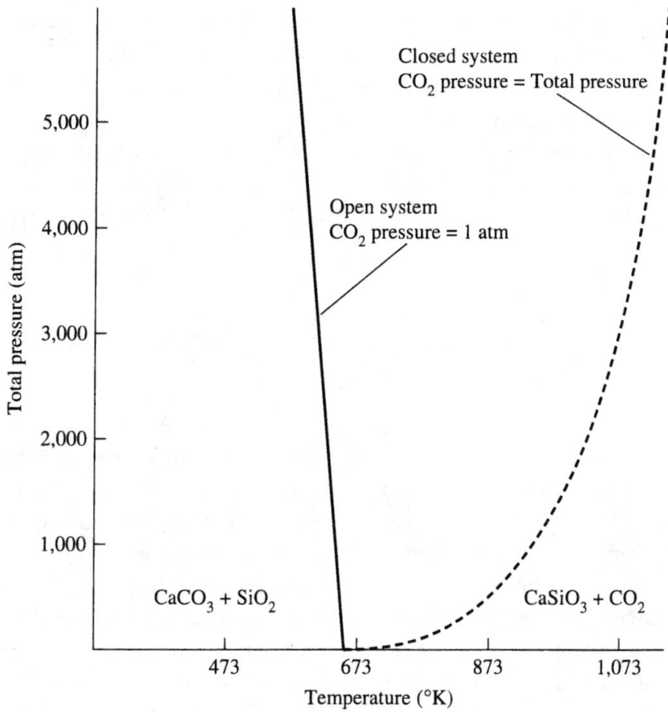

Figure 3-1 Pressure-temperature relationships for the reaction calcite + quartz ⇌ wollastonite + carbon dioxide. The dashed curve was first calculated by V. M. Goldschmidt in 1912 and is for a closed system. The solid line is for an open system from which carbon dioxide can freely escape. For most rocks the equilibrium curve will lie somewhere between the two curves shown. For an example of the field application of this diagram, see Greenwood (1967). (After T. F. W. Barth, 1962, *Theoretical Petrology*, 2nd ed., John Wiley & Sons, Inc., New York, p. 258.)

tropy. Thus entropy can be used to determine the direction that an adiabatic process will take. Additional information on heat exchange with the surroundings is needed to determine direction for a nonadiabatic process. The two values, entropy change and heat change, can be combined in the Gibbs function to give the direction of a nonadiabatic reaction as discussed previously. One other point should be mentioned. In developing the theory of the second law, it can be shown that entropy, like internal energy, is a function only of the state of a system. It does not matter how the system reached that state.

To understand how entropy can be used quantitatively in studying natural processes, we need to know how entropy values can be assigned to substances and how entropy changes can be calculated. The *third law of thermodynamics* states that every perfectly ordered, pure crystalline substance has the same entropy at 0 K (absolute zero). Thus absolute zero can be used as a reference level for zero entropy. The entropy of a substance at some other temperature can be calculated by use of a quantity known as the *heat capacity.* The heat capacity, C, of a substance is defined as

$$C = \frac{dQ}{dT} \tag{3-8}$$

where dQ is the heat absorbed by the substance when its temperature is raised the amount dT.

Heat capacity may be measured with the substance kept at constant volume or under constant pressure, with different values obtained for the two situations. Entropies of gases can also be calculated from spectroscopic data; the entropies of liquids and ions can be obtained by measuring heats of fusion and heats of solution.

Let us increase the temperature of a substance by a reversible process from a temperature T_1 to a temperature T_2. Using the equation relating entropy to absolute temperature and heat flow, expressing it for an infinitesimal change, and substituting for dQ from equation (3-8), we have

$$dS = \frac{dQ}{T} \tag{3-9}$$

$$\int_{S_1}^{S_2} dS = S_2 - S_1 = \int_{T_1}^{T_2} \frac{C}{T}\, dT \tag{3-10}$$

We see that the entropy change from one state to another can be determined by integration if the heat capacity is known. Although a reversible process is an ideal process, which cannot be carried out exactly in a laboratory, we can do it mathematically and thus determine entropy changes.* We do have to determine the heat capacity C experimentally. Also it must be assumed that C is a constant over the temperature range dealt with (a reasonable assumption if the temperature range is small). Otherwise, the heat capacity must be expressed as a function of temperature in equation (3-10). Changes that take place at a constant temperature, such as a phase change, can be treated more simply as follows. The absolute temperature of transition and the resulting heat flow are measured. Assuming that the change is a reversible process,

$$\Delta S = \frac{Q_r}{T} \tag{3-6}$$

Let us conclude our discussion of entropy with a return to geology. How do the entropies of different minerals compare? Graphite at 25°C is found to have a higher entropy than diamond. We can interpret this as being due to the larger volume of a unit mass of graphite as compared to the same mass of diamond. The larger volume allows more disorder of the carbon atoms and thus gives a higher entropy. In general, the greater the difference in specific or molar volume between two polymorphs, the greater the difference in entropy. If we compare the volumes of the common polymorphs of SiO_2, we find tridymite > cristobalite > quartz. The entropies at 25°C show the same relationship, with tridymite having the highest entropy. Since increasing temperature tends to decrease the disorder of atoms, we might interpret this to indicate that the stable high-temperature phase of SiO_2 should be tridymite. However, at the transition temperature of the two phases, cristobalite has a slightly higher entropy and is the stable high-temperature phase. Obviously, not only does the entropy of a substance change with temperature, it may also change more for one substance than for another. We can generalize the relationship between volume and entropy by saying that, for most minerals, the greater the volume, the greater the entropy. If we increase the pressure on a mineral, we tend to decrease its volume. It has also been found that in general the entropy of a mineral decreases as the pressure is increased.

The entropy of a given mineral is apparently related to a number of other factors in addition to volume. These include the chemical composition, crystal structure, degree of solid solu-

*We can also describe theoretically a reversible process for a state function such as entropy and then carry out the same change in state by an irreversible laboratory process. The amount of change for a state function is the same for both processes since it does not matter how the change is carried out.

tion (if any), and type of bonding. Consider the effect these factors have on the ordering of the atoms (ions) in a mineral. A mineral made up of a large number of different elements should have a larger entropy compared to another mineral with fewer components and thus more order. Different crystal structures represent different degrees of order and thus of entropy. The occurrence of solid solution in a mineral should increase the relative disorder and therefore the entropy. If the atoms (ions) of a mineral are tightly bonded to each other, they have little freedom of movement about their positions in the crystal structure. This would tend to produce a lower entropy than if they could move more freely. These qualitative interpretations have been supported by experimental measurements of the actual entropies of minerals. A review of our understanding of mineral entropies is given by Ulbrich and Waldbaum (1976).

GIBBS FUNCTION

We have now discussed the first and second laws of thermodynamics and introduced three properties of a system that depend only on the state of the system. These are internal energy, enthalpy, and entropy, which are often referred to as "state functions." The difference in value for any one of these properties between two states of a system can be calculated. Even though the calculation assumes an ideal process, which cannot be completely achieved experimentally, the difference found nevertheless represents a real difference between these two states. The same difference would be found, if it were possible to measure it, in any reversible process carried out in a laboratory in which the same change from one state to the other is achieved. Irreversible processes that produce the same change in states as that of a given reversible process will also result in the same difference in each of the properties.

We now proceed to combine the first and second laws and to develop two other functions of the state of a system. The basic theory of thermodynamics has been summarized above. The new functions presented here are related to those already discussed and are more useful in dealing with certain types of problems. For a reversible change in state, the first law can be expressed as

$$\Delta U = Q_r - W_r \tag{3-11}$$

Substituting the second law equation and assuming a closed system and that the work done involves a reversible volume change at constant pressure and temperature, we get

$$\Delta U = T\,\Delta S - P\,\Delta V \tag{3-12}$$

This equation thus combines the first and second laws of thermodynamics. It holds true only for reversible processes, since the terms $T\,\Delta S$ and $P\,\Delta V$ can be identified as heat absorbed and work done only in the case of a reversible process.* However, it is important to note that, for the change from a given initial state to a given final state, ΔU has a definite value that depends only on these states and not on how the change takes place.

If we wish to state the combination of the two laws in a form involving enthalpy, we add $\Delta(PV)$ to both sides of equation (3-12):

$$\begin{aligned} \Delta U + \Delta(PV) &= T\,\Delta S - P\,\Delta V + P\,\Delta V + V\,\Delta P \\ &= T\,\Delta S + V\,\Delta P \end{aligned} \tag{3-13}$$

*When there are no irreversible changes in composition, equation (3-12) can also be applied to an irreversible change. Such changes in composition could occur due to chemical exchange with the surroundings (an open system) or to an irreversible chemical reaction taking place inside an open or closed system.

which is equal to ΔH for a reversible process. Thus enthalpy is useful in studying systems in which pressure and entropy are variables. The internal energy is more useful when volume and entropy are variables. Suppose that we are carrying out an experiment in which we study a system by varying the temperature and volume of the system. Then in order to determine the relationships of interest we need a function that depends on these two properties as variables. We define the *Helmholtz function* as $A = U - TS$. Thus

$$dA = dU - T\,dS - S\,dT = T\,dS - P\,dV - T\,dS - S\,dT$$

$$= -S\,dT - P\,dV \text{(infinitesimal change)} \qquad (3\text{-}14)$$

and $$\Delta A = -S\,\Delta T - \Delta V \text{(finite change)}$$

The letter *F* is used in some textbooks instead of *A* for the Helmholtz function. It was named for the man who first applied the results of Joule's experiments to problems in physical chemistry.

In most chemical problems we are concerned with the variables pressure and temperature. To obtain a function for these problems, we define the *Gibbs function* (mentioned earlier in the discussion of entropy) as

$$G = U + PV - TS = H - TS = A + PV$$

Thus

$$dG = dA + P\,dV + V\,dP = -S\,dT - P\,dV + P\,dV + V\,dP$$

$$= V\,dP - S\,dT \quad \text{(infinitesimal change)} \qquad (3\text{-}15)$$

and $$\Delta G = V\,\Delta P - S\,\Delta T \quad \text{(finite change)}$$

Note that ΔG is also equal to $\Delta H - \Delta(TS)$ and that at constant temperature

$$\Delta G = \Delta H - T\,\Delta S \qquad (3\text{-}16)$$

The function *G* is called the Gibbs function after the man who developed the basic theory of thermodynamics. The letter *F* is used in some textbooks instead of *G* for the Gibbs function.

The Helmholtz function and the Gibbs function are also known as the Helmholtz free energy and the Gibbs free energy. Both of them depend only on the state of a system and not on how that state was reached (as is true of internal energy, enthalpy, and entropy). Which of the four functions (*U, H, A,* and *G*) we use depends on what changes in the system we want to study. They are all related to each other and all come from the same basic principles. Any of them can be used in applying these principles to natural processes.

Earlier we discussed enthalpy and the concepts of heat of reaction and standard heat of formation. Similarly, we can calculate the free energy of reaction and the standard free energy of formation (Table 3-2). These terms usually refer to the Gibbs function. The change in free energy accompanying a chemical reaction is called the *free energy of reaction.* It is defined as the sum of the free energies of the products minus the sum of the free energies of the reactants. To calculate this, we need to assign free-energy values to the substances taking part in the reaction. This is done by assigning values to compounds that are based on arbitrary values for the elements. The standard free energy of formation of a compound (ΔG_f^0) is defined as the free-energy change resulting from the formation of that compound from its constituent elements under chosen standard conditions (pressure of 1 atmosphere, 25°C). The standard free energy of formation of the elements is defined as zero for the form of the element stable under the standard conditions.

Once we have set up this system for assigning a free-energy value to each compound and element, we can calculate the free-energy change for any other conditions, and we can calculate the free energy of reaction for any chemical reaction. If the products and reactants are all in specified standard states, then we speak of the standard free energy of reaction (ΔG^0).* (The term *standard state* refers to a reference state defined by the amount and manner of behavior of material at a selected pressure and temperature; see bottom of Table 3-2.)

Let us consider two examples. First, we want to calculate the standard free energy of formation of galena. We use the formula $\Delta G_f^0 = \Delta H - T\,\Delta S$ for formation at a temperature of 25°C (298.15 K) and 1 atmosphere pressure. The standard heat of formation of galena for the reaction $Pb + S \rightarrow PbS$ is –23,353 calories per mole. The change in entropy for the reaction is equal to the entropy of galena (21.84 calories per mole-degree) minus the sum of the entropies of lead (15.55 calories per mole-degree) and sulfur (7.60 calories per mole-degree). Thus

$$\Delta G_f^0 = -23{,}353 - (298.15 \times -1.31) = -22{,}962 \text{ cal/mol}$$

This is the standard free energy of formation of galena. Values for free energies of formation of compounds given in various references are often slightly different for the same compound. This is due to the fact that such values can be determined in several different ways, and the results are usually not in exact agreement. Also, as experimental and mathematical techniques improve, more precise results become available for calculating free-energy values (Nordstrom and Munoz 1985, Chaps. 11 and 12). It should be noted that the precision of experimental techniques is such that the value actually given for the free energy of formation of galena is a figure such as –22,962 ± 200 calories per mole.

If a limestone containing quartz grains is subjected to metamorphism, the following reaction goes significantly to the right at about 650 K and 1 atmosphere pressure:

$$CaCO_3 \text{ (calcite)} + SiO_2 \text{ (quartz)} \rightleftharpoons CaSiO_3 \text{ (wollastonite)} + CO_2 \text{ (gas)}$$

First, let us calculate the standard free energy of reaction at 298.15 K (25°C) and 1 atmosphere pressure (using data from Table 3-2):

$$\begin{aligned}
\Delta G^0 = \Delta H - T\,\Delta S &= \Delta H_f^0 \text{ (woll.)} + \Delta H_f^0 \text{ (}CO_2\text{)} - [\Delta H_f^0 \text{ (cal.)} \\
&\quad + \Delta H_f^0 \text{ (qtz.)}] - 298.15 \times [S^0 \text{ (woll.)} + S^0 \text{ (}CO_2\text{)}] \\
&\quad + 298.15 \times [S^0 \text{ (cal.)} + S^0 \text{ (qtz.)}] \\
&= -390{,}640 - 94{,}051 - (-288{,}592 - 217{,}650) - 298.15 \\
&\quad \times (19.60 + 51.06) + 298.15 \times (22.15 + 9.88) \\
&= -484{,}691 + 506{,}242 - 21{,}067 + 9{,}550 \\
&= +10{,}034 \text{ cal}
\end{aligned}$$

To get the free-energy change at some other temperature, such as 650 K (total pressure still 1 atmosphere), we must determine ΔH and ΔS for that temperature. The change of enthalpy for

*The standard free-energy change of a reaction is the total free energy of the pure products in their stoichiometric proportions minus the total free energy of the pure reactants in their stoichiometric proportions under standard-state conditions.

each compound can be calculated as follows: $\Delta H = Q$ at constant pressure, and therefore $\Delta H = C_p\, \Delta T$, where C_p is heat capacity at constant pressure.* Thus, for each compound,

$$H_{650\text{K}} - H_{298.15\text{K}} = \int_{298.15}^{650} C_p\, dT$$

When we have an enthalpy value for each compound at 650 K, we can subtract the sum of the enthalpies of the reactants from the sum of the enthalpies of the products and get a value for ΔH of the reaction occurring in the limestone. A faster procedure would be to directly calculate ΔH for the reaction by using the equation

$$\Delta H_{650\text{K}} = \int_{298.15}^{650} \Delta C_p\, dT + \Delta H_{298.15\text{K}}$$

where $\Delta C_p = \sum C_p$ (products) $- \sum C_p$ (reactants).

We can do the same thing for the entropies of the substances since, for each compound, $\Delta H = Q = T\, \Delta S$ when the pressure is constant. Thus, for each compound,

$$\Delta S_{650\text{K}} - \Delta S_{298.15\text{K}} = \int_{298.15}^{650} \frac{C_p}{T}\, dT$$

The entropy change for the reaction at 650 K is

$$\Delta S_{650\text{K}} = \int_{298.15}^{650} \frac{\Delta C_p}{T}\, dT + \Delta S_{298.15\text{K}}$$

where $\Delta C_p = \sum C_p$ (products) $- \sum C_p$ (reactants).

To get the free energy of a reaction at any temperature, we determine ΔH and ΔS for the reaction at the specified temperature and then use $\Delta G = \Delta H - T\, \Delta S$. If we were to calculate the value of ΔG for the above calcite-quartz reaction at 650 K, we would find that it is very close to zero. Thus, at 298.15 K, ΔG is +10,034 calories, whereas at 650 K it is approximately 0 calories. The significance of these values will be pointed out shortly.

We can also calculate the free energy of a reaction at higher pressures. Looking at equation (3-15), we see that $\Delta G/\Delta P = V$ for a substance at constant temperature and this leads to the following expression for the change in ΔG of a reaction due to a change in pressure when temperature is held constant:

$$\left(\frac{\partial \Delta G}{\partial P}\right)_T = \Delta V \qquad (3\text{-}17)$$

Thus the free energy of reaction at a selected temperature and pressure is

$$\Delta G(T, P) = \Delta G(T, 1) + \int_1^P \Delta V\, dP \qquad (3\text{-}18)$$

where $\Delta G(T, 1)$ is the free energy of reaction at one atmosphere and the temperature of interest (reported above for the calcite-quartz reaction at 650 K) and ΔV is the difference between the

*Remember that heat capacity is a function of temperature. Often the heat capacity of a mineral is experimentally measured and then put in the form of an empirical power series: $C_p = a + bT + cT^2 + dT^3 + \ldots$, where a, b, c, d, etc. are constants.

total volume of the products and the total volume of the reactants. Because the effect of modest pressure and temperature changes on mineral volumes in the Earth's crust is generally insignificant, the value of ΔV for reactions involving only solids can be considered a constant and can be calculated using molar volumes measured at 298.15 K (25°C) and one atmosphere.

When gases are involved in a reaction, ΔV is definitely not a constant (i.e., it is a function of temperature and pressure) and the calculation of the free energy of reaction is more complicated. The simplest approach is to (1) ignore all condensed phases and consider that any volume difference between reactants and products is due to gases only, and (2) assume ideal gas behavior for the gases. When only one gas is involved in a reaction, the change in volume for the reaction is essentially equal to the volume of the gas. (Molar volumes of gases are much larger than the molar volumes of minerals. For example, at 298.15 K and one atmosphere, the molar volume of quartz is 22.69 cm^3 and the molar volume of carbon dioxide gas is 22,263 cm^3.)

Returning to the calcite-quartz reaction, the effect of changing pressure on the reaction is given by

$$\Delta G(T, P) = \Delta G(T, 1) + \int_1^P \Delta V\, dP = \Delta G(T, 1) + \int_1^P \frac{RT}{P}\, dP \qquad (3\text{-}19)$$

where the ideal gas law is assumed for carbon dioxide and the molar volume of the gas, $\overline{V} = RT/P$, is substituted for the ΔV of the reaction. At 650 K and 2000 atmospheres, the free energy of reaction is

$$\Delta G = 0 + RT \ln P = 82.06 \text{ cm}^3\text{-atm-deg}^{-1}\text{-mol}^{-1} \times 650 \text{ deg} \times \ln 2000$$
$$= 410{,}646.3 \times 10^6 \text{ dyne-centimeters} = 9814.4 \text{ calories}$$

Reactions between minerals at very high temperatures and pressures have ΔV values that are not independent of pressure and temperature. To determine the effect of changing pressure and temperature on the free energies of such reactions, we can use the thermal expansion α and the isothermal compressibility β of minerals,* where

$$\alpha = \frac{1}{V}\left(\frac{\Delta V}{\Delta T}\right)_P \quad \text{and} \quad \beta = -\frac{1}{V}\left(\frac{\Delta V}{\Delta P}\right)_T$$

Richardson and McSween (1989, 246) derive the following expression for the molar volume of a mineral at a given temperature and pressure:

$$\overline{V}(T, P) = \overline{V}(298.15, 1 \text{ atm})[1 + \alpha(T - 298.15) - \beta(P - 1)] \qquad (3\text{-}20)$$

*Both α and β are based on fractional changes in volume, dV/V, and are thus intensive properties. α can be written as $(\Delta V/V)/\Delta T$ and is the ratio of fractional volume change to the temperature change ΔT that caused it. Similarly, β can be written as $(\Delta V/V)/\Delta P$ and is the ratio of fractional volume change to the pressure change ΔP. Minerals have very small values of α and β. For example, quartz has $\alpha = 0.000034$ deg^{-1} at one atmosphere and 293 K and $\beta = -0.0000027$ bar^{-1} at one atmosphere and 298 K. Most minerals have similar values. Thus temperature changes of a few hundred degrees change the volume of most minerals by about one part in a hundred. Pressure changes on the order of several thousand bars change the volume of most minerals by about one percent. If a mineral is subjected to increases in both temperature and pressure, the change in volume is even smaller, since α is generally positive and β is always negative, that is, increase in temperature increases volume and increase in pressure decreases volume. Thus the molar volumes of minerals can be considered constant to within one part in a hundred in at least the upper ten kilometers of the Earth's crust.

Calculation of the molar volumes of the minerals involved in a reaction taking place under conditions of very high temperature and pressure gives us ΔV for the reaction.

We can combine the above equations to get a general equation for the dependence of the free energy of a reaction on temperature and pressure, starting at 298.15 K and one atmosphere,

$$\begin{aligned} \Delta G(T, P) &= \Delta H(298.15,\ 1\ \text{atm}) \\ &+ \int_{298.15}^{T} \Delta C_P\, dT - T\,\Delta S\,(298.15,\ 1\ \text{atm}) \\ &- T \int_{298.15}^{T} \frac{\Delta C_P}{T}\, dT + \int_{1}^{P} \Delta V\, dP \end{aligned} \tag{3-21}$$

where $\Delta C_p = \sum C_p$ (products) $- \sum C_p$ (reactants) and $\Delta V = \sum V$ (products) $- \sum V$ (reactants).

EQUILIBRIUM AND EQUILIBRIUM CONSTANTS

Earlier in our discussion of entropy we gave some general rules for determining whether or not a reaction will occur. We now proceed to make these rules quantitative. It was indicated that the direction and extent of a reaction depend on the balance between entropy (order) and heat flow. For reactions in a closed system (no material added or taken away) at a given temperature and pressure, this balance is represented by the change in Gibbs function, $\Delta G = \Delta H - T\,\Delta S$ [equation (3-16)], where the first term represents change due to heat flow and the second term represents change due to a change in order. (If we wanted to study a reaction occurring at constant volume and constant temperature, we would use the change in Helmholtz function rather than the change in Gibbs function.)

A rule stated earlier was that reactions that give off heat and produce less order (an increase in entropy) will always occur. For these reactions, then, ΔG will be negative (negative ΔH minus $T\,\Delta S$). Another rule stated that reactions that absorb heat and produce more order (a decrease in entropy) will not occur. For such reactions, ΔG will be positive (positive ΔH minus T times negative ΔS). In both these cases the two factors, heat flow and entropy change, work together. In the remaining two cases they do not. Reactions that give off heat and result in more order will occur only if the heat lost is greater than the increase in Gibbs function due to the decrease in entropy (negative ΔH greater in value than the positive value of the $T\,\Delta S$ term). Similarly, reactions that absorb heat and produce less order will occur only if the decrease in Gibbs function due to the increase in entropy is greater than the increase due to heat absorption (negative $T\,\Delta S$ term greater than positive ΔH term). In both of the last two cases, ΔG has to be negative for a reaction to occur. Thus, if we have values for ΔH and ΔS at a particular temperature and pressure for a reaction, we can calculate ΔG and predict whether that reaction will or will not occur. If we know the variation of heat capacities and molar volumes with temperature and pressure, we can make a similar prediction for other conditions.

The above discussion explains how the Gibbs function can be used as a criterion for predicting whether or not a reaction will occur. A derivation of this use of the Gibbs function can be done as follows. Consider the change of a closed system (no transfer of matter across the system boundary) at constant temperature and pressure between states 1 and 2. From the definition of the Gibbs function, we have

$$G_2 - G_1 = (U_2 - U_1) + (PV_2 - PV_1) - (TS_2 - TS_1)$$

Since $U_2 - U_1 = Q - W$ [equation (3-2)],

$$G_2 - G_1 = Q - W + P(V_2 - V_1) - T(S_2 - S_1)$$

From equation (3-7), we know that, for natural processes,

$$Q < T(S_2 - S_1)$$

Therefore,

$$-W + P(V_2 - V_1) < G_2 - G_1$$

The term W stands for the total work done by the system, whereas $P(V_2 - V_1)$ represents only work involving a volume change of the system. Other types of work, such as electrical work, could occur, and thus the total work is

$$W = P(V_2 - V_1) + W'$$

where W' represents other types of work. Substituting the above equation into the previous equation,

$$W' < -(G_2 - G_1) \quad \text{or for infinitesimal change} \quad dW' < -dG$$

Thus, for natural processes involving work and occurring at constant temperature and pressure, the work done, other than volume change, is less than the decrease of the Gibbs function. When $dW' = 0$ and the other conditions are the same,

$$dG < 0 \text{ (infinitesimal change)} \quad \text{or} \quad \Delta G < 0 \text{ (finite change)}$$

Under these conditions, the Gibbs function can only decrease (ΔG is negative) for natural processes.

In the previous section we calculated the standard free energy of reaction for calcite reacting with quartz at 298.15 K:

$$CaCO_3 \text{ (calcite)} + SiO_2 \text{ (quartz)} \rightleftharpoons CaSiO_3 \text{ (wollastonite)} + CO_2 \text{ (gas)}$$

The result was $\Delta G^0 = +10{,}034$ calories. Thus this reaction will not occur from left to right to any significant degree at this temperature. At about 650 K, ΔG is zero. There is a zero net change in the Gibbs function because the two opposite reactions occur in equal amounts. At this temperature, calcite and quartz are in equilibrium with wollastonite and carbon dioxide. If we were to calculate ΔG at a still higher temperature (keeping the pressure constant at 1 atmosphere), we would find the value of ΔG to be negative. So the reaction goes strongly to the right at higher temperatures. At 650 K and 2000 atmospheres we found $\Delta G = 9814$ calories and thus the reaction will not occur. By taking into account variations of ΔG with both temperature and pressure it is possible to construct a diagram showing the stability field of calcite plus quartz and the stability field of wollastonite plus carbon dioxide (Figure 3-1). A line dividing stability fields is an *equilibrium curve,* and ΔG equals zero along the curve. Note in Figure 3-1 that, when the composition of the system can change (open system), the free-energy relationships are different and thus the stability fields are different.

The magnitude of a ΔG value for a reaction is an indication of how far from equilibrium the reaction is under the given conditions. It is a measure of the extent to which the reactants will change into the products, or vice versa. At high temperatures and low pressures (large, negative ΔG), calcite and quartz in a closed system will react until one or the other is completely used up. For reactions that can be easily reversed (not true for many reactions of geological interest), we can say that these reactions generally behave in such a way that they do not go entirely to completion; that is, there is always a small amount of reactants present (and a large amount of products) after a reaction has reached equilibrium (no further net reaction). Remember that the above discussion does not take into account the question of kinetics. Some reactions go to completion at a very slow rate (such as the rusting of metal); others go to completion at a very fast rate (such as the dissolving of many solids in water).

We now discuss the relationship between the standard free energy of a reaction, ΔG^0, and the equilibrium constant, *K*, of a reaction. The mass-action law of chemical equilibrium states that, for a reaction $aA + bB \rightleftharpoons cC + dD$ at a given temperature,

$$K = \frac{(C)^c(D)^d}{(A)^a(B)^b} \tag{3-22}$$

where the quotient involves the concentrations of *A, B, C,* and *D*. The quantity *K* is a constant characteristic of the reaction whose value varies with temperature. In using the expression for *K* for real solutions, we have to substitute activity values for the concentration values (activities are modified concentrations defined in the next section; parentheses are normally used to indicate concentrations and brackets are used to indicate activities). It is possible to derive a relationship between the free-energy change of a reaction and the equilibrium constant. For the reaction $aA + bB \rightleftharpoons cC + dD$,

$$\Delta G_r = \Delta G^0 + RT\ln\frac{[C]^c[D]^d}{[A]^a[B]^b} \tag{3-23}$$

where ΔG_r is the free-energy change of the reaction, ΔG^0 is the free-energy change when the reactants and products are in a standard state of unit activity (i.e., 1 mole/kilogram for ions or 1 atmosphere pressure for gases), *R* is the gas constant, *T* is absolute temperature, and $[A]$ is the activity of *A*, $[B]$ the activity of *B*, etc.* At equilibrium, $\Delta G_r = 0$ and

$$\Delta G^0 = -RT\ \ln\frac{[C]^c[D]^d}{[A]^a[B]^b} = -RT\ \ln K \tag{3-24}$$

where $[A]$ is now the activity of *A* at equilibrium, $[B]$ now the activity of *B* at equilibrium, etc. The quantity ΔG^0 is the standard free energy of reaction discussed earlier, and *K* is the equilib-

*For any reaction ΔG^0 is the change in free energy when *a* moles of *A* and *b* moles of *B* are completely changed into *c* moles of *C* and *d* moles of *D* under standard-state conditions.

rium constant of the mass-action law. It should be emphasized that ΔG^0 refers to standard-state conditions and K to equilibrium conditions. These are usually *not* the same.* When they are the same, then $K = 1$ since the activity of each product and each reactant would be 1 for standard-state conditions. With $K = 1$, $\Delta G^0 = 0$. The amount that ΔG^0 differs from zero for a given reaction can be considered a measure of how far the standard-state activities (concentrations) differ from those to be found at equilibrium, whereas the equilibrium constant is based on the actual equilibrium values of the activities. Most calculations for reactions involve the standard free energy of reaction (ΔG^0) rather than the free energy of reaction for products and reactants occurring at nonstandard activities (ΔG). When the reactants and products are not at unit (standard) activity (and their activities are known and selected), ΔG can be calculated using the more general equation, equation (3-23).

It is possible to derive a relation expressing the dependence of the equilibrium constant K on temperature (van't Hoff equation). One form of this equation, at constant pressure, is

$$\frac{d \ln K}{dT} = \frac{\Delta H^0}{RT^2} \tag{3-25}$$

where ΔH^0 is the standard enthalpy change of the reaction. When ΔH^0 is independent of temperature, as it usually is,

$$\ln \frac{K_2}{K_1} = -\frac{\Delta H^0}{R}\left(\frac{1}{T_2} - \frac{1}{T_1}\right) \tag{3-26}$$

where K_2 is the equilibrium constant at T_2 and K_1 is the equilibrium constant at T_1. Thus if ΔH^0 and K are known at one temperature, the K at any other temperature can be easily calculated. When ΔH^0 is not independent of temperature, knowledge of the variation of ΔC_p with temperature can be used to express K in terms of ΔH^0 and temperature. Alternatively, an average change in enthalpy from T_1 to T_2 can be used in equation (3-26).

As an example of the use of the relationship between the standard free energy of reaction and the equilibrium constant, we shall consider the conditions under which the mineral malachite might precipitate from groundwater in the oxidized zone of a copper deposit:

$$2Cu^{2+} + CO_3^{2-} + 2OH^- \rightleftharpoons Cu_2(OH)_2CO_3 \text{ (malachite)}$$

Malachite is commonly found in the oxidized zone of ore deposits, particularly those associated with limestone. The standard free-energy change for the reaction, using ΔG_f^0 values from Table 3-2, is $\Delta G^0 = -46{,}020$ calories. (Note that ΔG^0 depends on the amounts of the substances taking part in the reaction. The free energy of formation values for cupric ion and hydroxide ion must be doubled in calculating ΔG^0.) We can now set $-RT \ln K = -46{,}020$, and K for the reaction at 25°C is thus approximately $10^{33.7}$. The large negative value for ΔG^0 and the large positive value for K indicate that the reaction will go far to the right when the ions are present at unit activities. (A large positive value for ΔG^0 would correspond to a very small value of K.)

*As mentioned earlier, a standard state is a reference state that is chosen arbitrarily. Because thermodynamic processes involve relative changes of various functions, it is necessary to use reference states. The selection of an appropriate standard state for various geochemical problems is discussed in detail in Nordstrom and Munoz (1985). The standard states used in this book are described at the bottom of Table 3-2.

Using the mass-action law,

$$[Cu^{2+}]^2[CO_3^{2-}][OH^-]^2 = 10^{-33.7}$$

where $[Cu^{2+}]$ is the activity of cupric ion, $[CO_3^{2-}]$ is the activity of carbonate ion, and $[OH^-]$ is the activity of hydroxide ion (the activity of a solid phase is set at 1 by convention). Assume that we want to know the concentration of cupric ion necessary to cause precipitation of malachite. In dilute solutions, activities of solutes are approximately equal to concentrations of solutes, and thus we can use concentration values measured in groundwater for this example. If our hypothetical copper deposit is in limestone, the pH of the water can be assumed to be about 8; thus $(OH^-) = 10^{-6}$ mole per liter. Under the conditions of near-surface groundwater and a pH of 8, the concentration of carbonate ion in the water would probably be about 10^{-6} mole per liter (most of the carbonate material in the water would occur in the form HCO_3^-; see Figure 4-12). Solving for (Cu^{2+}),

$$(Cu^{2+})^2 = \frac{10^{-33.7}}{(10^{-6})(10^{-6})^2} = 10^{-15.7}$$

$$Cu^{2+} = 10^{-7.85} \text{ mole/liter}$$

We can conclude that very little copper is needed in solution to cause the precipitation of malachite. Measurements of the copper content of groundwater draining sulfide ore deposits give values on the order of 10^{-3} mole per liter. Many of these deposits do not occur in limestone areas, and the water draining them would have smaller concentrations of carbonate ion and hydroxide ion.

CHEMICAL POTENTIAL, FUGACITY, ACTIVITY

So far we have considered only closed systems of constant composition. In his development of thermodynamics, J. W. Gibbs used a quantity now known as the chemical potential to deal with systems and phases of variable composition.

The chemical potential is particularly useful in establishing the conditions necessary for equilibrium and for spontaneous movement of components. In addition, two related quantities, fugacity and activity, are useful in dealing with the pressure of gases (fugacity) and with the concentrations of components in liquid and solid solutions (activity).

The *chemical potential* of a component in a system can be defined as

$$\mu_i = \left(\frac{\partial U}{\partial n_i}\right)_{S,V,n_j} \tag{3-27}$$

This is a partial differential that says that the chemical potential of component i is equal to the change in internal energy of a system when the number of moles of component $i\,(n_i)$ is changed at constant entropy and volume and for constant amounts of the other components present. It can be shown that the total change of internal energy for a system of variable composition is

$$dU = T\,dS - P\,dV + \sum_i \mu_i\,dn_i \text{ (infinitesimal change)}$$

or

$$\Delta U = T\,\Delta S - P\,\Delta V + \sum_i \mu_i\,\Delta n_i \text{ (finite change)} \tag{3-28}$$

The last term in equation (3-28) sums up the changes in amount of each component multiplied by the chemical potential of that component. Thus for such a system the internal energy can be changed in three ways: (1) by changing the entropy (amount of order), (2) through work done on or by the system, or (3) by changing the composition of the system.

Similar equations can be written for the other functions, such as G:

$$\mu_i = \left(\frac{\partial G}{\partial n_i}\right)_{T,P,n_j} \tag{3-29}$$

and

$$dG = V\,dP - S\,dT + \sum_i \mu_i\,dn_i \text{ (infinitesimal change)}$$

or

$$\Delta G = V\,\Delta P - S\,\Delta T + \sum_i \mu_i\,\Delta n_i \text{(finite change)} \tag{3-30}$$

The total Gibbs function of a system of one component in one phase is equal to the Gibbs function per mole of the pure component multiplied by the number of moles present. When two or more components are mixed together in a system, the Gibbs function of the system cannot be calculated by summing up the respective Gibbs function per mole values of the pure components since the process of mixing changes their contribution to the total Gibbs function of the system. Instead we have to deal with the *effective Gibbs function* per mole of each component, which is the chemical potential. Only for a system of one component is it equal to the Gibbs function per mole value of the pure component.

It can be shown that, for equilibrium in a given system, the chemical potential of a particular component must be the same in every phase in the system.* Consider a system consisting of granite. If it formed under equilibrium conditions, the chemical potential of potassium, for example, should be the same in each type of mineral found in the granite. It would be nice to relate the chemical potential of potassium in a mineral to the amount of potassium in the mineral. Then we could analyze the minerals of the granite (or of any other rock) and reach conclusions about whether or not equilibrium was reached in the formation of the rock. To relate the chemical potential of a component to its amount in a phase, we have to consider the two additional terms, fugacity and activity.

A general procedure in thermodynamics is to develop an idealized picture of a gas, liquid, or solid through various simple assumptions about the nature of the material. Mathematical relationships among the properties of such a material are then worked out. For example, one assumption made in describing an ideal gas is that there are no forces acting between the atoms of the gas. One resulting equation (an equation of state) for such a gas is

$$PV = nRT \tag{3-31}$$

where P is pressure, V is volume, n is number of moles of the component making up the gas, R is a constant depending in value upon the units used for the other terms (Table 3-1), and T is the

*A difference in the chemical potential of a component between two phases causes some of the component to tend to move from the phase of higher potential to the phase of lower potential. The direction of movement (higher to lower *potential*) is usually, but not always, the direction of decreasing *concentration.* The tendency of all natural processes is to minimize the chemical potentials of all components in all phases. Even though a given phase does not contain a particular component, we can still define a chemical potential for that component in that phase.

absolute (Kelvin) temperature. Similarly, the various thermodynamic functions can be expressed for an ideal gas. With respect to chemical potential, equation (3-31) can be used to show that for a gaseous mixture made up of several ideal gases (components), at a constant temperature,

$$\mu_i = \mu_i^* + RT \ln P_i \tag{3-32}$$

where μ_i^* is the chemical potential of pure component i at unit pressure, and P_i is the partial pressure of component i (P_i = mole fraction of component i times total gas pressure).

Most of the equations developed for idealized material are relatively simple and uncomplicated, and it would be helpful to be able to use them for real-life materials. It was for this purpose that G. N. Lewis introduced a fictitious pressure known as the *fugacity* (f). Essentially, the fugacity is the pressure value needed, at a given temperature, to make the properties of a real (nonideal) gas satisfy the equations of an ideal gas. For a real gas

$$f_i = \gamma_i P_i \tag{3-33}$$

where γ_i is the *fugacity coefficient* and P_i is the partial pressure for component i of the gaseous mixture. For an ideal gas, $\gamma_i = 1$; for real gases, the ratio f_i/P_i approaches 1 as total pressure approaches zero. Gaseous water at low temperatures and pressures closely approximates an ideal gas. Deviations from ideality increase as the pressure increases. Experimental study of gases allows calculation of the fugacity of a given gas at any temperature and pressure. Fugacity values can also be calculated from an equation of state for a gas. Such values are needed because, in applying thermodynamic formulas to real gases, it is necessary to substitute fugacity values wherever pressure P occurs in the equations.

One geologic use of experimental fugacity studies is in estimating the fugacities of gases such as O_2 and H_2O that existed at the time the minerals of a rock formed. This is done by experimentally determining relationships between gas fugacities and the composition of minerals that show solid solution in nature. By determining the chemical composition of the minerals of a rock, it is thus possible to estimate gas fugacities at the time of formation. The possible range of fugacity values is limited by the actual mineral assemblage present, so that some conclusions can be reached even without compositional data for the minerals. Work of this type has indicated that the fugacity of gases such as oxygen can be as important as temperature and total pressure in determining the mineralogy of igneous and metamorphic rocks (Eugster and Skippen 1967; Carmichael and Eugster 1987).

Several oxidation-reduction reactions are referred to as "oxygen buffers" because they fix the value of oxygen fugacity for a given temperature and pressure. For example, the following reaction could occur between iron-rich olivine (fayalite) and oxygen in a magma:

$$3Fe_2SiO_4 \text{ (fayalite)} + O_2 \text{ (gas)} \rightleftharpoons 2Fe_3O_4 \text{ (magnetite)} + 3SiO_2 \text{ (quartz)}$$

This reaction occurs under specific conditions of temperature and oxygen fugacity (FMQ curve in Figure 3-2). The amount of oxygen in this system can increase or decrease without changing the oxygen fugacity as long as the three mineral phases are present. The buffering capacity of the system is ended when one or more of the minerals is used up; for instance, by complete conversion of fayalite to magnetite and quartz. Oxygen buffers have been very useful in experimental studies of mineral reactions at high temperatures and pressures (Huebner 1971; Holloway and Wood 1988). The stability of any iron-bearing mineral in any environment is affected by oxygen fugacity (pressure). When free molecules of oxygen are present in very small amounts (as in many

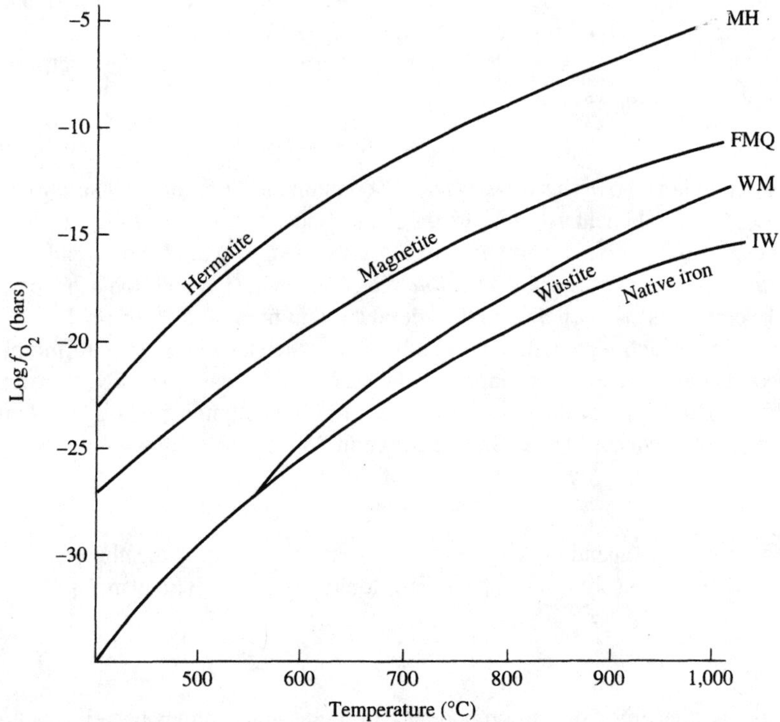

Figure 3-2 Oxygen buffer curves in the system Fe-Si-O at one bar total pressure. When any pair of phases, such as magnetite and hematite, are present, they act as a buffer for oxygen fugacity. Most oxygen fugacities for igneous and metamorphic rocks occur above the FMQ line. The curves are for the following reactions:

MH: $4Fe_3O_4$ (magnetite) + O_2 (gas) $\rightleftharpoons$ $6Fe_2O_3$ (hematite)

FMQ: $3Fe_2SiO_4$ (fayalite) + O_2 (gas) $\rightleftharpoons$ $2Fe_3O_4$ (magnetite) + $3SiO_2$ (quartz)

WM: $\frac{6}{4x-3}$ Fe_xO (wüstite) + O_2 (gas) $\rightleftharpoons$ $\frac{2x}{4x-3}$ Fe_3O_4 (magnetite)

IW: 2xFe (iron) + O_2 (gas) $\rightleftharpoons$ $2Fe_xO$ (wüstite)

Wüstite is a nonstoichiometric phase with an Fe/O ratio that depends on oxygen fugacity and temperature. After Eugster and Wones (1962).

igneous and metamorphic environments), other molecules such as H_2O and CO_2 can serve as sources or sinks of oxygen.

Solutions (both solid and liquid) can also be considered to be ideal or perfect, and the properties of such solutions are then related in a simple manner. These relationships are found to hold true in real cases for very dilute liquid solutions and for certain other solutions. The chemical potential for an ideal solution at constant temperature and pressure can be expressed as

$$\mu_i = \mu_i^* + RT \ln N_i \tag{3-34}$$

where N_i is the mole fraction of component i in the solution and μ_i^* is the chemical potential when $N_i = 1$ (pure i). Under most conditions, solutions, like gases, are nonideal; thus it is convenient to substitute a quantity for N_i such that the above equation (and other thermodynamic relationships involving N_i) will hold true for real-life (nonideal) solutions. This concept was also introduced by G. N. Lewis, who called the fictitious quantity the *activity* (a). The activity is related to the mole fraction as follows:

$$a_i = \gamma_i N_i \tag{3-35}$$

where γ_i is known as the *activity coefficient* of component i. For an ideal solution, $\gamma_i = 1$, and the activity of a component is equal to the mole fraction of that component in the solution. Activities can also be expressed in terms of molality as $a_i = \gamma_i m_i$.* For nonideal solutions, a_i can be considered the *effective concentration* of a component. The solution behaves as though the amount of i present is a_i, even though the actual amount present is N_i or m_i.

Activities and activity coefficients can be determined by a number of different experimental and theoretical procedures [see Garrels and Christ (1965, Chap. 2) and Nordstrom and Munoz (1985, Chaps. 6 and 7)]. For dilute electrolyte solutions a common method of calculating activity coefficients is use of the Debye-Huckel equation

$$\log \gamma_i = -Az_i^2\sqrt{I} \tag{3-36}$$

where A is a constant depending on pressure and temperature, z_i is the charge on ion i, and I is the ionic strength of the solution being studied. Ionic strength is defined as

$$I = \frac{1}{2}\sum m_i z_i^2 \tag{3-37}$$

where m_i is the concentration of component i in molal units (moles per kilogram of solvent). Modified versions of the above Debye-Huckel equation are used for ionic strengths greater than 10^{-3} molal.

Table 3-3 lists activity coefficients calculated using a Debye-Huckel equation for various ionic strengths. These agree well with experimental data. For ionic strengths above 0.1 molal (equivalent to about 5.8 g/kg of dissolved ions in a NaCl solution) other equations often must be used to get activity coefficients that agree with experimental results. For example, no form of the Debye-Huckel equation works for seawater activity coefficients, since the ionic strength of seawater is about 0.67 molal. As shown in Table 3-3, for solutions whose ionic strength is less than 10^{-4} (less than about 50 mg/kg of dissolved ions), activity coefficients for most ions are 0.95 or more. Thus, for these solutions, activity values are essentially equal to measured concentrations.

The importance of activities in geology is that, in dealing with reactions between minerals, between minerals and natural solutions, or within natural solutions, the use of equilibrium con-

*The activity of a substance is defined in terms of the fugacity of the substance, $a = f/f^*$, where a and f are the activity and the fugacity of the substance in a particular state, and f^* is the fugacity in a standard reference state. The theoretical definitions of activity and activity coefficient are such that activity is a dimensionless quantity, and activity coefficients are expressed in units of reciprocal concentration. The activity is directly related to the concentration (such as mole fraction) of a component by the activity coefficient and can be thought of as a concentration. It is incorrect to use a molarity scale rather than a molality scale when calculating activities, since the molarity of a given solution does not have the same value at two different temperatures or pressures (because of the change in density). This distinction is particularly important for high pressures and temperatures. Nordstrom and Munoz (1985, 132–135) point out that the definition of activity is dependent on the choice of standard-state chemical potential and that different standard states may be used in calculating activities in different geochemical problems.

TABLE 3-3 Calculated Ion Activity Coefficients

Ion	Ionic Strength (moles per kilogram)				
	10^{-4}	10^{-3}	10^{-2}	0.05	10^{-1}
H^+	0.99	0.97	0.91	0.86	0.83
Al^{3+}, Fe^{3+}, La^{3+}, Ce^{3+}	0.90	0.74	0.44	0.24	0.18
Mg^{2+}, Be^{2+}	0.96	0.87	0.69	0.52	0.45
Ca^{2+}, Zn^{2+}, Cu^{2+}, Sn^{2+}, Mn^{2+}, Fe^{2+}	0.96	0.87	0.68	0.48	0.40
Ba^{2+}, Sr^{2+}, Pb^{2+}, CO_3^{2-}	0.96	0.87	0.67	0.46	0.39
Na^+, HCO_3^-, $H_2PO_4^{3-}$, CH_3COO^-	0.99	0.96	0.90	0.81	0.77
SO_4^{2-}, HPO_4^{2-}	0.96	0.87	0.66	0.44	0.36
PO_4^{3-}	0.90	0.72	0.40	0.16	0.10
K^+, Ag^+, NH_4^+, OH^-, Cl^-, ClO_4^-, NO_3^-, HS^-	0.99	0.96	0.90	0.80	0.76

Sources: Data from Kielland (1937, 1675); also see Figure 4-2. After W. Stumm and J. J. Morgan, *Aquatic Chemistry*, 2nd ed. Copyright © 1981 by John Wiley & Sons. Reprinted by permission of John Wiley & Sons, Inc.

stants and other thermodynamic relationships involves activity values, whereas what is measured in minerals and natural solutions are usually concentrations (it is possible to measure activities by using specific ion electrodes). For example, suppose that we are interested in studying the chemistry of seawater. Although seawater does not have a large amount of dissolved material (about 35 grams of dissolved salts in 1,000 grams of seawater), it is a very complicated solution. This is due to the fact that the elements in seawater occur not only as simple ions, but also as complex ions (such as $CaHCO_3^+$) and as neutral molecules (such as $MgSO_4$). Furthermore, the interactions among these various species make seawater behavior very different from that of an ideal solution. In general, the greater the amount of dissolved material, the lower the activity coefficients of the various species present in the water. Garrels and Thompson (1962) calculated activities and activity coefficients for seawater by assuming certain properties for the dissolved materials (Table 3-4). Their theoretical values for the activity coefficients compare favorably with experimentally measured values. More sophisticated models of seawater are discussed by Pytkowicz (1983).

TABLE 3-4 Occurrence and Properties of Major Ions in Seawater

Major ions in average seawater	Concentration (moles/kilogram)	Amount of ion occurring as free ion (%)	Activity coefficient	Activity of free ion
Na^+	0.475	99	0.76	0.357
Mg^{2+}	0.054	87	0.36	0.017
Ca^{2+}	0.010	91	0.28	0.0025
K^+	0.010	99	0.64	0.0063
Cl^-	0.56	100	0.64	0.36
SO_4^{2-}	0.028	54	0.12	0.0018
HCO_3^-	0.0024	69	0.68	0.0011
CO_3^{2-}	0.0003	9	0.20	0.0000054

Source: After R. M. Garrels and M. E. Thompson, *American Journal of Science*, v. 260, 1962. Reprinted by permission of American Journal of Science.

Let us consider the solubility of calcite in seawater. To do this we need to use the experimentally determined equilibrium constant for the reaction,

$$CaCO_3 \text{ (calcite)} \rightleftharpoons Ca^{2+} + CO_3^{2-}$$

for which

$$K = [Ca^{2+}][CO_3^{2-}] = 0.45 \times 10^{-8}$$

For seawater, using the activity values from Table 3-4, we get

$$[Ca^{2+}][CO_3^{2}] = 1.35 \times 10^{-8}$$

Since this number is close to the value of the equilibrium constant, we can say that seawater is approximately saturated with calcium carbonate. Slight changes in conditions could lead to precipitation of calcite or to solution of any available solid calcite.*

For solutions that do not deviate greatly from ideality (such as stream water), concentrations can often be used instead of activities in solubility and other calculations without introducing large errors. This is commonly done for natural solutions, because the form, activities, and activity coefficients of the dissolved species are not well known for such solutions. The importance of knowing the chemistry of the solution being dealt with is illustrated by the occurrence of CO_3^{2-} in seawater (Table 3-4). The measured concentration of CO_3^{2-} is 0.0003 mole per kilogram, but, because of its low activity coefficient and because part of it occurs in other forms, its activity as a free ion (effective concentration) is only about 2 percent of its actual concentration.

We have pointed out that dilute liquid solutions can, to a first approximation, be treated as ideal solutions. Similarly, the occurrence of a trace element in a mineral can be considered as a dilute solid solution and thus approximately ideal. This makes it possible to apply thermodynamic laws to such problems as the distribution of a trace element between two coexisting minerals. Suppose that we are interested in the distribution of cobalt between biotite and hornblende in an igneous rock. At equilibrium the chemical potential of cobalt in biotite must be equal to the chemical potential of cobalt in hornblende:

$$\mu_{\text{Co}}^{\text{Bio}} = \mu_{\text{Co}}^{*\text{Bio}} + RT \ln a_{\text{Co}}^{\text{Bio}} = \mu_{\text{Co}}^{\text{Hbd}} = \mu_{\text{Co}}^{*\text{Hbd}} + RT \ln a_{\text{Co}}^{\text{Hbd}}$$

Since $\mu_{\text{Co}}^{*\text{Bio}}$ and $\mu_{\text{Co}}^{*\text{Hbd}}$ are by definition constants at any specified temperature and pressure, we can say that

$$\frac{a_{\text{Co}}^{\text{Bio}}}{a_{\text{Co}}^{\text{Hbd}}} = \text{constant } (K_D)$$

for that temperature and pressure. For ideal solid solution, $a_{\text{Co}} = N_{\text{Co}}$; thus the ratio of the mole fraction of cobalt in biotite to the mole fraction of cobalt in hornblende should be a constant if the rock in which the minerals occur formed under equilibrium conditions and if cobalt exhibits ideal behavior in these two minerals. Such ratios (K_D) are known as *distribution coefficients.* (They are also referred to as partition coefficients.) Often the concentrations of the trace elements are expressed as parts per million rather than as mole fractions.

*Most calcium carbonate is precipitated in seawater as a result of biological processes rather than inorganic chemical reactions. Also the solubility and precipitation of inorganic calcium carbonate is affected by the common substitution of magnesium ions for some of the calcium ions. Carbonate sediments are discussed in Chapter Seven.

Various studies of trace element distributions indicate that coexisting minerals do have a constant distribution coefficient for many trace elements (Figure 3-3). Ideally, such distributions could be used as a geothermometer, since the distribution coefficient should vary with temperature (and pressure). Various attempts to apply such a geothermometer have not been too successful because of several restrictions on the method. For instance, the overall composition of the minerals involved must remain essentially constant. Experimental determination of trace element distribution coefficients for sulfide minerals and problems in using the results to estimate temperatures of ore formation are discussed by Bethke and Barton (1971).

Other applications of distribution coefficients have involved studying the distribution of major elements among seawater and precipitating minerals and among coexisting minerals of metamorphic rocks. The most widely used geothermometers in metamorphic rocks involve measurement of iron-magnesium ratios in mineral pairs such as garnet-clinopyroxene and garnet-biotite (Essene 1982). In the case of garnet-biotite pairs (Figure 3-4), the distribution coefficient can be expressed as

$$K_D = \frac{(\mathrm{Fe/Mg})_{\mathrm{garnet}}}{(\mathrm{Fe/Mg})_{\mathrm{biotite}}}$$

Use of measured compositions (mole fractions or elemental ratios) to express values of K_D avoids the problem of converting compositions to activities, which requires determination of activity coefficients and is not easily done. Further discussion of these geothermometers can be found in Chapter Nine.

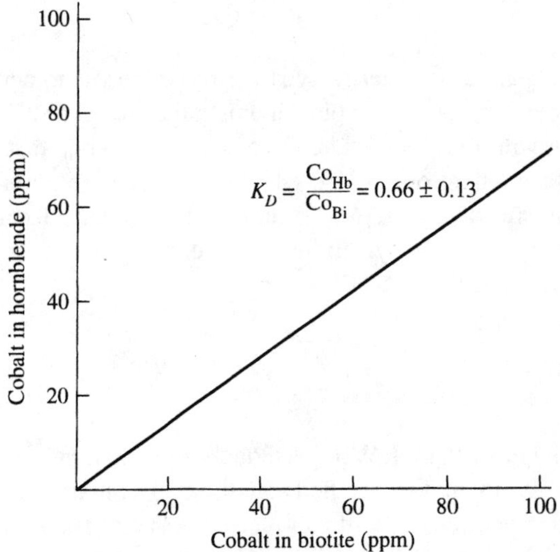

Figure 3-3 Plot of average distribution coefficient (K_D) for cobalt between coexisting biotite and hornblende. The line is based on analyses of 95 samples from igneous rocks of various compositions. It was concluded from the data that the distribution coefficient is unaffected by initial crystallization temperature, cooling history, or major element composition of the minerals or magmas. These data suggest that cobalt partitioning gives a useful estimate of the degree of equilibrium attained by a given rock. After Greenland et al. (1971).

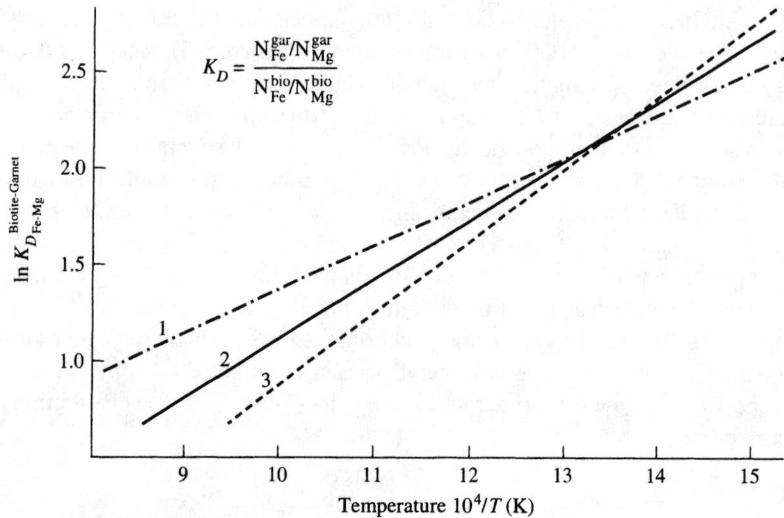

Figure 3-4 Three different calibrations of the biotite-garnet K_D thermometer. Curve one is based on laboratory experimentation [Ferry and Spear (1978)], curve two on field observations [Thompson (1976)], and curve three on comparison with isotopic thermometry [Goldman and Albee (1977)]. Solid solution of components other than iron and magnesium can affect the validity of this and similar geothermometers. N_{Fe}^{gar} is the mole fraction of iron in garnet and so forth. After Essene (1982).

Very little is known about the thermodynamic properties of minerals showing extensive solid solution. Data are available for many pure end members of such solutions, but such data are not adequate for dealing with these minerals. In particular, minerals that exhibit interstitial and omission types of solid solution (as opposed to substitutional solid solution) appear to depart drastically from the behavior expected of an ideal solution. For example, Toulmin and Barton (1964) have shown that the activity of FeS in pyrrhotite (which exhibits omission solid solution) is about 0.4 when the mole fraction of FeS is 0.9.

THE PHASE RULE

The phase rule, developed in 1875 by J. W. Gibbs, deals with the number of phases that can exist when a system is in equilibrium under specified conditions. The main use of the phase rule in geology is in the construction and interpretation of phase diagrams. These are discussed in the next section. To derive the rule we make a list of the number of variables that can affect a system and another list of the number or relationships involving these variables that must hold at equilibrium. The variables we are usually interested in are pressure, temperature, and the mole fractions of each component in each phase. If the total number of components is c and the total number of phases is p, the phase rule tells us how many of these $cp + 2$ variables we can vary and still keep our system in equilibrium.

At equilibrium, the chemical potential of each component in each phase must be equal.* This gives us $c(p-1)$ relationships of the type

$$\mu_1^1 = \mu_1^2 = \ldots \mu_1^p$$
$$\mu_2^1 = \mu_2^2 = \ldots \mu_2^p$$
$$\mu_c^1 = \mu_c^2 = \ldots \mu_c^p$$

For each phase the sum of the mole fractions of the components must equal 1. So there are p additional relationships from this requirement. Thus we have $cp + 2$ variables, and we have $c(p-1) + p$ relationships among these variables. We want to know how many of the variables we can arbitrarily change and still have equilibrium. This number (usually referred to as the *degrees of freedom*), represented here by f, should be equal to the total number of variables minus the number of conditions that fix a portion of these variables. Stating this mathematically,

$$f = cp + 2 - [c(p-1) + p] = c + 2 - p \tag{3-38}$$

The expression $f = c + 2 - p$ is the *Gibbs phase rule*. It states that, for a system in which the number of phases is equal to the number of components plus 2 (i.e., for $f = 0$), it is impossible to vary the temperature, pressure, or mole fraction in a particular phase of any of the components and still have the system in equilibrium (zero degrees of freedom). If f is equal to 1, then any one (and only one) of the variables can be arbitrarily changed and the system can remain in equilibrium. If f is equal to 2, then two variables can be changed, and so forth. It should be pointed out that the number 2 that appears in the phase rule applies only to a system with two variables, temperature and pressure, in addition to the compositional (mole fraction) variables. For a system with two different pressure variables (such as rocks containing groundwater, with solid phases under one pressure and the water under another), the rule should be $f = c + 3 - p$. For many changes at the surface of the Earth, the pressure can be considered a constant, and then $f = c + 1 - p$.

We need to discuss further the selection of a value for c, the number of components in a system. In general, any substances may be picked as components. The actual selection depends on the composition of the phases of the system of interest and on the possible reactions that can occur in the system. The total number of components should be the smallest number needed to describe all possible variations in the composition of all phases. For a system consisting of anhydrite, gypsum, and water the number of components is two, $CaSO_4$ and H_2O. The composition of each of the three phases can be specified in terms of these two components. Note that the occurrence of calcium and sulfate ions in the water as a result of solution of gypsum or anhydrite would not increase the number of components because the ratio Ca^{2+}/SO_4^{2-} would be constant. Thus components

*It should be noted that equation (3-34) tells us that the chemical potentials are a function of the mole fraction of each component in each phase as well as of temperature and pressure. The equality of the chemical potential of a given component between two phases is therefore a relation among the variables of the system. In deriving the phase rule Gibbs used a slightly different approach than the one given here. He employed the following equation, which describes the possible changes in each phase in terms of the variables temperature, pressure, and chemical potential of each component.

$$0 = S\,dT - V\,dP + \sum_i n_i d\mu_i$$

This equation, known as the Gibbs-Duhem equation, is a relationship among the variables of phases in internal, homogeneous equilibrium. A rigorous derivation of the equation is given by Tunell (1979).

should be picked on the basis of the kinds of variation that can occur within a system. For igneous and metamorphic rocks, simple oxides such as K_2O, Al_2O_3, etc., are usually selected as components. A nepheline syenite made up of nepheline, orthoclase, albite, muscovite, and corundum would have five components, SiO_2, Al_2O_3, Na_2O, K_2O, and H_2O. The same components would be listed for a gneiss consisting of quartz, orthoclase, albite, muscovite, and kyanite.

A geological form of the phase rule was suggested by V. M. Goldschmidt in his 1911 Ph.D. thesis. He pointed out that most rocks are subjected to a range of temperatures and pressures, and yet certain groupings of minerals are commonly found, apparently indicating formation under equilibrium conditions. Thus there must be at least two degrees of freedom—variation in temperature and in pressure—for most rocks. If f is equal to 2, then p equals c, and if f is greater than 2, then p is less than c. This is summarized in *Goldschmidt's mineralogical phase rule*, which states that the number of minerals that can coexist in equilibrium in a rock is equal to or less than the number of components making up the rock. Although this form of the phase rule is of little use in detailed studies of rocks, it does explain why most rocks consist of relatively few minerals. They usually contain five to ten major components, and their formation was governed by equilibrium requirements. Thus they ordinarily are made up of five to ten minerals. Think how complicated the study of rocks would be if each one were made up of forty or fifty different minerals!

Let us apply the phase rule to the system shown in Figure 3-1. The number of phases coexisting along the line for a closed system is four: quartz, wollastonite, calcite, and carbon dioxide. The number of components needed to describe these phases is three: SiO_2, CaO, and CO_2. Thus $f = c + 2 - p = 3 + 2 - 4 = 1$. There is one degree of freedom for the system. This means we can arbitrarily vary the temperature to a new value, and there will be a fixed pressure at which the four components are in equilibrium. In other words, we can stay on the equilibrium line by arbitrarily varying either temperature or pressure. But we cannot stay on the line if we arbitrarily select new values for *both* temperature and pressure.

The phase rule was further modified for geological applications by D. S. Korzhinsky in 1936. He suggested that the phase rule expression is different for an open system than it is for a closed system. Many geological systems have to be treated as open systems in which material can leave or be added to the system. This is particularly true of metamorphic processes. Korzhinsky classified components into two types, mobile and inert. *Mobile components* can move in and out of a system, whereas *inert components* cannot exchange with the surroundings. In the case of the limestone containing quartz grains (Figure 3-1), fractures in the rock could allow carbon dioxide to escape and then it would be a mobile component. Water is another component that is probably mobile during metamorphism. Korzhinsky suggested that the chemical potential of mobile components is determined (fixed) by the surroundings of a system, and therefore these components should not be counted in using the phase rule. In this case the mineralogical phase rule would be changed to state that the number of minerals in a rock is equal to or less than the number of *inert* components. Korzhinsky supported his suggestion with examples of metamorphic rocks that apparently formed as a result of metasomatism (exchange of material with migrating fluids). These rocks have a number of major components as such, but consist of three or fewer phases. Perhaps the best examples are the monomineralic and bimineralic reaction zones often found between ultramafic rocks and siliceous country rock (Figure 3-5). Other examples of sharp contacts with one or two phases in a zone include weathering crusts on rocks and the alteration zones surrounding ore veins.

Extensive discussions of the application of the phase rule to metamorphic and metasomatic processes have been presented by Thompson (1959, 1970, 1982). Other authors have given alternative views on the importance of mobile components such as H_2O and CO_2 in the formation

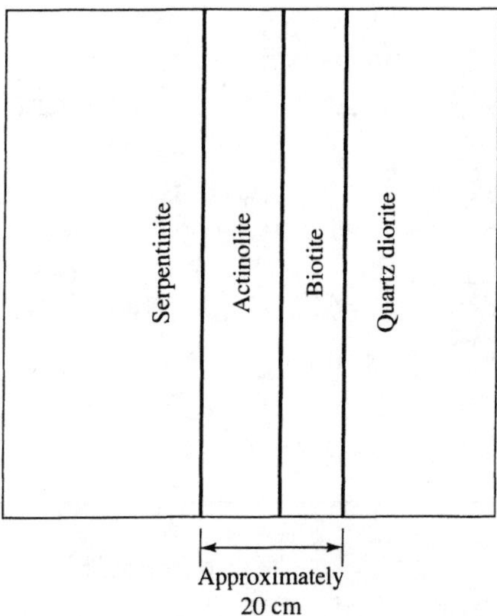

Figure 3-5 Monomineralic contact bands formed by reaction between serpentinite and quartz diorite at Westfield, Massachusetts. Variations found in the (Mg/Mg + Fe) ratios of biotite and actinolite samples indicate that complete chemical equilibrium was not reached during formation of the bands. Most of the components in the contact bands were probably mobile, thus limiting to one the number of phases that could form in each band. Data from Brownlow (1961).

of metamorphic rocks. Rice and Ferry (1982) state that there probably exists a continuous spectrum of possibilities from one extreme (an external reservoir of fluid controls the final mineralogy and mineral composition) to the other (final mineralogy and composition controlled by changes in a closed system involving initial pore fluids and fluids produced by devolatilization reactions). Examples of infiltration (open system), internal buffering (closed system), and control by combined infiltration and buffering are used by Rice and Ferry (1982) to illustrate that natural mineral assemblages have acted as buffers and thus regulated temperatures as well as fluid and solid compositions during metamorphism. They conclude that internal buffering is the most important control of fluid composition during metamorphism.

Zen (1963) has pointed out a number of problems involved in trying to apply the phase rule to geologic problems. For instance, there is no way to positively identify a particular component as inert or mobile. Furthermore, correct counting of the phases in a rock is complicated by the occurrence of zoned minerals and of intimate intergrowths of different composition (such as cryptoperthite). Some elements readily substitute for others in minerals. An example is the substitution of magnesium for iron in ferromagnesian minerals. Can we treat this as a case of one component, (Mg, Fe)O, or must the two elements be considered as independent components, MgO and FeO? Perhaps most important is the fact that the phase rule is a necessary but not sufficient test of equilibrium. (A sufficient test would be to show that any conceivable infinitesimal change in a system is reversible). Rocks that do not violate the phase rule may still have formed as a disequilibrium assemblage. Many rocks may show *local* equilibrium measured in centimeters. In this case different parts of a rock mass would not be in equilibrium with each other. At least we can say that rocks that violate the phase rule did not form under equilibrium conditions (assuming correct counting of components and phases).

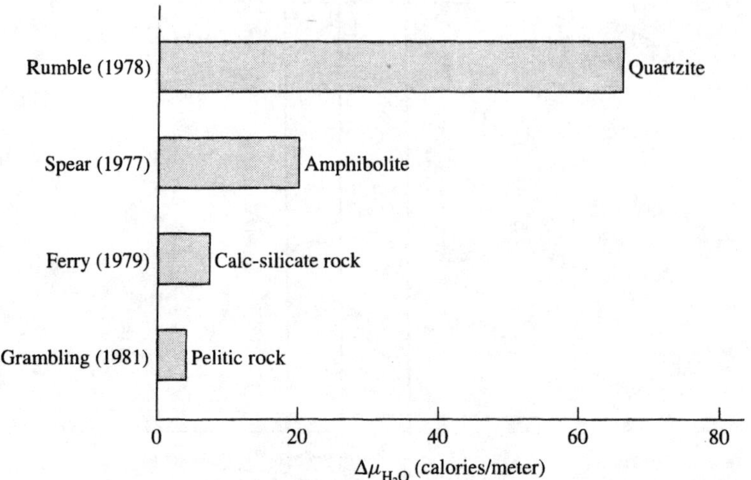

Figure 3-6 Plot of $\Delta\mu_{H_2O}$ for four different metamorphic rock types. Chemical analyses of minerals in the rocks were combined with thermodynamic equilibrium equations to determine values of $\Delta\mu_{H_2O}$ in individual samples at the time of metamorphism. This figure indicates the magnitude of the chemical potential gradient for water in each of the four rock types. These and other results show that the chemical potentials of volatiles such as H_2O vary during metamorphism over distances measured in meters. After Spear et al. (1982). Data from Spear (1977); Rumble (1978); Ferry (1979); and Grambling (1981).

Although there are problems in applying the phase rule as such to the study of rocks, an extension of this thermodynamic approach to phase equilibria provides information on variations in pressure, temperature, chemical potentials μ_i of components, and mole fractions N_i of components. Spear et al. (1982) describe a procedure for studying metamorphic rocks that can be summarized as follows:

1. Write a set of simultaneous equations describing the relationships among the parameters temperature, pressure, μ_i, and N_i for equilibrium conditions.
2. Solve the equations to relate compositions of minerals in an equilibrium assemblage to the variables temperature, pressure, μ_i, and N_i.

This procedure can be used to study the behavior and chemical potentials of H_2O, CO_2, and O_2 in metamorphic rocks (Figure 3-6).

PHASE DIAGRAMS

A phase diagram is a graphical representation of phase relationships. Those diagrams that are based on thermodynamic relationships can also be considered graphical expressions of the phase rule. The coordinates of phase diagrams are of two main types, intensive and extensive. *Intensive variables* or properties are those that are independent of the amount of material being considered. Examples are temperature, pressure, and density. *Extensive variables* are dependent on the amount of material involved. Volume, mass, and internal energy are extensive variables. Composition as a variable is obviously an extensive variable. Extensive properties can be changed to intensive properties by expressing them in units of per mole or per gram. Volume per mole is an intensive property. Similarly, chemical potential, fugacity, and activity are intensive properties. We shall limit our discussion to a few of the most important types of phase diagrams used in geology. Further discussion of selected diagrams can be found in Chapters Seven to Nine. Detailed reviews of phase diagrams are given by Garrels and Christ (1965), Ehlers (1972), Ernst (1976), and Richardson and McSween (1989).

The simplest type of phase diagram involves only one component, with temperature and pressure as variables (Figure 3-7). The phase rule tells us that when three phases are together in equilibrium (point A in Figure 3-7) then $f = 1 + 2 - 3 = 0$, and we cannot vary temperature or pressure and still maintain the three phases together in equilibrium. When two phases are together (along one of the lines AB, AC, or AD), $f = 1$, and we can vary either the pressure or the temperature and still have the possibility of the two phases in stable coexistence. If, for example, we have kyanite and sillimanite in equilibrium at about 8 kilobars and 725°C, and then arbitrarily change the temperature to 825°C, the pressure will have to change to about 10 kilobars to keep the two minerals in equilibrium with each other. When we have only one phase, then $f = 2$, and both pressure and temperature can vary arbitrarily without changing the situation unless we cross one of the lines AB, AC, or AD.

Thus, for Figure 3-7, the triple point A represents no degrees of freedom. This means that we should very rarely find a rock containing all three of the polymorphs of Al_2SiO_5, since they

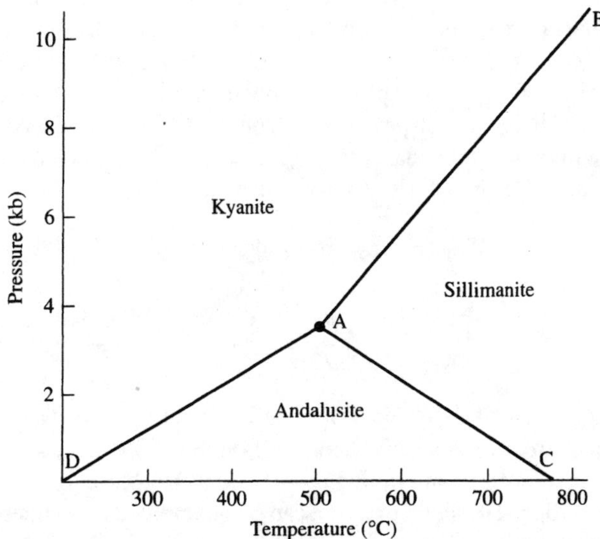

Figure 3-7 Phase diagram for the one-component system Al_2SiO_5. Despite intense study, the stability relations are still marked by some uncertainty. The most accepted diagram is shown here with the triple point at 501°C and 3.76 kbar. Other calibrations place the triple point at 620°C and 5.5 kbar (Richardson et al. 1969) and at 511°C and 3.87 kbar (Hemingway et al. 1991). This diagram is of great significance in the study of metamorphic rocks. For a field example of an apparent equilibrium occurrence of the Al_2SiO_5 triple point, see Grambling (1981). A detailed discussion of the Al_2SiO_5 polymorphs can be found in Kerrick (1990). (After M. H. Holdaway, *American Journal of Science*, vol. 271, 1971. Reprinted by permission of American Journal of Science.)

can coexist in equilibrium at only one specific value of temperature and one specific value of pressure. The occurrence of two of the phases together should be (and is) more common, since there is a range of conditions, represented by the phase boundary lines, under which they are stable. Finally, most likely is the occurrence of only one of the polymorphs in a rock, since any particular polymorph is stable under a wide variety of temperature and pressure conditions (represented by the regions between the phase boundary lines).

The experimental and theoretical work on the Al_2SiO_5 system is a good example of some of the different methods that are used to develop phase diagrams. At the same time this research illustrates some of the problems involved in preparing such diagrams. One method consists of subjecting a sample of the appropriate composition to a selected temperature and pressure until equilibrium is attained (hopefully), then quenching the material to room temperature and pressure for laboratory identification. Many runs at different temperatures and pressures are made and plotted on a diagram. To check the location of phase boundary lines, samples of kyanite, for example, can be taken to a higher temperature across the sillimanite boundary to change them to sillimanite. Similarly, sillimanite at high temperatures would be taken into the kyanite field. The location of the phase changes should agree when approaching the boundary curve from either side.

Another method involves location of the triple point by calculation of free-energy relations using data on the solubilities of andalusite, kyanite, and sillimanite in appropriate liquids. Still another method uses measurements of heats of formation and of heat capacities to determine the triple-point location. Because this system is particularly difficult to work with, the results of the various approaches do not agree. Pressure values found for the triple point vary from about 2.5 to 9 kilobars, and temperature values vary from about 400 to 700°C. Most petrologists use the triple point located by Holdaway (1971) at 501°C and 3.76 kilobars.

It is possible to calculate the slope of the univariant (one degree of freedom) phase boundary lines in Figure 3-7 using the Clapeyron equation, which can be derived as follows. For equilibrium between sillimanite and andalusite $G_{sil} - G_{and} = 0$. To maintain equilibrium when temperature and pressure are changed, dG_{sil} must stay equal to dG_{and}. Thus

$$\begin{aligned} V_{and}\,dP - S_{and}\,dT &= V_{sil}\,dP - S_{sil}\,dT \\ (V_{sil} - V_{and})\,dP &= (S_{sil} - S_{and})\,dT \\ \frac{dP}{dT} &= \frac{S_{sil} - S_{and}}{V_{sil} - V_{and}} = \frac{\Delta S}{\Delta V} \end{aligned} \tag{3–39}$$

This expression is one form of the Clapeyron equation. It can be applied to two phases of any one-component system or to any other univariant reaction. Since the units of $\Delta S/\Delta V$ are cal/cm^3-deg, the value obtained must be multiplied by 41.3 atm-cm^3 (the equivalent of 1 cal) to get the slope in terms of atm/deg. The slope of a line can be determined experimentally and used to put limits on $\Delta S/\Delta V$ for a reaction, or the slope can be calculated directly from previously measured entropy and volume data.

Other pairs of intensive variables can be used instead of pressure and temperature. Examples are fugacity-temperature diagrams (Figure 3-2) and diagrams with chemical potentials (Figure 9-11), fugacities (Figure 3-8), or activities (Figure 7-4) as variables. Other diagrams use as variables the partial pressure of a gas and the activity of hydrogen ion (pH) [Figure 3-9] or activities of dissolved aqueous species and pH (Figure 3-10). Figure 3-8 is an example of a system plotted with two different pairs of variables. The choice of which intensive variables to measure

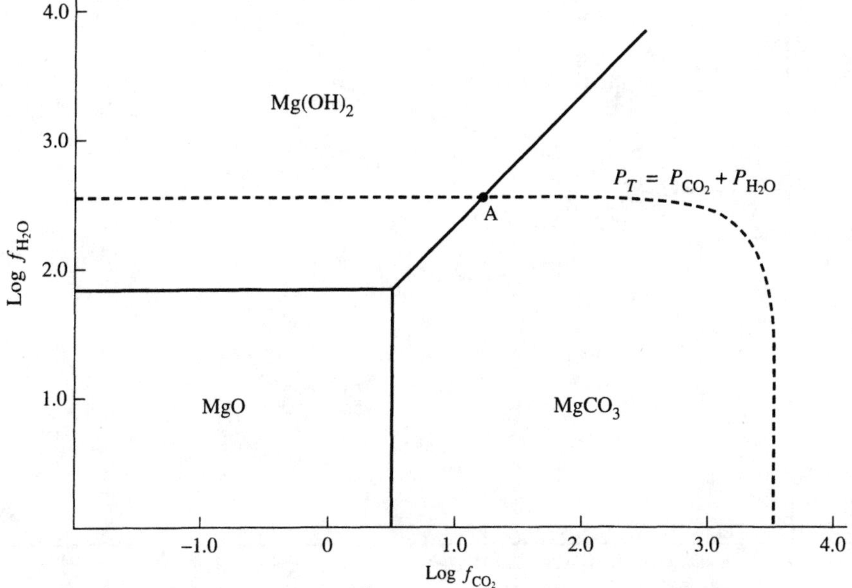

Figure 3-8a Fugacity diagram for the system MgO-H_2O-CO_2 at 700 K and a total pressure of 2 kbar. The dashed curve represents the unique locus of possible fugacities in the system when the total pressure is equal to the sum of the partial pressures of H_2O and CO_2. Under these conditions, likely for geologic systems, all other areas of the diagram are inaccessible. Point A is also plotted in Figure 3-8b. The minerals represented are brucite [$Mg(OH)_2$], magnesite [$MgCO_3$], and periclase [MgO]. (After D. K. Nordstrom and J. L. Munoz, *Geochemical Thermodynamics.* Copyright © 1985. Adapted by permission of Blackwell Science, Ltd., Oxford, England.)

or calculate (and thus the choice of which types of diagram to use) in a given research project depends on the nature and goals of the research.

All the diagrams discussed so far have involved intensive variables. Diagrams involving only composition (usually plotted as mole percent or weight percent) as a variable are commonly used to show stable assemblages of minerals (Figure 3-11). These assemblages are often based on field occurrence, and the lines on the diagrams are not quantitatively connected by thermodynamic relationships as is the case for most diagrams involving intensive variables. Figure 3-11 is an example of an *ACF diagram,* which is named for the components Al_2O_3, CaO, and (Mg, Fe)O. In this case it is assumed that excess SiO_2 (represented by quartz) is also present for all rocks of the system. Although the lines of Figure 3-11 have no thermodynamic significance, the diagram can be considered to exhibit the results of the phase rule. For a system of four components and with both pressure and temperature as variables (two degrees of freedom), we get

$$f = c + 2 - p \qquad 2 = 4 + 2 - p \qquad p = 4$$

Thus a rock in equilibrium should exhibit a maximum of four phases. The triangular sections of Figure 3-11 represent the three minerals that would occur with quartz for a given composition. For example, an aluminum-rich rock would consist of the four minerals andalusite, anorthite, cordierite,

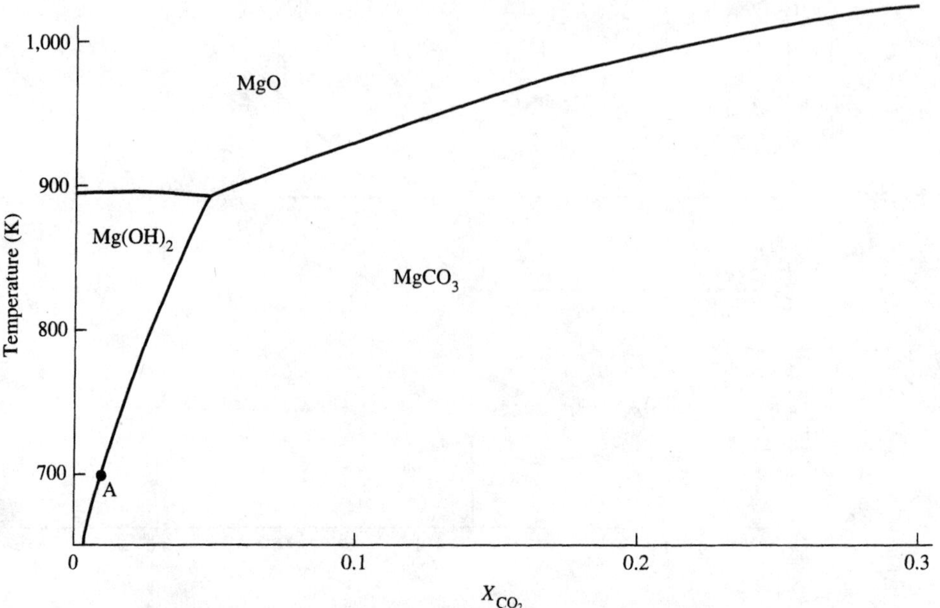

Figure 3-8b T-X_{CO_2} (mole fraction of CO_2) diagram for the system MgO-H_2O-CO_2 at a total pressure of 2 kbar. This type of diagram is more useful in petrologic studies than the fugacity diagram of Figure 3-8a. Point A is also plotted in Figure 3-8a. (After D. K. Nordstrom and J. L. Munoz, *Geochemical Thermodynamics*. Copyright © 1985. Adapted by permission of Blackwell Science, Ltd., Oxford, England.)

and quartz. As part of his 1911 Ph.D. thesis, V. M. Goldschmidt determined the chemical composition of contact metamorphic rocks in the Oslo region of Norway and compared the results with the mineralogy of the rocks. He found relationships between chemical composition and mineralogy, such as those shown in Figure 3-11. If another component, for example K_2O, was found in a rock, another mineral (phase) was also found (orthoclase in the case of K_2O). The regularity of the mineralogy led Goldschmidt to conclude that equilibrium was reached in the formation of the rocks.

Many phase diagrams combine extensive and intensive variables. The most popular diagrams are those using temperature and composition (Figure 3-12). Diagrams of this type can be constructed by direct experiment or by use of thermodynamic data and relationships. For the system shown in Figure 3-12, there are two components, $KAlSi_2O_6$ and SiO_2 (binary system). (All compositions along a composition line can be considered to be a mixture of the two end-point compositions.) Only one of the two components can be counted as an independent variable, since an arbitrary amount for one fixes the amount for the other. For the system represented by Figure 3-12, pressure is not a possible variable under the experimental conditions used to construct the diagram. Thus the phase rule in this case is $f = c + 1 - p$, and, when neither composition nor temperature can vary, $0 = 2 + 1 - p$ and $p = 3$. Three phases can occur together only at a point (constant temperature and constant composition), such as point E in Figure 3-12, where K-feldspar, tridymite, and liquid can coexist. Point E is an example of an *invariant point* (a point for which $f = 0$). If one of the two possible variables can arbitrarily vary, then $p = 2$. The curves ARE and BE represent this situation (univariant equilibrium). The coexistence of leucite and liquid can be maintained after arbitrarily changing one variable if the other is changed a fixed amount

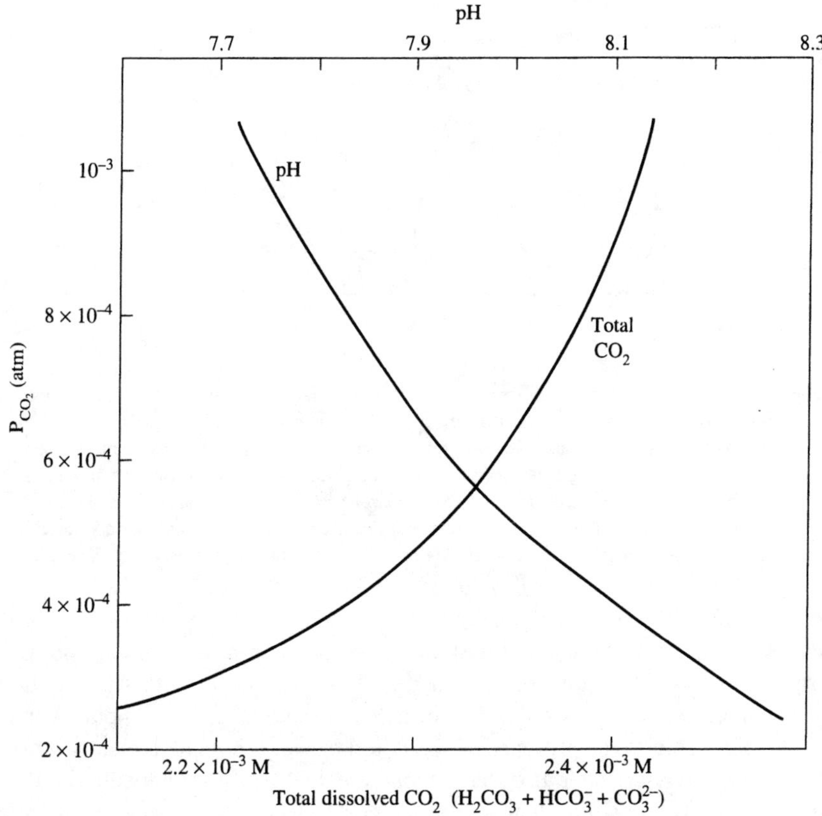

Figure 3-9 Effect of increasing CO_2 content of the atmosphere upon total dissolved CO_2 and pH of surface ocean water. The approximate present value of P_{CO_2} is 3.3×10^{-4} atm. A doubling of this value would increase total dissolved CO_2 by about 5 to 6% and would approximately double the hydrogen ion concentration (a pH change from about 8.16 to about 7.90). The control and limitation of the pH of ocean water by the aqueous carbonate system is referred to as buffering. (After W. Stumm and J. J. Morgan, *Aquatic Chemistry*, 2nd ed. Copyright © 1981 by John Wiley & Sons. Reprinted by permission of John Wiley & Sons, Inc.)

to compensate for the arbitrary change. (This will result in a change in the proportion of the two phases present, since we will move to a different location on the curve ARE. This curve defines the relative amounts of the two phases for any given temperature and overall composition as explained in a later paragraph). When both composition and temperature are arbitrarily varied,

$$f = c + 1 - p \qquad 2 = 2 + 1 - p \qquad \text{and} \qquad p = 1$$

This is represented by the liquid region of Figure 3-12.

Although both point E and point R of Figure 3-12 are invariant points, there is an important difference between them. Point E is a binary *eutectic point,* a point at which two solids are in equilibrium with the liquid phase. Cooling of the liquid phase at point E causes it to solidify completely, without change of composition or temperature, to a mixture of K-feldspar and tridymite. Similarly, if we were to heat a mechanical mixture of K-feldspar and tridymite to the

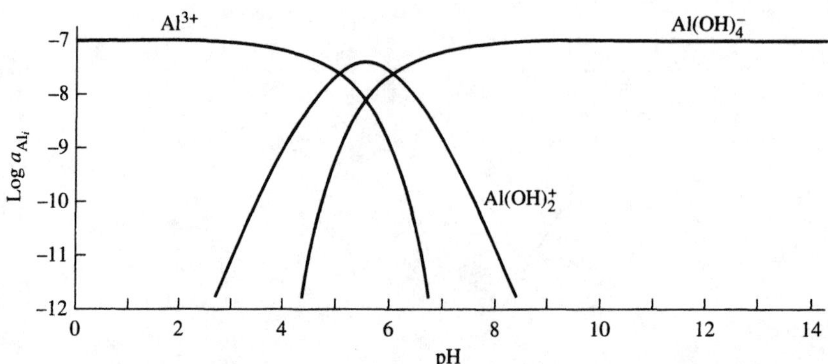

Figure 3-10 Activities of aqueous species of aluminum as a function of pH at 25°C. The total aluminum is assumed to be 10^{-7} mol/liter. The most abundant species for pH less than 5 is Al^{3+}, for pH 5 to 6 is $Al(OH)_2^+$, and for pH greater than 6 is $Al(OH)_4^-$. The diagram has been constructed by using the equilibrium constants for individual hydrolysis reactions of aluminum. This type of diagram has been referred to as a master-variable diagram because it shows the dependence of a variable (aluminum hydrolysis species) on an independent (master) variable (pH). (After D. K. Nordstrom and J. L. Munoz, *Geochemical Thermodynamics*. Copyright © 1985. Adapted by permission of Blackwell Science, Ltd., Oxford, England.)

eutectic temperature (about 990°C), a liquid of the composition represented by point E would start to form. Point R, on the other hand, is a *peritectic* (reaction) *point*. As a cooling liquid changes composition along line AR, leucite forms from the liquid. When point R is reached, leucite is no longer stable in the presence of the liquid phase. It begins to react with the liquid to form K-feldspar, and K-feldspar also starts to precipitate directly from the liquid. Depending on the original liquid composition, leucite may be completely used up (the result for original compositions to the right of the $KAlSi_3O_8$ composition point such as point I, or some leucite may remain when the liquid is used up (the result for original compositions to the left of the $KAlSi_3O_8$ composition point such as point J). If we start with K-feldspar by itself at a temperature below that of point R (1,150°C), and heat it to that temperature, the K-feldspar will begin to decompose into leucite and a liquid of a composition represented by point R (incongruent melting). Thus, although both point E and point R represent the occurrence together of three phases (invariant points with $f = 0$), the phases are stable together at point E but are not stable together at point R. Examples of the importance of peritectic points in the formation of rocks are given in Chapter Seven (for evaporite precipitation) and Chapter Eight (for crystallization of igneous rocks).

An important distinction should be made between the one-phase region of Figure 3-12 and the regions representing two phases in equilibrium. The curves ARE and BE separate the one-phase (liquid) region from the areas that represent equilibrium between a solid and a liquid. The other solid lines separate regions that represent two phases in equilibrium with each other. Each location in the one-phase region, such as point C, represents the actual state of the system in terms of phases, composition, and temperature. Point C corresponds to a liquid of composition 50 percent SiO_2, 50 percent $KAlSi_2O_6$ at a temperature of 1,600°C. A point such as point D or point F corresponds to the overall composition of two phases in equilibrium, but does not represent the phases in the system as such. For example, point D represents the equilibrium between leucite (of composition $KAlSi_2O_6$) and liquid of composition 30 percent SiO_2, 70 percent $KAlSi_2O_6$ (point

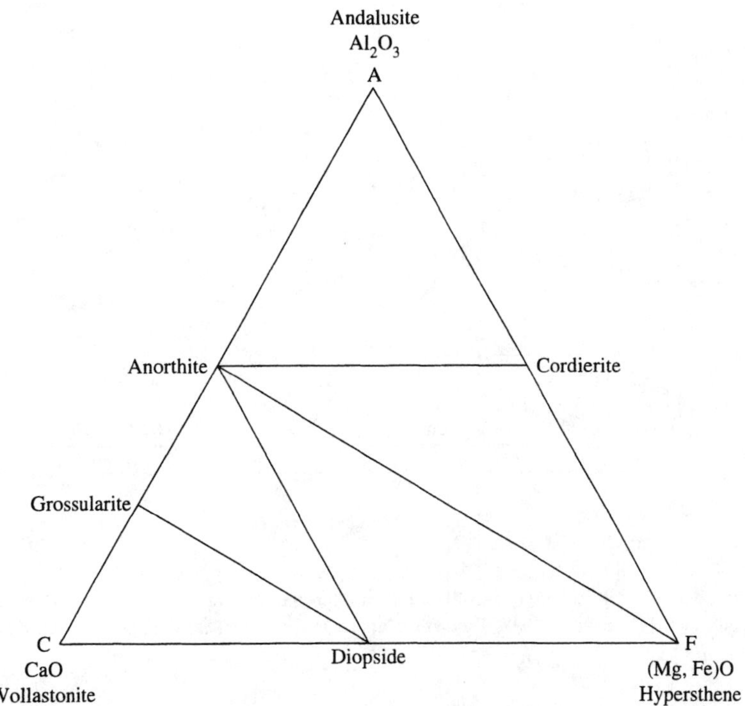

Figure 3-11 ACF diagram for contact metamorphic rocks near Oslo, Norway. Excess SiO_2 is assumed to have been present at the time of metamorphism, and quartz is a possible member of each assemblage. The three components, Al_2O_3, CaO, and (Mg, Fe)O, are plotted (in mole percent) for a rock to determine the equilibrium mineral assemblage. After Goldschmidt (1911).

H).* Similarly, point F does not represent the actual composition of a phase, but instead represents equilibrium between a certain proportion of the two solid phases, leucite and K-feldspar, at 1,000°C. No liquid would be present. (Point F represents the result produced by cooling liquid J to 1,000°C.)

The two regions of Figure 3-12 that represent two solid phases in equilibrium show what the final product of a cooling liquid of any given composition would be. For example, a liquid consisting of 70 percent SiO_2 would cool until it reached a temperature of about 1,280°C, at which point tridymite would begin to crystallize. With further cooling, the composition of the liquid would change along the curve BE (such curves are known as *liquidus lines*), and tridymite would continue to form. At point E, K-feldspar would start to form along with tridymite, and the liquid would stay at this temperature and composition until all the liquid was used up in the crystallization of K-feldspar and tridymite. The final product would be a mixture of K-feldspar and tridymite. For the liquid at point J, the first solid, leucite, would form when the liquid cooled to about 1,650°C. Leucite would continue to form as the liquid composition changed along AR. When point R is reached, leucite reacts with liquid to form K-feldspar, and also K-feldspar

*The relative amounts of leucite and liquid for point D are represented by the line lengths DH and DG according to the lever rule of phase diagrams. Thus DH is a measure of the relative amount of leucite, and DG is a measure of the relative amount of liquid. The percent of liquid present, for example, is equal to 100 × DG/GH.

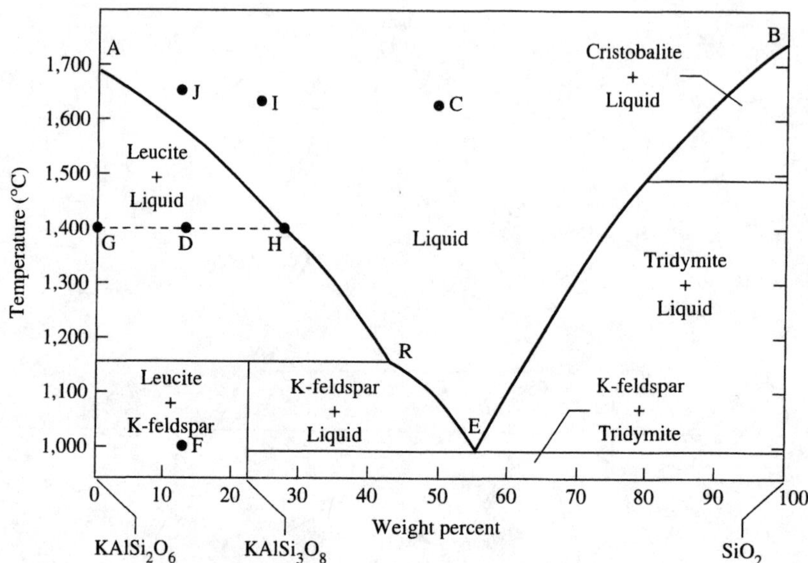

Figure 3-12 Temperature-composition diagram for the system leucite-silica. (After J. F. Schairer and N. L. Bowen, 1947, The system anorthite-leuctite-silica, *Bull. Soc. Geol. Finland*, v. 20, p. 74.)

precipitates directly from the liquid. This continues at a constant temperature of 1,150°C until the liquid is all used up, the final solid product being a mixture of leucite and K-feldspar.

The liquid represented by point I would have a similar early cooling history, with leucite first forming at about 1,425°C. However, at point R, the leucite would all be converted to K-feldspar before the liquid was entirely used up. The liquid would then change in composition and temperature along RE, with K-feldspar continuing to form. At point E, K-feldspar and tridymite would crystallize together, the final product being a mixture of these two minerals. (There would be much more K-feldspar than tridymite since the original composition I is much closer to the composition point of $KAlSi_3O_8$ than to the composition point of SiO_2.) Thus, for equilibrium crystallization, all liquids whose original composition is to the right of the $KAlSi_3O_8$ composition point will produce a solid consisting of K-feldspar and tridymite. All liquids to the left of this point will produce a solid consisting of K-feldspar and leucite.

EXPERIMENTAL GEOCHEMISTRY

The phase diagram and other data needed to apply thermodynamics to geologic problems come from laboratory measurements on minerals, rocks, melts, and fluids. A good introduction to experimental geochemistry is provided by Holloway and Wood (1988). They point out that temperatures from 0–2000°C and pressures from 1 atm to 50 kbar (equal to 150 km in depth) can be attained experimentally with high precision and accuracy. Even higher pressures can be produced, but with lower precision. Table 3-5 is a list of some of the machines commonly used in experimental geochemistry, along with typical applications to geologic problems.

TABLE 3-5 Machines Commonly Used in Experimental Geochemistry

Name	Typical Applications
Quench and gas mixing furnace	Liquidus relations in iron-bearing synthetic systems and rocks, synthesis of iron-bearing minerals and glasses, element partitioning experiments
Dickson rocking autoclave	Hydrothermal alteration of rocks, mineral solubilities, hydrothermal alteration of nuclear waste forms, kinetic studies
Barnes rocking autoclave	Studies in aqueous sulfide systems (measurement of sulfide mineral solubilities, determination of speciation in the fluid)
Cold-seal vessel ("Tuttle bomb")	Used in Tuttle and Bowen's (1958) classical experiments on melting of granites; used for metamorphic experiments, hydrothermal synthesis of minerals, diffusion experiments, and for studies of fluid phase chemistry
Internally heated gas vessel	Studies of melting relations in hydrous andesitic to rhyolitic melts, fluid/melt and fluid/crystal equilibria, and for synthesis of starting materials
Piston-cylinder	Studies of melting relations in basaltic and ultramafic systems and of the thermodynamic properties of pyroxenes and garnets
Diamond anvil	Exploration of mid- to lower-mantle phase equilibria, measurement of mineral compressibilities and thermal expansions at high pressure, exploration of the core/mantle boundary, studies of phase changes in the interiors of Jupiter and Saturn
Multiple anvil device	Phase transitions in the mid- to deep-mantle, equations of state of minerals, melting relations of mantle compositions to 300 kbar
Shock tube	Measuring densities of minerals and phase changes at very high pressure and temperature conditions of the lower mantle and core of the Earth
High temperature reaction calorimeter	Heat of dissolution of silicate in a solvent is compared with heats of constituent oxide dissolution to get heat of formation data
Drop calorimeter	Heat capacity measurements
Differential scanning calorimeter	Heat capacity measurements
Adiabatic calorimeter	Heat capacity measurements
Oxygen activity sensor	Determination of equilibrium oxygen activity (fugacity) for solid buffers; measurement of intrinsic oxygen fugacities of minerals and rocks
High temperature pH electrode	pH measurement at elevated temperature in hydrothermal solutions

Source: After Holloway and Wood (1988).

Measurements of thermodynamic quantities are also discussed in Chapters 11 and 12 of Nordstrom and Munoz (1985). They review calorimetric (heat) measurements, phase equilibrium studies, molar volume measurements, solubility determinations, conductivity measurements, electromotive force values, and the development and use of molecular data. Table 3-6 lists the types of data obtained from these measurements. In addition, Nordstrom and Munoz discuss

TABLE 3-6 Experimental Methods Used to Obtain Thermodynamic Data

Method	Data Obtained
Calorimetry	Heat capacities, third-law entropies, standard enthalpies and free energies, temperature effect on entropy, enthalpy, and free energy
Determination of temperature and pressure coordinates of univariant reactions	Free energies of reaction
Molar volume measurements	Effect of pressure on free energy, fugacities of gases and gas mixtures, partial molal volumes of electrolytes
Solubility measurements	Solubility-product constants, hydrolysis constants, activity coefficients
Conductivity measurements	Contributions to electrolyte theory, degree of association and/or ion pairing of electrolytes
Electromotive-force measurements	Free energies of reaction, dissociation constants, activities of ions, activity coefficients
Spectroscopic and spectrophotometric measurements	Equilibrium constants, heat capacities

Source: After Nordstrom and Munoz (1985).

techniques available for the estimation of thermodynamic values and methods for evaluation of the validity of published experimental data. They emphasize that many incorrect values can be found in the literature. Appendix F of Nordstrom and Munoz (1985) is an extensive list of sources of thermodynamic data.

An example of the uncertainties inherent in experimental measurements of thermodynamic data, and of the resulting uncertainties in calculated phase boundaries and equilibrium temperatures, has been given by Best (1982) for the aluminosilicate polymorphs (Figure 3-7 and Table 3-7). In many calculations of this type, we have to deal with small differences between large numbers. Small errors in the experimental data can lead to large errors in the calculated equilibrium relationships.

TABLE 3-7a Thermodynamic Data per Mole for the Aluminosilicate Polymorphs at 500°C and One Atmosphere with Estimated Maximum Uncertainties

Mineral	Volume (cm^3)	Entropy (joules per kelvin)	Enthalpy (megajoules)
Andalusite	52.29 ± 0.05	245.1 ± 0.4	-2.51515 ± 0.00021
Sillimanite	50.23 ± 0.05	246.9 ± 0.4	-2.51278 ± 0.00021
Kyanite	44.69 ± 0.05	236.0 ± 0.4	-2.51931 ± 0.00021

Sources: Data from Holdaway (1971, 97–131). From IGNEOUS AND METAMORPHIC PETROLOGY by Best. Copyright © 1982 by W. H. Freeman and Company. Used with permission.

TABLE 3-7b Slopes and Equilibrium Temperature Values at One Atmosphere for Reaction Lines Delineating Stability Field Boundaries of the Aluminosilicate Polymorphs Calculated from the Data in Table 3-7a. Note the Large Uncertainties in the Equilibrium Temperature Values

Reaction	Slope (megapascals per kelvin)	Equilibrium Temperature (°C)
Kyanite → andalusite	1.20 ± 0.12	184^{+95}_{-79}
Kyanite → sillimanite	1.98 ± 0.17	326^{+89}_{-77}
Andalusite → sillimanite	-0.87 ± 0.36	1044^{+1473}_{-567}

Source: From IGNEOUS AND METAMORPHIC PETROLOGY by Best. Copyright © 1982 by W. H. Freeman and Company. Used with permission.

SUMMARY

Thermodynamics deals with the effect of heat flow and work done on matter. The basic principles are summarized in the first and second laws of thermodynamics. The first law gives the relationship among internal energy, heat, and work. The second law defines entropy and can be used to determine the direction in which any given reaction will go. From the first law comes the concept of enthalpy and of the various types of enthalpy changes, such as heat of formation. By combining the first and second laws, a quantity known as the Gibbs function can be defined. The Gibbs function is the most useful thermodynamic property of a system since it allows, for systems in which temperature and pressure are variables, a quantitative determination of the direction in which a reaction will go.

Because thermodynamics can be applied to any material, geologists can use it to study all types of natural processes. Applications can be made to laboratory, field, and theoretical research. Information can be obtained on the stability of minerals under various conditions, on chemical reactions in natural waters, and on the relative importance for a given process of the numerous variables encountered in nature. In order to obtain this information, it is necessary to determine the thermodynamic properties of minerals, liquids, gases, and ions. Particularly useful are the heat capacities, molar volumes, heats of formation, entropies, and free energies of formation. The major limitation of the use of thermodynamics involves rates and paths of reactions. It tells us nothing about how fast, and by what mechanism, a given reaction takes place.

Entropy can be thought of as a measurement of the degree of order of a material or of a system. Whether or not a possible process will take place depends on the entropy change involved and on the heat (enthalpy) change involved. In general, processes that produce less order and result in a loss of heat are the ones that occur. The following expression states the relationship between these two factors and the change in Gibbs function: $\Delta G = \Delta H - T\,\Delta S$ (at constant temperature). All processes should go in the direction that produces a decrease in Gibbs function. Provided heat capacities, molar volumes, and related quantities are available for reactants and products, we can calculate ΔG of a reaction at any temperature and pressure. Calculations of ΔG, however, do not tell us the manner in which a change will occur or how fast it will occur.

Another important equation states the relationship of the standard free-energy change of a reaction and the equilibrium constant of the reaction: $\Delta G^0 = -RT \ln K$. This gives us a means of relating the activities (modified concentrations) of reactants and products to the free-energy change for a reaction. In using the thermodynamic relationships, it is necessary to work with activities instead of concentrations and with fugacities instead of pressures. Activities can be considered concentration values that would be obtained if solids and liquids behaved in an ideal manner, and fugacities can be considered pressure values that would be obtained if gases behaved in an ideal manner. The theoretical definition of the activity or fugacity of a component of a system is based on its relationship to the chemical potential of the component. The chemical potential of a given component is defined in terms of the change in a thermodynamic function produced when the amount of that component is changed.

At equilibrium in a system the chemical potentials of each component in each phase must be equal. This requirement allows derivation of the phase rule, which is a relationship among the phases, components, and variables of a system in equilibrium. It can be used to estimate whether or not a rock has formed under equilibrium conditions and is particularly useful in the construction and interpretation of phase diagrams. Such diagrams allow geologists to portray calculated relationships between materials, to represent field relationships, and to illustrate the results of laboratory studies. Various combinations of variables can be used, and many different types of diagrams can be constructed. A clear understanding of their method of construction is necessary in applying them to natural processes.

QUESTIONS

1. The heat of formation of carbon monoxide is hard to determine experimentally, since it is difficult to burn carbon without producing at least some carbon dioxide as well as carbon monoxide. Use equations for the formation of carbon dioxide from carbon and then from carbon monoxide to calculate the heat of formation of carbon monoxide. The heat given off in these two reactions is 94,051 and 67,635 calories respectively.

2. (a) What would you expect to be the general relationship among the hardness values and entropies of minerals?

(b) Using the entropy values in Table 3-2, predict which of the Al_2SiO_5 polymorphs would be stable at high temperatures. Check your answer with Figure 3-7.

3. (a) Calculate the standard free-energy change for the conversion of aragonite to calcite at $25^{\pm}$C and 1 atmosphere. Which is stable under these conditions?

(b) Which of the two polymorphs would you expect to be the stable form of $CaCO_3$ at high temperatures? Why?

(c) Given the molar volumes of calcite (36.93 cubic centimeters per mole) and aragonite (34.15 cubic centimeters per mole), which one would you expect to be stable at high pressures?

(d) Calculations for very low temperatures (below about 80 K) show that aragonite has a lower Gibbs function than calcite. Using only this information and the information from parts (a), (b), and (c), sketch a diagram showing the general stability fields of aragonite and calcite as a function of pressure and temperature.

4. (a) The change in the internal energy of an adiabatic system due to the performance of work is independent of (does not depend on) the way the work is done. True or False? Justify your answer.

(b) Under what conditions is the heat change between a system and its surroundings equal to the enthalpy change?

5. The following questions refer to Table 3-2.

(a) Why are the heat of formation and free energy of formation of O_2 zero, whereas they are not zero for S_2?

(b) Why do the gases listed in Table 3-2 have larger entropy values than the solid substances?

(c) If the free energies of formation of the Al_2SiO_5 polymorphs were given for formation from the oxides ($\Delta G_f^0 = 0$ for oxides), how would the values differ from those given in Table 3-2? Keep in mind that the values in Table 3-2 are based on $\Delta G_f^0 = 0$ for the elements in their standard states.

6. Calculate ΔG for the reaction

$$CaSO_4 \cdot 2H_2O \text{ (gypsum)} \rightleftharpoons CaSO_4\text{(anhydrite)} + H_2O \text{ (water)}$$

at 25 and 50°C, with a pressure of 1 atmosphere in both cases. Which of the two minerals is stable at 25°C and which at 50°C? The heat capacity C_p is $14.10 + 0.033T$ calories per degree-mole for anhydrite, $21.84 + 0.076T$ calories per degree-mole for gypsum, and 18.02 calories per degree-mole for water.

7. Is the following statement True or False? Explain your answer. The Gibbs free energy of a substance always increases with pressure when temperature is kept constant.

8. (a) What would be the change in molar volume of quartz if the pressure acting on it is changed from one bar to 10,000 bars? Assume the temperature is constant at 298.15 K. The molar volume of alpha quartz at 298.15 K and one bar is 22.69 cm^3.
   (b) What would be the change in Gibbs free energy of one mole of quartz undergoing this pressure increase?

9. Calculate ΔG for the reaction kyanite $\rightarrow$ sillimanite at 10 kb and 800 K. Given that C_p (kyanite) $= 45.32 + 2.34 \times 10^{-3}$ T, C_p (sillimanite) $= 40.69 + 5.86 \times 10^{-3}$ T, molar volume (kyanite) $= 44.69$ cm^3, and molar volume (sillimanite) $= 50.23$ cm^3. Which mineral is stable under these conditions? Compare your answer with Figure 3-7.

10. (a) Calculate the entropy change produced by heating 1 mole of water from 0 to 100°C at a constant pressure of 1 atmosphere. The heat capacity of water is $C_p = 18.02$ calories per degree-mole.
    (b) Calculate the entropy change produced by vaporizing 1 mole of water at 100°C and 1 atmosphere. The heat of vaporization of water is 540 calories per gram.

11. (a) What equation is used to calculate the slopes given in Table 3-7b? What equation can be used to calculate the equilibrium temperature values given in Table 3-7b?
    (b) What additional source of uncertainty should be considered in calculating the slopes and temperatures of Table 3-7b?

12. The rate at which a reaction occurs is expressed in kinetic theory as a rate constant *K*, which varies with temperature. A relationship between *K* and temperature is given by the Arrhenius equation

$$K = Ae^{(-\Delta E/RT)}$$

where *A* is the Arrhenius constant, ΔE is activation energy, *R* is the gas constant, and *T* is absolute temperature. Use the Arrhenius equation to calculate the change in *K* between 250°C and 300°C for the reaction

$$CaCO_3 \text{ (calcite)} + Mg^{2+} \rightarrow CaMg(CO_3)_2 \text{ (dolomite)} + Ca^{2+}$$

Use $A = 1.0 \times 10^{14}$ sec^{-1} and $\Delta E = 49$ kcal/mole [Katz and Matthews (1977, 297–308)].

13. (a) If a given reaction had a standard free-energy change that is positive and very large in value, what would you expect the value of the equilibrium constant for the reaction to be, very large or very small?
    (b) Which would be present at higher concentrations, the reactants or the products, when this reaction is complete?
    (c) What can be said about the initial activities of the reactants and products when the free-energy change of a reaction is equal to the standard free-energy change?

14. (a) Calculate the standard free energy of reaction and the equilibrium constant for the following reaction at 25°C and 1 atmosphere:

$$2CaCO_3 \text{ (calcite)} + Mg^{2+} \rightleftharpoons CaMg(CO_3)_2 \text{ (dolomite)} + Ca^{2+}$$

(b) Why can't the value of ΔG^0 be used as such to tell which way the reaction would go in seawater?

(c) Use the activity values given in Table 3-4 for Ca^{2+} and Mg^{2+} to predict the direction of the reaction in seawater.

15. **(a)** What concentration of Cu^{2+} would be necessary to precipitate malachite from seawater? Assume a hydroxide ion activity of 10^{-6} and use the carbonate ion activity value in Table 3-4.

(b) How does this value compare with that given for copper concentration in Table 1-11?

16. **(a)** Using Table 3-2, calculate the equilibrium constant for the following reaction at 25°C:

$$CaSO_4 \text{ (anhydrite)} \rightleftharpoons Ca^{2+} + SO_4^{2-}$$

(b) Considering the activity values in Table 3-4, would you say that seawater is approximately saturated with calcium sulfate?

17. **(a)** A mineral assemblage consisting of the ferrous iron biotite, annite, along with K-feldspar and magnetite, can serve as an oxygen buffer in a rock system. Write an equation for the breakdown of annite to K-feldspar and magnetite to represent this buffer. The formula of annite is $KFe_3AlSi_3O_{10}(OH)_2$, where the iron is all in the ferrous state.

(b) Would you expect annite to be stable under conditions in the lower part or upper part of Figure 3-2?

18. Given is a solution containing 0.2 mole per kilogram of Fe^{2+}. If 80 percent of the Fe^{2+} occurs as a free ion, and if the activity coefficient of Fe^{2+} in this solution is 0.6, what is the activity of the Fe^{2+} in the solution?

19. **(a)** Given is a rock containing calcite, dolomite, enstatite, wollastonite, and diopside. Assume a closed system and that none of the minerals has iron in it. Can this combination exist in equilibrium? If it can exist, how many degrees of freedom does it have? If it can exist at some one temperature, what would happen if the temperature were lowered a few degrees?

(b) Given is the same rock as in (a), but with the addition of a gas phase consisting of CO_2. Answer the same questions about this system as were asked in (a).

20. Complete the following table for pressure-temperature diagrams such as Figure 3-7.

Number of components	Geometrical form	Maximum number of phases possible
One	Point	
Two	Point	
Three	Point	
One	Line	
Two	Line	
Three	Line	
One	Region	
Two	Region	
Three	Region	

21. **(a)** Write an equation for the reaction of annite and pyrope garnet to form phlogopite and almandine garnet. The formulas are $KMg_3AlSi_3O_{10}(OH)_2$ (phlogopite), $KFe_3AlSi_3O_{10}(OH)_2$ (annite) $Mg_3Al_2Si_3O_{12}$ (pyrope), and $Fe_3Al_2Si_3O_{12}$ (almandine).

(b) Write an expression for a distribution coefficient K_D in terms of the mole fractions of iron and magnesium in biotite and garnet representing solid solutions of phlogopite-annite and pyrope-almandine.

(c) Given are the following chemical analyses (weight percent) of coexisting biotite and garnet in a metamorphic rock. Use Figure 3-4 to determine the temperature of formation of the rock.

	FeO	MgO
biotite	17.98	10.51
garnet	29.61	2.26

Data from Novak and Holdaway (1981).

(d) Name two sources of error for part (c) that are related to the compositions of biotite and garnet.

22. The following questions refer to Figure 3-12.

(a) How many phases can occur together at point H?

(b) Is point H an invariant point?

(c) What would be the first solid phase to separate from a cooling liquid with 20 percent SiO_2? with 50 percent SiO_2?

(d) What would be the final products for a cooling liquid with 20 percent SiO_2? with 50 percent SiO_2?

(e) What would be the composition of the last bit of cooling liquid for an initial composition of 20 percent SiO_2?

REFERENCES

BARTH, T. F. W. 1962. *Theoretical Petrology,* 2d ed. John Wiley & Sons, Inc., New York.

BERMAN, R. G. 1988. Internally consistent thermodynamic data for stoichiometric minerals in the system Na_2O-K_2O-CaO-MgO-FeO-Fe_2O_3-Al_2O_3-SiO_2-TiO_2-H_2O-CO_2. *J. Petrol.,* v. 29:445–522.

BEST, M. G. 1982. *Igneous and Metamorphic Petrology.* W. H. Freeman and Company, New York.

BETHKE, P. M., and P. B. BARTON, JR. 1971. Distribution of some minor elements between coexisting sulfide minerals. *Econ. Geol.,* v. 66:140–163.

BROWNLOW, A. H. 1961. Variation in composition of biotite and actinolite from monomineralic contact bands near Westfield, Massachusetts. *Am. J. Sci.,* v. 259:353–370.

CARMICHAEL, I. S. E., and H. P. EUGSTER, eds. 1987. *Thermodynamic Modeling of Geological Materials: Minerals, Fluids and Melts,* Rev. in Mineralogy, v. 17. Mineralogical Society of America, Washington, D.C.

COX, J. D., D. D. WAGMAN, and V. A. MEDVEDEV. 1989. *CODATA Key Values for Thermodynamics.* Hemisphere Publishing Corp., Bristol, PA.

EHLERS, E. G. 1972. *The Interpretation of Geological Phase Diagrams.* W. H. Freeman and Company, New York.

ERNST, W. G. 1976. *Petrologic Phase Equilibria.* W. H. Freeman and Company, New York.

ESSENE, E. J. 1982. Geologic thermometry and barometry. In *Characterization of Metamorphism through Mineral Equilibria,* Rev. in Mineralogy, v. 10, J. M. Ferry, ed. Mineralogical Society of America, Washington, D.C. 153–206.

EUGSTER, H. P., and G. B. SKIPPEN. 1967. Igneous and metamorphic reactions involving gas equilibria. In *Researches in Geochemistry,* v. 2, P. H. Abelson, ed. John Wiley & Sons, Inc., New York. 492–520.

EUGSTER, H. P., and D. R. WONES. 1962. Stability relations of the ferruginous biotite, annite. *J. Petrol.,* v. 3:82–125.

FERRY, J. M. 1979. A map of chemical potential differences within an outcrop. *Am. Mineralogist,* v. 64:966–985.

FERRY, J. M., and F. S. SPEAR. 1978. Experimental calibration of the partitioning of Fe and Mg between biotite and garnet. *Contrib. Mineral. Petrol.,* v. 66:113–117.

FLETCHER, P. 1993. *Chemical Thermodynamics for Earth Scientists.* John Wiley & Sons, Inc., New York.

FRASER, D. G., ed. 1977. *Thermodynamics in Geology.* D. Reidel Publishing Company, Boston.

GARRELS, R. M., and M. E. THOMPSON. 1962. A chemical model for seawater. *Am. J. Sci.,* v. 260:57–66.

GARRELS, R. M., and C. L. CHRIST. 1965. *Solutions, Minerals, and Equilibria.* Harper & Row, Inc., New York.

GOLDMAN, D. S., and A. L. ALBEE. 1977. Correlation of Mg/Fe partitioning between garnet and biotite with $^{18}O/^{16}O$ partitioning between quartz and megnetite. *Am. J. Sci.,* v. 277:750–767.

GOLDSCHMIDT, V. M. 1911. Die Kontaktmetamorphose im Kristianiagebiet. *Kristiania Vidensk. Skr., I, Math-Naturv. Klasse.*

GRAMBLING, J. A. 1981. Kyanite, andalusite, sillimanite, and related mineral assemblages in the Truchas Peaks region, New Mexico. *Am. Mineralogist,* v. 66:702–722.

GREENLAND, L. P., R. I. TILLING, and D. GOTTFRIED. 1971. Distribution of cobalt between coexisting biotite and hornblende in igneous rocks. *N. Jb. Miner, Mh.,* v. 1:33–42.

GREENWOOD, H. J. 1967. Wollastonite: stability in H_2O-CO_2 mixtures and occurrence in a contact-metamorphic aureole near Salmo, British Columbia, Canada. *Am. Mineralogist,* v. 52:1669–1680.

HELGESON, H. C. 1982. Errata: thermodynamics of minerals, reactions, and aqueous solutions at high temperatures and pressures. *Am. J. Sci.,* v. 282:1144–1149.

HELGESON, H. C. 1985. Errata II: thermodynamics of minerals, reactions, and aqueous solutions at high temperatures and pressures. *Am. J. Sci.,* v. 285:845–855.

HELGESON, H. C., J. M. DELANY, H. W. NESBITT, and D. K. BIRD. 1978. Summary and critique of the thermodynamic properties of rock-forming minerals. *Am. J. Sci.,* v. 278A:1–229.

HELGESON, H. C., and G. C. FLOWERS. 1982. Theoretical prediction of the thermodynamic behavior of aqueous electrolytes at high pressures and temperatures. IV. Calculation of activity coefficients, asmotic coefficients, and partial molal and standard and relative partial molal properties to 600°C and 5kb. *Am. J. Sci.,* v. 281:1249–1516.

HELGESON, H. C., D. H. KIRKHAM, and G. C. FLOWERS. 1981. Theoretical prediction of the thermodynamic behavior of aqueous electrolytes at high pressures and temperatures. IV. Calculation of activity coefficients, osmotic coefficients, and apparent molal and standard and relative partial molal properties to 600° C and 5kb. *Am. J. Sci.,* v. 281:1249–1516.

HEMINGWAY, B. S., R. A. ROBIE, H. T. EVANS, and D. M. KERRICK. 1991. Heat capacities and entropies of sillimanite, fibrolite, andalusite, kyanite, and quartz and the Al_2SiO_5 phase diagram. *Am. Mineralogist,* v. 76:1597–1613.

HOLDAWAY, M. J. 1971. Stability of andalusite and the aluminum silicate phase diagram. *Am. J. Sci.,* v. 271:97–131.

HOLLOWAY, J. R., and B. J. WOODS. 1988. *Simulating the Earth: Experimental Geochemistry.* Unwin Hyman, Boston.

HUEBNER, J. S. 1971. Buffering techniques for hydrostatic systems at elevated pressures. In *Research Techniques for High Pressure and High Temperature,* G. C. Ulmer, ed. Springer-Verlag, New York.

KATZ, A., and A. MATTHEWS. 1977. The dolomitization of $CaCO_3$: an experimental study at 252–295°C. *Geochim. et Cosmochim. Acta,* v. 41:297–308.

KERRICK, M. 1990. *The Al_2SiO_5 Polymorphs.* Rev. in Mineralogy, v. 22, Mineralogical Society of America, Washington, D.C.

KIELLAND, J. 1937. Individual activity coefficients of ions in aqueous solutions. *J. Amer. Chem. Soc.,* v. 59:1675–1678.

KRAUSKOPF, K. 1979. *Introduction to Geochemistry,* 2d ed. McGraw-Hill Book Company, New York.

LASAGA, A. C., and R. J. KIRKPATRICK. 1981. *Kinetics of Geochemical Processes,* Rev. in Mineralogy, v. 8, Mineralogical Society of America, Washington, D.C.

Morse, S. A. 1980. *Basalts and Phase Diagrams.* Springer-Verlag, New York.

Novak, J. M., and M. J. Holdaway. 1981. Metamorphic petrology, mineral equilibria and polymorphism in the Augusta Quadrangle, south-central Maine. *Am. Mineralogist,* v. 66:56–57.

Nordstrom, D. K., and J. L. Munoz. 1985. *Geochemical Thermodynamics.* The Benjamin/Cummings Publishing Company, Menlo Park, CA.

Nordstrom, D. K., and J. L. Munoz. 1994. *Geochemical Thermodynamics,* 2d ed. Blackwell Scientific Publications, Oxford, England.

Powell, R. 1978. *Equilibrium Thermodynamics in Petrology.* Harper & Row, New York.

Pytkowicz, R. M. 1983. *Equilibria, Nonequilibria, and Natural Waters,* v. 1. Wiley-Interscience, New York.

Rice, J. M., and J. M. Ferry. 1982. Buffering, infiltration, and the control of intensive variables during metamorphism. In *Characterization of Metamorphism through Mineral Equilibria,* Rev. in Mineralogy, v. 10, J. M. Ferry, ed. Mineralogical Society of America, Washington, D.C. 263–326.

Richardson, S. M., and H. Y. McSween, Jr. 1989. *Geochemistry: Pathways and Processes.* Prentice-Hall, Englewood Cliffs, NJ.

Robie, R. A., and D. R. Waldbaum. 1968. Thermodynamic properties of minerals and related substances at 298.15 K (25°C) and one atmosphere (1.013 bars) pressure and at high temperatures. U.S. Geol. Survey Bull. 1259.

Robie, R. A., B. S. Hemingway, and J. R. Fisher. 1978. Thermodynamic properties of minerals and related substances at 298.15 K and 1 bar (10^5 pascals) pressure and at higher temperatures. U.S. Geol. Survey Bull. 1452.

Rumble, D., III. 1978. Mineralogy, petrology and oxygen isotope geochemistry of the Clough Formation, Black Mountain, western New Hampshire, U.S.A. *J. Petrol.,* v. 19:317–340.

Schairer, J. F., and N. L. Bowen. 1947. The system anorthite-leucite-silica. *Bull. Soc. Geol. Finlande,* v. 20:74.

Spear, F. S. 1977. Carnegie Inst. Wash. Year Book 76. 613–619.

Spear, F. S., J. M. Ferry, and D. Rumble, III. 1982. Analytical formulation of phase equilibria: the Gibbs method. In *Characterization of Metamorphism through Mineral Equilibria,* Rev. in Mineralogy, v. 10, Mineralogical Society of America, Washington, D.C. 105–152.

Stumm, W., and J. J. Morgan. 1981. *Aquatic Chemistry,* 2d ed. Wiley-Interscience, New York.

Thompson, A. B. 1976. Mineral reactions in pelitic rocks. II. Calculation of some P-T-X (Fe-Mg) phase relations. *Am. J. Sci.,* v. 276:425–454.

Thompson, J. B., Jr. 1959. Local equilibrium in metasomatic processes. In *Researches in Geochemistry,* P. H. Abelson, ed. John Wiley & Sons, Inc., New York. 427–457.

Thompson, J. B., Jr. 1970. Geochemical reaction and open systems. *Geochim. Cosmochim. Acta,* v. 34:529–551.

Thompson, J. B., Jr. 1982. Reaction space: an algebraic and geometric approach. In *Characterization of Metamorphism through Mineral Equilibria,* Rev. in Mineralogy, v. 10, J. M. Ferry, ed. Mineralogical Society of America, Washington, D.C. 33–52.

Toulmin, P., III, and P. B. Barton, Jr. 1964. A thermodynamic study of pyrite and pyrrhotite. *Geochim. Cosmochim. Acta,* v. 28:641–671.

Tunell, G. 1979. On the mathematical derivation of Gibbs equation from experimentally determinable thermodynamic relations. Carnegie Inst. Washington Supp. Publ. No. 49.

Ulbrich, H. H., and D. R. Waldbaum. 1976. Structural and other contributions to the third-law entropies of silicates. *Geochim. Cosmochim. Acta,* v. 40:1–24.

Wagman, D. D., W. H. Evans, V. B. Parker, R. H. Schumm, I. Harlow, S. M. Bailey, K. L. Churney, and R. L. Nuttall. 1982. The NBS tables of chemical thermodynamic properties: selected values for inorganic and C_1 and C_2 organic substances in SI units. *J. Phys. Chem. Ref. Data 11* (Suppl. 2). 1–392.

Wood, B. J., and D. G. Fraser. 1976. *Elementary Thermodynamics for Geologists.* Oxford University Press, New York.

Woods, T. L., and R. M. Garrels. 1986. *Thermodynamic Values at Low Temperature for Natural Inorganic Materials: An Uncritical Summary.* Oxford University Press, New York.

Zen, E. 1963. Components, phases, and criteria of chemical equilibrium in rocks. *Am. J. Sci.,* v. 261: 929–942.

Water Chemistry

Water is an unusual substance. In particular, its physical and chemical properties are very different from those of most other liquids. Water, as the only common and abundant liquid on the Earth, has been a prime factor in the evolution of the Earth's surface and in the evolution of life on the Earth. The weathering and erosion of rocks is mainly carried out by processes involving water. Most sedimentary rocks form by precipitation from, or physical deposition in, water. Water also plays a role, although not as great, in the formation of igneous and metamorphic rocks. Water is a basic component of living matter, and most forms of life exist in water. Finally, the circulation of most of the common elements and compounds at the Earth's surface is brought about by the movement of water or its gaseous equivalent. Thus a knowledge of water chemistry is vital for any study of the Earth or its inhabitants.

PROPERTIES OF WATER

The unusual properties of water can be best explained in terms of the structure of the H_2O molecule and in terms of the intermolecular forces that hold water molecules together. These bonding forces are known as *hydrogen bonds* and are a direct consequence of the structure of the H_2O molecule. A representation of the structure of the molecule is shown in Figure 4-1. The location of two hydrogen atoms on one side of an oxygen atom results in a dipolar molecule with a positive charge on one side and a negative charge on the other side. The dipolar nature of the water molecule is the cause of its unusual properties, both as a pure substance and as a solution containing ions and molecules of various elements.

A hydrogen bond forms when a hydrogen atom (proton) of one water molecule interacts with the negatively charged side of another water molecule. A general description of the physical

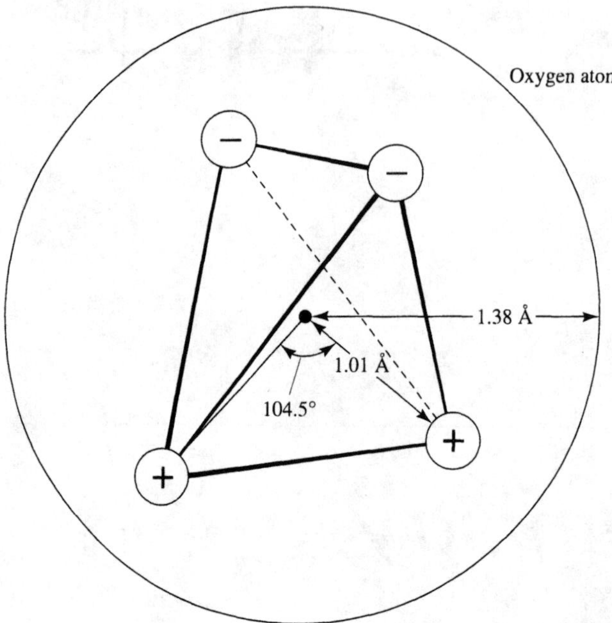

Figure 4-1 Distribution of charge on the water molecule. The two positively charged hydrogen atoms are embedded in one side of the oxygen atom, and the resulting charge distribution can be represented by a tetrahedron with two positively charged corners (location of the hydrogen atoms) and two negatively charged corners (resulting from the overall electron distribution). Although the molecule has an overall neutral charge, there is a net negative charge on one side and a net positive charge on the other. Bonding of a hydrogen atom (proton) of one water molecule to the negatively charged side of another water molecule produces a hydrogen bond. After Evans (1964, 269).

structure of water would thus be that it consists of H_2O molecules held together in three dimensions by hydrogen bonds. In detail, neither the structure of water nor the nature of hydrogen bonds is well understood. The structure of ice (a hexagonal arrangement of water molecules, each bonded to four neighbors by hydrogen bonds) is fairly well known, and various models have been proposed to explain what happens when ice melts. Any model must explain a number of unusual properties exhibited by water near its freezing point. For instance, when ice melts there is an increase in density (decrease in volume). One popular model suggests that water consists of regions with the four-coordination of ice separated by regions of unordered molecules. Such a model is viewed as having smaller and fewer void spaces than ice in the vicinity of the freezing point and thus a greater density. So far no structural model has been found that will satisfactorily explain in detail all, or even most, of the properties of liquid water over the temperature range from zero to 100°C. General explanations of these properties can, however, be given in terms of the dipolar nature of the water molecule.

Let us briefly review some of the important properties of water. We have already mentioned that ice is less dense than water near the freezing point. (The density of water reaches its maximum value at 4°C.) One important result of this property is that ice forms on the top of a lake, rather than on the bottom, allowing organisms to survive in the lake during cold weather. The boiling point of water is higher than that of most liquids of similar molecular weight, because the hydrogen bonds that hold liquid water together are stronger than the weak van der Waals bonds (owing to the interaction of the electrons of one molecule with the nuclei of other molecules) that hold together most liquids. As a result, a higher temperature must be reached to break the bonds of water and transform it to a gas. Similarly, the strong bonding of ice is a major factor in the high melting point of ice. The temperature range at the surface of the Earth is such that H_2O occurs mainly in the liquid state rather than as a solid or a gas. For this reason the history of the Earth has been quite different from that of those planets in the solar system where liquid H_2O cannot exist in abundance.

Probably the most important property of water from a geochemical point of view is its high *dielectric constant.* The dielectric constant of a substance is a measure of the attraction of opposite electric charges for each other in that substance. For liquid water at 18°C, the dielectric constant is 81. This means that two charges in water attract each other with a force only 1/81 as strong as in a vacuum. The combination of a high dielectric constant and the fact that water molecules tend to attach themselves to ions results in water having an extraordinary capacity to dissolve ionic substances. When a compound, such as sodium chloride, is dissolved in water, the attraction between the sodium ions and the chloride ions is weak, because the dipolar water molecules align their positive and negative sides in such a way that the attraction is partially neutralized. In addition, each positively charged sodium ion is surrounded by several water molecules with their negative sides attached to the ion. The positive sides of other water molecules attach themselves to the chloride ions. This formation of hydrated ions also weakens the attraction of the sodium and chloride ions for each other. Thus water can dissolve and hold large amounts of those compounds and elements that form ions in solution. Most organic compounds do not form ions in liquids and thus will not dissolve easily in water. They will dissolve extensively in liquids that have low dielectric constants, such as benzene.

SOLUTIONS AND SOLUBILITIES

In this section we review the solubility properties of water solutions. Such solutions consist of water (the *solvent*) and a wide variety of components known as *solutes.* Inorganic solutes consist of simple ions, complex ions, and uncharged molecules. Organic compounds also occur in natural waters, particularly those that are polluted or are strongly colored.

In field, laboratory, and theoretical studies of solutions, it is necessary to know and describe the concentrations of solutes. These concentrations are given in weight-per-weight units or in weight-per-volume units. An example of the former is parts per million (ppm). One part per million of a solute is equal to 1 gram of solute in 1 million grams of solution. This unit is often used for expressing the concentration of trace elements in water. Sophisticated analytical techniques now make it possible to detect and measure concentrations of components at the parts-per-billion level (ppb). Abundant solutes in seawater are often reported as parts per thousand (‰). Weight-per-volume units are usually based on amounts of solute in 1 liter of water. To convert such values to weight-per-weight units, it is assumed usually that 1 liter of water solution weighs exactly 1 kilogram (the weight of 1 liter of pure water at 4°C). This assumption does not ordinarily introduce any great error in concentration calculations because most natural waters are very dilute solutions (brines and hydrothermal and metamorphic waters are exceptions). Water analyses are commonly reported in units of either milligrams per liter (mg/liter) or parts per million, and these two are equivalent for most cases.

Other common ways to express concentration for individual ions or compounds are as moles per liter (*molarity*), moles per kilogram of solvent (*molality*), or as equivalent weights per liter (*normality*). Use of the first two terms allows comparison of elements and compounds on a basis that takes into account their differing molecular weights, while the last term takes into account differences in cation or anion charge.* A 1-molar (1 *M*) solution of a substance contains 1 mole (atomic or molecular weight in grams) of that substance per liter of solution and a 1-normal (1 *N*) solution contains one equivalent of the substance per liter of solution. An *equivalent* is the amount of a substance that reacts with one atomic weight of hydrogen. For example, a 1 *N* solution of HCl would contain 35.45 grams of Cl^- per liter or 1 *M* Cl^-. However, a 1 *N* solution of H_2SO_4 would contain 0.5 *M* SO_4^{2-}. The most common equivalent weight unit for water solutions is milliequivalents per liter (mEq/liter) and, for a given analysis of a water solution, the total milliequivalents per liter of cations should equal the total milliequivalents per liter of anions to preserve electrical neutrality. Table 4-1 contains an estimate of the mean composition of river waters of the world with the amounts of the constituents expressed in the common units of concentration. Note that the values are the same when expressed as parts per million or as milligrams per liter, and that, for singly charged ions, values in milliequivalents per liter are the same as values in millimoles per liter.

A major use of concentration measurements and calculations is in studying the solubility and precipitation of solutes. The *solubility* of a given substance is the maximum amount of that substance that can be dissolved in a solution that is in equilibrium with a solid source of the solute. Such a solution is called a *saturated solution.* The major tool for studying solubilities is a form of equilibrium constant known as the *solubility product.* Consider the reaction

$$AgCl \rightleftharpoons Ag^+ + Cl^- \tag{4-1}$$

In a solution saturated with silver chloride, the activities of Ag^+ and Cl^- obey the equation

$$[Ag^+][Cl^-] = K_{sp} = \frac{\text{products}}{\text{reactants}} \tag{4-2}$$

where K_{sp} is the solubility product constant (the activity of a solid, such as silver chloride, is conventionally set to 1).† The activities of the ions are expressed in moles per liter of solution. The solubility of AgCl in moles per liter is equal to the activity of Ag^+ or Cl^-, since 1 mole of AgCl dissolves in pure water to give 1 mole of each in solution.‡ The value of K_{sp} for silver chloride at 25°C is $10^{-9.8}$; thus the solubility of silver chloride in pure water at this temperature is

$$\text{solubility} = [Ag^+] = [Cl^-] = \sqrt{10^{-9.8}} = 10^{-4.9} \text{ mol/liter} \tag{4-3}$$

The product $[Ag^+][Cl^-]$ can have any value less than K_{sp} for an unsaturated solution. If the product is greater than K_{sp}, precipitation of silver chloride should occur.

*Significant differences between molar and molal units only occur at high temperatures (above 100°C) or when dealing with solutions containing large amounts of dissolved materials (more than 10,000 mg/l total dissolved solids).

†For the rest of this chapter we shall use "activities" in place of "concentrations." For very dilute solutions, such as river water, activities are essentially equal to concentrations. This means we can use the values of concentrations measured in natural waters for our solubility calculations. In concentrated solutions, the interactions of the various constituents result in concentration values not being equal to activity values, and it then becomes particularly important to use activity values in calculations involving solubility products. For example, in studying the solubility of NaCl (halite) in initially pure water, the concentrations of Na^+ and Cl^- are not equal to their activities because of the unusually high solubility of halite. See the discussion of seawater concentrations and activities in Chapter Three.

‡Note that the solubility of NaCl in pure water is equal to the concentration of Na^+ or Cl^-, but is not equal to the activity of Na^+ or Cl^-, because in this case we are no longer dealing with a very dilute solution.

TABLE 4-1 Mean Composition of River Waters of the World

Constituent	Milligrams per liter or parts per million (mg/l or ppm)	Milliequivalents per liter (mEq/l)	Millimoles per liter (mM/l)
HCO_3^-	58.4	0.958	0.958
SO_4^{2-}	11.2	0.233	0.117
Cl^-	7.8	0.220	0.220
NO_3^-	1.0	0.017	0.017
Total anions	78.4	1.428	1.312
Ca^{2+}	15.0	0.750	0.375
Mg^{2+}	4.1	0.342	0.171
Na^+	6.3	0.274	0.274
K^+	2.3	0.059	0.059
Total cations	27.7	1.425	0.879
Fe	0.7	—	0.013
SiO_2	13.1	—	0.218
Grand total	119.9	—	2.422

Sources: Estimated composition from Livingstone (1963, G41).
Also see Berner and Berner (1987, Table 5.6, 190–191.)

The values of individual solubility products can be obtained either by direct experiment or by calculations using experimentally determined free-energy values for the substances involved. Published values of solubility products often are in disagreement with each other because of experimental difficulties. A representative list of mineral solubility products for pure water is given in Table 4-2. It is extremely difficult to apply these values to natural solutions (for instance, to determine the extent of saturation of a compound such as $CaCO_3$), since to be precise it is necessary to know the chemical form (single ions, complex ions, or neutral molecules) and activity of all ionic species in a given solution. Thus most geochemical solubility calculations are only approximations and not nearly as accurate as we would like. Usually, estimated solubilities for natural solutions are less than true solubilities.

Suppose that we want to use the expression for the solubility of fluorite (CaF_2) in pure water. This can be written as

$$\begin{gathered} CaF_2 \rightleftharpoons Ca^{2+} + 2F^- \\ [Ca^{2+}][F^-]^2 = K_{sp} = [Ca^{2+}][2\,Ca^{2+}]^2 \end{gathered} \tag{4-4}$$

Students often ask why the number 2 occurs twice in the last part of this expression. First, 1 mole of CaF_2 will dissolve to give 2 moles of F^-. So the activity of F^- will be twice that of Ca^{2+}. But why does this value of 2 Ca^{2+} have to be squared? Let us label the Ca^{2+} ions as *A*, one-half of the F^- ions (which would together have the same activity as *A*) as *B*, and the other half of the F^- ions as *C* (again with the same activity as *A*). If we think of *A*, *B*, and *C* as red balls (*A*), black

TABLE 4-2 Mineral Solubility Products (25°C)

Mineral	K_{sp}
NaCl (halite)*	38 or $10^{1.58}$
$CaCO_3$ (calcite)	$10^{-8.35}$
$CaCO_3$ (aragonite)	$10^{-8.22}$
$FeCO_3$ (siderite)	$10^{-10.7}$
$CaSO_4$ (anhydrite)	$10^{-4.5}$
$CaSO_4 \cdot 2\,H_2O$ (gypsum)	$10^{-4.6}$
CaF_2 (fluorite)	$10^{-10.4}$
PbS (galena)	$10^{-27.5}$
ZnS (sphalerite)	$10^{-24.7}$
$Cu_2(OH)_2CO_3$ (malachite)	$10^{-33.8}$
$CaMg(CO_3)_2$ (dolomite)	$10^{-16.7}$
$Fe(OH)_3$ (amorphous)	$10^{-38.7}$
$MgCO_3$ (magnesite)	$10^{-7.50}$

Note: Solubility product values from different sources often disagree significantly because of experimental difficulties and because of differences in the physical state of the compounds being studied. Solubility products calculated from free-energy values also usually show differences from direct experimental values.

*The halite value has been calculated from free-energy data.

Sources: The values for dolomite and ferric hydroxide are from Stumm and Morgan (1981, 241–243). The other values are from Krauskopf (1979, 552–553).

balls (B), and green balls (C) in a box (containing large and equal amounts of each), what is the probability of taking three balls (one at a time) out of the box and forming the combination ABC (i.e., CaF_2)? The chances of getting a red ball on the first draw are one in three. The probability that the next ball will be a black ball is one in three and the probability of getting a green ball on the last draw is also one in three. (We do not have to get the three balls in any particular order.) Thus the probability of getting the combination ABC is

$$P = \left(\frac{1}{3}\right)\left(\frac{1}{3}\right)\left(\frac{1}{3}\right) = \frac{1}{27} \tag{4-5}$$

(from the multiplication theorem for probabilities), or, expressed as concentrations of A, B, and C,

$$P = (\text{conc. } A)(\text{conc. } B)(\text{conc. } C) \tag{4-6}$$

Since B and C both represent fluoride ion,

$$P = (\text{conc. } Ca^{2+})(\text{conc. } F^-)^2 \tag{4-7}$$

The reaction

$$Ca^{2+} + 2F^- \rightleftharpoons CaF_2 \tag{4-8}$$

can be viewed as depending on the probability of the three ions involved getting together to form CaF_2. This in turn depends upon their activities (concentrations), and it is not surprising that the expression for the solubility product is found *experimentally* to be

$$K_{sp} = [Ca^{2+}][F^-]^2 = [Ca^{2+}][2\ Ca^{2+}]^2 = 4[Ca^{2+}]^3 \tag{4-9}$$

Note that the solubility of fluorite is equal to the activity of calcium ion, but is equal only to one half the fluoride ion activity.

The solubility of an ionic salt depends only on the solubility product if no other similar ions are already in the solution. If we want to dissolve AgCl in a solution already containing some silver ion, the solubility of the AgCl will be less than the solubility in pure water. Once the solution becomes saturated by adding AgCl to it, the activities of the ions will be

$$[Ag^+_{total}] = [Ag^+_{initial}] + [Ag^+_{added}] \tag{4-10}$$

$$[Cl^-_{total}] = [Cl^-_{added}] \tag{4-11}$$

and thus

$$K_{sp} = [Ag^+_{initial} + Ag^+_{added}][Cl^-_{added}] \tag{4-12}$$

The solubility of the AgCl in such a solution will be equal to the activity of the chloride ion and will be less than the solubility of AgCl in pure water. This decrease in solubility is said to be due to the *common-ion effect.* An example of the common-ion effect would be the precipitation of various compounds downstream from the point where two streams of contrasting chemical composition come together. Precipitates of aluminous hydroxides have been observed, for example, below the confluence of two streams with very different hydroxide ion activities (Theobald et al. 1963).

Although the solubility product of a particular compound is a constant for all solutions, no matter what their composition, this constant value applies only to a specific temperature. For most compounds the value of the solubility product (and thus the solubility) increases with temperature. A few compounds, such as calcite and gypsum, have a lower solubility at higher temperatures. One way of predicting the effect of temperature is to determine whether the solubility reaction is exothermic (heat given off) or endothermic (heat absorbed). An increase in temperature will increase K_{sp} and the solubility (drive the reaction further to the right) if the reaction is endothermic, and will decrease K_{sp} and the solubility (drive the reaction further to the left) if the reaction is exothermic. This is a good example of the application of Le Chatelier's principle, named for the French chemist Henry Louis Le Chatelier (1850–1936). This principle can be stated as follows: changing the conditions of a system will result in reactions that tend to restore the original conditions. Thus, when we add heat to a solution by raising the temperature, an endothermic reaction will go further to completion in an attempt to absorb the additional heat to keep the temperature from increasing. The ΔH value for the reaction

$$NaCl \rightleftharpoons Na^+ + Cl^- \tag{4-13}$$

is positive (endothermic reaction), and thus the solubility of halite is greater at higher temperatures.

We have to qualify further our use of solubility products when we are dealing with compounds that, in dissolving, react with water (a *hydrolysis reaction*) or that form complex ions in solutions containing other ions. In other words, additional equilibria besides the specific solubility equilibrium may be involved. Solubility of compounds in natural waters thus depends on many other equilibrium constants besides the solubility products. The chemical weathering of silicate minerals involves hydrolysis reactions, and their solubilities cannot be determined from simple application of solubility products. As an example, the hydrolysis of enstatite can be represented by

$$MgSiO_3 \text{ (enstatite)} + 3\,H_2O \rightleftharpoons Mg^{2+} + H_4SiO_4 + 2\,OH^- \quad (4\text{-}14)$$

Reactions of this type depend on the pH (H^+ activity) of the solution involved. In equation (4-14), the SiO_3 "anion" is shown combining with hydrogen ion. Other hydrolysis reactions involve combination of the cation of the solid with hydroxide ion (an example is $MgOH^+$). Anion hydrolysis, which produces additional hydroxide ion, is usually more extensive in acid (high H^+ activity) solutions, and cation hydrolysis, which produces additional hydrogen ion, is usually more extensive in basic solutions (low H^+ activity). Both types of hydrolysis tend to decrease the activity of the simple ions of the dissolving solid, and thus they increase the solubility of the solid over that calculated from the solubility product alone. Further discussion of hydrolysis and of the weathering of silicate minerals can be found in Chapter Seven.

Another process that can increase the solubility of a solid is the formation of one or more complex ions by the solid's cation and/or anion. A *complex ion* can be defined as an ion containing a central metal cation bonded to one or more molecules or ions. Examples are $Cu(H_2O)_6^{2+}$ and $HgCl_4^{2-}$. Some dissolved species that otherwise fit this definition do not have a charge and, strictly speaking, should be labeled *complexes* rather than complex ions (e.g., $CuSO_4^0$).

Complexes and complex ions play an important role in the solution and precipitation of minerals during weathering and the formation of sedimentary rocks, and are also believed to be important in the transport of metal ions in hydrothermal solutions. The formation of complex ions tends to decrease the activity of the simple ions of a dissolving solid, and thus increases the solubility of the solid over that which would be found if no complex-ion formation had occurred. For example, if the mineral argentite (Ag_2S) were to dissolve in a hydrothermal solution with a high concentration of chloride ion, silver in the solution might occur as simple silver ions (Ag^+), as uncharged silver chloride molecules (AgCl), and as complex ions such as $AgCl_2^-$ and $AgCl_3^{2-}$. The total solubility of Ag_2S in such a solution would be

$$\text{solubility} = [Ag^+] + [AgCl] + [AgCl_2^-] + [AgCl_3^{2-}] \quad (4\text{-}15)$$

Since the extent of hydrolysis reactions in natural waters is not well known, and since many complex ions probably exist in natural solutions, it is easily seen why an earlier statement was made to the effect that estimated solubilities (particularly those calculated from solubility products) are usually less than true solubilities. A further complication occurs in solutions that have relatively high concentrations of solutes, such as seawater. In these solutions, activities of the various species differ from their concentrations and must be estimated (assuming the identity of the species can be determined), introducing additional uncertainties in our understanding of solution behavior (Figure 4-2).

Let us apply the above concepts to the evaporation of seawater. When seawater evaporates, the measured concentrations of the various ions increase in value. As mentioned above, because

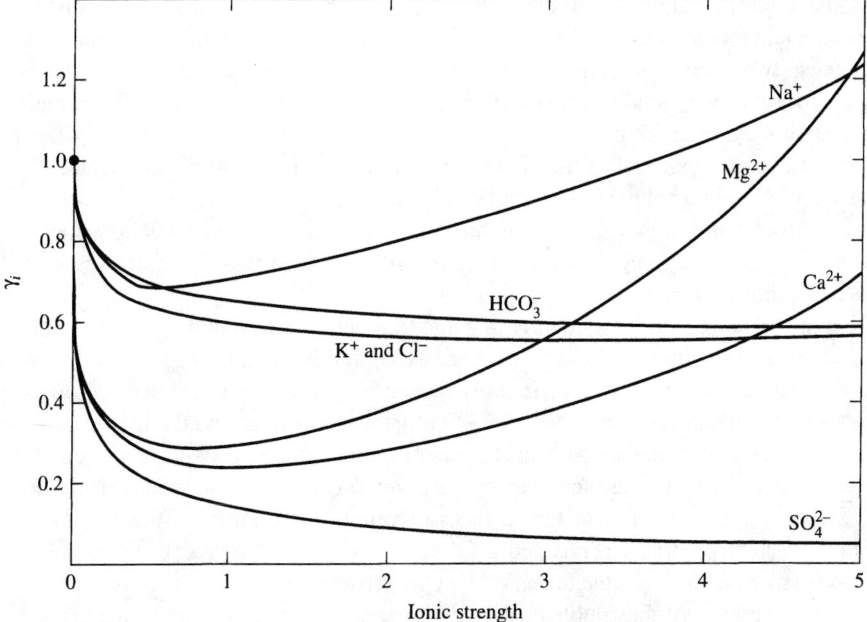

Figure 4-2 Activity coefficients for the common ions in natural waters up to 5 molal ionic strengths. Positive ions tend to go through a minimum and then increase markedly at high ionic strengths, whereas negative ions tend to be constant at high ionic strengths. Divalent ions show more deviation from ideality than univalent ions. Ionic strength $I = 1/2\ \Sigma m_i z_i^2$, where m_i = moles per kilogram of solvent for component i and z_i is the charge on ion i. When the activity coefficient γ of an ion is one, its activity value is equal to its concentration. The further the value of the activity coefficient is from one, the greater the difference between the activity (effective concentration) and the actual concentration. See discussion of activities in Chapter Three. (After D. K. Nordstrom and J. L. Munoz. *Geochemical Thermodynamics, Guide to Problems*, p. 43. Copyright © 1987. Adapted by permission of Blackwell Science, Ltd., Oxford, England.)

of the number and abundance of dissolved ions in seawater, the activities of the ions are not equal to their concentrations. Table 3-4 lists concentrations and estimated activities of the major ions in seawater, with the activities calculated by taking into account the effects of complex-ion formation and of interactions among ions. In Chapter Three we used these values to show that seawater is approximately saturated with calcium carbonate. What about the degree of saturation of the evaporite minerals, gypsum and halite? Using the activities in Table 3-4, we find the following (the activity of water is assumed equal to 1):

Mineral	K_{sp}	Calculated activity product
Halite	38	0.13
Gypsum	2.5×10^{-5}	4.5×10^{-6}

The values of the activity products are less than the K_{sp} values, and thus seawater is undersaturated with respect to both minerals. But halite is much further from saturation than is gypsum. This is why gypsum precipitates before halite from evaporating seawater.

We have implied in our discussion so far that a solute is either dissolved uniformly in a solution as ions and molecules or else it exists as a solid mass. However, there is a transitional state between these two conditions (Figure 4-3). Some materials, such as organic compounds in natural waters, occur as very small particles (diameters less than 10 micrometers) that remain in suspension indefinitely. Such particles are referred to as *colloidal particles,* and they grade into the atomic and molecular material of true solutions and into the larger suspended material that can physically settle out of a solution. A colloid can be defined as a stable electrostatic suspension of very small particles in a liquid. Chemical changes occurring at colloid particle surfaces, colloid formation, and colloid aggregation (flocculation) all affect the dissolved and suspended constituents of natural waters.

We can still talk about the solubility of a material that forms colloidal particles, but this solubility is controlled by the large number of processes that can affect colloid stability. Once a colloid has formed (as a result of weathering, organic activity, etc.), its continued stability depends on many factors. One factor is the frequency of collision of particles and the interaction of repulsive and attractive forces that accompanies collision. For example, clay particles generally have a net negative charge on their surfaces due to loss of interlayer ions or to lattice substitutions such as Al^{3+} for Si^{4+}. As a result, near the surfaces of water-borne particles, there is an excess of cations in the aqueous phase. The charged solid surface and adjacent layer of oppositely charged solution ions is known as a *double layer.* When two particles come close together, repulsion occurs due to the presence of the double layer and attraction occurs as a result of the van der Waals force (a weak electrostatic force that acts between molecules). The overall balance between attraction and repulsion determines whether or not particles come together and aggregate into large particles (flocculation).

This balance of forces is affected by various physical and chemical changes. One important control factor is the total ionic strength of the solution (polyvalent ions have a greater effect than monovalent ions). When river water flows into an estuary, total ionic strength increases (due to

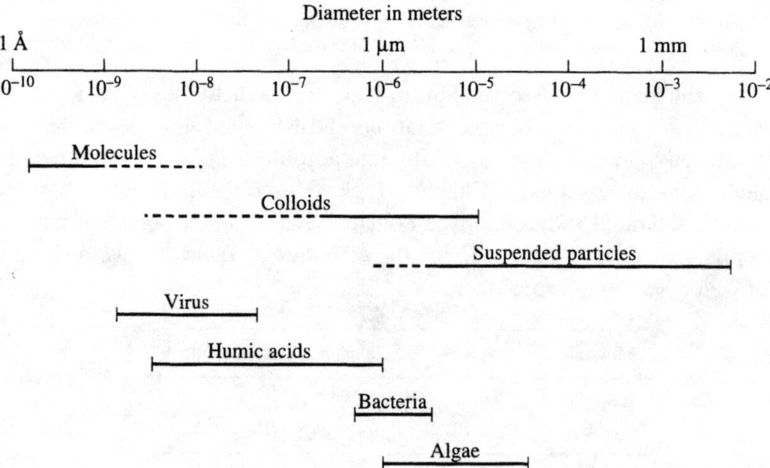

Figure 4-3 Size spectrum of water-borne particles. (After W. Stumm and J. J. Morgan. *Aquatic Chemistry*, 2d ed, p. 647. Copyright © 1981 by John Wiley & Sons. Reprinted by permission of John Wiley & Sons, Inc.)

mixing with seawater), the solution ion layer next to the surfaces of colloidal particles becomes thinner, and the result is flocculation as the attractive force exceeds the repulsive force. Thus one of the causes of high sediment deposition in estuaries is the flocculation of clay and iron oxide suspensions.

Another factor controlling stability of colloids involves chemical changes at the surface of colloid particles. For example, adsorption of ions by the particles can increase or decrease the surface charge and thus alter the repulsive force. Metallic ions and organic material often react chemically or are adsorbed at colloid surfaces. As with most chemical changes in natural waters, pH is a major factor in determining whether or not a given reaction takes place. However, no simple rules or equations can be given to explain colloid stability. They are an important factor in natural water chemistry and need to be better understood.

We close our discussion of solubility with a consideration of the solubility of gases in water. We can write an equation for the equilibrium between carbon dioxide gas of the atmosphere and dissolved carbon dioxide in river water as

$$CO_2\,(g) \rightleftharpoons CO_2\,(aq) \tag{4-16}$$

If we use $[CO_2\,(aq)]$ to represent the activity of aqueous CO_2 (it also represents the solubility of CO_2 when we have equilibrium), and $[CO_2\,(g)]$ to represent the activity of gaseous CO_2, then

$$K_1 = \frac{[CO_2\,(aq)]}{[CO_2\,(g)]} \tag{4-17}$$

[We could discuss the gaseous CO_2 in terms of its fugacity (ideal pressure) rather than in terms of its activity. However there would be no difference in the conclusions reached. See Chapter Three for a discussion of the relationship of fugacity to activity.] It can be shown that the activity of a gas at the temperature and pressures existing at the Earth's surface is essentially equal to its partial pressure. Substituting a pressure value for a gas activity is analogous to substituting a concentration value for a solute activity. Thus the equilibrium for the above reaction can be expressed as

$$K_2 = \frac{(CO_2\,(aq))}{P_{CO_2}} \tag{4-18}$$

Note that K_1 and K_2 are two different kinds of equilibrium constants in the sense that K_1 is dimensionless and K_2 would be expressed as moles per liter-atmosphere.

The expression for K_2 is one form of Henry's law, which states that, at a given temperature, the concentration of a species in a dilute solution is proportional to the partial pressure of that species in a gas phase in equilibrium with the solution. This is a limiting law, which becomes valid as the concentration and pressure approach zero. In applying Henry's law it should be pointed out that (1) we must use the partial pressure of the species in the gas phase and not the total pressure of the gas phase, and (2) only the amount of the species in the solution in the form of uncombined molecules or hydrated with water should be considered. For example, when studying the distribution of CO_2 between the atmosphere and river water, we would use 0.0003 atmosphere for the partial pressure of $CO_2\,(g)$. (The *partial pressure* of a gas species is equal to total gas pressure times the mole fraction of the species; the atmosphere contains 0.03 mole percent CO_2.) For the value of $[CO_2\,(aq)]$ in the Henry's law equation, we would use total dissolved CO_2 [equal to the concentration of CO_2 (aq) plus the concentration of H_2CO_3 (aq), since some of the dissolved carbon dioxide reacts with water to form H_2CO_3 (aq)].* The amount of CO_2 that can be dissolved in natural surface waters at 25°C is approximately 10^{-5} mole per liter, and thus the value of K_2 is 0.033 mole per liter-atmosphere at this temperature.

pH

In our earlier discussion of silicate solubilities it was mentioned that pH is another term for hydrogen ion activity. The original definition of pH (in 1909) was in terms of hydrogen ion concentration:

$$\text{pH} = -\log\,(\text{H}^+) \qquad (4\text{-}19)$$

Use of this term allowed reference to hydrogen ion concentration in terms of common, positive numbers (Figure 4-4). The modern definition is in terms of hydrogen ion activity:

$$\text{pH} = -\log\,[\text{H}^+] \qquad (4\text{-}20)$$

Instruments used to determine the pH of a solution actually measure hydrogen ion activity rather than hydrogen ion concentration.† In simple, dilute solutions there is essentially no difference between hydrogen ion activity and hydrogen ion concentration. In other words, the *effective hydrogen ion concentration* is very similar to the actual hydrogen ion concentration.

A hydrogen ion is nothing more than a proton. Rather than occurring by itself, it attaches itself to one or more water molecules, forming an ion such as the hydronium ion, H_3O^+. However, for convenience, we usually discuss the proton as though it occurred in the form H^+. Hydroxide ions, OH^-, also occur as hydrated ions, but, like H^+, they are written as though they were not hydrated. The ionization of water can thus be written

$$\text{H}_2\text{O} \rightleftharpoons \text{H}^+ + \text{OH}^- \qquad (4\text{-}21)$$

The equilibrium constant for this reaction at 25°C is

$$K = [\text{H}^+][\text{OH}^-] = 10^{-14} \qquad (4\text{-}22)$$

For pure water, $[H^+] = [OH^-]$ and pH = 7. Any solution with pH = 7 is by definition a neutral solution. No matter what other solutes occur in a given solution, the product of the hydrogen and hydroxide ion activities will always be 10^{-14} at 25°C. Note that the value of this equilibrium constant changes with temperature, as do all equilibrium constants. For this reaction at 230°C, $K = 10^{-11.4}$ and a neutral solution would have a pH of 5.7.

Geologists dealing with natural waters are interested in pH values of these solutions, because most of the reactions (organic and inorganic) that take place in such waters involve hydrogen ion or are related in some way to hydrogen ion abundance. Hydrogen ion is important even though it is often not as abundant as several other ions in a given solution. For example, river water with a pH of 7 has a hydrogen ion concentration of 10^{-7} mole per liter. Analyses of various river waters show that several ions occur commonly with concentrations significantly greater than 10^{-7} mole per liter (Table 4-1). The overall pH of a natural water is controlled by the various possible

*It is customary to represent CO_2 (aq) by the formula H_2CO_3 (aq), even though the actual concentration of CO_2 molecules in a given solution is much greater than the concentration of H_2CO_3 molecules.

†A typical pH meter is an electrochemical cell that uses two electrodes to measure the electromotive force of a solution. The electrodes are designed so that the voltage of the cell is a function only of the hydrogen ion activity. See the discussion of electrochemical cells in the next section.

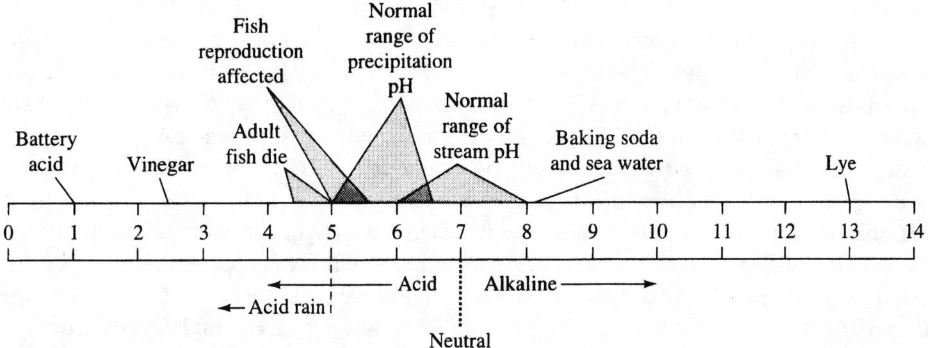

Figure 4-4 pH scale. The normal range of precipitation pH is from about 5.0 to 6.5. The pH of pure water in equilibrium with atmospheric carbon dioxide is 5.6. Normal rain is slightly acidic because of carbon dioxide (which forms the weak acid carbonic acid) and natural emissions of sulfur and nitrogen oxides and certain organic acids. Human activities produce more of these compounds and the result is "acid rain." The formation of strong acids such as sulfuric and nitric acids causes the pH of precipitation to be less than 5.0, producing acid rain. Occasional pH readings of well below 2.4, the acidity of vinegar, have been reported for rain and fog in large cities and highly industrialized areas.

reactions involving the major ions, other nonionic solutes, solids, and gases associated with the solution. Most natural solutions are buffered; that is, their pH cannot change greatly because of various reactions that work against such a change. The range of pH values for various natural materials is discussed in a later section of this chapter.

OXIDATION AND REDUCTION (Eh)

When a water solution contains ions, it can serve as an electrical conductor. If we pass a current of electricity through such a solution by placing two electrodes from a battery in the solution, chemical reactions occur at the two electrodes. Electrons enter the solution at the cathode and react with water to produce hydrogen gas and hydroxide ion. A current is carried across the solution by the ions present. At the anode, electrons leave the solution as a result of the breakdown of water molecules to oxygen gas and hydrogen ions. These reactions can occur no matter what kind of solute ions are in solution, and they even occur in pure water, although not very fast or very extensively because of the small number of hydrogen and hydroxide ions present. The reaction at the anode is known as an *oxidation reaction* and that at the cathode as a *reduction reaction.* Oxidation can be defined as the removal of electrons from an atom or atoms and reduction as the gain of electrons by an atom or atoms. (The less general, and older, definition of oxidation is any reaction involving combination with oxygen.)

For a particular chemical reaction, an *oxidizing agent* is any material that takes on electrons and a *reducing agent* is any material that gives up electrons. Consider the reaction in groundwater between ferric iron and hydrogen sulfide produced by decaying organic matter:

$$8\,Fe^{3+} + H_2S + 4\,H_2O \rightleftharpoons 8\,Fe^{2+} + SO_4^{2-} + 10\,H^+ \qquad (4\text{-}23)$$

In this reaction the oxidizing agent is ferric ion and the reducing agent is the hydrogen sulfide. In such reactions an oxidation step is balanced by a related reduction step to maintain electrical neutrality. In the above case the ferric ion is reduced to ferrous ion, and the sulfur of the hydrogen sulfide is oxidized to the higher oxidation number needed to form sulfate ion. The *oxidation number* of an element in a particular form is the electric charge that it appears to have on the basis of reactions in which it takes part. A set of rules has been developed by chemists for assigning oxidation states. The sulfur atom in hydrogen sulfide is assigned an oxidation number of –2, and the sulfur atom in sulfate ion is assigned an oxidation number of +6. Usually, experimental evidence allows assignment of oxidation numbers unambiguously.

If, instead of using a battery to decompose water, we want to construct an electric battery, we can do so by putting electrodes of zinc and copper in a solution of copper sulfate (Figure 4-5). This must be done in such a way that the oxidizing and reducing agents do not come into direct contact, so that electrons from the reducing agent can flow through an external circuit before joining the oxidizing agent. In this case the zinc is the reducing agent, and zinc ions will go into solution at the zinc electrode (the anode). Electrons will flow from the zinc electrode through a wire to the copper electrode (the cathode), where they will react with copper ions to deposit copper on the copper electrode. The total reaction is

$$Zn + Cu^{2+} \rightarrow Cu + Zn^{2+} \tag{4-24}$$

This reaction could be written as two *half-reactions,*

$$Zn \rightarrow Zn^{2+} + 2\,e^- \tag{4-25}$$

and

$$Cu^{2+} + 2\,e^- \rightarrow Cu \tag{4-26}$$

These can be added together to get the total reaction. The voltage produced by our battery can be measured and depends on the relative concentrations of the ions involved and the relative differ-

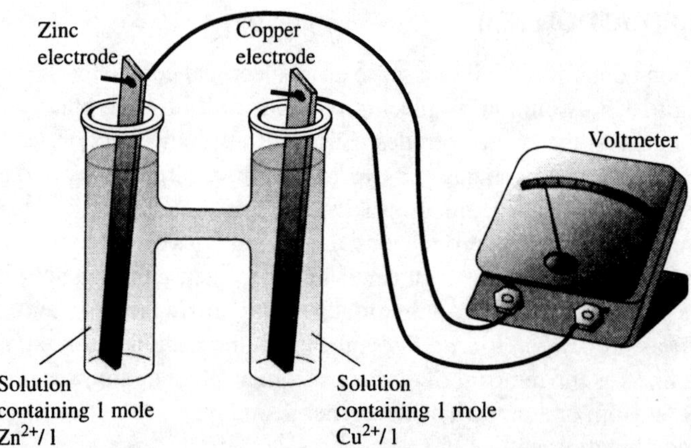

Figure 4-5 Cell representative of the copper-zinc battery. The term "cell" refers to a receptacle containing electrodes and an electrolyte (in this case dissolved copper sulfate), either for generating electricity by chemical action or for use in electrolysis (the electrical decomposition of a substance). In the above cell, sulfate ions migrate toward the zinc electrode (anode) and Zn^{2+} and Cu^{2+} migrate toward the copper electrode (cathode), that is, a current is carried across the solution by the ions. As zinc ions form at the anode, the electrons produced move through the wire to the voltmeter and then to the cathode, where Cu^{2+} is reduced to solid copper. (From *General Chemistry*, Second Edition, by Linus Pauling, p. 256. W. H. Freeman and Company. Copyright © 1953.)

ence between the tendencies of the metals to become ions. A voltage produced in this way is called an *electromotive force.* It can be considered a measure of the tendency of the zinc-copper reaction to take place.

The zinc-copper battery is an example of the fact that an element with a smaller affinity for its electrons (zinc) can give them up to an element (copper) whose affinity for electrons is greater. By carrying out a series of experiments, it is possible to list various materials in order of their relative oxidizing and reducing powers. The process is standardized by specifying for the experiments that all dissolved substances have a concentration of unit activity (essentially 1 mole per liter), and that all gases present also have unit activity (essentially 1 atmosphere pressure). The temperature is fixed at 25°C. The usual procedure in preparing a list of oxidizing and reducing agents is to tabulate half-reactions with oxidizing agents on one side and reducing agents on the other side. For example, the iron-sulfur and copper-zinc reactions previously discussed could be listed in this manner:

Reducing agents	Half-reaction	Oxidizing agents
Strong	$Zn \rightleftharpoons Zn^{2+} + 2\,e^-$	*Weak*
	$H_2S + 4\,H_2O \rightleftharpoons SO_4^{2-} + 10\,H^+ + 8\,e^-$	
	$Cu \rightleftharpoons Cu^{2+} + 2\,e^-$	
Weak	$Fe^{2+} \rightleftharpoons Fe^{3+} + e^-$	*Strong*

This list tells us that solid zinc is a relatively strong reducing agent, which can reduce sulfate ion to hydrogen sulfide, cupric ion to solid copper, and ferric ion to ferrous ion. Similarly, ferric ion is a relatively strong oxidizing agent, which can oxidize solid copper to cupric ion, hydrogen sulfide to sulfate ion, and solid zinc to ionic zinc. On the other hand, the list also tells us that cupric ion cannot oxidize ferrous ion. It should be pointed out that, strictly speaking, this is not quite true. Any reaction, such as

$$2\,Fe^{2+} + Cu^{2+} \rightleftharpoons Cu + 2\,Fe^{3+} \tag{4-27}$$

goes in both directions. What experimental work (and thus the list) tells us is that the above reaction goes far to the left. At equilibrium, the inverse of the activity product will be a large number. This equilibrium state could be approached from either side of the equation; that is, we could start out with ferrous ion, cupric ion, and no solid copper. In this case the reaction would go to the right until equilibrium was reached.

The standard reference half-reaction used in listing oxidation-reduction pairs is the hydrogen half-reaction

$$H_2 \rightleftharpoons 2\,H^+ + 2\,e^- \tag{4-28}$$

This is assigned a *standard potential* (E^0) of 0.00 volt. The values of E^0 for the other half-reactions represent the electromotive force between electrodes when the hydrogen half-reaction is combined with a given half-reaction. Thus the E^0 value for the half-reaction

$$Zn \rightleftharpoons Zn^{2+} + 2\,e^- \tag{4-25}$$

is determined by using a zinc electrode in a solution of zinc sulfate and a platinum electrode in a solution of sulfuric acid. (Hydrogen gas dissolves sufficiently in the platinum to allow the hydrogen half-reaction to occur.) It is found that solid zinc will reduce hydrogen ion, producing hydrogen gas. At the same time, zinc ion is produced by oxidation of the solid zinc. The measured potential (E^0) is 0.76 volt. The sign convention adopted here is such that this value is given a negative sign. If we do the same thing with the copper half-reaction, we find that cupric ion is reduced to solid copper by hydrogen gas, which is oxidized to hydrogen ion. Thus the current flows in the opposite direction to that found for the zinc half-reaction, and the measured potential (0.34 volt) is therefore given a positive sign. All potentials for half-reactions that cause hydrogen gas to be oxidized to hydrogen ion have positive values, and all half-reactions that produce hydrogen gas by reduction of hydrogen have negative values. (The reader is warned that some textbooks reverse these signs, giving strong reducing agents positive potentials and strong oxidizing agents negative potentials.) The potential values are referred to as electrode potentials, as oxidation potentials, or as oxidation-reduction potentials.

A list of selected half-reactions is given in Table 4-3. The utility of this list lies in the fact that the potential for any given reaction can be calculated (without doing an experiment) by subtracting one half-reaction from another. For example, to get the potential for the zinc-copper reaction, we subtract the copper half-reaction from the zinc half-reaction:

$$\begin{array}{lr} Zn \rightleftharpoons Zn^{2+} + 2\,e^- & -0.76 \text{ volt} \\ Cu \rightleftharpoons Cu^{2+} + 2\,e^- & +0.34 \text{ volt} \\ \hline Zn + Cu^{2+} \rightleftharpoons Zn^{2+} + Cu & -1.10 \text{ volts} \end{array}$$

We know from experiment that this reaction goes far to the right, and the result of the arbitrary assignment of positive and negative values to half-reaction potentials is that any reaction with a negative potential will go to the right. Any reaction with a positive potential will go to the left. When combining two half-reactions, the number of electrons must balance off; thus one or both half-reactions may have to be multiplied by a coefficient. The potential values are *not* multiplied since they represent standard conditions and do not change with the way an equation is written.

TABLE 4-3 Electrode Potentials for Selected Half-reactions

Half-reaction	E^0
$Zn(s) \rightleftharpoons Zn^{2+} + 2\,e^-$	−0.76
$H_2S\ (g) + 4\,H_2O \rightleftharpoons SO_4^{2-} + 10\,H^+ + 8\,e^-$	−0.34
$H_2\ (g) \rightleftharpoons 2\,H^+ + 2\,e^-$	0.00
$Cu^+ \rightleftharpoons Cu^{2+} + e^-$	+0.16
$FeCO_3$ (siderite) $+ H_2O \rightleftharpoons Fe_3O_4$ (magnetite) $+ 3\,CO_2\ (g) + 2\,H^+ + 2\,e^-$	+0.32
$Cu\ (s) \rightleftharpoons Cu^{2+} + 2\,e^-$	+0.34
$Fe^{2+} \rightleftharpoons Fe^{3+} + e^-$	+0.77
$Fe^{2+} + 3\,H_2O \rightleftharpoons Fe(OH)_3\ (s) + 3\,H^+ + e^-$	+1.06
$FeCO_3$ (siderite) $+ 3\,H_2O \rightleftharpoons Fe(OH)_3\ (s) + HCO_3^- + 2\,H^+ + e^-$	+1.08
$2\,H_2O \rightleftharpoons O_2(g) + 4\,H^+ + 4\,e^-$	+1.23
$Mn^{2+} \rightleftharpoons Mn^{3+} + e^-$	+1.51

Note: Values are for 25°C, 1 atmosphere, and unit activity of substances.

Sources: E^0 values for the two reactions involving siderite were calculated from free-energy data. The other E^0 values are measured values from Krauskopf (1979, App. 9) and values rearranged from Stumm and Morgan (1981, Tables 7-4 and 7-5, 447–449).

You will recall from Chapter Three that the sign of the Gibbs free-energy change (ΔG) of a reaction can be used to tell the direction in which the reaction will go, whereas the magnitude of ΔG indicates how far from equilibrium the reaction is under given conditions. This is just as true for oxidation-reduction reactions as it is for other kinds of reactions. Thus there should be a relationship between the standard potential of an oxidation-reduction reaction and the standard Gibbs free-energy change of the reaction. Furthermore, there should be a relationship between the standard potential and the equilibrium constant of a reaction, just as there is between the standard Gibbs free-energy change and the equilibrium constant.

Let us consider the electrical work done when a quantity of electricity Q (expressed in coulomb units) is transferred from an electrode of potential V_1 to an electrode of potential V_2, where the electromotive force E (expressed as volts) is equal to $V_2 - V_1$. We know from the principles of electricity that the amount of electrical work done is

$$W = (V_2 - V_1)Q = EQ \quad \text{(volt-coulombs or joules)} \tag{4-29}$$

A fundamental law of electrochemistry states that the passage of electricity at an electrode involves a chemical change that can be expressed in terms of chemical equivalents (defined earlier in this chapter), and that the number of chemical equivalents (N) for a given change is directly proportional to the amount of electricity passing through the electrode. If we denote by F the proportionality constant, the law (known as Faraday's law) can be expressed as

$$Q = NF \tag{4-30}$$

The constant F is unaffected by changes in conditions such as temperature or concentration. It represents the quantity of electricity that will produce a chemical change involving one equivalent. One equivalent in this case is the weight of a substance that is capable of oxidizing one atomic weight of hydrogen, or that has the same reducing power as one atomic weight of hydrogen. The number N can be thought of as the number of moles of electrons that move through an electrode. The value of F (known as a faraday) has been found to be 96,500 coulombs per mole of singly charged ion. A faraday represents the charge of 1 mole of electrons. The quantity of electricity flowing through an electrode in a given reaction is equal to NF coulombs [from equation (4-30)]; thus, using equation (4-29), the electrical work done in the reaction is

$$W = NEF \text{ volt-coulombs} \tag{4-31}$$

If we want to determine the Gibbs free-energy change for an oxidation-reduction reaction in an electrochemical cell, we measure the potential of the cell under standard conditions (including a constant temperature and pressure). The Helmholtz free-energy change (ΔA) was expressed in Chapter Three as

$$\Delta A = -S\,\Delta T - P\,\Delta V \tag{3-13}$$

When the temperature is constant, it can be shown that

$$\Delta A^0 = W = \text{work done for any change of state} \tag{4-32}$$

If our oxidation-reduction reaction involves a gas, such as the reaction

$$\tfrac{1}{2}\,H_2\,(g) + AgCl\,(s) \rightleftharpoons Ag\,(s) + HCl\,(aq) \tag{4-33}$$

then the work done is of two kinds, electrical work and work involving a volume change. Thus

$$\Delta A^0 = NE^0F - P\,\Delta V \tag{4-34}$$

This can be rewritten as

$$NE^0F = \Delta A^0 + P\,\Delta V = \Delta G^0 \text{ (constant pressure and temperature)} \tag{4-35}$$

Equation (4-35) gives us our relationship between Gibbs free-energy change and the potential of an oxidation-reduction reaction:

$$\Delta G^0 = NE^0F \tag{4-36}$$

If the conditions of the reaction are not those defined as standard conditions, then

$$\Delta G = NEF \tag{4-37}$$

(If ΔG is to be given in calories, then F has to be expressed in calories per volt rather than in coulombs. When F is given in coulombs, ΔG is expressed in volt-coulombs, or joules. See Table 3-1.) Finally, since from Chapter Three we know that

$$\Delta G^0 = -RT\ln K \tag{3-18}$$

the relationship of electrode potential to the equilibrium constant of a reaction is

$$NE^0F = -RT\ln K \tag{4-38}$$

Also given in Chapter Three was the expression for the Gibbs free-energy change under non-standard conditions,

$$\Delta G = \Delta G^0 + RT\ \ln\frac{[C][D]}{[A][B]} \tag{3-17}$$

for the reaction

$$A + B \rightleftharpoons C + D$$

In this case,

$$NEF = \Delta G^0 + RT\ \ln\frac{[C][D]}{[A][B]} \tag{4-39}$$

Having developed the principles of oxidation-reduction reactions, how can we apply them in nature? Since the activities of ions in natural solutions are usually not the standard values that can be imposed in the laboratory, what we are concerned with is E rather than E^0. The key equation is (4-39), which can be combined with equation (4-36) and written as

$$E = E^0 + \frac{RT}{NF}\ln\frac{[C][D]}{[A][B]} \tag{4-40}$$

This is called the *Nernst equation.* Since R (1.987 calories per degree-mole) and F (23,060 calories per volt) are constants, and setting the temperature at 25°C (298.15 K), we can use the equation in the form

$$E = E^0 + \frac{0.0257}{N}\ln\frac{[C][D]}{[A][B]} \tag{4-41}$$

or, changing from natural log to common log form by using the factor 2.303,

$$E = E^0 + \frac{0.059}{N}\log\frac{[C][D]}{[A][B]} \tag{4-42}$$

It is possible to measure the potential of a natural solution with respect to the standard hydrogen half-cell (the method of doing this is discussed later in this section); when this is done, the symbol Eh is used for the measured potential (in place of E in the above equations). This potential is known as the *oxidation potential* (the term we shall use here) or as the *redox potential.* The oxidation potential reflects a balancing of all the different possible reactions in a given solution (metastable equilibrium is assumed to exist in the solution, although this is not completely true).* It is a measure of the tendency of the solution to allow any particular reaction of interest to take place. (The term oxidation potential and the symbol Eh are also used for the potential of individual half-reactions under nonstandard conditions.)†

For example, suppose that we are interested in whether the copper dissolved in a stream is mainly in the form of Cu^+ or Cu^{2+}. The standard potential for the reaction

$$Cu^+ \rightleftharpoons Cu^{2+} + e^- \tag{4-43}$$

is +0.16 volt (Table 4-3). Assume that the measured potential of the stream is +0.52 volt. This is a more oxidizing potential than that of the copper half-reaction (Table 4-3). Therefore, we would expect Cu^{2+} to be the dominant ion. We can even determine the ratio Cu^{2+}/Cu^+ by applying equation (4-42) to the half-cell reaction as follows:

$$+0.52 = +0.16 + \frac{0.059}{1} \log \frac{[Cu^{2+}]}{[Cu^+]}$$

$$\log \frac{[Cu^{2+}]}{[Cu^+]} = \frac{0.36}{0.059} = 6.1$$

$$\frac{[Cu^{2+}]}{[Cu^+]} = 10^{6.1} \approx 1,259,000$$

Thus there are over a million times as many Cu^{2+} ions in our stream as there are Cu^+ ions. It should be pointed out that copper ions are most abundant in acid solutions, and that in basic solutions copper tends to precipitate out.

In doing the above calculation we are actually applying equation (4-42) to the reaction

$$H^+ + Cu^+ \rightleftharpoons Cu^{2+} + \tfrac{1}{2} H_2\,(g) \tag{4-44}$$

*The composition of a natural solution with Eh $\neq 0$ is not the same as it would be in the absence of a measured potential. The existence of an electrode potential is due to the fact that natural solutions are poor electronic conductors. Thus there can be a state of *metastable* equilibrium determined by the potential difference. True equilibrium between a solution and a standard hydrogen electrode occurs only when Eh $= 0$ and $\Delta G = 0$ at constant pressure and temperature. When Eh $= E^0$, this is not a statement of equilibrium, but only means that all products and all reactants are at unit activity.

†Some water chemists prefer to discuss oxidation-reduction equilibria in terms of pe instead of Eh, where

$$pe = -\log[e^-]$$

This is similar to the definition of pH and allows the use of numbers instead of voltages to express the oxidation-reduction environment (electron activity) of a solution. It can be shown that

$$pe = \frac{F}{2.3RT}\,Eh = \frac{Eh}{0.059\ V} \quad (25^\circ C \text{ and } 1 \text{ atm})$$

For a discussion of pe calculations and the use of pe in studying oxidation-reduction equilibria, see Chapter 7 of Stumm and Morgan (1981) and Thorstenson (1984). Since the pH of natural solutions is determined by measuring a voltage (as is Eh), we could express the hydrogen ion activity of a solution as a voltage rather than as a pH number. Most geologists express hydrogen ion activity as a number (pH) and electron activity as a voltage (Eh).

The E^0 value of +0.16 volt is the difference in potential between the copper half-cell under standard conditions and the standard hydrogen half-cell; the measured E value of +0.52 volt can be considered the difference between the copper half-cell under conditions existing in the stream and the standard hydrogen half-cell. Applying equation (4-42) to the full reaction,

$$+0.52 = +0.16 + \frac{0.059}{1} \log \frac{[Cu^{2+}][H_2]^{1/2}}{[Cu^+][H^+]}$$

Since the activities of hydrogen gas and hydrogen ion are set at 1 for standard conditions, the equation can be written in the form

$$+0.52 = +0.16 + \frac{0.059}{1} \log \frac{[Cu^{2+}]}{[Cu^+]}$$

as was done above. Thus the usual procedure is to apply equation (4-42) to the half-reaction of interest, rather than to the full reaction as such.

The oxidation potential of natural material can be measured by immersion of two electrodes into the material. One electrode is a reference electrode that is chemically maintained at a known voltage relative to a standard hydrogen electrode. The other electrode is an inert electrode usually made of platinum. This electrode will provide electrons to, or receive electrons from, the material being tested, depending on the relationship of the material's potential to the fixed electromotive force of the reference electrode. The overall electromotive force (potential) that is measured is equal to the difference between the potential of the material being tested and the reference electrode potential. This can then be used to determine the potential of the material relative to the standard hydrogen half-reaction. If the material contains a large amount of one or more reducing agents, a large, negative Eh value will be obtained. Similarly, if the material contains a large amount of one or more oxidizing agents, a large, positive Eh value will be obtained. Measurements of Eh have been made on soils and sediments as well as on natural water solutions. Soils and sediments are either measured directly (if they contain water) or a sample of the material is suspended in water. In either case, what is being measured is the potential of a water solution, which may or may not be directly related to the composition of the soil or sediment.

There are a number of problems in measuring Eh in natural materials. The equipment is sensitive, particularly to temperature changes. Inserting the electrodes into a material can change the chemical environment, particularly if air is introduced to material normally out of contact with the atmosphere. Contamination of electrodes can occur. In many cases the material being measured is not in internal (metastable) equilibrium. Some geochemists feel that calculating Eh values for individual oxidation-reduction couples is more practical than trying to measure an overall, meaningful Eh value for a natural water (Thorstenson 1984).* The main point here is that Eh measurements must be done carefully and must be evaluated carefully. They can be very useful provided the limitations of such measurements are understood and taken into account. A detailed discussion of pH and Eh measurement is given by Garrels and Christ (1965) and by Stumm and Morgan (1981). The use of special types of electrodes to measure individual ion activities is also briefly discussed in these two textbooks.

*Determination of the relative concentrations of the oxidized and reduced species in a couple allows an estimate of Eh for the couple. Agreement of Eh values from several couples in a multicomponent system suggests an approach to equilibrium, while lack of agreement suggests disequilibrium. In the latter case, an overall Eh value has no meaning.

It is important to note that Eh measurements in waters exposed to the atmosphere give lower values than would be expected for oxygen content of the waters. Apparently, the potential expected for oxygen reactions is not reached, and the oxygen acts as a relatively weak oxidizing agent. If we keep this fact in mind, along with the other problems mentioned above, we can still use measured Eh values to study natural environments.

Suppose that we want to know whether the conditions of a sediment in a swamp are *aerobic* (oxygen available) or *anaerobic* (oxygen not available). The presence or absence of oxygen is a critical environmental factor with respect to the kinds of inorganic reactions that take place in the sediment, the types of organisms present, and the direction and extent of organic decay. The dissolved-oxygen content of the water of the sediment (or of any solution) is a measure of the balance between oxygen-consuming and oxygen-producing processes. Assume that the Eh value found for our sediment is $+0.36$ volt. Assume that we have also measured the pH and obtained a value of 7. The pertinent half-reaction is

$$2\,H_2O \rightleftharpoons O_2\,(g) + 4\,H^+ + 4\,e^-, \quad E^0 = +1.23\,V \tag{4-45}$$

Substituting into equation (4-42),

$$Eh = E^0 + \frac{0.059}{N}\log\frac{[H^+]^4[O_2\,(g)]}{[H_2O]^2}$$

Because of the behavior of oxygen mentioned above, a more realistic E^0 value for natural waters is $+1.00$ volt. Since the activity of water is by convention set equal to 1, and substituting oxygen pressure in atmospheres for $[O_2\,(g)]$,

$$+0.36 = +1.00 + \frac{0.059}{4}\log\,[H^+]^4 p_{O_2}$$

$$-0.64 = \frac{0.059}{4}\log\,[H^+]^4 + \frac{0.059}{4}\log p_{O_2}$$

$$-0.64 = 0.059\log\,[H^+] + 0.015\log p_{O_2}$$

The term $\log\,[H^+]$ can be replaced by $-$pH, and thus

$$-0.64 = -0.059\,\text{pH} + 0.015\log p_{O_2}$$

Therefore,

$$\log p_{O_2} = -\frac{0.227}{0.015} = -15$$

and

$$p_{O_2} = 10^{-15}\text{ atm}$$

To get an indication of the amount (activity) of dissolved oxygen in the water of the sediment, we apply Henry's law (discussed earlier in the chapter):

$$O_2\,(g) \rightleftharpoons O_2\,(aq) \tag{4-46}$$

$$K = \frac{O_2\,(aq)}{p_{O_2}} \approx 10^{-2} \tag{4-47}$$

Substituting $p_{O_2} = 10^{-15}$, we find that the amount of dissolved O_2 is about 10^{-17} mole per liter. In other words, there is essentially no oxygen present and the environment is anaerobic.

Thus we find that a relatively high oxidation potential of +0.36 volt (at a pH of 7) does *not* mean that there is a significant amount of oxygen present in the environment. If we take water that is saturated with oxygen (Eh $\approx$ 0.45 volt for pH of 7) and reduce its oxygen content drastically, there would be a very small lowering of the Eh value of the water. Keep in mind that Eh is determined by the overall composition of a solution and not just by oxygen content. Further, an overall Eh has no meaning unless major redox couples are in equilibrium. Despite problems in obtaining and interpreting Eh values of natural waters, these values are still useful in general descriptions of natural environments as discussed in the next section. The Eh concept and the interpretation of Eh measurements in natural waters are rigorously discussed by Hostettler (1984).

Eh AND pH OF NATURAL ENVIRONMENTS

What ranges of Eh and pH values are actually found in nature? This question was dealt with in an important paper by Baas-Becking et al. (1960). They reviewed and evaluated 6,200 sets of measurements (their own data and that of others) for many different kinds of natural material (Table 4-4). It was found that Eh values varied from about +0.8 to –0.5 volt, and pH values varied from a little less than 1.0 to a little more than 10.0. The approximate positions of some natural environments in terms of Eh and pH are given in Figure 4-6. (Note that the upper part of the diagram represents oxidizing conditions, whereas in our list of half-reactions, Table 4-3, the most reducing materials are listed in the upper part of the table.) Baas-Becking et al. found that there is a progressive increase in the range of values for Eh and pH in the following order: rainwater, mine water, peat bogs, seawater, rivers and lakes, marine sediments, evaporites, and geothermal waters. Thus, if we plotted Eh and pH values on Figure 4-6 for rainwater, a smaller area would be outlined than would be found for geothermal waters.

The Eh and pH values of natural materials are controlled by various factors, some of which are listed in Table 4-4. The major controls on Eh and pH are (1) the organic processes of photosynthesis, respiration, and decay; (2) oxidation-reduction reactions involving iron, sulfur, nitrogen, and carbon; and (3) the balance between dissolved CO_2 and calcium carbonate in natural waters. Photosynthesis converts light energy (plus water and carbon dioxide) to chemical energy in the form of organic compounds (plus oxygen). The organic compounds plus oxygen have a high Gibbs function value and are thus unstable. Respiratory and decay processes reduce the value of the Gibbs function, use up oxygen, and produce carbon dioxide. Thus we can say that photosynthesis tends to produce disequilibrium and high oxidation potential, whereas respiration and decay tend to restore equilibrium and produce a low oxidation potential. The production of CO_2 in natural waters as a result of respiration and/or decay causes a lowering of pH, since CO_2 combines with water to form a weak (slightly ionized) acid, H_2CO_3:

$$H_2O + CO_2 \rightleftharpoons H_2CO_3 \rightleftharpoons HCO_3^- + H^+ \tag{4-48}$$

The uptake of CO_2 by photosynthesis would reduce the amount of H_2CO_3, reduce the amount of H^+, and thus raise the pH. Oxidation-reduction reactions involving carbon, nitrogen, and sulfur can be organic or inorganic or a combination of the two. Almost all of them at the Earth's surface are biologically mediated. Iron reactions are important, because iron is the most abundant of those elements in the Earth's crust that exhibit more than one oxidation state.

TABLE 4-4 Range and Controls of pH and Eh of Various Natural Materials

Material	pH range	Eh range (mV)	Controls
Meteoric waters	4–8	+800 to +300	Amount of dissolved CO_2, O_2
Peat bogs			
Depressions	7–8	+500 to negative values	Topography, organic reactions
Mounds	3–7	+500 to negative values	
Soils	2.8–10+	+750 to −350	Amount of water and kinds of minerals present; organic content
Groundwater	5–9	+500 to −100	Rock environment, organic reactions
Mine water			
Oxidized zone	2–9	+800 to +200	Oxidation of pyrite
Primary zone	6–9	+200 to −100	Rock environment
Freshwater rivers and lakes	4–10	+600 to −100	Amount and kind of dissolved material
Freshwater sediments	4–9	+600 to −200	Sediment composition (organic and inorganic)
Marginal marine sediments (in deltas, estuaries, etc.)	5–10	+500 to −400	Degree of isolation; kind and amount of flora and fauna; sediment composition
Seawater	6–10	+500 to −200	Algae; balance between dissolved CO_2 and $CaCO_3$
Open-sea sediments	6–9	+600 to −400	Algae; balance between dissolved CO_2 and $CaCO_3$; sediment composition
Evaporites	6–10	+600 to −500	Composition of evaporite water; organic content
Geothermal waters	0.85–9.5	+700 to −185	Numerous factors
Connate waters	5–8	+100 to −300	Isolation from air; rock environment

Source: Data from Baas-Becking et al. (1960, 243–284).

The control of pH in the oceans has long been believed to be the equilibrium involving CO_2 and calcium carbonate:

$$\begin{array}{c} H_2O + CO_2(g) \\ \downarrow\uparrow \\ CaCO_3(s) + H_2CO_3 \rightleftharpoons Ca^{2+} + 2\,HCO_3^- \end{array} \tag{4-49}$$

As pointed out in the previous paragraph, increasing CO_2 abundance will increase the amount of H_2CO_3 and the above equations show that this will in turn cause a decrease in the amount of

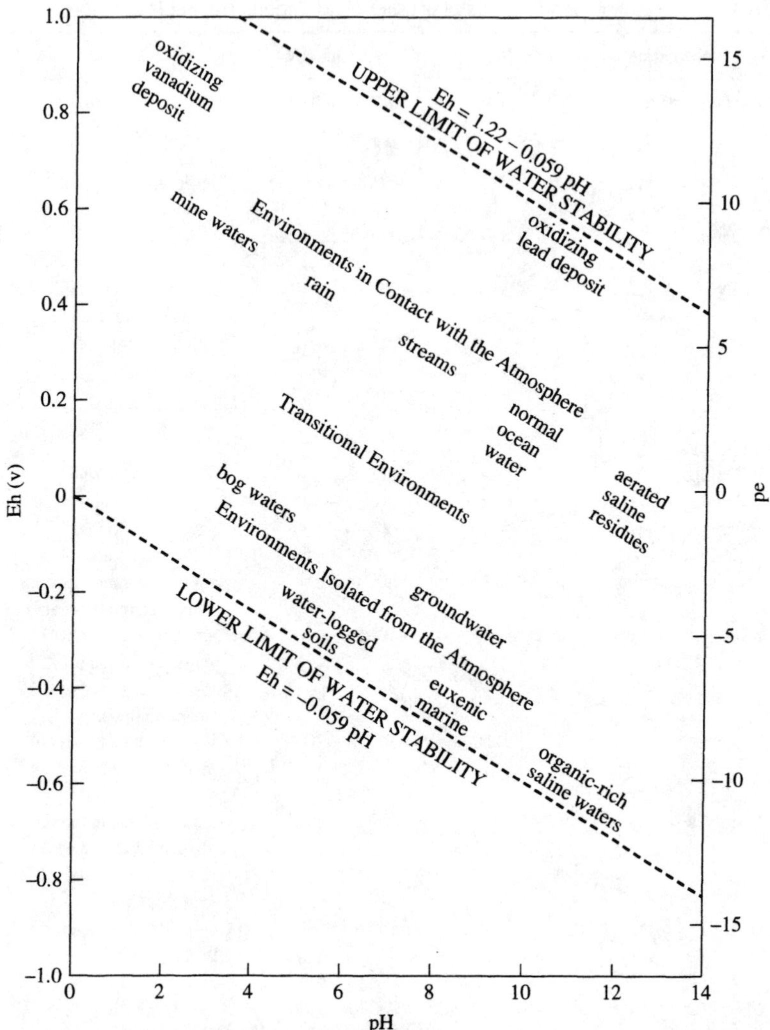

Figure 4-6 Approximate position of some natural environments as characterized by Eh and pH. pe units are shown on the right side of the diagram and in Figures 4-7 and 4-8. (After R. M. Garrels and C. L. Christ, *Solutions, Minerals, and Equilibria*, Freeman-Cooper, San Francisco, p. 381.)

calcium carbonate. Decreasing CO_2 abundance will result in precipitation of calcium carbonate.

The relationship between H_2CO_3 and HCO_3^- is such that water containing them acts as a *buffer*; that is, the water can take on or give up large amounts of H^+ without causing much change in the pH of the water. Consider the reaction

$$H_2CO_3 \rightleftharpoons H^+ + HCO_3^- \tag{4-50}$$

We can express the equilibrium as

$$\frac{K}{[H^+]} = \frac{[HCO_3^-]}{[H_2CO_3]} \tag{4-51}$$

where $K = 10^{-6.4}$. If we have equal amounts of HCO_3^- and H_2CO_3 (e.g., 10^{-2} mole per liter), then $[H^+] = 10^{-6.4}$ mole per liter and the pH is 6.4. In our calculation we assumed that $[HCO_3^-] = 10^{-2}$ mole per liter, and that $[H_2CO_3] = 10^{-2}$ mole per liter, giving a total for dissolved carbonate species of 2×10^{-2} mole per liter. When acid is added to the solution, equation (4-50) goes to the left and more H_2CO_3 is produced. Let us add enough acid to change the $[HCO_3^-]/[H_2CO_3]$ ratio to 1/3. This would require the addition of about 5×10^{-3} mole per liter of acid to our solution. We are thus adding about 10^4 times as much H^+ to the solution as was originally present. The new pH can be calculated as follows:

$$\frac{10^{-6.4}}{[H^+]} = \frac{1}{3} \qquad [H^+] = 3 \times 10^{-6.4} \approx 10^{-5.9}$$

$$\text{pH} = 5.9$$

So the pH has changed very little, even though we added much more hydrogen ion than was originally present in the solution. Most of the additional hydrogen ions occur in the form of H_2CO_3, and very little occurs as H^+.

The chemistry of seawater is much more complex than the simple considerations discussed above imply. However, the overall effect of the relationship between dissolved species, CO_2 in the atmosphere, and solid calcium carbonate is to produce a buffered solution whose pH is normally between 7.8 and 8.4.* Locally, organic activity or other causes can result in temporary values as low as 6 and as high as 10. Many years ago it was suggested that reactions involving clay minerals and other silicates are the ultimate control on the pH of seawater. The carbonate reactions discussed above are faster than silicate reactions and thus have a greater immediate effect on pH when any change occurs.

Eh-pH DIAGRAMS

We can calculate theoretical upper and lower limits of Eh in nature. The upper oxidizing limit of Eh is defined by the reaction previously discussed,

$$2\,H_2O \rightleftharpoons O_2\,(g) + 4\,H^+ + 4\,e^- \qquad E^0 = +1.23\text{ V} \tag{4-45}$$

since no stronger oxidizing material exists in nature that can oxidize water to produce oxygen gas. If we call the natural potential of this reaction Eh, then, from equation (4-42),

$$\text{Eh} = 1.23 + \frac{0.059}{4} \log\,[H^+]^4\, p_{O_2}$$

The partial pressure of oxygen in the atmosphere is 0.2 atmosphere; thus

$$\text{Eh} = 1.23 - 0.059\,\text{pH} + 0.015 \log\,(0.2)$$

*The pH of surface seawater is about 8.2. Deep water is generally below 8.0.

From this we get the following equation relating Eh and pH:

$$\text{Eh} = 1.22 - 0.059\ \text{pH} \tag{4-52}$$

This is the equation of the upper line shown in Figure 4-6.

The lower limit of Eh can be defined by the half-reaction

$$H_2 \rightleftharpoons 2\,H^+ + 2\,e^- \qquad E^0 = 0.00\ \text{V} \tag{4-53}$$

since very few natural reducing agents produce hydrogen gas from the hydrogen ion of water. Again using equation (4-42),

$$\begin{aligned}\text{Eh} &= 0.00 + \frac{0.059}{2}\log\frac{[H^+]^2}{[H_2]}\\ &= \frac{0.059}{2}\log[H^+]^2 - \frac{0.059}{2}\log[H_2]\\ &= -0.059\ \text{pH} - \frac{0.059}{2}\log p_{H_2}\end{aligned}$$

The lowest possible value for this Eh would be when hydrogen pressure equals atmospheric pressure (1 atmosphere); thus our equation for the lower limit of Eh on Figure 4-6 is

$$\text{Eh} = -0.059\ \text{pH} \tag{4-54}$$

We can think of Eh as reflecting the abundance of electrons in the environment. A large number of available electrons would give a reducing environment. An absence of available electrons would give an oxidizing environment. Similarly, we can think of pH as representing the abundance of protons. A large number of available protons would give an acid environment, and a scarcity of available protons would give a basic environment. Since protons and electrons have opposite charges, we might expect that when we have an abundance of one we would have a shortage of the other. In other words, we would expect that an oxidizing environment (high Eh) would tend to be acidic (low pH), and a reducing environment (low Eh) would tend to be basic (high pH). Note that both of the limiting lines on Figure 4-6 slope down to the right. These and other reaction lines plotted on Eh-pH diagrams usually slope to the right because the reactions occur at a lower Eh when the pH is raised.

We can use Eh-pH diagrams to illustrate three different kinds of reactions in nature. We shall illustrate these reactions by reviewing a system consisting of water and various forms of iron. One type of reaction involves relations between dissolved material only. Consider the half-reaction

$$Fe^{2+} \rightleftharpoons Fe^{3+} + e^- \qquad E^0 = +0.77\ \text{V} \tag{4-55}$$

Applying equation (4-42) and using Eh for the potential of the reaction in a natural solution,

$$\begin{aligned}\text{Eh} &= E^0 + \frac{0.059}{N}\log\frac{[Fe^{3+}]}{[Fe^{2+}]}\\ &= 0.77 + 0.059\log\frac{[Fe^{3+}]}{[Fe^{2+}]}\end{aligned} \tag{4-56}$$

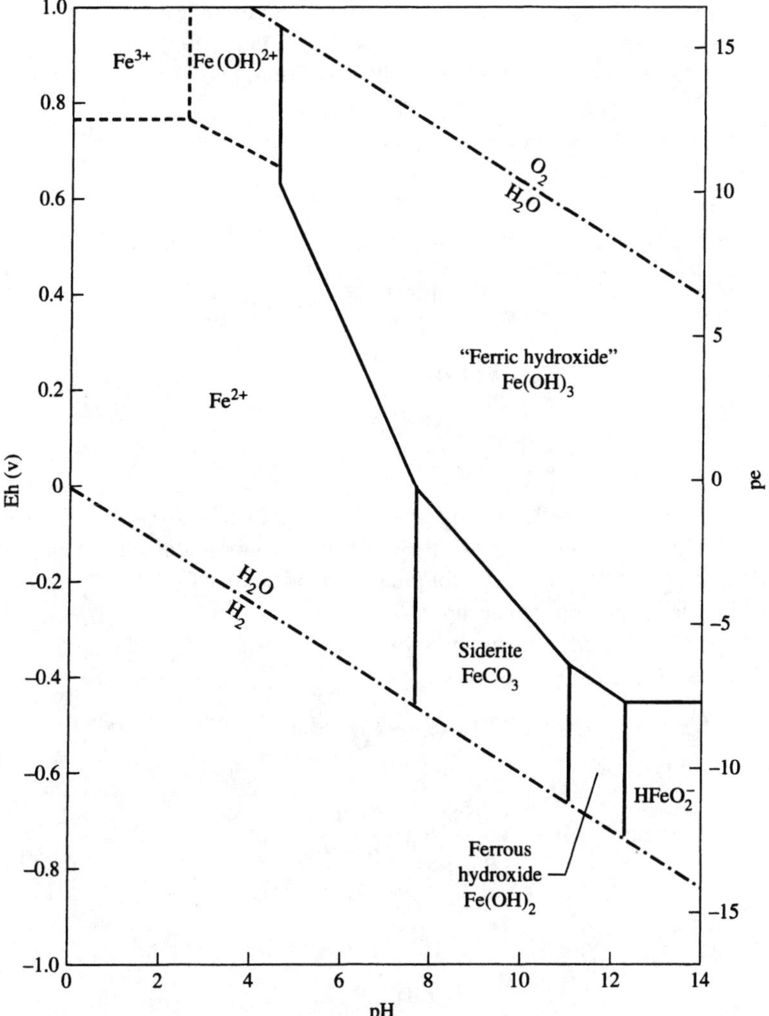

Figure 4-7 Diagram showing the relations among the metastable iron hydroxides and siderite at 25°C and 1 atm total pressure. Boundary between solids and ions at total activity of dissolved species $= 10^{-6}$. Total dissolved carbonate species $= 10^{-2}$. Dashed lines are boundaries between fields dominated by the labeled ion. (After R. M. Garrels and C. L. Christ, *Solutions, Minerals, and Equilibria*, Freeman-Cooper, San Francisco, p. 210.)

When the ratio of the two iron ions is equal to 1,

$$\text{Eh} = E^0 = 0.77 \text{ V}$$

This will plot as a straight line on our Eh-pH diagram (Figure 4-7). It is a horizontal line because the reaction as such is not affected by pH. We plot it only at very low pH values, since we know experimentally that ferric ion does not ordinarily occur in abundance at moderate and high values of pH. We mark the area above the line as the stability field of Fe^{3+}, since values higher than 0.77 volt will be given by equation (4-56) when the ratio $[Fe^{3+}]/[Fe^{2+}]$ is greater than 1. Similarly, we mark the field below the line as the stability field of Fe^{2+}, since Eh values will be less than 0.77 volt when $[Fe^{3+}]/[Fe^{2+}]$ is less than 1.

As an example of a reaction that is not affected by Eh, we have

$$Fe^{3+} + H_2O \rightleftharpoons Fe(OH)^{2+} + H^+ \qquad K = 10^{-3.05} \tag{4-57}$$

Nothing is oxidized or reduced. The equilibrium constant for this reaction is expressed as

$$K = \frac{[H^+][Fe(OH)^{2+}]}{[Fe^{3+}]} = 10^{-3.05} \tag{4-58}$$

When the ratio of the iron ions is 1, we have

$$[H^+] = 10^{-3.05}$$
$$\log [H^+] = -3.05 \quad \text{and} \quad pH = 3.05$$

Therefore, a vertical line on our Eh-pH diagram (Figure 4-7) at pH = 3.05 will separate the stability fields of Fe^{3+} and $Fe(OH)^{2+}$. When there is a greater amount of Fe^{3+} than $Fe(OH)^{2+}$, the calculated pH will be less than 3.05, and for greater abundance of $Fe(OH)^{2+}$ it will be more than 3.05. The stability fields are marked accordingly.

A second kind of Eh reaction involves dissolved material and solid material (minerals and other natural solids). Consider the equation

$$Fe^{2+} + 3\,H_2O \rightleftharpoons Fe(OH)_3\,(s) + 3\,H^+ + e^- \qquad E^0 = +1.06 \tag{4-59}$$

Using equation (4-42),

$$Eh = 1.06 + \frac{0.059}{N} \log \frac{[H^+]^3}{[Fe^{2+}]}$$
$$= 1.06 + 0.059 \times 3 \log[H^+] - 0.059 \log [Fe^{2+}]$$

If we assume a typical value for $[Fe^{2+}]$ of 10^{-6} mole per liter, we get the equation

$$Eh = 1.41 - 0.177\,pH \tag{4-60}$$

This is the equation of the line in Figure 4-7 separating the field where Fe^{2+} would be the dominant form of iron from the field where $Fe(OH)_3$ would be the dominant form. The "ferric hydroxide" that precipitates in nature is an unstable hydrous oxide of indeterminate water content. With time it converts to hematite or goethite. If we assumed a different value for $[Fe^{2+}]$, we would get the equation of a line that would be offset slightly from the line of Figure 4-7 and parallel to it.

We could use the equation

$$Fe^{2+} + 3H_2O \rightleftharpoons Fe(OH)_3\,(s) + 3\,H^+ + e^- \tag{4-59}$$

in a different way. Instead of locating a line on a theoretical Eh-pH diagram, we can determine the amount of Fe^{2+} in metastable equilibrium with solid $Fe(OH)_3$ in a particular natural solution. Assume a swampy area has precipitated "ferric hydroxide," a measured pH of 6, and a measured

Eh of +0.20 volt. Then we can write

$$\begin{aligned} Eh &= 1.06 + \frac{0.059}{N} \log \frac{[H^+]^3}{[Fe^{2+}]} \\ &= 1.06 - 0.177\ pH - 0.059 \log [Fe^{2+}] \end{aligned}$$

Substituting our values for Eh and pH,

$$\log [Fe^{2+}] = -3.4 \quad \text{and} \quad [Fe^{2+}] = 10^{-3.4}\ \text{mol/liter}$$

A third kind of Eh-pH reaction involves two or more solids. Consider the reaction

$$\begin{aligned} &FeCO_3\ \text{(siderite)} + 3\ H_2O \rightleftharpoons Fe(OH)_3\ (s) + HCO_3^- + 2\ H^+ + e^- \\ &E^0 = +1.08 \end{aligned} \tag{4-61}$$

We write the equation in this form because we know experimentally that siderite forms at pH values roughly in the range 7–10, and we also know that most of the dissolved carbon dioxide in solutions with these pH values is in the form of HCO_3^- (rather than H_2CO_3 or CO_3^{2-}). Writing our expression for Eh,

$$\begin{aligned} Eh &= 1.08 + \frac{0.059}{N} \log \frac{[H^+]^2[HCO_3^-]}{1} \\ &= 1.08 + 0.118 \log [H^+] + 0.059 \log [HCO_3^-] \\ &= 1.08 - 0.118\ pH + 0.059 \log [HCO_3^-] \end{aligned}$$

If we assume a typical value of 10^{-2} mole per liter for $[HCO_3^-]$, we get

$$Eh = 0.96 - 0.118\ pH$$

This is the equation of the line dividing the fields of siderite and "ferric hydroxide" in Figure 4-7. The other boundary lines of this diagram were derived in a manner similar to the preceding examples.

The same methods used to construct Eh-pH diagrams can be used to construct diagrams with other variables, such as gas activities (partial pressures) or ion activities. One example will be given here. If we are interested in studying siderite formation, an appropriate pair of variables might be Eh and the partial pressure of carbon dioxide (Figure 4-8). The line separating the magnetite and siderite fields in Figure 4-8 is defined by the following equation:

$$\begin{aligned} &3\ FeCO_3\ \text{(siderite)} + H_2O \rightleftharpoons Fe_3O_4\ \text{(magnetite)} + 3\ CO_2 + 2\ H^+ + 2\ e^- \\ &E^0 = +0.319 \end{aligned} \tag{4-62}$$

Writing our expression for Eh, and substituting p_{CO_2} for $[CO_2]$,

$$\begin{aligned} Eh &= 0.319 + \frac{0.059}{2} \log [H^+]^2[CO_2]^3 \\ &= 0.319 + 0.059 \log[H^+] + \tfrac{3}{2} \times 0.059 \log p_{CO_2} \\ &= 0.319 - 0.059\ pH + 0.089 \log p_{CO_2} \end{aligned} \tag{4-63}$$

We could assume a value for the pH, or we could use the equation (Berner 1971, 194)

$$pH = 6.17 - \frac{1}{2} \log p_{CO_2} \tag{4-64}$$

since it can be shown that pH is directly proportional to log p_{CO_2}. Substituting equation (4-64) into (4-63), we get

$$Eh = -0.046 + 0.118 \log p_{CO_2}$$

which is the equation of the siderite-magnetite line of Figure 4-8. The other lines of Figure 4-8 are obtained in a similar manner.

Figure 4-8 tells us that very special conditions are necessary for siderite formation. These conditions are low Eh, high p_{CO_2}, and negligible sulfide ion present. Such conditions are usually only developed in nonmarine environments, and siderite is normally found in nonmarine sedimentary rocks.

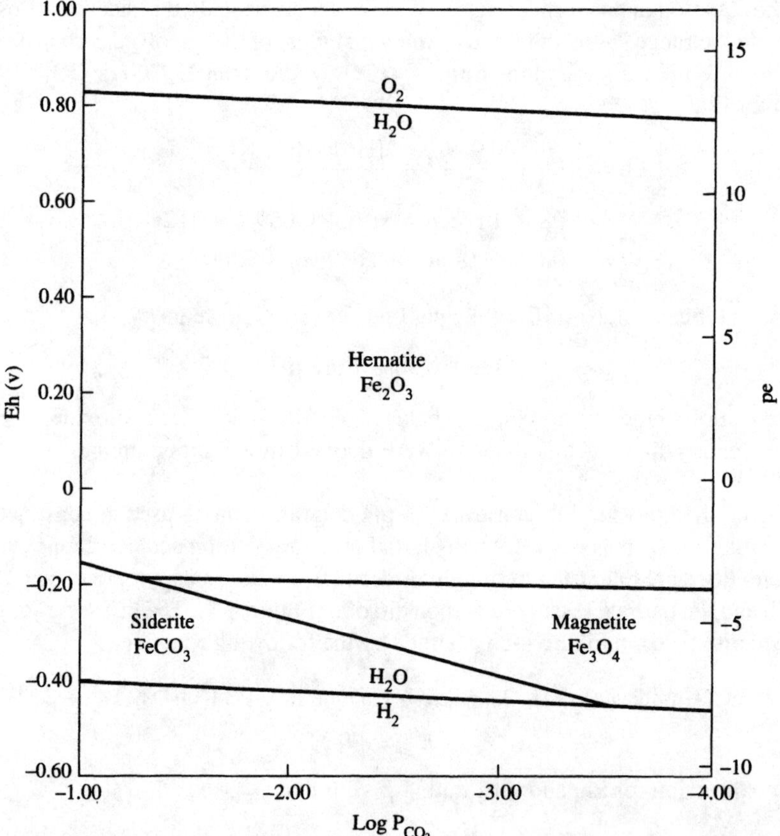

Figure 4-8 Eh-log P_{CO_2} diagram for hematite, magnetite, and siderite in marine sediments. $T = 25°C$, $P_{total} = 1$ atm, $a_{Ca^{2+}} = 10^{-2.58}$, equilibrium with calcite is assumed. The activity of sulfide ion is assumed to be so low that pyrite and pyrrhotite do not plot stably. (From *Principals of Chemical Sedimentology* by R. A. Berner, p. 196. Copyright 1971 by McGraw-Hill Book Company. Used with permission of McGraw-Hill Book Company.)

A final note of caution in using Eh-pH and similar diagrams. They are only as good as the data and assumptions used to construct them. Equilibrium is assumed (often not a valid assumption in natural systems), and minerals and other compounds are assumed to be pure and of known composition (which is not true for many natural materials such as the "ferric hydroxide" discussed earlier). We measure concentrations but use activities in the calculations. These are often not the same. There are many variables and many possible reactions, and a given diagram can consider only some of these. Other limitations could be mentioned. The important point is this: such diagrams are useful to a geologist if they seem to explain observed facts and if their limitations are used to qualify any conclusions reached. Brookins (1988) has published a reference book of Eh-pH diagrams for 75 elements.

WATER AT HIGH TEMPERATURES AND PRESSURES

Much of what we have said so far concerns water occurring under conditions existing at or near the surface of the Earth. In particular, temperatures for this environment are on the order of 25°C and pressure is about 1 atmosphere. However, many surface and subsurface geologic processes go on (and have gone on) under conditions of higher temperature values and/or higher pressure values. Water is a major factor determining the properties and crystallization of magmas. For example, water reduces melt viscosity, depresses crystallization temperatures, and affects the formation and stability of hydrous minerals such as micas and amphiboles. Many of the reactions that occur during metamorphism involve water. Also, water-rich solutions are responsible for the formation of hydrothermal ore deposits. So we need to know something about how water behaves at high temperatures and pressures. Unfortunately, because of experimental and other difficulties, we know much less about water and water solutions occurring under these conditions.

Let us first consider pure H_2O at high temperatures and pressures. Figure 4-9a is a phase diagram showing the stable form of H_2O over a range of temperatures and pressures. At low temperatures and pressures, H_2O occurs either as a liquid (disordered, close-packed molecules) or as a gas (disordered, widely spaced molecules). When the liquid is in equilibrium with its vapor, the pressure of the gas molecules is a certain value known as the *vapor pressure* of the liquid. This equilibrium vapor pressure varies with temperature, as shown in Figure 4-9a. Above a certain temperature and pressure, H_2O exhibits a continuity between the liquid and gaseous states. In other words, there is no longer a distinction between what is gas and what is liquid. This form of H_2O is known as a *supercritical fluid.* The temperature and pressure beyond which the distinction no longer exists is known as the *critical point* (Figure 4-9a). Other substances also have critical points, each at a distinctive pressure and temperature.

Also shown in Figure 4-9a is the variation in density of liquid H_2O with temperature. The density decreases significantly as the temperature is raised to the temperature of the critical point. At still higher temperatures, we will have either water vapor or supercritical fluid, depending on the pressure conditions. In this high-temperature region the density of the fluid or vapor depends mainly on the pressure (Figure 4-9b).

Of what significance are the H_2O critical point and the density of H_2O phases to geologists? We are not as interested in pure H_2O as in H_2O solutions containing various elements and compounds. It is generally believed that many such solutions in nature consist mainly of H_2O. Their critical point and density would then show relationships similar to those discussed above for pure H_2O. But why are we interested in these two features? The critical point is important be-

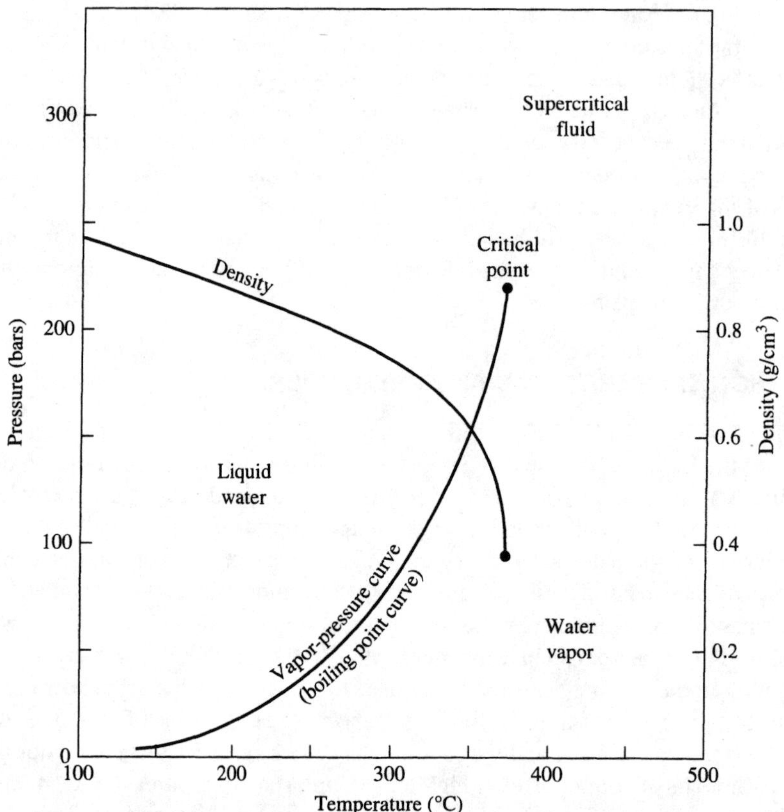

Figure 4-9a Variation of the density and vapor pressure of liquid water. At temperatures and pressures beyond the critical point (374.15°C and 221 bars), H_2O exists as a supercritical fluid. Under these conditions the transition from the gaseous state to the liquid state occurs without a discontinuity. The density of water vapor and of the supercritical fluid is variable, depending upon pressure. At high temperatures the density of a given phase of H_2O is a major control on the solubility of material in the phase. (Modified from J. Verhoogen, F. J. Turner, L. E. Weiss, C. Wahrhaftig, and W. S. Fyfe, *The Earth: An Introduction to Physical Geology,* p. 244. Copyright © 1970 by Holt, Rinehart and Winston. Reprinted by permission.)

cause, for example, the physical situation in a magma chamber may involve two or three phases, depending on the temperature and pressure. In one case there may be only silicate liquid and a water-rich, supercritical fluid that has separated from it. At lower temperatures and pressures, below the critical point of the fluid, there could be three phases over a period of time: silicate liquid, a water-rich gas, and, at lower temperatures, a water-rich liquid. A third possibility would be one phase only, silicate liquid, if the original magma contained very little water. In this case no water would separate as crystallization of silicates proceeded, since the residual magma could not become saturated with water. This would depend on the pressure, too, with high pressure preventing separation of a H_2O phase.

Krauskopf (1979) has discussed these three possibilities in some detail, and suggests that aplites (fine-grained, sugary-looking igneous rocks) form from a residual melt that crystallized without separation of a H_2O phase, whereas pegmatites (coarse-grained rocks with the same gen-

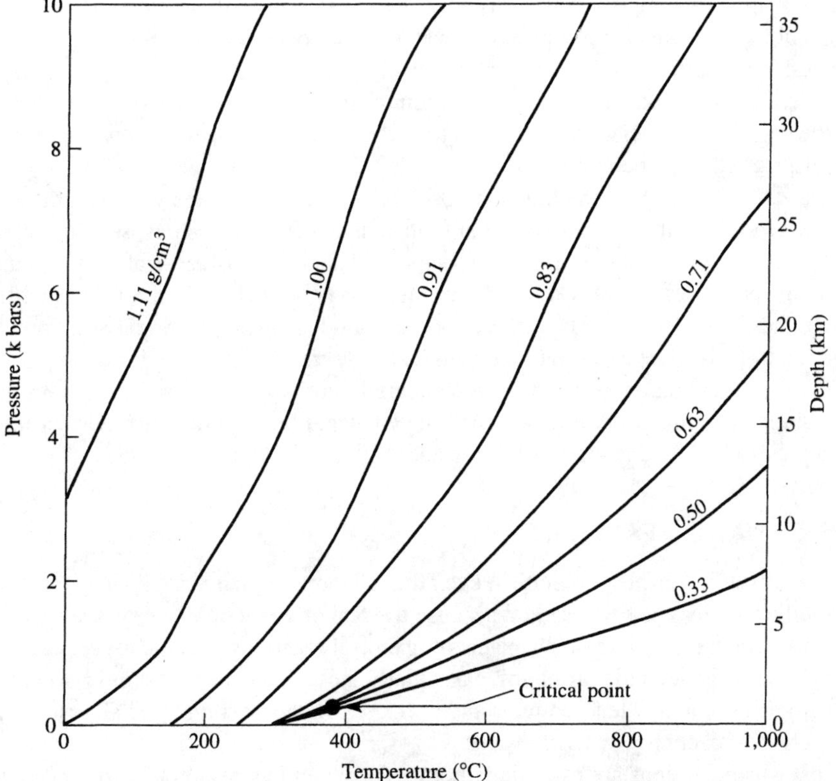

Figure 4-9b Variation of the density of supercritical water as a function of pressure and temperature. At high pressures the supercritical fluid has a density close to that of liquid water at the surface of the Earth (1 g/cm^3). At low pressures and high temperatures the density can deviate significantly from 1 g/cm^3. Low densities result in low solubility of ionic materials. Note the location of the critical point. From IGNEOUS AND METAMORPHIC PETROLOGY by Best. Copyright © 1982 by W. H. Freeman and Company. Used by permission. After Fyfe et al. (1978).

eral mineralogy as aplites) formed from residual melts whose temperature-pressure conditions did allow separation of a H_2O phase. The existence of the H_2O phase would cause slower cooling, permit more effective ion diffusion, and thus allow large crystals to grow. Experimental work by Jahns and Burnham (1969) supports this hypothesis. Krauskopf further suggests that the third possibility, three phases, may be related to the formation of hydrothermal vein deposits, with one of the two H_2O-rich phases migrating away from the magma chamber to eventually form vein minerals as precipitates from a low-temperature, liquid solution.

Having given an example of the possible significance of the critical point of a H_2O-rich solution, we now turn to the density of such a solution. The solubility of a given solute in a H_2O-rich solution may increase or decrease as temperature and/or pressure rise. Limited experimental work suggests that, in general, as temperature rises there is probably a greater solubility of compounds in the form of un-ionized molecules and a lower solubility of compounds in the form of ions. At very high temperatures and under any conditions where we are dealing with water vapor or supercritical fluid, the density of the material is a major factor affecting solubility, particularly ionic solubility. The dielectric constant of water, which determines its ability to dissolve ionic substances, is proportional to density and inversely proportional to temperature. Therefore high temperatures decrease the degree of dissociation of acids, bases, and salts and high pressures increase the degree of dissociation. The ionization of water increases strongly with pressure and, to a lesser degree, with temperature.

Thus at low densities (which would occur at low pressures) there would be little solubility, whereas extensive solubility could occur at high densities (high pressures; see Figure 4-9b). The pH and solubility properties of H_2O-rich, supercritical solutions associated with igneous and metamorphic processes probably depend very much on pressure conditions, more so than on temperature conditions. If the pressure exerted on a high-density, supercritical solution carrying a large amount of dissolved material were suddenly lowered by faulting, for example, the result could be rapid precipitation of the dissolved material. Similarly, the dissolving power of a supercritical solution passing through rocks would be greater at higher pressures. Further information on this topic can be found in articles by Eugster (1986) and Sverjensky (1987).

KINETICS OF WATER REACTIONS

It was pointed out in Chapter Three that, even though thermodynamics tells us a reaction should occur, equilibrium may not be reached because the rate of a reaction is very slow. Chemical kinetics deals with the rates and mechanisms of chemical reactions. Kinetic aspects are important in the study of many geologic problems, particularly those that occur (or should occur according to thermodynamics) at low temperatures and pressures. In this section we shall restrict ourselves to the kinetics of reactions involving water.

Most water solutions are essentially homogeneous. It has been found that many reactions involving only dissolved material and gases are rapid enough that equilibrium can be assumed to exist at any given time. There are important exceptions. In general, the rate of all reactions in a given body of water will depend on its size, rate of mixing, and general chemical and physical environment. Reactions would tend to occur more quickly in a river (probably hours or days for reaction completion time) as compared to the ocean (probably thousands to millions of years for completion of some reactions). Groundwaters would tend to have rates of reaction somewhat slower than those of rivers (because of less rapid mixing) and somewhat faster than those of oceans (because of much smaller volume).

Even though most inorganic reactions involving dissolved material have rates that can be expressed in seconds or hours (assuming small volume and rapid mixing), that is not true for a reaction such as

$$H^+ + SO_4^{2-} + NH_3\ (aq) \rightleftharpoons H_2S\ (g) + NO_3^- + H_2O \qquad (4\text{-}65)$$

This reaction might be used as an explanation for the formation of H_2S gas in anaerobic groundwater, where ammonia has been produced by the decay of organic matter. However, it is known from laboratory study that the inorganic reduction of sulfate ion is extremely slow. The reduction of sulfate ion does occur in nature and is in fact a major source of H_2S in sediments. The sulfur of this H_2S may later occur in rocks as native sulfur (in the cap rocks of salt domes, for example) or as pyrite in black shales containing abundant organic matter. The rapid reduction of sulfate ion in nature is due to the action of a certain type of bacteria, which serves as a catalyst and derives energy from the process. Organic carbon is oxidized and the sulfate ion is reduced by reactions such as

$$2\,C_6H_{12}O_6\ \text{(carbohydrate)} + 6\,SO_4^{2-} \rightleftharpoons 6\,H_2S + 12\,HCO_3^- \tag{4-66}$$

In discussing natural processes we can contrast reactions in solutions with reactions involving solids. The former are usually fast. The latter are often very slow. Equilibrium is generally reached in solution reactions, whereas it often is not reached in liquid-solid reactions. It is in the slow reactions that we are particularly concerned with the mechanisms by which the reactions occur. Frequently, they occur in a series of steps, with each step having its own rate of reaction. Laboratory studies of water-feldspar systems indicate that the alteration of feldspar minerals to clay minerals in chemical weathering probably occurs in a series of steps. As all geologists know, the overall rate of change of feldspar to clay mineral at the surface of the Earth is quite slow. A discussion of the kinetics of clay mineral formation is given in Chapter Seven.

Using experimental data we can describe the rate at which a given chemical reaction occurs. Actual concentrations are used in writing rate laws whereas activity values (effective or thermodynamic concentrations) are used in describing equilibrium relationships. Some reactions are independent of the concentration of the reactants. If the reaction

$$A + B \rightarrow C + D \tag{4-67}$$

is independent of the concentrations of A and B and therefore obeys the equations

$$-\frac{d(A)}{dt} = k_1 \quad \text{and} \quad -\frac{d(B)}{dt} = k_2 \tag{4-68}$$

then the reaction is said to be a zero-order reaction for both A and B where k_1 and k_2 are rate constants that can be determined experimentally. If the reaction depends on the concentrations of A and B and obeys the equations

$$-\frac{d(A)}{dt} = k_3(A) \quad \text{and} \quad -\frac{d(B)}{dt} = k_4(B) \tag{4-69}$$

then the reaction is first-order in A and first-order in B. Many reactions obey the equations

$$-\frac{d(A)}{dt} = k_5(A)(B) \quad \text{and} \quad -\frac{d(B)}{dt} = k_6(A)(B) \tag{4-70}$$

Such reactions are described as second-order overall (the disappearances of A and B are first-order in A and first-order in B). For the reaction

$$A + A \rightarrow A_2 \tag{4-71}$$

the rate equation is

$$-\frac{d(A)}{dt} = k_7(A)^2 \tag{4-72}$$

and the disappearance of A is said to be second-order in A.

Table 4-5 lists these three types of rate laws and the result of integrating each rate equation. A general form for these rate laws is

$$\text{rate} = k(A)^m(B)^n \ldots \tag{4-73}$$

TABLE 4-5 Types of Rate Laws

Zero-order	
$-\frac{d(A)}{dt} = k$	$A = A^\circ - kt$, where A° is initial concentration
First-order	
$-\frac{d(A)}{dt} = k(A)$	$A = A^\circ e^{-kt}$, where A° is initial concentration
Second-order	
$-\frac{d(A)}{dt} = k(A)(B)$	$kt = \frac{1}{(B^\circ - A^\circ)} \ln \frac{(A^\circ)(B)}{(B^\circ)(A)}$, where A° and B° are initial concentrations
or	
$-\frac{d(A)}{dt} = k(A)^2$	$kt - \frac{1}{A} - \frac{1}{A^\circ}$, where A° is initial concentration

where k is the rate constant, $A, B, \ldots$ are the reactants, and the overall order of the reactions is $m + n + \ldots$. Figure 4-10 is a plot of experimental data for a first-order reaction.

Temperature has an important effect on reaction rates. This dependence can be represented by the Arrhenius equation

$$k \text{ (rate constant)} = Ae^{-E/RT} \tag{4-74}$$

where A is an empirical constant related to the frequency of collision of molecules (the pre-exponential factor), E is the activation energy (an energy level that must be reached for the reaction to occur), R is the gas constant, and T is absolute temperature. The activation energy can be obtained by using the equation

$$E = -R\frac{d(\ln k)}{d(1/T)} \tag{4-75}$$

and plotting ln k versus $1/T$ (Figure 4-11).

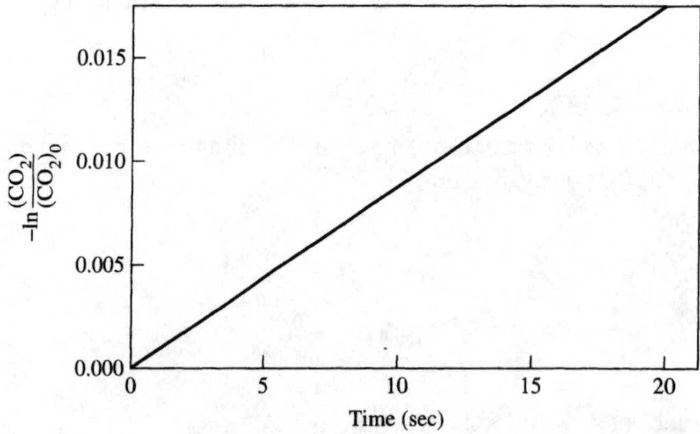

Figure 4-10 Plot of $-\ln[(CO_2)/(CO_2)_0]$ versus time for the first-order reaction CO_2 (aq) + H_2O (1) → H_2CO_3 (aq). Using the first-order rate law in Table 4-5 it can be shown that the slope of the line $= k = 2.0 \times 10^{-3}$ seconds^{-1} at 0°C. The value of k can be used to calculate the time required to consume half of the original amount of CO_2, $(CO_2)_0$. This half-life value is about 5.78 minutes. After Jones et al. (1964, 610–612) and Lasaga (1981, 24).

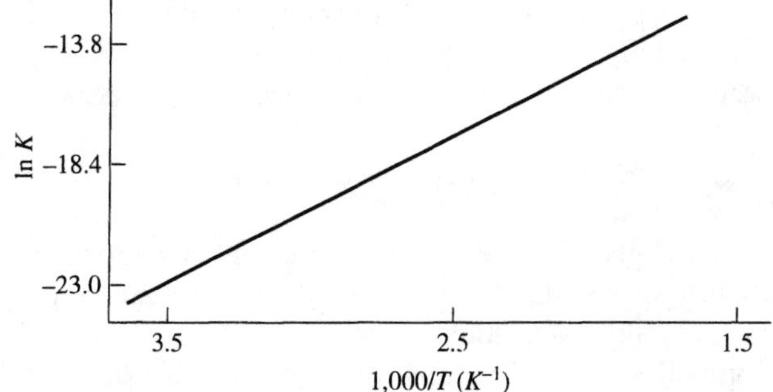

Figure 4-11 Arrhenius plot of experimentally determined precipitation rate constants for silica-water reactions. For the reaction $H_4SiO_4 \rightarrow SiO_2$ (quartz) $+ 2\,H_2O$, the activation energy, obtained from the slope of the line, is equal to 11.9 kcal/mole. After Rimstidt and Barnes (1980, 1683–1699) and Lasaga (1981, 32–33).

Readers interested in further information on the kinetics of geologic reactions are referred to Lasaga and Kirkpatrick (1981) and Stumm and Morgan (1981). The book edited by Lasaga and Kirkpatrick is a detailed review of the kinetics of geochemical processes, covering both low-temperature and high-temperature processes. Extensive discussion of rates of reaction of processes occurring in natural waters can be found in Stumm and Morgan.

EVALUATION OF WATER ANALYSES

It is very useful for geologists to be able to understand and evaluate chemical analyses of natural water. This can be done by applying the concepts and definitions discussed earlier in this chapter. In addition, we need to define a few more terms used by water analysts. We shall first briefly review the mode of occurrence of the constituents of natural waters. This will be followed by a review of the various measurements made for natural waters. Finally, an evaluation of some typical analyses will be given.

Inorganic materials can occur in natural waters as cations, anions, complex ions, and undissociated molecules. In addition, they occur in the form of colloids and suspended matter. A major decision that must be made when a water sample is to be analyzed is whether or not to filter out the suspended material. Knowledge of the composition of the suspended material is important for certain types of water research. However, most analyses are made on filtered solutions, and it is assumed that the constituents found occur dissolved in the solutions. This may or may not be true depending on the size of the filter used. Filters with pores as small as 0.45 micrometer (a commonly used size) will allow significant amounts of colloidal material to pass through them. Analyses of such samples will reflect the presence of the colloidal material. In addition to all these forms of inorganic material, a significant portion of some of the ions of a natural solution may occur adsorbed on suspended matter. Various reactions can occur in natural waters between solute ions and ions attached to sedimentary particles.

Many ions may occur as part of soluble organic complexes. It is particularly common for metal ions to occur in this form. Organic material can occur as suspended matter, as colloidal material, as adsorbed compounds, and in the form of dissolved molecules. The organic matter may be living or dead. Specific organic compounds are not usually identified in water analyses unless the possibility of organic pollution is being studied. The amount of dissolved organic matter in most waters is usually very low. The proportion of a given element occurring in any one form, inorganic or organic, changes with changes in the environment of the water, which would include changes in the amount and kind of suspended sediment, in the amount of organic activity, and in the overall chemical composition of the water. Thus a given element might behave differently in river water as compared to groundwater. In most of the following discussion we shall assume that the water has been filtered (the usual procedure), and thus most or all of the suspended matter, organic or inorganic, has been removed.

We now pass on to some of the general properties of natural waters that can be measured or identified. The properties known as Eh and pH were discussed earlier in this chapter. Much of the discussion that follows is based on material given by Hem (1985). The reader interested in more details is referred to that publication. [For specific details of analytical procedures, see Skougstad et al. (1979) and American Public Health Association and others (1985)].

The *color* of natural waters is due primarily to organic matter, which may occur mainly in the dissolved state, in colloidal suspension, or as coarse suspended matter. Color is a physical property, which is sometimes described in a chemical analysis by using an empirical number scale to express the intensity of the color. The color of a given water will depend on whether or not it has been filtered. *Turbidity* is a property that is measured before filtering (or when filtering is not done at all). This term refers to the reduction of transparency due to the presence of suspended particulate matter. The suspended matter causes light to be scattered and absorbed. As with color, turbidity is expressed by use of an empirical number scale set up for prepared standards.

A parameter of water that is directly related to its organic content is known as *biological oxygen demand* (BOD). Organic life produces oxygen in water by photosynthesis; respiration and the decay of organic matter (by organic or inorganic processes) use up oxygen. For waters in contact, or recently in contact, with the atmosphere, the oxygen content is also controlled by exchange with the atmosphere (Table 4-6). The amount of dissolved oxygen in a given water sample is determined mainly by temperature and by the balance between biological oxygen production and consumption. A BOD determination is made by adding oxygenated water to a sample and measuring the weight of oxygen used up over a five-day period (by oxidation of organic matter) per unit volume of sample. Not all the organic material present reacts with oxygen. Thus *total organic carbon* is often determined by measuring the amount of carbon dioxide produced when the dry residue of a water sample is burned. Another way of estimating how much oxidizable organic material is present in a sample involves wet chemical oxidation of the sample with a strong oxidizing agent. The amount of the agent used is expressed in equivalents of oxygen; this number is known as *chemical oxygen demand* (COD). Measurements of BOD, COD, and of total organic carbon are usually

TABLE 4-6 Solubility of Oxygen in Water in Equilibrium with Air at a Total Pressure of One Atmosphere

Temperature (°C)	Oxygen (mg/l)
0	14.16
5	12.37
10	10.92
15	9.76
20	8.84
25	8.11
30	7.53
35	7.04

Source: Drever, J. I., *The Geochemistry of Natural Waters*, p. 317, 2nd ed. Copyright © 1988. Adapted by permission of Prentice-Hall, Inc., Upper Saddle River, NJ.

TABLE 4-7 Five-day Biochemical Oxygen Demand (BOD) of Some Common Effluents

Effluent source	BOD at 20°C (mg/l)
Distilling	10,000–30,000
Pulp and paper processing	20–20,000
Wool scouring	200–10,000
Canning industry	400–4,000
Cattle barns and dairies	200–4,000
Meat packing plants	600–2,000
Feedlots	400–2,000
Sugar-beet mills	400–2,000
Dairy processing plants	200–2,000
Cotton mills	50–1,750
Breweries	500–1,250
Untreated domestic sewage	100–400
Combined sewer systems	50–400
Urban storm runoff	> 10

Note: For domestic sewage and many industrial wastes, about 70 to 80 percent of the total BOD is exerted within five days. For more resistant materials, the five-day BOD underestimates the full pollution load.

Source: From WATER IN ENVIRONMENTAL PLANNING by Dunne and Leopold. Copyright © 1978 by W. H. Freeman and Company. Used with permission.

made to determine the extent of organic pollution of natural waters by humankind (Table 4-7). The oxygen demand of polluted water and of sewage depends upon the content of organic material and of inorganic reducing agents that will react with dissolved oxygen. Not much is known about specific organic compounds in polluted water, with the exception of extensive work on pesticides, detergents, and trace organic pollutants such as trichloroethylene (TCE).

A commonly reported item in water analyses is *total dissolved solids.* This can be determined in several ways. Usually, the term refers to the weight of the dry residue produced by evaporation of a portion of the water sample. Another way of estimating the amount of dissolved solids is by measurement of the *specific conductance* of a sample. The conductance of a sample refers to its ability to conduct an electric current, and this depends on the amount of charged ions in solution. Thus the specific conductance is more a measure of ionic concentration than of total dissolved solids. However, there is often not much difference between the two values for water samples. For simple solutions, such as most river water, a direct relationship is found between specific conductance and total dissolved solids determined by evaporation. For a more complicated solution like seawater, this relationship does not hold. Another common way of determining total dissolved solids is to add up the amounts of cations and anions found by chemical analysis. Ideally, this computed value would agree with the value found by evaporation. Generally, it does not.

The significance of the value found for total dissolved solids lies in its indication of the past and present environment of the water. Extensive weathering of soluble minerals and rocks will produce high values for waters of an area. Water that occurs in a dry climate will also tend to have high values owing to little dilution by rainwater and to concentration by evaporation. For samples with high amounts of dissolved solids, it may become important to measure the *density* of the sample. A density value is needed if it is desired to convert analytical values from weight per sample volume to parts per million.

The values given in water analyses for the carbonate and bicarbonate ions come from a quantity known as the *alkalinity.* This is defined as the capacity of a solution to neutralize acid (react with hydrogen ions). Any ion in the solution that could react with acid added to the water is part of the total alkalinity. For natural waters the most important ions are the carbonate and bicarbonate ions, and alkalinity is usually reported in terms of concentrations of these two ions. However, the actual measurement (in units of milliequivalents of acid per liter) is by titration of the water sample, and other ions may sometimes contribute to the alkalinity value. Thus for some waters the concentration values given for the carbonate and bicarbonate ions may be significantly in error. The amount of dissolved carbon dioxide in the form of carbonate ion, bicarbonate ion, and dissolved carbonic acid varies with pH (Figure 4-12). We see from Figure 4-12 that for water with pH $= 5.0$, 95 percent of the total dissolved carbon dioxide is in the form of H_2CO_3 (aq) and 5 percent is in the form of HCO_3^-. Water with a pH of 9 will have about 95 percent of the total dissolved carbon dioxide as HCO_3^- and 5 percent as CO_3^{2-}. Thus in a chemical analysis the alkalinity may be given as a concentration of bicarbonate ion (pH values less than 8) or as a concentration of bicarbonate and carbonate ions (pH values greater than 8).

In addition to defining alkalinity in terms of the neutralization of acid (and measuring it by titration), we can also define it mathematically. For waters in which the alkalinity is due mainly to carbonate species (a common situation),

$$\begin{aligned} \text{carbonate alkalinity} &= (\text{conc. } HCO_3^-) + 2(\text{conc. } CO_3^{2-}) \\ &\approx \text{total alkalinity in eq/l} \end{aligned} \tag{4-76}$$

Under these conditions, it is possible to use a pH value and the titration results for a water sample to determine how much each of the two ions HCO_3^- and CO_3^{2-} contributed to the alkalinity value. In other words, the initial composition of the solution can be estimated, even though ac-

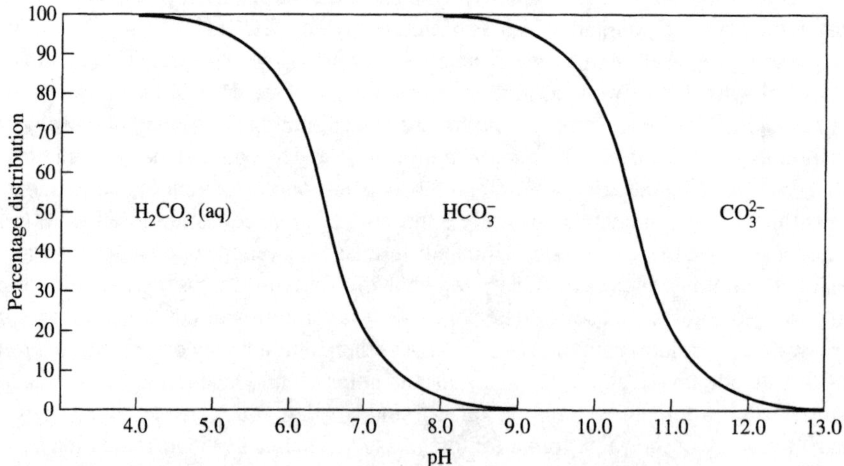

Figure 4-12 Percentages of carbonate species in solution as a function of pH. At a pH of 6.35 the activities (and concentrations) of H_2CO_3 (aq) and HCO_3^- are equal (50% each) and at a pH of 10.33 the activities (and concentrations) of HCO_3^- and CO_3^{2-} are similar. Temperature 25°C. Pressure 1 atm. Most of the material labeled H_2CO_3 (aq) is actually CO_2 (aq). By convention, all the CO_2 molecules in the liquid are assumed to be combined with H_2O. After Hem (1985, 107).

tual measurements of HCO_3^- and CO_3^{2-} were not made. However, it should be kept in mind that some natural waters (and many polluted waters) contain compounds such as ammonia and organic bases that contribute to total alkalinity values obtained by titration. Usually these other compounds are not measured and the total alkalinity is reported as concentrations of HCO_3^- and CO_3^{2-}.

Water with a high alkalinity has a high concentration of inorganic carbon. The alkalinity value is an indicator of the ability of the water to support aquatic life and thus is used by water biologists as an approximate measure of water fertility. It is also used to estimate the capacity of a water body to neutralize acidic wastes without disrupting its aquatic life. Note that a high pH value is not the same as a high alkalinity value. pH is an *intensity* measurement and alkalinity is a *capacity* measurement.

A term found in water analyses that is vaguely related to alkalinity is *hardness.* This is a very old term as applied to water; to the general public it refers to the way water behaves when it is boiled or when soap is added to it. Calcium, magnesium, and other less abundant ions in hard water react with soap to form insoluble compounds. In practical terms, soap is wasted since it will not cleanse or lather until the offending ions are precipitated. Heating hard water (in a boiler, for instance) results in the precipitation of a coating of calcium and magnesium carbonates, calcium sulfate, and other dissolved compounds. Thus hardness is a rather vague property referring to the overall effect of several dissolved constituents. The usual procedure in a chemical analysis is to add the values found for Ca^{2+} and Mg^{2+} (in milliequivalents per liter), multiply the result by 50 (half the atomic weight of $CaCO_3$), and list the value obtained as *hardness as* $CaCO_3$ in units of milligrams per liter. If this hardness value exceeds the alkalinity (in terms of $CaCO_3$), the excess is reported as *noncarbonate hardness.* The hardness values given in water analyses have no great geochemical significance; however, they are useful as an indicator of the way a given water will behave when used for domestic purposes.

The *acidity* reported in water analyses is defined as the capacity of a solution to react with hydroxyl ions. A number of different ions, such as H^+, Fe^3, and HSO_4^-, may contribute to an acidity value. An acidity value has no particular geochemical significance, and it is not reported in some water analyses. When it is reported, it may be expressed in terms of the equivalent amount of hydrogen ion or of sulfuric acid. Sometimes it is reported as the amount of calcium carbonate that would be required to neutralize the sample.

The major dissolved constituents found in most natural waters are listed in Table 4-8. Included in the table is an indication of the major source or sources for each constituent. A general range of concentration is also given for each constituent. Most of these constituents are those listed in standard chemical analyses, as in Table 4-9. In addition to the usual constituents reported in analyses, the table includes information on dissolved organic matter and on dissolved oxygen. A wide variety of other constituents can also occur in trace amounts in natural waters. It is important to note that the listing of constituents in chemical analyses such as those of Table 4-9 does not mean that those constituents actually occur in the form given. For example, part of the material reported in an analysis as NO_3^- may actually occur in the form of NO_2^- or NH_4^+.

Chemical analyses of natural waters are not as accurate as we would like because of various sources of error. A major source of error lies in the method of sampling. As with other types of investigations, results obtained are only good to the extent that the samples analyzed are representative of the water mass being studied. Often the water mass is not homogeneous, and often its composition changes or fluctuates fairly rapidly. Many water samples are analyzed in a laboratory sometime after collection. The longer the period of time between collection and analysis,

TABLE 4-8 Major Dissolved Constituents of Natural Waters

Element	Common dissolved species	Common range of concentration (mg/l)	Geologic source
Silicon	H_4SiO_4	1–30 (SiO_2)	Weathering of silicate minerals
Aluminum	Al^{3+}, $Al(OH)_4^-$	0–0.5 (Al)	Weathering of silicate minerals
Iron	Fe^{2+}, Fe^{3+}, colloidal ferric hydroxide	0–15	Iron silicate, iron oxide, and iron sulfide species
Calcium	Ca^{2+}, $CaSO_4$	4–400 (Ca^{2+})	Weathering of silicate and carbonate minerals
Magnesium	Mg^{2+}, $MgSO_4$	1–1,350 (Mg^{2+})	Weathering of silicate and carbonate minerals
Sodium	Na^+	0–10,500	Weathering of silicate minerals, sea salt
Potassium	K^+	0–380	Weathering of silicate minerals
Carbon	HCO_3^-, CO_3^{2-}, H_2CO_3, CO_2	0–1,000 (HCO_3^-)	Plants, animals, decaying organic matter
Sulfur	SO_4^{2-}, H_2S, HS^-	0–2,700 (SO_4^{2-})	Weathering of sulfide minerals
Chlorine	Cl^-	1–19,000	Weathering of sedimentary rocks, sea salt
Fluorine	F^-	0–10	Weathering of F-bearing minerals
Nitrogen	NO_2^-, NO_3^-, NH_4^+	0–15 (NO_3^-)	Atmosphere, organic waste, soil
Dissolved organic matter	Many species	0.1–10	Biological activity
Oxygen	O_2	0–12	Atmosphere, photosynthesis

Source: Data mainly from Hem (1985).

the more likely it is that unreliable values will be obtained. Analytical errors, under optimum conditions, will be small, but large errors are always possible. The overall composition of a given water sample will affect the amount of analytical error. The accuracy of a value for a given constituent will in general depend on the abundance of the constituent, with large errors possible for trace constituents. Hem (1985) states that values for solutes present in concentrations less than 1 milligram per liter probably will not, at best, be closer than ±10 percent to the actual value. For solutes that are more abundant, the values found will usually be within ±2 to ±10 percent of the actual value.

Table 4-9 contains four selected water analyses, which illustrate the relationship between water environment and water chemistry. The first analysis is from a creek in Pennsylvania. The analysis shows several unusual features, all of which are related. The pH is very low, the specific conductance is high, and the water has rather large amounts of aluminum, iron, and sulfate. This creek is located in the vicinity of coal mines, and the coal contains pyrite. Oxidation of the pyrite

TABLE 4-9 Selected Water Analyses

Constituent	1 (mg/l)	2 (mg/l)	3 (mg/l)	4 (mg/l)
Silica (SiO_2)	21	7.9	8.2	363
Aluminum (Al)	29	0.6	0	0.2
Iron (Fe)	15	11	0.04	0.06
Calcium (Ca)	119	8.4	40	0.8
Magnesium (Mg)	68	1.5	50	0
Sodium (Na)	17 (Na + K)	1.5	699	352
Potassium (K)		3.6	16	24
Carbonate (CO_3)	0	0	26	117*
Bicarbonate (HCO_3)	0	30	456	0
Sulfate (SO_4)	817	5.9	1,320	23
Chloride (Cl)	22	1.8	17	405
Fluoride (F)	0.1	0.1	1.0	25
Nitrate (NO_3)	0.4	0.4	1.9	1.8
Dissolved solids				
Calculated	1,260	47	2,400	1,310
Residue on evaporation	—	44	2,410	—
Hardness				
As $CaCO_3$	845	27	306	2
Noncarbonate	845	2	0	0
Specific conductance (microsiemens at 25°C)	1,780	68.8	3,140	1,790
pH	3.0	6.3	8.2	9.6

Notes: 1: Shamokin Creek, Weighscale, Pa. Affected by drainage from coal mines (p. 85).

2: City well, Fulton, Miss. Depth: 210 feet. Water from the Tuscaloosa Formation (p. 85).

3: Moreau River, Bixby, S. D. Drains Pierre Shale, Fox Hills Sandstone, and Hell Creek Formation (p. 102).

4: Spring, Upper Geyser Basin, Yellowstone National Park, Wyo. Also reported: 1.5 mg/l Br, 1.5 mg/l As, 5.2 mg/l Li, 0.3 mg/l I, 1.3 mg/l PO_4, and 2.6 mg/l H_2S. Temperature: 94°C (p. 70).

*According to Hem, the alkalinity represented by this number is mostly attributable to silicate.

Sources: All analyses from Hem (1985).

results in a high content of sulfuric acid in the water of the creek, thus the low pH and the high value for sulfate ion. As a result of the low pH, a large amount of iron from the general area remains in solution rather than precipitating out by formation of ferric hydroxide (Figure 4-7). Water with a pH below 4.0 can dissolve and hold large amounts of aluminum. In contrast, at pH values between 4 and 9, the solubility of aluminum is usually less than 1 milligram per liter. The acid condition of the water allows it to react with carbonate and other minerals and take into solution a large amount of various ions, indicated by the high specific conductance value and the calculated value for dissolved solids. Because of the low pH, none of the dissolved carbon dioxide occurs in the form of HCO_3^- or CO_3^{2-} (Figure 4-12). Therefore, the alkalinity as represented by carbonate and bicarbonate ions is zero. A hardness value is still reported as $CaCO_3$ (by convention), since the hardness is due to the cations Ca^{2+} and Mg^{2+} and not to anions such as HCO_3^- and CO_3^{2-}.

The groundwater reported from a well in Mississippi in the second analysis of Table 4-9 has a pH of 6.3. This water is probably somewhat depleted in oxygen and would give a reducing value if the Eh were measured. Such a conclusion is suggested by the fact that the iron content of the water is high for a pH of 6.3. The oxidizing conditions at the surface should cause the water to change from clear when taken out of the well to cloudy brown as ferric hydroxide precipitates. The aluminum value is very low since the pH is greater than 4. The pH value tells us that approximately equal amounts of H_2CO_3 (aq) and HCO_3^- occur in the water; only HCO_3^- is reported in the analysis. In comparison with the first analysis, this water has much less calcium, magnesium, sulfate, and chloride. Thus the total for dissolved solids is low, and the specific conductance value is low. An even more detailed explanation for the composition of the water could be given by using knowledge of the mineralogy and chemistry of the Tuscaloosa Formation (source of the water) at this location.

In the area of the Moreau River in western South Dakota, the country rock is fine grained and contains soluble material. Because of the rather dry climate, a high concentration of various salts builds up on the surface of the rocks and in the river water. The Moreau River (analysis 3, Table 4-9) has a high concentration of sodium and sulfate ions and a high content of dissolved solids. The pH of 8.2 results in essentially all iron being precipitated out of solution. Aluminum is also very insoluble at this pH. Alkalinity is represented by both bicarbonate and carbonate ion, since both occur at pH values higher than 8.0 (Figure 4-12). The pH of analysis 2 (6.3) is probably due mainly to dissolved carbon dioxide; the pH of analysis 3 suggests that it is in contact with solid calcium carbonate. Pure water in contact with the atmosphere at 25°C has a dissolved carbon dioxide content that produces a pH of 5.65. Pure water in contact with solid calcium carbonate has a pH of 8.4. It should be kept in mind that natural waters are not pure, and their pH values represent a balancing of a number of different reactions. A second factor that controls pH is alkalinity.* The lower pH of analysis 2 is due, in part, to its lower alkalinity.

Analysis 4 is for hot spring water from Yellowstone National Park. The two most unusual values in the analysis are the high silica value and the high pH value. Note that this water has a temperature of 94°C. It is known that temperature has a strong effect on the solubility of silica, with the solubility of amorphous silica about three times as great at 94°C as compared to 25°C. The water of analysis 4 precipitates silica when it cools. The high pH value for this water is probably due to the existence of the dissolved silica in forms such as $H_3SiO_4^-$. The equilibrium relations between SiO_2, $H_3SiO_4^-$, H^+, OH^-, and other pertinent species are such that the pH of otherwise pure water containing large amounts of silica tends to be in the range 9–11. When water reacts with silicate minerals (hydrolysis) as part of the weathering process, two products are hydroxide ions (giving a high pH) and dissolved silica. In the case of the spring water at Yellowstone National Park, extensive reaction has occurred because of the high temperature. Note that the alkalinity value in analysis 4 is given as carbonate ion, even though much of the alkalinity (capacity to neutralize acid) of this water is probably due to the high concentration of dissolved silica

*Drever (1988, 52–53) points out that alkalinity is independent of P_{CO} (and thus of dissolved CO_2). Although an increase in P_{CO} will cause an increase in HCO_3^-, the net effect on alkalinity is zero because the increase due to more HCO_3^- is balanced by a decrease in CO_3^{2-} and/or an increase in H^+.

rather than to the presence of carbonate ion. The relatively high values for chlorine and fluorine in analysis 4 are undoubtedly due to passage of the water through volcanic rocks. Various forms of chlorine and fluorine are commonly associated with volcanic and fumarolic gases.

CHARACTERISTICS OF NATURAL WATERS

In the following sections a brief summary is given of the main chemical characteristics of natural waters. The characteristics of interest are pH, Eh, and general composition in terms of the most abundant constituents. We shall not devote much attention to the solid particles in natural waters, since the amount and kind of such particles is extremely variable. These solid particles can, at times, have a great influence on the chemical composition and behavior of water bodies. Adsorption of ions on solids and various exchange reactions between solids and surrounding water are particularly important processes.

Rivers and Lakes

An extensive review of the chemical composition of rivers and lakes has been given by Livingstone (1963). Readers interested in further information are referred to this publication and to Berner and Berner (1987) and Drever (1988). The source of most of the water in rivers and lakes is rainwater. Rainwater generally contains very small amounts of various dissolved species (Berner and Berner 1987, Chap. 3). We should note that the overall composition of rainwater is quite variable, varying not only from one locality to another, but also varying with time at any one locality. Perhaps the most important characteristic of rainwater is its pH. This is mainly controlled by equilibrium with carbon dioxide in the atmosphere. The carbon dioxide of the atmosphere dissolves in the water, with the following reaction taking place:

$$CO_2 + H_2O \rightleftharpoons H_2CO_3 \rightleftharpoons H^+ + HCO_3^- \qquad (4\text{-}77)$$

The equilibrium relationships are such that hydrogen ions from this source give rainwater a pH of about 5.6. At equilibrium, the HCO_3^- formed cannot amount to more than about 0.15 parts per million.

As newly fallen water passes through soil and rock, it interacts with inorganic and organic matter. The water becomes less pure as various ions and compounds are taken into solution. Minerals contribute ions such as K^+, Mg^{2+}, and Ca^{2+}; organisms affect the content of oxygen, carbon dioxide, and nutrient elements such as phosphorus and nitrogen. The nutrients are taken out of the water by living organisms and are added to it when organisms die. The water is continually adjusting its composition as it encounters new conditions. Eventually, it enters a river or lake, either as surface runoff or as groundwater entering beneath the river or lake surface.

In general, the composition of the water of a river or lake depends on the type of soil and rock that the water has passed through and whether the main source of the water is runoff or groundwater. Because of its environment, groundwater usually contains more solutes than water from surface runoff. It has been found that the composition of rivers and lakes is quite variable, both with time and location. This is not surprising since many important factors vary with time and location. Examples are the relative contribution of runoff and groundwater, climate, the amount of organic matter present, the amount of physical mixing of the water, the amount of suspended matter, and so on. Thus a true understanding of the chemistry of a given river or lake will come only from many chemical analyses, carried out at many locations and over a long period of time.

Enough work has been done to allow an estimate of the mean composition of river waters of the world (Table 4-1). The most abundant constituents, in order of abundance, are HCO_3^-, Ca^{2+}, SiO_2, SO_4^{2-}, Cl^-, Na^+, Mg^{2+}, and K^+. Most river water, although less pure than rainwater, is still quite dilute in that the total amount of dissolved material is, on the average, only 120 parts per million (Figure 4-13). The effect of reactions between the water and various minerals is to increase the pH, because hydrogen ion in the water becomes tied up in reaction products such as H_4SiO_4 and HCO_3^-. The pH of river and lake water also depends on the carbon dioxide content of the atmosphere, unless there is no contact with the atmosphere as is the case for some deep lake water. The carbon dioxide content and pH of the water are also affected by production of carbon dioxide as a result of respiration and organic decay. The pH values of most river and lake waters vary from 6 to 8. Higher and lower values are not uncommon. Although the Eh of surface water is affected by the oxygen content of the atmosphere, it has been found that the Eh values of rivers and lakes represent a lower oxidizing state than would be expected if the water were in equilibrium with the atmosphere. Most river and lake waters have Eh values that fall in the upper (oxidizing) part of Figure 4-6. Table 4-4 gives the overall range of pH and Eh values found for rivers and lakes.

As an example of the variability of river water, we can return to the Moreau River discussed in connection with Table 4-9. Regular analyses of the river over a period of a year showed that the total ion content ranged from 160 to 3,400 parts per million, and the pH varied from 7.1 to 8.9 (Livingstone 1963, G4). During the same year the mean discharge of the river varied from 3.2 cubic feet per second to 7,200 cubic feet per second. As would be expected, the lower ion concentrations occurred at the times of highest discharge. The lower pH values also occurred at these times.

Lakes tend to be, overall, less chemically variable than rivers. However, those elements and compounds associated with biological activity can show large changes in concentration. Lake waters are often not homogeneous, showing a stratification of temperature and chemical composition when vertical mixing cannot occur at depth. The lower part of such lakes does not get a continuous supply of oxygen and tends to have a more reducing environment as a result of depletion of oxygen by respiration and decay of organic material. The pH of many lakes is also mainly controlled by biological processes (rather than by overall composition or by equilibration with the atmosphere), with photosynthesis reducing the carbon dioxide content and thus increasing the pH, while the respiration of plants and animals does the opposite. Since photosynthesis mainly occurs in the surface water, we would expect the pH to decrease with depth. The variation of Eh and pH from the top to the bottom of many lakes results in variations in the form and content of the various solutes. Iron, for example, would tend to occur as Fe^{3+} in the upper layers and as Fe^{2+} in the lower layers. If it precipitates out, it would do so as hydrous ferric hydroxide in the upper layers and as iron sulfide in the lower layers. As lakes get older, they tend to have more and more organic life. *Eutrophication* is a general name for this increase in organic activity. It is a natural process that can be greatly speeded up by pollution from human activities.

Wetlands and Estuaries

Wetlands represent a special type of aqueous surface environment. Included here are areas referred to as marshes, swamps, bogs, peatlands, and so forth. Although all wetlands have certain

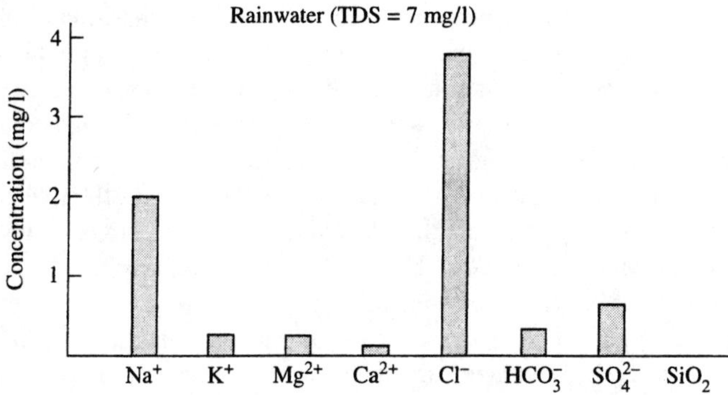

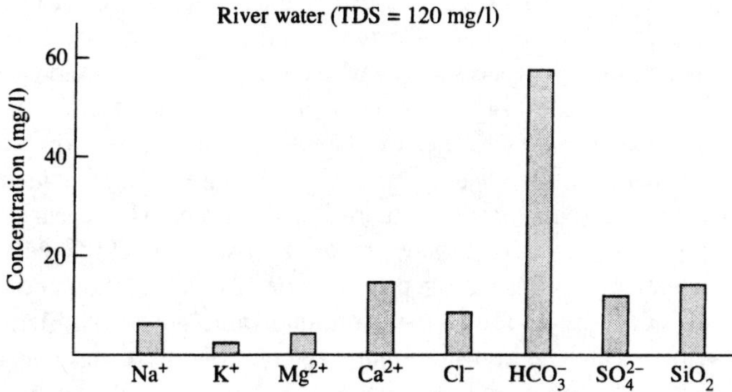

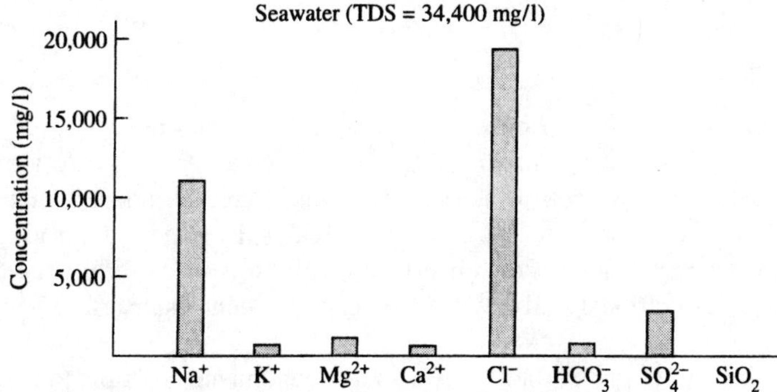

Figure 4-13 Average chemical compositions of rainwater, river water, and seawater. TDS = total dissolved solids. Note that total concentrations increase from rainwater to river water to seawater. Note also the change of scale from top to bottom. The SiO_2 concentration in rainwater and seawater is too low to appear. [After The Open University (1989, 101). Reprinted by permission of Butterworth-Heinemann Ltd.]

common characteristics, the composition of the aqueous phase varies over a wide range. Compositional variation results from the many sources of water (precipitation, runoff, groundwater influx, tides), large Eh variations, and the important role of vegetation in controlling the total chemistry of elements such as nitrogen and phosphorus. Wetlands often serve as sinks for chemicals transported to them. Various elements are removed by precipitation, adsorption, sediment-water exchange, and uptake by vegetation. Wetlands are effective in removing nitrogen from polluted waters flowing through them. Anaerobic bacteria break down the nitrate ion to obtain oxygen and nitrogen is released as N_2 or N_2O into the atmosphere. A detailed review of wetlands can be found in Mitsch and Gosselink (1986).

Estuaries represent another special type of aqueous surface environment. They occur at the contact zone between two very different water compositions, river water and seawater (Figure 4-13). They also have associated wetlands. Three important characteristics are variable salinity (vertically and horizontally), a high rate of photosynthesis, and the occurrence of regular tidal fluctuations. As with wetlands, estuaries serve as sinks of chemicals brought to them by rivers and other sources. Flocculation of clay minerals, organic matter, and various oxides and hydroxides is a major removal process, along with those processes mentioned above for wetlands.

One goal of geochemistry is to describe quantitatively the cycles followed by elements at the Earth's surface. The passage in water from land to the oceans is an important part of the cycles of many elements. A major question involves the conservative versus nonconservative behavior of elements as they pass through estuaries on their way to the oceans. A conservative element exhibits no loss or gain as river water mixes with seawater (Figure 4-14). Many elements behave nonconservatively, being removed in estuaries by flocculation and other processes, or, more rarely, being added by various organic and inorganic processes. Elements removed from water entering estuaries include iron, aluminum, and phosphorus. Conservative elements include sodium, potassium, magnesium, and calcium. Some elements, such as silicon (as dissolved silica), are conservative in some estuaries and removed or added in other estuaries. Further discussion of this topic can be found in Berner and Berner (1987).

Subsurface Waters

The chemical compositions of various types of subsurface water are reviewed by White et al. (1963). Most subsurface water is groundwater (i.e., water that occurs below the water table). White et al. discuss both dilute subsurface water (the most typical kind) and mineral water (characterized by dissolved material in concentrations greater than a few thousand milligrams per liter). About 300 chemical analyses of subsurface waters from many different environments are given by White et al. Additional analyses and further discussion of subsurface water can be found in Hem (1985).

As mentioned earlier, subsurface waters tend to contain more dissolved material than river water because of their more intimate and longer contact with organic material and with soil and rock particles. Subsurface waters tend to be less well mixed and thus less homogeneous than surface waters. Often there is a fairly direct relationship between the composition of a given subsurface water and its host rock (or soil). This does not always hold true, since a number of factors affect the composition of subsurface water. The porosity and permeability of the host rock are obviously important. A major factor is the abundance and type of organic matter present in the host rock and in the water. Often the anion composition of subsurface waters is controlled by organic reactions, whereas the cation composition is controlled by inorganic reactions. Water in contact

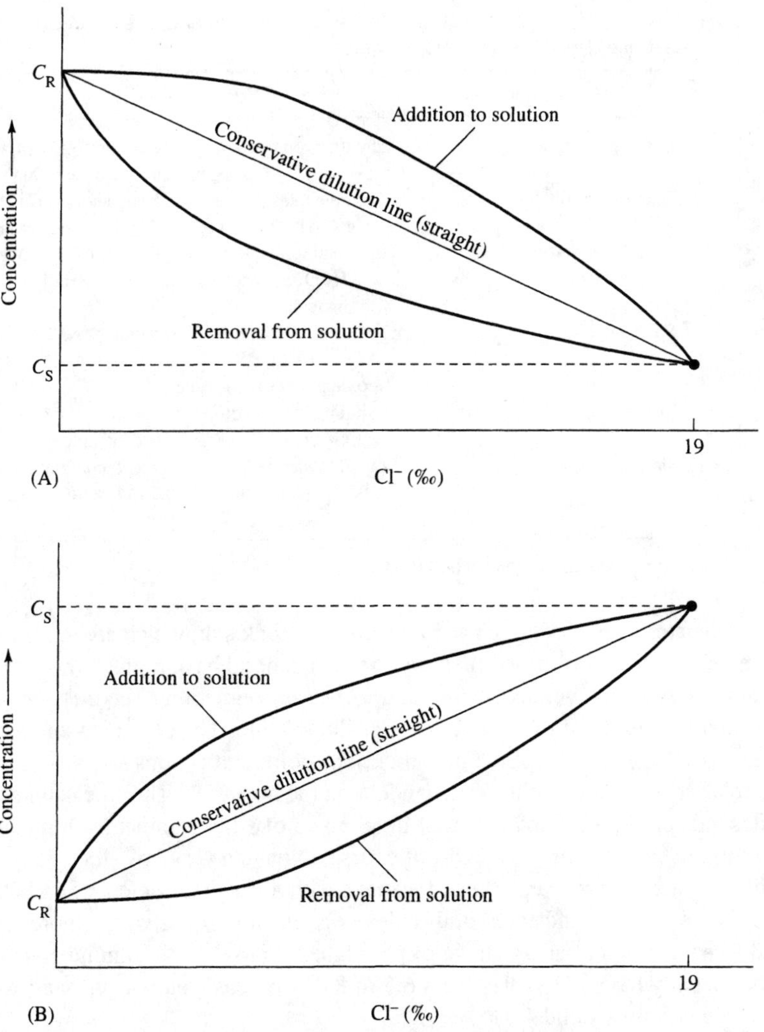

Figure 4-14 Idealized plots for estuarine water of the concentration of dissolved components versus chloride (which serves as a conservative measure of the degree of mixing between fresh water and seawater). C_R = concentration in river water; C_S = concentration in seawater. (A) Component whose concentration in fresh water is greater than it is in seawater (for example P, N, Si). (B) Component whose concentration in fresh water is less than it is in seawater (for example Ca, Mg, K). (After Berner, E. K. and Berner, R. A., *The Global Water Cycle—Geochemistry and Environment*, Copyright © 1987. p. 295. Adapted by permission of Prentice Hall, Inc., Upper Saddle River, NJ.) After Liss (1976, 95).

or recently in contact with the atmosphere will have a different chemistry from that of water which has been long out of contact with air.

Table 4-10 lists generalized characteristics of dilute subsurface waters whose compositions have been mainly affected by reactions with their host rocks. Water from rocks containing one or more chemically resistant minerals, such as felsic igneous rocks, tends to have a lower total ion

TABLE 4-10 Characteristics of Subsurface Waters of Low Mineral Content Associated with Common Rock Types

Rock types	Water characteristics
Granite, rhyolite	Low total ion content; dominant ions Na^+, HCO_3^-; pH 6.3 to 7.9; SiO_2 content moderate to high
Gabbro, basalt	Moderate total ion content; dominant ions Ca^{2+}, Mg^{2+}, HCO_3^-; pH 6.7 to 8.5; SiO_2 content high
Sandstone, arkose, graywacke	High total ion content; dominant ions Ca^{2+}, Mg^{2+} Na^+, HCO_3^-; pH 5.6 to 9.2; SiO_2 content low to moderate
Siltstone, clay, shale	High total ion content; dominant ions Na^+, Ca^{2+}, Mg^{2+}, HCO_3^-; SO_4^{2-}, Cl^-; pH 4.0 to 8.6; SiO_2 content low to moderate
Limestone, dolomite, marble	High total ion content; dominant ions Ca^{2+}, Mg^{2+}, HCO_3^-; pH 7.0 to 8.2; SiO_2 content low
Slate, schist, gneiss	Low to moderate total ion content; dominant ions HCO_3^-; Ca^{2+}, Na^+; pH 5.2 to 8.1; SiO_2 content low

Source: Data from White et al. (1963).

content than water from rocks such as limestone and rock salt, which are made up of chemically soluble minerals. Waters from sedimentary rocks in general have high total ion contents, because most sedimentary rocks were deposited in a salty environment and because cementing material and adsorbed ions in these rocks tend to be easily soluble. The dominant anion in most subsurface waters is HCO_3^- for all types of host rock. The dominant cations are Na^+, Ca^{2+}, and Mg^{2+}, with the relative abundance of these dependent on the host rock. The silica content of subsurface waters depends on the ease and extent of breakdown of silicate minerals in the host rock. Water in rocks without silicate minerals tends, of course, to have a very low silica content. The pH values of dilute waters show a large range for rocks with a variable content of carbonate and sulfide minerals. Examples are sandstones and shales. Waters from rocks with a more regular composition tend to have a more narrow range of pH values. Thus waters from igneous rocks and from limestones usually have pH values from 6.5 to 8.0, whereas waters from sandstones and shales can have values from around 4.0 to over 9.0.

In contrast to the meteoric, dilute waters discussed above, the connate waters (trapped during rock formation) from deeply buried sedimentary rocks tend to have a high content of dissolved material. Total ion content can go as high as 300,000 milligrams per liter. Such waters are known as brine, mineral, or saline waters. For these waters the dominant cation is usually Na^+, and the dominant anion is Cl^-. Spring water may be mainly made up of meteoric or connate water. Subsurface waters with special characteristics are found in areas of recent or active volcanism. These waters, often very hot, tend to have very high contents of silica as a result of intense reaction with host rocks. They usually contain high amounts of one or more of the following: hydrogen sulfide, boron, sulfate, lithium, fluoride, and arsenic. They often contain trace amounts of heavy metals such as copper and lead.

The Eh values of subsurface waters show a wide range, with near-surface waters tending to have positive (oxidizing) values and deep waters tending to have negative (reducing) values. Deep water generally has lost its dissolved oxygen by various oxygen-consuming reactions, such as those involved in the decay of organic matter. It contains anaerobic organisms that break down sulfate ion to hydrogen sulfide as part of the continued oxidation of organic matter (the organic carbon is the reducing agent). Another ion that serves as an oxidizing agent for organic matter is Fe^{3+}. The overall trend produced by the various reactions is a slow decrease in the Eh value of the water. Strong oxidizing agents are reduced, and weaker oxidizing agents such as carbon dioxide are formed.

Seawater

Seawater is briefly discussed in Chapter Three as an example of a solution that has a high enough concentration of solutes to cause the concentration and activity values of a given constituent to be significantly different. Millero (1974) gives a good summary of the thermodynamic properties of seawater. A brief review of the overall composition of seawater is given in Chapter One, and abundance values can be found in Table 1-11. Seawater is a dilute solution in the sense that the total amount of dissolved material (approximately 35,000 milligrams per liter) is low compared to the maximum amount that could be dissolved. On the other hand, it is a concentrated solution compared to river water and to dilute subsurface waters. The major ions of seawater are listed in Table 3-4. Sodium and chloride are the dominant ions. The most abundant ions, in order, are chloride, sodium, sulfate, magnesium, calcium, and potassium. Ninety-nine percent of all the dissolved material, excluding gases, is made up of these six components.

The concentrations of elements and compounds not involved in biological activity are essentially constant in seawater, varying only within a narrow range. Relative abundances are the same throughout the oceans, even when the total concentration of dissolved salts (the salinity) shows small variations. The major dissolved gases (N_2, O_2, CO_2, and H_2S) have concentrations that are directly related to organic activity. Biological activity in the oceans varies in intensity both horizontally and vertically, and so does the dissolved gas content. Most organic activity (particularly photosynthesis) is concentrated near the surface, with nutrient (fertilizing) elements such as phosphorus (from PO_4^{3-}) and nitrogen (from NO_3^-) being taken out of the water. Decay of organic material, particularly at depth, returns these elements to the water. Chemical variations in the composition of seawater are thus determined by organic activity and by the extent of mixing of surface and deep waters. When mixing occurs, the water brought from depth is often rich in nutrient elements, and an organism-rich surface zone results. When mixing does not occur, oxygen becomes depleted in the deep water, a reducing value for Eh develops, and sulfate ion is broken down by the same process discussed earlier for anaerobic subsurface water. The pH of surface seawater is about 8.2, which roughly corresponds to what would be found for equilibrium with calcium carbonate and with carbon dioxide in the atmosphere. Reactions involving carbon dioxide and calcium carbonate tend to keep the pH value constant; thus seawater is a buffered solution.

Any detailed study of marine chemistry has to be concerned with three types of exchanges: (1) between seawater and the atmosphere, (2) between seawater and bottom sediments, and (3) between seawater and living organisms. Oxygen and carbon dioxide are both exchanged with the atmosphere. In addition, they are produced by organic processes. In near-surface waters the dissolved oxygen concentration is about 5 milliliters per liter of seawater. The concentration of molecular carbon dioxide plus carbonic acid is about 0.2 milliliter per liter. Marine organisms add and subtract a number of elements, particularly carbon, oxygen, nitrogen, phosphorus, silicon, and various trace metals. Organic compounds occur dissolved in seawater, and small particles of living and dead organic material are common. We know very little about this organic material, much of which has not yet been identified. Revelle and Suess (1957) estimate that only about 0.3

percent of the organic carbon in the oceans is associated with living organisms. The total concentration of dissolved organic compounds ranges from about 0.1 to as much as 10 milligrams per liter. Further discussion of the organic geochemistry of seawater can be found in Chapter Six. Processes involving seawater and bottom sediments include compaction and cementation of sediment, ion exchange with clay minerals, diffusion of elements through sediment pore waters, chemical alteration ("weathering") of detrital rocks and minerals, and precipitation and dissolution of compounds such as calcium carbonate and manganese oxide. The narrow zone divided by the sediment-water interface is chemically complex, and a critical area in understanding sediment formation and diagenesis.

A number of chemical reactions occur at spreading centers where submarine volcanism is active (Von Damm et al. 1985; Von Damm 1990). Cold seawater penetrates the sea floor, reacts chemically, and is heated as it moves through ocean crust (mainly basalt). It emerges as hot, chemically active water at mid-ocean ridges and other spreading centers. The nature and extent of the chemical reactions below and above the sea floor depends on temperature and the water-rock ratio. Laboratory experiments have duplicated the hydrothermal alteration seen in rocks from ocean spreading centers and from geothermal areas on land. Elements removed from seawater by water-rock interaction include potassium at low temperatures ($<50°C$) and magnesium at high temperatures ($>50°C$). Some elements appear to be released to solution at high temperatures and taken up by basalt at low temperatures. Leaching of metals from the rock, and reduction of seawater SO_4^{2-} to H_2S, could result in the development of an ore-forming solution. The overall effect of these processes on seawater composition is not well known.

Since the proportions of the major constituents of seawater are essentially constant, it is possible to determine the *salinity* (total concentration of dissolved material) by analyzing only one constituent. Early marine chemists selected chlorine and defined the *chlorinity* of seawater as the equivalent amount of halogens (F, Cl, Br, I) as measured by volumetric titration with silver nitrate. Since the other halogens occur in very low concentrations, chlorinity can be defined as the number of grams of chloride in 1,000 grams of seawater. The relation of salinity to chlorinity is approximately

$$\text{salinity (ppt)} = 1.806 \text{ chlorinity (ppt)}$$

Salinity is usually expressed as parts per thousand by weight.

Another way to determine salinity is to measure the electrical conductivity of water samples. Instruments developed for this type of measurement (salinometers) have proved to be more convenient and more precise than the older titration method. Salinity can be measured to ± 0.003 parts per thousand. The concentration of all the major constituents of a given sample of seawater can be calculated from the salinity measurement. Note, however, that a salinity value determined from conductivity depends on temperature and pressure and is not exactly equal to the total dissolved salts in a sample. Determining abundances of trace elements and of dissolved gases requires special analytical techniques. Readers interested in further details on seawater are referred to Strickland and Parsons (1972) and to The Open University (1989).

Fluid Inclusions

Fluid inclusions in crystals represent a special type of natural fluid. They are actual samples of fluids that existed at the time rocks formed or at a later time when they were altered in some way. Data obtainable from fluid inclusions provide information on temperature and pressure at the time of rock formation or alteration, and density and composition of the fluids involved in a rock's history. Often data obtained from fluid inclusions cannot be obtained in any other way. A comprehensive review of fluid-inclusion work has been published by Roedder (1984). An update of this research is provided by the collection of articles published in the March, 1990 issue of *Geochimica et Cosmochimica Acta* (v. 54, no. 3).

MODELING OF WATER CHEMISTRY

A model of water chemistry is an attempt to describe quantitatively the reactions and species that control the chemical composition of a selected water body. Such models are used to explain the origin of present chemistry or to predict future changes in composition and properties. Most models are computer-based.

Some computer programs use a water analysis as input and calculate concentrations, activities, activity coefficients, and state of saturation (speciation-saturation programs). Others add the effects of processes such as precipitation, dissolution, adsorption, and oxidation-reduction (mass balance and reaction-path programs). The most active modeling research has been applied to hydrothermal solutions, contaminant transport in groundwater, major-element chemistry of groundwater, interstitial water in sediments, and solute transport in streams. Two examples of this type of research will be given here. Further information can be found in Melchior and Bassett (1989).

Rogers (1987) has modeled the chemical evolution of groundwater in stratified-drift and arkosic bedrock aquifers of north-central Connecticut. Observed changes in water chemistry (from chemical analyses of groundwater samples) were used to determine the quantities of minerals and gases that must enter or leave solution to produce the changes found between selected initial and final points. The type of program used is known as a mass balance or mass transfer program. In addition, various models using hypothetical sets of reactions were tested as possible simulants of the evolution of the water chemistry as the water moved through the two aquifers (Figure 4-15). These reaction-path models use speciation-saturation and mass balance calculations along with selected thermodynamic constraints (Plummer et al. 1983).

The four models illustrated in Figure 4-15 have the following characteristics:

1. Reaction path model RP-1 is for a system that is closed to CO_2 and the CO_2 pressure of the system is controlled internally by water-rock reactions.
2. In model RP-2 the system is open to CO_2 and the partial pressure of the gas is externally fixed at $10^{-2.35}$ atm.
3. For model RP-3 the system is partially open to CO_2 and the partial pressure of the gas decreases at a constant rate to a final pressure of $10^{-2.76}$ atm.
4. Model RP-4 is similar to RP-3 except that the CO_2 partial pressure decreases at a slower rate to a final value of $10^{-3.15}$ atm.

Comparison of the composition of water samples with reaction path lines such as those of Figure 4-15 resulted in the conclusion that model RP-4 gives the best simulation of the actual changes in water chemistry in the stratified drift. Rogers (1987) also concluded that openness of the system to CO_2 in the unsaturated zone is a major factor controlling water chemistry in the aquifer. It appears that the rate at which CO_2 is transported into the groundwater system and the partial pressure of the gas both decrease as water moves along flow paths.

It is possible to combine a mass balance model with a hydrologic transport model to study the change in shape and composition with time of a contamination plume in groundwater

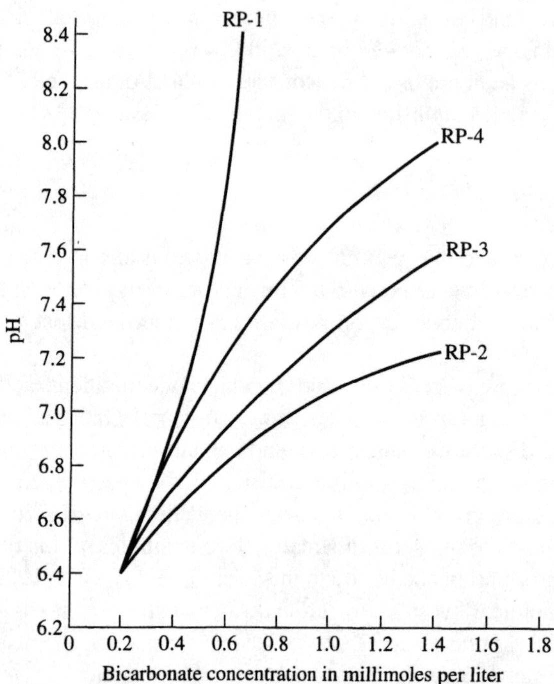

Figure 4-15 Simulated relationship between pH and HCO_3^- in stratified drift. Each line is a reaction path calculated for a specific model of the system. All four models start with a pH of 6.4 and a CO_2 pressure of $10^{-2.35}$ atm as determined from the least evolved water in the stratified drift. See text for details. (After R. J. Rogers, *Water Resources Research*, vol. 23, pp. 1531–1545, 1987, copyright by the American Geophysical Union.)

(Figure 4-16). Narasimhan et al. (1986) used a three-stage sequential approach to model the hydrology and geochemistry of a uranium mill tailings plume at Riverton, Wyoming. In stage one they used a fluid flow program to determine the quantity of tailings water that entered the groundwater system. Stage two involved calculation of the chemical results of mixing tailings water and groundwater (done with a combined mass transport–mass balance program). In the last stage the authors simulated the evolution of the plume for individual components such as sulfate ion using a mass transport program. They compared the outcome of each of the three stages with corresponding field measurements and found reasonably good agreement. Narasimhan et al. point out that this type of modeling can be used to study other types of surface and groundwater contamination and to study natural processes such as supergene sulfide enrichment.

SUMMARY

The unusual properties of water are due to the dipolar nature of the water molecule. The bonding of H_2O molecules is stronger than in most compounds, resulting in a high melting point for ice and a high boiling point for liquid water. Because of the uneven distribution of electric charge on water molecules, water has a high dielectric constant and a high capacity to dissolve substances that form ions in solution.

The maximum amount of a given compound that can dissolve in a solution is known as its solubility and is expressed in the form of an equilibrium constant known as the solubility product. The solubility product for a saturated solution is

$$K_{sp} = \frac{\text{products}}{\text{reactants}}$$

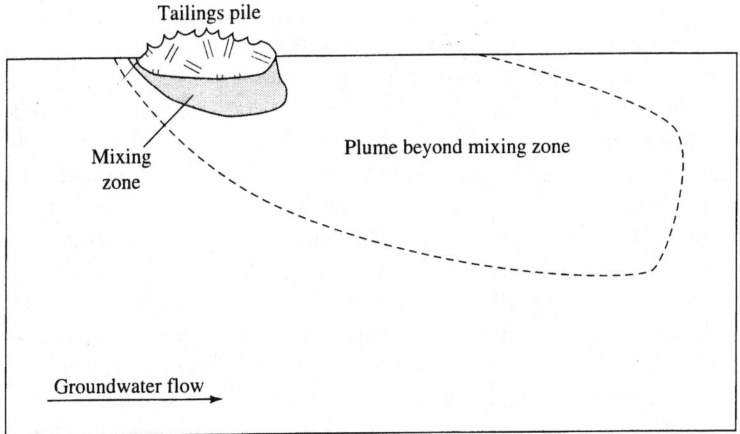

Figure 4-16 Migration of contaminants from a uranium mill tailings pile. At most sites precipitation of metal oxides and hydroxides occurs in the mixing zone due to reaction of the acidic tailings pore water with calcite in the soil. Such precipitation lowers the very high contaminant levels in the tailings water. A mixture of tailings water and groundwater containing lower levels of radioactive and trace metal contaminants spreads out in a plume below and beyond the mixing zone. The mixing zone is only a few feet thick while the plume may cover several hundred acres and go to a depth of at least 150 feet. (After Drever, J. I., *The Geochemistry of Natural Waters,* 2d ed. Copyright © 1988. p. 360. Adapted by permission of Prentice Hall, Inc., Upper Saddle River, NJ.)

where the species involved are expressed as activities. Solubility products are not of great use in the study of natural waters because we must consider the effects of one or more of the following: (1) other ions in solution, (2) hydrolysis reactions, (3) complex ion formation, and (4) development of colloidal solutions. Much of the needed information in a given problem is not known; thus it is difficult to correct solubility calculations for the above factors, which generally tend to increase the solubility over that found for pure water.

The two most important chemical properties of natural waters are pH and Eh, where pH is a measure of the hydrogen ion activity (abundance of protons) and Eh is a measure of the oxidizing power (electron activity or availability) of a solution. The oxidizing power represents a balancing (metastable equilibrium) of the activities of all the oxidizing and reducing agents in the solution. Oxidizing and reducing agents can be listed in the form of half-reactions, each with a unique potential relative to the reference hydrogen half-reaction. Two half-reactions can be combined to find the standard potential of a given reaction. By comparing this potential with the measured potential of a particular solution, it is possible to determine the direction in which the reaction will tend to go and also to determine the abundances of the species involved. The key equation relating measured potential to the standard potential of a reaction and to species abundance is

$$\mathrm{Eh} = E^0 + \frac{0.059}{N} \log \frac{\text{products}}{\text{reactants}}$$

Although it is difficult to measure accurately the Eh values of natural solutions and materials, such values in combination with measured pH values are very useful in explaining the chemical properties of these solutions and materials. Eh values have been found to vary from about +0.8 to –0.5 volt, and pH values to vary from less than 1.0 to more than 10.0. The major factors affecting Eh and pH are (1) the organic processes of photosynthesis, respiration, and decay; (2) oxidation-reduction reactions involving iron, sulfur, nitrogen, and carbon; and (3) the balance between dissolved carbon dioxide and calcium carbonate in natural waters. A useful way to represent the chemical conditions of natural environments is with an Eh-pH diagram. On such a diagram, calculated stability fields of various dissolved species and of solids can be outlined. Careful application of such diagrams to natural environments makes it possible to explain many of the observed chemical properties of these environments.

We have very limited knowledge of the properties of water solutions under conditions of high temperature and pressure. Pure water has a critical point, beyond which it exists as a supercritical fluid rather than as a liquid or a gas. Igneous and metamorphic processes involving water may differ greatly, depending upon whether or not the water is above its critical point. Prevailing pressure conditions determine the density of water vapor and supercritical water, and in turn the density is a major control on solubility, with extensive solubility occurring at high densities.

The kinetics of reactions in or involving water depend upon pressure, temperature, water volume, rate of mixing, and general chemical environment. Experimental data can be used to describe the rate at which a given chemical reaction occurs. Reactions involving only liquid material are usually rapid (seconds to hours); reactions in which solids take part are usually slow (years to thousands of years). For many geologic reactions, kinetics is the controlling factor rather than thermodynamics.

The evaluation of analyses of natural waters requires knowledge of the concepts and definitions presented in this chapter. Specifically, an understanding of the significance of the following terms is required: pH, Eh, dissolved solids, specific conductance, dissolved oxygen, and alkalinity. River and near-surface lake waters generally have a pH near 7, low dissolved solid content, and a high Eh. Seawater has a pH near 8, high dissolved solid content, and a variable Eh. Subsurface water has characteristics ranging from those of rivers and lakes to those of seawater. The following are average values in parts per million for total dissolved solids: rainwater, 7; rivers, 120; subsurface waters, 900; seawater, 34,400.

The most abundant constituents found in natural waters and formed as a result of inorganic reactions are HCO_3^-, Cl^-, Ca^{2+}, Na^+, Mg^{2+}, SiO_2, and SO_4^{2-}. Species whose abundance is related to organic reactions include O_2, CO_2, SO_4^{2-}, H_2S, PO_4^{3-}, and NO_3^-. Models of water masses can be used to explain the origin of present chemistry or to predict future changes.

QUESTIONS

1. (a) Calculate the solubility of siderite in pure water at 25°C. Express the solubility as moles per liter, as grams per liter, and as parts per million.

(b) Calculate the solubility of siderite in river water with the average composition given in Table 4-1. Assume that all dissolved iron is Fe^{2+}.

(c) If some of the river water passes through an area of carbonate rocks where CO_3^{2-} is added to the water, what concentration of CO_3^{2-} could lead to precipitation of siderite?

2. (a) What is the solubility of anhydrite in pure water at 25°C? Would the solubility in seawater be greater or less than this value?

(b) Write an expression for the equilibrium constant for the reaction gypsum $\rightleftharpoons$ anhydrite + water. Assume that the reaction involves evaporating seawater with H_2O activity less than 1.

(c) Use free-energy values from Table 3-2 to determine which mineral is stable at 25°C and 1 atmosphere pressure. (This was also calculated in Problem 6 of Chapter Three.)

(d) The equilibrium constant for the reaction of part (b) is found experimentally to be about 0.61 at 25°C and about 0.85 at 50°C (Hardie 1967, 171–200). Use these values to plot the line dividing stability fields of gypsum and anhydrite on a diagram with variables of water activity and temperature.

3. What is the difference between an *activity product* and a *solubility product*?

4. (a) Calculate the equilibrium constant for the following reaction at 25°C:

$$2\ CaCO_3\ \text{(calcite)} + Mg^{2+} \rightleftharpoons CaMg(CO_3)_2\ \text{(dolomite)} + Ca^{2+}$$

(b) Suppose that a solution with ten times as much Ca^{2+} as Mg^{2+} were to pass through some limestone. Would you expect dolomite to form in place of calcite?

5. Given: the solubility of SO_3 in water is 0.49 liter per liter of water at 25°C when the pressure of SO_3 is 2 atmospheres. What is the value of K for the following reaction?

$$SO_3\ (g) + H_2O \rightleftharpoons H_2SO_4$$

Assume that all the dissolved SO_3 forms sulfuric acid.

6. (a) What is the pH of a solution with a hydrogen ion concentration of 0.00002 mole per liter?

(b) Given is a solution with a pH of 0. What is the hydrogen ion concentration in moles per liter?

7. What is the Eh of the half-cell reaction $Cu \rightleftharpoons Cu^{2+} + 2e^-$ as measured against a standard hydrogen electrode if the activity of Cu^{2+} is $10^{-3.0}\ M$?

8. (a) Given is groundwater with a pH of 7 and an Eh of 0.6 volt. If this water passed through a lead ore deposit, how much Pb^{2+} in solution would be necessary to cause precipitation of plattnerite (PbO_2)? Use the following half-reaction:

$$Pb^{2+} + 2\ H_2O \rightleftharpoons PbO_2\ (s) + 4\ H^+ + 2e^- \qquad E^0 = +1.46\ V$$

(b) If the solution became more alkaline, would the amount of PbO_2 precipitated increase or decrease?

9. (a) Given is a stream with an Eh of 0.92 volt and a pH of 5.5. Which form of manganese, Mn^{2+} or Mn^{3+}, would be the most abundant in the stream?

(b) Calculate the ratio Mn^{2+}/Mn^{3+} for the stream.

10. (a) Write a half-reaction for the oxidation of ferrous iron to hematite.

(b) Calculate E^0 for this reaction.

(c) Plot a line representing this equation on an Eh-pH diagram. Assume a value of $10^{-6}\ M$ for Fe^{2+}.

(d) If stream water with a pH of 6 and an Eh of 0.47 volt passed over a supergene deposit of hematite, what concentration of Fe^{2+} in the water would be in metastable equilibrium with the hematite?

11. Indicate whether the oxidation potential was high or low when each of the following was formed.

(a) meteorites

(b) Chilean nitrate deposits

(c) coal

(d) sedimentary manganese deposits

(e) carnotite

(f) pyrite in shale

12. (a) Some rivers show a variation in pH with the seasons of the year. Would you expect a river to have a higher pH in the summer or in the winter? Explain your answer.

(b) Some lakes show a variation in pH over a 24-hour period. Would you expect a lake to have a higher pH during the day or at night? Explain your answer.

13. Assume halite dissolves in a supercritical fluid according to the following reaction:

$$NaCl\ \text{(halite)} = Na^+ + Cl^- \qquad K = [Na^+][Cl^-]$$

Using Figure 4-9b, plot a graph of log K versus temperature for supercritical water at a pressure of 2 kb and for the temperature range 400°C to 800°C . You do not need to calculate actual values for log K.

14. (a) Derive an expression for the half-life of a first-order reaction.

(b) If the rate constant for a first-order irreversible reaction is 10^{-3} sec^{-1}, what is the half-life of the reaction?

(c) If the term A in the Arrhenius equation is 10^{10} sec^{-1} at 25°C for the reaction in part (b), what is the activation energy for the reaction?

15. (a) Water draining coal mine areas is often acidic because of oxidation of pyrite in the coal to form sulfuric acid and ferrous ion. Such oxidation may occur by reaction of O_2 adsorbed on pyrite grain surfaces (oxygenation) or by reduction of ferric ion in solution by pyrite. Write equations for these two reactions.

(b) The ferrous ion formed in the above reactions can react with dissolved O_2 to form additional ferric ion. Write an equation for this reaction.

(c) The ferric ion formed in the above reaction may hydrolyze to form "ferric hydroxide." Write an equation for this reaction. Would the water tend to become more acid as a result of this reaction?

(d) Stumm and Morgan (1970, 540–542) propose that the oxidation of pyrite, once started, mainly occurs in two steps: (1) oxidation of ferrous ion to ferric ion, and (2) oxidation of pyrite by ferric ion. Singer and Stumm (1970, 1121–1123) report that the oxidation of pyrite is found experimentally to be independent of the surface area of pyrite grains. If this is true, which of these two steps would you guess would be slower and, therefore, the rate-determining step for the overall change of pyrite to sulfuric acid?

16. (a) Calculate the approximate fraction of total dissolved CO_2 existing as each of the three forms H_2CO_3, HCO_3^-, and CO_3^{2-} for a pH of 8.3. The pertinent reactions and equilibrium constants are

$$H_2CO_3\,(aq) \rightleftharpoons H^+ + HCO_3^- \qquad K = 10^{-6.4}$$

$$HCO_3^- \rightleftharpoons H^+ + CO_3^{2-} \qquad K = 10^{-10.3}$$

Compare your answer with Figure 4-12.

(b) Use Figure 4-12 to estimate the concentration of H_2CO_3, HCO_3^-, and CO_3^{2-} in a solution with a pH of 7.0 and with total dissolved CO_2 equal to 0.001 M.

17. Express the alkalinity of the water from the Moreau River listed in Table 4-9 in terms of (a) milliequivalents per liter and (b) milligrams per liter as $CaCO_3$.

18. Given the following chemical results for water samples from an estuary:

	Sample Number				
	1	2	3	4	5
Dissolved Iron (μm/l)	4.1	1.7	0.95	0.61	0.18
Salinity (‰)	0.5	4.9	10.2	14.8	19.7

Plot a curve for dissolved iron (vertical axis) versus salinity (horizontal axis). Suggest an explanation for the variation of dissolved iron with salinity.

19. (a) Some spring water in Arizona was chemically analyzed in the field, then stored in a bottle, and reanalyzed six months later. The new analysis showed a decrease in calcium from 93 to 22 mg/l, a decrease in HCO_3^- from 556 to 297 mg/l, and an increase in pH from 7.5 to 8.4. All other constituents gave the same results as in the first analysis. Explain the changes that occurred.

(b) Water from a well in Los Angeles County, California, showed a change with time in its chemical composition. Calcium went from 27 to 438 mg/l, magnesium from 11 to 418 mg/l, sodium from 82 to 1,865 mg/l, SO_4^{2-} from 40 to 565 mg/l, and Cl^- from 40 to 4,410 mg/l. Suggest an explanation for this change.

20. Compare models RP-1 and RP-2 in Figure 4-15. Note that the pH increases much more rapidly in RP-1 as compared to RP-2 and bicarbonate concentration increases much more rapidly in RP-2 as compared to RP-1. Explain these differences in the two models.

REFERENCES

American Public Health Association and others. 1985. *Standard Methods for the Examination of Water and Wastewater,* 16th ed. Washington, D.C.

Baas-Becking, L. G. M., I. R. Kaplan, and O. Moore. 1960. Limits of the natural environment in terms of pH and oxidation-reduction potentials. *J. Geol.,* v. 68:243–284.

Berner, E. K., and R. A. Berner. 1987. *The Global Water Cycle–Geochemistry and Environment.* Prentice-Hall, Inc., Englewood Cliffs, NJ.

Berner, R. A. 1971. *Principals of Chemical Sedimentology.* McGraw-Hill Book Company, New York.

Best, M. G. 1982. *Igneous and Metamorphic Petrology.* W. H. Freeman and Company, New York.

Brookins, D. G. 1988. *Eh-pH Diagrams for Geochemistry.* Springer-Verlag, Inc., New York.

Drever, J. I. 1988. *The Geochemistry of Natural Waters,* 2d ed. Prentice-Hall, Inc., Englewood Cliffs, NJ.

Dunne, T., and L. B. Leopold. 1978. *Water in Environmental Planning.* W. H. Freeman and Company, New York.

Eugster, H. P. 1986. Minerals in hot water. *Am. Min.,* v. 71, nos. 5 and 6:655–673.

Evans, R. C. 1964. *An Introduction to Crystal Chemistry,* 2d ed. Cambridge University Press, New York.

Fyfe, W. S., N. J. Price, and A. B. Thompson. 1978. *Fluids in the Earth's Crust.* Elsevier Publishing Company, New York.

Garrels, R. M., and C. L. Christ. 1965. *Solutions, Minerals, and Equilibria.* Harper & Row, Inc., New York.

Hardie, L. A. 1967. The gypsum-anhydrite equilibrium at one atmosphere pressure. *Am. Mineralogist,* v. 52:171–200.

Hem, J. D. 1985. *Study and Interpretation of the Chemical Characteristics of Natural Water,* 3d ed. U.S. Geol. Survey Water-Supply Paper 2254.

Hostettler, J. D. 1984. Electrode electrons, aqueous electrons, and redox potentials in natural waters. *Am. J. Sci.,* v. 284:734–759.

Jahns, R. H., and C. W. Burnham. 1969. Experimental studies of pegmatite genesis: I. A model for the derivation and crystallization of granitic pegmatites. *Econ. Geol.,* v. 64:843–864.

Jones, P., M. L. Haggett, and J. L. Longridge. 1964. The hydration of carbon dioxide. *J. Chem. Educ.,* v. 41:610–612.

Krauskopf, K. 1979. *Introduction to Geochemistry,* 2d ed. McGraw-Hill Book Company, New York.

Lasaga, A. C. 1981. Rate laws of chemical reactions. In *Kinetics of Geochemical Processes,* Rev. in Mineralogy, v. 8, A. C. Lasaga and R. J. Kirkpatrick, eds. Mineralogical Society of America, Washington, D.C. 1–68.

Lasaga, A. C., and R. J. Kirkpatrick, eds. 1981. *Kinetics of Geochemical Processes,* Rev. in Mineralogy, v. 8. Mineralogical Society of America, Washington, D.C.

LISS, P. S. 1976. Conservative and non-conservative behavior of dissolved constituents during estuarine mixing. In *Estuarine Chemistry,* J. D. Burton and P. S. Liss, eds. Academic Press, New York. 93–130.

LIVINGSTONE, D. A. 1963. *Chemical Composition of Rivers and Lakes.* U.S. Geol. Survey Profess. paper 440–G.

MELCHIOR, D. C., and R. L. BASSETT, eds. 1989. *Chemical Modeling of Aqueous Systems II.* Symposium Series no. 416, American Chemical Society, Washington, D.C.

MILLERO, F. J. 1974. The physical chemistry of seawater. In *Ann. Rev. of Earth Planet. Sci.,* v. 2. F. A. Donath, ed. 101–150.

MITSCH, W. J., and J. G. GOSSELINK. 1986. *Wetlands.* Van Nostrand Reinhold, New York.

NARASIMHAN, T. N., A. F. WHITE, and T. TOKUNAGA. 1986. Groundwater contamination from an inactive uranium mill tailings pile, 2, Application of a dynamic mixing model. *Water Resources Research,* v. 22, no. 13:1820–1834.

NORDSTROM, D. K. and J. L. MUNOZ. 1987. *Geochemical Thermodynamics, Guide to Problems.* Blackwell Scientific Publications, Palo Alto, CA.

OPEN UNIVERSITY. 1989. *Seawater: Its Composition, Properties, and Behaviour.* Pergamon Press, New York.

PLUMMER, L. N., D. L. PARKHURST, and D. C. THORSTENSEN. 1983. Development of reaction models for ground-water systems. *Geochim. et Cosmochim. Acta,* v. 47, no. 4:665–685.

REVELLE, R., and H. E. SUESS. 1957. Carbon dioxide exchange between atmosphere and ocean and the question of an increase of atmospheric CO_2 during the past decades. *Tellus,* v. 9:18–32.

RIMSTIDT, J. D., and H. L. BARNES. 1980. The kinetics of silica-water reactions. *Geochim. Cosmochim. Acta,* v. 44:1683–1699.

ROEDDER, E. 1984. *Fluid Inclusions.* Rev. in Mineralogy, v. 12. Mineralogical Society of America, Washington, D.C.

ROGERS, R. J. 1987. Geochemical evolution of groundwater in stratified-drift and arkosic bedrock aquifers in north central Connecticut. *Water Resources Research,* v. 23, no. 8:1531–1545.

SINGER, P. C. and W. STUMM. 1970. Acidic mine drainage: the rate determining step. *Science,* v. 167:1121–1123.

SKOUGSTAD, M. W., N. J. FISHMAN, L. C. FRIEDMAN, D. E. ERDMANN, and S. S. DUNCAN, eds. 1979. Methods for determination of inorganic substances in water and fluvial sediments. Techniques of Water-Resources Investigations. U.S. Geol. Survey, Book 5, Chapter A1.

STRICKLAND, J. D. H., and T. R. PARSONS. 1972. *A Practical Handbook of Seawater Analysis,* 2d ed. Bulletin 167, Fisheries Research Board of Canada, Ottawa.

STUMM, W., and J. J. MORGAN. 1981. *Aquatic Chemistry,* 2d ed. Wiley-Interscience, New York.

SVERJENSKY, D. A. 1987. Calculation of the thermodynamic properties of aqueous species and the solubilities of minerals in supercritical electrolyte solutions. In *Reviews in Mineralogy,* v. 17, I. S. E. Carmichael and H. P. Eugster, eds. Mineralogical Society of America, Washington, D.C. 177–209.

THEOBALD, P. K., H. W. LAKIN, and D. B. HAWKINS. 1963. The precipitation of aluminum, iron, and manganese at the junction of Deer Creek with the Snake River in Summit County, Colorado. *Geochim. Cosmochim. Acta,* v. 27:121–132.

THORSTENSON, D. C. 1984. The concept of electron activity and its relation to redox potentials in aqueous geochemical systems. U.S. Geol. Survey Open-File Report 84–072.

VERHOOGEN, J., F. J. TURNER, L. E. WEISS, C. WAHRHAFTIG, and W. S. FYFE. 1970. *The Earth: An Introduction to Physical Geology.* Holt, Rinehart and Winston, 244.

VON DAMM, K. L. 1990. Seafloor hydrothermal activity: black smoker chemistry and chimneys. In *Ann. Rev. Earth Planet. Sci.,* v. 18, G. W. Wetherill, ed. 173–204.

Von Damm, K. L., J. M. Edmond, B. Grant, C. I. Measures, B. Walden, and R. F. Weiss. 1985. Chemistry of submarine hydrothermal solutions at 21° N, East Pacific Rise. *Geochim. Cosmochim. Acta,* v. 49:2197–2220.

White, D. E., J. D. Hem, and G. A. Waring. 1963. Chemical Composition of Subsurface Waters. U.S. Geol. Survey Profess. Paper 440–F.

Crystal Chemistry

Crystal chemistry deals with the relationships between the architecture of solids and their chemical composition and physical properties. We shall be concerned here with minerals, some of which exhibit simple crystal chemical relationships, but most of which are very complex solids. By studying the crystal chemistry of minerals, we can estimate the conditions under which they formed and the processes that produced them. The chemistry of minerals was first studied systematically in the nineteenth century. A giant step forward came with the discovery of the diffraction of X rays by crystals. It then became possible to determine the internal structure of minerals. A number of other techniques, involving the interaction of crystals with various types of radiation, have since furthered our knowledge of both the composition and the internal structure of minerals.

NATURE OF SOLIDS

We generally think of solids as having certain properties, such as hardness, that distinguish them from substances we call liquids or gases. However, a more specific and more useful definition would be as follows: *solids* are materials that exhibit a long-range regularity of atomic or molecular arrangement. Atoms or molecules repeat at regular intervals along parallel lines. Such materials are said to occur in the crystalline state. *Liquids* exhibit short-range order (involving a few atoms or molecules), and *gases* show little or no order. The requirement of long-range order for a solid means that amorphous (noncrystalline) materials such as volcanic glass have to be considered as very viscous liquids. Such materials are thermodynamically unstable and may eventually become crystalline.

Crystals grow from a liquid or gas by first forming a tiny nucleus of their own composition (owing to chance collisions of atoms) or by adding material to a foreign particle (a seed crystal). Minerals forming in a cooling magma would represent the first process; an example of the latter process would be the precipitation in the ocean of inorganic calcium carbonate on a shell fragment. When a small nucleus of a compound is first formed, it is thermodynamically unstable (Figure 5-1). This is because small crystals have a large surface area in comparison to their size, and this gives them a large surface free energy (similar to that due to the surface tension of a liquid). As a nucleus grows, there is a decrease in surface free energy relative to volume energy and thus in total crystal free energy. This implies that large crystals are more stable than small crystals. Larger crystals grow by addition of individual units, representing the crystal composition, to the crystal nucleus. If the solution from which the crystal is growing is highly supersaturated, such building up of a crystal is easily done by continually adding layers of atoms or molecules. However, growth of a crystal in a solution that is only slightly supersaturated requires the existence of imperfections (such as slight bumps or irregular steps) on the outside of the crystal. These imperfections (known as *dislocations*) allow atoms and molecules to be easily added to the crystal at the site of the imperfections without formation of a new, extensive layer of atoms or molecules on the outside of the crystal. The dislocations tend to be self-perpetuating; thus they can control the growth pattern of a crystal.

The previous discussion involves nucleation in a liquid or gas. Nucleation in solids, such as new mineral formation in metamorphic rocks, is more complicated. There is always an activation energy barrier to change. This is the minimum energy required to initiate nucleation. The amount of free energy needed to create a critical nucleus is affected by temperature, pressure, and the presence or absence of defects in the host solid. The more the temperature differs from (overshoots) the transformation temperature ($\Delta G_r = 0$) of the reaction, the lower the value of the activation energy of the reaction. The presence of defects also reduces the amount of activation energy needed. Thus a large ΔT (ΔT = actual T – transformation T) and the existence of grain boundaries, cracks, etc. in the host material both increase the possibility for nucleation.*

The overall free energy change, ΔG_r, involved in forming a nucleus A in a solid matrix B can be expressed as follows:

$$\Delta G_r = \Delta G_{\text{A–B}} + \Delta G_{\text{surface}} + \Delta G_{\text{strain}} \tag{5-1}$$

The $\Delta G_{\text{A–B}}$ term represents the general free energy difference between A and B while $\Delta G_{\text{surface}}$ refers to the surface free energy discussed above and in Figure 5-1. The final term, ΔG_{strain}, represents the energy needed by A to make room for itself. For a crystal nucleus below the critical size (Figure 5-1), the sum of $\Delta G_{\text{surface}}$ plus ΔG_{strain} is positive and greater than the negative value of $\Delta G_{\text{A–B}}$. Thus ΔG_r is positive and A does not form. If $\Delta G_{\text{A–B}}$ is larger than the other two terms (a negative value for ΔG_r), crystal growth occurs.

Crystalline solids, no matter how they form, consist of a regular internal arrangement of their constituent particles. Crystals show symmetry on the outside because the atoms and molecules on the inside are arranged in a symmetrical pattern. The distances between the planes of atoms in a crystal are extremely small (about 10^{-8} centimeter). X rays have wavelengths of about the same dimension, and thus crystals will diffract X rays; that is, the radiation is scattered and

*Many physical and chemical properties of minerals are affected by the existence of defects *within* minerals. Most common are defects due to the presence of impurities or to displacement of atoms from their normal positions in a structure. These "point defects" are discussed in Schock (1985) and a review of all types of defects is given by Putnis (1992).

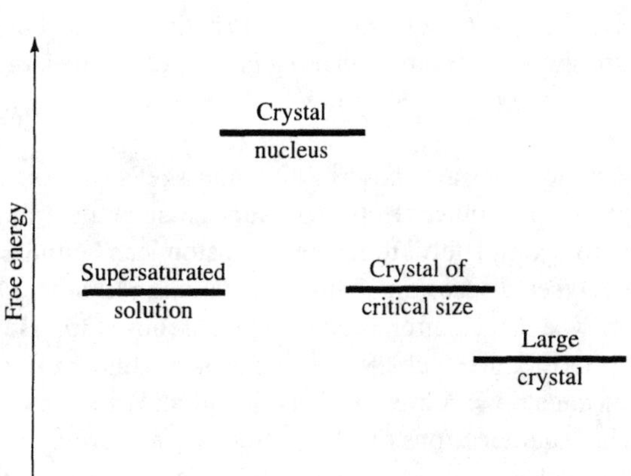

Figure 5-1 Relations among the free energy of a supersaturated solution, a crystal nucleus, a small stable crystal, and a large crystal. The crystal nucleus has a high surface energy (due to its small size) and is unstable relative to the other states. The size of the critical nucleus is a function of the amount of supersaturation of the solution, with nuclei forming more easily as the degree of supersaturation increases. Thus a high degree of supersaturation leads to many small crystals, whereas slight supersaturation will result in the formation of a few large crystals. (From W. S. Fyfe, *Geochemistry of Solids*, p. 172. Copyright 1964 by McGraw-Hill Book Company. Used with permission of McGraw-Hill Book Company.)

constructive interference of some of the radiation causes it to appear to be reflected from the planes of the atoms. It is possible to use diffraction data to determine the size and shape of unit cells, as well as the arrangement of atoms within the cells. These unit cells, made up of a number of atoms, represent the basic building blocks of the crystals. A *unit cell* is a representative unit of a crystal, from which all crystal properties can be derived.* For minerals, this arrangement can be considered an orderly stacking of spherical atoms. Molecules, as such, do not exist in most minerals.

As indicated above, one experimental method for studying minerals involves the diffraction of X rays by crystalline solids. Another type of diffraction study uses the scattering of neutrons. These experiments give information about the regularity of internal structure and about energy relationships. Knowledge of energy levels and of the structure of atoms and molecules comes also from emission and absorption spectroscopic studies. Such techniques use radiation from throughout the electromagnetic spectrum (Putnis 1992). Examples include Raman, infrared, X-ray, and microwave spectroscopy and ultraviolet-visible spectrophotometry. Two other useful techniques are nuclear magnetic resonance (NMR) and electron spin resonance (ESR). Application of these techniques to crystal chemical studies are discussed in Kieffer and Navrotsky (1985), Berry and Vaughan (1985), Tossell and Vaughan (1992), and Navrotsky (1994).

BONDING FORCES

The forces that hold the atoms of a solid together are electrical in nature and are related to the electron structure of the atoms. These forces can be described in terms of two ideal bond types. The pure *ionic bond* is due to the attraction of oppositely charged ions; the pure *covalent bond* is due to mutual interaction of equally shared electrons. In the case of ionic bonding, one or more electrons are transferred completely from one atom to another to produce charged ions. In covalent bonding, electrons are not transferred at all. Instead, the electron clouds of atoms overlap and outer electrons are shared equally, resulting in attractive forces between the atoms. Bonds

*The periodicity of internal structure of every crystal can be related to a regular array of identical points known as a *space lattice.* A unit cell is formed by joining these points to form a parallelepiped. For each type of crystal, the periodicity of the space lattice is unique, but the shape of the unit cell is arbitrary in many cases (i.e., there is more than one way to connect the points of a space lattice in order to form a complete unit of pattern).

representing transition states somewhere between the two extreme types can also exist, with electrons shared unequally between atoms. The type of bond that exists in a given solid depends upon the potential and kinetic energy relationships of the atoms making up the solid. The structure that tends to form from a liquid or a gas is the structure with the lowest possible total energy, and thus the maximum stability.

For bonding, the most important electrons are those in the outer shells of atoms. Energy changes occur when a neutral atom gives up an outer electron or gains an outer electron. The energy required to remove an outer electron completely from a gaseous atom can be measured and is known as the *first ionization potential* (see discussion of atom excitation in Chapter One). The *electron affinity* is the energy given off when an electron is added to a gaseous atom. How easy it is to add or subtract electrons from atoms of a given element depends on a number of factors, including the size and nuclear charge of the atoms. A high nuclear charge makes it easy to add an electron and hard to take one away. The outer electrons of a large atom are relatively far from the nucleus and are not strongly attracted to it. They can be removed without expending much energy and not much energy is given off when electrons are added. The electrons between the nucleus and the outer electrons can have a screening effect, which results in a lower attraction of the nucleus for the outer electrons. The general tendency for most atoms is to add or lose electrons in such a way that they achieve an outer shell that has eight electrons (inert-gas structure).

By combining ionization potential and electron affinity values for any particular element, we can get a measure of the degree to which its atoms in a molecule or a crystal tend to attract electrons. This property is known as the *electronegativity* of the element. It can be considered an average of the effects of the first ionization potential and the electron affinity.* Study of first ionization potentials has shown that they decrease from right to left in rows of the periodic table, and also decrease from top to bottom of columns of the table. Electron-affinity values can be used with ionization potentials to calculate electronegativities. There are several other ways in which electronegativity numbers can be calculated. However this is done, it is found that the values generally decrease from right to left in rows of the periodic table; they also tend to decrease from top to bottom of columns of the periodic table (see Table 5-1). In other words, elements on the right side of the periodic table have a strong attraction for electrons, and those on the left side have a weak attraction for electrons. The large atoms at the bottom of the periodic table have a large number of electron shells and thus have a weak attraction for electrons, whereas the small atoms at the top of the periodic table have fewer shells and display a strong attraction for electrons.

We can use electronegativity values to predict the type of bond that will form between two different elements (Figure 5-2). Elements with a large difference in electronegativity (for example, 3.0) will tend to form ions and thus develop ionic bonding (the more electronegative element forms the negative ions and the other element forms positive ions). Elements with similar electronegativities have equal attractions for electrons and will therefore share electrons in a covalent bond.†

*Note that electron affinity refers to an isolated atom's attraction for electrons, while electronegativity refers to this attraction when there is a chemical bond with another atom.

†A special type of covalent bond occurs when *both* electrons in such a bond come from *one* of the two participating atoms instead of one electron coming from each of the two bonding atoms. This type of bond is known as a coordinate or dative bond. Examples include sulfate and nitrate minerals, with coordinate bonds occurring between the central sulfur or nitrogen atom and some of the surrounding oxygen atoms in the NO_3^- and SO_4^{2-} groups. Oxygen atoms donate electron pairs to form the coordinate bonds. To describe an ion such as the nitrate ion we actually need three different "resonance structures," since no one location for a coordinate bond (among the three possibilities) is adequate to represent the structure. The term "resonance" refers to the joint use of more than one structure or model to represent a given molecule.

TABLE 5-1 Pauling's Electronegativity Scale

Li	**Be**	**B**											**C**	**N**	**O**	**F**
1.0	1.5	2.0											2.5	3.0	3.5	4.0
Na	**Mg**	**Al**											**Si**	**P**	**S**	**Cl**
0.9	1.2	1.5											1.8	2.1	2.5	3.0
K	**Ca**	**Sc**	**Ti**	**V**	**Cr**	**Mn**	**Fe**	**Co**	**Ni**	**Cu**	**Zn**	**Ga**	**Ge**	**As**	**Se**	**Br**
0.8	1.0	1.3	1.5	1.6	1.6	1.5	1.8	1.8	1.8	1.9	1.6	1.6	1.8	2.0	2.4	2.8
Rb	**Sr**	**Y**	**Zr**	**Nb**	**Mo**	**Tc**	**Ru**	**Rh**	**Pd**	**Ag**	**Cd**	**In**	**Sn**	**Sb**	**Te**	**I**
0.8	1.0	1.2	1.4	1.6	1.8	1.9	2.2	2.2	2.2	1.9	1.7	1.7	1.8	1.9	2.1	2.5
Cs	**Ba**	**La–Lu**	**Hf**	**Ta**	**W**	**Re**	**Os**	**Ir**	**Pt**	**Au**	**Hg**	**Tl**	**Pb**	**Bi**	**Po**	**At**
0.7	0.9	1.1–1.2	1.3	1.5	1.7	1.9	2.2	2.2	2.2	2.4	1.9	1.8	1.8	1.9	2.0	2.2
Fr	**Ra**	**Ac**	**Th**	**Pa**	**U**	**Np–No**										
0.7	0.9	1.1	1.3	1.5	1.7	1.3										

Note: The values given in the table refer to the common oxidation states of the elements. For some elements, significant variation of the electronegativity with oxidation number is observed: for example, Fe^{2+}, 1.8, Fe^{3+}, 1.9; Cu^{+}, 1.9, Cu^{2+}, 2.0; Sn^{2+}, 1.8, Sn^{4+}, 1.9. The numbers are relative to hydrogen, which is given the value 2.1.

Source: After Linus Pauling, *The Nature of the Chemical Bond, Third Edition.* Table 3-8, p. 93. Copyright 1939 and 1940, third edition ©1960, by Cornell University. Used by permission of the publisher, Cornell University Press.

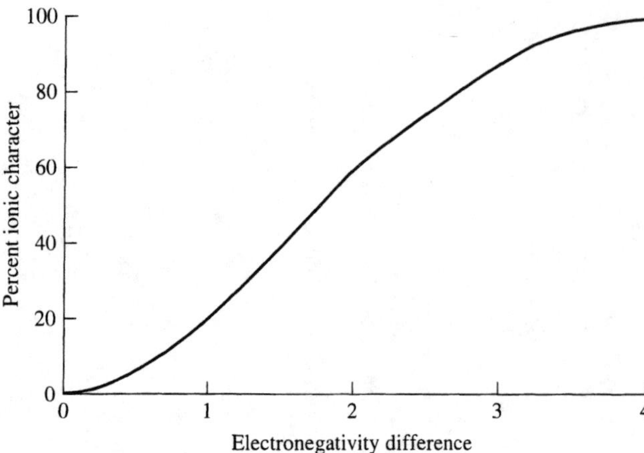

Figure 5-2 Relation between electronegativity difference and percent ionic character. The largest difference there can be in a bond is 3.3, which is the difference between the most electronegative element F (4.0), and the least electronegative element Cs (0.7). (After R. Chang, *General Chemistry*. Copyright 1986 by McGraw-Hill, Inc. Used with permission of McGraw-Hill, Inc.)

Elements with differences in electronegativity of about 1.7 will tend to form a bond that is half ionic and half covalent. A mixed bond can be viewed as consisting of anions with deformed electron clouds due to the attraction of highly charged cations. Anions exert a weaker pull on their outer electrons than do cations because, in terms of nuclear charge, anions have extra electrons, whereas cations have a deficiency of electrons. Thus cations can deform (polarize) the electron cloud of anions.

Some solids, and parts of some minerals, are held together by a much weaker bond than the two described above. This bond, known as the *van der Waals bond,* can be considered a special case of the ionic bond. In the true ionic bond, oppositely charged ions are held together by electrostatic attraction. The van der Waals bond acts between neutral molecules. Although neutral overall, these molecules can have a little more negative charge at one end and a little more positive charge at the other end. This leads to a very weak electrostatic attraction (bonding) between the molecules. The binding force is about one-hundredth that of an ionic or covalent bond. An example is the interlayer bonding that holds together the strong, covalently bonded sheets found in graphite. The extreme softness and greasy feel of graphite is due to the weak van der Waals bonds between sheets.

The *metallic bond* found in solids such as native copper can be considered a special case of the covalent bond. Each atom shares its electrons with all its neighboring atoms in turn, rather than with just one neighbor or a few neighbors. The shared electrons must be able to move about. Thus metals are viewed as consisting of positive ions immersed in a "sea" of electrons. When more than one kind of atom is involved, it is possible to have a mixed bond that is partially covalent and partially metallic. In this case the electrons are not freely shared and can show only limited movement.

The standard example of a purely ionic compound is halite (Figure 5-3a), and a good example of a covalent compound is orthorhombic native sulfur (Figure 5-3b). Many minerals have bonds that are partially ionic and partially covalent. As a result, they display some properties characteristic of purely ionic compounds and other properties characteristic of purely covalent compounds. Quartz is a good example. It does not have discrete molecules and has a high melting point, both indicating ionic bonding. It is not soluble in solvents of high dielectric constant, and, when melted, it does not conduct electricity. These two properties indicate covalent bonding. The various bonds in a given mineral can show differences in character. In silicate minerals the Si-O

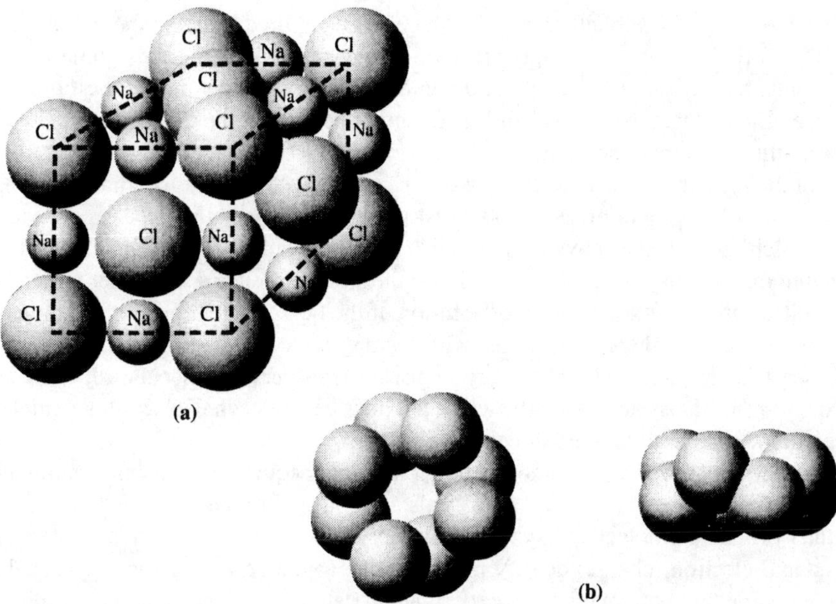

Figure 5-3 (a) Halite crystal structure. No NaCl molecule as such is present. Each sodium ion is surrounded by six chlorine ions and each chlorine ion by six sodium ions. Each chlorine ion can be considered to be neutralized by six $+\frac{1}{6}$ charges received from the adjacent sodium ions. As a result of electron transfer, sodium ions and chlorine ions both have eight electrons in their outer shell. The unit cell of halite is outlined by the dashed lines. When unit cells are stacked together, corner ions are shared among eight cells, ions at cube-face centers are shared between two cells, and ions at the centers of the edges are shared among four cells. Thus there is a total of four sodium ions and four chlorine ions in each unit cell. This gives four formula units in the unit cell. (b) The S_8 ring molecule, viewed parallel and perpendicular to the axis of the ring. Orthorhombic sulfur consists of these covalently bonded molecules held together by van der Waals bonds. Each sulfur atom is bonded to two other sulfur atoms, because the electron structure of a sulfur atom has two orbitals in the *p* subshell that each contain a single electron. Each of these two electrons forms a covalent bond with an electron of a neighboring atom. As a result, each sulfur atom can be considered to have eight electrons in its outer shell, six of its own and two provided by other atoms. The angle between the two bonds of an atom is 108°, resulting in four of the atoms in a ring being slightly offset from the other four. There are 128 atoms (equivalent to 16 rings) in the unit cell of sulfur. Because adjacent rings are held together by weak van der Waals bonds, this form of sulfur has a low hardness (about two on the Mohs scale) and a low melting point (112.8°C). (a) After Frye (1974, 55); (b) after Bragg and Claringbull (1965, 36).

bond is roughly 40 percent ionic and 60 percent covalent. The other bonds, such as Al-O and K-O, are 60–75 percent ionic.

The sulfide minerals have bonds that tend to be a mixture of ionic, covalent, and metallic bonding. For most sulfides, the bonds seem to be largely covalent, with lesser metallic character and a small amount of ionic character. Although the crystal structure and most properties of sulfides, such as pyrite, indicate covalent bonding, the minerals also exhibit properties typical of metallic bonding (such as metallic luster). The metallic bonding in these minerals would be between metal atoms, whereas metal-sulfur bonds and sulfur-sulfur bonds would be covalent.

The strength of these bond types (i.e., the energy required to break them) can be roughly described as follows. Covalent bonds are the strongest, with ionic bonds somewhat weaker. Metallic bonds have even less strength and van der Waals bonds are the weakest of all. Falling between metallic and van der Waals bonds is the hydrogen bond (see Chapter Four), responsible for the bonding in water and ice.

All of the preceding discussion can be described as part of the classic mineralogy approach to the problem of bonding in minerals. Navrotsky (1994) describes two other, more modern approaches, which she calls the "physics approach" and the "chemistry approach." These two methods are based on applying quantum mechanics to the study of bonding and other mineral properties. They differ in the method chosen for solution of the Schrödinger equation (the fundamental equation of quantum mechanics that deals with the probability of finding an electron in a given volume—see Chapter One). The chemistry approach considers a molecular cluster of a small number of atoms as the system of study, while the physics approach deals with a complete crystal (represented by an infinite three-dimensional lattice).

In the physics approach a solution to the Schrödinger equation gives information about allowable energy bands (band structure) in crystals as compared to the more narrow energy levels of molecular orbitals in molecules. Symmetry and crystallographic direction are factors that are considered and electron (charge) density maps are one of the products. The theoretical density maps can be compared with maps from experimental data to better understand bonding in minerals (Figure 5-4). For example, Cohen (1991) studied electron density and band structure in stishovite and concluded that a mixture of ionic and covalent bonding occurs in that mineral. He found that the ionicity of O^{2-} and Si^{4+} is reduced to roughly $O^{1.4-}$ and $Si^{2.8+}$ in stishovite.

The chemistry approach uses molecular orbital theory. This theory assumes that a new set of orbitals (molecular orbitals) forms when atoms interact during bond formation (Figure 5-5). The molecular orbitals define the region where there is a high probability of finding a particular electron. When a pair of *s* atomic orbitals interact during bond formation, two different types of molecular orbitals form: sigma bonding orbitals with lower energy than either parent atomic orbital, and sigma* antibonding orbitals with higher energy than either parent orbital. For *p* atomic orbitals, two different types of interaction can occur, forming either sigma orbitals or pi orbitals (Figure 5-5). The difference in energy between two atomic orbitals and a bonding molecular orbital formed from them is related to the amount of interaction between the two atomic orbitals. Extensive interaction produces a large energy difference and a strong chemical bond, while little interaction results in little energy difference and a weak bond. Symmetrical molecular orbitals (symmetrical, homogeneous distribution of electrons) produce a covalent bond. A lack of symmetry means that some electrons are more likely to be found near one nucleus than the other (partial ionic bonding).

Gibbs (1982) and others have studied the structure of minerals from a molecular orbital viewpoint and have used their results to explain and predict the properties of minerals. Among the parameters that can be calculated or determined are bond types (ionic or covalent), strengths, force constants, lengths, and angles. With this kind of information it should be possible to explain known mineral structures and predict unknown structures if chemical composition is given. Navrotsky (1994, 255) points out that, in some cases, the type of structure can be related to just two parameters, one related to atom size and one related to bond type. Figure 5-6 is a Mooser-Pearson diagram plotting electronegativity difference (related to bond type) versus average principal quantum number of valence electrons (related to atom radius). The diagram shows that the structure type of AB compounds depends on the values of these two parameters.

Figure 5-4 A map of the static electron density distribution obtained from X-ray diffraction data in a pseudoatom model refinement of coesite, calculated through the plane of the Si(1)-O(5)-Si(2) group. The circle centered on O(5) and labeled A defines the outermost limit of the O atom as defined by its atomic radius (0.6 Å). The circle labeled I defines the outermost limits of the oxygen ion as defined by its ionic radius (1.32 Å). In terms of the electron density distribution, the oxygen atomic radius is too small and the oxygen ionic radius is too large. A "bonded radius" can be derived by measuring, along the path between the Si and O atoms, the distance from the center of the oxygen atom to the point of minimum electron density. Such a measurement yields a radius of about 0.96 Å. The electron density distribution of the SiOSi group is compressed along both SiO bonds and is expanded into the empty regions of the coesite structure, rendering a nonspherical distribution for O(5). Coesite is a crystalline form of silica with fourfold-coordinate Si. After Gibbs et al. (1992).

The application of molecular quantum chemistry and theoretical solid-state physics to minerals and related materials, briefly described above, has been named "quantum geochemistry" by Tossell and Vaughan (1992). Other names, such as mineral physics, have also been applied. Whatever title is used, it is clear that quantum mechanical studies of minerals will be a major contributor in the future to our understanding of bonding and other features of crystal chemistry. Readers interested in further information are referred to Tossell and Vaughan (1992) and Navrotsky (1994).

COVALENT AND IONIC RADII

To describe the arrangement of atoms in a given crystal structure, it is helpful to know the relative sizes of the different types of atoms that occur in the structure. One factor that determines the

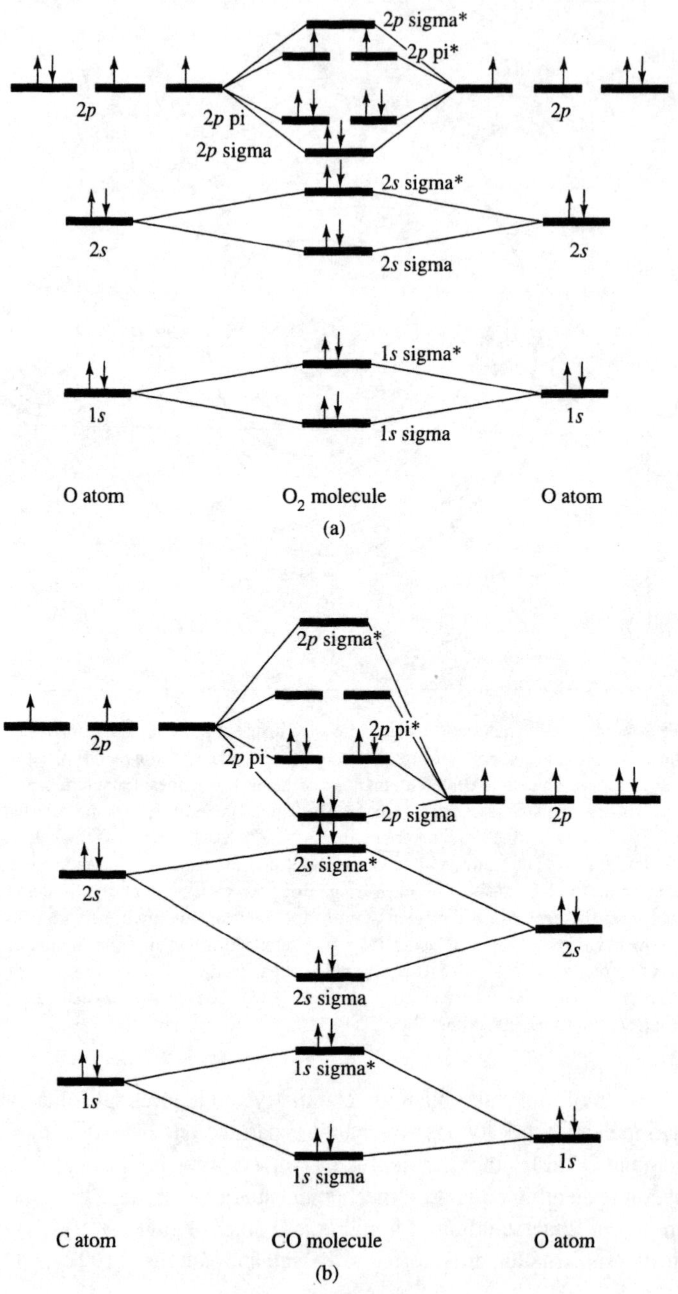

Figure 5-5 Molecular orbitals for (a) O_2 and (b) CO. The electron density distribution is identical around the two oxygen atoms in O_2, while the electron density distribution is greater near the oxygen atom than near the carbon atom in CO. The bond in O_2 is covalent while the bond in CO is partially ionic. (After A. Navrotsky, *Physics and Chemistry of Earth Materials*, Copyright 1994. Reprinted with the permission of Cambridge University Press.)

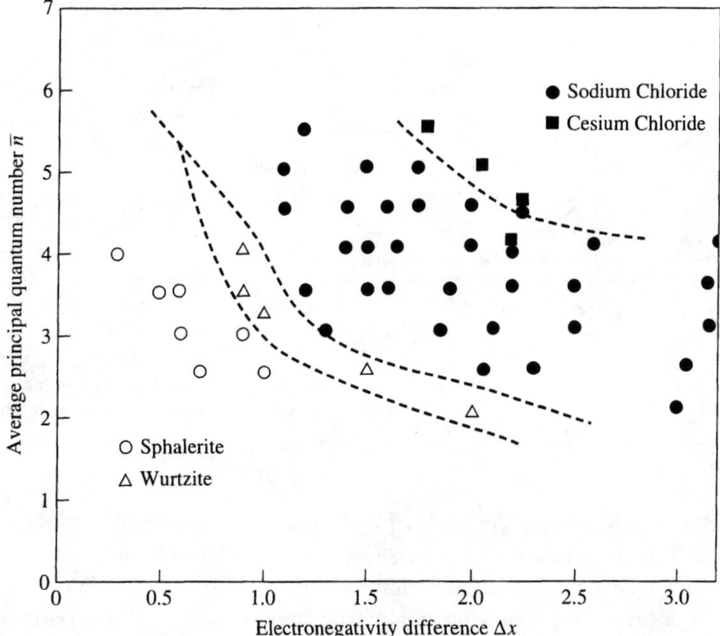

Figure 5-6 Mooser-Pearson diagram. The diagram shows separation of structure types of AB compounds according to electronegativity and average principal quantum number of valence electrons. After Pearson (1972). (After A. Navrotsky, *Physics and Chemistry of Earth Materials*, Copyright 1994. Reprinted with the permission of Cambridge University Press.)

size of the atoms of any particular element is the electron structure of that element. Atoms of high atomic number have more electron shells and are obviously going to be bigger than atoms of low-atomic-number elements. You will recall from Chapter One that we cannot determine the exact locations of the electrons of an atom. We can only describe the probability of finding electrons at various distances from the nucleus. This probability is highest at certain distances from the nucleus (and in certain zones at these distances), and these are the distances used to specify the location of the electron shells, subshells, and orbitals of the atoms. However, in most cases we cannot determine an *absolute* value for the radius of an atom or an ion in a solid.

For two atoms that are bonded together, an important type of measurement that can be made is the distance from the center of one atom to the center of the other atom. This distance is known as the *bond length.* For most elements, and for most combinations of bonding and structure, atoms can be considered as spheres that are just touching or that have slight interpenetration of their electron shells. The atoms arrange themselves such that the bonding forces that hold them together are just balanced by the repulsive forces between positively charged nuclei.

One way in which bond length can be determined involves the interaction of X rays with the atoms of crystals. When a beam of monochromatic (characteristic) X radiation strikes a crystal, the radiation is scattered as a result of interaction with the electron shells of all the atoms in the crystal. Some of the randomly scattered radiation combines (*constructive interference*) to form a diffracted beam, which leaves the crystal at the same angle as that of the incident beam. Although this diffraction is due to *all* the atoms in a crystal, the diffracted beam can be explained in terms of parallel planes of atoms within the crystal (*lattice planes*). There is a relationship,

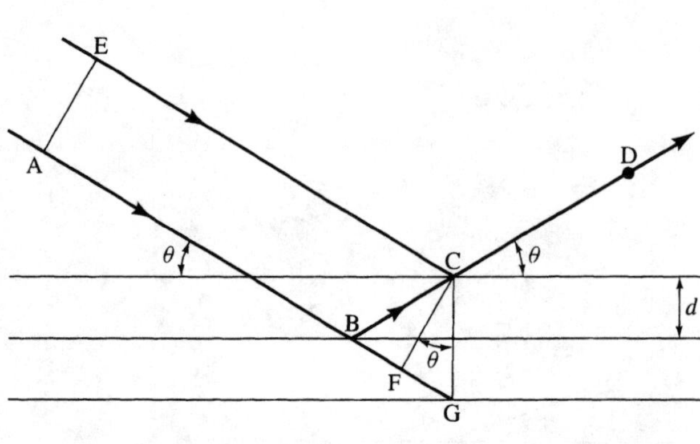

Figure 5-7 Diffraction by a set of parallel planes. Incident X rays AB and EC interact with the rows of atoms (planes) to cause emission of secondary X rays. Some of the emitted rays combine to produce a reinforced ray CD, whose angle of travel relative to the planes (θ) is similar to that of the incident rays. For reinforcement to occur, the path difference between ABD and ECD must be a multiple of the wavelength (λ) of the incident rays: $n\lambda = \text{ABD} - \text{ECD} = \text{AB} + \text{BC} - \text{EC} = \text{AF} - \text{FG} - \text{EC} = \text{FG} = 2d \sin \theta$, where n is an integer, λ the wavelength of the incident rays, d the interplanar spacing, and θ the angle between the rows of atoms and the incident rays. The equation $n\lambda = 2d \sin \theta$ is known as Bragg's law.

known as *Bragg's law,* among the angle of incidence, the wavelength of the X-ray beam, and the perpendicular distance between the planes (Figure 5-7). Thus we can determine the distance between planes of atoms in a certain direction by finding the specific angle of incidence producing diffraction of X rays of a given wavelength. The angle found is characteristic of a particular set of planes, and a series of measurements is made to study all the sets of planes in a crystal.

Measurement of the distance between planes of atoms (*d* spacing) can give us the sum of the radii (bond length) of the atoms in any two of the planes. However, knowledge of the bond length does not tell us the value of the radius of an atom in either plane unless the atoms of both planes are of the same element. The bond length in a diamond is found to be 1.54 Å (1 angstrom $= 10^{-8}$ centimeter). Therefore, the radius of an individual carbon atom is 0.77 Å.* This value for the carbon atom is often referred to as its *atomic* radius. It is also referred to as its *covalent* radius, since it was determined in a covalent compound and can only be used for structures in which carbon is covalently bonded.

It is possible to determine in a similar manner the covalent radius of certain other elements, for example, nitrogen in the N_2 molecule. If the measurements are correct, the bond length found for a carbon-nitrogen bond in a covalently bonded solid containing both elements should be equal to the sum of the two radii found in compounds consisting only of the individual elements. The agreement is usually quite good. With knowledge of the covalent radii of some elements, the radii of others can be found by measuring bond lengths in compounds containing an element of known radius and another element whose radius is to be determined. A list of covalent radii is given in Table 5-2. Atoms of elements such as copper occur metallically bonded to each other, and bond lengths from compounds such as native copper give metallic radii for these elements (also in Table 5-2). These radii are also often referred to as *atomic* radii.

It should be pointed out that the distance between two atoms that are not bonded together at all, or that are bonded only by van der Waals bonds, will be longer than the sum of the two co-

*In organic compounds, two carbon atoms can form a multiple covalent bond in which two or more pairs of electrons are shared by two atoms. In this case, the bond length would be different than that found for diamond, in which one pair of electrons is shared (single bond) between any two atoms. Multiple bonds are stronger than a single bond, and thus the covalent radius is smaller.

valent radii. For example, the covalent radius of chlorine in a Cl_2 molecule is 0.99 Å. The individual Cl_2 molecules of solid chlorine are held together by van der Waals bonds, and the distance between the adjacent atoms of two molecules is about 3.6 Å.

A further complication when dealing with covalent bonds is the formation of *hybrid* bonds. Such bonds are due to electrons from two or more different orbitals getting together to form a new or hybrid orbital, which has a mixture of the properties of the original orbitals. Carbon bonds are an example of hybrid bonds. It was mentioned earlier that a covalent bond between two carbon atoms may be a single bond or a multiple bond. Each type of bond, including the single bond of diamond, is actually a hybrid bond, and the carbon covalent radius is different for each type. Hybrid bonds are formed by a number of elements, and the resultant radii may or may not be different from their normal covalent radii. Fortunately, this effect is not important in most minerals.

The gain or loss of outer electrons to form ions should change the size of an atom. It has been found that large changes in size do occur, not only in going from an atom to an ion, but also in going from one valence to another. Ionic radii cannot be determined in absolute values with the procedure used for covalent radii, since no purely ionic compound can consist of only one element. Thus there is a major problem in deciding how much of a measured bond length belongs to the cation and how much belongs to the anion. Although ionic radii are often used to reproduce measured bond lengths in various materials, Gibbs (1995, personal communication) points out that covalent (atomic) radii can be used to generate bond length data for metals, molecules, and ionic and covalently bonded crystals equally well (Slater 1964).

Early attempts at estimating ionic radii were carried out by assuming a value for the radius of O^{2-}. The oxygen anion is an important constituent of many compounds, including the silicate minerals, and its large size ($\sim$ 1.32 Å) is a major control on the structure of such compounds. Following the early estimates of radii, new data were obtained on the relationship of ionic radii to various properties such as unit cell dimensions and ionization potentials. Some current values are given in Table 5-2, which lists the common ions of each element and the typical coordination numbers of each ion.

A major factor affecting ionic radii is the *coordination number* of a given cation or anion. This is the number of ions (neighbors) packed around a given ion. Early estimates of ionic radii assumed octahedral coordination for both cations and anions, that is, six neighboring ions around a given ion (Figure 5-8). However, if the number of neighboring ions is changed, there is a change in the effective size of the central ion due to a different interaction between the electrons of the central ion and the electrons of the new number of neighbors. For example, the estimated radius of Si^{4+} for octahedral coordination by oxygen ions is about 0.48 Å, whereas the radius is about 0.34 Å for tetrahedral (fourfold) coordination. In general, the greater the number of oxygen ions around a given cation, the greater is the effective radius of the cation. Additional neighbors have the effect of "pulling out" the outer electrons of a cation. Anion radii show a smaller change in magnitude with coordination number than do cation radii (see oxygen in Table 5-2).

Another factor that affects the ionic radii of some transitional metals is the electronic spin state of their outer *d*-orbital electrons. In certain situations the *d*-orbital electrons separate into two sets of unequal potential energy (Figure 5-9). The *high-spin state* occurs when the difference in energy of the two sets is small, and the *low-spin state* when the difference is large. In the high-spin state, each orbital is half-filled (has one electron) before any orbital becomes completely filled. In the low-spin state, the large difference in energy of the two sets causes the orbitals in the set of lower energy to become completely filled before any electrons enter the higher-energy set

TABLE 5-2 Selected Covalent, Metallic, and Ionic Radii

Atom or ion	Coordination number	Radius (Å) [covalent (c) or metallic (m) for atoms; ionic for ions]	Atom or ion	Coordination number	Radius (Å) [covalent (c) or metallic (m) for atoms; ionic for ions]
Ag^{+}	6	1.23	K^{+}	12	1.68
Al	12	1.43 (m)	La^{3+}	6	1.13
Al	4	1.26 (c)	Li^{+}	6	0.82
Al^{3+}	4	0.47	Lu^{3+}	6	0.94
Al^{3+}	6	0.61	Mg^{2+}	6	0.80
As^{5+}	6	0.58	Mn^{2+}	6	0.75 (LS)
Au^{3+}	4	0.78	Mn^{2+}	6	0.91 (HS)
B^{3+}	4	0.20	Mn^{3+}	6	0.66 (LS)
Ba^{2+}	8	1.50	Mn^{3+}	6	0.73 (HS)
Be^{2+}	4	0.35	Mn^{4+}	6	0.62
Bi^{3+}	6	1.10	Mo^{6+}	6	0.68
C	4	0.77 (c)	Na^{+}	6	1.10
Ca^{2+}	6	1.08	Na^{+}	8	1.24
Ca^{2+}	8	1.20	Nb^{5+}	6	0.72
Cd^{2+}	6	1.03	Nd^{3+}	6	1.06
Ce^{3+}	6	1.09	Ni^{2+}	6	0.77
Cl	4	0.99 (c)	O	4	0.73 (c)
Cl^{-}	6	1.72	O^{2-}	4	1.30
Co^{2+}	6	0.73 (LS)	O^{2-}	6	1.32
Co^{2+}	6	0.83 (HS)	O^{2-}	8	1.34
Cr^{3+}	6	0.70	P^{5+}	4	0.25
Cs^{+}	12	1.96	Pb	12	1.75 (m)
Cu	12	1.28 (m)	Pb^{2+}	8	1.37
Cu^{+}	2	0.54	Pm^{3+}	6	1.04
Cu^{2+}	6	0.81	Pr^{3+}	6	1.08
Dy^{3+}	6	0.99	Rb^{+}	8	1.68
Er^{3+}	6	0.97	Rb^{+}	12	1.81
Eu^{3+}	6	1.03	Re^{4+}	6	0.71
F^{-}	4	1.23	S	4	1.04 (c)
Fe	12	1.26 (m)	S^{2-}	6	1.72
Fe^{2+}	4	0.71 (HS)	S^{6+}	4	0.20
Fe^{2+}	6	0.69 (LS)	Sb^{5+}	6	0.69
Fe^{2+}	6	0.86 (HS)	Sc^{3+}	6	0.83
Fe^{3+}	4	0.57 (HS)	Se^{2-}	4	1.88
Fe^{3+}	6	0.63 (LS)	Se^{6+}	4	0.37
Fe^{3+}	6	0.73 (HS)	Si^{4+}	4	0.34
Ga^{3+}	4	0.55	Si^{4+}	6	0.48
Gd^{3+}	6	1.02	Sm^{3+}	6	1.04
Ge^{4+}	4	0.48	Sn^{4+}	6	0.77
Hf^{4+}	8	0.91	Sr^{2+}	8	1.33
Hg^{2+}	6	1.10	Ta^{5+}	6	0.72
Ho^{3+}	6	0.98	Tb^{3+}	6	1.00
In^{3+}	6	0.88	Th^{4+}	8	1.12
K^{+}	8	1.59	Ti^{3+}	6	0.75

TABLE 5-2 *(cont.)*

Atom or ion	Coordination number	Radius (Å) [covalent (c) or metallic (m) for atoms; ionic for ions]	Atom or ion	Coordination number	Radius (Å) [covalent (c) or metallic (m) for atoms; ionic for ions]
Ti^{4+}	6	0.69	W^{4+}	6	0.73
Tl^{3+}	6	0.97	W^{6+}	6	0.68
Tm^{3+}	6	0.96	Y^{3+}	8	1.10
U^{4+}	8	1.08	Yb^{3+}	6	0.95
U^{6+}	6	0.81	Zn	12	1.39 (m)
V^{2+}	6	0.87	Zn	4	1.31 (c)
V^{3+}	6	0.72	Zn^{2+}	4	0.68
V^{4+}	6	0.67	Zn^{2+}	6	0.83
V^{5+}	6	0.62	Zr^{4+}	8	0.92

Note: HS, high-spin state (*d* electrons unpaired); LS, low-spin (*d* electrons paired).

Sources: For covalent and metallic radii, Zhdanov (1965, 192–193); for ionic radii, Whittaker and Muntus (1970, 952–953).

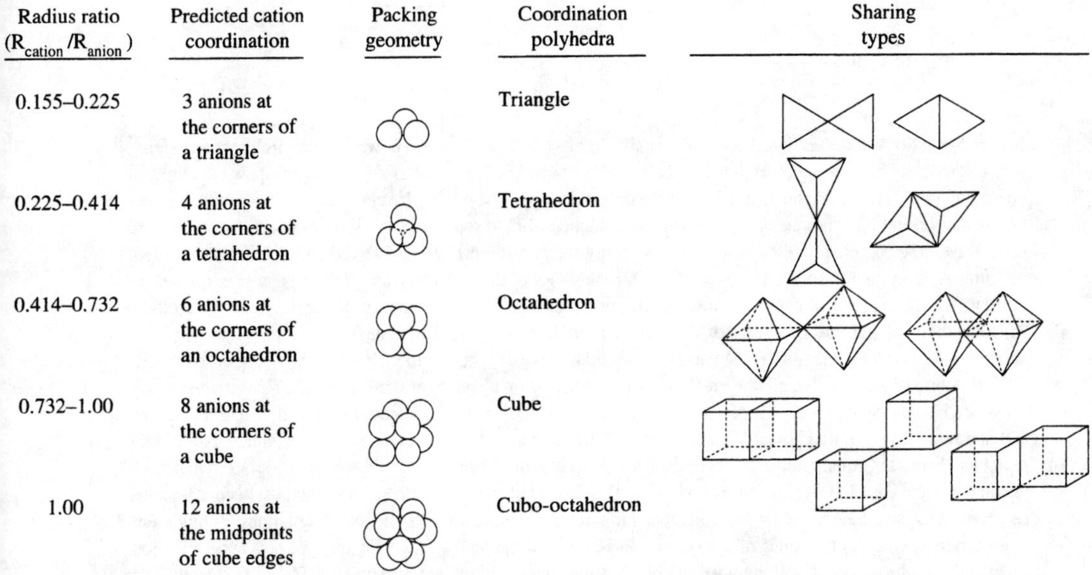

Radius ratio (R_{cation}/R_{anion})	Predicted cation coordination	Packing geometry	Coordination polyhedra	Sharing types
0.155–0.225	3 anions at the corners of a triangle		Triangle	
0.225–0.414	4 anions at the corners of a tetrahedron		Tetrahedron	
0.414–0.732	6 anions at the corners of an octahedron		Octahedron	
0.732–1.00	8 anions at the corners of a cube		Cube	
1.00	12 anions at the midpoints of cube edges		Cubo-octahedron	

Figure 5-8 Relationship of relative ionic size to cation coordination. Coordination polyhedra may be visualized in terms of the geometric figure formed by imaginary lines that connect the nuclei of the anions. If one anion is common to two coordination polyhedra, the polyhedra share corners. If two anions are common, edges are shared. If three or four anions are common, faces are shared. Coordination polyhedra in real crystal structures tend to share as few anions as possible. (After William H. Dennen, *Principles of Mineralogy,* Rev. Ptg.© 1960, The Ronald Press Company, New York, Table 2-4.)

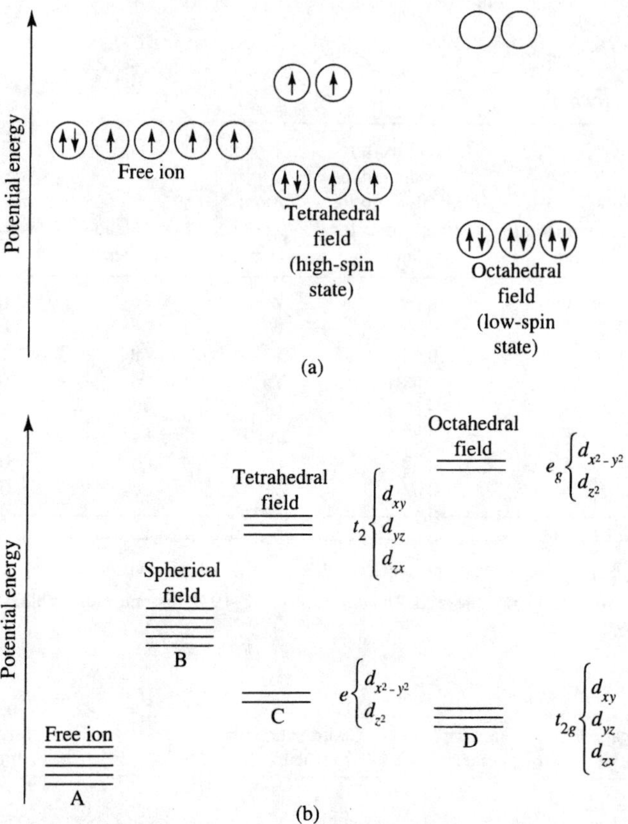

Figure 5-9 (a) A possible distribution of the six $3d$ electrons of Fe^{2+} is shown by the circles, representing orbitals, with each arrow representing an electron. A maximum of two electrons, each with opposite spin, can occupy an orbital (see Chapter One). The tetrahedral field represents the high-spin state (each orbital half-filled before any orbital becomes completely filled); the octahedral field represents the low-spin state (lower-energy orbitals become completely filled before any electrons enter the higher energy orbitals). In the octahedral field, the difference in energy of the two sets of orbitals is so great that the electron-electron repulsion caused by putting two electrons in the same orbital is compensated by the lower energy of the low-energy orbitals. (b) Schematic of the energy levels of the five d orbitals (represented by lines) for a transition-metal ion according to whether the ion is (A) a free ion that lacks neighbors like the one in the gaseous state, (B) surrounded by a negative charge uniformly disposed as if on a spherical shell around it, (C) surrounded by four neighbors (ligands) at the corners of a perfect tetrahedron centered around the metal ion, or (D) surrounded by six neighbors that are located at the corners of a perfect octahedron centered around the metal ion. From the equal-energy or degenerate states shown in (A) and (B), the d orbitals split into two sets of unequal energy, e and t_2 in (C) and e_g and t_{2g} in (D). The splitting in octahedral coordination occurs because set e_g orbitals have lobes of their electron clouds that are directed toward the surrounding anions, giving them higher energy, whereas the lobes of set t_{2g} extend between these anions. (See Figure 1-2 for a representation of the shapes of the orbitals.) In tetrahedral coordination the e orbitals are more distant from the surrounding anions, and thus of lower energy levels, than those of the t_2 orbitals. Energy differences can also arise between individual orbitals in a set if the distribution of anions is distorted from that of a perfect octahedron or tetrahedron. If the octahedron or tetrahedron is located in a crystal structure, all energy levels in (C) and (D) are lowered so that even the higher energy levels, that is, t_2 in (C) and e_g in (D), are below the levels in (A) because of electrostatic interactions with other ions. [(a) From W. S. Fyfe, *Geochemistry of Solids*, p. 165, Copyright 1964 by McGraw-Hill Book Company. Used with permission of McGraw-Hill Book Company. (b) After *Crystallography and Crystal Chemistry: An Introduction* by F. Donald Bloss, p. 444. Copyright © 1971 by Holt, Rinehart and Winston, Inc. Reprinted by permission of Holt, Rinehart and Winston.]

of orbitals. (This is an example of the aufbau principle mentioned in Chapter One.) Shannon and Prewitt (1969) have shown that ions of transition metals have larger radii when they are in the high-spin state. The state that the ions reach in a given crystal structure is related to the kind and number of neighboring ions. The terms "high spin" and "low spin" are used because the quantum number indicating which of two possible ways an electron is spinning is known as the spin quantum number (see discussion of quantum numbers in Chapter One).

Most minerals have bonds that are mainly ionic in character. Thus ionic radii are the most useful in studying the structures of minerals. Two modern tables of ionic radii are those of Shannon and Prewitt (1969, 1970) and of Whittaker and Muntus (1970).* Shannon and Prewitt base their radius values on experimentally determined interatomic distances (bond lengths) for oxides and fluorides and on a relationship between ionic volume and unit cell volume in oxides and fluorides. They take into account electronic spin state for transition elements and coordination number for both cations and anions. Earlier tables of radii did not consider electron spin state and anion coordination. Shannon and Prewitt give two different sets of values, one (IR or ionic radius values) based on the assumption that O^{2-} has a radius of 1.40 Å (octahedral coordination), and the other (CR or crystal radius values) on the assumption of a fluorine ion radius of 1.19 Å (octahedral coordination). Whittaker and Muntus have modified Shannon and Prewitt's values by considering the *radius ratio* principles of crystal chemistry. Their radii are given in Table 5-2.

The importance of the ratio, radius (cation)/radius (anion), in determining crystal structures was realized soon after the initial values for ionic radii were developed. Ions tend to pack together in such a way that they are as close together as possible. This puts them in a low-energy, and therefore stable, state. The geometry of different-sized spheres is such that the most stable arrangements can be classified in terms of the ratio of cation radius to anion radius (Figure 5-8). Since oxygen is the major anion in most minerals, we are particularly interested in the radius ratio of various cations with respect to oxygen. There is fairly good agreement between coordination numbers predicted on the basis of geometry and the actual values found in minerals.

Gibbs et al. (1992) point out that radius ratio arguments fail to predict the observed coordination numbers for many compounds. They feel that "an effort should be made to purge from textbooks the notion that the coordination number of a cation in a coordinated structure is governed by the relative sizes of the cation and anion that make up the coordinated polyhedra" (p. 748). Thus, although they have been utilized for many years, there is some question as to the usefulness of radius ratio principles. As a general guide they are useful, but they must be employed with caution. Gibbs et al. also suggest that it is unreasonable to assume a unique radius for anions like O^{2-}. Now that it is possible to map the electron density distribution of atoms in some structures, we know that the distribution around a nucleus is not spherical (implied by the term radius). A bonded radius (Figure 5-4) refers to the distance to the minimum electron density in the direction between two nuclei. The outer extent of an atom in other directions is usually different (the atom is not spherical) because of a different distribution of electrons (Figure 5-4).

Since the radius-ratio approach to explaining crystal structures has been somewhat successful, Whittaker and Muntus (1970) proposed adjusting the Shannon-Prewitt values on the basis of an oxygen anion radius of 1.32 Å, since values obtained in this way give a better conformity with known coordination numbers. These new values are intermediate between the two sets of values suggested by Shannon and Prewitt. It should be emphasized that all proposed ionic radii

*Some of the Shannon and Prewitt values were revised by Shannon (1976).

represent relative rather than absolute values, and that for a given solid discrepancies can occur between calculated and observed bond lengths. Ionic radii are particularly useful in comparing the geochemical behavior of different elements, since relative sizes rather than absolute sizes are most important for such comparisons. Figure 5-10 illustrates the relative sizes of the most common ions. It is also important to realize that the distance between a particular cation-anion pair in a specific configuration in a mineral can vary in different samples of the mineral, and ionic radii such as those in Table 5-2 are necessarily based on average interatomic distances.

A comparison of the ionic and atomic (covalent and metallic) radii in Table 5-2 indicates that the radius of a cation is always less than the atomic radius of the atom, whereas the radius of an anion is always larger than the atomic radius. This is to be expected, since in one case we are subtracting electrons and in the other case we are adding electrons. Carrying this reasoning further, we would expect Fe^{3+} to be smaller than Fe^{2+}, since Fe^{3+} represents the loss of an additional electron. On the other hand, Mg^{2+} has the same number of electrons as Na^{+}, but is smaller for two reasons: (1) the higher charge of the magnesium nucleus pulls the electrons closer to the nucleus, and (2) the higher charge also pulls surrounding anions closer to the magnesium ion. For anions with the same number of electrons, such as F^{-} and O^{2-}, the smaller nuclear charge of the oxygen nucleus exerts a smaller pull on the electrons and on surrounding cations, and thus the oxygen anion has a larger radius.

When ionic radii are related to the periodic table (Figure 5-10), we find that the radii increase from top to bottom in a column, and that there is a tendency for the radii to decrease going from left to right in a period. The vertical change is due to addition of electron shells in going from period to period; the horizontal decrease is due to increasing nuclear charge acting on the same number of electron shells. In the period beginning with potassium, the effect of the 10 additional transition elements is to cause an additional decrease in radius (from $K^{+} = 1.59$ Å to $Ga^{3+} = 0.55$ Å), as compared to the decrease in the previous period (from $Na^{+} = 1.24$ Å to $Al^{3+} = 0.47$ Å). In the fifth period, the decrease in radius is even greater as a result of the addition of both 10 transition elements and 14 lanthanide elements. This decrease is so great that it offsets the increase that occurs in going to the sixth period with its additional electron shell. As a result, hafnium, with a much higher atomic number than zirconium, has similar atomic and ionic radii and therefore similar chemical behavior. Other pairs showing this similarity are niobium-tantalum and cadmium-mercury. The contraction found in going from the fifth to the sixth period is known as the *lanthanide contraction.* The lanthanide elements themselves show only a small (~ 0.21 Å) decrease in radius with increasing atomic number. Thus all of them have very similar radii and they tend to occur together in nature. For these elements, electrons are added to the third shell in when going from lower to higher atomic number. As a result, contraction due to increasing nuclear charge is not as great as it is for addition of electrons to a less crowded outer shell.

Keep in mind that, in the above discussion of ionic radii and the periodic table, we have not taken into account the effect on radii of coordination number and of the possibility of an increase in the number of electron shells within a period when ions rather than atoms are being compared. For example, the ionic radius of Pb^{2+} can vary from 1.02 Å (four coordination) to 1.57 Å (twelve coordination). The divalent lead ion has one more electron shell than its neighbor in the same period, Tl^{3+}, and thus is much larger than the thallium ion. We have also ignored the complications of the transition elements, in which more than one valence can occur and in which the number of

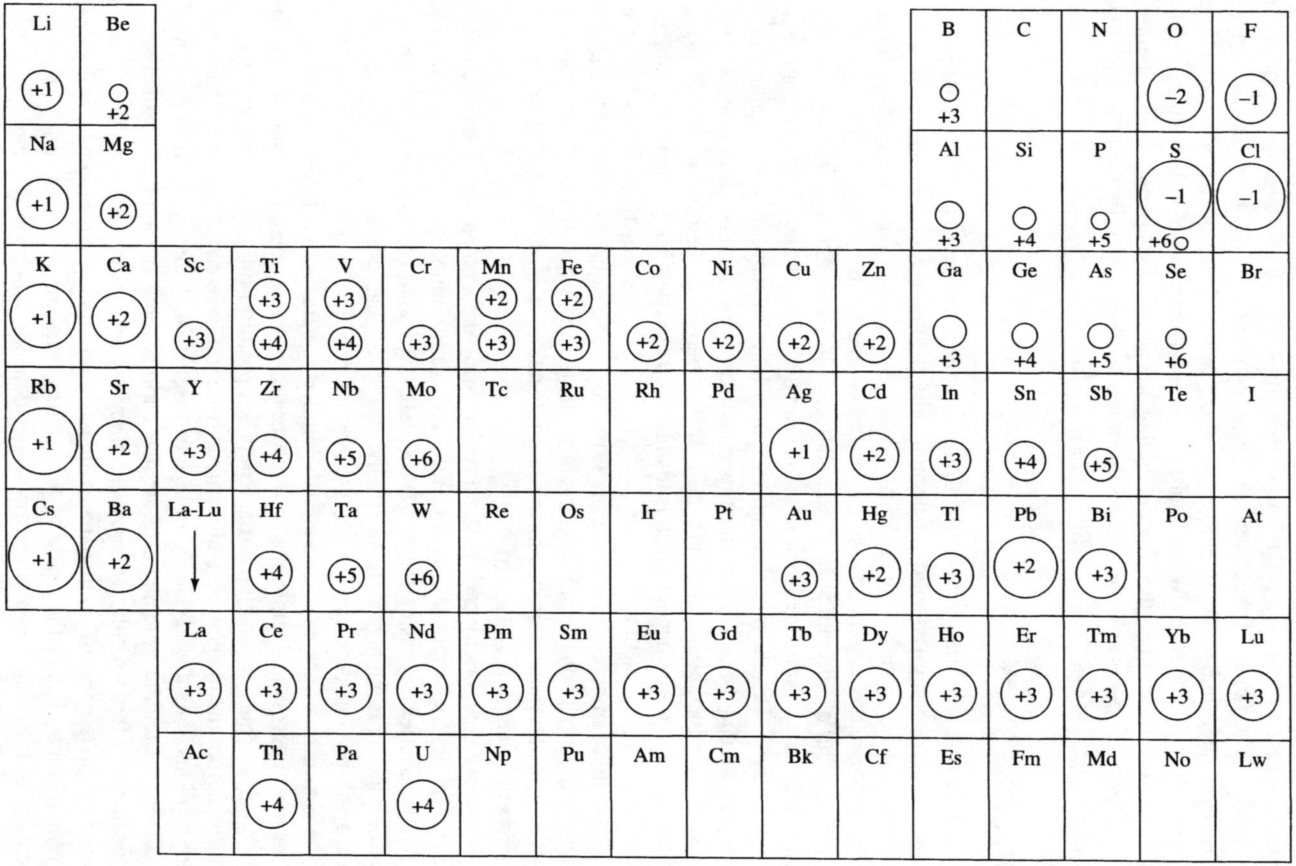

Li +1	Be +2											B +3	C	N	O −2	F −1
Na +1	Mg +2											Al +3	Si +4	P +5	S −1, +6	Cl −1
K +1	Ca +2	Sc +3	Ti +3, +4	V +3, +4	Cr +3	Mn +2, +3	Fe +2, +3	Co +2	Ni +2	Cu +2	Zn +2	Ga +3	Ge +4	As +5	Se +6	Br
Rb +1	Sr +2	Y +3	Zr +4	Nb +5	Mo +6	Tc	Ru	Rh	Pd	Ag +1	Cd +2	In +3	Sn +4	Sb +5	Te	I
Cs +1	Ba +2	La-Lu ↓	Hf +4	Ta +5	W +6	Re	Os	Ir	Pt	Au +3	Hg +2	Tl +3	Pb +2	Bi +3	Po	At
		La +3	Ce +3	Pr +3	Nd +3	Pm +3	Sm +3	Eu +3	Gd +3	Tb +3	Dy +3	Ho +3	Er +3	Tm +3	Yb +3	Lu +3
		Ac	Th +4	Pa	U +4	Np	Pu	Am	Cm	Bk	Cf	Es	Fm	Md	No	Lw

Figure 5-10 Relative ionic sizes. Radii (see Table 5-2) are for common valences and coordination numbers. Transition metals in high-spin state. (Scale: 1 Å = 3 mm)

outer-shell electrons can differ greatly among ions of elements relatively close together in the periodic table. Vanadium can occur with valences of +2, +3, +4, and +5. For octahedral coordination, the radius of the divalent ion is 0.87 Å, whereas for the same coordination the radius of the +5 ion is 0.62 Å. The radius of Ag^+ is much larger than that of Mo^{6+} because there are 10 more electrons in the outer shell of the silver ion, even though silver is only five places to the right of molybdenum in the periodic table. When coordination number, transition-element complications, and other factors are constant, the effect of adding a proton and keeping the same number of electrons is to decrease the radius. This is clearly shown in Figure 5-10 by the following sets of ions: Na^+ to S^{6+}, Rb^+ to Mo^{6+}, Ag^+ to Sb^{5+}, and Cs^+ to W^{6+}.

CRYSTAL STRUCTURES

The crystal structure of a solid is the configuration of its constituent particles (atoms, ions, and molecules). The structure of a given solid is determined by the relative sizes of its particles, the relative abundance of these particles, the chemical nature of the particles, and the types of bonding forces acting between the particles. These factors are all interrelated.

There are a number of different ways used to describe and illustrate the structures of minerals. The basic shape of any structure is the *unit cell,* and descriptions and illustrations are usually in terms of unit cells. The chemical formula of a mineral represents the relative abundance of the elements present, and the unit cell of the mineral contains one or more formula units. Thus one way of classifying mineral structures is to group them in terms of their formulas (AX structures, AX_2 structures, etc.). Another approach is to classify them chemically (elements, halides, oxides, silicates, etc.). A third method uses bond type to differentiate structures. Because of the variety and complexity of mineral structures, there is no easy and unique way to classify and describe them. Detailed descriptions of mineral structures are given by Smyth and Bish (1988). A series of books on specific mineral groups has been published by the Mineralogical Society of America (Reviews in Mineralogy, 1974 on).

In this section we shall briefly review some typical structures, emphasizing the interrelationship of chemistry and structure. Many minerals have fairly simple structures marked by mainly ionic bonding. After discussing these in some detail, we shall contrast them with structures in which the bonding is mainly covalent. Finally, structures characterized by both ionic and covalent bonding will be considered.

One of the controlling features of an ionically bonded mineral is the size of the anions of the mineral. For most common minerals, the anion is the oxygen ion, sometimes supplemented by the hydroxide ion. Because the small hydrogen ion (i.e., proton) is effectively inside the oxygen of the hydroxide ion, the oxygen ion and the hydroxide ion have essentially the same ionic radius. The larger anions of a mineral usually arrange themselves such that they can be viewed as touching each other in a regular grouping. The smaller cations occur in the interstices between the anions. Thus there is often a regular arrangement of anions, known as a *coordination polyhedron,* around each cation (Figure 5-8). The number of anions around a given cation generally depends on the ratio of the radius of the cation to the radius of the anion (*radius ratio*). In other words, the geometry of the arrangement of spherical ions depends upon their relative sizes, with the anions packing together in the closest manner possible. Thus there are more anions around a big cation than around a small cation.

A regular grouping of cations around a given anion is not likely in most minerals for two reasons. First, most minerals contain more than one kind of cation, each with a different size. Second, the cations usually do not touch each other because they occur in holes whose size is determined by the packing of the larger anions. Thus a good way to describe ionically bonded minerals is in terms of cation coordination polyhedra outlined by the location of anions around cations. Since all minerals have a regular arrangement of their constituents, coordination polyhedra can also be used to describe minerals that are not ionically bonded.

An example of a simple ionic solid is the mineral halite. The halite unit cell and the packing of its ions are shown in Figure 5-3a. The relationship of the unit cell to cation coordination is shown in Figure 5-11. The chlorine ions form octahedrons around the sodium ions, and each octahedron shares every one of its 12 edges with other octahedrons.

Ionically bonded solids, particularly simple structures such as halite, show certain regular features, which were first summarized by Pauling in 1928. They are discussed in detail in his book on chemical bonding (Pauling 1960). Pauling's rules are as follows:

1. A coordination polyhedron of anions is formed about each cation, the cation-anion distance being determined by the sum of the two radii, and the coordination number of the cation being determined by the radius ratio.
2. In a stable ionic structure, the valence of each anion, with changed sign, is exactly or nearly equal to the sum of the strengths of the electrostatic bonds to it from adjacent cations. [This rule states that crystals should be electrostatically neutral over very short distances. Each of the six sodium cations around a chlorine anion in halite can be considered to have a bond strength of $\frac{1}{6}$ (cation charge over number of neighbors) with the anion, the sum of these equaling the unit valence of the anion.]
3. The presence of shared edges and especially of shared faces (see Figure 5-8) in a coordinated structure decreases its stability; this effect is large for cations with high valency and small coordination number. (Essentially, this says that any two cations tend to be as far from each other as possible. When polyhedra do share edges or faces, the cations have to be fairly close together, and the polyhedra are thus distorted due to the mutual repulsion of the positively charged cations.)

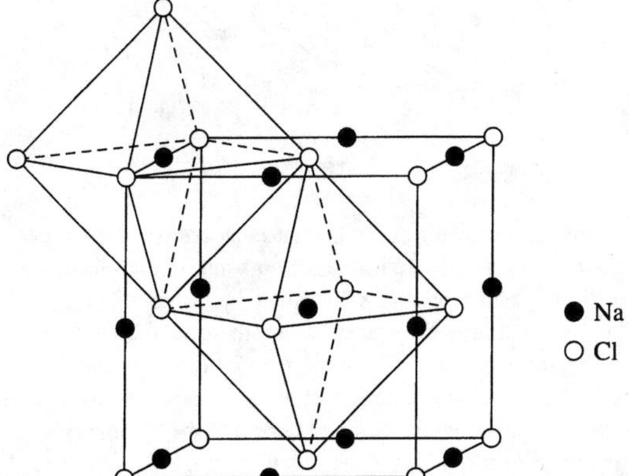

Figure 5-11 Cubic structure of sodium chloride, showing the coordinating octahedra of anions around the cations. After Evans (1964, 33).

4. In a crystal containing different cations, those with large valence and small coordination number tend not to share polyhedron corners, edges, or faces with each other. (This also reflects the dislike of cations for each other.)
5. The number of essentially different kinds of constituents in a crystal tends to be small. (This rule can be restated as follows. The structural environment of a given cation or anion tends to be the same throughout a crystal; that is, only one type of coordination is likely for a given element. This rule is not useful for many minerals. The oxygen ions in silicate minerals, for example, have different structural environments.)

A modern reinterpretation of Pauling's rules has been given by Burdett and McLarnan (1984). Their work generally supports Pauling's predictions and adds some modifications that take into account research since Pauling's original publication of his rules. Burnham (1990) points out that the ionic model, first developed in the early part of the twentieth century, is still very useful in understanding mineral structures, even those structures in which the ionic character of the bonding is not high.

As an example of a more complicated ionic solid, let us look at the perovskite structure (Figure 5-12). The unit cell of $CaTiO_3$ is very close to cubic symmetry but is actually orthorhombic. Another compound with the perovskite structure, $SrTiO_3$, is cubic. Because the ionic radii differ, Ti has a different coordination number from those of Ca and Sr. The coordination numbers of Ca, Sr, and Ti are, respectively, 8, 12, and 6. The structure of these two compounds, and of others with the perovskite structure, can be shown in terms of TiO_6 octahedra sharing all their corners (Figure 5-12).

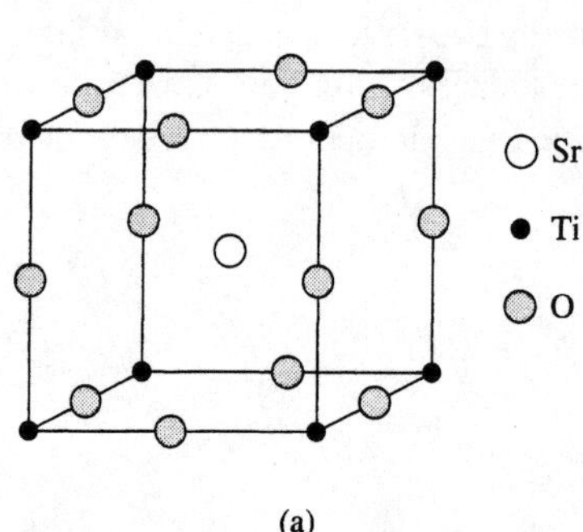

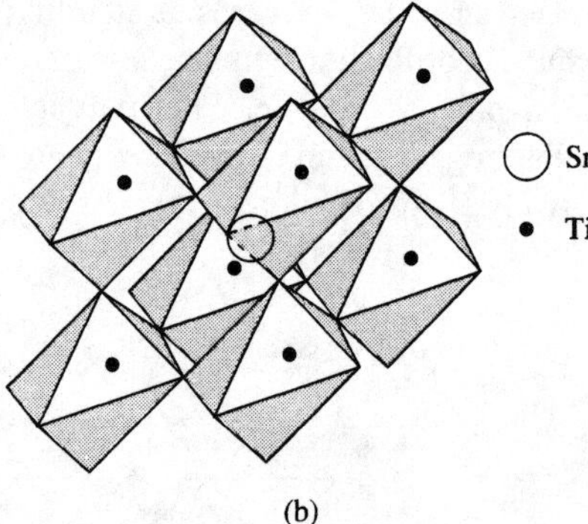

Figure 5-12 Perovskite structure. (a) Unit cell showing Ti^{4+} ions at each corner, O^{2-} ions along each cell edge, and Sr^{2+} ion at the cube center. There is one formula unit in the unit cell. (b) Each Ti^{4+} ion is at the center of a TiO_6^{8-} octahedron (six coordination), which shares all corners with adjacent octahedra. The Sr^{2+} ions occupy the spaces among the octahedra and are in twelvefold coordination. The nearest neighbors of each oxygen ion are two titanium ions and four strontium ions slightly farther away. The structure is named after the mineral perovskite ($CaTiO_3$). However, the unit cell of perovskite is a slightly distorted cube and the symmetry of the mineral is orthorhombic rather than cubic. Several different distortions of this structure have been identified. These distortions allow a much wider range of cation sizes to fit into the perovskite structure than is possible with the ideal cubic structure. After Frye (1974, 63).

Applying Pauling's second rule, we calculate the titanium-oxygen bond strengths to be 4/6 (cation charge over number of neighbors). Similarly, the strontium-oxygen bond strength is 2/12 or 1/6. The total of the bond strengths reaching a given oxygen is thus 2 times 4/6 from its two titanium neighbors plus 4 times 1/6 from its four strontium neighbors, giving a total of 12/6 or 2. Calculating the bond strengths of minerals is helpful in understanding their physical properties. Cleavage, for example, tends to occur along planes that pass through weak bonds. When studying unknown structures, Pauling's second rule is useful in figuring out the most likely structure.

Many minerals with the formula ABX_3 have the perovskite structure or a slight variation from it. The *A* ions must be large cations, and the sum of the valences of *A* and *B* must be six in order to give electrical neutrality. If the *A* and *X* ions (the large ions of the structure that determine its form) are exactly the same size, then a unit cell that is a perfect cube results. Such a cube is shown in Figure 5-12a. However, most *A* ions, such as the calcium ion of perovskite, are smaller than the oxygen ion, and this results in unit cells that are slightly deformed from a perfect cube. The crystal system of any given mineral depends on the symmetry of its unit cell. Halite, which has a cubic unit cell, belongs to the isometric system; perovskite, which is almost but not quite cubic, is orthorhombic. In minerals where fairly severe distortion of the ideal perovskite structure occurs, it is necessary to select a larger unit cell containing several formula units in order to represent the symmetry and structure of the minerals.

We now pass on to minerals dominated by covalent bonding. The sulfide minerals are the most important group here, and pyrite (FeS_2) is an interesting example (Figure 5-13). The pyrite structure is similar to that of halite, with sodium replaced by iron and chlorine replaced by S_2 groups. The sulfur atoms thus occur in pairs, and the bond distance between them is 2.10 Å. This is much less than twice the ionic radius of a sulfur ion (about 3.12 Å for four coordination) and indicates that each pair is covalently bonded. The bonding between iron and sulfur atoms is an example of hybrid bonding. The bonds are d^2sp^3 hybrids formed by two $3d$, one $4s$, and three $4p$ orbitals. These directed covalent bonds form an octahedron around the iron atoms, with sulfur atoms at the corners of the octahedron. The FeS_6 octahedra share faces, another indication of covalent bonding, since the polyhedra of ionic compounds almost never share faces. In addition to the covalent bonding, a certain amount of metallic bonding may occur between the iron atoms.

Naturally occurring organic material is made up of compounds whose structures are dominated by covalent bonding. Like the orthorhombic sulfur shown in Figure 5-3b, solid organic compounds are made up of discrete molecules held together by covalent bonds, with the

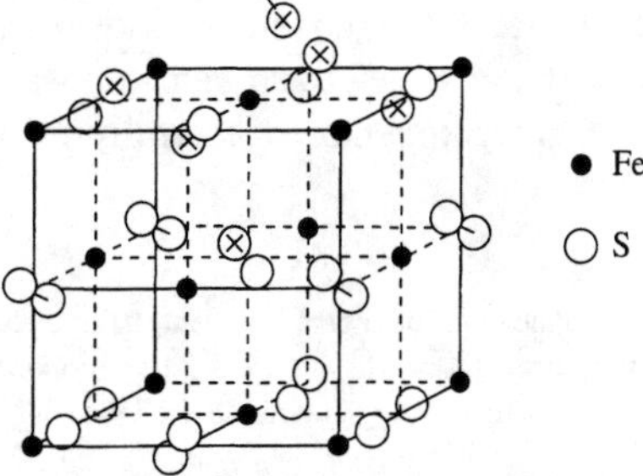

Figure 5-13 Diagrammatic location of atoms in the cubic structure of pyrite. Each iron atom is surrounded by six sulfur atoms, and each sulfur atom is linked to another sulfur atom and to three iron atoms. The unit cell is a cube containing four units of FeS_2. The six sulfur atoms marked with an × form an octahedral coordination about the iron atom at the center of the top face. Note that some of the sulfur atoms shown are outside the unit cell. After Ernst (1969, 56).

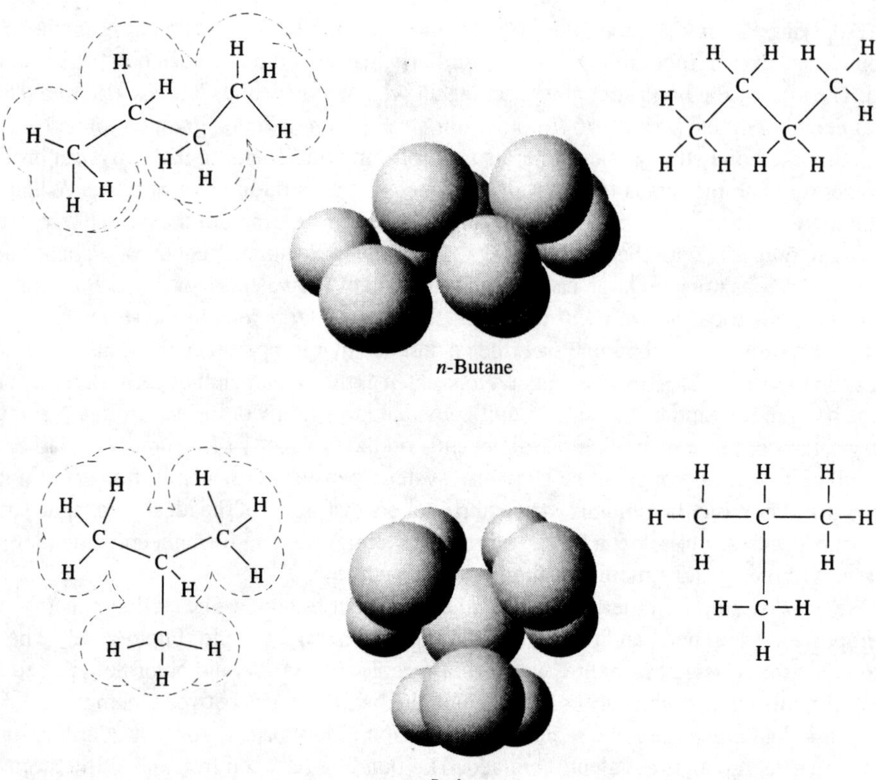

Figure 5-14 Structures of the isomers *n*-butane and isobutane. These compounds belong to the methane series of hydrocarbons and occur in natural gas and petroleum. The C_4H_{10} molecules shown here can occur in solid, liquid, and gaseous form. (After *General Chemistry, Second Edition,* by Linus Pauling, p. 571. W. H. Freeman and Company. Copyright © 1953.)

molecules attached to each other by weaker bonding (van der Waals or hydrogen bonds).* Solid organic compounds tend to melt by breaking the weaker bonds, and the characteristic molecules usually do not break down. Organic gases are also composed of these molecules. The structures of the solids are determined by the irregularly shaped molecules rather than by the spherical anions or atoms of minerals. The hydrocarbons are the simplest organic compounds. Examples of hydrocarbons are the gases *n*-butane and isobutane (Figure 5-14). Because the carbon atom has four electrons in its outer shell, it can form a total of four covalent bonds. In *n*-butane and isobutane each carbon atom forms single bonds with individual hydrogen atoms and also single bonds with other carbon atoms. These atoms can group together structurally in two different ways, forming either *n*-butane or isobutane (Figure 5-14).

*When a hydrogen atom is covalently bonded to the atom of another element, its one electron mainly occurs on the side of the hydrogen atom where the second atom occurs. This results in a weak attraction, on the opposite side, of the positively charged hydrogen nucleus for negative ions, giving a hydrogen bond. See Figure 4-1, which shows the cause of hydrogen bonding between water molecules.

The occurrence of different structures with the same composition (polymorphism in minerals) is known in organic chemistry as *isomerism* and is a common feature of organic compounds. Normal butane (*n*-butane) has a linear chain of zigzag carbons like a crooked railroad track, whereas isobutane has a branch line attached to its straight track. This difference is illustrated in the structural formulas on the right side of Figure 5-14. In addition to the common occurrence of isomerism, another complicating factor in the study of organic compounds is the occurrence of double and triple covalent bonds between two carbon atoms or between carbon and other atoms of a structure. Such multiple bonding has a strong effect on bond lengths and on the chemical properties of the compounds in which it occurs.

Next we consider structures characterized by the existence of some bonds that are mainly ionic and other bonds that are mainly covalent. Most of these minerals can be thought of as ionic structures having complex anions held together by simple cations. The bonding within the complex anions is primarily covalent. An example is the mineral calcite (Figure 5-15). The carbonate ion in this and other minerals consists of a planar grouping of three oxygen atoms around a carbon atom. These ions, and the oxygen atoms themselves, occur in six coordination around each calcium ion. The carbonate ion exists as a discrete unit, because the bonding between carbon and oxygen atoms, which is mainly covalent, is much stronger than the ionic bonding between

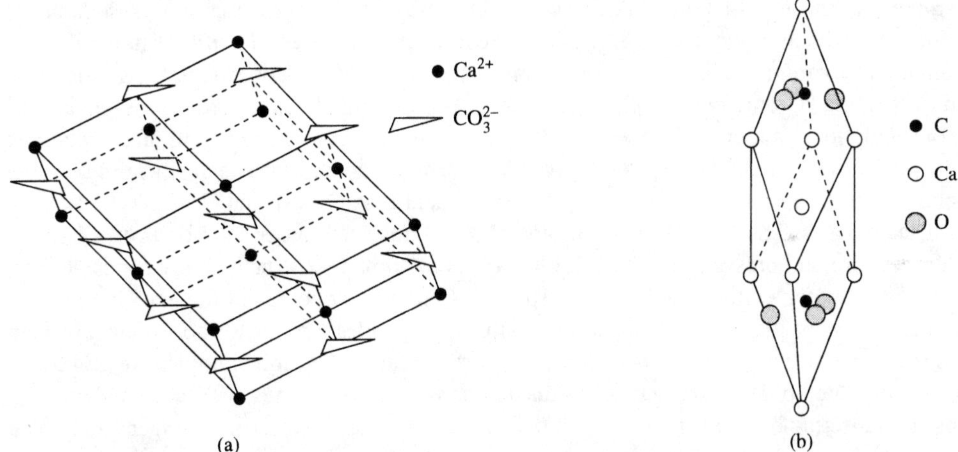

Figure 5-15 Structure of calcite ($CaCO_3$). (a) The cleavage rhombohedron can be thought of as a distorted sodium chloride cube with each chlorine ion replaced by a calcium ion and each sodium ion replaced by a planar carbonate ion group. The rhombohedron, which contains four formula units of $CaCO_3$, does not represent the calcite unit cell. (b) The true unit cell has calcium ions at the corners and in the center, with two CO_3^{2-} groups that are the inverse of each other on the vertical threefold axis. The true unit cell has two formula units of $CaCO_3$. The two very different refractive indexes of calcite are controlled by the relationship between the direction of light travel and the plane of the CO_3^{2-} groups. The high index occurs when a light ray passes through the calcite parallel to the plane of the CO_3^{2-} groups; the low index occurs when a light ray travels perpendicular to the plane of the CO_3^{2-} groups. Most of the refractivity is due to interaction of light waves with the oxygen atoms. [(a) From W. S. Fyfe, *Geochemistry of Solids,* p. 113. Copyright 1964 by McGraw-Hill Book Company. Used with permission of McGraw-Hill Book Company.] (b) After Bragg and Claringbull (1965, 127–134).

calcium and oxygen atoms. When calcite dissolves in water, the carbonate ion continues to exist as a distinct entity. Even though perovskite ($CaTiO_3$) and calcite ($CaCO_3$) both have formulas of the ABX_3 type, the structure of the two minerals is very different. This is due to the difference in bonding that occurs for the titanium and carbon atoms. In turn, the difference in bonding is due to the difference in electronegativity values for the two elements. Titanium has a value (1.5) that is much lower than that of oxygen (3.5), thus allowing a dominantly ionic bond to form between them. Carbon's electronegativity value (2.5) is much closer to that of oxygen and results in a bond that is mainly covalent.

An important example of a complex anion occurring in minerals is the SiO_4^{4-} ion. In some silicate minerals such anions occur as discrete ions in a manner similar to that of the carbonate ion. The difference in electronegativity between silicon and oxygen (1.7) results in bonding between them that is about 60 percent covalent and 40 percent ionic. The oxygen atoms form a tetrahedron around a silicon atom, and minerals with isolated tetrahedrons can be described in terms of the relationship of the tetrahedrons to the cations of the mineral. In most silicates, however, more complex structures occur in which the tetrahedrons are linked together, by sharing corner oxygen atoms, into various units of structure (Figure 5-16). These silicates are classified on the basis of the type of tetrahedral linkage that occurs (Table 5-3). No mineral is known in which two fully occupied silicate tetrahedrons share an edge or face, in agreement with Pauling's third rule. In detail, silicate structures are extremely complicated, and some silicate minerals do not fit into the simple classification of Table 5-3. The common substitution of one cation for another, and of aluminum for silicon in the tetrahedrons, further complicates the study of silicate minerals. The coordination of various cations can often be predicted from radius-ratio calculations. Such calculations assume an ionic radius for oxygen, which is not as reliable a procedure for silicates as it is for purely ionic minerals. Further details on the structure, composition, and phase relations of some of the major silicate minerals are given later in this chapter. Reviews of silicate structures and classifications are given in Liebau (1985), Smyth and Bish (1988), and Griffen (1992).

The above discussion assumes that mineral structures exist under conditions of relatively low temperature and pressure. We have known for some time that structural variations accompany changes in temperature and pressure. To understand the properties of the deep interiors of the Earth and other planets we need to have simplifying rules similar to those developed by Pauling for near-surface conditions. In recent years a great deal of mineralogical research has dealt with the study of structural variations due to changing temperatures, pressures, and compositions. This discipline is known as comparative crystal chemistry (Hazen and Finger 1982). The results of this research allow prediction of the structure, properties, and stability of minerals at high temperatures and pressures. For example, Knittle and Jeanloz (1987) obtained experimental data for (Mg, Fe)SiO_3 perovskite under pressure conditions representative of the lower mantle. Their data suggest that this dense, perovskite-structured mineral can exist throughout the pressure range of the lower mantle and may therefore be the most abundant mineral in the Earth.

It has been found that, for many common minerals, pressure and temperature have an inverse relationship. An increase in temperature has the same effect as a decrease in pressure and vice versa. The alkali feldspars show this relationship (Figure 5-17). Changes in pressure and temperature cause significant changes in the sodium-oxygen and potassium-oxygen bond lengths and have little effect on the aluminum-oxygen and silicon-oxygen bond lengths within the tetrahedra. In addition to pressure and temperature, compositional variation also affects the structure. The substitution of smaller sodium for larger potassium atoms has the same effect as an increase in pressure or a decrease in temperature (Figure 5-17).

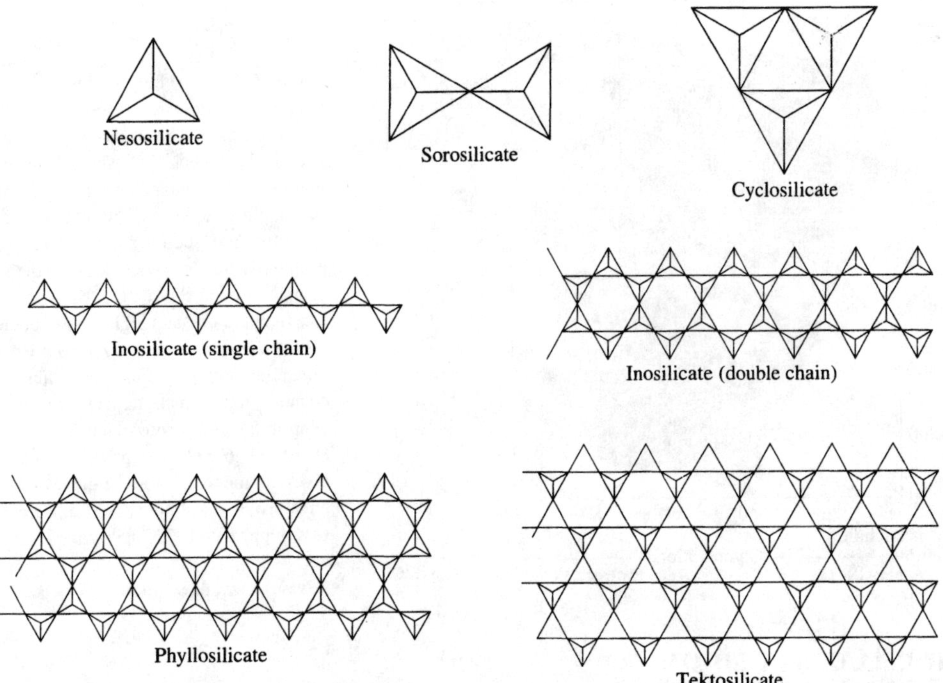

Figure 5-16 Examples of the structural units of the silicate classes listed in Table 5-3. The basic unit of all classes is the SiO_4^{4-} tetrahedron, consisting of four oxygen atoms around a silicon atom. The apexes of all the tetrahedra shown here are pointing upward for all classes except the tektosilicate example. This particular tektosilicate structure consists of layers of hexagonal rings of tetrahedra with alternate rows of apexes pointing in opposite directions. The upward-pointing apexes of a given layer coincide with the downward-pointing apexes of the layer above it. Thus all four corners of each tetrahedron are shared.

TABLE 5-3 Silicate Classification

Class	Arrangement of tetrahedra	Shared corners	Repeat unit	Si : O	Example
Nesosilicates	Independent tetrahedra	0	SiO_4^{4-}	1 : 4	Olivine
Sorosilicates	Pair of tetrahedra sharing corner	1	$Si_2O_7^{6-}$	$1 : 3\frac{1}{2}$	Hemimorphite
Cyclosilicates	Closed rings of tetrahedra each sharing two corners	2	SiO_3^{2-}	1 : 3	Tourmaline
Inosilicates	Infinite single chains of tetrahedra each sharing two corners	2	SiO_3^{2-}	1 : 3	Pyroxenes
	Infinite double chains of tetrahedra alternately sharing two and three corners	$2\frac{1}{2}$	$Si_4O_{11}^{6-}$	$1 : 2\frac{3}{4}$	Amphiboles
Phyllosilicates	Infinite sheets of tetrahedra each sharing three corners	3	$Si_2O_5^{2-}$	$1 : 2\frac{1}{2}$	Micas
Tektosilicates	Unbounded framework of tetrahedra each sharing four corners	4	SiO_2	1 : 2	Quartz, feldspars

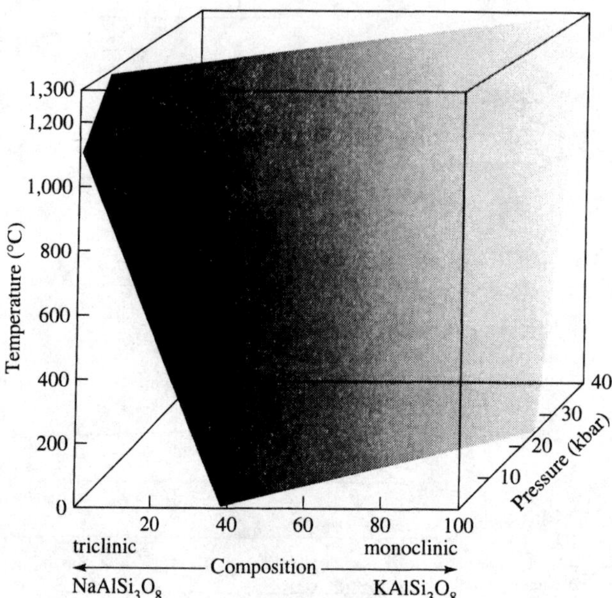

Figure 5-17 The cube represents temperature-pressure-composition space for alkali feldspars. The surface within the cube represents conditions of constant molar volume and constant structure. A decrease in temperature, an increase in pressure, or a substitution of sodium for potassium, could each individually produce exactly the same change in mineral structure. High pressure, low temperature, and high sodium content result in triclinic symmetry. Low pressure, high temperature, and high potassium content produce monoclinic symmetry. (Reprinted with permission from R. M. Hazen, 1976, Sanidine: predicted and observed monclinic-to-triclinic reversible transformations at high pressure, *Science*, v. 194, pp. 105–107. Copyright 1976 by the American Association for the Advancement of Science.)

UNIT CELL COMPOSITION

The diffraction of X rays by a crystal makes it possible to determine both the location of atoms in a unit cell and the volume of the unit cell. Knowledge of the volume, along with knowledge of the molecular weight and density of the mineral, allows us to calculate the number of formula units (Z) in the unit cell:

$$\begin{aligned}\text{density (g)/cm}^3) &= \frac{\text{weight (g) of unit cell}}{\text{volume in cm}^3\text{ of unit cell}}\\ &= \frac{(\text{molecular weight in grams}) \times Z}{(\text{number of formula units in 1 mol}) \times (\text{volume in cm}^3)}\\ &= \frac{M \times Z}{6.023 \times 10^{23} \times V(\text{Å}) \times 10^{-24}}\\ &= \frac{M \times Z \times 1.66}{V(\text{Å})}\end{aligned} \tag{5-2}$$

As an example, consider the mineral pyrite (FeS_2), with a density of 5.01 grams per cubic centimeter, a molecular weight of 119.97, and a cubic unit cell with an edge length of 5.42 Å. Substituting into the above formula,

$$5.01 = \frac{119.97 \times Z \times 1.66}{(5.42)^3} = \frac{199.05\,Z}{159.24}$$

$$Z = \frac{797.79}{199.05} = 4$$

Thus each unit cell of pyrite contains four iron atoms and eight sulfur atoms.

Because of substitution of atoms and ions for each other, different samples of a given mineral can have different compositions. The chemical analysis of a particular sample can be used to find the number of atoms of each element in the unit cell of the sample. An example of the procedure is given below for a sample of the mineral biotite. The unit cell volume of this monoclinic mineral is given by the formula $V = abc \sin\beta$, where a (5.3 Å), b (9.2 Å), and c (10.2 Å) are the lengths of the sides of the unit cell, and β (100°) is the angle between sides c and a. For our sample, the volume is 489.89 Å^3 and the density is 3.04 grams per cubic centimeter. Since the chemical analysis is given in oxides, we need to calculate the oxide contents per unit cell. The weight (W) of an oxide such as SiO_2 in the unit cell is

$$W = \frac{\text{total cell weight (volume} \times \text{density)} \times \text{oxide percent } (P)}{100} \tag{5-3}$$

The number of oxide units per unit cell is

$$\frac{W}{\text{oxide molecular weight } (OW) \times 1.66 \times 10^{-24}}$$

The number 1.66×10^{-24} is the weight in grams of a hypothetical atom of atomic weight 1. It is equal to $(1/6.023 \times 10^{23})$, and has to be used since the atomic weights of the elements represent the weight of 6.023×10^{23} atoms of an element. Putting the above equations together, the number of oxide units per unit cell is

$$\begin{aligned} \frac{W}{OW \times 1.66 \times 10^{-24}} &= \frac{V(\text{Å}) \times D \times P \times 10^{-24}}{100 \times OW \times 1.66 \times 10^{-24}} \\ &= \frac{V \times D}{166} \times \text{molecular proportion } \frac{P}{(OW)} \end{aligned} \tag{5-4}$$

The following table shows the calculation of atoms per unit cell for a biotite sample:

	1	2	3
SiO_2	38.28	0.637	5.71
Al_2O_3	12.17	0.119	1.07
FeO	21.03	0.293	2.63
MgO	14.51	0.360	3.23
K_2O	10.18	0.108	0.97
H_2O	3.81	0.212	1.90
Total	99.98		

1: Chemical analysis.

2: Molecular proportions obtained by dividing weight percent by molecular weight of the oxide.

3: Oxide contents per unit cell obtained by multiplying molecular proportions times 8.97, which is the value of $V \times (D/166)$.

The calculations show that the number of atoms per unit cell are Si, 5.71; Al, 2.14; Fe, 2.63; Mg, 3.23; K, 1.94; H, 3.8; and O, 23.36 $(11.42 + 3.21 + 2.63 + 3.23 + 0.97 + 1.90)$. Ideally, the number of oxygens should be 24, corresponding to the formula

$$K_2(Mg,\ Fe,\ Al)_6(Al_{2\text{–}3}\ Si_{6\text{–}5})O_{20}(OH)_4$$

Slight errors in the chemical analysis and in the measurement of the density will cause deviations from the ideal total of 24 oxygens. Adjusting the numbers to a basis of 24 oxygens, we get the following formula for our sample:

$$K_{2.00}Mg_{3.33}Fe_{2.71}Al_{2.20}Si_{5.89}O_{20.09}(OH)_{3.91}$$

Note that in biotite aluminum can substitute for both silicon (fourfold coordination) and for magnesium-iron (sixfold coordination). In this particular sample, there is apparently no substitution for magnesium-iron, since the total number of atoms of these two elements is 6.04, and the ideal formula requires six atoms in sixfold coordination. The silicon and aluminum add up to 8.09 atoms, and the ideal formula calls for a total of eight atoms. Fluorine and chlorine, which often substitute for hydroxide ion, were not included in the chemical analysis, and this may explain why the value found for hydroxide ions is below the ideal number of four ions. Many other trace elements can also substitute for the cations in the biotite structure. If we were to calculate Z for this or any other biotite sample, we would find $Z = 1$; that is, there is one formula unit of the above formula in the unit cell. The general formula for biotite is often given as $K(Mg, Fe)_3AlSi_3O_{10}(OH)_2$ and, in this case, $Z = 2$.

SOLID SOLUTION

A dictionary definition of the word "solution" might read as follows: a homogeneous mixture formed by combining two or more substances. With this definition, a solution could be a gas, a liquid, or a solid. Most rock-forming minerals have a variable composition in terms of two or more of their components. The mineral biotite, for example, varies in its magnesium-iron ratio and, to a lesser extent, in its aluminum-silicon ratio. We are not considering here variation in trace element content, which also occurs, but which usually does not significantly affect the formula of a sample. To a first approximation, any given sample of a mineral can be considered to be homogeneous and thus we can talk about minerals as crystalline solutions. In detail, the distribution of a particular type of atom or ion may not be perfectly uniform throughout a sample.

The term *solid solution* applies only to crystalline solids. Any mineral, different samples of which have different amounts of one or more major or minor components, is said to show solid solution. This term should not be confused with the terms isotypism, isostructuralism, and isomorphism. Two minerals that have exactly the same structures are called *isotypes.* An example is the pair halite-galena. Each atom in one of the minerals has a counterpart in the other. *Isostructural* minerals have very similar structures, but not a one-to-one relationship between their atoms. For example, coesite (SiO_2) and albite ($NaAlSi_3O_8$) have similar structures, which can be illustrated in this manner:

Coesite		Si	Si	Si_2O_8
Albite	Na	Al	Si	Si_2O_8

However, there is no counterpart of the albite sodium atom in coesite. The albite structure can be described in terms of the coesite structure, with the sodium atom of albite stuffed into spaces of the structure. The most general term for similarity of mineral structures is *isomorphism.* The pyroxene group of minerals can be considered isomorphous, but in detail their structures have significant differences. The term isomorphism is often used to describe the whole range of similarity of structures, from isotypic structures to structures with only a gross similarity.

A given mineral may or may not exhibit solid solution involving atoms of a companion isomorphous mineral. Because the magnesium and iron ions have similar sizes, complete solid solution occurs between the isotypic minerals forsterite and fayalite, giving the mineral olivine. On the other hand, the cation and anion of a given mineral *AB* may be much larger than the cation and anion of another mineral *CD*. However, if the ratio *A/B* is very similar to the ratio *C/D*, the two minerals may have closely related structures (isostructuralism) or generally similar structures (isomorphism). Because of the differences in size, *A* could not substitute for *C* and *B* could not substitute for *D*. Thus no solid solution could occur with respect to these atoms.

Now let us assume that *A* and *C* are roughly the same size as *B* and *D*. In this case, some *A* may substitute for *C* in *CD*, and some *B* may substitute for *D* in *CD*. Also, *C* and *D* may substitute for *A* and *B* in *AB*. These are examples of the most common type of solid solution, known as *substitutional* solid solution. This substitution may be complete across an entire range of composition as in olivine, which varies from Fe_2SiO_4 to Mg_2SiO_4. Often the substitution is not complete, an example being the limited substitution of sodium for potassium in low-temperature alkali feldspars. The occurrence of substitutional solid solution is mainly governed by geometrical considerations. Since the structures of Fe_2SiO_4 and Mg_2SiO_4 are exactly the same, the only requirement for magnesium and iron to substitute for each other is that they have approximately the same radii. In plagioclase feldspar, the radii of sodium and calcium are similar, and substitution occurs even though the ions have different valences (since charge balance must be maintained, Na^+-Ca^{2+} substitution is accompanied by Si^{4+}-Al^{3+} substitution). Solid solution can even occur between two end members with very different structures and/or with very different chemical compositions. In these cases partial substitution is all that occurs (and is all that is possible).

Another type of solid solution is known as *interstitial* solid solution. In this case, foreign atoms or ions occur in openings between the regular constituents of the structure. An example is the occurrence of aluminum and sodium in tridymite, SiO_2. Aluminum can replace a small amount of silicon in this structure, and, in order to maintain electrical neutrality, sodium ions are accommodated in interstitial areas. The occurrence of foreign ions in tridymite makes the structure more rigid and extends its stability field to lower temperatures as compared to the stability field of pure tridymite. *Omission* solid solution is a third type. In this case some ions are missing from sites in a structure. In the mineral pyrrhotite, $Fe_{1-x}S$, as many as one out of every five iron sites may be vacant. Iron ions have both +2 and +3 valences to provide electrical neutrality.

Interstitial and omission solid solutions as discussed above are two examples of defects that occur more or less throughout the structures of minerals. Local chemical defects can also occur in some samples of a given mineral but not in other samples. Examples would be (1) missing atoms at some structural sites, (2) atoms displaced from their normal site, and (3) atoms at sites of a different coordination from that of most atoms of the element in question. There are several kinds of local defects that are physical defects only. The composition of a "faulty" mineral sample is not changed as a result of this kind of defect. An example would be what is known as an *edge dislocation.* This refers to a plane of atoms in a structure that stops partway through the structure, with the parallel planes on either side joining each other beyond the point of stoppage. This disturbance of the structure makes the edge dislocation a likely place for displacement of atoms to occur when the structure is put under stress. Both the physical and chemical properties of a given mineral sample may be strongly affected by the occurrence of one or more defects in its structure. Many synthetic solids used in industry owe their unique properties to the deliberate development of defects, chemical or physical, in their structures.

THERMODYNAMICS OF CRYSTALLINE SOLUTIONS

Most applications of thermodynamics to geologic problems have dealt with heterogeneous equilibria (equilibria involving a number of different phases) or with homogeneous equilibria in fluids (equilibria involving a number of different components in one phase). Starting about 1960, a small number of geologists have used thermodynamics to study homogeneous equilibria inside minerals (Saxena 1973; Ganguly and Saxena 1987). In this section we shall briefly discuss the basic theory and give a few examples of how it can be applied to minerals. Readers interested in further information are referred to Ganguly and Saxena (1987) and Navrotsky (1994).

In studying crystalline solutions various models are used. One such model, the ideal solution, is mentioned in Chapter Three in the discussion of activities, activity coefficients, and chemical potentials. An ideal solution, whether it is a gas, liquid, or solid, has to satisfy the following relationship:

$$\mu_i = \mu_i^* + RT \ln N_i \qquad (5\text{-}5)$$

where μ_i is the chemical potential of component i in the solution, N_i the mole fraction of component i in the solution, and μ_i^* the chemical potential at a given temperature and pressure when $N_i = 1$ (pure i). In an ideal solution the activity of component i, a_i, is equal to N_i.

How would a mineral behave if it were ideal? Consider olivine, $(Mg, Fe)_2SiO_4$. The two variable components can be chosen to be the extreme or end members of the solution, forsterite, Mg_2SiO_4, and fayalite, Fe_2SiO_4. If this mineral exhibits ideal behavior, substitution of magnesium ions for iron ions should cause a regular change in the properties of the mineral that is proportional to the change in mole fraction of forsterite. For example, the molar volume of an olivine sample should satisfy the equation

$$\bar{V}_{\text{olivine}} = \frac{n_1}{n_1 + n_2}\bar{V}_{\text{forsterite}} + \frac{n_2}{n_1 + n_2}\bar{V}_{\text{fayalite}} \qquad (5\text{-}6)$$

where n_1 is moles of forsterite in the sample, n_2 moles of fayalite in the sample, $n_1/(n_1 + n_2)$ the mole fraction of forsterite, and $n_2/(n_1 + n_2)$ the mole fraction of fayalite; that is, the volume of mixing equals zero. The molar internal energy and molar enthalpy will also be proportional to the mole fraction of each component.

There are other requirements that must be satisfied for a crystalline solution to be considered an ideal solution. The heat of formation (ΔH_f) of an ideal solution from its end members is zero; that is, the heat of mixing equals zero. For this to be so, the structure of the two end members, and of any mixture of the two, must be the same; thus the ions that replace each other in the mixture must do so without causing any change in the structure. Furthermore, the ions should not be ordered in the structure, but should instead be randomly distributed. Formation of the mixture does cause a difference in the entropy of the mixture as compared to that of the two end members before mixing. By forming the mixture, we have increased the amount of disorder and thus increased the entropy. The increase is known as the *entropy of mixing*. Since $\Delta G = \Delta H - T\,\Delta S$, the Gibbs function of the mixture is less than that of the pure separated components.

For nonideal solutions

$$\mu_i = \mu_i^* + RT \ln a_i \qquad (5\text{-}7)$$

with $a_i = \gamma_i N_i$, where γ_i is the activity coefficient of component i. The value of the activity coefficient for an ideal solution is 1, and for nonideal solutions the activity coefficient can be thought of as a correction factor that has to be used because of the deviation from ideal behavior. One procedure in studying a particular crystalline solution is to calculate mole fractions for given conditions using experimental data for end members and theoretical relationships involving chemical potentials, activities, Gibbs functions, etc. These calculated relationships for the mixture can then be compared with experimental studies of the mixture under the given conditions. For example, the calculated locations of the liquidus and solidus curves in the melting diagram for olivine (Figure 5-18) are very similar to the locations found experimentally by Bowen and Schairer (1935). Other experimental evidence also indicates that olivine is close to being an ideal crystalline solution. A regular variation in heats of solution with change in composition has been found (Sahama and Torgeson 1949). Studies of the distribution of iron and magnesium between olivine and coexisting pyroxene give a straight-line relationship between the activity of fayalite and the mole fraction of fayalite (Saxena 1973, 116). Wood and Kleppa (1981) report a small positive heat of mixing for olivine.

The alkali feldspars are an example of a crystalline solution that deviates significantly from ideality. The two end members, $KAlSi_3O_8$ and $NaAlSi_3O_8$, exhibit unmixing upon cooling, and the heat of formation of any given mixture of the two is not zero. Figure 5-19 shows how molar volumes and heats of solution deviate from ideality. Despite the strong deviation from ideality, it is possible to carry out thermodynamic calculations assuming a particular, nonideal model for the feldspar solution. These calculations, along with experimental data such as that in Figure 5-19, lead to an equation of state (an equation that relates certain properties of a system) for the feldspars; this equation can be used to calculate the relationships of the alkali feldspars at various temperatures and pressures.

The statement that $a_i = N_i$ is known as Raoult's Law. In this case the mixing of two different components in a solid solution is ideal and thus the activities of the components are equal to their mole fractions. Olivine, discussed above, exhibits Raoult's Law behavior. Another kind of mixing behavior is represented by Henry's Law, which is expressed as $a_i = h_i N_i$, where h_i is a proportionality constant determined by pressure and temperature. Henry's Law is obeyed when

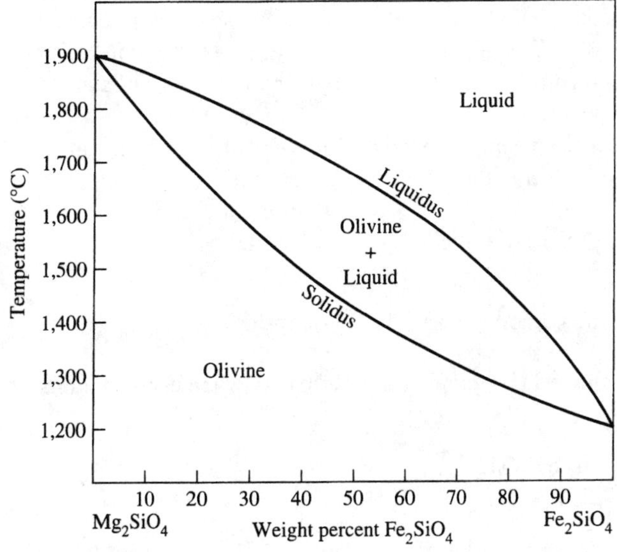

Figure 5-18 Phase diagram for the system forsterite-fayalite. After Bowen and Schairer (1935).

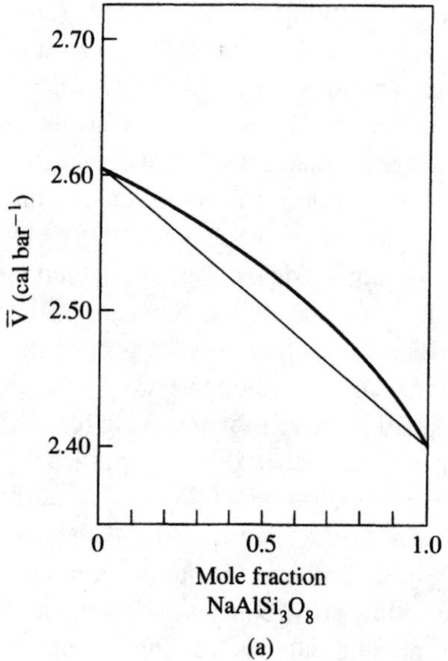

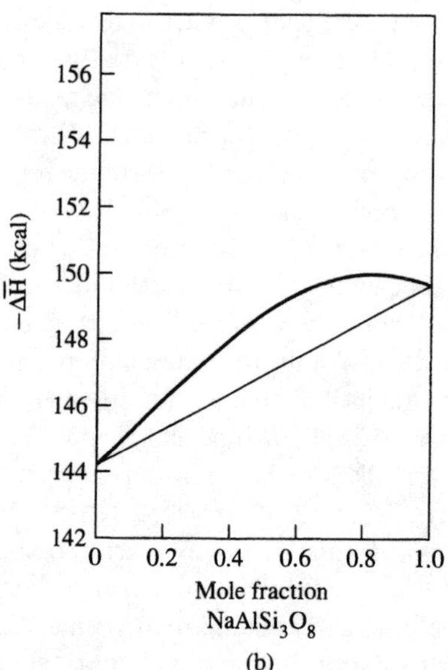

Figure 5-19 (a) Molar volumes of sanidine-high albite crystalline solutions. The straight line represents values to be expected for an ideal solution. One cal bar^{-1} = 41.840 cm^3. (b) Heats of solution for microcline-low albite crystalline solutions. The straight line represents values to be expected for an ideal solution. (a) After Waldbaum and Thompson (1968). (b) After Waldbaum (1966).

component i is a solute and N_i is very small. The substitution of a trace element, such as nickel, in olivine follows Henry's Law.

Solutions can exhibit three types of behavior, depending on the amount of component i present (Figure 5-20). At the extremes are behavior represented by Raoult's Law and Henry's Law. Between these two extremes is behavior such as that shown by the alkali feldspars. For nonideal behavior like that of the feldspars, $a_i = \gamma_i N_i$. The activity coefficient, γ_i, is a correction factor that takes into account interactions between component i and all the other components of a given solution.

The distribution of elements between coexisting minerals exhibiting solid solution can be used as a geothermometer. As an example, consider the following equilibrium:

$$\frac{1}{3}\ Mg_3Al_2Si_3O_{12}\text{ (pyrope garnet)} + CaFeSi_2O_6\text{(hedenbergite)}$$

$$\rightleftharpoons \frac{1}{3}\ Fe_3Al_2Si_3O_{12}\text{ (almandine garnet)} + CaMgSi_2O_6\text{ (diopside)} \tag{5-8}$$

The distribution coefficient, K_D, for the mineral pair garnet-clinopyroxene, when each contains iron and magnesium, is defined as

$$K_D = \frac{(Fe^{2+}/Mg)^{1/3}_{\text{garnet}}}{(Fe^{2+}/Mg)_{\text{clinopyroxene}}} \tag{5-9}$$

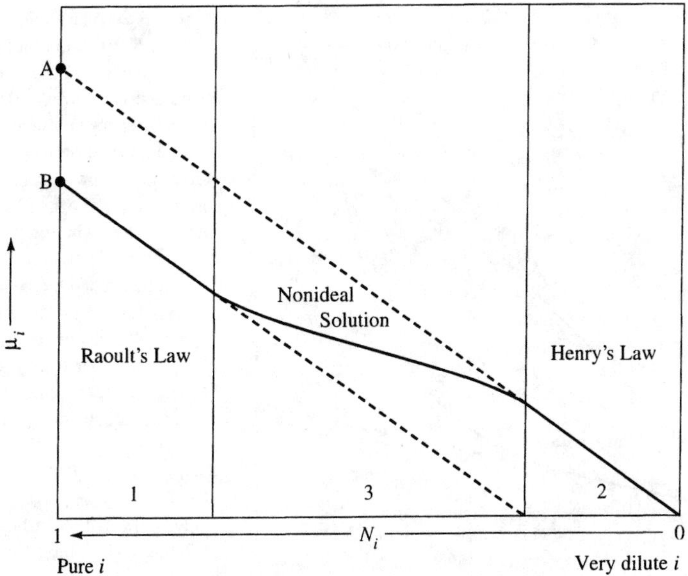

Figure 5-20a Schematic plot of the chemical potential μ_i of component i in solution as a function of mole fraction N_i. The plot is divided into three areas of solution behavior. Area one is when component i is a solvent and N_i is close to unity. Because of the large amount of i present, interactions between i and the other species present are insignificant. As a result, the activity of i, a_i, is equal to N_i. Area two represents the opposite situation; component i is a solute and N_i is very small. Because of the small amount of i present, interactions between i and the other species present are insignificant. In this case $a_i = h_i N_i$ where h_i is a proportionality constant dependent on temperature and pressure. The third area represents behavior between the two extremes. In this situation the behavior of component i depends on the other components present as well as on temperature and pressure and must be determined experimentally or theoretically for each solution. When ideal mixing occurs at all compositions, area three disappears and areas one and two connect along a straight line. In the Raoult's Law region the standard state (reference value) of component i is μ_i^*, which is the chemical potential of pure i at a selected temperature and pressure (point B). In this region $\mu_i = \mu_i^* + RT \ln N_i$. In the Henry's Law region $\mu_i = \mu_i^* + RT \ln h_i N_i$ where h_i is a proportionality constant. In the intermediate region (area three) $\mu_i = \mu_i^* + RT \ln \gamma_i N_i$ where γ_i is the activity coefficient of component i. For the significance of point A see Figure 5-20b. (After D. K. Nordstrom and J. L. Munoz, *Geochemical Thermodynamics*. Copyright © 1985 by Blackwell Science, Ltd. Reprinted by permission of Blackwell Science, Ltd., Oxford, England.)

For ideal solutions (Raoult's Law obeyed) and constant pressure, it can be shown that

$$\ln K_D = A + B/T \qquad (5\text{-}10)$$

where A and B are constants and T is temperature.

It should be noted that distribution coefficients, although functions of pressure and temperature, do not behave exactly like equilibrium constants. The relationship between the two is

$$K_{\text{equil}} = K_D K_\lambda \qquad (5\text{-}11)$$

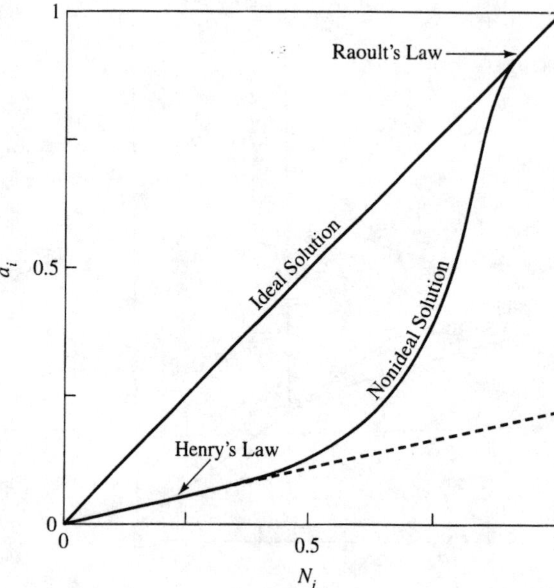

Figure 5-20b Relations between activity of component i and mole fraction of component i for the three types of behavior discussed in Figure 5-20a. The relationship shown here for nonideal behavior assumes a standard state of μ_i^*, the chemical potential of pure i at a selected temperature and pressure. The standard state could have been defined to be the *hypothetical* pure state at the selected temperature and pressure obtained by extrapolation of Henry's Law along the dashed line in Figure 5-20a to $N_i = 1$ (point A). The choice of standard state depends on the type of geochemical problem being studied. See discussion of standard states in Nordstrom and Munoz (1985, 128–135). (After Richardson, S. M. and McSween, H. Y., Jr., *Geochemistry—Pathways and Processes.* Copyright © 1989. Adapted by permission of Prentice Hall, Inc., Upper Saddle River, NJ.)

or, for the garnet-clinopyroxene equilibrium above,

$$K_{\text{equil}} = \left[\frac{(N_{\text{Fe}}^{\text{gar}} / N_{\text{Mg}}^{\text{gar}})^{1/3}}{(N_{\text{Fe}}^{\text{pyr}} / N_{\text{Mg}}^{\text{pyr}})} \right] \left[\frac{(\lambda_{\text{Fe}}^{\text{gar}} / \lambda_{\text{Mg}}^{\text{gar}})^{1/3}}{(\lambda_{\text{Fe}}^{\text{pyr}} / \lambda_{\text{Mg}}^{\text{pyr}})} \right] \tag{5-12}$$

where $N_{\text{Fe}}^{\text{gar}}$ is the mole fraction of Fe^{2+} in garnet and so forth and $\lambda_{\text{Fe}}^{\text{gar}}$ is the activity coefficient of Fe^{2+} in garnet and so forth. If both minerals exhibit ideal behavior, then the activity coefficients are all 1.0 and $K_D = K_{\text{equil}}$. K_λ takes into account compositional variation (other than the actual exchange reaction) in the coexisting minerals. For example, variation in the calcium content of garnet-clinopyroxene pairs affects the K_D value for the distribution of ferrous iron and magnesium between sites in the two minerals. In the case of the exchange of a single trace element between two phases, Henry's Law may be obeyed and K_D is then referred to as the Nernst distribution coefficient. Compositional variation in the host minerals also affects trace-element K_D values. For more information on distribution coefficients see Ganguly and Saxena (1987).

Equation (5-10) indicates that measurement of K_D for a coexisting mineral pair can be used to estimate the temperature at which a rock reached equilibrium. For garnet-clinopyroxene pairs it is necessary to take into account the effect of calcium on the distribution of iron and magnesium. Ellis and Green (1979) used experimental data to derive the following expression for temperature for garnet-clinopyroxene pairs:

$$T\text{ (K)} = \frac{3104\, N_{\text{Ca}}^{\text{gar}} + 3030 + 10.86\text{ P(kb)}}{\ln K_D + 1.9039} \tag{5-13}$$

where $N_{\text{Ca}}^{\text{garnet}}$ refers to the mole fraction of calcium in garnet. Note that the effect of pressure is small and thus only a rough estimate of pressure is needed to calculate temperature. An independently obtained value for pressure will of course allow a more exact determination of temperature. Other expressions have been developed for this thermometer using thermodynamic data and by making various assumptions about the mixing behavior of the two minerals. An example of the use of this thermometer to obtain temperatures for high-grade metamorphic rocks is given by Dahl (1980).

As indicated above, distribution coefficients are functions of pressure as well as of temperature (Figure 5-21). Thus experimental or thermodynamic data can also be used to obtain equations that will allow determination of pressure (geobarometers). An independent estimate of temperature may be needed to get a precise value for pressure. A useful geobarometer is based on the reaction

$$3CaAl_2Si_2O_8 \text{ (anorthite)} \rightleftharpoons Ca_3Al_2Si_3O_{12} \text{ (grossularite garnet)}$$

$$+ 2Al_2SiO_5 \text{ (kyanite or sillimanite)} + SiO_2 \text{ (quartz)} \quad (5\text{-}14)$$

Ganguly and Saxena (1987, 245–247) discuss the derivation of the necessary equations for this reaction. For a reaction involving sillimanite (Koziol and Newton 1988)

$$P(\text{bar}) = 23.41T\ (^\circ\text{C}) - 25.0 - 0.0001872T^2 \quad (5\text{-}15)$$

and for a reaction involving kyanite

$$P(\text{bar}) = 22.80T\,(^\circ\text{C}) - 1093 \quad (5\text{-}16)$$

Equation (5-16) was determined experimentally by Koziol and Newton (1988) and equation (5-15) was calculated by combining their results with experimental results for the kyanite-sillimanite reaction (Holdaway 1971). Martignole and Nantel (1982) used a different sillimanite equation to estimate equilibrium pressures of metapelites in Quebec:

$$P(\text{bar}) = 23.8T\,(^\circ\text{C}) + 113.0 \quad (5\text{-}17)$$

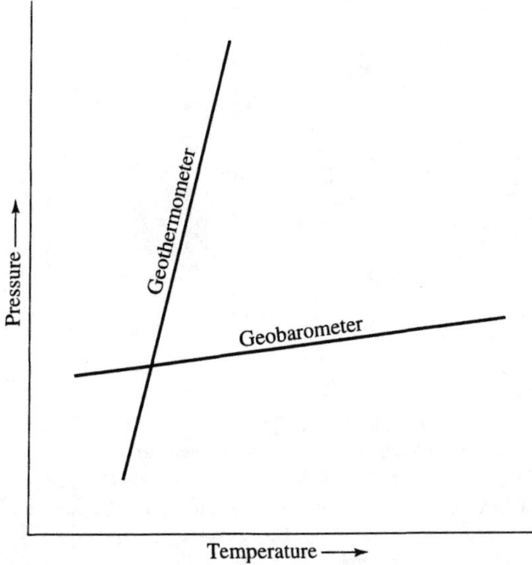

Figure 5-21 Equilibrium curves for reactions that would make a good geothermometer and a good geobarometer. A good geothermometer will be sensitive to temperature change and insensitive to change of pressure. Conversely, a good geobarometer will be sensitive to pressure change and will not be affected significantly by temperature change. The position of each of these curves is fixed at a given value of the distribution coefficient K_D. The geothermometer curve will be translated to the left, parallel to its present position, with increasing K_D. The geobarometer curve will be translated upward with increasing K_D, but it too will retain the same slope. (After Richardson, S. M., and McSween, H. Y., Jr., *Geochemistry—Pathways and Processes*. Copyright © 1989. Adapted by permission of Prentice Hall, Inc., Upper Saddle River, NJ.)

A discussion of the geobarometry of high-grade metamorphic rocks has been given by Newton (1983) and Essene (1982, 1989) reviews both geologic thermometry and barometry. Further discussion of this topic is provided in Chapter Nine.

ORDER-DISORDER

In our earlier discussion of a solution, we required that the solution be homogeneous. For a crystal structure to be completely homogeneous, it is necessary that every crystallographically equivalent site (sites that are the same in terms of symmetry) be occupied by only one kind of atom. Thus all minerals that exhibit solid solution are not homogeneous, since variation in composition occurs as a result of the substitution of one kind of atom for another at particular sites in the structures. A sample of olivine is homogeneous in the sense that cation positions between the silica tetrahedra are only occupied by magnesium or iron (ignoring trace elements). Ions of these two elements have similar radii; thus the structures of minerals of the olivine series are essentially all the same. However, detailed study of the olivine structure has shown that there are two crystallographically distinct positions occupied by the magnesium and iron ions (Figure 5-22). The oxygen coordination polyhedra about the two different sites are slightly different in shape. Crystallographers designate the two positions as M_1 and M_2. The filling of these positions could occur by a random distribution of magnesium and iron ions in both sites (typical of an ideal so-

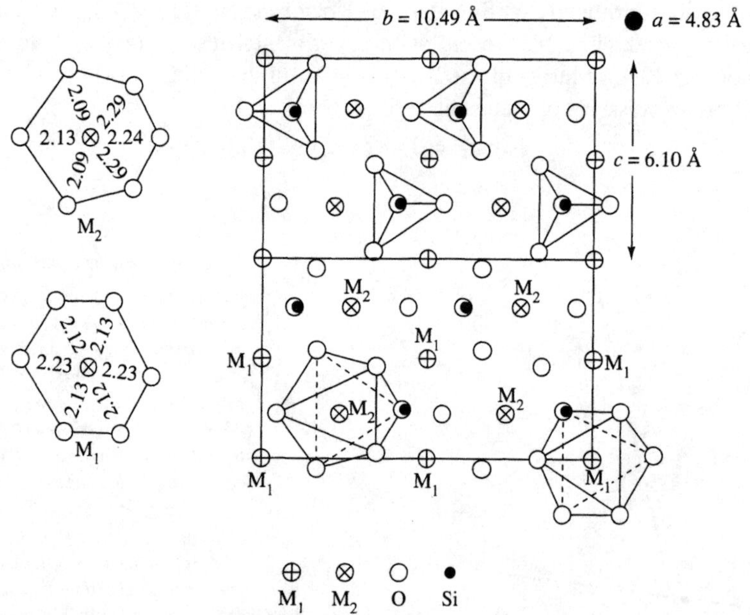

Figure 5-22 Olivine crystal structure. The structure is shown here projected onto (100). Two unit cells are represented, with silica tetrahedra outlined in the upper cell and oxygen coordination polyhedra about the two different metal positions (M_1 and M_2) outlined in the lower cell. The metal-oxygen distances in each oxygen polyhedron vary (as shown on the left): thus each polyhedron is distorted from octahedral symmetry. The average metal-oxygen distances in forsterite, Fa_{10}, are M_1 site, 2.10 Å, and M_2 site, 2.14 Å. In fayalite, Fa_{100}, these distances are M_1 site, 2.16 Å, and M_2 site, 2.19 Å. After Burns (1970, 82).

lution), or there could be a complete or partial ordering of the ions, with magnesium occurring completely or mainly in one site and iron completely or mainly in the other site. Note that complete ordering could only occur in a crystal with the composition $MgFeSiO_4$. All other compositions would involve at least one element occurring in both sites.

The distribution of ions in a mineral can be found by a variety of techniques. These site population measurements involve either diffraction studies (X-ray, electron, and neutron diffraction) or measurement of absorption spectra (Mössbauer, infrared, and other spectra). Development of the electron microscope has allowed study of order-disorder and other phenomena in crystals at a scale of a few angstroms. In silicate minerals, most of which exhibit solid solution, examples have been found that run the gamut from almost complete *order* to almost complete *disorder.* Knowledge of order-disorder relationships in minerals allows geologists to better understand the conditions and processes involved in mineral formation. For example, experimental results on mineral stability can be related to the amount of ordering in minerals. Another important result of order-disorder studies is to provide new knowledge on the thermodynamic properties of minerals.* Two examples involving order-disorder in silicate minerals will be given here.

It was indicated in the previous section that experimental data show olivine to be almost ideal in its behavior. Studies of possible cation ordering in olivine also suggest that it is close to an ideal solution, since little or no ordering has been found. Orthopyroxene, $(Mg, Fe^{2+})SiO_3$, is another mineral that has distinct M_1 and M_2 positions for its cations. Cations in the M_1 sites are each bonded to oxygens, which are also bonded to single silicon ions. In the M_2 sites, four of the oxygens are bonded to single silicon ions and two are each bonded to two silicon ions. Significant ordering does occur in this mineral. The amount of disorder increases with temperature. Chatillon-Colinet et al. (1983) carried out an experimental study of $(Fe^{2+}, Mg)SiO_3$ that suggests that, despite appreciable ordering, thermodynamic departures from ideality are very small in the temperature range in which orthopyroxene is formed in the Earth. In this mineral, calcium can also substitute in the M_1 and M_2 sites. If more than two kinds of ions can occupy a given site, it is much more difficult to determine site occupancies. In the case of orthopyroxene and other pyroxenes, calcium seems to occur only in the M_2 site, and this helps in estimating cation distribution.

An important example of order-disorder relationships is found in the K-feldspars. In this case we are concerned with the distribution of aluminum and silicon in the silicon-oxygen framework of the three polymorphs: sanidine, microcline, and orthoclase. They all have the same composition, but they have different stability fields in terms of temperature. The differences among them lie in the degree of ordering of aluminum and silicon. There are two different structural sites, known as T_1 and T_2, in the tetrahedrons of sanidine and orthoclase. High sanidine is the disordered, high-temperature form of $KAlSi_3O_8$, with random distribution of the aluminum and silicon ions. Since there are three silicons for every aluminum, this means that, on the average, one out of every four adjacent T_1 or T_2 sites will be occupied by an aluminum ion. Orthoclase is the partially ordered, medium-temperature member of the trio, with aluminum tending to be concentrated in the T_1 sites. In other words, more than one out of every four T_1 sites will contain an aluminum ion in any given sample of orthoclase. In microcline, two of four adjacent T_1 sites differ slightly from the other two T_1 sites in terms of symmetry. Thus they become structurally different

*Navrotsky (1987, 1994) points out that ordering phenomena make it very difficult to understand the thermodynamic behavior of mineral solid solutions. The extent of ordering depends on temperature in a complex fashion and is often kinetically controlled. Thermodynamic models for solid solutions must take into account the significant effects caused by mixing and ordering of ions. A good review of the effects of order-disorder transformations in mineral solid solutions can be found in Carpenter (1985).

sites. A similar situation is found for four adjacent T_2 sites, which become two different pairs. Microcline, the low-temperature K-feldspar, shows the greatest amount of ordering, with aluminum tending to concentrate in one pair of the T_1 sites. In what is known as maximum microcline, aluminum occurs only in these two sites. Thus the differences in symmetry, unit-cell dimensions, refractive indexes, twinning, and other properties of the K-feldspars are mainly due to variations in the ordering of their aluminum and silicon ions. This variation appears to be a continuous function of temperature. Albite ($NaAlSi_3O_8$) shows similar degrees of Al-Si ordering, with high-temperature albite showing the greatest disorder and low-temperature albite the greatest order (Goldsmith and Jenkins 1985).

The previous discussion has dealt with the pure end members of the $KAlSi_3O_8$-$NaAlSi_3O_8$ solid solution. This solution clearly does not behave as an ideal solution at all temperatures and compositions. As a result of free energy relationships, a process known as exsolution occurs (Figure 5-23). At high temperatures a single homogeneous phase is stable at all compositions. This situation can be represented by the curve labeled G_{ideal} in Figure 5-23a. A mechanical mixture of the two endmembers has the free energy of the curve labeled $G_{mixture}$ in Figure 5-23a. At all temperatures the actual free energy of solutions is affected by the excess free energy, which can be defined as

$$G_{excess} = G_{actual} - G_{ideal} \qquad (5\text{-}18)$$

G_{excess} is determined by repulsive and attractive interactions between mixing components and can be positive or negative. For an ideal solution G_{excess} is zero.

Figure 5-23b shows that, at some temperatures, G_{actual} has a maximum near the center of the curve. In this central area a mixture of two separate phases (not of endmember compositions) has a lower free energy than does a homogeneous solution of equivalent composition. The range of composition of the central area is given by A_1 and A_2 in Figure 5-23b. Such exsolution intergrowths formed in feldspars are known as perthites. The solvus curve in Figure 5-23c represents the compositions of coexisting phases for all temperatures below the temperatures at which a single homogeneous phase should not exist. However, for compositions between the solvus curve and the spinodal curve, the free energy relationships are such that the creation of two phases from a single phase (due to a lowering of temperature) is not favored unless kinetic factors allow phase formation at nucleation sites. Inside the spinodal curve two separate phases are always more stable and can separate more easily than outside the curve.* Further discussion of the kinetics of exsolution can be found in Richardson and McSween (1989). A review of the thermodynamics of solutions is provided in Nordstrom and Munoz (1994, Chap. 5).

*Spinodal decomposition is a process involving continuous breakdown of a homogeneous solid solution without phase formation at specific nucleation sites. In the case of the alkali feldspars, diffusion of the potassium and sodium atoms forms compositionally different regions, but the aluminosilicate framework is continuous across these regions. For compositions between the spinodal curve and the solvus, formation of two phases involves development of new phases at nucleation sites located at points of structural defects, such as dislocations and missing atoms. Thermodynamic theory tells us that the difference between the area inside the spinodal curve and the area between the two curves is as follows. Inside the spinodal curve a homogeneous phase can lower its free energy by spontaneously developing local areas of slightly different composition. Between the two curves formation of such areas initially causes a slight increase in free energy. Thus the existence of nucleation sites is necessary to make formation of separate phases energetically possible. Kinetic theory tells us that the spinodal curve is a kinetic boundary, with spontaneous decomposition possible inside the curve (one phase unstable) and decomposition between the two curves occurring only under special circumstances (one phase metastable).

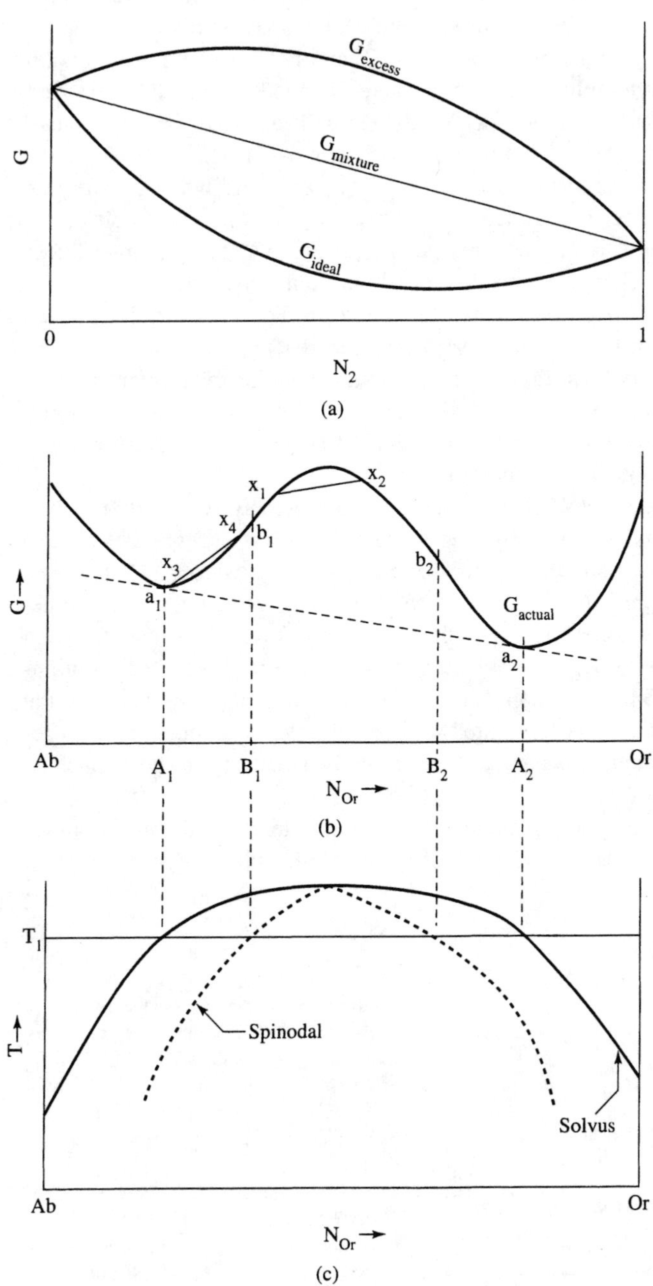

Figure 5-23 (a) Schematic isothermal, isobaric $\bar{G}$-N plot for a real solution composed of components 1 and 2 showing contributions from mechanical mixing, ideal mixing, and excess mixing. G_{excess} can be positive (shown here) or negative. (After D. K. Nordstrom and J. L. Munoz, *Geochemical Thermodynamics*. Copyright © 1985. Adapted by permission of Blackwell Science, Ltd., Oxford, England.) (b) Schematic G-N diagram for the alkali feldspar system at some temperature T_1 below the crest of the solvus. $G_{actual} = G_{ideal} + G_{excess}$. Tangent points a_1 and a_2 define the locations of solvus limbs, and inflection points b_1 and b_2 define spinodal limbs at this temperature. If any feldspar having an overall composition between B_1 and B_2 contains small local fluctuations in composition (as, for example, x_1 and x_2), its free energy will be less than that for a homogeneous feldspar. This situation results in unmixing and the process is called *spinodal decomposition.* Within the limits of the spinodal (b_1 to b_2), coexisting compositions like x_1 and x_2 are stable relative to a homogeneous phase. Outside the spinodal a homogeneous phase is stable with respect to exsolved phases like x_3 and x_4 according to thermodynamic reasoning. However, Nordstrom and Munoz (1985, 143) point out that random shifts in the positions of atoms, which cannot be predicted by classical thermodynamics, may result in local areas of compositional inhomogeneity, leading to macroscopic phase separation. (c) Idealized T-N diagram of the subsolidus portion of the alkali feldspar system showing locations of the experimentally determined solvus and the calculated spinodal (Waldbaum and Thompson 1969). The solvus is the line separating a higher temperature field containing a single homogeneous solid solution from a lower temperature field in which two phases may form by exsolution from the solid solution. The temperature T_1 defines the location of points a_1, b_1, a_2 and b_2 in (b). See discussion in text. [(b) and (c) after Richardson, S. M., and McSween, H. Y., Jr., *Geochemistry—Pathways and Processes*. Copyright © 1989. Adapted by permission of Prentice Hall, Inc., Upper Saddle River, NJ.]

POLYMORPHISM

Polymorphism refers to the occurrence of a given element or compound in two or more different structural forms. The difference between the various structural forms may involve bonding, site arrangement, cation coordination, or degree of ion ordering. The only important mineral example of bond difference between polymorphs is the pair graphite-diamond. Diamond has only covalent bonds; graphite has both covalent and van der Waals bonds. The change from low quartz to high quartz involves only a slight change in ion location (Figure 5-24). This type of change, known as a *displacive transformation,* occurs quickly and can be reversed easily in laboratory experiments. There is very little difference in the energy of the two polymorphs. A larger difference in energy is found when comparing tridymite with high quartz, because there is a bigger difference between the two structures. To change one to the other, bonds must be broken, whereas this is not necessary in changing low quartz to high quartz, or vice versa. The change of high quartz to tridymite is an example of a *reconstructive transformation.* Such changes may or may not result in a change of coordination number. For the pairs high quartz-tridymite and tridymite-cristobalite, there is no change in coordination number. For the pair high quartz-stishovite, the coordination of silicon changes from four in quartz to six in stishovite. Reconstructive transformations occur slowly and are difficult to reverse in the laboratory.

The K-feldspars discussed in the last section are examples of polymorphs that differ in their degree of ordering. Such polymorphs show a continuous change in their properties as they change into each other. These transformations are referred to as *second-order transitions,* whereas displacive and reconstructive transformations, which involve sharp changes in volume and other properties, are known as *first-order transitions.* For first-order transitions the Gibbs function shows a continuous change in value through the transition point; however, there are discontinuities in the first derivatives of the Gibbs function. Entropy, volume, and enthalpy change at the equilibrium transformation temperature and pressure. For second-order transitions the first derivatives of the Gibbs function are continuous and the second derivatives, such as heat capacity, are discontinuous.

Several different examples of polymorphism were discussed in terms of thermodynamics in Chapter Three. We shall discuss the thermodynamics of Figure 5-24, which is a phase diagram

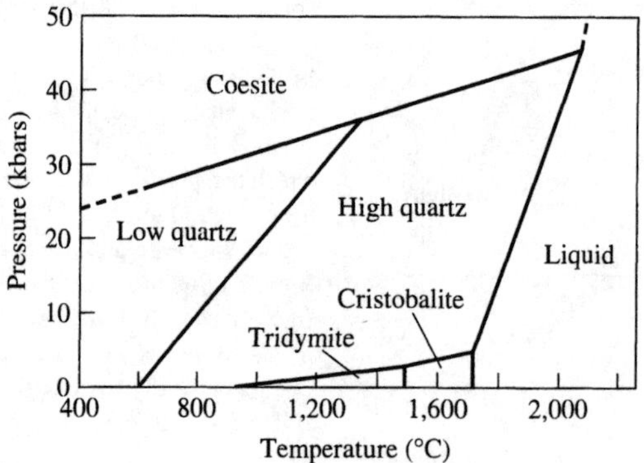

Figure 5-24 Phase diagram for SiO_2 showing the stability fields of the polymorphic forms of SiO_2 and liquid SiO_2. Stishovite forms at higher pressures (above 100 kbars) than shown in the diagram. Silica glass is the only phase that never possesses the lowest relative free energy. It is metastable at all temperatures and pressures. (After W. S. Broecker and V. M. Oversby, *Chemical Equilibria in the Earth,* p. 187. Copyright 1971 by McGraw-Hill Book Company. Used with permission of McGraw-Hill Book Company.)

for the SiO_2 system. The stable phase at any point in the diagram is the phase with the lowest Gibbs function value. We use equation (3-15) in the form

$$dG = V\,dP - S\,dT \tag{5-19}$$

and state the change in free energy with respect to pressure at constant temperature as

$$\left(\frac{\partial G}{\partial P}\right)_T = V \tag{5-20}$$

This tells us that the change in the Gibbs function of a phase as pressure increases, at constant temperature, is proportional to the volume of the phase. Let us compare low quartz (molar volume 22.69 cubic centimeters) and coesite (molar volume 20.64 cubic centimeters). If we start in the stability field of low quartz and increase the pressure, equation (5-20) tells us that the Gibbs function of low quartz will increase faster, because of its larger volume, than will the Gibbs function of coesite. As a result, the Gibbs function of low quartz eventually becomes greater than that of coesite, and then coesite becomes the stable form of SiO_2.

In a similar manner, for the change of Gibbs function with temperature at constant pressure, we can write

$$\left(\frac{\partial G}{\partial T}\right)_P = -S \tag{5-21}$$

The entropy of low quartz at 25°C is 9.88 calories per mole-degree, and that of coesite is 9.30 calories per mole-degree. If we start in the stability field of coesite (Figure 5-24) and increase the temperature, at constant pressure, equation (5-21) tells us that the Gibbs function of low quartz will decrease faster, because of its higher entropy, than will that of coesite. Thus at some point the Gibbs function of low quartz will become less than that of coesite, and low quartz will be the stable form of SiO_2. It was pointed out in Chapter Three that the entropy of minerals changes with temperature. This has to be kept in mind in using the above reasoning. For example, tridymite has a higher entropy than cristobalite at 25°C, and this could be taken to indicate that, with increasing temperature, cristobalite should change to tridymite. However, at higher temperatures cristobalite has the higher entropy (and the lower Gibbs function value) and is the stable phase. The phase diagram shown in Figure 5-24 was constructed by determining the relative Gibbs functions of the silica phases at different temperatures and pressures.

The slope of the phase boundaries in Figure 5-24 can be expressed in terms of entropy and volume as

$$\frac{dP}{dT} = \frac{\Delta S}{\Delta V} \tag{5-22}$$

This relation is one form of the *Clapeyron equation,* which can be applied to any phase boundary representing one degree of freedom (see discussion of phase diagrams in Chapter Three). All the boundaries in Figure 5-24 are of this type, as shown by applying the phase rule

$$f = c + 2 - p = 1 + 2 - 2 = 1$$

For example, the slope of the low quartz-coesite boundary is positive and can be calculated as follows (see Table 3-1 for conversion factors):

$$\frac{\bar{S}_c - \bar{S}_q}{\bar{V}_c - \bar{V}_q} = \frac{0.58 \text{ cal/mol-deg}}{2.05 \text{ cm}^3\text{/mol}} = \frac{0.58 \text{ cal}}{2.05 \text{ deg-cm}^3}$$

$$= \frac{0.58 \times 4.184 \text{ J}}{2.05 \text{ deg-cm}^3} = \frac{0.58 \times 4.184 \times 10^7 \text{ ergs}}{2.05 \text{ deg-cm}^3}$$

$$= \frac{0.58 \times 4.184 \times 10^7 \text{ dyne-cm}}{2.05 \text{ deg-cm}^3} = \frac{0.58 \times 4.184 \times 10^7 \text{ dynes}}{2.05 \text{ deg-cm}^2}$$

$$= \frac{0.58 \times 4.184 \times 10^7 \times 10^{-6} \text{ bar}}{2.05 \text{ deg}} = \frac{0.58 \times 4.184 \times 10^{-2} \text{ kbar}}{2.05 \text{ deg}}$$

$$= \frac{0.58}{2.05} \times 0.04184 \text{ kbar/deg} = 0.0117 \text{ kbar/deg}$$

This is the slope plotted in Figure 5-24. The slope of most polymorphic phase boundaries between solids is positive, because solids with larger volumes tend to have larger entropies (more randomness).

A special kind of polymorphism, known as *polytypism,* occurs in minerals with a layered structure. Polytypes are formed by varying the manner in which identical layers are superimposed. The mica minerals, for instance, are made up of sheets of linked silica tetrahedra that are held together vertically by cations between the sheets. Composite layers composed of silica sheets and interstitial cations have hexagonal symmetry, and there are six simple ways of stacking individual layers in an ordered manner (Table 5-4). More complex forms are also possible. Although each mica mineral tends to occur mainly in one or two of the six simple forms, individual samples may show other forms. In addition, many samples show various degrees of disordered stacking of the mica layers. The occurrence of polytypism, along with other characteristic complications, makes the micas, clay minerals, and other layered structures particularly difficult to study.

Advances in theory and research instrumentation in recent years have made it possible to study minerals at the atomic level. The clay minerals are an example of a silicate group that has

TABLE 5-4 Simple Mica Polymorphs

Polymorph	Symmetry	Number of composite layers in unit cell	Unit cell a Å	b Å	c Å	β	Common example*
1M	Monoclinic	1	5.3	9.2	10	100°	Biotite
$2M_1$	Monoclinic	2	5.3	9.2	20	95°	Muscovite
$2M_2$	Monoclinic	2	9.2	5.3	20	98°	Lepidolite
2O	Orthorhombic	2	5.3	9.2	20	90°	Anandite
3T	Trigonal	3	5.3	—	30	—	Glauconite
6H	Hexagonal	6	5.3	—	60	—	Not found in nature

*Any given mica may form more than one of these polymorphs. The 2O structure has been found only at one locality in the only known occurrence of anandite.

Source: After Smith and Yoder (1956).

been the object of much atomic-level research. Bleam (1993) reviews four different types of investigation of the clay minerals: quantum chemistry, statistical mechanics, electrostatic theory, and atomic-level crystal chemistry. One use of quantum chemistry techniques is to quantitatively compute bond properties and the effect of cation substitutions on bonding (Bleam and Hoffman 1988). Another use allows us to better understand adsorption and other mineral-surface reactions (Lasaga 1992). Statistical-mechanical (probabilistic) simulations have been constructed to study the properties of clay-water-ion systems by modeling the surfaces of clay minerals (Skipper et al. 1991). Computations of the electrostatic energy of clay minerals can be used to study interlayer bonding of clay minerals and micas (Giese 1984). Recent atomic-level crystal chemistry research has mainly dealt with mineral surface chemistry and with processes occurring at mineral-water interfaces (Hochella and White 1990).

All of these studies can be combined to give us a better understanding of the layered silicates which, because of their complex properties and behavior, are not easy to understand. The clay minerals are discussed further in Chapter Seven.

TRACE ELEMENTS IN MINERALS

We use the term *trace element* here to refer to any element that is not a significant component of a given mineral; that is, it does not occur in the mineral's formula. (In talking about trace elements in general, a common definition is as follows: those elements whose concentration in the crust is less than 0.1 percent by weight.) The development by V. M. Goldschmidt of the principles of trace element geochemistry [for distribution of these elements in both minerals and the Earth as a whole (Goldschmidt 1937)] is one of the reasons he is considered the father of modern geochemistry. (Other reasons are discussed in Chapters One and Three.) These principles were a result of Goldschmidt's earlier work on the structure and composition of minerals. In his classic 1937 paper, Goldschmidt suggested that the varying abundance of trace elements in minerals is mainly due to differences in ionic size, with ionic charge sometimes also playing a role. He used data on the ionic size and mineral abundance of various trace elements to postulate a set of rules for the distribution of trace elements during magmatic crystallization:

1. If two ions have the same radius and the same charge, they will enter into a crystallizing mineral with equal ease. (If one element is a trace element and the other is a major element, the trace element is said to be *camouflaged* in the structure. Two or more elements that can occupy the same site in a structure are said to be *diadochic* in that structure.)
2. If two ions have similar but not equal radii and the same charge, the smaller ion will be preferentially concentrated in early formed samples of a crystallizing mineral. (The smaller ion forms a stronger bond. Extensive substitution does not generally occur between elements whose radii differ by more than 15 percent.)
3. If two ions have similar radii but different charges, the ion with the higher charge will be preferentially concentrated in early formed samples of a crystallizing mineral. (A trace element with higher charge than a major element is said to be *captured* by the structure, and a trace element with a lower charge is said to be *admitted* by the structure. The element with higher charge forms a stronger bond.)

Goldschmidt's rules can be used to explain a number of trace element occurrences in minerals of igneous rocks. An example of his first rule is the camouflaging of hafnium (ionic radius

0.79 Å for six coordination) in zircon, where it substitutes for zirconium (ionic radius 0.80 Å for six coordination). Rubidium is commonly concentrated relative to potassium in late-formed samples of K-feldspar, as predicted by the second rule. The rubidium radius for the nine-coordination position in feldspar is 1.71 Å; that of potassium is 1.63 Å. Examples of the third rule are the capture of barium (+2, 1.55 Å) in place of potassium (+1, 1.63 Å) by K-feldspar, and the admittance of lithium (+1, 0.82 Å, six coordination) into biotite as a replacement for magnesium (+2, 0.80 Å, six coordination).

Over the years it has become necessary to modify Goldschmidt's rules to explain the numerous exceptions that have been found. The first major modification, by Ringwood (1955), dealt with Goldschmidt's assumption of ionic bonding in developing his rules. As we have discussed earlier, there is a strong covalent component in the bonding of silicate minerals, and these are the minerals to which Goldschmidt's rules are most commonly applied. Ringwood suggested that, by taking into account the electronegativities of substituting elements, corrections could be made for differences in the strengths of the bonds formed by the elements. He amended the rules by adding the following:

4. If two elements have similar radii and the same charge, the one with the lower electronegativity will be preferentially concentrated in early formed samples of a crystallizing mineral. (The element with the lower electronegativity will form a stronger bond. This rule seems to be useful for elements that differ in electronegativity by more than 0.1.)

Even with Ringwood's modification of the rules, many exceptions were found, mainly involving the transition elements. An approach for dealing with the transition elements has been outlined by Burns and Fyfe (1967). They describe the behavior of the transition elements in terms of crystal-field theory, which deals with the effects of surrounding anions on the energy levels of the electrons of a transition-metal ion in a crystal structure. The outer electrons of many of the transition-metal ions can have more than one configuration (see the earlier discussion of high-spin and low-spin states with respect to ionic radii). Using crystal-field theory, it is possible to determine the relative energies of the incompletely filled inner orbitals of the transition elements in various coordinations (Figure 5-9). This in turn allows prediction of the relative affinity of an ion in a magma for a particular site in a crystallizing silicate mineral. Burns and Fyfe also emphasize that any rules on trace elements in minerals forming from a magma must take into account the properties of the magma.

A thorough discussion of the use of crystal-field theory in mineralogy can be found in Burns (1993). An example of its use is the explanation of the enrichment of nickel in early formed minerals of fractionally crystallized magmas, such as that of the Skaergaard intrusion of Greenland. According to the Goldschmidt rules, nickel, which substitutes for magnesium, should not be concentrated in either the early or the late formed minerals, because the ionic radii of the two elements are very similar [using the radii of Whittaker and Muntus (1970) in Table 5-2]. (Earlier estimates of the radii gave nickel the larger radius, and it was believed that nickel should be concentrated in late formed minerals according to the Goldschmidt rules.) Nickel has a higher electronegativity than magnesium and, using Ringwood's addition to the rules, should therefore be concentrated in late forming minerals. Burns explains the early extraction of nickel from a magma as due to a high crystal field stabilization energy, which it develops in early formed minerals relative to magma. The sequence of removal from a magma of various transition elements after nickel, as predicted by Burns, agrees well with that found in the Skaergaard intrusion.

It is obvious that the Goldschmidt rules have only limited use, even with various modifications. Burns (1973, 1993) has listed a number of problems in using the rules. Some of these are as follows:

1. Ionic radii are based on average interatomic distances within a particular coordination polyhedron. The majority of coordination polyhedra are distorted and have metal-oxygen distances that may vary by up to 0.5 Å. For example, the metal-oxygen distances for the M_2 site in orthopyroxene, discussed earlier with respect to order-disorder, vary from 2.04 to 2.52 Å. Thus small differences in ionic radii must be used carefully in predicting or explaining trace element distributions.
2. Geochemists have tended to interpret analytical data for bulk rock samples without reference to a specific mineral structure. Since the geometries and dimensions of coordination polyhedra of an individual major cation may differ from one mineral to another and also within a single mineral structure, the ionic-radius criterion is too simple to explain trends in bulk rock analyses.
3. Goldschmidt's rules assume random distribution of ions throughout a crystal structure. We now know this is not true for cations in several common minerals. Trace elements may tend to be ordered in that they prefer certain sites in a structure. Analytical data on minerals cannot be used to prove a diadochic relationship between two elements in minerals with more than one site for substitution.
4. Most of the discussions of element distribution and fractionation during mineral formation have considered bonding forces of ions in crystalline phases only. The stabilities of ions in both the mineral and the phase from which the mineral crystallized have to be considered. Most of the necessary information for magmas and other fluids is not available.

Crystal field theory, discussed above, is a relatively simple approach to explaining the chemical behavior of minerals. It considers only the electrostatic interactions between a central ion and its neighbors (ligands), thought of as point charges. A more general approach, ligand field theory, takes into account the covalent nature of bonds and the symmetry of ligand orbitals. When orbitals are considered as associated with molecules rather than atoms the resulting theory is known as molecular orbital theory. This theory also deals with electronic transitions between atoms, in contrast to the other two theories, which consider only transitions within an ion. Crystal field theory and ligand field theory can be considered special cases of molecular orbital theory. Figure 5-9 is an example of the use of crystal field theory.

As an example of the use of ligand field theory, let us consider the cause of color in ruby. Pure corundum, Al_2O_3, is colorless and has Al-O bonds that are about 60 percent ionic and 40 percent covalent. The cause of the red color in ruby is the substitution of Cr^{3+} for Al^{3+} in trace amounts (about one out of every hundred aluminum atoms is replaced by a chromium atom). The electron structure of Cr^{3+} is such that it has three unpaired electrons in the unfilled $3d$ orbital (the first and second shells are filled, as are the $3s$ and $3p$ orbitals). When a chromium ion substitutes for an aluminum ion in corundum, splitting of the five $3d$ orbitals occurs in a similar manner to that shown in Figure 5-9b.

When white light passes into a ruby, absorption of parts of the visible spectrum occurs as a result of changes in the electron energy levels of the chromium ions (Figure 5-25). As the radiation interacts with the ruby, wavelengths whose energies correspond to the energy differences between the orbital energy levels of chromium are absorbed and electrons are excited from a lower

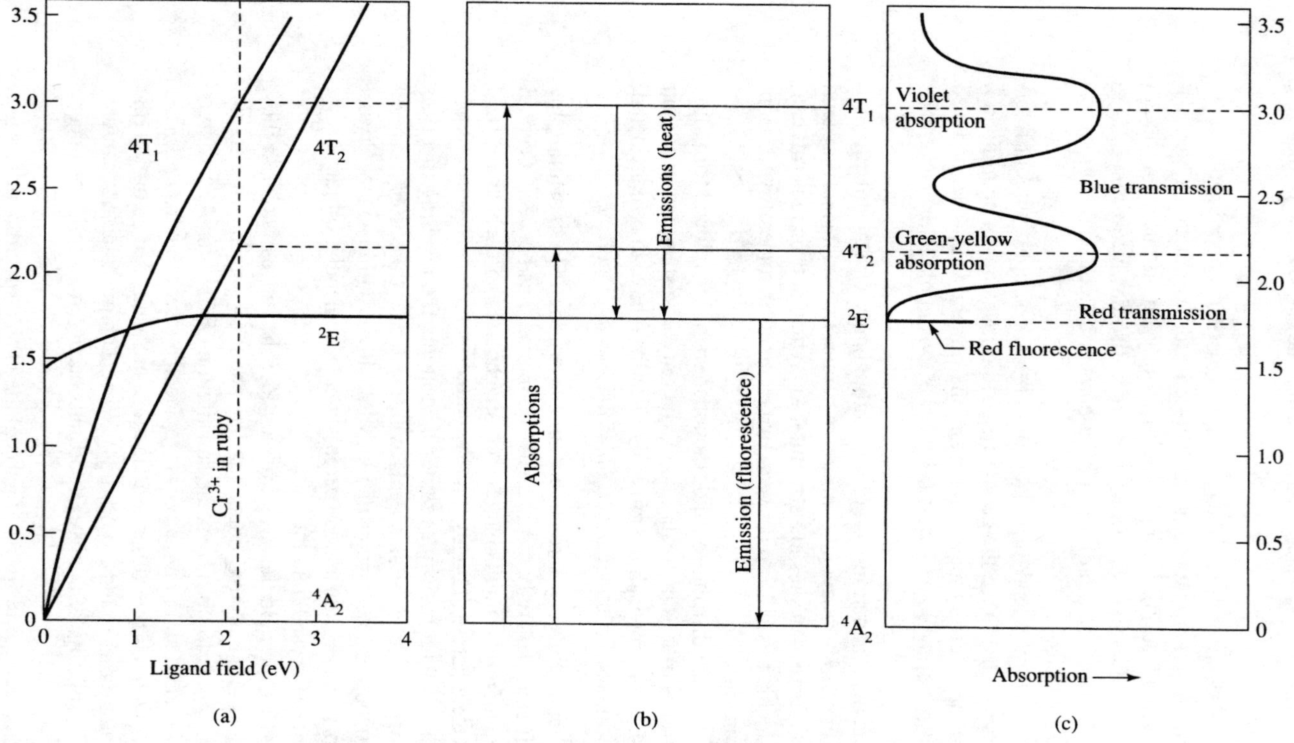

Figure 5-25 Explanation of the color and fluorescence of ruby. (a) The energy state of Cr^{3+} in a distorted octahedral ligand field. The four lowest electron energy levels are shown, ground level (4_{A_2}) and three higher levels ^2E, $4T_2$, and $4T_1$. The field of Cr^{3+} in ruby is shown by the vertical dashed line, which determines the energy levels for ruby shown in (b). Note that energy level ^2E changes very little with the strength of the ligand field while levels $4T_2$ and $4T_1$ change significantly. (b) The energy levels and transitions in ruby. Increases in energy levels result from absorption of various wavelengths and decreases produce heat and result in emission of visible red light (fluorescence). Fluorescence requires incident ultraviolet radiation to produce emission of visible light. Energy is absorbed as electrons reach energy level ^2E and is re-emitted as the electrons fall to ground level. (c) The absorption spectrum and fluorescence of ruby. The color of ruby is determined by the wavelengths that are least absorbed, which are in the red portion of the visible spectrum. In addition to strong red transmission at about 1.8 eV, there is some blue transmission at about 2.5 eV. Green-yellow absorption and violet absorption occur in the vicinity of 2.2 eV and 3.0 eV, respectively. Red fluorescence occurs at energy levels of 1.788 and 1.791 eV. (After K. Nassau, *The Physics and Chemistry of Color*. Copyright © 1983 by John Wiley & Sons. Reprinted by permission of John Wiley & Sons, Inc.)

level to a higher level. Return of the electrons to intermediate levels causes production of heat in a ruby and return to the ground state results in fluorescence (Figure 5-25). Thus ligand field theory explains quite well the color and fluorescence of ruby.*

The exact color resulting from splitting of electronic energies depends on the element involved and its oxidation state, both of which determine the number of *d* electrons present. The geometry of the coordination site is very important, as is the type of bonding. The presence of certain elements can cause color when the element is a major constituent as well as when it is a trace constituent. For example, iron is the cause of the color in olivine and copper is the cause of the color in turquoise. Transition metals such as chromium, iron, and copper cause color in many minerals and other materials and such colors are often explained by crystal field and ligand field theory. There are a number of other causes of color in solids. An excellent reference on this topic is Nassau (1983).

Another important aspect of trace element geochemistry is the study of the distribution of these elements between coexisting minerals. Earlier we discussed the distribution of major elements between coexisting minerals. Trace element distribution is also used to obtain information about rock history. Knowledge of these distributions can provide information about the attainment of equilibrium during mineral formation and, in some cases, can indicate temperatures of formation. A number of different geothermometers using trace elements have been proposed involving dependence of the distribution coefficient, K_D, on temperature. These coefficients have been discussed in Chapter Three. The example there dealt with an ideal, dilute solid solution involving cobalt in biotite and hornblende, for which

$$K_D = \frac{a_{\mathrm{Co}}^{\mathrm{Bio}}}{a_{\mathrm{Co}}^{\mathrm{Hbd}}} = \frac{N_{\mathrm{Co}}^{\mathrm{Bio}}}{N_{\mathrm{Co}}^{\mathrm{Hbd}}} \tag{5-23}$$

where $a_{\mathrm{Co}}^{\mathrm{Bio}}$ is the activity of cobalt in biotite, $a_{\mathrm{Co}}^{\mathrm{Hbd}}$ the activity of cobalt in hornblende, $N_{\mathrm{Co}}^{\mathrm{Bio}}$ the mole fraction of cobalt in biotite, and $N_{\mathrm{Co}}^{\mathrm{Hbd}}$ the mole fraction of cobalt in hornblende. The usual procedure in dealing with trace elements is to use a ratio of concentrations in parts per million instead of a ratio of mole fractions. It is assumed that the trace element obeys Henry's Law (discussed earlier). In Figure 3-3, K_D is shown to be a constant for cobalt distribution between biotite and hornblende in igneous rocks. This suggests attainment of equilibrium in these rocks for this distribution, and further suggests that differences in temperature of formation for the various rocks had little effect on K_D.

Theoretically, K_D should vary with temperature according to the relation (O'Nions and Powell 1977)

$$\left(\frac{\partial \ln K_D}{\partial T}\right)_P = \frac{\Delta H^0}{RT^2} \tag{5-24}$$

or

$$\ln K_D = -\frac{\Delta H^0}{RT} + B$$

*Chromium also causes the green color of emerald, a gem form of beryl. In both ruby and emerald, Cr^{3+} substitutes for Al^{3+} and the site geometry is similar. The different colors for the two gemstones are a result of a difference in bonding (the oxygen-cation bonding in corundum is more ionic than in beryl). The difference in bonding causes a difference in the crystal field splitting of the *d* orbitals of chromium and thus a difference in the wavelengths of visible light that are absorbed. Green wavelengths are absorbed in ruby and transmitted in emerald.

where B is an integration constant and ΔH represents the difference between the heats of solution for the trace element in the two minerals of interest. In some cases it has been found that K_D varies sufficiently with temperature to have potential use as a geothermometer. An example is the study by Häkli and Wright (1967) on nickel distribution in a lava lake in Hawaii (Figure 5-26). They suggest that under favorable conditions crystallization temperatures can be predicted within an accuracy of 10 to 20°C.

More recent work has established that nickel partitioning is *not* a strong function of temperature because it is highly dependent on major element composition (Hart and Davis 1978). We will discuss in detail the behavior of trace elements during magmatic crystallization in Chapter Eight. It has been found that trace element concentrations vary greatly during magmatic processes and thus can be used as monitors of igneous evolution.

The value of K_D should also change with pressure according to the relation (O'Nions and Powell 1977)

$$\left(\frac{\partial \ln K_D}{\partial P}\right)_T = \frac{-\Delta V^0}{RT} \tag{5-25}$$

In some cases, for example metamorphic rock formation, pressure may have a greater effect on K_D than temperature. When this is true, trace element distribution can be used as a geobarometer. However, little is known about the effect of pressure, if any, on the K_D values of trace elements in coexisting minerals. Also, as mentioned above for nickel, changes in major element concentrations generally have a significant effect on the K_D values of trace element distributions, thus limiting their use as geothermometers and geobarometers.

As discussed earlier in this chapter, K_D values for the exchange of major components of coexisting minerals can be used as both geothermometers and geobarometers (Figure 5-21). Equations 5-24 and 5-25 show that reactions with a large ΔH^0 and a small ΔV^0 can be used as geothermometers and those with a small ΔH^0 and a large ΔV^0 can be used as geobarometers.

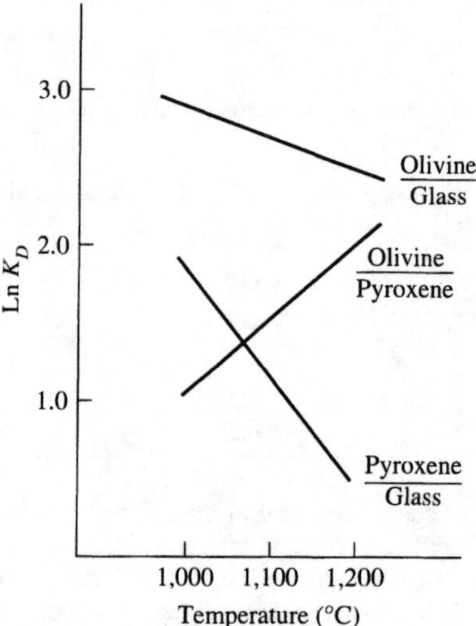

Figure 5-26 Relation between temperature and the natural log of nickel-distribution coefficients for the pairs olivine-glass, clinopyroxene-glass, and olivine-clinopyroxene in the Makaopuhi lava lake, Hawaii. The straight lines indicate that the three phases are in equilibrium with respect to nickel distribution and that they obey the thermodynamic relationship given in equation (5-24). Measurement of nickel distribution between any two of these phases in a sample should give the temperature of formation of the sample. However, this geothermometer must be used with caution, because of the strong effect on K_D of changes in the composition of the phases. After Häkli and Wright (1967).

SUMMARY

This chapter reviews the chemistry of solids that exhibit a long-range regularity of atomic or molecular arrangement. The bonding forces of these solids can be described in terms of the extent to which electrons are transferred from an atom to one or more of its neighbors. Complete transfer results in an ionic bond due to electrostatic attraction between oppositely charged ions. Partial transfer results in a sharing of electrons, a lower total energy for the atoms involved, and a bond known as a covalent bond. Gradational situations between complete transfer and even sharing can occur. The metallic bond is an extreme type of a covalent bond, with a random sharing of electrons occurring. The weak van der Waals bond is similar to the ionic bond, but involves electrostatic attraction between molecules rather than between ions. Two or more bond types, with different strengths, are found in many minerals. The type of bonding that occurs in a given mineral depends on the kinds of elements making up the mineral and on the electron structure of these elements.

Crystal structures can be described in terms of the sizes and arrangement of the approximately spherical atoms and/or ions from which they are built. The arrangement of the spheres depends on their relative sizes and on the kinds of bonding that develop. Absolute covalent and metallic radii can be obtained by use of X-ray diffraction. Only relative radii values can be measured for ionically bonded elements. The bonding of most minerals is chiefly ionic in character, and the major anion found is oxygen. Thus the ionic radii of cations are usually related to an assumed value for the oxygen anion. The ionic radius of a particular ion in a specific mineral depends on the nature of the element (as indicated by its location in the periodic table), on the coordination number (number of nearest neighbors) of the ion in the mineral, and on the number and arrangement of the ion's outer electrons. A particularly large variation in ionic size is shown by the elements in the central portion of the periodic table, since these elements can exhibit more than one valence, can occur with several different coordination numbers (as can most of the elements), and their outer electrons can occur in several different energy configurations.

The basic shape of any structure is the unit cell, which represents a complete unit of the pattern of the structure. However, an outline of the unit cell of a mineral does not always give a clear picture of the packing geometry. Many minerals consist of one large anion and one or more smaller cations. The larger anion controls the structure and tends to exhibit a regular grouping around the cations. Thus, in describing and visualizing the structure of a mineral, it is often convenient to think in terms of the arrangement of the anion around the cations. This geometrical form is known as a coordination polyhedron. The number of anion spheres in a polyhedron and its shape depend on the radius ratio of cation to anion. The relationship among radius ratio, cation coordination, and the shape and packing of coordination polyhedra is shown in Figure 5-8. By using radius ratios, geochemists can predict the correct coordination numbers for some compounds but not others. Thus the coordination numbers of cations and anions are not always determined by relative size. The silicate minerals can be described in terms of the arrangement of oxygen tetrahedra, which may share one, two, three, or four corners with each other, or which may share none at all. The tetrahedra do not share edges or faces, since this would bring the central cations of the tetrahedra (silicon or aluminum) too close together.

The study of the silicate minerals, and of certain other common minerals, is complicated by the occurrence of solid-solution and order-disorder phenomena. A mineral that exhibits a variation in content of two or more of its major elements is said to show solid solution. This is usually due to the partial substitution of one element for another in the mineral structure (substitutional solid solution). Two elements substituting for each other may be randomly distributed throughout the available sites in the mineral (disorder), or various degrees of segregation (ordering) of the elements may occur. Some mineral solid solutions are ideal or nearly ideal solutions in terms of thermodynamics; others are not. Mixing of two components in a solid solution can be described in terms of three types of behavior. Two of these are represented by Raoult's Law and Henry's Law while details of the third (nonideal behavior) must be determined experimentally or theoretically for each individual case.

Extensive research has been done on the alkali feldspars, which at high temperature exhibit complete solid solution between $KAlSi_3O_8$ and $NaAlSi_3O_8$. The K-feldspars also exhibit a wide range of ordering of aluminum and silicon in tetrahedral sites. This variation in order distinguishes the polymorphs sanidine, orthoclase, and microcline. Other examples of polymorphism are due mainly to variations in site arrangement, type of bonding, or cation coordination. The common SiO_2 polymorphs result from small (low quartz–high quartz) or large (high quartz–tridymite) changes in site arrangement. The rare form stishovite differs principally in having a silicon coordination number of six rather than four.

The distribution of major and trace elements between coexisting minerals can be used as a geothermometer or a geobarometer. The occurrence and abundance of trace elements in minerals is generally controlled by their availability at the time of mineral formation, by ionic size and charge, and by the nature of the host mineral structure. If trace elements are available for substitution in a crystallizing mineral, they tend to substitute for major elements of similar size and charge. Using the factors size and charge, Goldschmidt postulated a set of rules in the 1930s to explain trace element distribution in minerals. These rules are now known to have limited use in explaining trace element geochemistry. Other factors, such as the nature of the liquid from which a mineral forms, the type of bonding in the mineral, and the energy relationships between an ion and its structural neighbors, must be considered.

QUESTIONS

1. For each of the minerals listed below, indicate which one of the following types of bonding you would expect: (1) all bonds essentially ionic, (2) all bonds essentially covalent, or (3) some bonds mainly ionic and other bonds mainly covalent.
   (a) Apatite
   (b) Chalcopyrite
   (c) Fluorite
   (d) Native arsenic
   (e) Spinel
2. **(a)** The ionic radius of iron in aegirine ($NaFeSi_2O_6$) is not the same as the ionic radius of iron in hypersthene ($FeSiO_3$). Suggest an explanation for this.
   (b) The ionic radius of aluminum in orthoclase ($KAlSi_3O_8$) is not the same as the ionic radius of aluminum in spodumene ($LiAlSi_2O_6$). Suggest an explanation for this.
3. **(a)** What is the most common coordination number for metallically bonded atoms?
   (b) What is the most common coordination number for ionically bonded elements?
   (c) Why isn't coordination number an important factor for covalently bonded atoms?

4. **(a)** Using the radii given in Table 5-2, what coordination number would you expect to be most common for each of the following elements when packed around oxygen anions? Assume that the oxygen anion has a radius of 1.32 Å.
   (A) potassium; (B) rubidium; (C) calcium; (D) sodium; (E) boron.
   (b) Do any of these elements ever occur in minerals with a different coordination number than that predicted by you in part (a)?
5. Why is the coordination number of potassium in silicate minerals often higher than the coordination number of sodium in silicate minerals?
6. In which one of the following situations for a transition element would you expect it to have the largest ionic radius?
   (a) +2 charge, coordination number 4, low-spin state.
   (b) +2 charge, coordination number 4, high-spin state.
   (c) +2 charge, coordination number 6, low-spin state.
   (d) +2 charge, coordination number 6, high-spin state.
   (e) +3 charge, coordination number 4, low-spin state.
   (f) +3 charge, coordination number 4, high-spin state.
   (g) +3 charge, coordination number 6, low-spin state.
   (h) +3 charge, coordination number 6, high-spin state.
7. For each of the following pairs, select the ion that you would expect to have the larger ionic radius. Assume that coordination number has no effect on the ionic radii.
   (a) Se^{2-} or Se^{6+}
   (b) Cu^{+} or Cu^{2+}
   (c) Nb^{5+} or Ta^{5+}
   (d) Zn^{2+} or Cd^{2+}
   (e) Mn^{2+} (high-spin state) or Mn^{2+} (low-spin state)
   (f) N^{5+} or B^{3+}
8. Figure 5-22 shows two unit cells of olivine. Each metal ion shown at the corner of a cell is shared among eight cells. Each metal ion shown at the midpoint of a horizontal cell edge actually occurs in the center of a cell face and is shared between two cells. Each metal ion shown at the midpoint of a vertical cell edge occurs on that edge and is shared among four cells. All the other metal ions shown within a cell belong only to that cell. With this information, and using Figure 5-22, calculate how many formula units are in the unit cell of olivine.
9. To which structural class of the silicates (see Table 5-3) does each of the following minerals belong?
   (a) Jadeite: $NaAlSi_2O_6$
   (b) Leucite: $KAlSi_2O_6$
   (c) Nepheline: $NaAlSiO_4$
   (d) Sillimanite: Al_2SiO_5
   (e) Vesuvianite: $Ca_{10}(Mg, Fe)_2Al_4Si_9O_{34}(OH, F)_4$
   (f) Cordierite: $(Mg, Fe)_2Al_4Si_5O_{18}$
10. The unit cell dimensions of the monoclinic mineral paragonite [$NaAl_3Si_3O_{10}(OH)_2$] are $a = 5.13$, $b = 8.89$, and $c = 19.00$ Å, and $\beta = 95°$. The density is 2.9 grams per cubic centimeter. How many atoms of aluminum are in the unit cell of this mineral?
11. **(a)** Given the following chemical analysis of a sample of olivine: MgO = 6.23 percent, FeO = 62.86 percent, and SiO_2 = 30.91 percent. What is the *exact* chemical formula of this sample?
    (b) What are the mole fractions of Fa (fayalite) and Fo (forsterite) for this sample?
    (c) Using the molar volumes of fayalite (46.39 cubic centimeters) and forsterite (43.67 cubic centimeters), and assuming ideal solid solution, calculate the molar volume of the sample.

12. Dahl (1980) reports the following results for a garnet-clinopyroxene pair (RMK-13) in a metamorphic rock from the Ruby Range, southwestern Montana.

$K_D(\text{Fe/Mg}) = 6.48$
mole fraction of calcium in the garnet = 0.255
estimated pressure of metamorphism = 7200 bars

(a) Use equation (5-13) to estimate the temperature of metamorphism.
(b) If the pressure estimate has an uncertainty of ± 1200 bars, what is the uncertainty for the temperature estimate?

13. **(a)** What are the expressions for the activity coefficients γ_i and γ_j of components *i* and *j* in a two-component solid solution that obeys Raoult's Law throughout the entire compositional range of the solid solution? What are the numerical values for the two activity coefficients?
(b) Assuming component *i* of a multi-component solid solution obeys Henry's Law over a certain range of composition, what is the relationship between the activity coefficient γ_i, the mol fraction N_i, and the activity a_i?

14. Answer the following questions about the two polymorphs, sanidine and microcline.
(a) Which unit cell would you expect to have the larger volume? Why?
(b) Which mineral would you expect to have the greater specific gravity? Why?
(c) Which mineral would you expect to have the greater hardness? Why?

15. **(a)** Given are two perthites (alkali feldspars) from two different igneous rocks. The two feldspars making up the perthite in sample A have compositions $Or_{95}Ab_5$ and Or_5Ab_{95}. Sample B has compositions $Or_{80}Ab_{20}$ and $Or_{20}Ab_{80}$. Which sample would you guess formed at a higher temperature?
(b) What can we say about the chemical potential of $NaAlSi_3O_8$ in the two phases making up a perthite?

16. **(a)** Using the relationships shown in Figure 5-24, which mineral would you expect to have a higher molar volume, coesite or high quartz?
(b) Using the relationships shown in Figure 5-24, which mineral would you expect to have a higher entropy, coesite or high quartz?

17. **(a)** In mica minerals would you expect a greater decrease in volume parallel to, or perpendicular to, the structural layers as a result of an increase in pressure?
(b) In mica minerals would you expect a greater increase in volume parallel to, or perpendicular to, the structural layers as a result of an increase in temperature?

18. For each of the following trace elements found in silicate minerals, name a major element that the trace element commonly replaces in silicate structures. (a) Rb; (b) Sr; (c) Ga; (d) Ti; (e) Li; (f) Ba; (g) Ge; (h) rare earths; (i) Pb; (j) Ni; (k) Mn.

19. **(a)** In the spinel group of minerals, Cr^{3+} and Fe^{3+} can substitute for each other in octahedral sites. Calculations based on crystal-field theory lead to the prediction that chromium ion rather than ferric ion would be concentrated in the early formed spinels of a crystallizing magma. Using Goldschmidt's rules, would you make the same prediction?
(b) The bonding in spinels may be partially covalent; thus Ringwood's modification of the Goldschmidt rules should perhaps be used to predict which of the two ions would be enriched in early formed crystals. If you use Ringwood's modification of the rules, which ion would be enriched in these crystals?

REFERENCES

BERRY, F. J., and D. J. VAUGHAN, eds. 1985. *Chemical Bonding and Spectroscopy in Mineral Chemistry.* Chapman and Hall, London.

BLEAM, W. F. 1993. Atomic theories of phyllosilicates: quantum chemistry, statistical mechanics, electrostatic theory, and crystal chemistry. *Rev. of Geophysics,* v. 31:51–73.

BLEAM, W. F., and R. HOFFMAN. 1988. Isomorphous substitution in phyllosilicates as an electronegativity perturbation: Its effect on bonding and charge distribution. *Inorg. Chem.,* v. 27:3180–3186.

BLOSS, F. D. 1971. *Crystallography and Crystal Chemistry: An Introduction.* Holt, Rinehart and Winston, Inc., New York.

BOWEN, N. L., and J. F. SCHAIRER. 1935. The system MgO-FeO-SiO_2, *Am. J. Sci.,* v. 29:151–217.

BRAGG, L., and G. F. CLARINGBULL. 1965. *Crystal Structures of Minerals.* G. Bell & Sons Ltd., London.

BROECKER, W. S., and V. M. OVERSBY. 1971. *Chemical Equilibria in the Earth.* McGraw-Hill Book Co., New York.

BURDETT, J. K., and T. J. MCLARNAN. 1984. An orbital interpretation of Pauling's rules. *Am. Mineralogist,* v. 69:601–621.

BURNHAM, C. W. 1990. The ionic model: perceptions and realities in mineralogy. *Am. Mineralogist,* v. 75:443–463.

BURNS, R. G. 1970. *Mineralogical Applications of Crystal Field Theory.* Cambridge University Press, New York.

BURNS, R. G. 1973. The partitioning of trace transition elements in crystal structures: A provocative review with applications to mantle geochemistry. *Geochim. Cosmochim. Acta,* v. 37:2395–2403.

BURNS, R. G. 1993. *Mineralogical Applications of Crystal Field Theory,* 2d ed. Cambridge University Press, New York.

BURNS, R. G., and W. S. FYFE. 1967. Crystal-field theory and the geochemistry of transition elements. In *Researches in Geochemistry,* vol. 2, P. H. Abelson, ed. John Wiley & Sons, Inc., New York. 259–285.

CARPENTER, M. A. 1985. Order-disorder transformations in mineral solid solutions. In *Microscopic to Macroscopic—Atomic Environments to Mineral Thermodynamics,* Rev. in Mineralogy, v. 14, S. W. Kieffer and A. Navrotsky, eds. Mineralogical Soc. Am., Washington, D.C. 187–223.

CHANG, R. 1986. *General Chemistry.* Random House, New York.

CHATILLON-COLINET, C., R. C. NEWTON, D. PERKINS, III, and O. J. KLEPPA. 1983. Thermochemistry of $(Fe^{++}, Mg)SiO_3$ orthopyroxene, *Geochim. Cosmochim. Acta,* v. 47:1597–1603.

COHEN, R. E. 1991. Bonding and elasticity of stishovite SiO_2 at high pressure: Linearized augmented plane wave calculations. *Am. Mineralogist,* v. 76:733–742.

DENNEN, W. H. 1960. *Principles of Mineralogy.* The Ronald Press Company, New York.

DAHL, P. S. 1980. The thermal-compositional dependence of Fe^{++}-Mg distributions between coexisting garnet and pyroxene: applications to geothermometry. *Am. Mineralogist,* v. 65:852–866.

ELLIS, D. J., and D. H. GREEN. 1979. An experimental study of the effect of Ca upon garnet-clinopyroxene Fe-Mg exchange equilibria. *Contrib. Min. Pet.,* v. 71:13–22.

ERNST, W. G. 1969. *Earth Materials.* Prentice-Hall, Inc., Englewood Cliffs, NJ.

ESSENE, E. J. 1982. Geologic thermometry and barometry. In *Characterization of Metamorphism through Mineral Equilibria,* Rev. in Mineralogy, v. 10, J. M. Ferry, ed. Mineralogical Soc. Am., Washington, D.C. 153–206.

ESSENE, E. J. 1989. The current status of thermobarometry in metamorphic rocks. In *Evolution of Metamorphic Belts,* J. S. Daly, R. A. Cliff, and B. W. D. Yardley, eds. The Geological Society, London. 1–44.

EVANS, B. W., and T. L. WRIGHT. 1972. Composition of liquidus chromite from the 1959 (Kilauea Iki) and 1965 (Makaopuhi) eruptions of Kilauea volcano, Hawaii. *Am. Mineralogist,* v. 57:217–230.

EVANS, R. C. 1964. *Crystal Chemistry,* 2d ed. Cambridge University Press, New York.

FRYE, K. 1974. *Modern Mineralogy.* Prentice-Hall, Inc., Englewood Cliffs, NJ.

Fyfe, W. S. 1964. *Geochemistry of Solids.* McGraw-Hill Book Co., New York.

Ganguly, J., and S. K. Saxena. 1987. *Mixtures and Mineral Reactions.* Springer-Verlag, Berlin.

Gibbs, G. V. 1982. Molecules as models for bonding in silicates. *Am. Mineralogist,* v. 67:421–450.

Gibbs, G. V., M. A. Spackman, and M. B. Boisen, Jr. 1992. Bonded and promolecule radii for molecules and crystals. *Am. Mineralogist,* v. 77:741–750.

Giese, R. F. 1984. Electrostatic energy models of micas. In *Micas,* Rev. in Mineralogy, v. 13, S. W. Bailey, ed. Mineralogical Soc. Am., Washington, D.C. 105–144.

Goldschmidt, V. M. 1937. The principles of distribution of chemical elements in minerals and rocks. *J. Chem. Soc.,* 655–672.

Goldsmith, J. R., and D. M. Jenkins. 1985. The high-low albite relations revealed by reversal of degree of order at high pressures. *Am. Mineralogist,* v. 70:911–923.

Griffen, D. T. 1992. *Silicate Crystal Chemistry.* Oxford University Press, New York.

Häkli, T. A., and T. L. Wright. 1967. The fractionation of nickel between olivine and augite as a geothermometer. *Geochim. Cosmochim. Acta,* v. 31:877–884.

Hart, S. R., and K. E. Davis. 1978. Nickel partitioning between olivine and silicate melt. *Earth Planet. Sci. Lett.,* v. 40:203–219.

Hazen, R. M. 1976. Sanidine: predicted and observed monoclinic-to-triclinic reversible transformations at high pressure. *Science,* v. 194:105–107.

Hazen, R. M., and L. W. Finger. 1982. *Comparative Crystal Chemistry.* Wiley-Interscience, New York.

Hochella, M. F., Jr., and A. F. White, eds. 1990. *Mineral-Water Interface Geochemistry,* Rev. in Mineralogy, v. 23. Mineralogical Soc. Am., Washington, D.C.

Holdaway, M. J. 1971. Stability of andalusite and the aluminum silicate phase diagram. *Am. J. Sci.,* v. 271:97–131.

Kieffer, S. W., and A. Navrotsky, eds. 1985. *Microscopic to Macroscopic—Atomic Environments to Mineral Thermodynamics,* Rev. in Mineralogy, v. 14. Mineralogical Soc. Am., Washington, D.C.

Knittle, E., and R. Jeanloz. 1987. Synthesis and equation of state of (Mg, Fe)SiO_3 perovskite to over 100 gigapascals. *Science,* v. 235:668–670.

Koziol, A. M., and R. C. Newton. 1988. Redetermination of the anorthite breakdown reaction and improvement of the plagioclase-garnet-Al_2SiO_5-quartz geobarometer. *Am. Mineralogist,* v. 73:216–223.

Lasaga, A. C. 1992. Ab initio methods in mineral surface reactions. *Rev. of Geophysics,* v. 30:269–303.

Liebau, F. 1985. *Structural Chemistry of Silicates.* Springer-Verlag, New York.

Martignole, J., and S. Nantel. 1982. Geothermobarometry of cordierite-bearing metapelites near the Morin anorthosite complex, Grenville province, Quebec. *Can. Mineralogist,* v. 20:307–318.

Nassau, K. 1983. *The Physics and Chemistry of Color.* Wiley-Interscience, New York.

Navrotsky, A. 1987. Models of crystalline solutions. In *Thermodynamic Modeling of Geological Materials: Minerals, Fluids, and Melts,* Rev. in Mineralogy, v. 17, I. S. E. Carmichael, and H. P. Eugster, eds. Mineralogical Soc. Am., Washington, D.C. 35–69.

Navrotsky, A. 1994. *Physics and Chemistry of Earth Materials.* Cambridge University Press, Cambridge, England.

Newton, R. C. 1983. Geobarometry of high-grade metamorphic rocks. *Am. J. Sci.,* v. 283–A:1–28.

Nordstrom, D. K., and J. L. Munoz. 1985. *Geochemical Thermodynamics.* The Benjamin Cummings Publishing Company, Inc., Menlo Park, CA.

Nordstrom, D. K., and J. L. Munoz. 1994. *Geochemical Thermodynamics,* 2d ed. Blackwell Scientific Publications, Oxford, England.

O'Nions, R. K., and R. Powell. 1977. The thermodynamics of trace element distribution. In *Thermodynamics in Geology,* D. G. Fraser, ed. D. Reidel Publishing Company, Boston.

Pauling, L. 1953. *General Chemistry, Second Edition.* W. H. Freeman and Company, New York.

Pauling, L. 1960. *The Nature of the Chemical Bond,* 3d ed. Cornell University Press, Ithaca, NY.

Pearson, W. B. 1972. *The Crystal Chemistry and Physics of Metals and Alloys.* Wiley-Interscience, New York.

Putnis, A. 1992. *Introduction to Mineral Sciences.* Cambridge University Press, New York.

Reviews in Mineralogy. 1974 (on). Mineralogical Soc. Am., Washington, D.C.

Richardson, S. M., and H. Y. McSween, Jr. 1989. *Geochemistry—Pathways and Processes.* Prentice-Hall, Englewood Cliffs, NJ.

Ringwood, A. E. 1955. The principles governing trace element distribution during magmatic crystallization. *Geochim. Cosmochim. Acta,* v. 7:189–202, 242–254.

Sahama, T. G., and D. R. Torgeson. 1949. Some examples of the application of thermochemistry to petrology. *J. Geol.,* v. 57:255–262.

Saxena, S. K. 1973. *Thermodynamics of Rock-forming Crystalline Solutions.* Springer-Verlag, Berlin.

Schock, R. N., ed. 1985. *Point Defects in Minerals,* Geophys. Monogr. Ser., v. 31. Min. Phys. Ser., v. 1. Am. Geophys. Union, Washington, D.C.

Shannon, R. D. 1976. Revised effective ionic radii and systematic studies of interatomic distances in halides and chalcogenides. *Acta Cryst.,* v. A32:751–767.

Shannon, R. D., and C. T. Prewitt. 1969. Effective crystal radii in oxides and fluorides. *Acta Cryst.,* v. B25:925–946.

Shannon, R. D., and C. T. Prewitt. 1970. Revised values of effective ionic radii. *Acta Cryst.,* v. B26:1046–1048.

Skipper, N. T., K. Refson, and J. D. C. McConnell. 1991. Computer simulation of interlayer water in 2:1 clays. *J. Chem. Phys.,* v. 94:7434–7445.

Slater, J. C. 1964. Atomic radii in crystals. *J. Chem. Physics,* v. 41:3199–3204.

Smith, J. V., and H. S. Yoder. 1956. Experimental and theoretical studies of the mica polymorphs. *Mineral. Mag.,* v. 31:209–235.

Smyth, J. R., and D. L. Bish. 1988. *Crystal Structures and Cation Sites of the Rock-Forming Minerals.* Allan & Unwin, Inc., Winchester, MA.

Tossell, J. A., and D. J. Vaughan. 1992. *Theoretical Geochemistry: Application of Quantum Mechanics in the Earth and Mineral Sciences.* Oxford University Press, New York.

Waldbaum, D. R. 1966. Calorimetric investigation of alkali feldspars. Ph.D. thesis, Harvard University, Cambridge, MA.

Waldbaum, D. R., and J. B. Thompson, Jr. 1968. Mixing properties of sanidine crystalline solutions: II. Calculations based on volume data. *Am. Mineralogist,* v. 53:2000–2017.

Waldbaum, D. R., and J. B. Thompson, Jr. 1969. Mixing properties of sanidine crystalline solutions, IV: phase diagrams from equations of state. *Am. Mineralogist,* v. 54:1274–1298.

Whittaker, E. J. W., and R. Muntus. 1970. Ionic radii for use in geochemistry. *Geochim. Cosmochim. Acta,* v. 34:945–956.

Wood, B. J., and O. J. Kleppa. 1981. Thermochemistry of forsterite-fayalite olivine solutions. *Geochim. Cosmochim. Acta,* v. 45:569–581.

Zhdanov, G. S. 1965. *Crystal Physics.* Academic Press, Inc., New York.

Organic Geochemistry

All geologists dealing with water, soils, sediments, and sedimentary rocks need to have some knowledge of organic chemistry and of the role of organisms in natural processes. An enormous amount of organic matter occurs disseminated in sedimentary materials. In the form of coal, petroleum, and organic-rich rock (oil shale), it is of great economic importance. Organic chemistry is obviously crucial in studying the conversion of living organisms to fossils and the effects of diagenetic processes on organic material. Organic compounds are significantly involved in the weathering of minerals and rocks. Although only trace amounts of organic compounds occur in most meteorites and metamorphic rocks, they can play an important role in attempts to decipher the history of these rocks. Study of the transport and fate of hazardous organic chemicals in natural environments is an important area of active research. Many environmental problems, involving both organic and inorganic pollutants, require for their solution a knowledge of organic chemistry.

We shall begin with a brief review of the classification and composition of organic matter, followed by a general summary of the forms of naturally occurring organic matter on the Earth at the present time. The general features of the carbon cycle and certain other biologically important cycles are briefly reviewed. Then a more detailed discussion will outline the organic chemistry of seawater and marine sediments and contrast this with the organic chemistry of the nonmarine environment. Next the nature and abundance of the organic material of sedimentary rocks will be summarized, followed by a discussion of our rather limited knowledge of the origin and chemistry of coal and petroleum. The chapter concludes with a brief review of research on the origin and early evolution of life on our planet.

ORGANIC CHEMISTRY

Early in the nineteenth century, chemists concluded that there was a fundamental difference between substances produced by organisms (organic) and all other substances (inorganic). At that time they knew organic substances could be converted into inorganic substances, but these processes did not seem to be reversible. Eventually, it was discovered that many organic compounds could be produced by conversion of inorganic substances and that they could also be formed by synthesis from the elements. We now know that chemists can not only form organic compounds similar to those produced by organisms, but they can also prepare compounds that are not known in nature but that have organic properties. Thus a definition of organic chemistry must be based on more than the distinction between living and nonliving matter. Since we now know that all organic substances contain one or more atoms of carbon, organic chemistry is generally defined as the chemistry of carbon compounds. Even this definition is not an ideal one, since some carbon-bearing compounds, such as those containing the carbonate ion (CO_3^{2-}), have properties that are more characteristic of inorganic compounds than of organic compounds. With due regard for these constraints, it seems reasonable to define *organic geochemistry* as the study of all naturally occurring carbonaceous (carbon-containing) materials.* A recent update of research in all areas of organic geochemistry can be found in Engel and Macko (1993).

A major reason for the differences in the properties of organic and inorganic compounds is a general difference in chemical structure. As pointed out in Chapter Five, most inorganic compounds have relatively simple structures that can be explained in terms of unit cells containing a small number of atoms. Most organic compounds, on the other hand, are made up of large, complex macromolecules each containing as many as 100,000 atoms. The structure and behavior of organic solids depends on these irregularly shaped molecules with their strong internal bonding (covalent bonds) and relatively weak external bonding (van der Waals bonds). This difference, and the great variety and complexity of organic materials, is mainly due to the bonding properties of the carbon atom. A carbon atom can form single (one pair of electrons shared) or multiple (two or more pairs shared) covalent bonds with other carbon atoms and with the atoms of several other elements. This not only results in different compositions for organic compounds, but also results in the common occurrence of two or more compounds with the same formula but different molecular structure (*isomerism*). Thus we find that organic compounds exhibit an almost infinite variety of compositions and molecular structures. At present there are over seven million known and characterized organic compounds.

NATURALLY OCCURRING ORGANIC MATTER

All organic matter is composed primarily of carbon, oxygen, and hydrogen. Table 6-1 gives average compositions for various organic materials. To discuss naturally occurring organic matter, we need to review briefly a general classification of the material that makes up plants and animals. This material can be divided into five groups: (1) proteins, (2) lipids, (3) carbohydrates, (4) pigments, and (5) lignins. Table 6-2 gives typical abundances for these groups in plants and animals.

*The subdivision of organic chemistry that deals with chemical processes that go on in living organisms is known as biochemistry. An understanding of the effects of pollutants on living organisms requires knowledge of the fundamentals of biochemistry.

TABLE 6-1 Typical Compositions of Organic Material

Organic material	C	O	H
Proteins	51*	22	7
Lipids (fats)	69	18	10
Carbohydrates	44	49	6
Lignin	53	27	5
Organic substances in recent marine sediments	56	30	8
Organic substances in sedimentary rocks	64	23	9
Petroleum	79–89	—	9–15
Peat	50–65	28–45	6–7
Lignite	65–78	16–28	5–6
Bituminous coal	78–87	5–16	5–6
Anthracite coal	87–91	2–5	4–5

*All compositions in weight percent.

Source: Modified from *Geochemistry* by Karl Hans Wedepohl. Copyright © 1971 by Holt, Rinehart and Winston, Inc. Reprinted by permission.

TABLE 6-2 Composition of Living Matter

	Weight percent of major constituents (dry, ash-free basis)			
Substance	Proteins	Carbohydrates	Lipids	Lignin
Plants:				
Spruce wood	1	66	4	29
Oak leaves	6	52	5	37
Scots-pine needles	8	47	28	17
Phytoplankton	23	66	11	0
Diatoms	29	63	8	0
Lycopodium	8	42	50	0
Animals:				
Zooplankton (mixed)	60	22	18	0
Copepods	65	25	10	0
Oysters	55	33	12	0
Higher invertebrates	70	20	10	0

Note: Plants contain predominantly carbohydrates, with the higher forms having lignin for strength. Animals are predominantly protein. Marine organisms such as corals and sponges also have mostly protein in their $CaCO_3$ matrix. Pigments are not abundant enough to include in the table. There is great variability for different species of each organism. Lipid variability is partly due to nutrient availability and health of the organisms. If food supply is limited, or there is crowding during growth, the organism increases its lipids. *Chlorella* (unicellular green alga) grown in a favorable environment will contain 20 percent lipids, in an unfavorable environment, 60 percent lipids.

Source: From PETROLEUM GEOCHEMISTRY AND GEOLOGY by Hunt. Copyright © 1979 by W. H. Freeman and Company. Used with permission.

Proteins make up structural elements, such as skin and muscles, in organisms. They also have a number of other important roles, including serving as essential catalysts of biological reactions (the enzymes). Proteins are composed of complex chains of an important group of compounds known as *amino acids* (Figure 6-1). Amino acids are substances characterized by the general structure

$$\begin{array}{c} NH_2 \\ | \\ R—C—COOH \\ | \\ H \end{array}$$

in which R represents various combinations of carbon, oxygen, hydrogen, and nitrogen atoms. The term protein is arbitrarily limited to linkages of amino acids that have high molecular weights. Combinations of amino acids that have molecular weights less than 10,000 are referred to as *peptides.*

The term *lipid* has no specific chemical connotation. It refers to a group of related substances that are insoluble in water but soluble in certain organic solvents, such as ether and benzene. They serve as a means of energy storage in organisms and are often referred to by the general term *fats.* Lipids occur in all living cells. Waxes, used as a protective coating by organisms, are also made up of lipid compounds. Lipids have a strong tendency to form complexes with proteins, and such compounds are important components of cell membranes.

The most important role of *carbohydrates* is as an energy source and a food reserve. The simpler carbohydrates are soluble in water and serve as a quick source of energy; the insoluble lipids provide long-term energy storage. Most carbohydrates have the molecular formula $C_x(H_2O)_y$ (Figure 6-1). Examples of carbohydrate compounds that are energy sources include sugars, starches, celluloses, and pectins. Chitin, which serves as a structural material in certain plants and animals, also belongs to the carbohydrate group.

Pigments are colored organic compounds found in small amounts in most organisms. The best-known pigments are the chlorophylls, which are a group of closely related green pigments occurring in plants and a few bacteria. The chlorophylls play an important role as catalysts for photosynthesis, the process by which the sun's energy is transformed to potential energy in carbohydrates. The general chlorophyll structure is shown in Figure 6-1. Two important groups of pigments are the *porphyrins* (to which the chlorophylls belong) and the *carotenoids* (the most widespread group of pigments). They differ in their compositions, structures, and their resistance to breakdown as organic residues.

Lignin is a major constituent, along with cellulose, of plant cell walls and helps to bind individual cells together. It makes up about 25 percent of the dry weight of wood. The structure of lignin is a complex combination of units, such as those shown in Figure 6-1. The units in lignin are bound together by very strong bonds; thus lignin is very resistant to chemical treatment and decay. Its overall composition is similar to that of carbohydrates.

Geochemists are mainly interested in what happens to organic matter after the death of the organisms that produced it. A related concern is the abundance and chemical role of organic debris in natural waters, sediments, and rocks. Essentially, what happens is that all organic matter breaks down in a series of steps, through intermediate compounds, to final products that are stable in natural environments. The basic reason for this breakdown is that the organic compounds

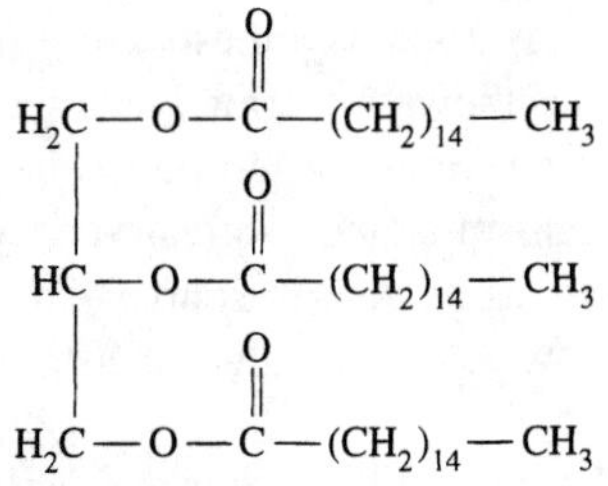

$$\begin{array}{l} H_2C-O-\overset{O}{\overset{\|}{C}}-(CH_2)_{14}-CH_3 \\ HC-O-\overset{O}{\overset{\|}{C}}-(CH_2)_{14}-CH_3 \\ H_2C-O-\overset{O}{\overset{\|}{C}}-(CH_2)_{14}-CH_3 \end{array}$$

Lipid (glyceryl palmitate)

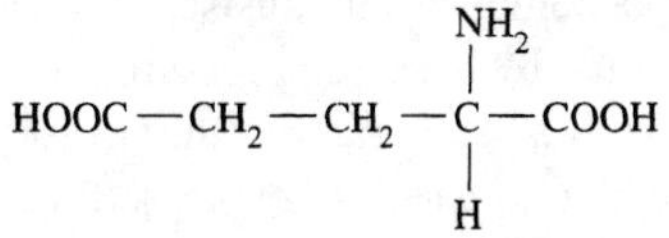

$$HOOC-CH_2-CH_2-\overset{NH_2}{\underset{H}{C}}-COOH$$

Amino acid (glutamic acid)

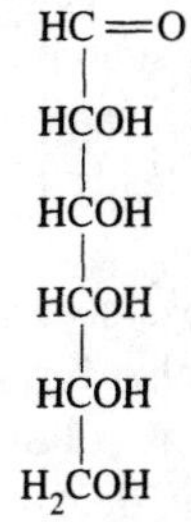

$$\begin{array}{l} HC=O \\ HCOH \\ HCOH \\ HCOH \\ HCOH \\ H_2COH \end{array}$$

Carbohydrate (glucose)

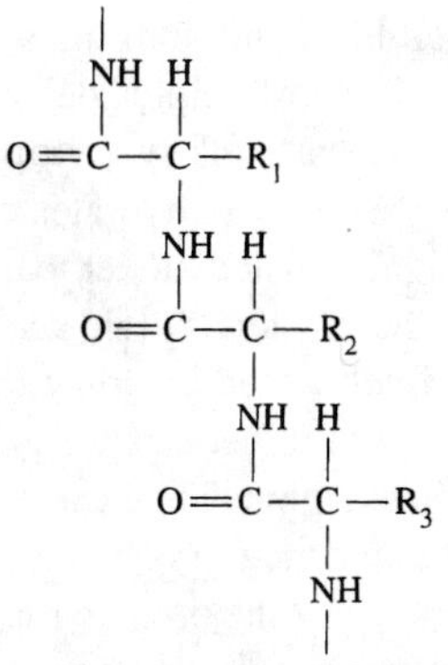

NH H | O=C—C—R₁ | NH H | O=C—C—R₂ | NH H | O=C—C—R₃ | NH

Protein

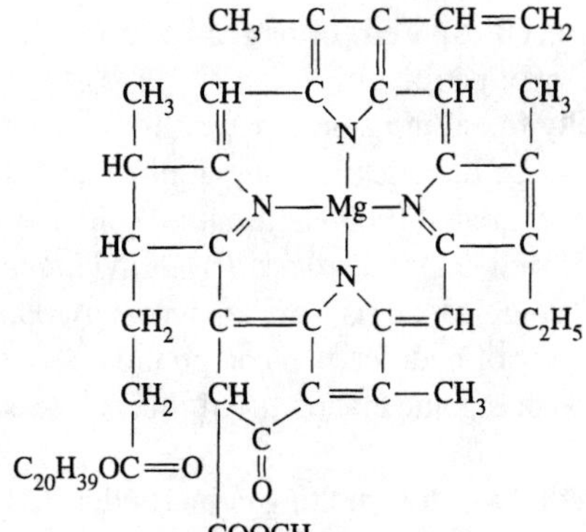

CH₃—C—C—CH=CH₂; CH₃ CH—C C—CH CH₃; HC—C N C—C; N—Mg—N; HC—C N C—C; CH₂ C=C C=CH C₂H₅; CH₂ CH C=C—CH₃; C₂₀H₃₉OC=O C=O; COOCH₃

Pigment (chlorophyll)

$$HO-C_6H_3(OCH_3)-CH=CH-CH_2OH$$

(coniferyl alcohol)

$$HO-C_6H_2(OCH_3)_2-CH=CH-CH_2OH$$

(sinapyl alcohol)

$$HO-C_6H_4-CH=CH-CH_2OH$$

(coumaryl alcohol)

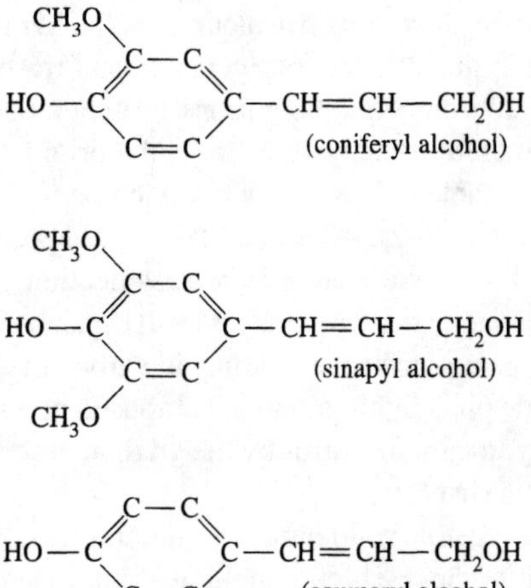

Components of lignin

Figure 6-1 Examples of common constituents of plants and animals. Not shown here is the water molecule, although water makes up two-thirds or more of the total weight of most organisms. The lines between letters represent single or double bonds. In the protein example R_1, R_2, and R_3 are nonspecific symbols representing parts of the molecule that are not needed to show the general structure.

of plants and animals represent states of high energy (photosynthesis concentrates energy by converting light energy to chemical energy), and thus they are thermodynamically unstable. The actual decomposition consists of various chemical reactions caused by oxidizing and reducing conditions, by heat and pressure, and by microorganism attack. A great variety of microorganisms, such as bacteria and fungi, use organic debris as an energy source. These organisms are most active in early stages of sediment diagenesis, with later changes caused by nonbiological *maturation*.

Table 6-3 lists the life substances discussed above and some of the decomposition intermediates and final products resulting from degradation of these substances. Proteins break down relatively easily into amino acids, which are much more stable over geological periods of time under anaerobic conditions in sediments. A wide variety of other products can also result (Table 6-3). Similarly, lipids decompose into a variety of products, of which fatty acids and hydrocarbons are particularly important. Lipids may serve as a source material for petroleum hydrocarbons. Many carbohydrates change to simple sugars such as ribose and xylose. The breakdown of lignin (listed under carbohydrates in Table 6-3) leads to the formation of humic substances (humic acid, fulvic acid, and humin; the three groups behave differently when humic material is fractionated by analytical procedures). Humic substances are extremely complicated mixtures of high-molecular-weight compounds, which can also form from other organic substances besides lignin. They can occur as dissolved material, as colloids, and as solids. Because of lignin's abundance in plants and its relative stability, lignin degradation products make up a large percentage of the organic matter in soils, sediments, and rocks.

Since organic pigments are not abundant in plants and animals, their decomposition products are not abundant in organic debris. The porphyrins are an important group of pigment products that form from chlorophyll. Porphyrins occur in trace amounts in a wide variety of organic deposits, including petroleum, coal, and oil shale. Because of their composition and stability, the porphyrins can serve as geochemical tracers to identify parent organic matter. They also reflect the environmental conditions to which they have been exposed. In particular, porphyrins have been used extensively to study the origin and history of petroleum.

Table 6-3 lists only organic materials typically found in natural waters (for concentrations see Figure 6-2). Additional terms are needed to describe the organic material of sediments, soils, and rocks. We shall use a classification given by Degens (1965) and outlined in Figure 6-3. *Bituminous substance* refers to all organic material in sediments and rocks. The term *bitumen* refers to organic substances soluble in carbon disulfide; insoluble materials are known as *pyrobitumens*. Crude oil is liquid bitumen composed of a wide variety of hydrocarbon compounds. This mixture of hydrocarbons forms by the gradual decomposition of organic matter. Its origin is discussed later in this chapter.

Fusible solid bitumen (mineral wax and asphalt) has a low melting point (under 230°F); asphaltites have a higher melting point. In addition to hydrogen and carbon, these bitumens contain significant amounts of oxygen, nitrogen, and sulphur. A wide variety of compounds are found in solid bitumens. Pyrobitumens decompose before melting. They can be subdivided into those with a low oxygen content and those with a high oxygen content. The latter group includes the abundant materials peat, coal, and kerogen. *Peat* forms from decaying plant material, and consists of original plant material such as cellulose and lignin and of intermediate breakdown products such as humic acids. *Coal* is an extremely complex material and represents more extensive change of original plant material. The formation and composition of coal are discussed later in this chapter.

TABLE 6-3 Decomposition Products of Life Substances Found in Natural Waters

Life substances	Decomposition intermediates	Intermediates and products typically found in nonpolluted natural waters
Proteins:	Polypeptides → $RCH(NH_2)COOH$ (amino acids) → {$RCOOH$, $RCH_2OHCOOH$, RCH_2OH, RCH_3, RCH_2NH_2}	NH_4^+, CO_2, HS^-, CH_4, HPO_4^{2-}, peptides, amino acids, urea, phenols, indole, fatty acids, mercaptans
Polynucleotides:	Nucleotides → purine and pyrimidine bases	
Lipids:		
{Fats, Waxes, Oils}	RCH_2CH_2COOH (fatty acids) + $CH_2OHCHOHCH_2OH$ (glycerol) → {RCH_2OH, $RCOOH$, shorter chain acids, RCH_3, RH}	CO_2, CH_4, aliphatic acids, acetic, lactic, citric, glycolic, malic, palmitic, stearic, oleic acids, carbohydrates, hydrocarbons
Carbohydrates:		
{Cellulose, Starch, Hemicellulose}	$C_x(H_2O)_y$ → {monosaccharides, oligosaccharides, chitin} → {hexoses, pentoses, glucosamine}	HPO_4^{2-}, CO_2, CH_4, glucose, fructose, galactose, arabinose, ribose, xylose
Lignin	$(C_2H_2O)_x$ → unsaturated aromatic alcohols → polyhydroxy carboxylic acids	
Porphyrines and Plant Pigments:		
{Chlorophyll, Hemin, Carotenes, Xantophylls}	Chlorin → pheophytin → hydrocarbons	Pristane, carotenoids
Complex Substances Formed from Breakdown Intermediates, e.g.:		
	Phenols + quinones + amino compounds →	Melanins, melanoidin, gelbshoffe
	Amino compounds + breakdown products of carbohydrates →	Humic acids, fulvic acids, "tannic" substances

Source: After Stumm and Morgan (1981).

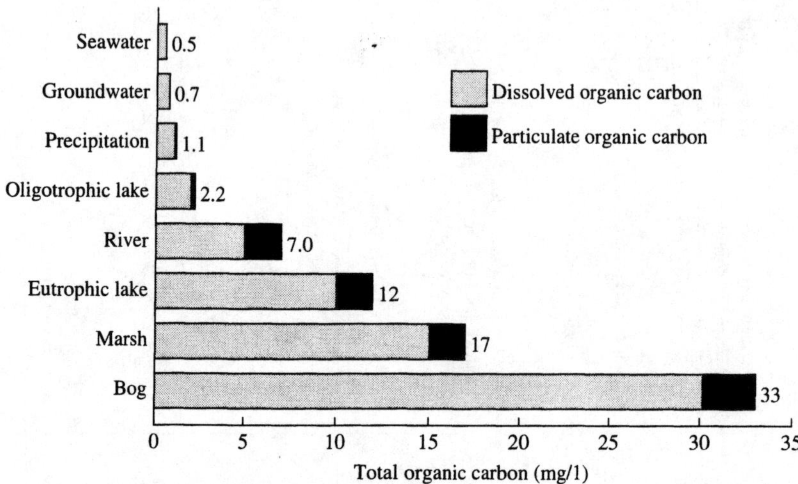

Figure 6-2 Approximate concentrations of dissolved and particulate organic carbon in natural waters. [After Thurman (1985, 8). Reprinted by permission of Kluwer Academic Publishers.]

The term *kerogen* refers to all high-oxygen pyrobitumens other than peat and coal that occur in sediments and sedimentary rocks.* Kerogen is the most common organic material on the Earth and is concentrated in the rocks known as oil shales. The major elements in kerogen are carbon, hydrogen, and oxygen. It is very difficult to obtain identifiable compounds from kerogen and its

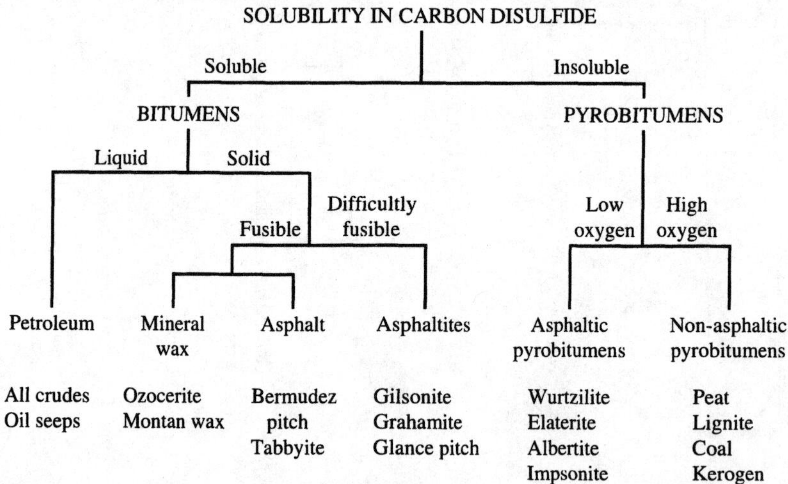

Figure 6-3 Terminology and classification of naturally occurring bituminous substances. Going across the figure from left to right, the carbon/hydrogen ratio and fusion point increase and the solubility in carbon disulfide decreases. After Degens (1965, 259). Adapted from Abraham (1960) and Hunt (1963).

*Hayes et al. (1983, 99) state: "The term 'kerogen' has about the same level of chemical specificity as the term 'metal.' 'Kerogen' can indicate materials with very different elemental compositions and, equally important, it is applied to materials with very different levels of maturity." Some classifications list coal and peat as a subcategory of kerogen.

structure is not well understood. Some kerogens have properties similar to those of coal; others do not (Waples 1985). Kerogen and bitumen are the two main components of the total organic carbon in any rock, with kerogen usually making up more than 90 percent of the total. In natural waters and unconsolidated sediments, precursors of kerogen such as humic and fulvic acids are the abundant components of organic matter.

THE CARBON CYCLE

Carbon is the element that is most extensively involved in both the organic and inorganic systems of the Earth. It occurs in a gaseous form in the atmosphere (CO_2), as part of various ions in the hydrosphere (HCO_3^-, H_2CO_3, etc.), as the major element of living and dead organic matter (proteins, carbohydrates, humic acids, hydrocarbons, calcium carbonate, etc.), and as a major component of carbonate sediments and sedimentary rocks (calcite, aragonite, dolomite). It is also a major component of the igneous rock carbonatite and the metamorphic rock marble. These different occurrences of carbon can be considered as reservoirs between which carbon exchanges as a result of inorganic and organic processes. Figure 6-4 illustrates the carbon cycle. It is very difficult to estimate the amounts of carbon in each reservoir and the rates of exchange between reservoirs. The numbers given in Figure 6-4 are rough but useful approximations.

At the present time, and in terms of yearly turnover, the major movements of carbon are a result of photosynthesis and respiration, with exchange occurring between the biosphere and atmosphere and between the biosphere and hydrosphere. The other important movement is the inorganic exchange that occurs between the atmosphere and hydrosphere. Early studies assumed the carbon cycle was at a steady state throughout geologic time, that is, there have been no changes with time in the amounts of carbon in the various reservoirs of Figure 6-4 or in the values of the fluxes between them. More recent research suggests that natural changes over long periods of time in the CO_2 content of the atmosphere have temporarily perturbed this steady state (Sundquist and Broecker 1985). Studies of the effects of past perturbations can help us to understand the relationship between changes in atmospheric CO_2 and climate.

A current concern is the possibility of global warming (the greenhouse effect) due to the burning of fossil fuels and other activities by modern humans.* Since 1850 about 200 billion tons of carbon dioxide (approximately 45×10^{15} moles of carbon) have been added to the atmosphere by burning fossil carbon. Since the carbon cycle is not well known, and since parts of it operate at rates involving hundreds of years, it is currently impossible to precisely predict the effects of this disturbance of the cycle (Post et al. 1990). We do know that this artificial release of CO_2 to the atmosphere is much faster than any natural release. Despite our uncertain knowledge of the carbon cycle, some scientists have estimated that a doubling of carbon dioxide in the atmosphere could cause a global warming of between 1.5 and 4.5°C. Since 1850 the concentration of CO_2 in the atmosphere has increased from about 285 microliters per liter to about 350 microliters per liter (Post et al. 1990).

If we think of the various exchanges that have occurred over geologic time, the most important are those processes, inorganic and organic, that have added carbon to the lithosphere. The rates of exchange for these processes, in terms of moles per year, are small. However, over millions of years the amount of carbon that has been concentrated in sedimentary rocks is estimated to be at least 600 times the present amount in the atmosphere, hydrosphere, and biosphere. This

*Other gases besides CO_2 contribute to the greenhouse effect. The total effect of synthetic methane, nitrous oxide, and other trace gases may equal that of the human-created increase in CO_2.

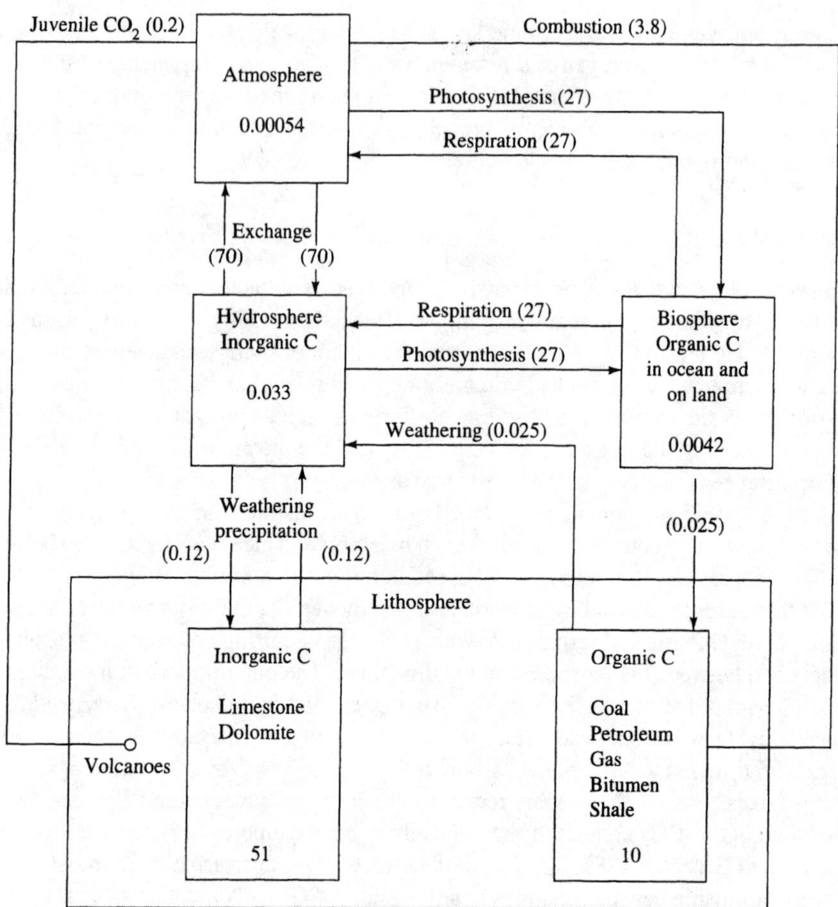

Figure 6-4 The carbon cycle. The circulation of carbon in nature is determined primarily by biochemical reactions. Amounts of carbon on the Earth's surface in 10^{20} mol (geomoles). Parentheses indicate rates of turnover of carbon in 10^{14} mol year^{-1} (microgeomol year^{-1}). Although the estimates given for the annual turnover are quite uncertain, they clearly illustrate that most of the carbon in the Earth's surface has been cycled through organisms. Coal and kerogen in shale account for about 95 percent of all organic matter. Over 99 percent of all carbon on the Earth occurs in the form of carbonates (mostly in sedimentary rocks) and as coal and kerogen. (After W. Stumm and J. J. Morgan, *Aquatic Chemistry*, 2d ed., p. 506. Copyright © 1981 by John Wiley & Sons. Reprinted by permission of John Wiley & Sons, Inc.)

carbon probably once existed in the atmosphere as carbon dioxide. About three-fourths of it now occurs as limestone and dolomite, and the other one-fourth as organic deposits, particularly as coal and disseminated kerogen in shales.

A useful way to study the carbon cycle, in terms of both present and past variations, involves measurements of the stable carbon isotopes ^{12}C and ^{13}C (see Chapter Two). It has been found that biological processes significantly fractionate carbon isotopes, and that the different biochemical compounds, such as proteins and carbohydrates, may show different degrees of fractionation (Figure 2-27). This opens up the possibility of understanding the details of biological

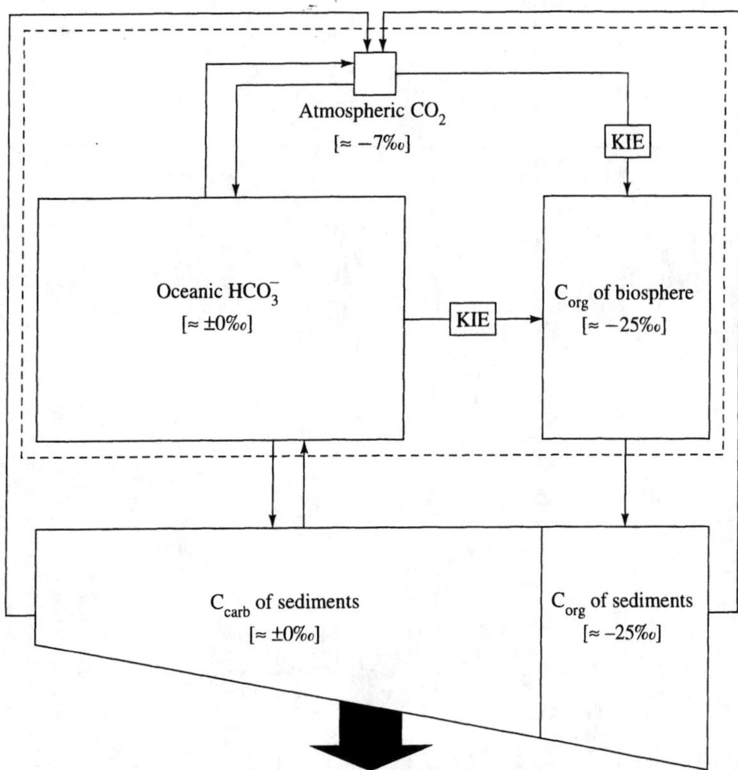

Figure 6-5 Isotopic expression of the carbon cycle. Numbers are $\delta\,^{13}C$ values. Enzymatically controlled kinetic isotope effects (KIE) are imposed on the two pathways leading from CO_2 and HCO_3^- to living matter (C_{org} of biosphere), the resulting fractionation being retained in subsequent geochemical cycling of this carbon species. Isotope fractionations in the CO_2-HCO_3^--CO_3^{2-} system are governed by equilibrium effects (double arrows) and are thus reversed in back reactions (a fractionation of about 1 between bicarbonate and sedimentary carbonate has been omitted for simplicity). (After Schidlowski, M., Hayes, J. M., and Kaplan I. R., Isotopic inferences of ancient biochemistries: carbon, sulfur, hydrogen, and nitrogen. In *Earth's Earliest Biosphere—Its Origin and Evolution*. Copyright © 1983. Reprinted by permission of Princeton University Press.)

reactions by isolating and analyzing discrete biochemicals. Carbon isotopes are also fractionated by inorganic processes; thus the inorganic part of the carbon cycle can be isotopically investigated. An isotopic expression of the carbon cycle is shown in Figure 6-5. A summary of the biogeochemistry of the stable carbon isotopes is given by Schidlowski et al. (1983). Figure 6-6 is an example of the use of carbon and oxygen isotopes to study the Earth's history.

NUTRIENT CYCLING

Elements needed by living things are known as nutrients. Their movement through various reservoirs is known as biogeochemical cycling. The most important nutrients for plants are listed in Table 6-4. An understanding of nutrient cycling is important in studying biological processes,

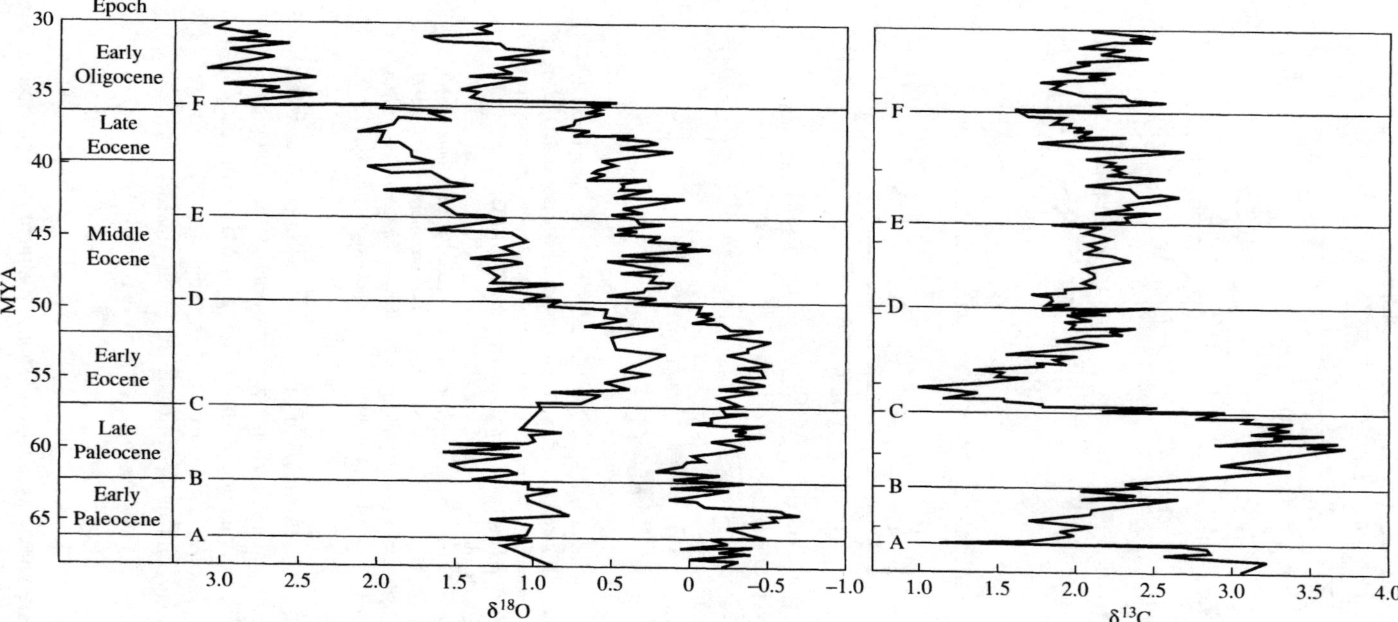

Figure 6-6 Isotopic history of Cenozoic ocean water. Shown here are ratios of stable isotopes extracted from the calcite skeletons of marine microfossils and plotted through the Paleocene, Eocene, and early Oligocene epochs. Ratios of oxygen 18 to oxygen 16 and of carbon 13 to carbon 12 are expressed as $\delta^{18}O$ and $\delta^{13}C$. The two oxygen curves give signals from the surface (right) and bottom waters (left) of the ocean, so that parallel changes through time are a good indication that the whole ocean is changing. A decrease in $\delta^{18}O$ indicates warming of the water in which biomineralization occurred (especially bottom water above level C); sharp increases in $\delta^{18}O$ are major cooling events (levels D and F). The carbon signal derives from the preferential uptake of carbon 12 during photosynthesis. Thus a marked increase in $\delta^{13}C$ suggests increased productivity, that is, removal of carbon 12. Note the sudden decrease in $\delta^{13}C$ at level A, suggesting a catastrophic collapse in photosynthesis and planktonic production at the Cretaceous-Tertiary boundary, and a large positive spike in $\delta^{13}C$ between levels B and C. Data from Shackleton (1986). After McGowran (1990).

TABLE 6-4 Elements Essential for Nutrition and Growth of Plants (Nutrients)

Element	Adequate concentration (% dry wt. of tissue)
Carbon	45
Oxygen	45
Hydrogen	6
Nitrogen	1.5
Potassium	1.0
Calcium	0.5
Phosphorus	0.2
Magnesium	0.2
Sulfur	0.1
Chlorine	0.01
Iron	0.01
Maganese	0.005
Zinc	0.002
Boron	0.002
Copper	0.0006
Molybdenum	0.00001

Sources: After Berner, E. K., and Berner R. A., *The Global Water Cycle.* Copyright © 1987 by Prentice Hall, Inc., Upper Saddle River, NJ. After Zinke (1977).

natural water chemistry, and pollution effects such as the accelerated eutrophication of lakes. Research on nutrient cycling involves determining a budget for an area or a body of water such as a lake (Table 6-5).

Berner and Berner (1987) discuss biogeochemical cycling in forests. They point out the importance of nutrient cycling on runoff composition. The composition of the stream water leaving an area is dependent on both rock weathering and biomass storage (accumulation of elements in living and dead vegetation) as shown in Table 6-5. For forested areas it is important to know if the forest is at a steady state, that is, not growing and storing nutrients. Also seasonal changes can have an effect on runoff composition.

According to Berner and Berner (1987), the most biologically affected elements are phosphorus and nitrogen.* These two elements play a major role in the pollution of water bodies by increasing the amount of plant growth. They come from sources such as sewage and fertilizers. Either element may be the controlling (limiting) nutrient (the nutrient element that limits growth because of a restricted supply). For most lakes phosphorus tends to be the limiting nutrient and reduction or limitation of its input is needed to reduce lake eutrophication. In many estuaries research has shown that nitrogen is the limiting nutrient. In addition to the rapid growth of algae (algal blooms) and other plants, another undesirable effect of excess nutrient addition to a body of water is the development of anaerobic conditions in the bottom water (Figure 6-7).

*In making a list of biologically active elements, Berner and Berner define biological activity as the ratio of the amount of an element stored annually in living and dead vegetation to the annual loss from the area in streamflow. Carbon is not included in the list.

TABLE 6-5 Element Budget in Percent for Watershed Ecosystems of the Hubbard Brook Experimental Forest

	Ca	K	Mg	Na	N	S
Source						
Precipitation input	9	11	15	22	31	65
Net gas or aerosol input	—	—	—	—	69	31
Weathering release	91	89	85	78	—	4
Storage or loss						
Biomass accumulation						
Vegetation	35	68	17	2	43	6
Forest floor	6	4	5	<1	37	4
Streamflow						
Dissolved substances	59	22	74	95	19	90
Particulate matter	<1	6	5	3	<1	<1

A large percentage (greater than 40 percent) of potassium, nitrogen, and calcium input from meteorologic sources and released by weathering is stored annually by living and dead biomass within ecosystems. In contrast, only about 2 percent of the sodium and 10 percent of the sulfur added to ecosystems is stored in the biomass, and the remainder, more than 90 percent, is lost in streamflow. Budget studies indicate that the undisturbed forest ecosystems show absolute gains in C, N, S, P, and Cl and absolute losses in Si, Ca, Na, Al, Mg, and K. Losses of the latter substances from the intrasystem nutrient cycle are made up by weathering primary minerals and the losses therefore imply a decrease in weathering substrate. Weathering is the major source of Ca, K, Mg, and Na in the ecosystem. Additional data are in Table 7-8.

Source: After Likens et al., *Biochemistry of a Forested Ecosystem.* Copyright 1977. Reprinted by permission of Springer-Verlag, New York.

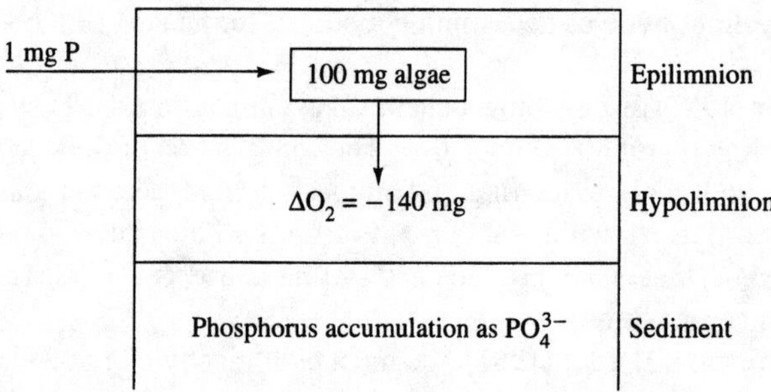

Figure 6-7 Transformation of phosphorus input to a stratified lake. Introduction of one milligram of phosphorus may lead to the synthesis of 100 milligrams of algae (dry mass), which eventually decays and accumulates in the hypolimnion, causing oxygen consumption of 140 milligrams. Some of the PO_4^{3-} formed by this process eventually accumulates in the lake sediment by adsorption on sediment particles. Excessive algal growth and decay leads to depletion of oxygen in the bottom water. The water-sediment contact area then becomes anaerobic and phosphorus is released back into the lake water from the sediment. In such a situation a decrease in the phosphorus input to the lake (such as a diversion of treated sewage) may not result in lower algal growth, since the lake sediment may become an important internal source of phosphorus. Phosphorus released from the bottom sediment can reach the surface waters by diffusion. In addition to this inorganic release of phosphorus, plants growing in the bottom sediment may remove phosphorus from the sediment with their roots and release PO_4^{3-} into the water from their leaves. (After W. Stumm and J. J. Morgan, *Aquatic Chemistry*, 2d ed., p. 706. Copyright © 1981 by John Wiley & Sons. Reprinted by permission of John Wiley & Sons., Inc.)

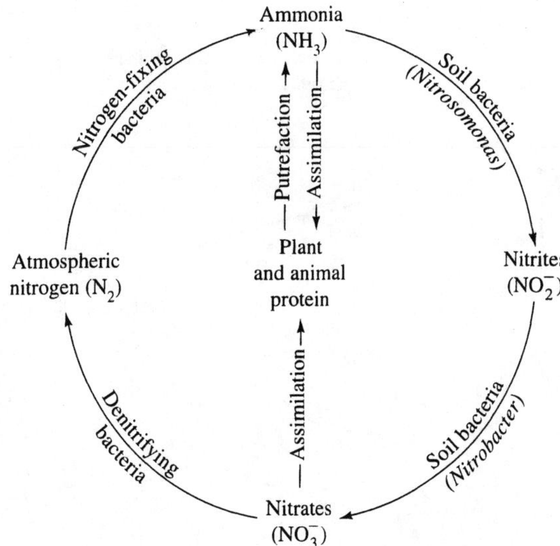

Figure 6-8 The nitrogen cycle. The processes of nitrogen fixation (N_2 to NH_3), nitrification (NH_3 to NO_3^-), and denitrification (NO_3^- to N_2) are all mediated by bacteria. Assimilation refers to the incorporation of nitrogen into plant and animal organic matter. Putrefaction refers to the decomposition of organic matter and also involves the actions of microorganisms. (After J. Postgate, *Microbes and Man.* Copyright 1992. Reprinted with permission of Cambridge University Press.)

Microorganisms (bacteria, fungi, algae) play a major role in the biogeochemical cycling of elements. An example is the role of bacteria in the nitrogen cycle (Figure 6-8). They mediate several of the major chemical reactions in the nitrogen cycle. In other words, they obtain chemical energy from reactions they cause to occur. The reaction may be an oxidation (NO_2^- to NO_3^-) or a reduction (NO_3^- to N_2). A detailed discussion of the vital role of microorganisms in the biogeochemical cycles of both abundant (C, H, O, N, P, and S) and minor elements of living organisms is provided by Atlas and Bartha (1987). Schlesinger (1991) has written a basic textbook on biogeochemical cycling of the elements.

TRACE ELEMENT CYCLES

All elements have natural cycles, including those that occur only in trace amounts in our environment. The biological part of a cycle is often extremely important for trace elements. The abundance and distribution of many of these elements are controlled by metabolic processes of microorganisms such as bacteria and algae. There has been increasing interest in trace element cycles in recent years, because it is now clear that many of these elements play a major role in human health and disease, particularly if the natural balance is disturbed by environmental pollution.

A general cycle for trace elements is shown in Figure 6-9a. Weathering of rocks puts these elements into the hydrosphere, where they are taken up by bacteria and planktonic organisms. Often they are more concentrated in these organisms than in the water in which the organisms live. They then pass on up the food chain through fish and eventually to humans, becoming increasingly concentrated. Along the way, the elements also pass back to the hydrosphere and eventually to the lithosphere as a result of organic decay. Disruptions of a cycle occur when industrial and/or other human activities introduce a large amount of an element, or a new form of an element, at some point in the natural cycle.

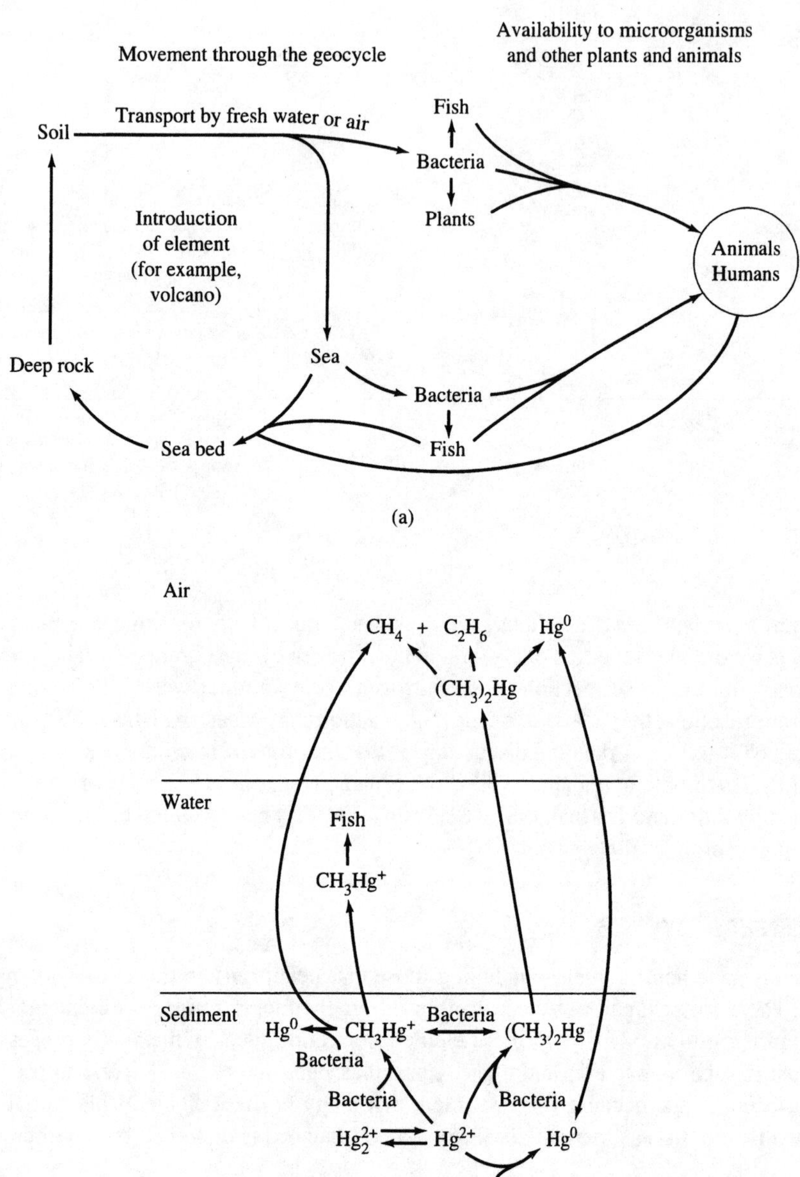

Figure 6-9 (a) Movement of toxic elements through the geocycle and their availability to microorganisms and other plants and animals; (b) the biological cycle for mercury. Methyl mercury (CH_3Hg^+) and dimethyl mercury [$(CH_3)_2Hg$] are toxic compounds, which are both produced and broken down by microorganisms. (After J. M. Wood. Copyright 1974 by the American Association for the Advancement of Science.)

TABLE 6-6 Classification of Elements* According to Their Toxicity**

Noncritical			Very toxic and relatively accessible			Toxic but very insoluble or very rare	
Na	C	F	Be	As	Au	Ti	Ga
K	P	Li	Co	Se	Hg	Hf	La
Mg	Fe	Rb	Ni	Te		Zr	Os
Ca	S	Sr	Cu	Pd	Pb	W	Rh
H	Cl	Al	Zn	Ag	Sb	Nb	Ir
O	Br	Si	Sn	Cd	Bi	Ta	Ru
N				Pt		Re	Ba

*Elements omitted from this table should not be neglected in the environmental sense. For example, iodine and manganese are important elements, but they fit more than one category for the above classification.

**Toxicity is a function of concentration. Many elements are essential in humans at very low concentrations, but become toxic at higher concentrations. Arsenic is an example.

Source: After J. M. Wood. Copyright 1974 by the American Association for the Advancement of Science.

Mercury is a good example of a toxic element whose natural cycle is fairly well known. In the form of organic mercury compounds, it damages the central nervous system of humans. Inhalation of mercury vapor affects the lungs. Some forms of mercury, such as liquid mercury, are not ordinarily toxic. Mercury occurs in trace amounts in the lithosphere (less than 1 part per million) and hydrosphere (about 1 part per billion). Organisms, however, tend to concentrate mercury.* It is also concentrated by humans for use in the chemical and electrical industries and in agriculture. Introduction of mercury into the atmosphere and hydrosphere by industrial processes leads to excess amounts of methyl mercury (CH_3Hg^+), the most injurious form of mercury. Methyl mercury is produced by bacteria from Hg^{2+} (Figure 6-9b). It is also formed by industrial synthesis. Under the right conditions, certain bacteria break down methyl mercury to Hg^0 (also toxic to humans) plus methane.

A good review of the biological cycles of toxic elements is given by Wood (1974). Wood points out that some trace elements, such as tin and platinum, behave in a manner similar to that of mercury, whereas others (lead, cadmium, zinc) show a different biochemical behavior. It is very important to know biochemical behavior in some detail in order to evaluate environmental health problems. Table 6-6 is a classification of elements according to their toxicity. The elements whose natural cycles need to be better understood from a health standpoint, and which should be monitored carefully, are those classified as very toxic and relatively accessible. For further information, the reader is referred to Freedman (1975), where the relation of trace elements to health and disease are discussed.

We now pass on to a detailed discussion of the organic chemistry of the hydrosphere and lithosphere. There are significant differences between the organic matter of marine and nonmarine waters and sediments. Thus we shall discuss these two environments separately.

*Fish in water containing 0.01 parts per billion mercury have been found with over 300 parts per billion mercury in their bodies, a biomagnification factor of 30,000.

ORGANIC GEOCHEMISTRY OF FRESHWATER AND SOILS

The organic carbon of natural waters is usually divided into two types, dissolved organic carbon (DOC) and particulate organic carbon (POC). DOC is the organic carbon passing through a 0.45 micrometer filter; POC is retained by the filter. The sum of DOC and POC is known as total organic carbon (TOC). For most natural waters 90 percent or more of total organic carbon consists of dissolved organic carbon. A chemical analysis of water could also measure dissolved, particulate, and total organic matter (DOM, POM, and TOM) where organic matter refers to organic molecules that contain other elements as well as carbon. According to Thurman (1985), DOM, POM, and TOM are generally equal to twice the DOC, POC, and TOC.

The organic matter of freshwater varies greatly in abundance and has a different makeup from that of seawater (Table 6-7). Fresh surface water usually has more dissolved organic matter, particularly if it has been polluted or occurs in a swampy area. An overall average value for dissolved organic carbon for freshwater is perhaps 4 mg/l, compared to 1 mg/l or less for seawater. Lake water, particularly in eutrophic lakes, usually has a higher concentration of organic carbon than river water. The carbon results from both photosynthesis within a water body and inflow from the surrounding land. Organic compounds due to decay of terrestrial plant material such as lignin are more abundant in freshwater than in seawater. An example of the abundance and type of organic matter in freshwaters has been given by Reeder et al. (1972) and Peake et al. (1972). They reported the following range of concentrations for the surface waters of the Mackenzie River drainage basin of Canada: (1) total organic carbon less than 2 to 42 parts per million; (2) amino acids, 12 to 299 parts per billion (bound to suspended matter and dissolved); (3) hydrocarbons, not detected to about 0.25 part per billion; and (4) pigments, 0.00003 to 0.1 part per billion.

There is a relatively delicate balance in freshwaters between the processes of photosynthesis, respiration, and organic decay. The rate of production of organic matter is more or less equal to the rate of destruction. In general, excess photosynthetic production of organic matter leads to an imbalance of organic life (algal blooms are an example) and to mass extinctions. It also raises the local pH as a result of carbon dioxide removal and maintains a high Eh by production of oxygen (see discussion in Chapter Four). The high rate of organic decay that usually follows excessive photosynthesis results in a lack of dissolved oxygen, reduction of inorganic ions, and pro-

TABLE 6-7 Abundance of Dissolved Organic Carbon in Natural Waters

Water	Mean (mg/l)	Normal range (mg/l)
Seawater	0.7	0.2–2.0
Groundwater	0.7	0.2–15.0
Rainwater	1.0	0.5–1.5
Estuary water	3.0	1.0–5.0
Lake water	5.0	1.0–30.0
River water	5.0	1.0–30.0
Soil interstitial water	7.0	2.0–30.0
Sediment interstitial water (aerobic)	9.0	4.0–20.0
Wetland water	25.0	10.0–400.0
Sediment interstitial water (anaerobic)	80.0	10.0–390.0

Source: After Thurman (1985).

duction of gases such as hydrogen sulfide and methane. The local pH has neutral to low values and Eh is low. When the natural balance is extensively upset, such as by addition of sewage wastes, a body of water may no longer be able to purify itself, and its chemical properties will be completely changed. In other cases the effect is to change the rate of natural processes. For example, addition of nutrients (such as phosphorus) to a lake greatly speeds up the development of life (eutrophication). Whether or not natural waters are polluted, their chemical composition is obviously controlled to a great extent by biological activity (Figure 6-7). A good discussion of the role of microorganisms in natural and polluted waters is given by Atlas and Bartha (1987). Thurman (1985) describes in detail the organic geochemistry of natural waters.

The organic matter of soils can be divided into (1) material of indefinite character (humic substances), and (2) other material that can be separated and identified (such as proteins, carbohydrates, lignins, lipids, pigments, and hydrocarbons). Most soil organic matter belongs to the first group and is produced as a result of nonphotosynthetic microorganisms using primary organic matter as an energy source. Humic substances are important for a variety of reasons. They affect the organic life of the soil, help in the breakdown of mineral particles, and probably control the abundance and distribution of trace elements in soil. One of their most important characteristics is their ability to adsorb metal ions and organic pollutants. Formation of humic material can occur by direct breakdown of the lignin of plants or by combination of the breakdown products of other precursors, such as proteins and carbohydrates. The kerogen of oil shales and other sedimentary rocks probably forms by alteration of humic substances. Schnitzer (1984) gives a good review of soil organic matter. The overall chemistry of soils is presented by Sposito (1989).

The organic matter of lake and stream sediments is very similar to that of soils. The greatest abundance of humic substances (as much as 50 percent by weight) occur in peats and organic muds. The abundances of the various nonhumic substances of soils and sediments tend to be in the range of 1–100,000 parts per million. Some of the organic material in sediments is introduced and may have been transported through several different chemical environments. Other material has accumulated from the environment in which it found. Thus a wide range of organic compounds is found, from practically unaltered to extremely stable degradation products.

ORGANIC GEOCHEMISTRY OF SEAWATER AND MARINE SEDIMENTS

The overall composition of seawater is given in Table 1-11 and the major ions are listed in Table 3-4. It was pointed out in Chapter Four that organic material can occur in natural waters as suspended particles, as colloidal material, as adsorbed compounds, and in the form of dissolved molecules. The total amount of colloidal and dissolved organic carbon in seawater averages about 1 part per million by weight. Particulate organic carbon, including planktonic (floating) organisms, averages about one-tenth of a part per million. These occurrences of carbon vary quite a bit in abundance vertically and laterally in the oceans. It is very difficult to estimate the amount of adsorbed organic matter. Some of it occurs in association with aggregates of inorganic and organic particles. These aggregates apparently are the result of the collection of dissolved organic matter on the surfaces of air bubbles, followed by some process of aggregate growth. The inorganic part of the aggregates consists of clay particles and other debris. The aggregates range in diameter from a few microns to several millimeters.

The organic matter of seawater mainly results from excretion by living organisms and from decomposition of dead organisms. Most of the production of the basic organic material for all forms of marine life is carried out by phytoplankton, floating, single-celled green plants (examples are algae, diatoms, and flagellates). The phytoplankton synthesize organic compounds by photosynthesis in the lighted surface waters of the oceans. Light energy is used to combine water and carbon dioxide into carbohydrates and oxygen:

$$6\,CO_2 + 6\,H_2O \rightarrow C_6H_{12}O_6 + 6\,O_2$$

Organic decay and respiration can be considered, in a general way, as the reverse of this reaction. Both photosynthesis and organic decay are actually very complicated processes involving distinct subprocesses.

The organic matter represented at the ocean surface by phytoplankton and organic-bearing aggregates serves as food for small planktonic animals (examples are copepods and euphausiids). These animals in turn serve as food for larger animals, and thus the marine food chain is built up. The other part of the marine food web is supplied by organic debris settling down from the surface layers. This debris is the most stable material left after attack by predators and marine bacteria. The abundance and distribution of phytoplankton vary seasonally and geographically in the oceans. This results in great variation in marine productivity (grams of organic carbon fixed by photosynthesis per unit of time) and in the abundance of organic life.

The identity and abundance of the various dissolved organic compounds produced by animal and plant metabolism and by decay are not well known. Major types found include amino acids, proteins, carbohydrates, lipids, organic acids, vitamins, and toxins. Although present in amounts measured in parts per million or parts per billion, these compounds are geochemically very important. Besides playing a vital role in the marine life cycle, they strongly affect the abundance and distribution of trace elements such as manganese. They are also a factor in the dissolution and precipitation of inorganic compounds such as calcium carbonate. Dissolved organic compounds are known to form organic coatings on carbonate mineral surfaces. These coatings are probably responsible for calcite supersaturation (disequilibrium) in seawater (see discussion in Chapter Seven).

The organic matter that accumulates in marine sediments consists of biochemically stable residual materials and particulate remains of organisms. After initial accumulation, further changes and additions result from bacterial, fungal, and animal processes in the bottom sediments. Organic material in marine sediments thus includes (1) shell particles and other essentially unaltered remains; (2) proteins and their decomposition products, particularly amino acids; (3) carbohydrates and their decomposition products (which include carbon dioxide and water); (4) stable lignins (mainly derived from land sources); (5) pigments and their decomposition products; (6) lipids and their decomposition products, including fatty acids and hydrocarbons; and (7) other organic matter of unknown history, including both decomposition products (such as humic acids) and contributions from the metabolism of organisms living in sediments. This complex of organic matter undergoes numerous changes after initial decomposition and through the stage of diagenesis. After burial, organic activity decreases, and a slower stage of nonbiological alteration begins and continues even after the sediment becomes rock. A detailed discussion of the accumulation and transformation of organic substances in marine sediments is given by Berner (1980).

The sources and history of organic matter in water and sediments can be studied by use of chemical and isotopic techniques. An example is the work by Fogel et al. (1989). They studied the organic matter in a Georgia salt marsh as follows. First, sources of organic matter were isotopically characterized and then a suite of relict muds was analyzed to document the progress of diagenesis of organic matter with time. The isotopic composition of present-day suspended particulate matter was measured and compared with source materials to determine the relative contributions of the major sources (Figure 6-10). Fogel et al. concluded that (1) the particulate organic matter in the salt marsh consisted primarily of microbially-altered material having an intermediate isotopic and chemical composition between fresh algae and Spartina plant matter and (2) total organic matter in marsh sediments has a generally constant chemical and isotopic composition over time on a macro-scale, but is comprised of a heterogeneous mixture of substances on a meso- to micro-scale.

The abundance of organic matter in typical marine sediments varies from practically none to several weight percent. It mainly occurs as very small, disseminated material associated with fine-grained sediment. Thus there is generally a direct correlation between sediment particle size and amount of contained organic matter. Table 6-8 is a specific example of organic matter in marine sediments (Aizenshtat et al. 1973). The 15 sediment samples listed in the table are from five drill sites and were collected by the Glomar *Challenger* as part of the JOIDES

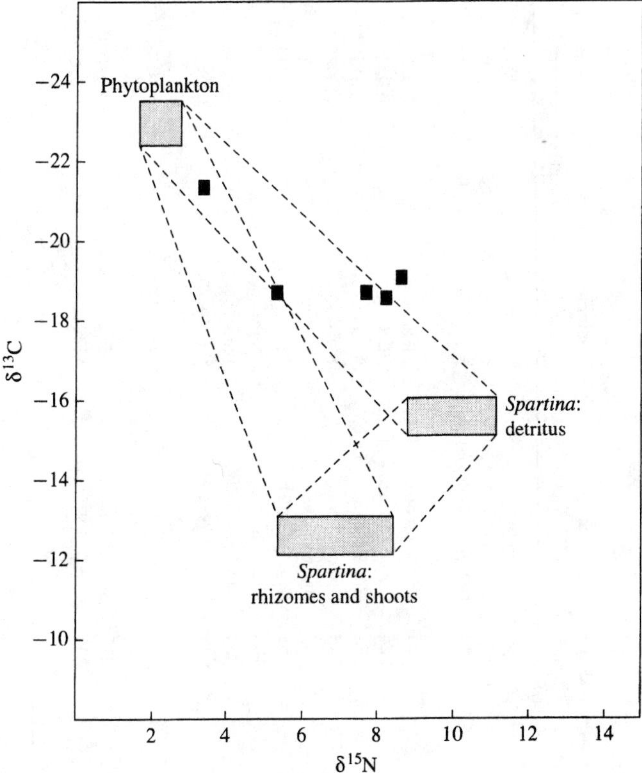

Figure 6-10 Stable carbon and nitrogen isotopic compositions of modeled endmembers in the water column of a salt marsh. The $\delta^{15}N$ and $\delta^{13}C$ values of phytoplankton are estimates; the size of the box in this diagram is 1 per mil for both C and N and covers the range of uncertainty in the estimation. Squares indicate suspended particulate matter collected in the waters of St. Catherines Island, Georgia. After Fogel et al. (1989).

TABLE 6-8 Organic Compounds in JOIDES Sediment from Gulf of Mexico and Western Atlantic

Sample location	Water depth (m)	Hole depth (m)	Sample description	Sample age	Total organic carbon (percent)	Hydro-carbons (percent of total organic carbon)	Fatty acids (percent of total organic carbon)	Alcohols (percent of total organic carbon)	Pigments (percent of total organic carbon)	Humic acids, kerogen, fulvic acid, etc. (percent of total organic carbon)
23°27.3′ N	3,572	20	Calcareous mud	Pleistocene	0.45	0.3216	0.01004	~0.0013	0.0004	99.67
92°35.2′ W		103	Coccolith ooze	Pliocene	0.38	0.0903	0.0014	~0.0015	0.0001	99.91
23°01.8′ N 92°02.6′ W	3,747	34	Calcareous silt and clay	Pleistocene	1.11	0.0714	0.0063	0.0017	0.0011	99.92
		209	Coccolith ooze and clay	Pleistocene	0.82	0.0061	0.0028	0.0005	0.0005	99.99
		324	Coccolith ooze and clay	Pliocene	0.47	0.0206	0.0060	0.0005	0.0003	99.97
		534	Grayish-green silty clay	Miocene	0.47	0.0185	0.0068	0.0016	0.0005	99.97
30°50.30′ N	5,125	15	Brown clay	Pleistocene	0.13	0.0085	<0.0030	n.d.*	n.d.	99.99
67°38.86′ W		43	Brown clay	Pliocene	0.16	0.0128	<0.0030	n.d.	n.d.	99.98
		153	Gray-green clay	Eocene	0.09	0.0114	<0.0030	n.d.	n.d.	99.99
10°53.55′ W	5,168	100	Gray silty clay	Pleistocene	0.87	0.0207	0.0263	0.0219	0.0049	99.93
44°02.57′ W		230	Olive-gray clay	Pleistocene	1.00	0.0238	0.0194	0.0332	0.0011	99.92
		478	Dark olive-gray silty claystone	Pleistocene	0.51	0.0568	0.0110	0.0096	0.0004	99.92
15°51.39′ N 56°52.76′ W	5,258	143	Light olive-gray clay	Miocene	0.27	0.0038	0.0009	0.0027	0.0001	99.99
		237	Stiff green-gray clay	Miocene	0.58	0.0119	0.0335	0.0191	0.0002	99.94
		249	Green-yellow mottled clay	Miocene	0.18	0.0151	0.0121	0.0018	0.0004	99.97

*n.d., not detected.

Source: Aizenshtat et al. (1973).

program of deep-sea drilling. The total organic carbon content of the sediments varies from 0.09 to 1.1 weight percent. The hydrocarbons, fatty acids, alcohols, and pigments represent the organic-extractable (benzene-soluble) fraction of the total organic matter. Note that for all but one of the samples this fraction makes up less than 0.01 percent of the total organic carbon (a concentration of less than 1 part per million for a total content of 1 percent carbon or 10,000 parts per million). The remainder of the carbon is difficult to study and belongs to such general groups as humic acids, fulvic acid, and kerogen. Aizenshtat et al. reached the following conclusions as a result of their study: (1) biogenic-derived matter in sediments rapidly transforms from original components to new, more stable organic compounds; (2) most of the carbon in unlithified sediment occurs in the form of humic acids and kerogen; and (3) the two major factors determining the amount and character of organic matter in older sediments are oxidizing conditions at the time of deposition and nature of the initial organic matter.

Organic processes that occur in marine sediments are generally different from those that take place in freshwater sediments. A good example of this is provided by Whiticar et al. (1986). They studied bacterial methane generation in anoxic sediments of both freshwater and marine environments. Biogenic methane is a product of microbially mediated reactions of organic molecules in anoxic sediments. This generally occurs by two different pathways: fermentation of acetate in freshwater sediments and reduction of carbon dioxide in marine sediments. Whiticar et al. used isotopic measurements of methane and coexisting carbon dioxide and water to study methane formation by these two pathways. They found that methane from freshwater and marine environments can clearly be distinguished using carbon and hydrogen isotopes and, further, that both can be distinguished from methane formed by thermal alteration of organic matter (Figure 6-11).

An interesting occurrence of methane in marine sediments is in the form of a gas hydrate, a crystalline solid that consists of a gas molecule surrounded by water molecules. Methane gas hydrates are abundant in marine sediments and also occur in permafrost areas. It has been estimated that the total carbon in gas hydrates is twice the amount of carbon in all fossil fuels on Earth. Thus this methane may become a future energy source. It also may play a role in the global greenhouse effect, since adding methane to the atmosphere produces global warming more effectively than adding carbon dioxide. We do not currently understand the natural processes and controls involved in the addition of methane to, or release from, the gas hydrate reservoir (Kennicutt II et al. 1993).

SEDIMENTARY ROCKS

The organic matter of all sedimentary rocks can be divided into the two types mentioned earlier: the small percentage that is extractable in organic solvents, and the much more abundant, unextractable material known as kerogen. Shales tend to have two or more times as much organic matter as associated sandstones and carbonate rocks. The organic compounds that occur in marine sediments (Table 6-8) are the same compounds usually found in sedimentary rocks of all ages. The major difference is the lower abundance of the various components in rocks. Table 6-9 gives concentration ranges for the organic matter of common sedimentary rocks. A sediment may contain several thousand parts per million of amino acids, whereas most rocks have less than 100 parts per million. Rocks tend to have very small amounts of original material, such as proteins,

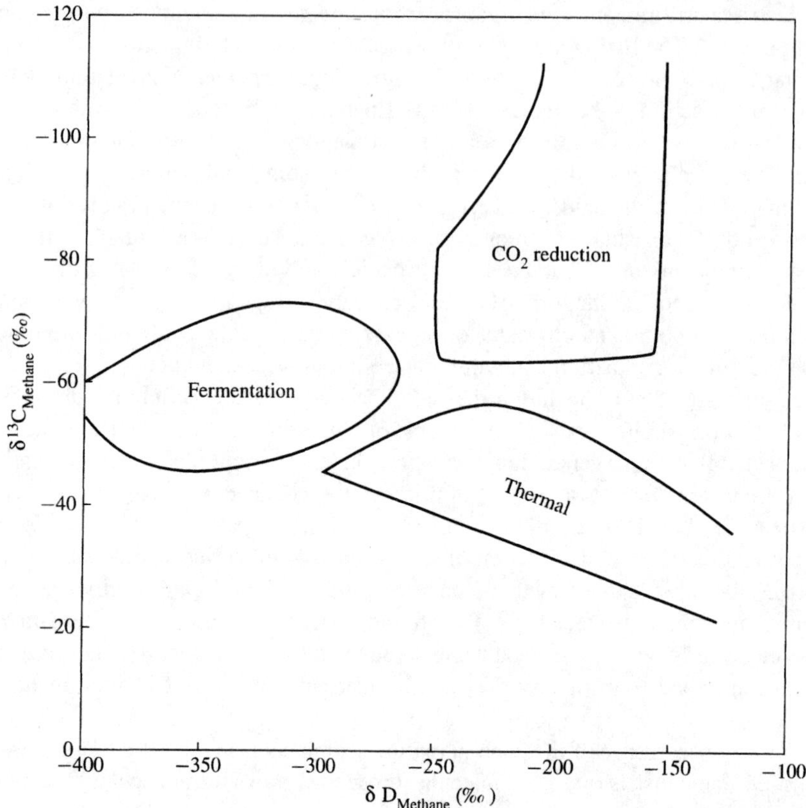

Figure 6-11 Natural gas genetic classification diagram using $\delta\,^{13}C$ and δ D of methane. The biogenic methane strongly depleted in deuterium through acetate-type fermentation can be distinguished from methanogenesis by CO_2 reduction and from thermogenic methane (responsible for much of the natural gas in oil-gas reservoirs). The fermentation process takes place in terrestrial freshwater sediments and the CO_2-reduction process takes place in marine sediments. (Reprinted from *Geochimica et Cosmochimica Acta,* v. 50, M. J. Whiticar, E. Faber, and M. Schoell, Biogenic methane formation in marine and freshwater environments: CO_2 vs. acetate fermentation—Isotope evidence, pp. 693–709. Copyright 1986, with kind permission from Elsevier Science Ltd., The Boulevard, Langford Lane, Kidlington OX5 1GB, UK.)

and proportionately larger amounts of their degradation products. Clearly, much of the organic matter of surface sediments is destroyed, or strongly altered, during diagenesis and lithification.

Gehman (1962) has provided extensive data on the abundances of organic matter in limestones and shales (Table 6-10). A mean organic content of 1.14 weight percent was found for shales from all over the world and of all ages from Cambrian to Tertiary. Limestones were found to have a mean organic content of 0.24 percent. Gehman compares these results with more restricted data for clay and lime sediments, and notes that limestones apparently lose a large part of their organic content during lithification. He suggests that limestones cannot hold and protect their organic matter, whereas shales retain this material as a result of adsorption on clay particles and low rock permeability. Gehman also points out that limestones may not lose their hydrocarbons

TABLE 6-9 Concentration Ranges of Organic Compounds in Common Sedimentary Rocks

Component	Range of concentration
Total organic matter	0–2 weight percent
Kerogen	0–2 weight percent
Hydrocarbons	0–4,000 ppm
Carbohydrates	0–700 ppm
Fatty acids	0–100 ppm
Amino acids	0–100 ppm
Pigments	0–5 ppm

during lithification, since the hydrocarbon contents of lime sediments and limestones are roughly equal, even though the total organic content of limestones is much lower (Table 6-10). Since hydrocarbons are relatively stable organic compounds, they may be enriched relative to other organic material that is partially or completely destroyed during lithification and the subsequent history of a rock. It is also possible that hydrocarbons have migrated into some limestones and thus do not represent original material.

Jones (1984) points out that the original grain size of rocks is directly related to organic content. Fine-grained carbonate rocks have significantly more organic matter than coarse-grained carbonates and thus they can serve as source rocks for petroleum. Most source rocks are believed to have formed in anoxic environments and to have original organic carbon contents greater than one weight percent. Hunt (1979) suggests that the quantity of organic matter converted to oil in source rocks ranges from a few percent to about 15 percent. The efficiency of hydrocarbon generation may be different for carbonates as compared to shales, since organic matter in shales is mainly wood material, while in carbonates it is mostly made up of amorphous algal remains.

Note that the values for organic matter in the sediments of Table 6-8 are given in terms of total organic carbon and thus are not directly comparable with the values of Table 6-10. The values for total organic matter in Table 6-10 were obtained by analyzing for total carbon and then

TABLE 6-10 Organic and Hydrocarbon Contents of Clay Sediments, Shales, Lime Sediments, and Limestones

	Mean total organic content (% of rock)	Mean hydrocarbon content (ppm of rock)	Mean hydrocarbon content (% of total organic content)
Clay sediments (74 samples)	1.0	60	0.6
Shales (1,066 samples)	1.14	96	0.9
Lime sediments (64 samples)	1.2	85	0.7
Limestones (346 samples)	0.24	98	4.1

Source: Data from Gehman (1962).

multiplying these values by the factor 1.22. This conversion factor is an arbitrary number, since the relationship between total organic carbon and total organic matter depends on the types and compositions of the organic materials being studied. Thus different investigators have used different numbers to convert total carbon to total organic matter. Because carbon content increases as organic matter matures, the numbers used have varied from 1.0 (very old rocks) to 1.8 (recent sediments).

An example of the regional and vertical variation of organic carbon in one sedimentary unit has been given by Schrayer and Zarrella (1963). They collected and analyzed samples from 24 stratigraphic sections of the Lower Cretaceous Mowry shale of Wyoming. The thickness of this marine shale varies from 150 to 450 feet over an area of 35,000 square miles. The average organic carbon content of the shale varies from 0.36 to 4.53 percent in one vertical section and shows significant variations in all sections. In general, the middle of the formation contains more organic carbon than the upper or lower parts. Regionally, organic carbon varies from 0.19 to 4.95 percent. A regional trend in the carbon content can be related to the depositional environment of the shale, with organic matter increasing in a seaward direction.

As indicated above, most sedimentary rocks have less than 2 weight percent organic matter. The major exceptions are the rocks known as *oil shales*. Most of these rocks are actually organic-rich marlstones with a significant carbonate component. When heated, the rocks yield hydrocarbons up to a maximum of about 100 gallons of oil per ton of rock. Substantial amounts of hydrocarbons form from kerogen of the shales at temperatures as low as 200°C. The types of hydrocarbons produced from oil shale are different from those that occur in petroleum. The sediments of oil shales accumulated under anaerobic conditions in either marine or nonmarine environments that had abundant organic life. In addition to kerogen, oil shales contain small amounts of most of the bitumens listed in Figure 6-3. The properties of the kerogen vary from rock to rock, and the types and abundances of the bitumens also vary.

The major elements making up kerogen are carbon (65–90 percent), hydrogen (5–12 percent), and oxygen (5–25 percent). Sulfur occurs organically bound in kerogen and also in associated pyrite. Hunt (1979) has classified kerogen into two types, *sapropelic* and *humic*. The sapropelic kerogen is amorphous and forms mostly from plankton. Humic kerogen is derived from higher plant remains (particularly lignin) from continental sources. The humic material is deposited in continental clastic sediments; the sapropelic material is most abundant in nonclastic sediments. There are chemical differences between the two types, with the marine-derived sapropelic kerogen having a hydrogen/carbon ratio of 1.6 compared to values of 1.0 or less for continental, humic kerogen. Furthermore, there are significant differences in the types of organic structures that make up sapropelic kerogen as compared to humic kerogen. These differences will be discussed further in the following sections on coal and oil.

COAL

Most of the organic matter of the Earth occurs in two forms: as kerogen in sedimentary rocks and as the organic rock coal. Some kerogen is similar to coal in organic structure and behavior; other kerogen is not (Figure 6-12). Coal is an extremely complex rock, which varies in its composition, structure, and properties. As with kerogen, coal can be classified into two general types, sapropelic and humic (Hunt 1979). Sapropelic coals (which are represented by coals known as cannel

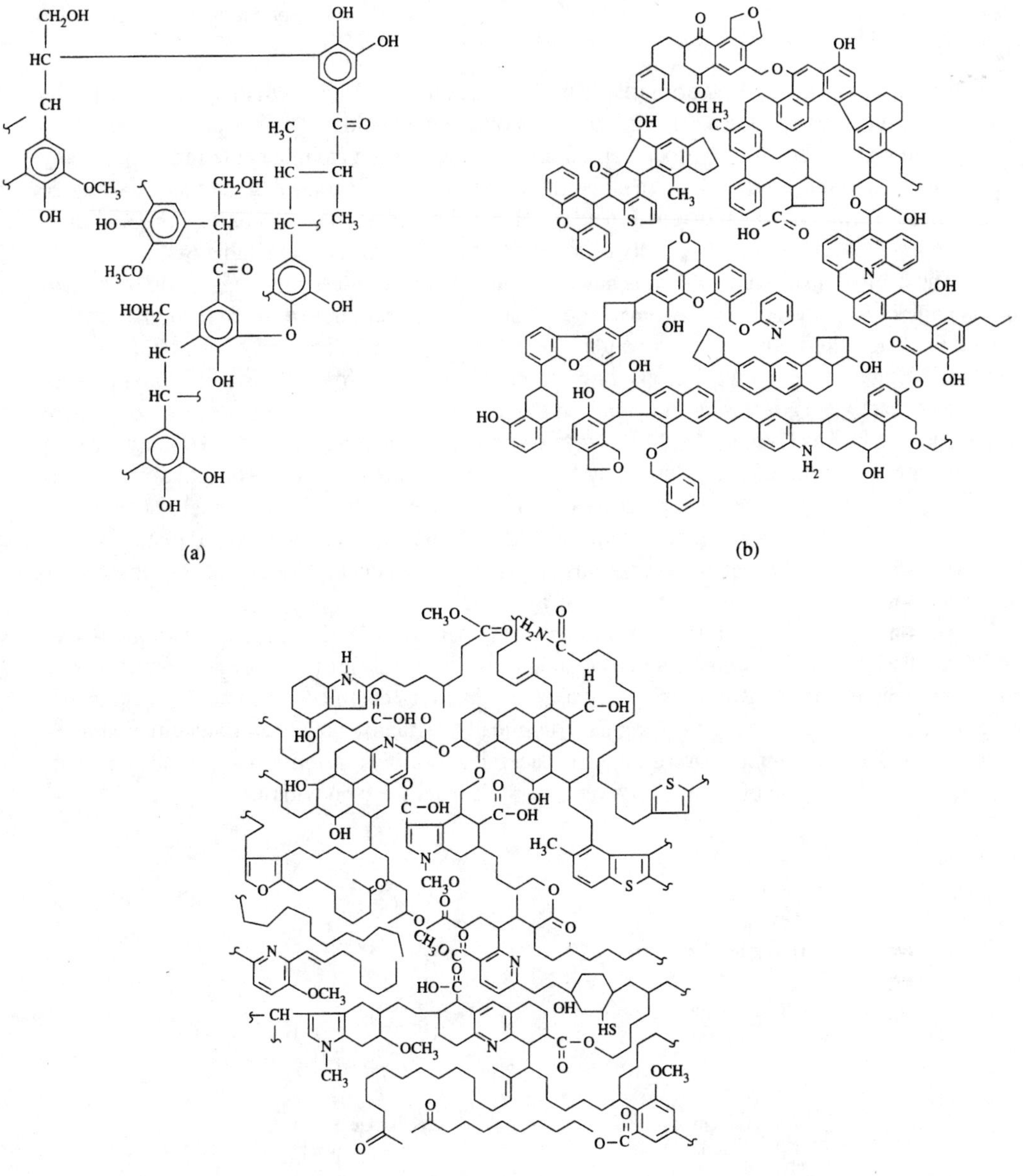

(a)

(b)

(c)

Figure 6-12 Structural models of coal and kerogen. (a) Lignite coal. (Reprinted from *Organic Geochemistry*, v. 16, P. G. Hatcher, Structural model of lignite coal, p. 959. Copyright 1990, with kind permission from Elsevier Science Ltd, The Boulevard, Langford Lane, Kidlington OX5 1GB, UK.) (b) Bituminous coal. (Reprinted from *Fuel*, v. 63, J. H. Shinn, Structural model of bituminous coal, p. 1187. Copyright 1984, with kind permission from Butterworth-Heinemann journals, Elsevier Science Ltd, The Boulevard, Langford Lane, Kidlington OX5 1GB, UK.) (c) Oil shale kerogen. (From C. G. Scouten et al. Preprints, *Am. Chem. Soc. Div. Pet. Chem.*, v. 34, p. 43, 1989.) (Reprinted with permission from M. Siskin and A. R. Katritzky, 1991, Reactivity of organic compounds in hot water: geochemical and technological implications, *Science*, v. 254, pp. 231–237. Copyright 1991 by the American Association for the Advancement of Science.)

and boghead coals) are derived primarily from spores, pollen, and algal remains; humic coals (the most common type) are derived from lignin and other compounds of higher plants. The differences mentioned above for sapropelic and humic kerogens, such as in the H/C ratio, also apply to sapropelic and humic coals. The most common way to classify coals is in terms of the degree of chemical change from the original material. Typical compositions for coals (lignite, bituminous, anthracite) and for peat (partially decayed plant matter) are given in Table 6-1.

The chemical structure of coal is not well known, but we can say that it is made up of *aromatic* and *aliphatic* compounds. Aromatic compounds have structures in which carbon atoms are linked in closed rings; aliphatic compounds have carbon atoms in the form of open chains (Figure 6-13). Some of these structures can have *functional groups* (groups that confer a characteristic type of reactivity) attached to those carbon atoms that are common to only one ring or that are at the end of a chain (Figure 6-13). An example of a functional group is —COOH, known as the carboxyl group. The functional groups play an important role in the reactivity of coal during metamorphism and during the various processes involved in the economic use of coal (such as carbonization, gasification, and liquefaction). Very little is known about the mode of occurrence of nitrogen and sulfur in the organic compounds of coal. They can occur in functional groups. Sulfur also occurs in inorganic forms in coal, mainly as pyrite.

The physical makeup of the organic portion of coal can be described in terms of *macerals,* which are the equivalent of minerals in inorganic rocks. Macerals are probably not homogeneous substances, but instead consist of irregular mixtures of various chemical compounds. Despite this inhomogeneity, macerals can be distinguished optically by their behavior in transmitted and reflected light. There are three major groups of macerals: *vitrinite, inertinite,* and *exinite.* The bulk of most coals is made up of vitrinite, which is derived from the woody tissues of plants. The in-

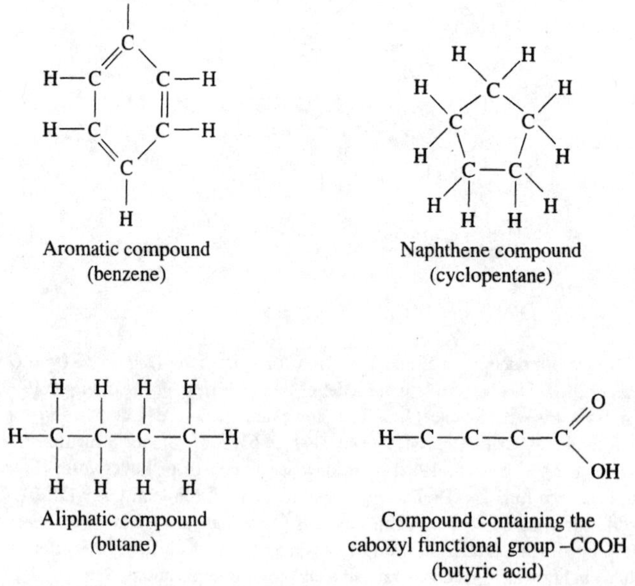

Aromatic compound
(benzene)

Naphthene compound
(cyclopentane)

Aliphatic compound
(butane)

Compound containing the
caboxyl functional group –COOH
(butyric acid)

Figure 6-13 Examples of organic compounds.

ertinite group includes the common macerals *micrinite* (origin uncertain) and *fusinite* (formed from charcoal). The various macerals of the exinite group form from cuticles (cutinite), resins (resinite), spores (sporinite), and algal bodies (alginite). The inertinite macerals have a high content of carbon, the exinite macerals are hydrogen rich, and vitrinites have carbon and hydrogen contents intermediate between the other two groups.

With this brief description of coal, we can now consider its origin. Coal results from the alteration of plant debris. The first step is the formation of peat. This stage of alteration is referred to as the *biochemical stage,* since microorganisms, such as fungi and bacteria, seem to play a major role. During the initial decay of plant material, the lignin and cellulose of plant tissues are broken down into smaller compounds. Peat contains, in addition to this decomposing organic matter, a wide variety of organic compounds (including amino acids) produced by microorganism activity. Metal-organic complexes form by concentrating trace elements from swamp water. All this material in peat is a mixture of aromatic and aliphatic compounds. With increased alteration, there is a steady decrease in the lignin-cellulose content and an increase in humic substances, which have a lower aliphatic fraction and a higher aromatic fraction. The overall properties and composition of a given peat depend on the environmental conditions that existed during initial accumulation of plant material and on the conditions that existed during the alteration period after burial. For instance, plants with a different biochemistry are found in freshwater environments as compared to marine environments, and they decay to different products. Whether or not oxygen is available during the decay period has an important influence on the type of compounds produced. For further details, see a group of papers dealing with past and present peat-forming environments (Cobb and Cecil 1993).

The second stage of coal formation is a *metamorphic stage* involving physical and chemical changes in peat brought about by the factors of time, increasing temperature, and increasing pressure. These changes are often referred to collectively as the *coalification process.* With increasing coalification, carbon content rises due to loss of water, oxygen (in the form of carbon dioxide), and hydrogen (in the form of methane). A useful way to show these changes is in the form of a plot of the atomic ratios H/C and O/C (Figure 6-14). The various components of peat and the various macerals of coal have different compositions and structures and thus react differently during coalification. For example, exinites contain a large proportion of aliphatic structures, whereas the structures of vitrinites and inertinites are mainly aromatic. The most abundant coal maceral group, vitrinite, forms from the major components of peat and thus has an initial composition similar to that of peat (Figure 6-14). Figure 6-14 shows that, with increasing coalification, vitrinite's composition changes along a path that parallels the change in composition from lignite to bituminous coal to anthracite. Other macerals follow different paths, with the paths coming together at the anthracite stage of coalification. The chemical components of the macerals become more aromatic as coalification increases. The coaly type of kerogen in sedimentary rocks (which is more aromatic than the noncoaly type) goes through similar chemical and structural changes.

PETROLEUM

The term *petroleum* is commonly used to refer to naturally occurring liquid hydrocarbons. To be more precise, the term should be used for all naturally occurring mixtures of hydrocarbon compounds, whether they are in the solid, liquid, or gaseous form. The solid forms include

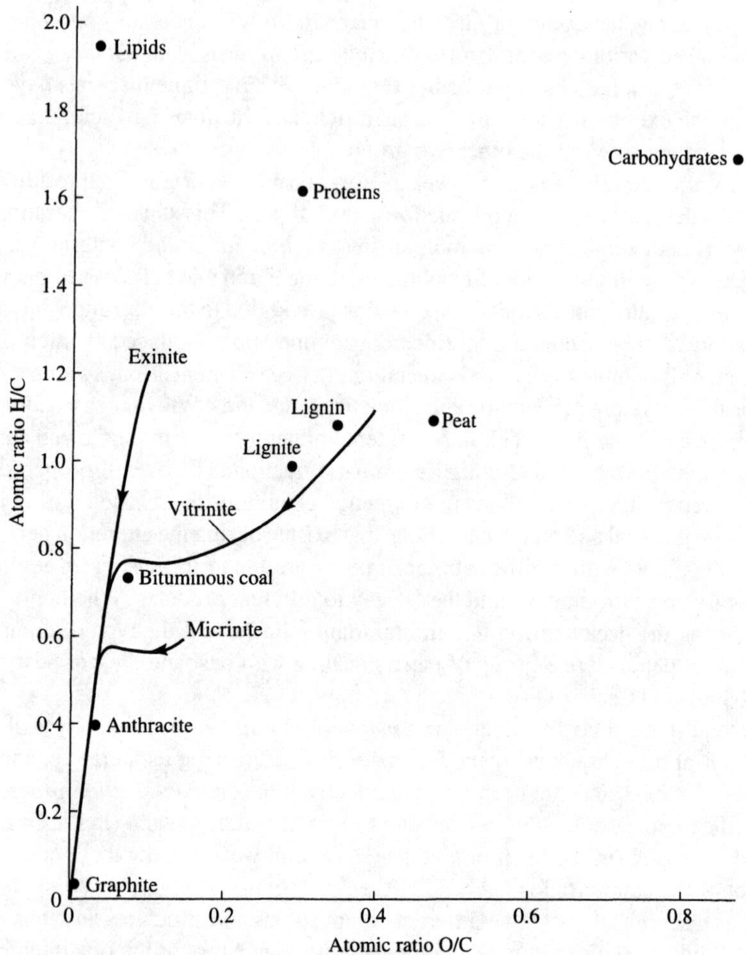

Figure 6-14 Composition of organic materials and coalification paths of macerals. After Degens (1965) and van Krevelen (1963).

asphalt and asphaltites (Figure 6-2); the gaseous material is known as natural gas. Crude oil is the proper term for liquid petroleum, and this section deals mainly with the composition and origin of crude oil.

The elemental composition of crude oil varies from about 79 to 89 percent carbon and from about 9 to 15 percent hydrogen. Small amounts of oxygen, nitrogen, and sulfur and trace amounts of metallic elements such as vanadium occur as part of various nonhydrocarbon compounds. Most of the hydrocarbons found in crude oil can be classified into three major groups. Two of these groups were mentioned earlier as components of coal. Aliphatic hydrocarbons (also referred to as paraffin or alkane hydrocarbons) have straight and branched chain structures in which all carbon valences are used in bonding to carbon and hydrogen atoms. The bond between each pair of atoms is a single covalent bond and there are no unshared electrons. Thus these hydrocarbons tend to be chemically inert, and they are referred to as *saturated* hydrocarbons. Butane is an example of this group of hydrocarbons (Figures 6-13 and 5-11).

The other group of hydrocarbons found in both coal and crude oil, aromatic hydrocarbons, contains compounds that are *unsaturated.* These compounds do not have enough hydrogen atoms to form single bonds with all the carbon atoms. Thus some of the carbon atoms have double bonds, and all aromatic hydrocarbons contain at least one six-carbon ring structure with alternate single and double bonds (see benzene in Figure 6-13). The remaining group of hydrocarbons found in crude oil is known as the *naphthenes* or cycloparaffins. They are similar to aliphatic hydrocarbons in being saturated, but have ring structures instead of open chains (see cyclopentane in Figure 6-13). There is a fourth group of hydrocarbons, the *olefins,* that are similar to aromatic hydrocarbons in being unsaturated, but they have straight and branching chains rather than ring structures. They are very rare in natural environments, even though they are the most abundant hydrocarbons in organisms. Very small quantities of olefins have been found in some crude oils. In addition to aliphatics, aromatics, and naphthenes, there are other fairly abundant hydrocarbons in crude oil that have mixed structures containing aromatic and naphthenic rings along with aliphatic chains. They are known as naphtheno-aromatics. The abundances of the various hydrocarbon groups in a typical crude oil are shown in Table 6-11. About 600 different hydrocarbon compounds have been identified in crude oils.

Organic compounds such as crude oil hydrocarbons are commonly characterized by their boiling points. Figure 6-15 is a plot of the boiling points of crude oil hydrocarbons in terms of the carbon numbers of the compounds (number of carbon atoms in each compound). Also shown are the boiling ranges of the various fractions that can be separated from crude oil. The lower boiling fractions of crude oils have been extensively studied; the higher boiling fractions are more complex and are not well understood. Table 6-11 lists the abundances of the major fractions in a typical crude oil.

TABLE 6-11 Composition of a Typical Crude Oil

Molecular size	Volume percent
Gasoline (C_5 to C_{10})	27
Kerosine (C_{11} to C_{13})	13
Diesel fuel (C_{14} to C_{18})	12
Heavy gas oil (C_{19} to C_{25})	10
Lubricating oil (C_{26} to C_{40})	20
Residuum ($>C_{40}$)	18
Total	100
Molecular type	**Weight percent**
Paraffins (aliphatics)	25
Naphthenes	50
Aromatics	17
Nonhydrocarbons	8
Total	100

Source: From PETROLEUM GEOCHEMISTRY AND GEOLOGY by Hunt. Copyright ©1979 by W. H. Freeman and Company. Used with permission.

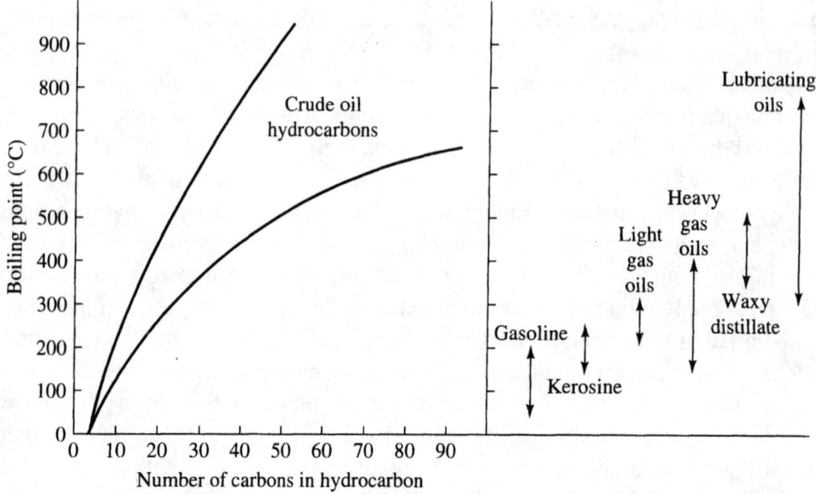

Figure 6-15 Boiling range of crude oil hydrocarbons. All known crude oil hydrocarbon compounds plot within the two lines. Approximate boiling ranges of the principal crude oil fractions are indicated on the right side of the figure. Gasoline, for example, consists of hydrocarbons with low carbon numbers and low boiling points. Natural gases associated with crude oils consist of hydrocarbons with very low carbon numbers and very low boiling points (such as methane, CH_4) along with nonhydrocarbons (such as N_2 and CO_2). Also see Table 6-11. Modified from Speers and Whitehead (1969).

The nonhydrocarbon compounds of crude oils usually make up 5 to 10 percent of the total liquid but may, in some heavy crude oils, be more abundant than the hydrocarbons. They are usually classified as sulfur compounds, oxygen compounds, nitrogen compounds, and metallic compounds. The sulfur content of crude oils can be as much as 10 percent by weight, whereas oxygen is usually less than 3 percent and nitrogen less than 1 percent. Sulfur compounds that have been found in crude oils include sulfides, disulfides, thiols, and thiophenes. Oxygen occurs mainly in organic acids and phenols; nitrogen occurs in a wide variety of compounds, many of which are not well known. The metal content of crude oils is commonly less than 100 parts per million. The metals of greatest abundance are usually vanadium and nickel. The most important metal compounds are the porphyrins. The occurrence of metalloporphyrins in crude oils provides support for the biological origin of petroleum, since these compounds probably formed from pigments found in organisms.

Porphyrins are an example of an important group of compounds found in crude oil known as biological markers, or *biomarkers* (Yen and Moldowan 1988; Waples and Machihara 1991). They are particular molecules formed by living organisms, or resulting from minor alteration of such molecules, and found in sediments and sedimentary rocks as well as in crude oil. Biomarkers can be thought of as geochemical fossils. They are useful in studying the origin and migration of petroleum and in source rock evaluation. Study of the types of alterations of precursor molecules provides information about pathways and processes involved in the formation of petroleum from sedimentary organic matter. An example of the combined use of biomarkers, isotope measurements, and geologic data to identify oil source rocks can be found in the article by Peters et al.

(1989). Biomarkers are also used in coal research. The organic geochemistry of biomarkers is reviewed by Philp and Lewis (1987).

As a result of a great deal of research, the origin and history of petroleum is now fairly well understood. There is general agreement that petroleum has a biogenic origin. Evidence for the origin of petroleum from organic matter includes the following:

1. It is rarely found in rocks that formed before life developed on Earth.
2. It contains compounds (metalloporphyrins) that can be related to pigments of living organisms.
3. It is enriched in ^{12}C relative to ^{13}C, as is all organic matter relative to inorganic matter.
4. Hydrocarbons in petroleum show optical activity (rotate the plane of polarized light) as do hydrocarbons and other compounds produced biologically.
5. The structures of many petroleum compounds are similar to those of lipids and other compounds found in living organisms and thus could be formed from them.

Any hypothesis put forward for the origin of petroleum must explain the following additional facts:

1. Petroleum is always found in or near sedimentary rocks that are dominantly, but not exclusively, marine in origin.
2. Although petroleum changes somewhat in composition with time, different samples generally contain the same hydrocarbon compounds (with variation in the relative amounts present).
3. Olefin hydrocarbons are the most abundant type of hydrocarbons in organisms and the least abundant in petroleum.
4. The relative abundances of the hydrocarbon groups are very different in rocks and in associated crude oils.
5. Cenozoic and older rocks contain higher concentrations of hydrocarbons than do Recent sediments.
6. Petroleum contains abundant low-molecular-weight hydrocarbons, whereas Recent sediments and living organisms have only a few of these hydrocarbons.
7. Paraffin (aliphatic) hydrocarbons of Recent sediments have carbon numbers that are predominantly odd, whereas the paraffins of petroleum show equal amounts of odd and even numbers.
8. Approximately 900 different compounds are known to occur in petroleum, and about one third of these are nonhydrocarbons.
9. Some compounds in petroleum are unstable above 300°C; thus much, if not all, of the history of petroleum occurred at low temperatures.
10. Low-molecular-weight hydrocarbons of petroleum are enriched in ^{12}C relative to petroleum as a whole.

We shall now briefly review three different hypotheses that have been suggested for the origin of petroleum. Some researchers have proposed that petroleum represents an accumulation of hydrocarbons that have been produced by living organisms. Although it is true that certain types of hydrocarbons occur in living organisms, these hydrocarbons are not abundant, and they are a different mixture of compounds than the mixture found in petroleum. For example, low-molecular-weight hydrocarbons are not found in organisms or Recent sediments, but they are abundant in petroleum. Thus, whereas some petroleum hydrocarbons may have originated directly from organisms, the history of petroleum formation must also involve generation of hydrocarbons from more abundant organic compounds such as lipids.

A second hypothesis proposes direct alteration of lipids and other complex organic compounds to the various hydrocarbons and nonhydrocarbons found in petroleum (Tissot and Welte 1984). This alteration occurs during diagenesis, and the final formation of petroleum occurs in the source rock. Philippi (1974) points out that a dominant part of all lipid molecules consists of hydrocarbon skeletons, and thus the chemical conversion of lipids into petroleum hydrocarbons is relatively simple. He gives a specific example of the possible conversion of fatty acids into paraffin hydrocarbons. The variation in composition of crude oils is explained by Philippi as a result of varying mixtures of land-derived material and marine organic debris. Terrestrial plant remains are dominantly made up of lignin and carbohydrates; marine remains, from plankton and other organisms, consist mostly of lipids and proteins. Thus petroleum derived mainly from marine organic matter (sapropelic kerogen) would have a different composition and a different mixture of hydrocarbon groups than would petroleum from organic matter with a significant amount of terrestrial plant material (humic kerogen).

Hunt (1977, 1979) describes a combination of the first two hypotheses (Table 6-12). He postulates three stages in the breakdown of organic matter: (1) diagenesis (biological and nonbiological alteration at temperatures below 60°C), (2) catagenesis (thermal alteration in the temperature range 60–200°C), and (3) metamorphism (final breakdown to methane and graphite due to heat and pressure at temperatures above 200°C). Hunt (1979) suggests that about 9 percent of the total hydrocarbons in sediments and sedimentary rocks form during diagenesis. This material is a mixture of hydrocarbons formed by the nonbiological breakdown of organic debris and hydrocarbons formed by living organisms. The other 91 percent of the total hydrocarbons forms by the thermal alteration of deeply buried organic matter (catagenesis stage). This consists of the alteration of kerogen formed during diagenesis to hydrocarbons with clay minerals generally acting as catalysts. An example of this process, which has been carried out experimentally, is the heating of clay organic complexes to 200°C, yielding a variety of hydrocarbons. Hunt (1979) concludes that the possible occurrence of intense oil generation in a given sediment or rock depends on temperature (generally controlled by depth), the abundance and composition of the kerogen (humic kerogens form more gas and less oil compared to sapropelic kerogens), and the nature of the mineral matrix (carbonate minerals are not as good catalysts as clay minerals). Time is also a factor in that at low temperatures long times are required to generate significant amounts of hydrocarbons (Connan 1974). Rocks in which oil has formed, but which later reach the metamorphic stage, have their oil destroyed.

It probably takes millions of years for petroleum formation to occur. Simoneit (1990) discusses "natural laboratories" for studying petroleum generation "instantaneously" in geological time. In recent years it has been found that hydrothermal conversion of immature organic debris to petroleum-like products occurs along oceanic rift systems with active venting of hydrothermal water. This process has been found occurring in the Gulf of California (Guaymas Basin), the East Pacific Rise, the Mid-Atlantic Ridge, the Red Sea, and other marine rift areas. In these locations hydrothermal pyrolysis results in "instantaneous" diagenesis and catagenesis.* Study of these

*Laboratory pyrolysis also produces crude oil. Putting small quantities of rock containing organic material into high-pressure vessels and cooking at high pressures (3,000 pounds per square inch) and high temperatures (400°C) can produce crude oil in about 72 hours.

TABLE 6-12 Maturation of Organic Matter

Depth (m)	T (°C)	Organic stage	Genesis of petroleum	Vitrinite reflectance	Kerogen color	Kerogen (wt. %C)	Coal rank
		Diagenesis		0.2		65	
1,000	47				Yellow	70	Lignite
				0.4			
2,000	75					75	Subbituminous
			Oil	0.6	Orange	80	High-volume bituminous
3,000	102						
		Catagenesis	Wet	1.0		85	Medium-volume bituminous
4,000	129		gas		Light brown		
			Dry	1.35		90	Low-volume bituminous
5,000	157		gas				
				2.0		92	Semianthracite
6,000	184				Dark brown,		
				3.0	black	94	Anthracite
7,000	211						
		Metamorphism		4.0			
8,000	238				Black		Metaanthracite
						96	
9,000	266						

Note: Genesis data are for an Eocene mixed-kerogen, fine-grained sedimentary rock. A temperature gradient of 2.7°C/100 m is assumed. The minor amount of oil and dry gas (CH_4) that forms at the surface is not shown. Vitrinite reflectance refers to the percentage of incident light reflected by a polished surface of the maceral vitrinite. Vitrinite particles commonly occur in sedimentary rocks. The occurrence of petroleum can be related to the degree of metamorphism of associated coals. The threshold of intense oil generation is approximately equivalent to the subbituminous coal rank. Measurement of vitrinite reflectance and of the maturation of kerogen (color, percent carbon) in sedimentary rocks can be used as a means of identifying possible petroleum source rocks.

Source: After PETROLEUM GEOCHEMISTRY AND GEOLOGY by Hunt. Copyright ©1979 by W. H. Freeman and Company. Used with permission.

"natural laboratories" will lead to a better understanding of petroleum formation and migration. The maturation, migration, and biogeochemistry of organic matter in hydrothermal systems are reviewed in a group of articles in the journal *Applied Geochemistry* (v. 5, no. 1/2, Jan./March, 1990).

A number of techniques are now used to identify petroleum source rocks (rocks that have a certain minimum amount of kerogen and that have reached the catagenesis state). Biomarker use was mentioned earlier. Other techniques are shown in Table 6-12. An unsolved problem involves understanding the mode of migration of crude oil from source rocks to reservoirs. It may travel in globules as a separate oil phase. Small amounts of organic material could also travel in solution or as a colloidal suspension in interstitial waters, with accumulation due to the different physical conditions of a reservoir rock. Methane in the presence of water can serve as a gas-phase carrier for oil.

We know that changes in the composition occur during migration but details of the changes are not well understood. Price and Clayton (1992) demonstrate that extractable hydrocarbons are not homogeneously distributed throughout source rocks and that primary migration processes can cause as much of a change in maturity indices as does increasing maturation rank itself. They suggest that maturity indices and source-rock indicators may not be as useful as previously thought.

Hunt (1979) gives a thorough review of the origin of oil and natural gas and of the nature of petroleum source rocks. Source and migration processes and geochemical evaluation techniques are reviewed in Merrill (1991). Emery and Robinson (1993) discuss in detail the most important inorganic geochemical methods used in petroleum geology. A history of the development of ideas on petroleum formation and migration is contained in Dott and Reynolds (1969).

ORIGIN OF LIFE

To conclude this chapter we shall review research on the origin and early evolution of life on our planet (Mason 1991). There are three major approaches used to study this topic. One approach combines chemical principles and laboratory experiments to explain the development of organic molecules and processes. This study of chemical evolution is known as *paleobiochemistry.* Another approach involves studying the organic remains of sedimentary rocks. These remains consist of various organic compounds, some of which may be little changed from their original composition and structure. Unaltered or slightly altered organic molecules serve as "chemical fossils," which give information on the life forms that produced them. Work on this material is part of the field of organic geochemistry. The third approach consists of identifying and studying Precambrian fossils. In this case the fossils, and the rocks in which they occur, are used to determine the evolution history of early life in terms of morphology and ecology. This field of study is known as *Precambrian paleontology* or *paleobiology.* Since about 1950, important advances have occurred in all three areas, and the study of the origin of life has become, after a long period of neglect, an active and productive field. We shall briefly review current knowledge and speculation in the three areas described above.

Calvin (1975) has summarized the time sequence of organic evolution, discussing in particular the chemical evolution that took place between original formation of the elements and the later biological evolution of living organisms (Figure 6-16). The chemical evolution of organic (carbon-containing) matter depended very much on the Earth's early environment. Atmospheric composition was particularly important. The composition of the primitive atmosphere may have been strongly reducing and CH_4-rich or it may have been mildly reducing and CO_2-rich. The

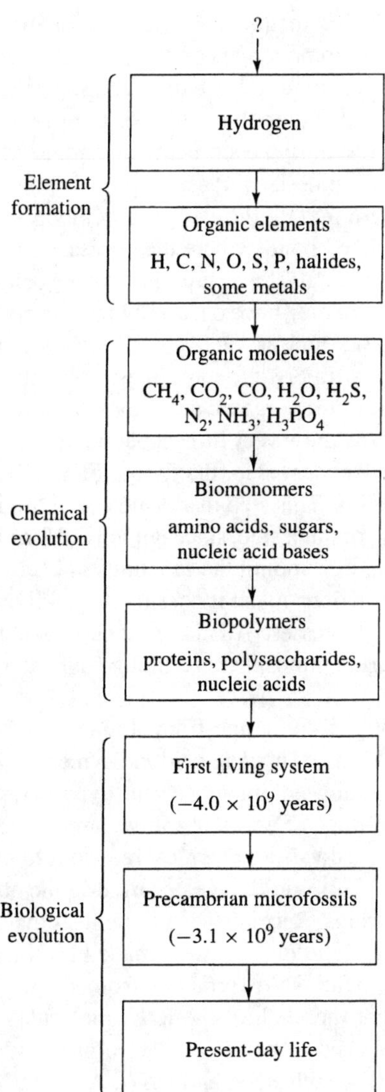

Figure 6-16 Time sequence of evolution from the formation of the elements to the present. [After Calvin (1975). Reprinted by permission, *American Scientist,* Journal of Sigma Xi, The Scientific Research Society of North America.]

original composition was determined by the source of the Earth's atmospheric gases. The volatiles could have been a result of the Earth's formation as a product of solar system evolution, mainly a result of later volcanic degassing of the Earth's interior, or some combination of the two (Chang et al. 1983).

Organic chemical evolution began when molecules like CO_2 and CH_4 were changed into simple organic compounds (biomonomers) such as amino acids. This conversion has been carried out by chemists in a number of different ways since amino acids were first produced in 1953 by exposing methane, ammonia, and hydrogen to a high-energy spark. Other mixtures and other energy sources (such as ultraviolet light and radioactivity) have been successfully used to form most of the basic building blocks of living matter. Thus it seems likely that special conditions were not necessary for the formation of biomonomers on the primitive Earth.

Special conditions may have been required for the next step, the combination of biomonomers into the structurally complicated biopolymers, such as proteins. Only relatively simple biopolymers have been formed in laboratory experiments, and none of the extremely complex polymers of living organisms has been synthesized. It seems probable that fairly special (but not necessarily unusual) conditions were required for the evolution of biopolymers. For instance, this evolution may have taken place in isolated ponds where the necessary biomonomers were concentrated by evaporation and a chemical catalyst was present to make certain reactions occur efficiently. On the other hand, this evolution may have occurred in the oceans, where the clay minerals could have served as concentrators and as catalysts. We know that clay minerals have chemically active surfaces and interact with organic molecules. Laboratory research has shown that clay minerals can bring together different organic molecules and can stabilize amino acids. All this, however, is pure speculation. We know very little about the formation of biopolymers on the Earth by nonbiological processes. The next step, the formation of a living thing, is also not understood in terms of chemical processes. This step marks the actual origin of life and was followed by biological evolution. It was a complex step, since it involved forming not only the structures of huge macromolecules, but also developing the functions of catalysis, information storage, and information transfer. These functions make it possible for all living organisms to carry out their metabolic and reproductive processes. All these steps undoubtedly occurred under anoxic conditions, with an oxygen-bearing atmosphere resulting from later biological photosynthesis.

Information on the beginning of life can come from studies of life's genetic processes. Genetic information is stored on two kinds of nucleic acid molecules, DNA (deoxyribonucleic acid) and RNA (ribonucleic acid). As indicated above, current hypotheses on the origin of life generally assume that the composition of the Earth's primitive atmosphere, along with the resulting surface-water environment, allowed various chemical reactions to produce biomonomers such as amino acids and nucleic acids. This could have occurred in the atmosphere, in surface lakes or ponds, or in deep-ocean hot springs. Recent research suggests that the first molecules to store information and catalyze their own reproduction were made of RNA (Oró et al. 1990).* A possible scenario would be as follows. After RNA formation from simpler organic compounds, RNA molecules synthesize proteins that then help some RNA molecules to evolve into DNA (DNA is the molecule that cells today need to build complex organic molecules). Simple life forms then evolve by processes involving both RNA and DNA. A major step was the development of membrane-bound polymolecular systems (Oró et al. 1990).

One problem in studying the organic matter of Precambrian rocks is determining whether a given group of compounds is biogenic or abiogenic. In other words, did the molecules form by biological processes, thus being potentially useful as chemical fossils, or did they form by nonbiological processes? Several criteria can be used to identify biological substances. One involves determining the relative abundances of the different organic molecules. Organisms contain characteristic patterns of various carbon molecules. It is unlikely that nonbiological synthesis would produce the same patterns, since there are millions of possible patterns. Another criterion involves

*One definition of "life" is that it is characterized by two processes: replication, the capacity to self-copy, and mutation, the ability to change and evolve.

measuring the optical activity of the material. Biological carbon compounds, because of their structure, rotate the plane of polarized light of a polarizing microscope and are said to be *optically active.* Nonbiological carbon compounds do not have this property. A third test for biological origin involves measurement of the isotope ratio $^{13}C/^{12}C$. For a given sample, all the various fractions of organic matter should have approximately the same ratio if they all came from the same biologic source and had the same diagenetic history. Measurements of the isotopic composition of the kerogen of some Precambrian rocks has shown that its carbon ratio is similar to that of younger, biologic organic matter, and different (enriched in ^{12}C) from that of coexisting and younger, inorganic carbonate carbon. This suggests a biologic origin for the kerogen. The isotopic evidence indicates that biological activity began on the Earth at least 3.5 billion years ago (Schidlowski et al. 1983).

Another problem with Precambrian organic matter is determining whether or not it is indigenous to the rock in which it is found. Movement of organic compounds such as hydrocarbons can and does occur during geologic time. Varying ratios of $^{13}C/^{12}C$ for different fractions of a sample would suggest possible seepage of some material into the rock (i.e., natural contamination). A final problem, because of the small amounts of organic matter in most Precambrian rocks (up to about 0.40 percent by weight), is contamination during sampling and laboratory analysis.

The organic matter of Precambrian rocks can be divided into two kinds: (1) that which is extractable in organic solvents (amino acids, hydrocarbons, pigments, etc.), and (2) the nonextractable material known as kerogen. Kerogen is by far the more abundant of the two (often greater than 99 percent of the total) and tells us little about its previous history. The extractable material can, ideally, give us information about the organisms that produced it. This material is also most likely to result from geologic or human contamination.

The first use of chemical fossils occurred in 1934 when it was shown that shales and crude oils contained organic molecules that were probably derivatives of the pigments chlorophyll and heme. This not only suggested that petroleum is of biological origin, but also put temperature and chemical limits on the types of environments in which petroleum could form and exist. It turns out that the hydrocarbons of sedimentary rocks are particularly useful as chemical fossils. They have formed by breakdown of the lipid and pigment portions of organisms. The types of hydrocarbons found in a given rock can sometimes indicate whether plant or animal debris was the major source of the organic matter. In the case of Precambrian rocks, the types of hydrocarbons in the oldest rocks can perhaps tell us approximately when the first living organisms developed and when photosynthetic life evolved. It may be that the remains of the biopolymers of chemical evolution are distinctively different from the remains of the macromolecules of early biological evolution. This type of evidence may also outline some of the details of early biological evolution. Readers interested in more information on the use of chemical fossils are referred to Hayes et al. (1983).

A major problem in working out the details of early biological evolution involves the relationship between the two major types of living systems, the *prokaryotes* and *eukaryotes.** Prokaryotes (bacteria) divide by fission and their cells do not have a nucleus. Members of the plant, animal, and other kingdoms are eukaryotes, with nucleated cells that divide by *mitosis*

*Many biologists use a five kingdom arrangement with two superkingdoms (Prokaryotae and Eukaryotae). The Prokaryotae consists solely of prokaryotes and is made up of one kingdom for bacteria (Monera). The Eukaryotae superkingdom contains four kingdoms (Protoctista, Fungi, Animalia, and Plantae). For further information see Margulis and Schwartz (1982).

(production of two new cells with the same number and kinds of chromosomes as in the original parent cell). Both groups existed in Precambrian time, but it seems likely that eukaryotes developed from a prokaryotic ancestor. Eukaryotes, with a few exceptions, require free oxygen to exist and thus probably did not develop until after the oxygen content of the atmosphere was built up by other, photosynthetic organisms. This probably occurred about 2,000 million years ago when deposition of banded iron formations stopped, and terrestrial red beds containing ferric oxides began to form. Fossil evidence indicates that eukaryotes were definitely present at least 1700 million years ago. Further discussion of eukaryote evolution can be found in Knoll (1992).

We now pass on to the most direct evidence of the origin and early history of life on Earth, the Precambrian fossil record. An excellent summary of the Precambrian record is given by Schopf (1983). Most Precambrian fossil occurrences belong to one of two types, *microfossil assemblages* and *stromatolites* (Table 6-13). Assemblages of microfossils, representing bacteria and other prim-

TABLE 6-13 Precambrian Fossils 2,000 Million or More Years in Age

Type	Approximate age (m.y.)	Geologic unit	Locality
Microfossil assemblage	2,000	Kasegalik Formation	Southern Hudson Bay, Canada
Microfossil assemblage	2,000	Pokegama Quartzite	Northeastern Minnesota
Stromatolite	2,000	Wyloo Group	West-Central Western Australia
Stromatolite	2,050	Hamersley Group	West-Central Western Australia
Stromatolite	2,100	Transvaal Dolomite	Eastern Transvaal, South Africa
Stromatolite	2,250	Olifants River Group	Northern Cape Africa Province, South Africa
Microfossil assemblage	2,250	Transvaal Dolomite	Eastern Transvaal, South Africa
Stromatolite	2,275	Wolkberg Group	Eastern Transvaal, South Africa
Stromatolite	2,600	Ventersdorp Group	Orange Free State, South Africa
Stromatolite	2,600	Steeprock Group	Steeprock Lake, Ontario
Stromatolite	2,600	Bulawayan Group	Zimbabwe
Stromatolite	2,750	Fortescue Group	West-Central Western Australia
Stromatolite	3,000	Insuzi Group	South Africa
Fossil-like microstructures	3,500	Swaziland Supergroup	Eastern Transvaal, South Africa
Stromatolite and microfossil assemblage	3,500	Warrawoona Group	Western Australia

Sources: After Schopf (1975); Walter (1983); and Schopf and Walter (1983).

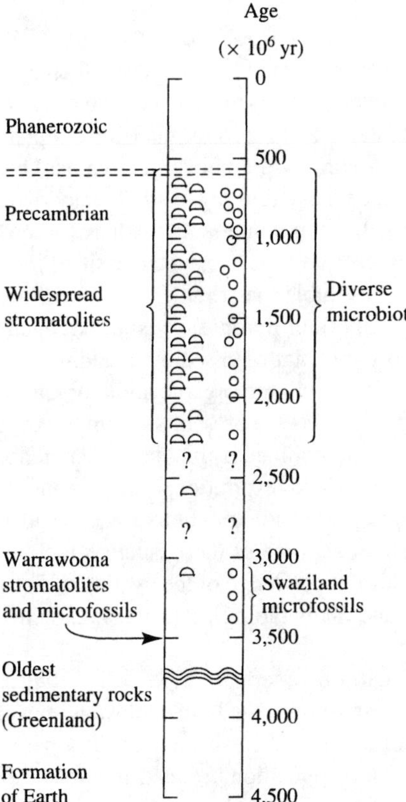

Figure 6-17 Geologic column summarizing the temporal distribution of stromatolites, microbiotas, and fossil-like microstructures now known from sediments of Archean (3760 < 2500 m.y.) and Proterozoic (2500 < 600 m.y.) age. After Schopf (1975); Walter (1983); and Schopf and Walter (1983).

itive life forms, have been found in cherts of North America, Africa, and Australia. These are often associated with stromatolitic carbonates, which may be partially silicified. Stromatolites (layered domal deposits of calcium carbonate), although they do not represent actual organic remains as do the chert microfossils, are definite evidence of the existence of organic life. Precambrian stromatolites are generally believed to have been formed by some type of cyanobacteria (also known as blue-green algae), since they are produced today by these primitive organisms.

The oldest known fossils have an age of about 3,500 million years (Table 6-13). Microfossil-like objects composed of organic matter have been found in the Swaziland Supergroup of South Africa, which has an age range from 3,100 to 3,500 million years. Similar objects and stromatolites occur in western Australia and are around 3,500 million years in age. The oldest known sedimentary rocks on the Earth have an age of about 3,800 million years.* Thus the origin of life probably occurred less than 1 billion years after the formation of the Earth. New discoveries of Precambrian fossils and more details on the origin and early evolution of life will undoubtedly result from the current interest in the beginnings of Earth history. Figure 6-17 summarizes the present Precambrian fossil record.

*The oldest known material consists of zircon crystals from Australia with an age (determined by radioactive dating) of 4,200 million years.

SUMMARY

Carbon is the characteristic element of organic compounds, which also contain large amounts of oxygen and hydrogen. Organic geochemistry is the study of naturally occurring carbonaceous material. The organic matter of sediments and rocks originates from plants and animals, which are composed of proteins, lipids, carbohydrates, pigments, and lignins. These substances are complex compounds, which, after the death of the parent organism, break down to a wide variety of other complex compounds, as well as to simpler compounds such as carbon dioxide and methane.

The organic matter of sediments and rocks can be classified into bitumens (soluble in carbon disulfide) and pyrobitumens (insoluble in carbon disulfide). Crude oil is liquid bitumen; asphalt is sold bitumen. The most important pyrobitumens are kerogen and coal, which together make up about 95 percent of all organic matter in sediment and rocks.

Carbon is a major component of both the organic and inorganic cycles of the Earth. However, the amount of carbon that circulates at the present time among the atmosphere, hydrosphere, biosphere, and lithosphere is mainly determined by biochemical reactions. The major movements of carbon are a result of photosynthesis and respiration. A large amount of carbon has been stored in sedimentary rocks over geologic time. Biological processes are also important in controlling the circulation of many trace elements. Some of these, such as mercury, are concentrated in organisms and can affect human health. Knowledge of the cycling of nutrients such as nitrogen and phosphorus is required to understand many biological processes and in dealing with the pollution of water bodies.

Organic material occurs in natural waters as suspended particles, as colloidal material, as adsorbed compounds, and in the form of dissolved molecules. In seawater dissolved organic carbon averages about 1 part per million and in freshwater about 4 parts per million. The organic matter of sediments includes both decomposition products and material contributed by sediment organic life. It varies in abundance from practically none to several weight percent. The nature of the source material and the environment of marine sediments are significantly different from those of freshwater sediments, resulting in organic debris of different characteristics.

Precambrian and younger sedimentary rocks contain the same types of organic matter as are found in present-day sediments. However, the rocks have smaller amounts, particularly of original material (such as proteins and carbohydrates), than the sediments. Most of the organic matter is kerogen, which occurs in amounts of up to 2 weight percent in shales. Other organic compounds occur in the parts per million range. Some of these compounds, such as hydrocarbons, are more abundant in rocks than in present-day sediments. Oil shales are kerogen-rich marlstones, which, when heated, yield up to 100 gallons of oil per ton of rock.

Coal forms from accumulated plant debris and is made up of 65 percent or more carbon, with significant amounts of hydrogen and oxygen. The organic compounds occurring in coal are not well known. Physically, coal consists of macerals, which have distinctive optical properties. Coal formation starts with peat developing as a result of microorganism alteration of the lignin and cellulose of plant tissues. This biochemical stage is followed by a metamorphic stage in which temperature, and other factors, cause further alteration to lignite, bituminous coal, and anthracite. With increasing coalification, aromatic organic compounds become more abundant.

Petroleum (natural gas, crude oil, and solid hydrocarbons) also forms from organic precursors. Crude oil is chiefly made up of several hundred different hydrocarbon compounds and contains 79 percent or more carbon. The three major groups of hydrocarbons are aromatics (unsaturated ring structures), aliphatics (saturated chain structures), and naphthenes (saturated ring structures). Nonhydrocarbon compounds containing oxygen, sulfur, and nitrogen, in addition to carbon and hydrogen, are also found in crude oil. The mode of origin of petroleum is now understood. It forms during sediment diagenesis and particularly during the catagenesis stage as a result of thermal alteration of kerogen. Factors determining whether or not petroleum forms, how much, and what type, are (1) temperature, (2) abundance and type of kerogen, (3) nature of the mineral matrix, and (4) time.

Evidence contributing to our understanding of the origin of life on the Earth comes from the study of possible chemical processes, of organic material in the Precambrian rocks, and of Precambrian fossils. Following original formation of the elements, molecules such as methane formed; these were later combined into simple organic compounds such as amino acids. The next step, formation of complex compounds such as proteins, has not been accomplished in laboratory research. The first molecules to catalyze their own reproduction may have been RNA molecules, some of which later could have evolved into DNA molecules. Eventually, the first organisms formed by some completely unknown process. These changes all occurred inorganically and were followed by biological evolution. Most of the organic matter in Precambrian rocks is kerogen, which is a complex and little understood material; thus we know little about its origin and history. The oldest Precambrian fossils are remains of primitive life forms such as algae, found mainly in cherts, and stromatolites, found in carbonate rocks and probably formed by algae or bacteria. The oldest remains that are generally agreed to represent organic life have been dated as 3,500 million years in age.

QUESTIONS

1. A carbon atom has four electrons in its outer shell with which to form covalent bonds and is said to have a valence of four. Similarly, hydrogen has a valence of one. By knowing the valence of these two elements and by knowing the carbon/hydrogen ratio of a compound, it is possible to predict the structure of an organic compound and the number of single and double carbon bonds in the compound.
   (a) What is the maximum number of double bonds that could occur in the compound C_5H_{10}?
   (b) What is the minimum number of double bonds that could occur in the compound C_5H_{10}?
   (c) What are the valences of oxygen and nitrogen in the protein structure of Figure 6-1?
2. Certain bacteria serve as catalysts in the decomposition of organic compounds in order to obtain energy for their metabolic needs. One of the products of bacterial decomposition of organic matter in sediments is methane. The methane can in turn be destroyed by sulfate-reducing bacteria in a reaction that results in H_2S and CO_2. Write a balanced equation for the reaction of CH_4 and SO_4^{2-} to produce H_2S and CO_2.
3. **(a)** According to Figure 6-4, is the carbon content of the atmosphere increasing or decreasing at the present time?
   (b) Assuming that the atmosphere has never had a large concentration of CO_2, what is a likely primary source for the immense amount of carbon that now occurs in sedimentary rocks?
4. The major carbon compounds involved in the carbon cycle (Figure 6-4) are CO_2, CH_4, CO_3^{2-}, HCO_3^-, and various organic compounds. Name three inorganic, non-mineral sulfur compounds that you would expect to play important roles in the sulfur cycle.

5. Use Table 6-5 to answer the following questions.
 (a) Which element has the largest percentage of its source material stored as biomass accumulation?
 (b) Which element has the largest percentage of its source material lost in streamflow?
 (c) Write an equation for the annual budget for an element in this ecosystem.
 (d) If the overall change in water composition in going from rainwater to stream water is due solely to rock weathering, what condition has to hold for the forest biomass?

6. You have the following data for a salt marsh. The isotopic composition of Spartina detritus is $\delta^{13}C = -15.5$ and $\delta^{15}N = +10$, and the isotopic composition of associated phytoplankton is $\delta^{13}C = -23$ and $\delta^{15}N = +2$ (from Figure 6-10). If an analysis of associated suspended particulate matter gives $\delta^{13}C = -18.6$ and $\delta^{15}N = +7.4$, and assuming this matter is composed of a mixture of Spartina detritus and phytoplankton, what percent of the particulate matter is made up of Spartina detritus and what percent is made up of phytoplankton?

7. **(a)** If the average concentration of mercury in rainwater is 100 ppb and the average concentration of mercury in soil organic matter (humus) is 5 ppm, what is the biomagnification (concentration) factor for soil organic matter?
 (b) Given is the following data:

average stream concentration of dissolved mercury: 7×10^{-5} ppm;
average stream concentration of particulate mercury: 8×10^{-5} ppm.

If annual stream flow is 10^{20} grams/year, how much mercury is carried to the oceans each year by streams?
 (c) If the cycle of mercury on the Earth's surface is considered to consist of four reservoirs (atmosphere, lithosphere, ocean, and sediments), which reservoir would you expect to have the shortest residence time for mercury? Which one would you expect to have the longest residence time for mercury?

8. **(a)** Is dissolved and colloidal organic material generally more abundant in seawater or in freshwater?
 (b) Are the breakdown products of lignin more abundant in marine sediments or nonmarine sediments?
 (c) Which is generally more abundant in sediments and sedimentary rocks, hydrocarbons or fatty acids?

9. **(a)** In aerobic (oxygen-bearing) waters and sediments, carbon dioxide is produced by the addition of free oxygen to the carbon in organic matter. This process can be represented by the reaction of carbohydrate ($C_6H_{12}O_6$) with oxygen to produce carbon dioxide and water. Write a balanced equation for this reaction.
 (b) Use the data in Table 3-2 to calculate ΔG^0 for this reaction. Assume the standard free energy of formation of the carbohydrate is –217,718 calories per mole.
 (c) In anaerobic (oxygen-free) waters and sediments, carbon dioxide is produced by fermentation reactions involving oxygen and carbon in organic matter. Based on your knowledge of the occurrence of organic matter in sediments, would you expect the free-energy change for such a fermentation reaction to be greater or less than that in part (b)?
 (d) In addition to carbon dioxide, fermentation reactions in anaerobic sediments produce methane. Write a balanced equation for the change of carbohydrate to carbon dioxide and methane.

10. How could ^{14}C measurements of dissolved organic carbon in surface waters be used to estimate the relative amounts of carbon due to (1) petrochemical industrial wastes, and (2) sewage wastes and naturally occurring animal and plant matter?

11. The occurrence of organic matter in sediments and sedimentary rocks produces a reducing environment. Suggest a mineral that is a possible product of reduction in each of the following cases:
 (a) Groundwater containing sulfate ion passing through a bituminous sandstone.
 (b) Swamp water with a high content of cuprous ion.

(c) Groundwater containing dissolved uranium passing a buried log.
(d) Anhydrite in a salt dome that has associated crude oil.

12. Would you expect the conversion factor used to calculate total organic matter from total carbon content (obtained by chemical analysis) to be higher or lower for metamorphic rocks as compared to sedimentary rocks? Explain your answer.

13. **(a)** What is the mean total carbon content for the clay sediments listed in Table 6-10?
(b) Calculate the mean hydrocarbon content of the clay sediments as percent of total organic carbon.
(c) Is the value found in part (b) higher or lower than the values reported for hydrocarbon content in the ocean sediment of Table 6-8?

14. To which hydrocarbon group (aliphatics, aromatics, naphthenes, and olefins) does each of the following compounds belong?

(a)
```
        H   H
         \ /
          C
   H    /   \    H
    \ C       C /
   H /|       |\ H
   H \|       |/ H
      C       C
   H /  \   /  \ H
          C
         / \
        H   H
```

(b)
```
          H
          |
          C
        /   \\
   H—C         C—OH
      ||        |
   H—C         C—H
        \    //
          C
          |
          H
```

(c)
```
      H   H
      |   |
   H—C—C—H
      |   |
      H   H
```

(d)
```
        H   H
   H    |   |    H
    \   |   |   /
     C=C—C=C
    /           \
   H             H
```

(e)
```
           H H H
            \|/
             C
   H         |         H
   H—C—C—C—H
   H         |         H
             C
            /|\
           H H H
```

15. Carbon isotope studies have been made on coals of various ages and degrees of coalification. The results show that there is no correlation between isotopic composition, degree of coalification, and age. There is very little difference between the $\delta^{13}C$ of land plants, peats, and coals. How would you interpret these results?

16. At what stage of the coalification process does the greatest loss of hydrogen occur?

17. An environmental problem associated with the use of coal as an energy source is air pollution. Given bituminous coal containing 80 percent carbon and 1 percent sulfur. What volumes of CO_2 and SO_2 will be formed from 100 grams of the coal under conditions of 1 atmosphere and 25°C?

18. Given the following four types of sedimentary basins. Which of them would be likely places to prospect for oil?
(a) Young, low-temperature basins.
(b) Young, high-temperature basins.
(c) Old, low-temperature basins.
(d) Old, high-temperature basins.

19. Major compositional changes occur in going from the original organic matter of sedimentary rocks to petroleum. These changes may occur mainly in the source rock or during migration to the reservoir rock. Some of the changes are due to chemical reactions and others may be due to physical separation during migration. Given a shale source rock and a sandstone reservoir rock (containing a large mass of petroleum) adjacent to the shale. Analyses of the organic matter in the shale give the following results:

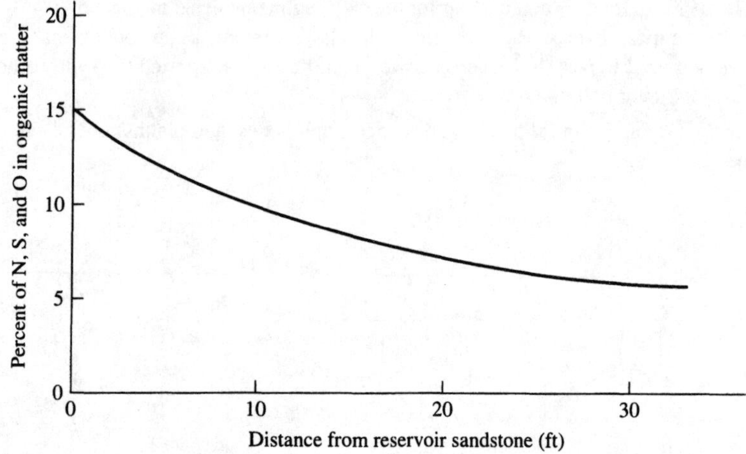

How would you interpret these data?

20. Given is the following diagram from Chang et al. (1983) for a reaction going from reactant A to product B ($A \rightarrow B$). The diagram shows the carbon isotopic composition of A and B versus the percentage of conversion of A to B. How can this diagram be used to explain two of the reaction pathways in Figure 6-5?

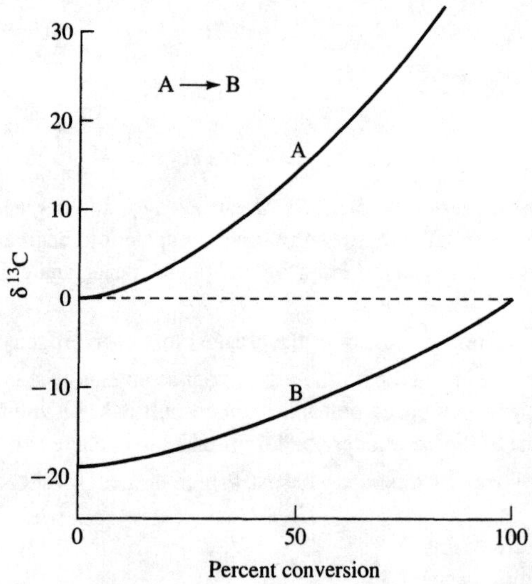

21. Detrital uraninite (UO_2) grains are found in Precambrian rocks older than 2 billion years but not in younger rocks. What does this fact suggest about the composition of the Earth's atmosphere before and after 2 billion years?

REFERENCES

Abraham, H. 1960. *Asphalts and Allied Substances,* 6th ed. Van Nostrand Reinhold Company, New York.

Aizenshtat, Z., M. J. Baedecker, and I. R. Kaplan. 1973. Distribution and diagenesis of organic compounds in JOIDES sediment from Gulf of Mexico and western Atlantic. *Geochim. Cosmochim. Acta,* v. 37:1881–1898.

Atlas, R. M., and R. Bartha. 1987. *Microbial Ecology,* 2d ed. The Benjamin/Cummings Publishing Company, Menlo Park, CA.

Berner, E. K., and R. A. Berner. 1987. *The Global Water Cycle.* Prentice-Hall, Englewood Cliffs, NJ.

Berner, R. A. 1980. *Early Diagenesis.* Princeton University Press, Princeton, NJ.

Calvin, M. 1975. Chemical evolution. *Am. Scientist,* v. 63:169–177.

Chang, S., D. DesMarais, R. Mack, S. L. Miller, and G. E. Strathearn. 1983. Prebiotic organic syntheses and the origin of life. In *Earth's Earliest Biosphere,* J. W. Schopf, ed. Princeton University Press, Princeton, NJ.

Cobb, J. C., and C. B. Cecil, eds. 1993. *Modern and Ancient Coal-Forming Environments.* Special Paper 286, The Geol. Soc. Am., Boulder, Colorado.

Connan, J. 1974. Time-temperature relation in oil genesis. *Bull. Am. Assoc. Petrol. Geol.,* v. 58:2516–2521.

Degens, E. T. 1965. *Geochemistry of Sediments.* Prentice-Hall, Englewood Cliffs, NJ.

Dott, R. H., Sr., and M. J. Reynolds, compilers. 1969. *Sourcebook for Petroleum Geology.* Am. Assoc. of Petrol. Geol., Tulsa, OK.

Emery, D., and A. Robinson. 1993. *Inorganic Geochemistry: Applications to Petroleum Geology.* Blackwell Scientific Publications, Oxford, England.

Engel, M. H., and S. A. Macko, eds. 1993. *Organic Geochemistry.* Plenum Publishing Corporation, New York.

Fogel, M. L., E. K. Sprague, A. P. Gize, and R. W. Frey. 1989. Diagenesis of organic matter in Georgia salt marshes. *Estuarine, Coastal and Shelf Science,* v. 28:211–230.

Freedman, J., ed. 1975. *Trace Element Geochemistry in Health and Disease.* Special Paper 155, Geol. Soc. Am., Boulder, CO.

Gehman, H. M., Jr. 1962. Organic matter in limestone. *Geochim. Cosmochim. Acta,* v. 26:885–897.

Hayes, J. M., I. R. Kaplan, and K. W. Wedeking. 1983. Precambrian organic geochemistry, preservation of the record. In *Earth's Earliest Biosphere,* J. W. Schopf, ed. Princeton University Press, Princeton, NJ.

Hunt, J. M. 1963. Composition and origin of the Unita Basin bitumens, *Oil and Gas Possibilities of Utah, Re-evaluated.* Utah Geol. Miner. Survey Bull., v. 54:249–273.

Hunt, J. M. 1977. Distribution of carbon as hydrocarbons and asphaltic compounds in sedimentary rocks. *Bull. Am. Assoc. Petrol. Geol.,* v. 61:100–104.

Hunt, J. M. 1979. *Petroleum Geochemistry and Geology.* W. H. Freeman and Company, San Francisco.

Jones, R. W. 1984. Comparison of carbonate and shale source rocks. In *Petroleum Geochemistry and Source Rock Potential of Carbonate Rocks,* Studies in Geology No. 18, J. G. Palacas, ed. Am. Assoc. Petrol. Geol., Tulsa, OK. 163–180.

Kennicutt II, M. C., J. M. Brooks, and H. B. Cox. 1993. The origin and distribution of gas hydrates in marine sediments. In *Organic Geochemistry,* M. H. Engel and S. A. Macko, eds. Plenum Press, New York.

Knoll, A. H. 1992. The early evolution of eukaryotes: a geological perspective. *Science,* v. 256:622–627.

Likens, G. E., F. H. Bormann, R. S. Pierce, J. S. Eaton, and N. M. Johnson. 1977. *Biogeochemistry of a Forested Ecosystem.* Springer-Verlag, New York.

Margulis, L., and K. V. Schwartz. 1982. *Five Kingdoms.* W. H. Freeman and Company, San Francisco.

Mason, S. F. 1991. *Chemical Evolution: Origins of the Elements, Molecules, and Living Systems.* Oxford University Press, New York.

McGowran, B. 1990. Fifty million years ago. *Am. Scientist,* v. 78:30–39.

Merrill, R. K., ed. 1991. *Source and Migration Processes and Evaluation Techniques.* Treatise of Petroleum Geology, Am. Assoc. Petrol. Geol., Tulsa, OK.

Oró, J., S. L. Miller, and A. Lazcano. 1990. The origin and early evolution of life on earth. In *Annual Review of Earth and Planetary Sciences,* v. 18, Annual Reviews, G. W. Wetherill, A. L. Albee, and F. G. Stehli, eds. Palo Alto, California. 317–356.

Peake, E., B. L. Baker, and G. W. Hodgson. 1972. Hydrogeochemistry of the surface waters of the Mackenzie River drainage basin, Canada. II. The contribution of amino acids, hydrocarbons, and chlorine to the Beaufort Sea by the Mackenzie River system. *Geochim. Cosmochim. Acta,* v. 36: 867–883.

Peters, K. E., J. M. Moldowan, A. R. Driscole, and G. J. Demaison. 1989. Origin of Beatrice oil by co-sourcing from Devonian and Middle Jurassic source rocks, Inner Moray Firth, United Kingdom. *Bull. Am. Assoc. Petrol. Geol.,* v. 73:455–471.

Philippi, G. T. 1974. The influence of marine and terrestrial source material on the composition of petroleum. *Geochim. Cosmochim. Acta,* v. 38:947–966.

Philp, R. P., and C. A. Lewis. 1987. Organic geochemistry of biomarkers. In *Annual Review of Earth and Planetary Sciences,* v. 15, Annual Reviews, G. W. Wetherill, A. L. Albee, an F. G. Stehli, eds. Palo Alto, CA. 363–395.

Post, W. M., T-H. Peng, W. R. Emanuel, A. W. King, V. H. Dale, and D. L. DeAngelis. 1990. The global carbon cycle. *Am. Scientist,* v. 78:310–326.

Postgate, J. 1992. *Microbes and Man,* Cambridge University Press, New York.

Price, L. C., and J. L. Clayton. 1992. Extraction of whole versus ground source rocks: Fundamental petroleum geochemical implications including oil-source rock correlations. *Geochim. Cosmochim. Acta,* v. 56:1213–1222.

Reeder, S. W., B. Hitchon, and A. A. Levinson. 1972. Hydrogeochemistry of the surface waters of the Mackenzie River drainage basin, Canada. I. Factors controlling inorganic composition. *Geochim. Cosmochim. Acta,* v. 36:825–865.

Schidlowski, M., J. M. Hayes, and I. R. Kaplan. 1983. Isotopic inferences of ancient biochemistries: carbon, sulfur, hydrogen, and nitrogen. In *Earth's Earliest Biosphere: Its Origin and Evolution,* J. W. Schopf, ed. Princeton University Press, Princeton, NJ. 149–186.

Schlesinger, W. H. 1991. *Biogeochemistry: An Analysis of Global Change.* Academic Press, San Diego, CA.

Schnitzer, M. 1984. Soil organic matter: its role in the environment. In *Short Course in Environmental Geochemistry,* Short Course Handbook, v. 10, M. E. Fleet, ed. Mineralogical Assoc. Can., London, Ontario. 237–267.

Schopf, J. W. 1975. Precambrian paleobiology: problems and perspectives. In *Annual Review of Earth and Planetary Sciences,* v. 3, Annual Reviews, F. A. Donath, ed. Palo Alto, CA. 213–249.

Schopf, J. W., ed. 1983. *Earth's Earliest Biosphere: Its Origin and Evolution.* Princeton University Press, Princeton, NJ.

Schopf, J. W., and M. R. Walter. 1983. Archean microfossils: new evidence of ancient microbes. In *Earth's Earliest Biosphere: Its Origin and Evolution,* J. W. Schopf, ed. Princeton University Press, Princeton, NJ. 214–239.

Schrayer, G. J., and W. M. Zarrella. 1963. Organic geochemistry of shale. I. Distribution of organic matter in the siliceous Mowry shale of Wyoming. *Geochim. Cosmochim. Acta,* v. 27:1033–1046.

Shackleton, N. J. 1986. Paleogene stable isotope events. *Palaeogeog. Palaeoclimatol. Palaeoecol.,* v. 57:91–102.

Simoneit, B. R. T. 1990. Petroleum generation, an easy and widespread process in hydrothermal systems: an overview. *Applied Geochemistry,* v. 5:3–15.

Siskin, M., and A. R. Katritzky. 1991. Reactivity of organic compounds in hot water: geochemical and technological implications. *Science,* v. 254:231–237.

Speers, G. C., and E. V. Whitehead. 1969. Crude petroleum. In *Organic Geochemistry,* G. Eglington and M. T. J. Murphy, eds. Springer-Verlag, New York. 638–675.

Sposito, G. 1989. *The Chemistry of Soils.* Oxford University Press, New York.

Stumm, W., and J. J. Morgan. 1981. *Aquatic Chemistry,* 2d ed. John Wiley & Sons, New York.

Sundquist, E. T., and W. S. Broecker, eds. 1985. *The Carbon Cycle and Atmospheric CO_2—Natural Variations Archean to Present.* Monograph 32, Am. Geophysical Union, Washington, D.C.

Thurman, E. M. 1985. *Organic Geochemistry of Natural Waters.* Martinus Nijhoff/Dr. W. Junk Publishers, Dordrecht, The Netherlands.

Tissot, B. P., and D. H. Welte. 1984. *Petroleum Formation and Occurrence,* 2d ed. Springer-Verlag, New York.

Van Krevelen, D. W. 1963. Geochemistry of Coal. In *Organic Geochemistry,* I. A. Breger, ed. Macmillan Publishing Co., Inc., New York. 183–247.

Walter, M. R. 1983. Archean stromatolites: evidence of the earth's earliest benthos. In *Earth's Earliest Biosphere: Its Origin and Evolution,* J. W. Schopf, ed. Princeton University Press, Princeton NJ. 187–213.

Waples, D. W. 1985. *Geochemistry in Petroleum Exploration.* International Human Resources Development Corporation, Boston, MA.

Waples, D. W., and T. Machihara. 1991. *Biomarkers for Geologists.* Am. Assoc. Petrol. Geol., Tulsa, OK.

Wedepohl, K. H. 1971. *Geochemistry.* Holt, Rinehart, and Winston, Inc., New York.

Whiticar, M. J., E. Faber, and M. Schoell. 1986. Biogenic methane formation in marine and fresh-water environments: CO_2 vs. acetate fermentation—Isotope evidence. *Geochim, Cosmochim. Acta,* v. 50:693–709.

Wood, J. M. 1974. Biological cycles for toxic elements in the environment. *Science,* v. 183:1049–1052.

Yen, T. F., and J. M. Moldowan, eds. 1988. *Geochemical Biomarkers,* Harwood Academic Publishers.

Zinke, P. J. 1977. Man's activities and their effect upon the limiting nutrients of primary productivity in marine and terrestrial ecosystems. In *Global Cycles and their Alterations by Man,* W. Stumm, ed. Dahlem Konferenzen, Berlin. 89–98.

Sedimentary Rocks

Geochemical studies can contribute to our understanding of both clastic and chemical sedimentary rocks. We have to consider all the chemical changes involved in weathering, transportation, deposition, diagenesis, and lithification. Postlithification changes can also be significant. This chapter begins with a review of the composition and mineralogy of sedimentary rocks. Next comes a detailed discussion of weathering processes and products. The clay minerals are major products of weathering and major components of sedimentary rocks. Their structure, composition, and properties are outlined. Next is a review of sediments containing the other major group of sedimentary minerals, the carbonates. The sediments resulting from weathering often undergo important changes after deposition, and chemical aspects of these diagenetic changes are discussed. The final section deals with the formation of evaporite rocks and with the relationships between evaporites, brine waters, and ore deposits. Throughout the chapter emphasis is placed on the processes leading to the compositional variability of soils, sediments, and sedimentary rocks.

ROCK COMPOSITION AND MINERALOGY

The minerals of the common sedimentary rocks can be divided in terms of origin into four major groups: (1) minerals that survive weathering and transportation (*detrital* minerals), (2) new minerals formed during weathering and transportation (*secondary* minerals), (3) minerals that form directly from solutions chemically or biochemically (*precipitated* minerals), and (4) minerals formed in sediments during and after deposition (*authigenic* minerals). The major detrital minerals are quartz, orthoclase, microcline, and plagioclase (Figure 7-1a). The clay minerals make up the bulk of the secondary minerals, as do calcite and aragonite for the precipitated minerals. All

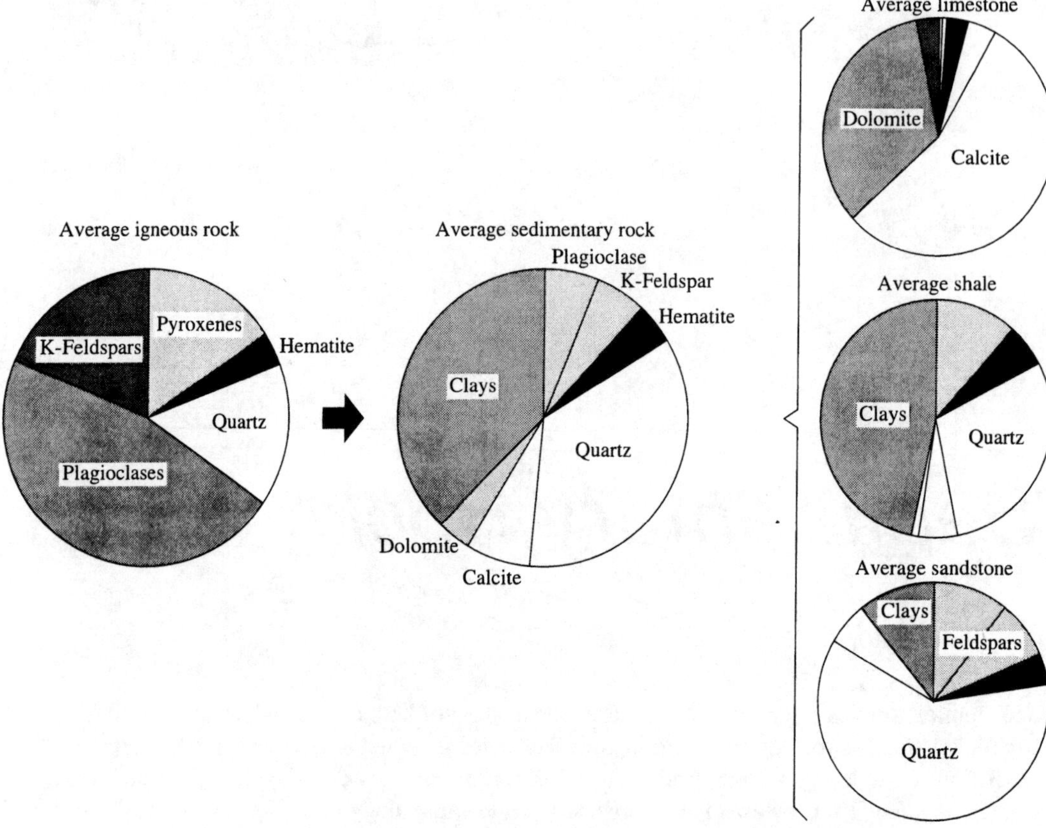

Figure 7-1a Normative mineral content of igneous and sedimentary rocks. The mineral contents are calculated from the average chemical compositions of the rocks. The weathering of igneous rocks is responsible for the conversion of most feldspars and mafic minerals to the clays which, with residual quartz, are the main components of the average sedimentary rock. The diagrams for the three major sedimentary rock types demonstrate that most sedimentary rocks are not monomineralic. Data from Garrels and Mackenzie (1971). After Railsback (1993).

these minerals can be authigenic minerals. Dolomite is a special case (discussed in detail later), which mainly forms as an authigenic mineral.

Sandstones can be considered to represent a physical segregation of quartz and thus a concentration of SiO_2. Similarly, the limestones represent a chemical segregation of CaO. Most limestones have a very low content of detrital and secondary silicate minerals. The shales, on the other hand, are more complicated. They are a combination of detrital, secondary, and precipitated minerals. All three rock types may undergo significant changes in mineralogy due to the formation of authigenic minerals. Since several of the major minerals of sedimentary rocks are relatively simple compounds (Table 7-1), it is fairly easy to relate variations in the composition of these rocks to their mineralogy. We shall first consider the two extremes, sandstones and limestones, and then discuss the more complex shales.

TABLE 7-1 Compositions of the Major Minerals of Common Sedimentary Rocks

Mineral	Composition
Quartz	SiO_2
Opal	$SiO_2 \cdot nH_2O$
Orthoclase	$KAlSi_3O_8$
Microcline	$KAlSi_3O_8$
Plagioclase	$NaAlSi_3O_8-CaAl_2Si_2O_8$
Muscovite	$KAl_3Si_3O_{10}(OH)_2$
Biotite	$K(Mg, Fe)_3AlSi_3O_{10}(OH)_2$
Kaolinite	$Al_4Si_4O_{10}(OH)_8$
Smectite	$(Al, Mg, Fe)_4(Al, Si)_8O_{20}(OH)_4 \cdot nH_2O$
Illite	$K_{0\text{-}2}Al_4(Al, Si)_8O_{20}(OH)_4$
Chlorite	$(Mg, Fe, Al)_3(Al, Si)_4O_{10}(OH)_2 \cdot (Mg, Al)_3(OH)_6$
Calcite	$CaCO_3$
Aragonite	$CaCO_3$
Dolomite	$CaMg(CO_3)_2$

Sandstones

Orthoquartzites (also known as quartz arenites) are composed almost entirely of quartz grains and chert (microcrystalline quartz) fragments. Their cement is usually silica or calcium carbonate. The orthoquartzite analysis in Table 7-2 is for a very "pure" rock cemented by silica, and thus its SiO_2 content is greater than 99 percent. Other sandstones contain both detrital quartz and detrital feldspars. These arkose sandstones have significant amounts of Al_2O_3, K_2O, Na_2O, and CaO (analysis 2, Table 7-2). In addition to quartz and one or more feldspars, many sandstones contain clay minerals, a variety of other minerals, and rock fragments. Terms that have been applied to these rocks include lithic sandstone, graywacke, and subgraywacke. The variety of possible physical constituents leads to a variety of chemical compositions involving, in addition to the oxides of orthoquartzites and arkoses, MgO, Fe_2O_3, and FeO (analyses 3 and 4, Table 7-2). Magnesium is contributed by smectite, carbonates, and chlorite. Iron is in clays, carbonates, sulfides, and oxides. Both elements are commonly found in the rock fragments of sandstones. Rocks with a calcite cement, such as the subgraywacke of Table 7-2, have large values for CaO and CO_2 in their analyses. Trace oxides such as TiO_2, P_2O_5, and MnO in a chemical analysis may come from clay minerals, from precipitation of minerals such as celestite during diagenesis, or from accessory detrital minerals such as rutile, zircon, monazite, ilmenite, titanite, apatite, and tourmaline. Table 7-3 lists background values for trace elements in sandstones.

Despite the relationship between mineralogy and composition, interpretation of sandstones using only chemical analyses is not realistic. Petrographic analyses are necessary to determine which minerals are detrital, which are authigenic, and which are cements. Most of the oxides can be contributed by several different minerals, and microscopic study must be done to determine which minerals are actually present. Comparison of the chemical analyses of different rocks should be done with care, since the chemical composition of a sandstone or shale is related to its grain size. It has been found that, with a decrease in grain size, SiO_2 tends to decrease, and Al_2O_3, K_2O, and iron oxides tend to increase. These changes are usually due to a decrease in

TABLE 7-2 Chemical Compositions of Common Sedimentary Rocks

	1 Ortho- quartzite	2 Arkose	3 Graywacke	4 Sub- graywacke	5 Lithographic limestone	6 Fossiliferous limestone
SiO_2	99.54	72.21	68.84	65.00	1.15	0.29
TiO_2	0.03	0.22	0.25	—	—	—
Al_2O_3	0.35	10.69	14.54	9.57	0.45	0.26
Fe_2O_3	0.09	0.80	0.62	1.59	—	0.11
FeO	—	0.72	2.47	1.08	0.26	—
MnO	—	0.22	—	—	—	0.01
MgO	0.06	1.47	1.94	0.40	0.56	0.70
CaO	0.19	3.85	2.23	10.10	53.80	55.53
Na_2O	—	2.30	3.88	2.14	0.07 }	0.07
K_2O	—	3.32	2.68	1.43		0.02
P_2O_5	—	0.10	0.15	—	—	0.05
H_2O^+	0.25	1.46	1.60	0.82	0.69	—
H_2O^-	—	0.08	0.35	0.23	0.23	—
CO_2	—	2.66	0.14	6.90	42.69	43.42
SO_3	—	—	0.15	0.04	—	0.15
Total:	100.51	100.10	99.93	99.54	99.90	100.61

Sources:

1: Tuscarora Quartzite, Silurian, Hyndman, Pa.; data from C. R. Fettke, 1981, *Pa. Topograph. Geol. Surv. Rept.* no. 12, p. 263.

2: Arkose, Oligocene, Switzerland; data from P. Niggli et al., 1930, *Beitr. Geol. Schweiz, Geotech. Ser. no. 14*, p. 262.

3: Graywacke, Jurassic and Cretaceous (Franciscan Formation), Piedmont, Calif.; data from E. F. Davis, 1918, *Bull. 11*, Calif. Univ. Dept. Geol., p. 22; analysis includes BaO = 0.04 and $ZrO_2 = 0.05$.

4: Calcareous subgraywacke, Oligocene (Frio Formation), Jim Wells and Kleberg Counties, Tex.; data from R. H. Hanz, 1954, *Bull. Am. Assoc. Petrol. Geol.*, v. 38, p. 114; analysis includes S = 0.16.

5: Lithographic limestone, Solenhofen, Bavaria; data from F. W. Clarke, 1924, *U.S. Geol. Surv. Bull. 770*, p. 564.

6: Fossiliferous limestone made up of crinoidal and bryozoan fragments, Salem Formation, Mississippian, Madison County, Ill.; data from D. L. Graf, 1960, *Ill. State Geol. Surv. Circ., 308*, p. 31.

quartz abundance and an increase in clay mineral content. Thus comparisons should only be made among rocks of similar grain size.

Limestones

All limestones are composed primarily of $CaCO_3$ in the form of calcite and aragonite. The $CaCO_3$ occurs as (1) discrete fragments and particles of either inorganic or organic origin, (2) microcrystalline material formed from a carbonate mud, and (3) coarsely or finely crystalline cementing material formed by inorganic or biochemical precipitation. A wide array of textures results from variations in the relative abundance of the three types and from variations in the nature of the discrete fragments and particles. No matter what the texture, the chemical composition of the rocks is quite uniform (analyses 5 through 7 of Table 7-2). The lithographic limestone of Table 7-2 formed from carbonate mud. The fossiliferous limestone in the table is composed of fossil fragments cemented by coarsely crystalline calcite. Cemented, inorganic or

TABLE 7-2 *(cont.)*

	7 Oolitic limestone	8 Dolomite	9 Siliceous shale	10 Potassic shale	11 Calcareous shale	12 Carbonaceous shale
SiO_2	0.75	3.24	84.14	56.29	25.05	51.03
TiO_2	0.03	—	0.22	0.64	—	—
Al_2O_3	0.25	0.17	5.79	19.22	8.28	13.47
Fe_2O_3	0.64	0.17	1.21	4.39 (Fe₂O₃ + FeO)	0.27	8.06
FeO	—	0.06	—		2.41	—
MnO	—	—	—	—	4.11	—
MgO	2.14	20.84	0.41	1.65	2.61	1.15
CaO	52.30	29.58	0.31	0.09	27.87	0.78
Na_2O	0.01	—	0.99	0.19	—	0.41
K_2O	0.04	—	0.50	10.85	—	3.16
P_2O_5	—	—	—	—	0.08	0.31
H_2O^+	0.50 (H_2O^+ + H_2O^-)	0.30 (H_2O^+ + H_2O^-)	5.56 (H_2O^+ + H_2O^-)	2.04	2.86	0.81 (H_2O^+ + H_2O^-)
H_2O^-				3.54	1.44	
CO_2	43.54	45.54	—	—	24.20	—
SO_3	—	—	—	0.72	—	—
Total:	100.20	99.90	100.03	99.62	99.18	102.90

Sources:

7: Oolitic limestone, Tertiary, eastern Styria, Austria; data from D. L. Graf, *ibid.*, p. 30.

8: Knox Dolomite, Cambro-Ordovician, Morrisville, Ala.; data from F. J. Pettijohn, 1957, *Sedimentary Rocks*, 2nd ed., Harper & Row, New York, p. 418.

9: Mowry Shale, Cretaceous, Black Hills, S.D.; data from F. J. Pettijohn, *ibid.*, p. 364.

10: Glenwood Shale, Ordovician, Minnesota; data from F. J. Pettijohn, *ibid.*, p. 370.

11: Calcareous shale, Cretaceous, Mount Diablo, Calif.; data from F. W. Clarke, *op. cit.*, p. 552.

12: Black shale, Devonian, Walker County, Ga.; data from F. W. Clarke, *ibid.*; analysis includes S = 7.29 and C = 13.11; total corrected for sulfur content is 100.17.

biochemical rounded masses (ooliths) make up the oolitic limestone. For all three rocks, despite their differences in texture, CaO and CO_2 are the only major components. Because calcite and aragonite are never compositionally pure, small amounts of magnesium, iron, manganese, barium, and strontium may be found in a particular analysis. Magnesium is the most abundant of these, since some calcitic skeletal material may contain over 10 percent $MgCO_3$. A value of MgO greater than 1 percent in a limestone analysis suggests that the mineral dolomite is present. Other oxides reported in limestone analyses, such as K_2O and Al_2O_3, are due to silicate impurities. Silicon dioxide is contributed by both silicate minerals and by chert nodules. Background values for trace elements in carbonate rocks are given in Table 7-3.

Many carbonate rocks, particularly Paleozoic and Precambrian rocks, have a significant amount of the mineral dolomite. Any rock containing more than 50 percent of the mineral is known as the rock dolomite (or dolostone). These rocks also represent a segregation of material, with the only major components being CaO, MgO, and CO_2 (analysis 8, Table 7-2). Most carbonate rocks are either limestones with less than 3 percent MgO or dolomites with more than 19 percent MgO. In some rocks the mineral dolomite clearly has a secondary rather than a primary origin, as shown by replacement textures and bedding relationships. In other rocks there is no evidence for either primary or secondary origin. In present-day environments, dolomite does not

TABLE 7-3 Compositional Variation of Rocks, Soils, and Plant Ash from Various Areas of the United States

	Rocks			Soils		Plant ash	
Element	Sandstone	Shale	Carbonate	Unculti-vated	Cultivated	Cultivated plants	Native species
Al (%)	0.43–3.0*	4.4–9.2	0.17–2.0	1.1–6.5	0.9–5.2	0.02–0.40	0.10–3.9
As (ppm)	1.1–4.3	6.4–9.0	0.74–2.5	6.7–13.0	5.5–12.0	No data	No data
Ba (ppm)	38–170	220–510	5.6–160	86–740	63–810	15–450	270–11,000
Be (ppm)	0.80	1.1–1.7	No data	0.76–1.3	1.0–1.2	No data	0.64–2.0
B (ppm)	18–36	43–110	29–31	18–63	21–41	37–540	140–600
Cd (ppm)	No data	No data	No data	No data	No data	0.37–2.3	0.95–20
Ca (%)	0.09–0.22	0.13–1.1	No data	0.07–1.7	0.08–0.66	0.29–20	13–35
C (%)							
Carbonate	0.01	0.06–0.16	No data	0.046–0.055	0.0075	No data	No data
Organic	0.30–0.35	0.27–0.32	0.10–0.28	0.70–2.8	0.91–2.2	No data	No data
Ce (ppm)	No data	No data	No data	50–110	120	No data	350
Cr (ppm)	2.0–39	62–130	2.7–29	11–78	15–70	0.42–6.6	2.2–22
Co (ppm)	1.6–7.4	4.8–13	1.3–7.1	1.0–14	1.3–10	0.50–6.2	0.65–400
Cu (ppm)	1.2–8.4	13–130	0.84–12	8.7–33	9.9–39	21–230	50–270
F (ppm)	9.8–120	700	38–100	160–480	160–440	0.43–0.49	0.50–1.6
Ga (ppm)	1.5–10	15–30	2.2–10	1.9–29	1.5–20	No data	1.5–2.8
I (ppm)	No data	No data	No data	No data	No data	4.6–13	2.8–5.4
Fe (%)	0.09–1.9	1.8–4.5	0.11–2.1	0.47–4.3	1.4–2.8	0.06–0.27	0.08–0.93
La (ppm)	6–36	29–67	24	26–45	18–49	No data	14–270
Pb (ppm)	5–17	11–24	4–18	2.6–25	2.6–27	7.1–87	24–480
Li (ppm)	2.1–17	25–79	0.78–2.6	15–32	15–24	No data	4.0–15
Mg (%)	0.09–0.21	0.61–1.6	No data	0.03–0.84	0.03–0.38	1.5–13	1.6–10
Mn (ppm)	29–300	65–420	83–910	61–1,100	99–740	96–810	470–14,000

TABLE 7-3 *(cont.)*

	Rocks			Soils		Plant ash	
Element	Sandstone	Shale	Carbonate	Unculti-vated	Cultivated	Cultivated plants	Native species
Hg (ppb)	7.9–16	45–340	22–30	45–160	30–69	No data	No data
Mo (ppm)	No data	No data	0.79	No data	No data	2.5–20	0.76–7.6
Nd (ppm)	No data	No data	No data	9.2–61	63	No data	No data
Ni (ppm)	1.2–18	21–110	2.3–16	4.4–23	1.8–18	2.7–130	0.81–130
Nb (ppm)	8.8	7.7	No data	5.8–19	6.6–16	No data	No data
P (%)	0.01–0.10	0.03–0.07	0.004–0.06	0.004–0.08	0.02–0.08	1.2–22	0.71–3.1
K (%)	0.08–0.66	1.8–5.4	0.12–0.56	0.07–2.6	0.04–1.7	18–41	2.9–23
Sc (ppm)	2.1–7.2	8.2–18	6.1–9.0	2.1–13	2.8–9.0	No data	No data
Se (ppm)	0.09–0.11	0.46–0.64	0.16–0.31	0.27–0.73	0.28–0.74	0.04–0.17	0.01–0.42
Ag (ppm)	No data	0.18	No data	No data	No data	No data	No data
Na (%)	0.01-0.19	0.09–0.50	0.01–0.17	0.02–0.62	0.45–0.79	0.0025–0.0039	0.02–0.31
Sr (ppm)	13–99	90–200	100–990	5.7–160	3.6–150	14–880	320–5,300
Ti (ppm)	83–2,200	2,300–5,700	31–810	1,700–6,600	1,700–4,000	4.7–250	69–1,200
V (ppm)	5.3–38	74–400	3.9–40	15–110	20–93	No data	2.6–23
Yb (ppm)	1.9	2.3–3.8	No data	1.8–28	1.5–3.8	No data	1.1–1.8
Y (ppm)	9–22	25–38	8–20	17–39	15–32	No data	2.1–47
Zn (ppm)	5.2–31	55–82	6.3–24	25–67	37–68	180–1,900	170–1,800
Zr (ppm)	22–170	95–230	6.5–42	120–460	140–360	No data	2.4–8.5

*Range of means.

Note: The data represent background concentrations for selected regions of the United States. Given here are the lowest mean value and the highest mean value found for the various regions studied. Where only one value is given, the element was studied in only one area. In most cases the means are based on 100 or more samples. Geometric means (antilog of the arithmetic mean of the logarithmic values) are used rather than arithmetic means because they are believed to give a better measure of central tendency.

Sources: Data from Connor and Shacklette, 1975. Also see Shacklette and Boerngen, 1984.

form by direct precipitation from ordinary seawater. Dolomite does replace calcite or aragonite particles in certain modern sediments and in carbonate rocks associated with these sediments. Apparently, dolomite can form as a metasomatic mineral at any time in the history of a carbonate rock, with the likelihood of replacement increasing with the age of the rock. Dolomite formation is discussed later in this chapter in the section on diagenesis.

Shales

The major minerals of shales are the clay minerals (kaolinite, smectite, illite, and chlorite) and quartz. Thus the most abundant oxides in an analysis are SiO_2 and Al_2O_3. A particularly high content of silica (analysis 9, Table 7-2) results when silica-bearing shells of microplankton such as diatoms or radiolaria or volcanic ash occur in a rock along with detrital quartz. A high K_2O content (analysis 10, Table 7-2) may be due to detrital feldspar, authigenic feldspar, detrital muscovite, illite, authigenic glauconite, or potassium adsorbed by clay minerals. The presence of calcite as fragments and/or cement is shown by the high values of CaO and CO_2 found for some shales (analysis 11, Table 7-2). Shales formed under anaerobic conditions retain some organic matter and may have as much as 13 percent carbon (see footnote for analysis 12, Table 7-2). Such shales often also have a high sulfur content. Other shales may be unusually rich in Al_2O_3 (from gibbsite), P_2O_5 (from apatite), MgO (from dolomite), and iron oxide (from limonite, pyrite, siderite, and glauconite).

Shales clearly have higher amounts of most trace elements when compared with sandstones and carbonate rocks. Table 7-3 shows that the following elements are enriched in shales: As, Ba, B, Cr, Cu, F, Ga, Li, Hg, Ni, Se, Ti, U, Y, and Zn. There are a number of reasons for this. Some of the trace elements occur in the same accessory detrital minerals mentioned above for sandstones. Some of the elements are concentrated owing to substitution for a major element in the various dominant minerals of shales. An example would be barium substituting for potassium in illite. Other elements are concentrated in the organic matter of shales or are part of new minerals formed by diagenetic reactions involving organic matter. Vanadium can be an original component of organic matter, and vanadium compounds also form from circulating groundwater owing to the reducing ability of organic matter. The majority of the trace elements in shales are concentrated there because of adsorption or ion exchange occurring on clay mineral surfaces. Fluorine, for example, replaces hydroxide ions on the surfaces and edges of clay particles.

Shales, like limestones, tend to change in mineralogy and chemical composition with age. Kaolinite and smectite become less abundant, and illite and particularly chlorite become more abundant with increasing age. Some of these changes, especially for Cenozoic and Mesozoic rocks, can be attributed to authigenic formation of illite and chlorite. The great abundance of illite in Paleozoic rocks may represent differences in weathering conditions and in the abundance of vegetation during the Paleozoic Era. The high illite content of Paleozoic shales results in a higher K_2O content for these rocks. They also tend to have more FeO and less Fe_2O_3 than younger shales. This may be due to oxidation of organic material and reduction of Fe_2O_3 as a rock grows older.

Variations in the compositions of the three major sedimentary rock types are illustrated in Figure 7-1b. The ordinate in this figure is SiO_2/Al_2O_3, and both sandstones and carbonate rocks show more variation in this parameter than in the abscissa, $(CaO + Na_2O)/K_2O$. The variation is primarily due to differences in the amount of quartz, chert, and clay minerals in sandstones and carbonates. Carbonates have high values of CaO and low values of Na_2O and K_2O, while shales and sandstones are low in Na_2O and CaO and high in K_2O. As pointed out earlier, sandstones

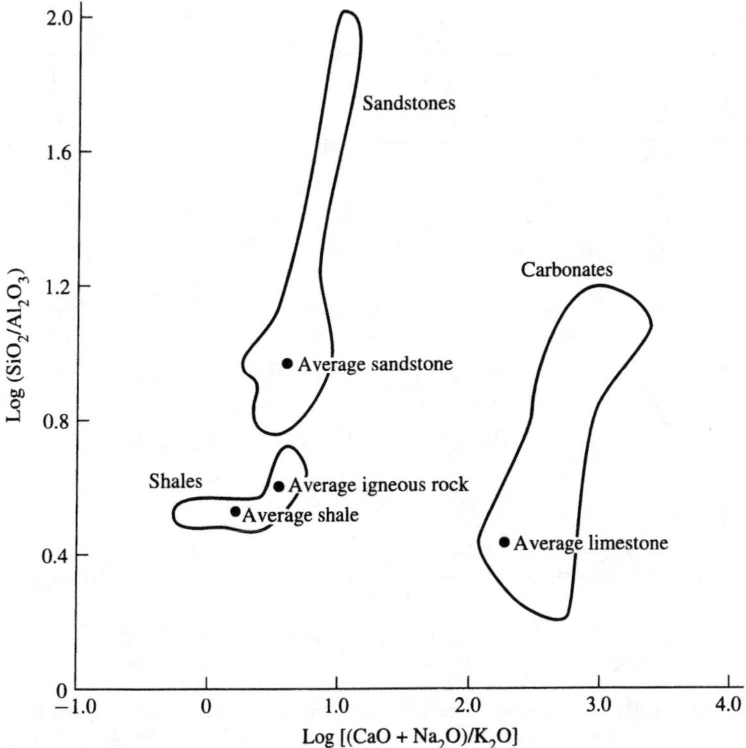

Figure 7-1b Variation in composition of sandstones, shales, and carbonate rocks. Rock compositions are in weight percent. From Garrels and MacKenzie (1971).

and carbonates represent segregations of the oxides SiO_2 and CaO, respectively. Variations in the extent of segregation are reflected in the large overall compositional variation shown for these two types in Figure 7-1b. Since shales represent less segregation of SiO_2 and CaO during weathering processes, and since the great majority of sedimentary rocks are shales, the average composition for shales is close to the average composition for igneous rocks (Figure 7-1b).

Further details on the chemical compositions, textures, and classification of sedimentary rocks can be found in Blatt (1992) and Boggs (1992).

CHEMICAL WEATHERING

The formation of sedimentary rocks begins with the physical and chemical breakdown of pre-existing rocks. Although the disintegration and decomposition of rocks are intimately related, we shall restrict ourselves here to a review of the chemical aspects of weathering. Chemical weathering results from the adjustment of thermodynamically unstable minerals to the surface environment of abundant water and atmospheric gases. Four different types of chemical processes are particularly important in the weathering of rocks: (1) dissolution, (2) hydrolysis, (3) adsorption, and (4) oxidation-reduction.

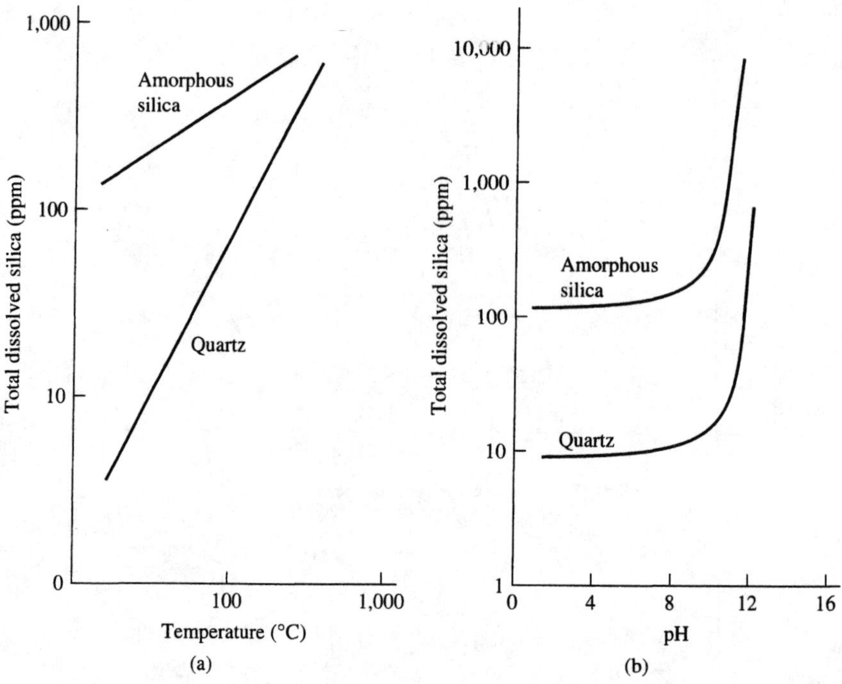

Figure 7-2 Solubility of silica as a function of temperature and pH: (a) solubility as a function of temperature at pH < 7; (b) solubility as a function of pH at 25°C. [(a) After Siever. © 1962 by the University of Chicago. All rights reserved. (b) After Garrels and MacKenzie (1971, 149). From Krauskopf, K., 1967, *Introduction to Geochemistry,* New York: McGraw-Hill, Inc.]

Dissolution refers to the dissolving of a mineral to form ions, aqueous molecules, and the dispersed molecular units of colloids. As pointed out in Chapter Four, the solubility products of most minerals are very small numbers, and thus most common minerals do not dissolve to any great extent in pure water or in natural waters. The most soluble common mineral is halite. Gypsum has a solubility about one-fortieth that of halite, and anhydrite has an even lower solubility. Calcite has a low solubility in *pure* water. The extensive solution of calcite in nature is due to the occurrence of dissolved carbon dioxide in water, giving carbonic acid that reacts with calcite:

$$CaCO_3 \text{ (calcite)} + H_2CO_3 \rightleftharpoons 2\ HCO_3^- + Ca^{2+} \tag{7-1}$$

The carbon dioxide in natural waters comes from the atmosphere and from the oxidation of surface organic matter. The more carbon dioxide there is in water, the more calcite it will dissolve. Calcite, aragonite, magnesite, and dolomite dissolve in a similar manner and have similar solubilities in natural waters, although magnesite and dolomite dissolve more slowly than aragonite and calcite. Quartz dissolves by a hydration reaction:

$$SiO_2 \text{ (quartz)} + 2\ H_2O \rightleftharpoons H_4SiO_4 \tag{7-2}$$

At low values of pH and temperature, the solubility of quartz is about 10 parts per million (Figure 7-2). At higher values of pH (above 9), silicic acid (a very weak acid) dissociates slightly,

$$H_4SiO_4 \rightleftharpoons H^+ + H_3SiO_4^- \tag{7-3}$$

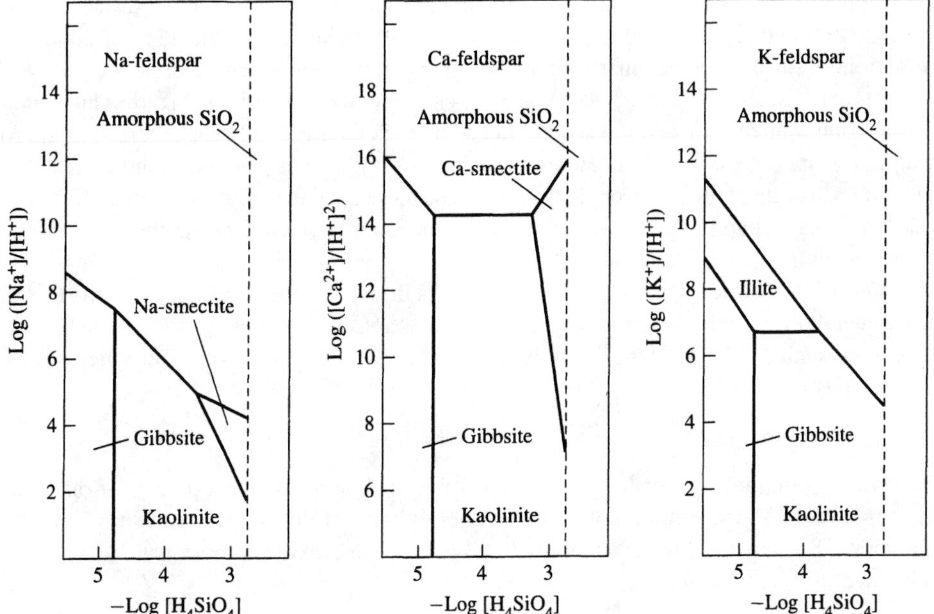

Figure 7-3 Activity-activity diagrams showing stability relations in the systems Na_2O-Al_2O_3-SiO_2-H_2O, CaO-Al_2O_3-SiO_2-H_2O, and K_2O-Al_2O_3-SiO_2-H_2O. The diagrams have been plotted using equilibrium constants for the various weathering reactions that can occur in the systems. A low abundance of Na^+, Ca^{2+}, and K^+ and a high abundance of H^+ favors formation of clay minerals from feldspars. When they form there is an increase in the abundance of Na^+, Ca^{2+}, and K^+ and a decrease in the abundance of H^+. An increasing amount of dissolved silica (H_4SiO_4) favors conversion of kaolinite to smectite. At concentrations above 192 ppm H_4SiO_4, amorphous silica precipitates. Gibbsite forms at very low concentrations of silica. Analyses of most soil waters, stream waters, and groundwaters plot in the kaolinite field of each diagram, indicating that this mineral is the primary product of chemical weathering of silicate minerals. After Stumm and Morgan (1981, p. 547).

and the solubility of quartz increases. Amorphous silica has a somewhat greater solubility (Figure 7-2).* The concentration of silica in continental surface and subsurface waters is usually less than the solubility of amorphous silica and more than the solubility of quartz. Most of the dissolved silica probably comes from other weathering reactions rather than from the solution of quartz.

The dissolutions of quartz and calcite are examples of congruent solubility. These reactions produce no new solid phase. Kaolinite is said to dissolve incongruently in water because a new solid phase, gibbsite, is produced:

$$Al_2Si_2O_5(OH)_4 \text{ (kaolinite)} + 5\,H_2O \rightleftharpoons Al_2O_3 \cdot 3\,H_2O \text{ (gibbsite)} + 2\,H_4SiO_4 \qquad (7\text{-}4)$$

The solubility of kaolinite is very small; however it will convert to gibbsite if brought in contact with water containing very little H_4SiO_4 (Figure 7-3). Many weathering dissolution reactions are incongruent.

Dissolution reactions of minerals can be *diffusion controlled* or *surface-reaction controlled.* Stumm and Wollast (1990) describe two main mechanisms of dissolution reactions: (1) transport

*Opal has often been described as amorphous silica containing some water. Recent research has shown that it diffracts X rays and consists of a mixture of small crystal blades of cristobalite and tridymite.

(diffusion) of a reactant or a weathering product through a leached layer at the surface of a weathering mineral and (2) chemical reactions at the surface involving the breaking of bonds and the formation of surface species that ultimately detach from the surface or break down into other products that detach from the surface (Figure 7-4). A number of weathering studies have indicated that, in natural environments, the second mechanism is the main process controlling the weathering rate of most oxides and silicates. In other words, the rate at which dissolution occurs is dependent on the rate of the surface reactions because these are generally slower than the rate of diffusion of reactants and products through the thin reaction layer that separates the mineral from the surrounding bulk solution.

One of the surface reactions that occurs in the dissolution of silicate minerals is *hydrolysis*. As a general term, hydrolysis refers to all those reactions of compounds with water in which a weak acid (such as H_2CO_3) or a weak base (such as NH_4OH) is formed. For silicate minerals, silicic acid is formed:

$$Mg_2SiO_4 \text{ (forsterite)} + 4\,H_2O \rightleftharpoons 2Mg^{2+} + 4\,OH^- + H_4SiO_4 \qquad (7\text{-}5)$$

The resulting solution is more basic since the silicic acid molecule ties up some of the hydrogen ion of the water. The weathering of silicate minerals can be represented by this type of hydrolysis reaction. Since natural waters contain dissolved carbon dioxide, a more realistic way to write such a hydrolysis reaction is

$$Mg_2SiO_4 \text{ (forsterite)} + 4\,H_2CO_3 \rightleftharpoons 2\,Mg^{2+} + 4\,HCO_3^- + H_4SiO_4 \qquad (7\text{-}6)$$

When an aluminosilicate is involved, an additional product is clay mineral:

$$2\,NaAlSi_3O_8 \text{ (albite)} + 2\,H_2CO_3 + 9\,H_2O$$

$$\rightleftharpoons 2\,Na^+ + 4\,H_4SiO_4 + Al_2Si_2O_5(OH)_4 \text{ (kaolinite)} + 2\,HCO_3^- \qquad (7\text{-}7)$$

Some of the silica set free in hydrolysis reactions may go into solution as colloidal silica rather than as the H_4SiO_4 molecule. As shown in the above reactions involving dissolved carbon dioxide, a normal product is bicarbonate ion (HCO_3^-), and this is why the bicarbonate ion is the dominant ion in nonmarine waters.

The term hydrolysis also refers to specific reactions occurring on an atomic scale at the surface of a weathering silicate mineral. Casey et al. (1988) report experimental work on a type of hydrolysis reaction known as depolymerization. The process of depolymerization for feldspar minerals involves the breaking of silicon-oxygen and aluminum-oxygen framework bonds and thus the breakdown of the aluminosilicate polymer.* The depolymerization process results in the formation of separated Si-OH and Al-OH groups that break down further and eventually detach from the weathering surface into the aqueous phase. Final products in the experiments of Casey et al. were H_4SiO_4 and Al^{3+}. Depending on pH, aluminum may occur as Al^{3+}, $Al(OH)_2$, or $Al(OH)_4^-$.

Field studies of plagioclase weathering indicate that some of the dissolved constituents precipitate as clay minerals close to the original site of dissolution. For instance, kaolinite formation can be represented by this reaction:

$$2\,Al(OH)_2^+ + 2\,H_4SiO_4 \rightarrow Al_2Si_2O_5(OH)_4 \text{ (kaolinite)} + 2\,H^+ + 3\,H_2O \qquad (7\text{-}8)$$

*A polymer or macromolecule is a large molecule made up of hundreds or thousands of atoms.

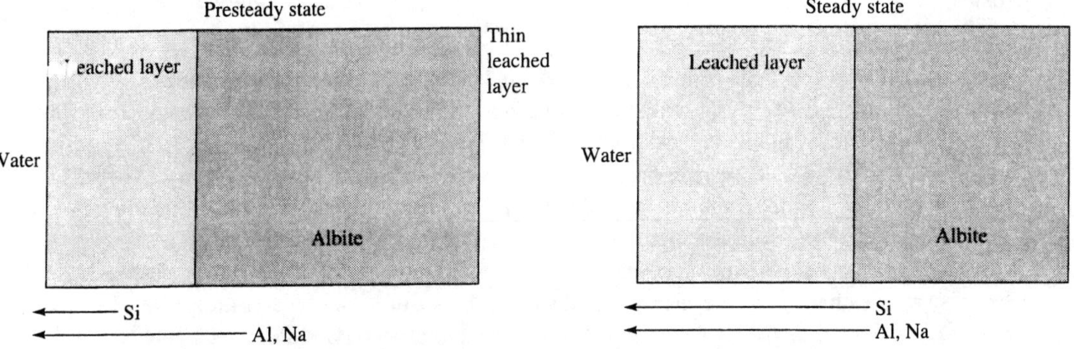

Figure 7-4a A simplified representation of leached layer formation on albite surfaces during experimental dissolution. Initially sodium and aluminum are preferentially leached with respect to silicon because Na-O and Al-O bonds are broken more rapidly than Si-O bonds. The thickness of the leached layer, which is a porous and open structure, increases during this initial period. Eventually a steady state is reached with detachment reactions occurring at equal stoichiometric rates for all three bond types. The dissolution rate of the albite is surface-reaction controlled (reactions at the solid-liquid interface control the rate) rather than diffusion controlled (diffusion of reaction products through the leached layer controls the rate). Experimental results show that the thickness of the leached layer is a function of pH, with the steady state thickness ranging from 10 to 900 Å. The data suggest that dissolution occurs nonuniformly with greater leaching at dislocations, microcracks, and other defects in the albite crystals. (Reprinted from *Geochimica et Cosmochimica Acta*, v. 54, Hellman et al., The formation of leached layers on albite surfaces during dissolution under hydrothermal conditions, pp. 1267–1281. Copyright 1990, with kind permission from Elsevier Science Ltd, The Boulevard, Langford Lane, Kidlington OX5 1GB, U.K.)

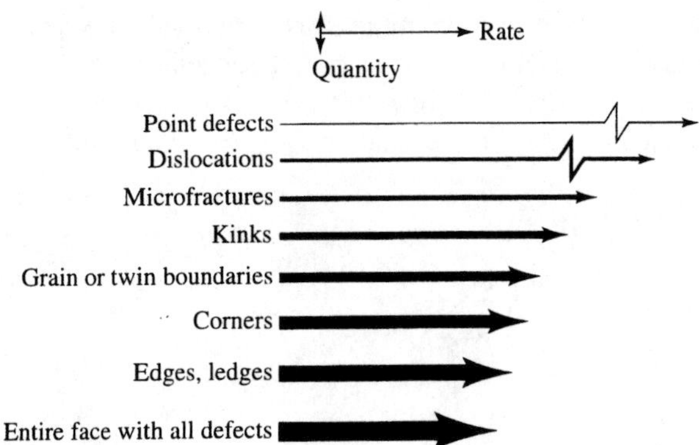

Figure 7-4b Schematic illustration of the location of dissolution processes involved in crystal dissolution. The horizontal length of each arrow indicates the relative rate of processes at each location (actual rates can differ by many orders of magnitude). The vertical thickness of each arrow represents the relative quantity of material dissolved and delivered to aqueous solution at that location. Thus, while dissolution occurs most rapidly at point and linear defects, they deliver less dissolved material to solution than slower dissolution of faces and pits occurring at edges, ledges, and corners. Dissolution rates are normally measured by analyzing the change in solution concentration with time. Those processes that deliver the greatest quantity of dissolved material will therefore control the measured rate. The slower dissolution occurring at edges, ledges, and corners determines overall reaction rate because these locations deliver much more material to solution. (Reprinted from *Geochimica et Cosmochimica Acta*, v. 53, Schott et al., Dissolution kinetics of strained calcite, pp. 378–383. Copyright 1989, with kind permission from Elsevier Science Ltd, The Boulevard, Langford Lane, Kidlington OX5 1GB, U.K.)

Also occurring at the same time as hydrolysis and clay formation is the exchange of hydrogen ions for cations in the feldspar, with cations such as K^+ and Na^+ released to the aqueous phase.

Two general conclusions result from studies [such as that of Casey et al. (1988)] of dissolution and hydrolysis weathering processes. First, weathering rates of minerals are directly dependent on their atomic structure and bond strengths. Second, these rates also depend on the nature of the weathering solution, particularly the abundance of hydrogen ions (pH).

Adsorption refers to the attachment of a dissolved ion or molecule to the surface of a solid. In studying weathering we are mainly interested in the adsorption of water molecules. For water molecules the bonding force of adsorption comes from hydrogen bonding (see Chapter Four). The first step in the hydrolysis process discussed above is the adsorption of water molecules at the sites of Si-O bonds. This leads to the breaking of the Si-O bonds and to the formation of other bonds involving the atoms of the water molecule as depolymerization proceeds (Figure 7-5). Adsorption by itself is important in the dissolution of minerals held together by ionic bonds linking covalently bonded polymers. Interaction of water molecules with the ions between polymers leads to a breakup into polymeric fragments. Lasaga (1990) reviews atom interactions at the mineral-water interface and discusses adsorption, hydrolysis, and dissolution processes.

Oxidation-reduction reactions play an important role in many areas of geochemistry. An example is their role in natural water chemistry as discussed in Chapter Four. Weathering can start with the occurrence of these reactions at mineral-water interfaces with mineral dissolution as a final result. Hering and Stumm (1990) discuss the oxidative and reductive dissolution of minerals. Examples are the reductive dissolution of ferric oxides such as magnetite and the oxidation reaction involved in the dissolution of ferrous silicates such as biotite. These reactions involve electron transfer between surface sites on minerals and multivalent aqueous species. In the case of an oxide like magnetite, transfer of electrons from a reductant such as organic matter to Fe^{3+} ions in the mineral results in a Fe^{2+}-O bond that is of lower strength than the original Fe^{3+}-O bond. As a result, surface material is more easily detached and put into solution.

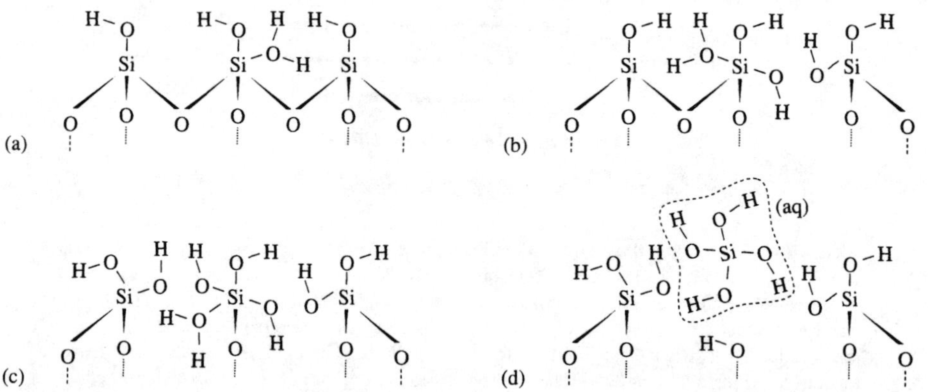

Figure 7-5 Sequence of steps in a proposed mechanism for the overall reaction in the dissolution of quartz. (a) Adsorption of H_2O molecule on middle Si-O bond; (b) hydrolysis reaction breaks Si-O bond; (c) further adsorption and bond breaking; (d) H_4SiO_4 molecule forms and goes into solution. (Reprinted from *Geochimica et Cosmochimica Acta*, v. 54, Dove and Crerar, Kinetics of quartz dissolution in electrolyte solutions using a hydrothermal mixed flow reactor, pp. 955–969. Copyright 1990, with kind permission from Elsevier Science Ltd, The Boulevard, Langford Lane, Kidlington OX5 1GB, UK.)

The dissolution of biotite is more complicated because several processes are occurring. White (1990) suggests that initially both K^+ ions and electrons are transferred from the mineral to solution. The electron transfer forms Fe^{3+} ions at the surface of the mineral. However, it appears from experimental work that the principal species released by dissolution is Fe^{2+} (White and Yee 1985). Thus, a further reaction involving Fe^{3+} ions in the solid and reducible species in the solution must cause the actual dissolution. The solid state oxidation reaction is followed by a reductive dissolution reaction that releases Fe^{2+} to solution. If no reducible species are present in solution, a reaction involving exchange of hydrogen ions for K^+ and Fe^{3+} can cause removal and reduction of Fe^{3+} in the mineral structure (White 1990). Similar weathering behavior is shown by other iron-containing silicates, such as pyroxenes and amphiboles.

Berner and Berner (1987, Chap. 4) point out that most weathering is brought about, directly or indirectly, by biological activity. Soil microorganisms (bacteria, fungi, algae, etc.) continually produce inorganic and organic acids that attack and break down minerals (Eckhardt 1985; Berthelin 1988). Carbonic acid (H_2CO_3) is produced by the oxidation of organic matter to CO_2, which combines with water to form H_2CO_3 and which in turn partly dissociates to H^+ and HCO_3^-. This oxidation of organic carbon is caused by microorganisms that act as catalysts and use the released energy of the reaction. In addition, organic acids such as humic acid and fulvic acid are formed by the alteration of organic matter. These acids are also generated at the root tips of plants. Berner and Berner (1987, 157) point out that, even though the actual acid attacking a mineral is organic, the overall reaction can be represented as though the only attacking acid is H_2CO_3 (equations 7-6 and 7-7). As mentioned above, hydrogen ions of whatever source are intimately involved in the breakdown of minerals, both as replacements for cations at mineral surfaces and, more generally, in the control of many reactions by pH.

An example of direct weathering by microorganisms has been reported by Hiebert and Bennett (1992). They studied an aquifer in Minnesota that was contaminated with crude oil in 1979 when a pipeline burst. The crude oil provided food for the subsurface bacteria and accelerated bacterial weathering, which probably also occurs in surface environments but at a much slower rate. Hiebert and Bennett found localized mineral etching in micro-reaction zones at bacteria-mineral interfaces where high concentrations of organic acids, formed during metabolism of hydrocarbon, selectively removed silica and aluminum from feldspar. Both quartz and feldspar grains in the sand and gravel aquifer were colonized by indigenous bacteria and chemically weathered at a rate faster than would be predicted for equilibrium conditions assuming only an inorganic environment.

Besides carbon, only four elements with more than one oxidation state are abundant enough to take part in a major way in oxidation-reduction reactions in surface environments. One of these, oxygen, plays a significant role in many oxidation processes. Another element, iron, forms compounds that are the most colorful products of weathering. Ferrous iron from weathering of iron-bearing silicates is oxidized by dissolved oxygen to form hematite:

$$2\,Fe^{2+} + 4\,HCO_3^- + \tfrac{1}{2}O_2 \rightleftharpoons Fe_2O_3\ \text{(hematite)} + 4\,CO_2 + 2\,H_2O \qquad (7\text{-}9)$$

Even though hematite is a stable iron compound under most weathering conditions, ferric iron often occurs as goethite ($HFeO_2$, a major constituent of the material known as limonite). The relationship between hematite and goethite can be represented by the equation

$$Fe_2O_3\ \text{(hematite)} + H_2O \rightleftharpoons 2\,HFeO_2\ \text{(goethite)} \qquad (7\text{-}10)$$

Laboratory studies show that the free-energy change of this reaction is very small, indicating that either mineral may be formed under surface conditions.

Two other important elements undergo valence changes during weathering. Sulfur has a valence of -2 in sulfide minerals, and bacterially-catalyzed oxidation during weathering produces SO_4^{2-} where its valence is $+6$. A common oxidation reaction involves pyrite:

$$2\ FeS_2\ \text{(pyrite)} + 2\ H_2O + 7\ O_2 \rightleftharpoons 2\ Fe^{2+} + 4\ H^+ + 4\ SO_4^{2-} \qquad (7\text{-}11)$$

Sulfuric acid is a strong acid that is involved in weathering wherever it occurs. Oxidation-reduction reactions involving sulfur compounds play an important role in natural water chemistry and in diagenesis and we will discuss some of these reactions later in the chapter. Manganese behaves in a manner similar to that of iron, with the major oxidation products being pyrolusite (MnO_2) and manganite ($Mn_2O_3 \cdot H_2O$). Mn^{2+} is the predominant manganese ion in the minerals of igneous rocks, while Mn^{3+} and Mn^{4+} predominate in compounds formed by weathering.

As suggested above, it is now possible, because of advances in analytical techniques, to directly study in situ geochemical processes that occur on mineral surfaces. Mogk (1990) reviews the application of Auger electron spectroscopy (AES) to studies of chemical weathering. One of the capabilities of this technique is its ability to provide direct measurement of the thickness of the reaction layer at the surface of a mineral. A sample depth profile is shown in Figure 7-6. Other ca-

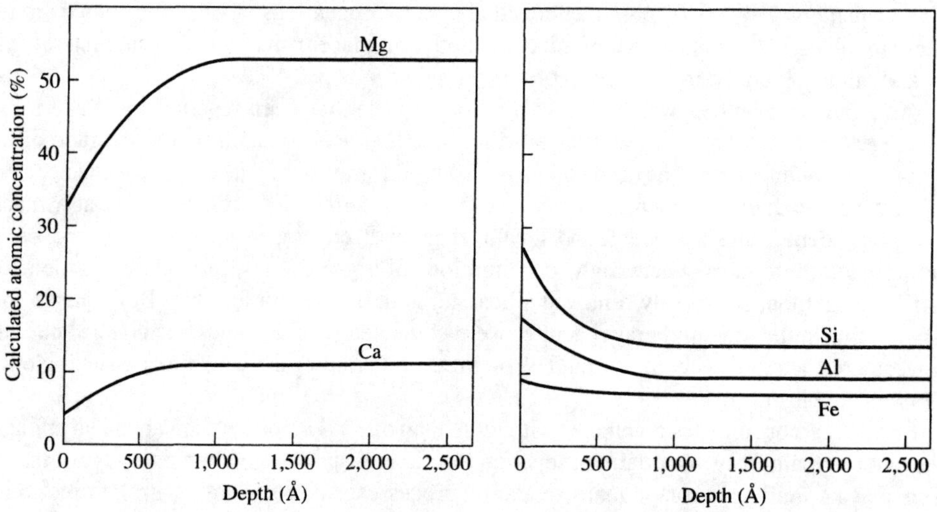

Figure 7-6 Depth profile of naturally weathered hornblende using Auger electron spectroscopy. The hornblende was collected from the B horizon of a soil on glacial outwash in Colorado. A regular change in relative atomic concentration as a function of depth occurs up to about 1200 Å depth. Silicon, aluminum, and iron show relative enrichment at the surface while magnesium and calcium are relatively depleted. If aluminum is assumed to be immobile, then relative to aluminum all cations were leached to a depth of 1200 Å. Cation depletion is greatest, although not complete, at the surface and decreases with depth. Dissolution has been incongruent, with significantly more depletion of calcium and magnesium than silicon and iron. Results of this and similar studies suggest that chemical weathering processes cannot be adequately explained solely by surface-controlled reactions, but must also involve cation depletion by volume diffusion processes. After Mogk (1990) and Mogk and Locke (1988).

pabilities of AES include chemical analyses of the near surfaces of minerals for all elements except hydrogen and helium and high lateral spatial resolution (>0.1 μm) to determine the composition of discrete domains on mineral surfaces. In addition to his discussion of AES, Mogk also reviews the capabilities and limitations of other surface-sensitive analytical techniques.

The discussion above deals with weathering of the surfaces of individual grains of minerals. At the other extreme, we can look at the weathering of large masses of a particular rock type or large areas containing several rock types. Several of the chapters in Colman and Dethier (1986) deal with this aspect of weathering. In the first chapter, the editors point out that the rate of chemical weathering of the rocks of an area depends on both the amount and chemistry of the weathering fluids (determined by variables such as climate and organic activity) and on the textures and mineral composition of the rocks.

Various types of field studies can be used to assess chemical weathering rates and products. The physical properties and mineral composition of soil profiles can be evaluated (Mahaney and Halvorson 1986). Another approach involves research on the geochemical mass balance of a watershed (Velbel 1986). It is possible to transform the results of mass-balance calculations on natural weathering rates to the units in which laboratory weathering rates are presented. Table 7-4 shows that there is relatively good agreement between the two very different methods of assessing weathering rates. As Table 7-4 shows, laboratory-determined rates for mineral breakdown are higher than rates determined in the study of natural areas, but the difference is only one order of magnitude. For the natural area studied by Velbel (1985), the rate at which the weathering front penetrated into the fresh rock is 3.8 cm/1000 years.

The products of the various weathering processes can be put into four groups: (1) soluble constituents removed from the weathering site, (2) residual primary minerals unaffected by weathering reactions, (3) new, stable minerals produced by these reactions, and (4) organic compounds resulting from organic decay reactions (Table 7-5). It should be emphasized that Table 7-5 is a very general summary, which does not take into account variations in the extent of weathering and in such other factors as climate and source rock. Many further changes occur after the initial weathering reactions are complete.

Berner (1971, 173) has emphasized the importance of water flow as a controlling factor in weathering. At moderate flow rates, albite is changed to kaolinite as shown in equation (7-7). For high rates of flow, silicic acid is removed fast enough to allow gibbsite to form instead of kaolinite:

$$\begin{aligned}&NaAlSi_3O_8\text{ (albite)} + H_2CO_3 + 7\,H_2O\\&\quad\rightleftharpoons Na^+ + 3\,H_4SiO_4 + Al(OH)_3\text{ (gibbsite)} + HCO_3^-\end{aligned} \tag{7-12}$$

TABLE 7-4 Comparison of Natural and Laboratory Weathering Rates

Mineral	Silica release rate (mole/meter^2-second)	Reference
Albite feldspar	3.1×10^{-12} (laboratory)	Nickel (1973)
Albite feldspar	9.6×10^{-12} (laboratory)	Busenberg and Clemency (1976)
Oligoclase feldspar	8.7×10^{-13} (natural)	Velbel (1985)
Almandine garnet	1.1×10^{-11} (laboratory)	Nickel (1973)
Almandine garnet	3.8×10^{-12} (natural)	Velbel (1985)

Source: After Velbel (1986). Reprinted by permission of Academic Press, Inc.

TABLE 7-5 Common Products of Chemical Weathering Processes

Soluble constituents	Na^+, Ca^{2+}, K^+, Mg^{2+}, H_4SiO_4, HCO_3^-, SO_4^{2-}, Cl^-
Residual primary minerals	Quartz, zircon, magnetite, ilmenite, rutile, garnet, titanite, tourmaline, monazite
New minerals	Kaolinite, smectite, illite, chlorite, hematite, geothite, gibbsite, boehmite, diaspore, amorphous silica, pyrolusite
Organic compounds	Organic acids, humic substances, kerogen

When flow rates are very slow, material is removed very slowly from the weathering site and, provided Mg^{2+} is available, the product is smectite

$$3\ NaAlSi_3O_8\ (\text{albite}) + Mg^{2+} + 4\ H_2O \rightleftharpoons 2\ Na_{0.5}Al_{1.5}Mg_{0.5}Si_4O_{10}(OH)_2\ (\text{smectite}) + 2\ Na^+ + H_4SiO_4 \quad (7\text{-}13)$$

These equations are an example of the control of climate and relief on weathering products. Bauxite minerals such as gibbsite tend to form in areas of high rainfall and high relief, since these areas would have a high flow rate for water passing through soils and bedrock. Smectite, on the other hand, is usually formed in areas with low rainfall and low relief, and thus a low flow rate. Intermediate conditions favor the formation of kaolinite, and kaolinite is the most common weathering product of feldspar. Figure 7-7 shows that all three products, kaolinite, smectite, and gibbsite, can form from the same rock type in areas with a large variation in annual rainfall.

The other two major factors determining the nature of the clays found in a particular area are the mineralogy of the rock being weathered and the chemical composition of the weathering solutions. The most common clay minerals formed initially by weathering are kaolinite and smectite. Feldspars tend to change to kaolinite, whereas ferromagnesian minerals provide magnesium for the formation of smectite. In certain weathering environments, aluminosilicates are changed to gibbsite [$Al(OH)_3$] with concurrent removal of all of the original silicon in the primary minerals. The amount of H_4SiO_4 in a weathering solution in contact with a primary aluminosilicate would tend to determine which of these three secondary minerals would be most abundant (Figure 7-3). Very little H_4SiO_4 would favor gibbsite formation, since it contains no silica. A larger amount of H_4SiO_4 would favor kaolinite formation and very large amounts should lead to formation of smectite, since it has a higher Si/Al ratio than kaolinite (Table 7-1). The occurrence of sodium or calcium in the weathering solution would favor smectite formation, since it contains one or both of these cations in its structure.

The stability relations of the clay minerals, the feldspars, and gibbsite can be calculated using the equilibrium constants of reactions such as that of equation (7-7). The diagrams of Figure 7-3 have been obtained by a process similar to that explained in Chapter Four for Eh-pH diagrams. Although Figure 7-3 represents estimates based on limited data, it gives an indication of the relationships of the minerals. With increasing activity (concentration) of dissolved silica, kaolinite is more stable than gibbsite, and smectite is more stable than kaolinite. At high silica activities, amorphous SiO_2 precipitates. Increasing activities for Na^+, Ca^{2+}, and K^+ along with decreasing hydrogen ion activity (higher pH) make smectite, illite, and the feldspars stable relative to kaolinite. These general relationships agree with those found in nature. One use of such diagrams is to plot analyses of various natural waters on the diagrams. A surface or subsurface water whose analysis plots in a particular stability field should be in equilibrium with the phase

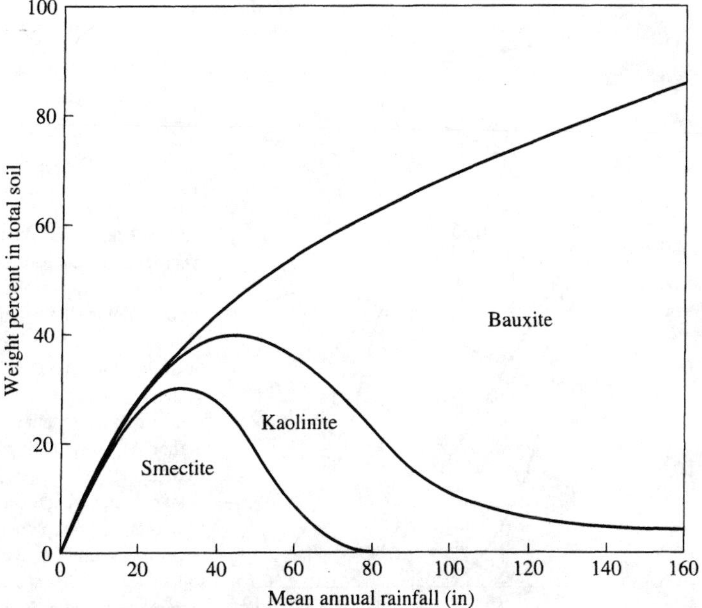

Figure 7-7 Plot of weight percent in total soil of bauxite, kaolinite, and smectite group minerals versus mean annual rainfall, island of Hawaii. Vertical distances between successive curves represent concentrations of each mineral group. Note the increase in total weathering products (top curve) with increase in rainfall. After Sherman (1952) and Berner (1971).

occupying that field. Most waters plot in the stability field of kaolinite, which is the major clay mineral produced by chemical weathering.

COMPOSITIONAL CHANGES PRODUCED BY WEATHERING

There are a number of ways to plot or calculate compositional changes during the course of chemical weathering. These approaches are usually applied to residual soils that have formed from underlying bedrock.

We will use here a graphic device developed by Reiche (1950). Figure 7-8 is a plot of the change in mole percent of each oxide through the various stages in the weathering of a granite gneiss. Since aluminum tends to stay at a weathering site in the form of clay minerals, the stages of weathering can be represented by the increasing percent of Al_2O_3 in the weathering products. Figure 7-8 shows that initial changes include a rapid drop in CaO and Na_2O and a rapid increase in H_2O. As weathering progresses, K_2O eventually shows a rapid decrease. Combined SiO_2 (all SiO_2 except that in quartz) and MgO exhibit a slow decrease throughout the weathering process, whereas Fe_2O_3 shows a steady increase. The final product is a soil composed mainly of SiO_2, Al_2O_3, H_2O, and Fe_2O_3 (analysis 1, Table 7-6). These chemical changes are caused by the mineralogical changes shown in Figure 7-9. Plagioclase and biotite break down rapidly; K-feldspar is more resistant. The difference in the stabilities of the two feldspars is the reason why K_2O in Figure 7-8 drops off after Na_2O and CaO. The decrease in combined SiO_2 is due to

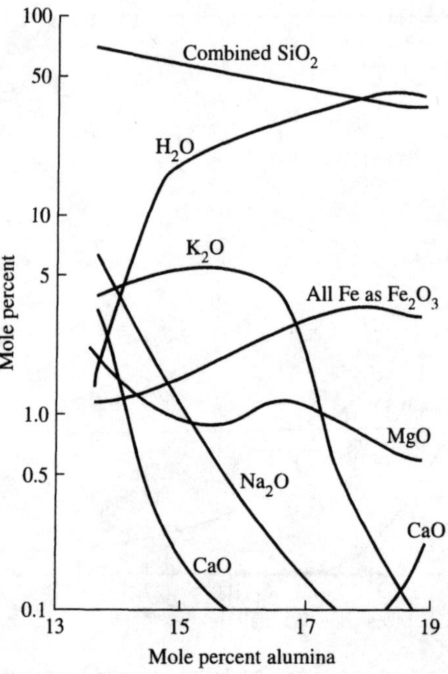

Figure 7-8 Variation diagram for the weathering of the Morton granite gneiss, Minnesota. A logarithmic vertical scale is used to plot mole percent (weight percent from chemical analysis divided by molecular weight) for each oxide, and mole percent of Al_2O_3 is used on the horizontal scale as a measure of the extent of weathering, with higher Al_2O_3 reflecting a greater degree of weathering. Combined SiO_2 refers to SiO_2 in silicate minerals other than quartz. Silica from quartz is not plotted. Fresh rock composition is plotted on the left, intermediate soils in the middle, and the final soil at the extreme right. After Reiche (1950); data from Goldich (1938).

removal of some of the silica from the breakdown of feldspars and biotite. Most of the silica stays at the weathering site in the form of kaolinite. The volume of kaolinite increases steadily, with the final soil containing 66 percent kaolinite. The rest of the soil consists of 25 percent residual quartz and 6 percent residual and secondary oxides such as magnetite, ilmenite, and leucoxene.

The residual soil discussed above is typical of the type of soil formed from granitic rocks in areas of temperate climate and moderate water-flow rates. In the case of rocks such as basalt and amphibolite, which contain larger amounts of ferromagnesian minerals, an abundance of smectite is often found in the overlying soil [equation (7-13)]. Since these rocks contain less SiO_2 and more MgO and iron oxide than granitic rocks, the soils formed from them have lower SiO_2 and higher MgO and Fe_2O_3 contents than granitic soils (compare analyses 1 and 2, Table 7-6). The magnesium occurs in smectite; the iron occurs mostly in goethite. The most intense chemical weathering occurs in some tropical areas with high water-flow rates. Soils known as *laterites* are formed by reactions such as that in equation (7-12). The major minerals in these soils are gibbsite, diaspore, boehmite, hematite, and goethite. Chemically, they consist almost entirely of Al_2O_3, Fe_2O_3, and H_2O. Anayses 2 and 3, Table 7-6, show the extreme differences between soil formed from a mafic igneous rock in a temperate climate and soil formed from the same kind of rock in a tropical climate.

Clay-rich soils may be made up of clay minerals that are residual rather than secondary minerals. Soils formed by the weathering of limestone and dolomite are usually made up of clay minerals, quartz, and iron oxides that were dispersed within the original carbonate rock. The types of clay minerals present, and thus the composition of the soil, depend mainly on the environmental conditions at the time of carbonate formation. However, the environment of weathering may cause some changes in clay mineralogy and composition. An example of a residual clay formed

TABLE 7-6 Chemical Compositions of Typical Soils and Sediments

	1	2	3	4	5	6	7	8	9
SiO_2	55.07	47.0	0.7	55.90	69.96	55.6	53.93	67.36	8.21
TiO_2	1.03	1.8	0.4	0.20	0.59	0.8	0.96	0.59	0.14
Al_2O_3	26.14	18.5	50.5	19.92	10.52	15.5	17.46	11.33	4.14
Fe_2O_3	3.72	14.6	23.4	7.30	3.47 (Fe₂O₃ + FeO)	7.4	8.53	3.40	4.09
FeO	2.53	—	—	0.39		—	0.45	1.42	—
MnO	—	—	—	—	0.06	0.2	0.78	0.19	0.54
MgO	0.33	5.2	—	1.18	1.41	2.2	4.56	2.29	0.83
CaO	0.16	1.5	—	0.50	2.17	1.0	1.56	1.74	44.55
Na_2O	0.05	0.3	—	0.23	1.51	2.2	1.27	1.64	1.32
K_2O	0.14	2.5	—	4.79	2.30	2.3	3.65	2.15	0.78
P_2O_5	—	0.7	—	0.10	0.18	0.1	0.09	0.10	0.26
H_2O^+	9.75	7.2 (H_2O^+ + H_2O^-)	25.0 (H_2O^+ + H_2O^-)	6.52	1.96	10.3 (H_2O^+ + H_2O^-)	6.30 (H_2O^+ + H_2O^-)	6.33 (H_2O^+ + H_2O^-)	—
H_2O^-	0.64			2.54	3.78				—
CO_2	0.36	—	—	0.38	1.40	1.4	0.40	1.30	34.75
Total:	99.92	99.3	100.0	99.95	100.62	100.0	100.09	99.90	100.00

Sources:

1: Residual clay from granite gneiss, Morton, Minn.; data from S. S. Goldich, 1938, p. 22.

2: Clay formed from dolerite, Staffordshire, England; data from H. E. Hawkes and J. S. Webb, 1962, *Geochemistry in Mineral Exploration*, Harper & Row, New York, p. 84.

3: Laterite formed from dolerite, Bombay, India; data from H. E. Hawkes and J. S. Webb, *ibid.*, p. 84.

4: Residual clay from limestone, Staunton, Va.; data from F. W. Clarke, 1924, *U.S. Geol. Surv. Bull. 770*, p. 512.

5: Composite analysis of 235 samples of Mississippi delta sediment; data from F. W. Clarke, *ibid.*, p. 509. Analysis includes Cl, 0.30; organic, 0.66; SO_3, 0.03; miscellaneous, 0.32.

6: Nearshore marine clay, Gulf of Paria; data from D. M. Hirst, 1962, *Geochim. Cosmochim. Acta*, v. 26, p. 317.

7: Deep-sea red clay, Pacific Ocean; data from S. K. El Wakeel and J. P. Riley, 1962, *Geochim. Cosmochim. Acta*, v. 25, p. 118. Analysis includes organic carbon, 0.13, and organic nitrogen, 0.016.

8: Deep-sea siliceous ooze, Atlantic Ocean; data from S. K. El Wakeel and J. P. Riley, *ibid.*, p. 119. Analysis includes organic carbon, 0.26.

9: Deep-sea Globigerina ooze, average for 66 stations, Pacific and Atlantic oceans; data from Z. L. Sujkowski, 1952, *Am. J. Sci.*, v. 250, p. 373. Analysis includes Cl, 0.31, and ZrO_2, 0.08.

from limestone is given in analysis 4, Table 7-6. The soil has a high amount of K_2O, suggesting the presence of illite, the clay mineral that is most common in marine sediments.

Comparison of chemical analyses of a parent rock and its residual soil allows calculation of percentage losses for the various elements as an indication of weathering mobility. A study by Marchand (1974) of a granitic igneous rock found the following sequence in order of decreasing mobility: Na, K, Mg > Ca > Si > Al > Fe. This type of study can be broadened to the observation of an entire drainage basin, comparing the average compositions of surface waters with that of soils and bedrock. An example is the study by Stallard (1985) of river chemistry, geology, geomorphology, and soils in the Amazon and Orinoco basins of northern South America. In this study changes in soil chemistry are shown to be directly related to the chemistry of the dissolved and solid loads of the rivers.

Studies of both local and regional weathering processes generally agree that the elements removed most quickly from a weathering site are sodium, calcium, and magnesium (Figure 7-8). Potassium is generally less mobile, followed by silicon, and then iron and aluminum. The relative

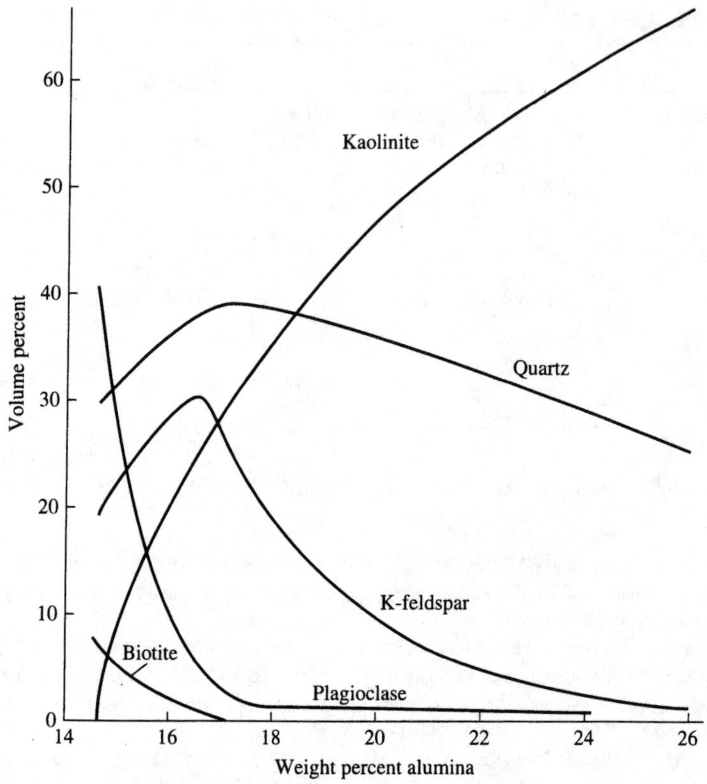

Figure 7-9 Mineral-variation diagram for the weathering of the Morton granite gneiss, Minnesota. Weight percent Al_2O_3 is used as a measure of the extent of weathering. Mineralogy of the fresh rock is plotted on the left and mineralogy of the final soil at the extreme right. After Goldich (1938).

mobility of elements is a reflection of the relative stability of primary minerals at weathering sites. For igneous rocks, as originally pointed out by Goldich (1938), the susceptibility to weathering is the reverse of the general order of crystallization from igneous melts (Table 7-7).

The mobility of trace elements is particularly hard to predict, since a given element released by weathering of a primary mineral may (1) become a structural part of a new secondary mineral; (2) be adsorbed by clay minerals, organic matter, or iron and manganese oxides; (3) be taken up by plants or other organisms; or (4) be removed in solution as simple or complex ions or as adsorbed ions on colloidal particles. The strong tendency for plants to concentrate trace elements is shown by the data for plant ash in Table 7-3.

A useful approach to investigating the weathering chemistry of a region is the calculation of an annual weathering budget. An example is given in Table 7-8 for the Hubbard Brook watershed of New Hampshire (Likens et al. 1977). The budget approach treats the watershed as a single reservoir with inputs from weathering and precipitation and output occurring as stream runoff (Velbel 1986). Because the forest system of the area is growing, it is necessary to include a term for the incorporation of elements into the biomass within the reservoir. The transfer of elements to soil minerals can be ignored in studying change over the span of one year. Thus elements lost

TABLE 7-7 Mineral Stability Series in Weathering

	Mafic minerals	Felsic minerals
↑ Increasing susceptibility to weathering	Olivine	Calcic plagioclase
	Pyroxene	Intermediate plagioclase
	Amphibole	Sodic plagioclase
	Biotite	K-feldspar
		Muscovite
		Quartz

Source: After Goldich (1938).

TABLE 7-8 Annual Weathering Budget for the Hubbard Brook Watershed

Species	Transported in runoff (T/km^2)	Incorporated in biomass (T/km^2)	Total (T/km^2)	Abundance in bedrock (wt. %)	Relative rate of release (%)
Ca^{2+}	1.16	0.95	2.11	1.4	100
Mg^{2+}	0.26	0.09	0.35	1.1	21
Na^+	0.56	0.02	0.58	1.6	24
K^+	0.10	0.61	0.71	2.9	16
Al^{3+}	0.19	0	0.19	8.3	2
SiO_2	3.85	0	3.85	66	2

T = metric tons. The relative rate of release (last column) is the percentage released by weathering, normalized to Ca^{2+} = 100. Additional data are in Table 6-5.

Sources: After Likens et al. (1977). (From Drever, J. I., *The Geochemistry of Natural Waters*, 2d ed., Copyright ©1988 by Prentice Hall, Inc. Adapted by permission of Prentice Hall, Inc., Upper Saddle River, NJ.)

by weathering of bedrock (granite and metamorphic rocks) and introduced by precipitation are either taken up by the biomass or removed in runoff.

Likens et al. (1977) estimated that cations incorporated into the biomass were equivalent to about 57 percent of the cations being removed by runoff. As shown in Table 7-8, the order of release of the elements by weathering is calcium, sodium, magnesium, potassium, and silicon-aluminum. This sequence is similar to that reported in the classic study by Goldich (1938) shown in Figure 7-7.

CLAY MINERALS

As pointed out in the preceding sections, the clay minerals are a major product of weathering processes and a major component of most sedimentary rocks. We now take a closer look at this interesting group of minerals. The clay minerals are usually classified on the basis of their crystal chemistry (Table 7-9). They are all hydrous silicates with continuous sheet structures (see discussion of phyllosilicates in Chapter Five). There are two types of sheets, *tetrahedral* sheets and *octahedral* sheets. The tetrahedral sheet consists of linked silica tetrahedra similar to that of other phyllosilicates such as the micas (Figure 7-10). The octahedral sheet is composed of linked octahedra, each of which consists of oxygen and hydroxide ions around an aluminum or magnesium ion (Figure 7-10). When aluminum ions are in the centers of the octahedra, only two-thirds of the

TABLE 7-9 Classification of Principal Clay Mineral Groups

Sheets	Type of octahedral sheet	Group	Example species	Ideal formula	Ion-exchange capacity (mEq/100 g)
Two	Dioctahedral	Kaolin	Kaolinite	$Al_2Si_2O_5(OH)_4$	1–18
	Trioctahedral	Serpentine	Chrysotile	$Mg_3Si_2O_5(OH_4)$	0
Three	Dioctahedral	Smectite	Montmorillonite	$Al_2Si_4O_{10}(OH)_2 \cdot nH_2O$	60–150
	Trioctahedral	Smectite	Saponite	$Mg_3Si_4O_{10}(OH)_2 \cdot nH_2O$	60–150
Three	Dioctahedral	Vermiculite	Vermiculite	$Al_2Si_4O_{10}(OH)_2 \cdot nH_2O$	120–200
	Trioctahedral	Vermiculite	Vermiculite	$Mg_3Si_4O_{10}(OH)_2 \cdot nH_2O$	120–200
Three	Dioctahedral	Mica	Illite	$KAl_2(AlSi_3)O_{10}(OH)_2$	10–40
	Trioctahedral	Mica	Phlogopite	$KMg_3(AlSi_3)O_{10}(OH)_2$	0
Three + One	Dioctahedral	Chlorite	Donbassite	$Al_2Si_4O_{10}(OH)_2 \cdot Mg_3(OH)_6$	1–5
	Trioctahedral	Chlorite	Clinochlore	$Mg_3Si_4O_{10}(OH)_2 \cdot Mg_3(OH)_6$	1–5

mEq = milliequivalents of +1 charge.

octahedral centers are filled (dioctahedral population or gibbsite sheet). When magnesium ions are in the centers, all the octahedra centers are filled (trioctahedral population or brucite sheet).

As shown in Table 7-9, the clay minerals can be subdivided into several groups on the basis of the number of sheets making up their basic structure (unit cell). These groups are further subdivided on the basis of the composition of the octahedral sheet. A final subdivision (not shown in

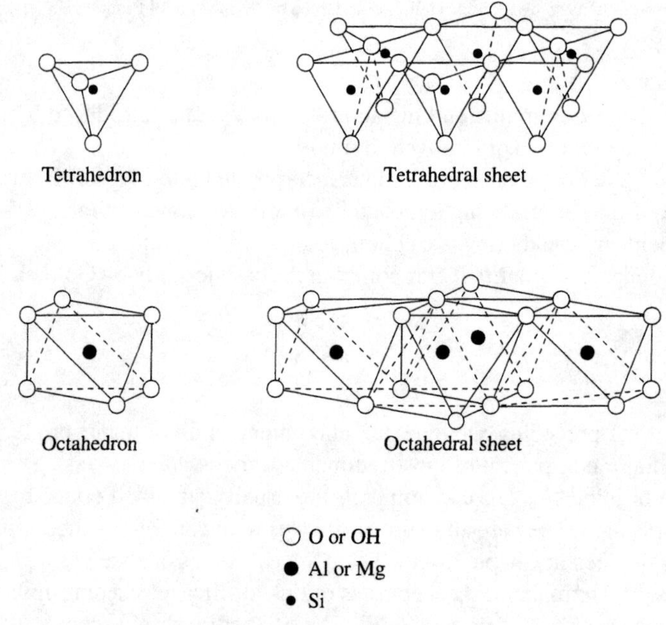

Figure 7-10 Basic components of clay mineral structures.

Table 7-9) is based on the occurrence of substitutions of various cations inside the sheets and on the occurrence of cations and water between the sheets.

The kaolin group consists of one tetrahedral sheet and one octahedral sheet of the gibbsite type, giving the formula $Al_2Si_2O_5(OH)_4$. Very little substitution occurs for the aluminum and silicon ions of the sheets. Kaolinite is the most important member of this group and is a common constituent of soils, sediments, and shales. The serpentine group has a two-sheet structure like kaolinite but with a brucite sheet substituted for the gibbsite sheet. Members of this group occur mainly in igneous and metamorphic rocks.

The term smectite is used for three-sheet clays having either a gibbsite sheet between two tetrahedral sheets (example: montmorillonite) or a brucite sheet between two tetrahedral sheets (example: saponite). Composition variations are common for the smectite clays with substitutions possible in both the tetrahedral (mainly Al for Si) and octahedral sheets (substitutions of H^+, Ca^{2+}, Na^+, Fe^{2+}, Fe^{3+}, Mg^{2+}, Al^{3+}, and other cations). Because of the substitutions, smectites often have a deficiency of positive charge, and this causes extensive adsorption of cations on the surfaces and edges of clay sheets. These ions can easily be replaced since they are not tightly held, and this is why smectite, in contrast to kaolinite, shows a high degree of ion exchange with adjacent solutions (Table 7-9). Interlayer cations and water can occur between the three-sheet groups (layers) of smectites. The interlayer cations are also exchangeable. The possible occurrence of interlayer water means that some smectites exhibit shrinking and swelling behavior.

Minerals of the vermiculite group are similar in composition and structure to those of the smectite group. The two groups differ in their expansion, contraction, and adsorption properties. These properties vary with the excess positive or negative charge occurring in the individual sheets as a result of ion substitution and with the amount and type of interlayer ions. There is a continuous gradation of properties between smectite and vermiculite, with vermiculite having more ion-exchange capacity and less ability to expand and contract. Smectite is much more abundant than vermiculite.

The mica group has a three-sheet structure similar to that of the smectite and vermiculite groups. However, substitution of aluminum for silicon in tetrahedral sheets is balanced by the occurrence of tightly held potassium ions between the three-sheet units. Because of the presence of the potassium ions, water molecules cannot slip between the units and the mica minerals are non-expandable. These minerals also have little or no ion exchange capacity (Table 7-9).

The common mica muscovite has the ideal formula $KAl_2(AlSi_3)O_{10}(OH)_2$, with little substitution of other ions. When brucite sheets rather than gibbsite sheets occur in the three-sheet mica structure, the minerals phlogopite and biotite result. Illite is a common clay mineral that has the same ideal formula as muscovite. However, ion substitution generally occurs in the illite structure. Most illites contain less potassium and aluminum and more silicon than muscovites. Many illite samples exhibit a mixed-layer structure, with up to 20 percent of the layers consisting of smectite.* Glauconite is a variety of illite in which Fe^{3+} replaces aluminum of the octahedral sheet.

The chlorite group, representing the third major structural type of clay mineral, has a three-sheet unit combined with a separate brucite sheet. When the octahedral sheet of the three-sheet

*The term mixed-layer clay is used to describe material that consists of random or regularly alternating layers of two or more different minerals. Since practically any combination of minerals can occur, a clay sediment may have almost any composition between the extremes represented by the individual clay minerals. Some clay material is poorly crystallized or amorphous and is given the general name allophane.

unit is also a brucite sheet, the most common type of chlorite mineral is formed. Dioctahedral chlorites are much less abundant than trioctahedral chlorites. A wide range of ionic substitutions occurs in the chlorite group and many names have been applied to members of the group. Chlorite minerals often occur in mixed-layer clay samples with minerals of the kaolin, smectite, or vermiculite groups.

Table 7-10 gives chemical analyses of samples of the various clay minerals. The major differences in composition can be related to (1) the type and number of sheets in each mineral, (2) the amount of intersheet water, and (3) the amount of substitution of one ion for another. Kaolinite has an Al/Si ratio of 1:1; montmorillonite has an additional tetrahedral sheet and an Al/Si ratio of 1:2. Thus montmorillonite has a higher SiO_2 value and a lower Al_2O_3 value in its analysis. Vermiculite and chlorite have brucite octahedral sheets rather than gibbsite sheets and thus have a high MgO content. Because vermiculite, saponite, and montmorillonite are swelling clays, they have the highest water contents of the analyses in Table 7-10. Although glauconite is a variety of illite, it has a much higher Fe_2O_3 value and a much lower Al_2O_3 value due to substitution of Fe^{3+} for Al^{3+} in the octahedral sheets. Trace amounts of an element in a clay mineral may be due to ionic substitution or to adsorption of ions on clay surfaces. Clay minerals may also have interlayer organic molecules.

The most common clay minerals in soils, sediments, and sedimentary rocks are kaolinite, illite, smectite clays, and chlorite clays. Mixed-layer clays are also very common. The origin of the clay minerals in a given material may be detrital or diagenetic, or a combination of the two. Table 7-11 contains examples of the clay-mineral composition of selected sediments and shales.

In addition to substitution of ions in their structural units, most of the clay minerals have the ability to exchange ions adsorbed on their surfaces, particularly cations. An approximate measure of this ability is known as the cation-exchange capacity (CEC). Table 7-9 gives ranges for the CEC values of the clay minerals. Exchange of adsorbed ions can occur when a clay particle in water moves into a new chemical environment or when water flows past a stationary clay particle. An example of the first type of exchange is the movement of river water carrying clay particles into seawater (Table 7-12). The main exchange shown in Table 7-12 is the release of calcium from the clay and its replacement by the abundant sodium of seawater.

An example of the second type of exchange involves the passage of water through soils. A cation-exchange capacity for a soil can be measured and depends on soil organic matter as well as clay mineral content. Exchange reactions involving the cations Na^+, Ca^{2+}, and Mg^{2+} are of great importance in determining the properties of soils, such as permeability and aggregate structure (Sposito 1989). For example, the presence of adsorbed Na^+ on smectite particles can prevent flocculation and thus reduce permeability because of the clogging effect of swelling clay particles dispersed in the soil solution.

In studying the origin and properties of most sedimentary rocks, transported material is more important than residual material. The sediments presently forming on the surface of the Earth can be broadly divided into three groups: (1) material accumulating in shallow-water regions of the continental shelves (deltas, bays, offshore basins, etc.) after river transport from the land, (2) carbonate sediments forming in shallow water on the shelves, and (3) deep-sea deposits of the ocean floor. Examples of the first group are the sediments of the Mississippi delta (analysis 5, Table 7-6) and of the Gulf of Paria offshore basin (analysis 6, Table 7-6). The composition of these terrigenous sediments is controlled by their distance from land and by their clay mineralogy. In general, the amount of quartz present decreases outward from the land, and more distant sediments have less SiO_2 in their analyses (with the exception of deep sea biogenic siliceous

TABLE 7-10 Chemical Analyses of Clay Minerals

	1	2	3	4	5	6	7	8
	Kaolinite	Montmorillonite	Saponite	Vermiculite	Illite	Mixed-layer illite-smectite	Glauconite	Chlorite
SiO_2	45.44	51.14	50.01	35.92	49.26	52.44	52.64	26.68
Al_2O_3	38.52	19.76	3.89	10.68	28.97	26.38	5.78	25.20
Fe_2O_3	0.80	0.83	0.21	10.94	2.27	0.31	17.88	—
FeO	—	—	—	0.82	0.57	—	3.85	8.70
MgO	0.08	3.22	25.61	22.00	1.32	3.57	3.43	26.96
CaO	0.08	1.62	1.31	0.44	0.67	0.66	0.12	0.28
Na_2O	0.66	0.04	—	—	0.13	0.16	0.18	—
K_2O	0.14	0.11	—	—	7.47	7.85	7.42	—
H_2O^+	13.60	7.99	12.02	19.84	6.03	4.78	5.86	11.70
H_2O^-	0.60	14.81	7.28	—	3.22	3.07	2.83	—
Total	99.92	99.52	100.33	100.64	99.91	99.32	99.99	99.52

Sources:

1: Roseland, Va.; data from C. S. Ross and P. F. Kerr, 1930, U.S. Geol. Survey Profess. Paper 165-E, p. 163.
2: Montmorillon, France; data from C. S. Ross and S. B. Hendricks, 1946, U.S. Geol. Survey Profess. Paper 105-B, p. 23.
3: Milford, Utah; data from H. P. Cahoon, 1954, *Am. Mineralogist*, v. 39, p. 226.
4: Pilot, Md.; data from C. S. Ross et al., 1928, *Econ. Geol.*, v. 23, p. 542.
5: Ballater, Scotland; data from R. C. Mackenzie et al., 1949, *Mineral. Mag.*, v. 28, p. 704.
6: Salina, Pa.; data from C. E. Weaver, 1953, *Geol. Soc. Amer. Bull.*, v. 64, p. 932.
7: New Zealand; data from C. O. Hutton and F. T. Seelye, 1941, *Am. Mineralogist*, v. 26, p. 596.
8: Ducktown, Tenn.; data from R. C. McMurchy, 1934, *Z. Krist.*, v. 88, p. 420.

TABLE 7-11 Average Clay-Mineral Composition of Some Sediments and Shales

	1	2	3	4	5	6	7	8	9	10
Kaolinite	5.7	—	—	65.4	26.5	14.1	6.2	—	12.1	14.4
Smectite	—	2.0	13.0	—	—	2.5	37.6	40.8	—	10.7
Illite	20.0	37.0	27.0	21.5	32.0	62.0	47.5	26.8	34.3	34.2
Chlorite	trace	—	—	5.7	26.5	21.5	8.9	—	11.1	8.2
Mixed-layer illite-smectite	74.4	—	—	—	—	—	—	—	42.5	13.0
Kaolinite-chlorite	—	61.0	60.0	—	—	—	—	32.2	—	17.0
Montmorillonite-mixed-layer clays	—	—	—	7.1	13.5	—	—	—	—	2.3

Sources: After Boggs, S., Jr., *Petrology of Sedimentary Rocks*. Copyright ©1989 by Prentice Hall, Inc. Adapted by permission of Prentice Hall, Inc., Upper Saddle River, NJ.

1: Cretaceous-Eocene shale; data from M. J. Johnson and R. C. Reynolds, 1986, *J. Sediment. Petrol.*, v. 56, p. 501–509.

2: Shelf sediments, NE Gulf of Alaska; data from B. F. Molnia and J. R. Hein, 1982, *J. Sediment. Petrol.*, v. 52, p. 515–527.

3: Miocene-Quaternary shale; data from B. F. Molnia and J. R. Hein, 1982, *J. Sediment. Petrol.*, v. 52, p. 515–527.

4: Cretaceous shale; data from T. P. Lonnie, 1982, *J. Sediment. Petrol.*, v. 52, p. 529–536.

5: Arctic Ocean sediments; data from D. A. Darby, 1975, *J. Sediment. Petrol.*, v. 45, p. 272–275.

6: Beauford Sea sediments; data from A. S. Naidu et al., 1971, *J. Sediment. Petrol.* v. 41, p. 691–694.

7: Black Sea sediments; data from P. Stoffers and G. Muller, 1972, *Sedimentology*, v. 18, p. 113–121.

8: Shelf sediments, California continental shelf; data from G. B. Griggs and J. R. Hein, 1980, *J. Geol.*, v. 88, p. 541–566.

9: Pennsylvanian shales; data from O. B. Raup, 1966, *Am. Assoc. Petroleum Geologists Bull.*, v. 50, p. 251–268.

10: Average of values reported in columns 1 through 9.

oozes). The clay mineralogy of the sediments is determined mainly by the weathering conditions in the source area, since major changes in clay mineralogy apparently do not occur until after deep burial or lithification. Thus any mixture of kaolinite, smectite, illite, and chlorite may be found. Because of the similarity in composition of the clay minerals, the varying mixtures produce little change in overall composition.

Shallow-water carbonate sediments tend to accumulate where land-derived material is scarce. Their compositions show little variation and are similar to those of nondolomitic carbonate rocks (Table 7-2). Deep-sea sediments may be largely siliceous, calcareous, or aluminous. The first two represent accumulation of organic oozes; the third consists of terrestrial clay material that has reached the deep ocean. Chemical analyses of these three types are given in Table 7-6, analyses 7 to 9. Trace elements in both shallow-water and deep-sea regions occur mainly in three forms: (1) combined structurally in the crystal structures of clay minerals, (2) as adsorbed ions on clay minerals, and (3) associated with concretions and nodules of hydrous iron and manganese oxides.

CARBONATE SEDIMENTS

The clay minerals of sedimentary rocks form mostly on the land, whereas the carbonate minerals of these rocks form mostly in the oceans. We now need to take a closer look at the carbonate sediments of the oceans. The two most important carbonate minerals found in ocean sediments are calcite and aragonite. To understand the abundance and distribution of these minerals we need to consider their solubility relationships in water.

TABLE 7-12 Cation Exchange on Smectite upon Transfer from River Water to Seawater

	Surface concentration (mEq/100 g dry wt.)		
Ion	Equilibrated with river water	Equilibrated with seawater	Change on clay
Ca^{2+}	57.6	15.7	−41.9
Mg^{2+}	18.4	18.1	−0.3
Na^+	2.3	44.3	+42.0
K^+	0.6	2.7	+2.1
H^+	2.0	0	−2.0
Total (CEC)	80.9	80.8	

Values are based on smectite reacting for seven days with average seawater after being previously equilibrated for seven days with synthetic average river water.

mEq = milliequivalents of +1 charge

CEC = cation-exchange capacity

Sources: Sayles and Mangelsdorf (1977, 951–960). (After Berner, E. K., and Berner, R. A., *The Global Water Cycle*, Copyright ©1987 by Prentice Hall, Inc. Adapted by permission of Prentice Hall, Inc., Upper Saddle River, NJ.)

It was pointed out in Chapter Three that seawater is approximately saturated with calcite. Let us assume that solid calcite is in equilibrium with water that is in contact with the atmosphere with $P_{CO_2} = 10^{-3.5}$ atm. What would be the values of pH, $[Ca^{2+}]$, and $[CO_3^{2-}]$? The necessary equations and equilibrium constants are given in Table 7-13. In addition we will use a charge balance equation expressing electrical neutrality in molal units:

$$2m_{Ca^{2+}} + m_{H^+} = 2m_{CO_3^{2-}} + m_{HCO_3^-} + m_{OH^-}$$

Using the first equation in Table 7-13, we have

$$[H_2CO_3] = 10^{-3.50} \times 10^{-1.46} = 10^{-4.96}$$

Using other equations in Table 7-13 gives the following:

$$[H^+][HCO_3^-] = 10^{-4.96} \times 10^{-6.35} = 10^{-11.31}$$

$$[HCO_3^-] = \frac{10^{-11.31}}{[H^+]} \tag{7-14}$$

$$[CO_3^{2-}] = \frac{10^{-11.31} \times 10^{-10.33}}{[H^+] \times [H^+]} = \frac{10^{-21.64}}{[H^+]^2} \tag{7-15}$$

$$[Ca^{2+}] = \frac{10^{-8.35}}{[CO_3^{2-}]} = \frac{[H^+]^2 \times 10^{-8.35}}{10^{-21.64}} = 10^{13.29}[H^+]^2 \tag{7-16}$$

$$[OH^-] = \frac{10^{-14.0}}{[H^+]} \tag{7-17}$$

TABLE 7-13 Equilibrium Constants for the Carbonate System

Reaction	Mass law equation	−Log K (25°C)
$CO_2(g) + H_2O = H_2CO_3$	$K_{CO_2} = \frac{[H_2CO_3]}{P_{CO_2}}$	1.46
$H_2CO_3 = H^+ + HCO_3^-$	$K_{H_2CO_3} = \frac{[H^+][HCO_3^-]}{[H_2CO_3]}$	6.35
$HCO_3^- = H^+ + CO_3^{2-}$	$K_{HCO_3^-} = \frac{[H^+][CO_3^{2-}]}{[HCO_3^-]}$	10.33
$H_2O = H^+ + OH^-$	$K_{H_2O} = [H^+][OH^-]$	14.0
$CaCO_3(\text{calcite}) = Ca^{2+} + CO_3^{2-}$	$K_{calcite} = [Ca^{2+}][CO_3^{2-}]$	8.35
$CaCO_3(\text{aragonite}) = Ca^{2+} + CO_3^{2-}$	$K_{aragonite} = [Ca^{2+}][CO_3^{2-}]$	8.22

Note that the first reaction in this table is a combination of the following two reactions: $CO_2(g) = CO_2(aq)$ and $CO_2(aq) + H_2O = H_2CO_3$. This is a common approach even though most dissolved carbon dioxide exists as $CO_2(aq)$ and not as H_2CO_3. The equilibrium relationships are not changed by combining the two reactions.

Substituting equations (7-14) through (7-17) into the charge balance equation given above,

$$2 \times 10^{13.29}[H^+]^2 + [H^+] = 2 \times \frac{10^{-21.64}}{[H^+]^2} + \frac{10^{-11.31}}{[H^+]} + \frac{10^{-14.0}}{[H^+]}$$

Multiplying through by $[H^+]^2$ and changing the number 2 to $10^{0.3}$,

$$10^{13.59}[H^+]^4 + [H^+]^3 = 10^{-21.34} + 10^{-11.31}[H^+] + 10^{-14.0}[H^+]$$

This equation can be solved by trial and error and gives

$$[H^+] = 10^{-8.4}$$

Putting the value for $[H^+]$ into equations (7-15) and (7-16) gives

$$[CO_3^{2-}] = 10^{-4.84} \quad \text{and} \quad [Ca^{2+}] = 10^{-3.51}$$

Our calculations give a pH of 8.4 for water in equilibrium with calcite and a CO_2 pressure fixed by contact with the atmosphere. The measured pH of the oceans varies from about 7.8 to 8.4. For the ion activity product (see Chapter Four) of calcium carbonate we get

$$[Ca^{2+}][CO_3^{2-}] = 10^{-3.51} \times 10^{-4.84} = 10^{-8.35}$$

which is the value of K_{sp} for calcite in Table 7-13 (and Table 4-2). Various combinations of laboratory experiments with synthetic seawater, field research, and theoretical studies all indicate that surface seawater is either at saturation or is supesaturated with respect to calcite. The K_{sp} value for aragonite (Table 7-13) is $10^{-8.22}$ and this suggests that calcite should always form instead of its polymorph aragonite. In other words, at a pressure of one atm and 25°C, calcite is less soluble than aragonite and is the stable form according to the equilibrium laws of thermodynamics.

There are a number of factors that influence the mineral composition and abundance of carbonate sediments in the oceans. First, in the above calculation involving calcite equilibrium with water, we have assumed that the activity coefficients of all ionic species are equal to one (see discussion of activity coefficients in Chapter Three). Actually, because seawater has an ionic strength of about 0.67 molal, the activity coefficients of some species are far from unity. For instance, Table 3-4 shows $\delta_{Ca^{2+}} = 0.28$ and $\delta_{CO_3^{2-}} = 0.20$. However, other factors are more important in affecting actual carbonate mineral behavior in seawater.*

It was also assumed in the above calculation that the other chemical species in seawater do not affect carbonate equilibria. Here we are talking about effects that are in addition to those of low activity coefficients resulting from a high total ionic strength. For example, various laboratory studies have shown that the rate of nucleation and crystal growth of calcite is greatly retarded by the presence of magnesium ion (Berner 1975; Mucci and Morse 1983). Dissolution of calcite can also be retarded. As a result of surface adsorption processes, organic matter occurs in thin layers on carbonate mineral surfaces. These layers can prevent dissolution of particles in undersaturated water (Suess 1973; Zullig and Morse 1988). Thus we can say that the extent of over- and undersaturation of calcium carbonate in seawater is strongly influenced by kinetic factors.

Most calcite in shalllow-marine environments is high-magnesium calcite containing 11 to 19 mole percent $MgCO_3$. This form of calcite appears to be less stable (more soluble) than pure calcite. Its solubility may depend more on kinetics than on thermodynamic equilibrium. Another form of calcite, known as low-magnesium calcite (up to 4 mole percent $MgCO_3$), is a minor constituent of shallow-ocean sediments and a major component of deep-sea, carbonate-rich sediments. It appears to be more stable thermodynamically than pure calcite in seawater. It should be noted that aragonite has a different relationship with magnesium as compared to calcite. Aragonite growth in seawater is not seriously affected by the presence of Mg^{2+} ions and aragonite does not take these ions into its structure to form a solid solution (Berner 1975).

It was mentioned above that the presence of Mg^{2+} in a solution can affect calcite nucleation and growth. The presence of an additional ion such as Mg^{2+} can affect the reaction rates of growth and dissolution in two ways. By forming complexes with Ca^{2+} or CO_3^{2-}, the additional ion can change the saturation state of the solution. Secondly, adsorption of the additional ion on the surface of the reacting solid tends to occur at sites that otherwise would be the most likely locations for growth or dissolution. In the case of Mg^{2+}, adsorption can also lead to the formation of magnesian calcite instead of pure calcite. Other ions that can affect calcium carbonate reaction kinetics are SO_4^{2-} and PO_4^{3+} (Walter and Burton 1986; Mucci et al. 1989), both of which are very abundant in seawater.

Morse and Mackenzie (1990, 72–85) review carbonate dissolution and precipitation kinetics. They point out that the degree of disequilibrium (super- or undersaturation) is one of the primary factors controlling the rate of reaction of carbonate minerals in aqueous solutions. Generally the rate of reaction increases with increasing disequilibrium. Most research has dealt with the kinetics of carbonate dissolution (Chou et al. 1989). This work indicates that a decrease in pH or an increase in temperature or CO_2 pressure will increase the rate of dissolution (Figure 7-11).

Three factors that directly affect calcium carbonate solubility are CO_2 pressure, temperature, and total pressure. In surface waters P_{CO_2} can be greater than the atmospheric value of $10^{-3.5}$ atm. A change from $10^{-3.5}$ atm to $10^{-2.5}$ atm would result in the dissolving of solid calcite and a

*Stumm and Morgan (1981, 408–416) discuss the difficult problem of determining activity coefficients in concentrated solutions such as seawater. They indicate that it is impossible to know unambiguously the relevant activity coefficients. Various models of seawater have been developed since the original model of Garrels and Thompson (1962); there is no general agreement on which model is the most accurate representation. Stumm and Morgan also point out that there is a problem in defining the pH of seawater. The problem arises because the high ionic strength results in the measured pH of seawater not corresponding accurately to $-\log[H^+]$.

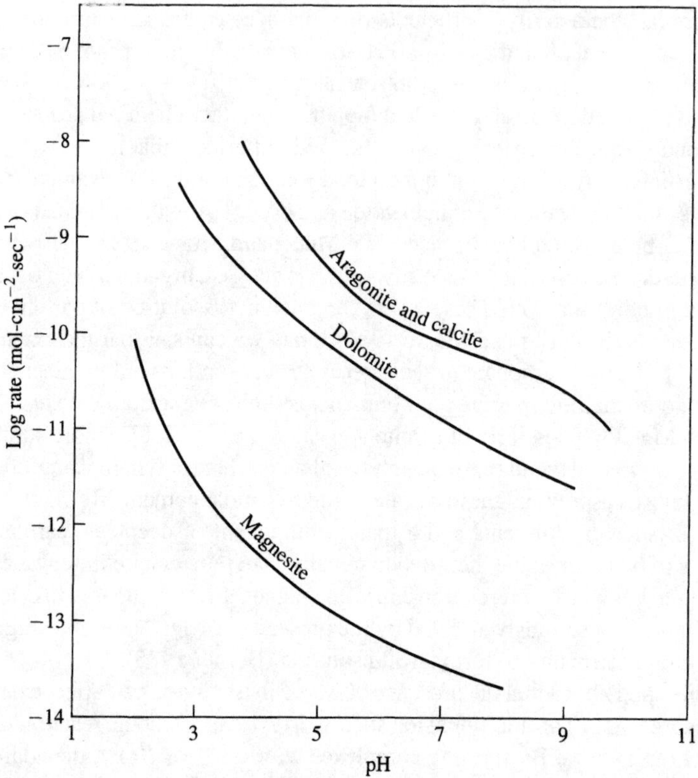

Figure 7-11 Dissolution rates as a function of pH for calcite, aragonite, dolomite, and magnesite. (After J. W. Morse and F. T. Mackenzie, *Geochemistry of Sedimentary Carbonates.* Copyright © 1990. Reprinted by permission of Elsevier Science B.V., Amsterdam.) Data from Chou et al. (1989).

change of pH from 8.4 to 7.6 (as shown by a calculation similar to the one above that gave us the pH value of 8.4). Biological production of CO_2 at shallow depths in ocean waters produces a lower pH, a lower activity of carbonate ion, and a trend toward undersaturation of calcium carbonate. Addition of CO_2 to deep water through the oxidation of organic matter gives similar results. Removal of CO_2 by photosynthesis produces the opposite result (increases the state of saturation and causes precipitation). The following double equation illustrates the relationship between gaseous CO_2 and calcium carbonate precipitation and dissolution:

$$\begin{array}{c} H_2O + CO_2\,(g) \\ \upharpoonleft\downharpoonright \\ CaCO_3\,(s) + H_2CO_3 \rightleftharpoons Ca^{2+} + 2HCO_3^- \end{array} \tag{7-18}$$

With increasing depth in the oceans, temperature decreases and pressure increases. Decreasing temperature causes K_{sp} of calcite to increase; thus the solubility tends to increase. Lower temperatures also produce higher activity coefficients, higher activities, and a tendency for calcite solubility to decrease. The effect on K_{sp} is the more important effect, and the overall result for temperature is an increased solubility in the cooler parts of the ocean. At lower temperatures, carbon dioxide is more soluble in seawater, and this also leads to increased solubility, as shown by equation (7-18). Increasing pressure increases K_{sp} and calcite solubility. Pressure has little effect on the activity coefficient of calcium ion, but increasing pressure increases the activity coefficient of carbonate ion, therefore tending to decrease solubility. The major effect is the change in K_{sp}; thus, in the deep ocean, the effect of pressure causes a significant increase in the solubility of calcite compared to surface zones. Table 7-14 shows the resultant effect of these various factors for seawater in the northern Pacific Ocean.

Figure 7-12 shows the effect of pressure and temperature (depth) on the saturation values of CO_3^{2-} for calcite and aragonite along with measured concentrations of carbonate ion at various depths in the South Atlantic Ocean. For the oceans in general, water at the surface is supersaturated with calcium carbonate (i.e., the activity product is much greater than the solubility product). As depth increases, the solubilities of both calcite and aragonite increase due to the effects of changing pressure and temperature as described above. At some depth, mainly because of the pressure effect, seawater becomes undersaturated (activity product less than solubility product).

Studies of deep-sea sediments show that there is practically no calcium carbonate in sediments accumulating below a water depth of about 5,000 meters. Thus at this depth the rate of carbonate formation must equal the rate of carbonate dissolution. This depth, known as the *carbonate compensation depth,* does not represent the depth at which the water becomes undersaturated with calcium carbonate (200 meters for the northern Pacific Ocean in Table 7-14). Instead, it represents a depth that is controlled by the kinetics of carbonate dissolution. The rate of dissolution is affected by the degree of undersaturation, by the occurrence of organic matter on carbonate grain surfaces, and by the adsorption of hydrated magnesium and other ions on grain surfaces. Somewhere above the compensation depth, biological and inorganic conditions affect the surfaces of the grains enough to allow dissolution to begin to take place at a significant rate.

Morse and Mackenzie (1990) provide a table of calculated values for the calcite saturation state of seawater under varying conditions of temperature, pressure, and partial pressure of CO_2 (Table 7-15). Five different cases are detailed in Table 7-15. Case one is typical of surface, subtropical, Atlantic seawater in equilibrium with the atmosphere. The seawater is supersaturated with respect to calcite. Case two illustrates lowering the temperature to approximately that of bottom seawater and keeping total CO_2 content constant. The lowering of temperature reduces the saturation by only about 10 percent. Case three represents an ocean depth of about 5000 m

TABLE 7-14 Carbonate Chemistry of Northern Pacific Seawater

Depth (m)	Approximate pressure (atm)	Temperature (°C)	pH	Calcite solubility product $[Ca^{2+}][CO_3^{2-}]$	Activity product $[Ca^{2+}][CO_3^{2-}]$	Activity product / solubility product
0	1	20	8.25	4.3×10^{-9}	16.6×10^{-9}	3.85
200	20	10	7.60	5.2×10^{-9}	3.8×10^{-9}	0.73
600	60	6	7.65	6.0×10^{-9}	3.6×10^{-9}	0.60
1,000	100	3	7.65	7.0×10^{-9}	3.6×10^{-9}	0.51
2,000	200	2	7.75	9.4×10^{-9}	5.1×10^{-9}	0.54
4,000	400	2	7.90	15.7×10^{-9}	9.1×10^{-9}	0.58
6,000	600	2	7.85	25.9×10^{-9}	10.5×10^{-9}	0.41

Source: After Berner (1971).

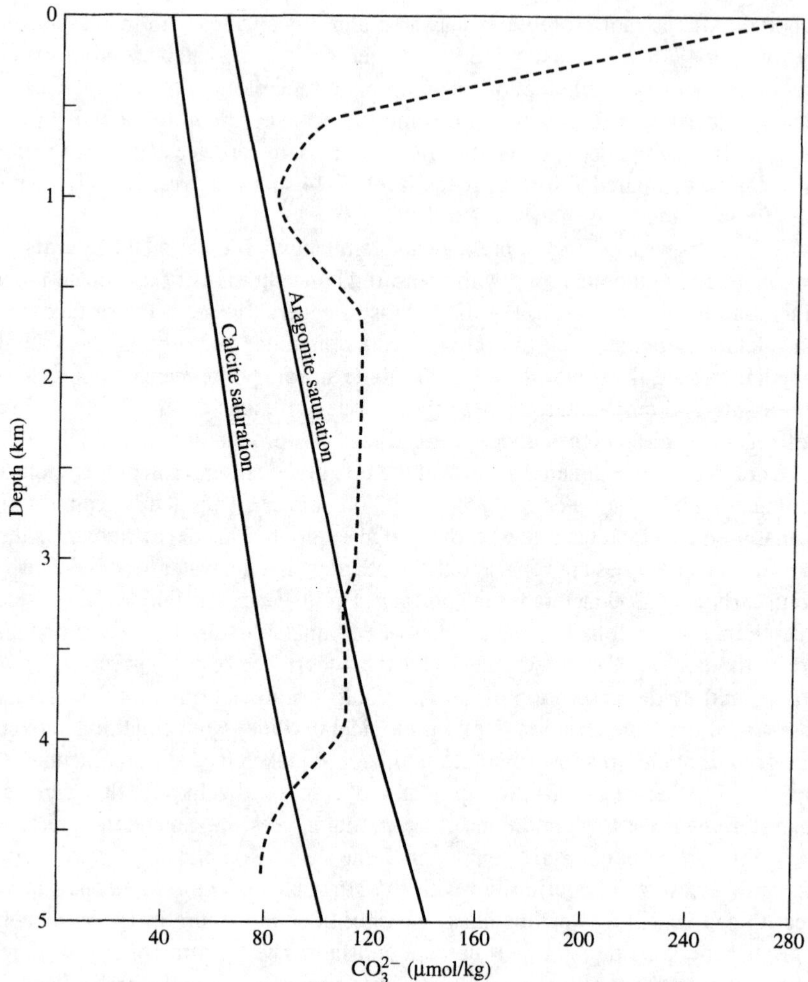

Figure 7-12 Carbonate ion variation (dashed line) with depth, South Atlantic Ocean. To the right of either solid line, seawater is supersaturated with respect to a carbonate mineral; to the left, it is undersaturated. As indicated by the dashed line, near-surface water is strongly supersaturated with calcium carbonate. An aragonite particle falling from surface water should dissolve below about 3300 meters and a calcite particle should dissolve below about 4300 meters. Studies in the Pacific Ocean indicate that supersaturation does not extend to as great depths as found in the Atlantic Ocean. See Table 7-13. (After Richardson, S. M., and McSween, H. Y., Jr., *Geochemistry—Pathways and Processes*. Copyright © 1989 by Prentice Hall, Inc. Adapted by permission of Prentice Hall, Inc., Upper Saddle River, NJ.)

and a pressure of 500 atm. Total CO_2 is the same as in the first two cases. This pressure change has a greater effect than the temperature change of case two; the saturation state drops to less than half the value of case one. Case four combines the lowered temperature and higher pressure; the saturation state drops by another quarter to about one-third the value of case one. The water is still supersaturated.

TABLE 7-15 Variation of Carbonate System Components and Calcite Saturation in Seawater Under Different Conditions

Case	T (°C)	P (atm)	P_{CO_2} (μatm)	CO_2(aq) (μmol/kg)	ΣCO_2 (mmol/kg)	HCO_3^- (μmol/kg)	CO_3^{2-} (μmol/kg)	pH	Ω
1	27	1	330	9	2.03	1729	293	8.28	6.70
2	2.2	1	116	7	2.03	1765	258	8.59	6.10
3	27	500	599	8	2.03	1755	266	8.12	2.99
4	2.2	500	215	6	2.03	1785	238	8.40	2.26
5	2.2	500	800	22	2.28	2167	94	7.92	0.90

Ω is the ratio $[(m_{Ca^{2+}})(m_{CO_3^{2-}})]$/calcite saturation constant. Activities and activity coefficients are not used to calculate this number. If Ω is less than one, the solution is undersaturated. If Ω is greater than one, the solution is supersaturated. See text for discussion of individual cases.

Source: After J. W. Morse and F. T. Mackenzie, *Geochemistry of Sedimentary Carbonates*. Copyright © 1990. Reprinted by permission of Elsevier Science B.V., Amsterdam.

Morse and Mackenzie (1990) point out that seawater at a depth of 5000 m in both the Atlantic and Pacific Oceans is substantially undersaturated with respect to both calcite and aragonite. Case 4 of Table 7-15 suggests that this undersaturation must come about through compositional changes. Morse and Mackenzie believe that the main change is an increase in CO_2 due to oxidation of organic matter and addition of CO_2 prior to deep-water formation. For case five in Table 7-15, the P_{CO_2} value has been increased to 800 μatm; as a result, total CO_2 is increased to 2.28 mmol/kg and CO_2 (aq) goes to the higher value of 22 μmol/kg. The result is undersaturation as indicated by the value of 0.90 in the last column. Note the drop in carbonate ion concentration to 94 μmol/kg for case five.

Most shallow-water sedimentary carbonate minerals are formed as a result of biological (metabolic) processes.* They can be metastable or unstable at the time of formation. Often they are structurally disordered and chemically heterogeneous. The three most abundant materials are low-magnesium calcite, high-magnesium calcite, and aragonite. Organisms can cause precipitation of $CaCO_3$ in water that is not saturated, but we know that most organic precipitation occurs in shallow water that is saturated or supersaturated with calcium carbonate. Apparently, direct precipitation by organisms to form skeletal material only takes place in water that is close to saturation. Since the solubility product for aragonite is similar to that of calcite, and since there is very little difference in the Gibbs energy of the two polymorphs, we might expect that either polymorph could result from organic activity. This is indeed the case, with some organisms forming hard parts from calcite and others preferring aragonite (Table 7-16). Both polymorphs also form inorganically, and it is difficult to tell whether fine-grained material has resulted from organic or inorganic processes. A major part of modern marine sediments is carbonate mud composed predominantly of clay-sized, needle-shaped aragonite crystals. Possible origins for this material include inorganic precipitation, organic precipitation, and abrasion of larger material of inorganic or organic origin.

As mentioned earlier, low-magnesium calcite has a lower solubility than high-magnesium calcite and is the more stable form of calcite. Aragonite has a solubility between that of the two

*The traditional definition of a mineral includes the phrase "formed by inorganic processes." Klein and Hurlbut (1993, 2) include in their definition "those organically produced compounds that answer all the other requirements of a mineral." They point out that the shells of mollusks, for example, are composed of material that is identical to inorganically formed calcium carbonate.

TABLE 7-16 Chemical and Mineralogical Composition of Modern Marine Invertebrates

Type of organism	Most common mineralogy	$CaCO_3$ (range, %)	$MgCO_3$ (range, %)	$SrCO_3$ (mean, %)
Foraminifera	Calcite	77–90	1–16	0.363
Calcareous sponges	Calcite	71–85	5–14	0.116
Madreporarian corals	Aragonite	98–99	0.1–0.8	1.33
Alcyonarian corals	Calcite	73–99	0.3–16	0.385
Echinoids	Calcite	78–92	4–16	0.334
Crinoids	Calcite	83–92	7–16	0.244
Asteroids	Calcite	84–91	9–16	0.225
Ophiuroids	Calcite	83–91	9–17	0.257
Bryozoa	Calcite, aragonite	63–97	0.2–11	0.383
Calcareous brachiopods	Calcite	89–99	0.5–9	0.190
Phosphatic brachiopods	Chitinophosphatic	?–8	2–7	—
Annelid worms	Calcite, aragonite	83–94	6–17	0.707
Pelecypods	Calcite, aragonite	98.6–99.8	0–3	0.258
Gastropods	Calcite, aragonite	96.6–99.9	0–2	0.234
Cephalopods	Aragonite	93.8–99.5	Trace–0.3	0.492
Crustaceans	Calcite, calcium phosphate	29–83	1–16	0.573
Calcareous algae	Calcite, aragonite	65–88	7–29	0.322

Source: After Garrels and MacKenzie (1971, 214).

types of calcite. Benthic skeletal debris occurs mainly in shallow water and consists of aragonite and high-magnesium calcite (corals, echinoderms, etc.). The planktonic skeletal debris of deep-sea sediments is made up mostly of low-magnesium calcite (formas, coccoliths, etc.). So we have a situation in shallow seawater where (1) the thermodynamically most stable form of calcium carbonate, calcite, is not as abundant as the less stable form, aragonite, and (2) when calcite does precipitate, it usually does so as the less stable form, high-magnesium calcite, rather than the more stable form, low-magnesium calcite. Inorganic and organic aragonites generally contain very small amounts of magnesium, since the aragonite structure favors substitution for calcium by ions with radii greater than that of Ca^{2+} (such as Sr^{2+}), whereas the calcite structure favors ions with radii less than that of Ca^{2+} (such as Mg^{2+} and Fe^{2+}). The amount and kind of trace element substitution in calcite and aragonite formed by organisms depends on many factors, including the type of organism involved, the form of calcium carbonate precipitated, particle surface properties, water temperature, and water composition. Table 7-16 gives magnesium and strontium values for the hard parts of various organisms.

In summary, we can say that modern marine carbonate sediments form mainly as a result of direct organic precipitation in shallow water. The majority of the material in carbonate mud is aragonite; most coarser material consists of calcite and aragonite fragments from skeletal components of organisms. Deep-sea carbonate sediments are accumulations of shells of floating and swimming organisms such as forams (calcite), coccoliths (calcite), and pteropods (aragonite). Shallow-water sediments consist mostly of aragonite and high-magnesium calcite while deep-sea sediments generally contain low-magnesium calcite. Because of a difference in the compensation depth, calcium carbonate ooze is more abundant in the Atlantic Ocean as compared to the Pacific

Ocean. No carbonate sediments occur below a depth of about 5,000 meters because of complete dissolution of $CaCO_3$ particles falling to the sea floor.

DIAGENESIS

Diagenesis is a general term for all the processes that take place after deposition of a sediment, excluding weathering and metamorphic processes. Thus it includes changes occurring while a sediment is in contact with overlying water and also changes occurring after the sediment has been buried and thus removed from direct contact with that water. Biological, physical, and chemical diagenetic processes alter the texture, structure, and mineralogy of sediments and lead to the eventual formation of indurated rocks. The chemical processes represent adjustments toward thermodynamic equilibrium brought about by an incompatible mineral assemblage occurring in an extremely variable (in time and space) chemical environment. The major chemical diagenetic processes can be grouped into five categories: oxidation-reduction, authigenesis (enlargement of initial minerals and formation of new minerals), dissolution, diffusion (element segregation), and chemical compaction. We shall first review the properties of the interstitial (pore) waters of sediments and then discuss each of the five categories. A final section describes the use of stable isotopes in the study of diagenesis.

In studying diagenetic processes in a particular sediment, it is important to have knowledge of the chemistry of the interstitial water of the sediment. This water is actively involved in the chemical changes occurring in the sediment, and its composition indicates the environment in which the changes take place. This composition changes with both time and depth. The major factors affecting composition change with depth are rate of sedimentation and amount of organic matter present. (See Chapter Six for a review of the organic matter of sediments.) These two are interrelated. When the rate of sedimentation is low, the organic carbon content of the sediment tends to be low as a result of oxidation of organic matter by the action of bacteria and other organisms. For low rates of sedimentation, diffusional exchange can occur between overlying water and material at great depths in the sediment. This results in relatively small changes in the composition of the interstitial water from the water-sediment contact to these depths.

When the rate of sedimentation is high, the organic carbon content of the sediment tends to be high, since organic matter is being rapidly buried and there is less contact with the dissolved oxygen of the overlying water. Various organisms use up most of the dissolved oxygen from the interstitial water to break down the abundant organic matter. Under these conditions, oxygen is then obtained by microorganisms from the nitrate and sulfate ions of the interstitial water. Extreme reducing conditions can be reached in which methane is produced, sulfate ion is completely depleted, and pyrite is formed in the sediment. The interstitial water may show large compositional changes with depth. As a generalization, deep-sea sediments tend to have low organic contents (slow depositional rates) and sediments deposited near the continents tend to have high organic contents (rapid depositional rates).

Thus many ocean sediments show a characteristic vertical and time-related redox reaction sequence. Near the sediment-water interface, aerobic organisms use dissolved oxygen from overlying and interstitital water to break down organic remains. In the uppermost layer of sediment, depletion of dissolved oxygen results in NO_3^- reduction followed at greater depths by oxygen removal from MnO_2, then Fe_2O_3, and then SO_4^{2-}. If enough organic matter is present, the final overall process is the formation of methane. According to Berner (1980), for a near-shore site,

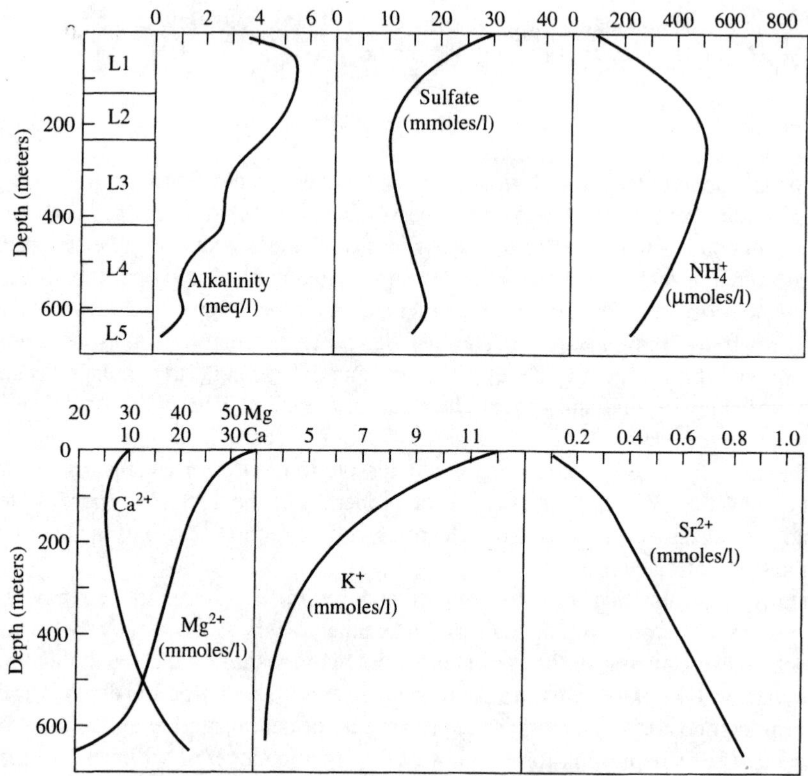

Figure 7-13 Composition of interstitial waters of Indian Ocean sediment. Location 15°50.65′S, 41°49.23′E. L1, gray foram-rich nanno-ooze; L2, gray clay and foram-bearing nanno-ooze; L3, gray clay and foram-bearing nannochalk; L4 and L5, brown clay and nannochalk separated by acoustic reflector. [After Gieskes (1975). Reproduced, with permission, from *Annual Reviews of Earth and Planetary Sciences,* volume 3. Copyright © 1975 by Annual Reviews, Inc. All rights reserved.]

nitrate reduction typically occurs in the top few centimeters, sulfate reduction over the next meter, and methane production below the depth where sulfate disappears. The zone of sulfate reduction is much thicker than that for nitrate reduction because sulfate is much more abundant in seawater than nitrate. The locations and behavior of the MnO_2 and Fe_2O_3 reduction zones are not well known. Presumably the reactions occur in order of their free-energy yield (given above) with microorganisms using each oxidation-reduction reaction as a source of energy for their metabolic processes.

Figure 7-13 is an example of the types of variations that can occur in the composition of interstitial waters. The deep-sea site of the sediment of Figure 7-13 is characterized by a fairly high rate of sedimentation (up to 5 centimeters per 1,000 years), and significant variations occur in the chemistry of the interstitial waters over the 600 meters that were sampled. The increase in alkalinity (capacity to neutralize acid; see Chapter Four) down to about 100 meters depth is due to bacterial reduction of sulfate ion as part of the oxidation of organic matter. This process results in an increase in the amount of HCO_3^- (and thus alkalinity), as represented by the following equation:

$$\frac{1}{53}\,(CH_2O)_{106}(NH_3)_{16}H_3PO_4 + SO_4^{2-} + \frac{15}{53}\,H^+ \quad \text{(organic matter)}$$
$$\rightarrow 2\,HCO_3^- + \frac{16}{35}\,NH_4^+ + H_2S + \frac{1}{53}\,H_2PO_4^- \tag{7-19}$$

As a result of the increase in HCO_3^-, there is also an increase in CO_3^{2-} (due to ionization of HCO_3^-). Thus the amount of Ca^{2+} in the interstitial water decreases, because the solubility product of $CaCO_3$ is exceeded and calcite or aragonite is precipitated. Below about 150 meters, little oxidation occurs and the alkalinity decreases steadily. Also below 150 meters, the amount of Ca^{2+} increases as a result of dissolution of calcium carbonate particles. Since strontium is an important trace element in aragonite and calcite, the strontium content of interstitial waters tends to parallel that of calcium, and Figure 7-13 shows a regular increase with depth for Sr^{2+}.

The decrease in sulfate ion in the upper part of the stratigraphic section is due to the bacterial breakdown of sulfate ion discussed above. The increase beginning at about 200 meters is apparently due to a lower rate of sedimentation for the deeper parts of the section. With a lower rate of accumulation, more oxygen is available for organic oxidation, and also sulfate ion can be replaced by diffusion from overlying seawater. The increase for NH_4^+ in the surface sediments is undoubtedly due to oxidation of organic matter as represented by equation (7-19). The regular decrease in the Mg^{2+} content of the interstitial water may be due to ion-exchange involving clay minerals or to reactions involving carbonate minerals. The decrease with depth of potassium is probably due to adsorption by clay minerals and the formation of illite from these clay minerals.

It should be emphasized that the changes shown in Figure 7-13 do not necessarily hold true for all sediments. Each sediment environment has its own set of characteristics, and the compositions of interstitial waters will vary accordingly. A comparision of two near-shore sediments with similar mineralogy and rates of sedimentation, but different Eh conditions, has been made by Sholkovitz (1973). The seawater above one sediment is almost anaerobic (0.05 to 0.1 milliliters per liter of O_2), while the other sediment is overlain by more oxygenated seawater (0.3 to 0.4 milliliters per liter of O_2). Over depths of 60 centimeters, and relative to seawater, the low-oxygen sediment shows large depletions in SO_4^{2-} and Ca^{2+} and large enrichments in alkalinity and NH_4^+. These changes are similar to those occurring in the top 100 meters of the deep-sea sediment of Figure 7-13. In contrast, the high-oxygen sediment shows only slight depletions in SO_4^{2-} and Ca^{2+} and very small increases in alkalinity and NH_4^+. Sholkovitz concludes that the interstitial water chemistry and the preservation of the calcareous microfossil record are extremely sensitive to small changes in the oxygen content of the seawater under which deposition occurs.

The composition of sediment interstitial waters of an anaerobic environment has been reported by Nissenbaum et al. (1972). For some core samples, they found a large decrease (up to 75 percent relative to seawater) in Ca^{2+} over a depth of 185 centimeters. Sulfate ion and total dissolved sulfur decrease with depth to a complete disappearance of all sulfur species from the interstitial water. Total dissolved CO_2 and NH_3 both increase strongly with depth up to 30 times their concentrations in overlying seawater. Thus it appears that the amount of change in the composition of interstitial waters with depth is high for an environment of low oxidation potential, high organic matter content in the sediment, and high rate of sedimentation. Conversely, there is little change with depth when dissolved oxygen is abundant, organic matter is scarce, and sediments accumulate slowly.

Changes in the chemical composition of overlying and interstitial waters of sediments has a direct effect on trace elements such as the transition metals. Shaw et al. (1990) studied the cycling of transition metals during early diagenesis in sediments from five different depositional environments off the coast of California. The study sites ranged from near-shore basins (low oxygen concentrations and high input of sediment) to the base of an escarpment with near deep-ocean conditions (high oxygen concentration and low sediment input). Figure 7-14 is a plot of observed pore water metal profiles at the escarpment site. Shaw et al. contrast the behavior of nickel and cobalt with that of vanadium and chromium. Nickel and cobalt associated with solid phase manganese oxides in the oxic zone are released to the pore water in the manganese reduction zone (Figure 7-14). Thus nickel and cobalt are remobilized from reducing sediments and enriched in oxic sediments. Vanadium and chromium are apparently transported to the sediments as reduced species in biogenic

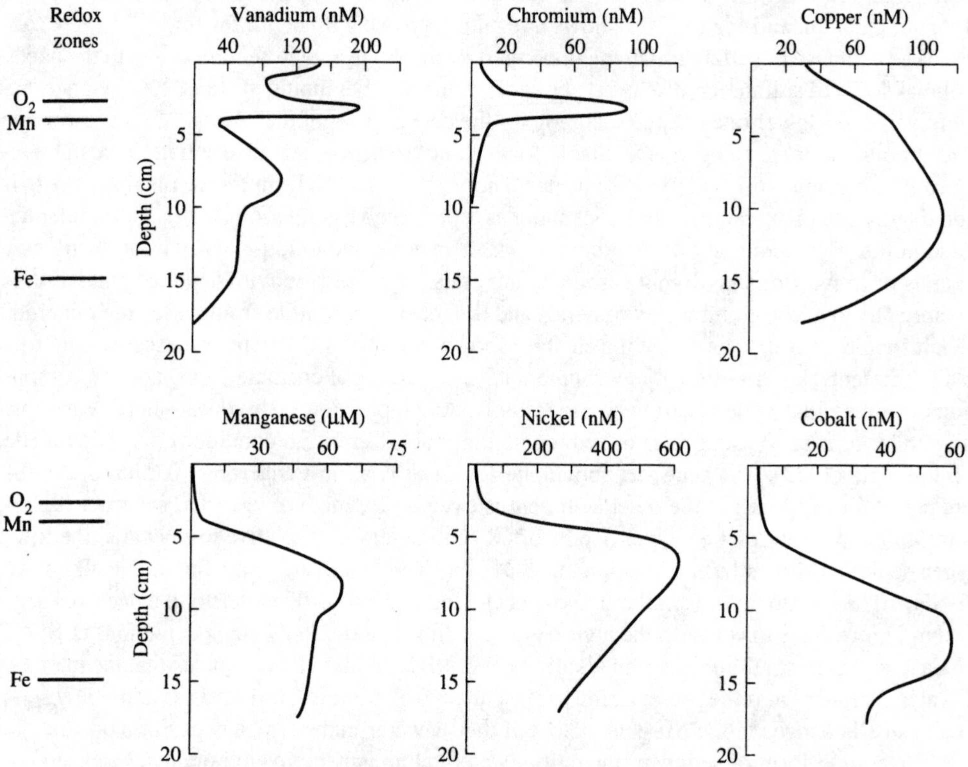

Figure 7-14 Observed metal abundances in interstitial water of sediments, Patton escarpment, southern California. Near-shore basin sites also studied show a redox reaction sequence similar to that in the Patton escarpment but with the reaction zones compressed toward the sediment-water interface. The sequence of reaction zones becomes more compressed with decreasing bottom water oxygen concentration. The redox zones shown represent the disappearance of dissolved oxygen followed at depth by the reduction of MnO_2 and then Fe_2O_3. For the Patton site, oxygen is no longer available as an oxidant at a depth of 2.5 cm, manganese is reduced starting at a depth of about 4.0 cm, and iron reduction starts at approximately 15.0 cm. For the basins, typical depth locations are 0.2 cm (oxygen), 1.0 cm (manganese), and 2.0 cm (iron). The Patton escarpment is located about 240 km off the coast of southern California and the near-shore basin sites are located from 30 to 120 km off the coast. (Reprinted from *Geochimica et Cosmochimica Acta*, v. 54, Shaw et. al., Early diagenesis in differing depositional environments: The response of transition metals in pore water. Copyright 1990, with kind permission from Elsevier Science Ltd, The Boulevard, Langford Lane, Kidlington OX5 1GB, UK.)

material and are released to the pore water in the near-surface oxic zone (Figure 7-14). They are removed from the pore water under reducing conditions below the oxic zone (probably by sediment adsorption of the reduced forms) and thus would tend to accumulate in sediments forming under reducing conditions. This behavior is opposite to that of nickel and cobalt. In contrast to these four metals, copper appears to be less sensitive to redox conditions (Figure 7-14).

It is clear from the above review of interstitial waters that oxidation-reduction reactions are an extremely important part of diagenesis. A large section of Chapter Four deals with oxidation-reduction reactions, including calculation and plotting of Eh-pH diagrams. Berner (1971, 118) has used such diagrams to predict the following order of reduction of dissolved species as Eh drops: O_2, NO_3^-, SO_4^{2-}, N_2, and finally HCO_3^-. The dissolved species that should be abundant in the most reduced waters are CH_4, HS^-, NH_4^+, and H_2. The oxidation-reduction reactions that are of greatest importance during diagenesis are those involving organic matter, iron- and manganese-bearing detrital minerals, and the dissolved species listed above. Some of these reactions have been discussed in Chapters Four and Six. The major minerals resulting from anaerobic reduction are pyrite and marcasite, which form from unstable iron sulfides (such as mackinawite) precipitated by reaction between dissolved sulfide (from bacterial sulfate reduction) and detrital iron-bearing minerals (such as goethite and chlorite). At the other extreme, environments with practically unlimited oxygen result in sediments with essentially no organic matter and containing minerals such as gypsum and hematite. As previously mentioned, minerals such as calcite can form indirectly as a result of oxidation-reduction reactions.

Authigenesis is a general name for the growth of minerals in a sediment. It includes both growth of minerals initially present in the sediment and formation of new minerals. An example of the former is the development of euhedral overgrowths of quartz on detrital quartz grains. As shown in Figure 7-2, quartz has a very low solubility. Most interstitial waters of modern sediments are supersaturated with respect to amorphous silica as a result of dissolution of diatoms and radiolaria. The buildup of the SiO_2 concentration can eventually lead to local precipitation in sediments of quartz and of opaline silica. Replacement of other minerals by silica is also a common occurrence. Direct precipitation of the rock chert (which is made up of microcrystalline quartz) apparently does not occur in modern sediments, and bedded cherts are believed to form during diagenesis from initial accumulations of siliceous organic remains.

Cementation is obviously an important authigenic process. Most sandstones and most carbonate rocks become lithified by precipitation of silica, calcium carbonate, or both. There are several possible sources for the dissolved silica that forms cement consisting of quartz, cristobalite, or opal. One source is the alteration of silicate minerals by circulating solutions, with reactions occurring similar to those described earlier for the weathering of silicates. Alteration of volcanic glass particles is also a source of silica when they are present. Another source is silica put into solution by the effect of pressure on buried quartz grains. Dissolution occurs at grain contacts, and the dissolved silica may precipitate as a local cement or migrate out of the source area. Some dissolved silica comes from the dissolution of diatom and radiolarian tests; this is the dominant source of silica in siliceous oozes.

We have already discussed the factors governing precipitation and dissolution of calcite and aragonite. For shallow-water sediments, formation of calcite or aragonite cement is apparently due to the same types of processes that cause formation of inorganic carbonate sediments. In deep-sea sediments, calcareous oozes are cemented more slowly by calcite precipitation after dissolution owing to increasing grain contact (pressure solution) and to decay of organic films on carbonate grains.

A large number of new minerals can form during diagenesis by replacement or alteration of pre-existing minerals. We have previously discussed the clay minerals and their ability to adsorb and exchange cations. Detrital clay minerals are thus able to undergo a number of changes in sediments. An example is the formation of illite from smectite by adsorption of potassium ions. Another example is represented by the alteration of clay minerals to micas. Other common authigenic minerals include feldspars (K-feldspar, albite), iron oxides (limonite, hematite), sulfates (gypsum, anhydrite), and sulfides (pyrite, marcasite).

The new carbonate mineral that most often forms during diagenesis is low-magnesium calcite. Morse and Mackenzie (1990, 293–295) list the following findings from experimental and other work on the transformation of aragonite to low-magnesium calcite:

1. The activation energy for this reaction is sufficiently large to make the kinetics of the polymorphic transition too slow to be of importance below temperatures of 300° to 500°C.
2. The rate of reaction is dependent on the nucleation and growth rates of calcite, not the dissolution rate of aragonite.
3. Mg^{2+}, SO_4^{2-}, and some organic compounds can inhibit the transformation.
4. The reaction is not isochemical.
5. Sedimentary aragonite persists for hundreds of millions of years under conditions where exposure to aqueous solutions is limited.

This information along with other research suggests that the diagenetic transformation of aragonite to calcite is *not* a polymorphic inversion, but rather a dissolution-precipitation process. This process consists of the simultaneous solution of the thermodynamically less stable aragonite and precipitation of the more stable calcite.

An important carbonate mineral, in addition to low-magnesium calcite, that forms during diagenesis is dolomite. A comparison of experimental measurements of the solubility product of dolomite with a calculation of the activity product using activity values (see Chapters Three and Four) indicates that seawater is supersaturated with dolomite (Table 7-17). However, dolomite, which has a strong ordering of calcium and magnesium ions in alternating crystallographic planes, apparently cannot form in seawater because the kinetics of the ordering of the cations requires special conditions for formation.

TABLE 7-17 Data Used for Calculation of the Activity Product of Dolomite in Seawater

Ion	Molality in seawater	Percent of ion occurring as free ion	Activity coefficient	Activity of free ion
Ca^{2+}	0.010	91	0.28	0.0025
Mg^{2+}	0.054	87	0.36	0.017
CO_3^{2-}	0.0003	9	0.20	0.000005

$$K = [Ca^{2+}][Mg^{2+}][CO_3^{2-}]^2 = (0.0025)(0.017)(0.000005)^2$$
$$= 1.06 \times 10^{-15}$$

The experimentally measured value of K_{sp} is on the order of 10^{-17}.

Source: Data from Table 3-4.

Dolomite forms under extreme conditions in both laboratory experiments (from solutions with unusual compositions) and field environments (hot springs, certain types of salt lagoons). It does not form in the laboratory from seawater at temperatures found in typical sedimentary environments and has not been found forming in normal sedimentary environments. Thus it appears that most dolomite forms during diagenesis of carbonate sediments or after formation of limestones.

The chemical change can be expressed as

$$2\ CaCO_3\ \text{(calcite)} + Mg^{2+} \rightarrow CaMg(CO_3)_2\ \text{(dolomite)} + Ca^{2+} \qquad (7\text{-}20)$$

Using free-energy values from Table 3-2, the standard free energy of this reaction is $\Delta G^0 = -2{,}436$ calories, and the equilibrium constant is

$$K = \frac{[Ca^{2+}]}{[Mg^{2+}]} = 10^{1.8}$$

In seawater the ratio of the activities of calcium and magnesium is less than this value (about 10^{-1}), and thus dolomite should form from calcite in normal seawater. However, even though dolomitization is favored thermodynamically, it does not occur in modern sediments except in special cases of high salinity (brine waters of supratidal flats and saline lakes), where dolomite does replace calcium carbonate particles. These special cases are not sufficient to explain the large volumes of dolomite found in the sedimentary rocks of the world, and the mode of formation of these rocks is still a major unsolved problem (Hardie 1987).

Groundwaters in dolomite aquifers generally have higher calcium and lower magnesium concentrations relative to those found in seawater. This may be due to dolomitization of pre-existing calcite. Buried sediments over time tend to reach higher temperatures and the increase of temperature favors dolomitization for two reasons: (1) water molecules attached to Mg^{2+} ions, which prevent reaction with other ions, are removed more easily at higher temperatures; and (2) the maximum ratio of Ca^{2+}/Mg^{2+} that will cause reaction (7-20) to go to the right increases and thus makes possible dolomite formation in calcium-rich waters (Figure 7-15). Recent research suggests that extensive dolomitization occurs in the deep subsurface at higher temperatures than those found at shallow depths (Machel and Mountjoy 1987).

The process of dissolution is extremely important in determining the composition of natural waters, soils, sediments, and sedimentary rocks. This process is discussed earlier in this chapter in the section on chemical weathering and has been mentioned in the preceding material dealing with oxidation-reduction and authigenesis. These earlier discussions indicate that often more than one chemical process has to be considered in studying precipitation and dissolution in natural systems such as sediments. For diagenetic systems, dissolution is often intimately related to oxidation-reduction reactions. Growth of new minerals (authigenesis) is commonly the final result of the oxidative and reductive dissolution of minerals.

Hering and Stumm (1990) point out that changes in the oxidation state of metals in minerals affect mineral solubilities. Some minerals, such as pyrite, become more soluble upon oxidation while others, such as hematite, become more soluble as a result of reduction. In both cases, metal-anion bonds are weakened by the redox process. Often oxidizable organic ligands are part of the process (Figure 7-16). In the case of the dissolution of ferric oxide or hydroxide shown in Figure 7-16, an important result, in addition to the dissolution of an iron-bearing compound, is the release of adsorbed phosphate ions and trace metals from the dissolving solid. The release of phosphate ions in lake sediments due to reducing conditions on the lake bottom increases the flux

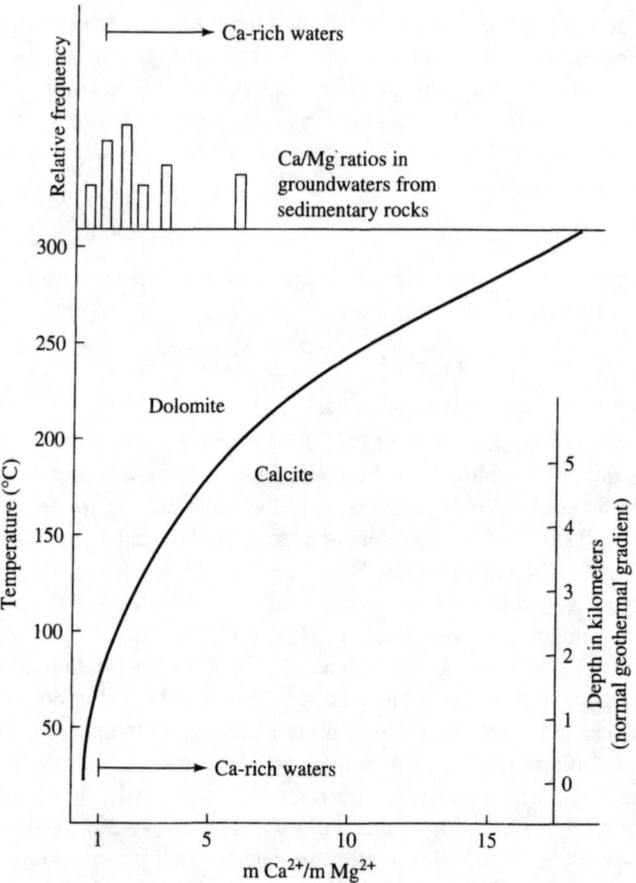

Figure 7-15 Plot of the expected stability field of dolomite as a function of temperature and molar Ca/Mg ratio in 1M chloride brines. At the top of the plot is a histogram showing the Ca/Mg ratios of natural groundwaters in contact with a variety of sedimentary rocks. According to this diagram, at 25°C a molar ratio for Ca^{2+}/Mg^{2+} of slightly less than one (or smaller) is needed to convert calcite to dolomite. At temperatures exceeding about 70°C, dolomite can form in calcium-rich waters (Ca^{2+}/Mg^{2+} ratio greater than one). In other words, at the higher temperatures found in buried sediments and sedimentary rocks, dolomite could form from calcite by reaction with circulating waters with very low Mg/Ca ratios. Since kinetic inhibitors of the calcite-to-dolomite reaction tend to be less important at higher temperatures, dolomite formation is favored as temperature increases by both thermodynamic and kinetic factors (Machel and Mountjoy 1986). After Boggs (1992). From Hardie (1987).

of phosphorus to the overlying water and can lead to an increase in the rate of eutrophication of the lake.

Molecular diffusion (net motion of atoms, ions, or molecules) in a sediment can take place through the interstitial water. Fick's First Law states that the mass flux of a given component *a* is proportional to the gradient of the concentration of the component

$$\text{flux}_a = -D_a\left(\frac{\partial C_a}{\partial x}\right) \tag{7-21}$$

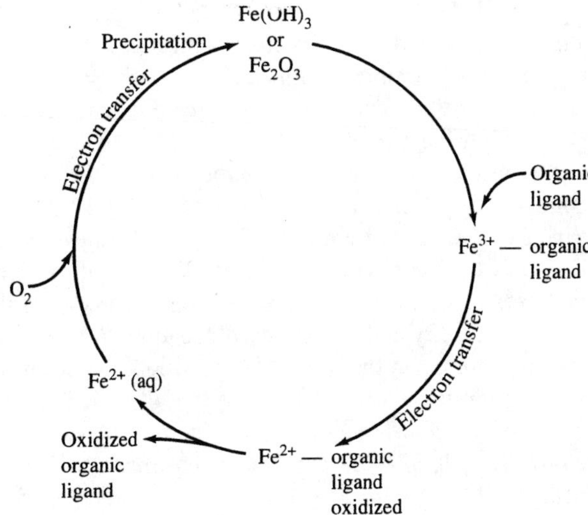

Figure 7-16a Iron cycle and redox processes typically observed in soils, sediments, and natural waters. The dissolution of ferric oxide or hydroxide by surface attachment of oxidizable organic ligands and the subsequent reoxidation of Fe^{2+} by O_2 causes relatively rapid cycling of electrons and of reactive elements (e.g., organic carbon, oxygen, trace metals, and phosphate) at the oxic-anoxic boundary of sediments. The source of the oxidizable organic ligand is usually biodegradable biogenic material. This cycle can also occur, photochemically induced, in oxic surface waters. On overall balance, the iron cycle mediates the progressive oxidation of organic matter by oxygen. Modified from Hering and Stumm (1990).

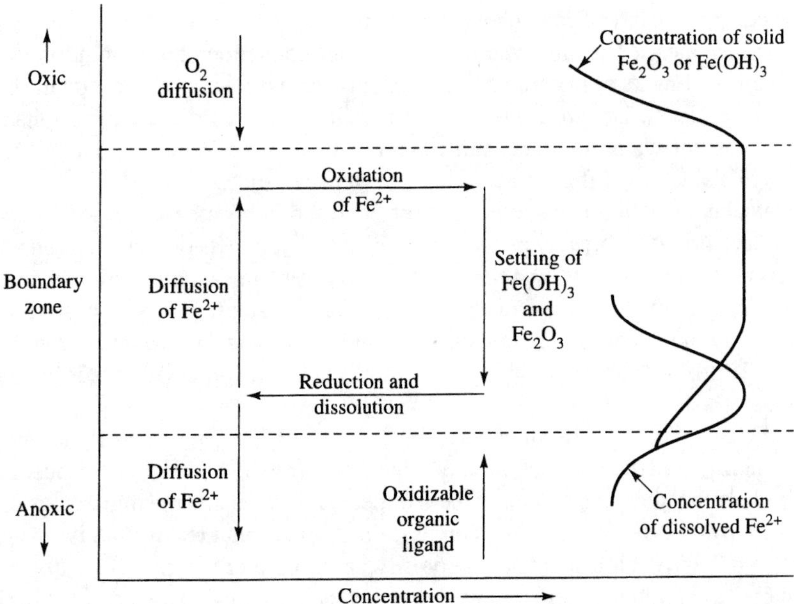

Figure 7-16b Transformations of iron at an oxic-anoxic (redox) boundary in a water column or sediment. Peaks in the concentrations of solid Fe_2O_3 and $Fe(OH)_3$ and of dissolved Fe^{2+} coincide with the depth of maximum Fe^{3+} and Fe^{2+} production. Reduction of Fe^{3+} in the solid Fe_2O_3 or $Fe(OH)_3$ by an oxidizable organic ligand leads to dissolution of the solid and production of Fe^{2+} (aq). Some Fe^{2+} diffuses upward and is oxidized to produce new solid Fe_2O_3 or $Fe(OH)_3$. Under some circumstances (flux of O_2 downward exceeds flux of Fe^{2+} upward) redox fronts may progess downward into sediments. Such non-steady state diagenesis can produce iron-rich layers in the sediments. After Davison (1985) and Hering and Stumm (1990).

where D_a = diffusion coefficient (proportionality constant) of a in $(\text{distance})^2$ per unit time, C_a = concentration of a in mass per unit volume, x = direction of maximum concentration gradient (perpendicular to the sediment-water interface for most sediments), and flux_a = diffusion flux of component a in mass per unit area per unit time. The movement of component a is from the point of high concentration to the point of low concentration, as indicated by the minus sign in equation (7-21). Diffusion coefficients of ions and molecules can be measured in the interstitial waters of sediments and tend to be on the order of 10^2 cm^2yr^{-1} (Li and Gregory 1974).

An example of the use of Fick's First Law has been provided by Trefry and Presley (1982), who studied the interstitial water chemistry of Mississippi Delta sediments. Because of reducing conditions in the sediments, significant fluxes of dissolved manganese (Mn^{2+}) pass from the Delta sediments to the overlying water column. Trefry and Presley used equation (7-21) to calculate the diffusional flux of Mn^{2+}. First they corrected the molecular diffusion coefficent for Mn^{2+} in the water for sediment porosity and tortuosity (due to the presence of solid particles that prevent an ion from diffusing freely in all directions). The corrected value for D_a is 107.7 cm^2yr^{-1}, and the value used for dC_a/dx is 7.8 μgcm^{-4} (from chemical analyses of Delta interstitial water at various depths). The calculated flux across the sediment-water interface is 840 $\mu gcm^{-2}yr^{-1}$. This manganese, transported by water movement from the Delta area seaward, is believed to contribute to the large amount of manganese found in deep-sea sediments and in manganese nodules.

Molecular diffusion of ions plays an important role in many chemical processes taking place during diagenesis. It is involved in the oxidation-reduction and dissolution reactions discussed previously. Fluxes of material from sediments can have a major effect on the composition of the overlying waters. The growth of monomineralic layers in sediments, precipitation of minerals at the sediment-water interface, and certain features of sedimentary rocks such as concretions can be explained by diffusion and related diagenetic processes.

Chemical compaction is the term used for the solution of minerals in sediments and rocks at grain contacts brought about by pressure. It is also referred to as pressure solution and is most common in carbonate sediments and rocks. After early mechanical compaction of sediments, the individual grains are in contact and further stress due to loading from above or to tectonic forces can cause dissolution at contact points. Ions pass into solution and move away from stress points by diffusion. This is another example of two or more diagenetic processes occurring together and simultaneously.

In carbonate masses the result of chemical compaction is often suture-like seams called stylolites. Commonly, constituents such as clay minerals with low solubility are concentrated in the seams. It is likely that the composition of the sediment pore water and the mineralogy of the grains affects the depth and extent of chemical compaction. An important result in all types of sediments is a reduction in bulk volume and thus in porosity. In the case of solution of quartz grains, some of the silica released into solution is reprecipitated locally as quartz overgrowths, further reducing porosity. Note, however, that diagenetic reactions can generate secondary porosity, as well as result in a decrease of porosity.

Stable isotopes have been used extensively to study diagenetic processes (Longstaffe 1987; Arthur et al. 1983). Burial-history curves can be produced for clastic sediments by combining radiogenic age-dating of mineral overgrowths with paleotemperatures obtained from oxygen isotope data (see Chapter Two). Information on pore-water evolution can be obtained using oxygen and other stable isotopes (Figure 7-17). Carbon, sulfur, nitrogen, and other isotopes are used to study the effect of diagenetic processes on organic matter; such research allows us to better un-

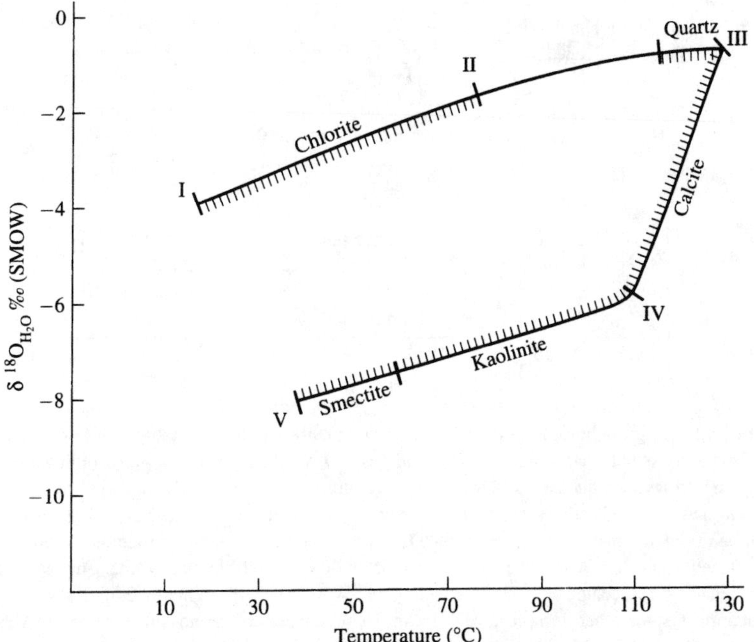

Figure 7-17 A model for the oxygen isotope evolution of pore water in the basal Belly River sandstone, determined from the paragenetic sequence and the $\delta\ ^{18}O$ values of diagenetic minerals. The solid line represents the variation in pore water $\delta\ ^{18}O$ values and temperature throughout diagenesis. The five stages of diagenesis marked on the line are as follows: stage I—early (shallow) diagenesis; stage II—burial diagenesis; stage III—maximum burial and uplift; stages IV and V—erosion to present conditions. The hatched areas along the line represent intervals of crystallization for the indicated diagenetic minerals. Mixed illite-smectite also forms in the interval labeled smectite. After Longstaffe (1987).

derstand the formation of kerogen, coal, and petroleum. Carbon and oxygen isotopes provide useful information on carbonate diagenesis (Figure 7-18). Further information on isotopic study of diagenesis can be found in Faure (1986), Arthur et al. (1983), and Longstaffe (1987).

EVAPORITES AND BRINES

It was pointed out in Chapter Four that the solubilities of compounds in water are governed by their solubility products. The solubility product constant for halite at 25°C is

$$K_{sp} = [Na^+][Cl^-] = 38$$

and the solubility is found as follows [assuming activity coefficents of 0.73 for Na^+ and 0.63 for Cl^- in seawater, as calculated by Garrels and Christ (1965, 63)]:

$$K_{sp} = \gamma_{Na^+}(Na^+)\gamma_{Cl^-}(Cl^-) = 38$$

$$0.46(Na^+)^2 = 38$$

$$(Na^+) = \text{solubility} = \sqrt{\frac{38}{0.46}} = 9.09 \text{ mol/l}$$

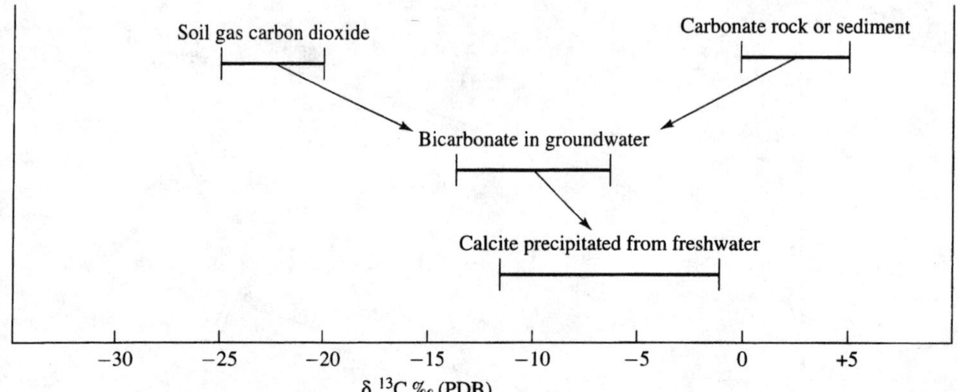

Figure 7-18 Diagram illustrating the effect of mixing carbon from metastable marine carbonate sediments or rocks with carbon derived from soil-gas CO_2 on the $\delta\,^{13}C$ composition of limestones undergoing freshwater diagenesis. This process occurs in Bermuda, where a large volume of limestone has been subaerially exposed for much of its existence. Soil-gas CO_2 derived from oxidation of organic matter and depleted in ^{13}C is produced in the vadose (unsaturated) zone and mixes in subsurface water with "heavier" carbon from dissolution of marine calcite. The resulting water contains bicarbonate ion of intermediate $\delta\,^{13}C$. When calcite precipitates from this water as a cement or in some other form, it is depleted in ^{13}C compared to marine carbonate. After Allan and Matthews (1982). (After J. W. Morse and F. T. Mackenzie, *Geochemistry of Sedimentary Carbonates*. Copyright © 1990. Reprinted by permission of Elsevier Science B.V., Amsterdam.)

Seawater contains Na^+ and Cl^- as two of its major components (Table 3-3), and evaporation of seawater leads to precipitation of halite when the product $[Na^+][Cl^-]$ becomes greater than K_{sp}. The other major ions in seawater are Mg^{2+}, Ca^{2+}, K^+, and SO_4^{2-}. Thus complete evaporation of seawater results in precipitation of a variety of minerals made up of these components (Table 7-18). Gypsum precipitates before halite because the value of its solubility product is much lower than that of halite. Other minerals, such as polyhalite, form after halite because their component ions have a relatively low concentration in seawater compared to that of Na^+ and Cl^-.

TABLE 7-18 Compositions of Minerals of Marine Evaporites

Gypsum	$CaSO_4 \cdot 2\,H_2O$
Anhydrite	$CaSO_4$
Halite	$NaCl$
Sylvite	KCl
Polyhalite	$K_2Ca_2Mg(SO_4)_4 \cdot 2\,H_2O$
Bloedite	$Na_2Mg(SO_4)_2 \cdot 4\,H_2O$
Epsomite	$MgSO_4 \cdot 7\,H_2O$
Kainite	$KMg(SO_4)Cl \cdot 3\,H_2O$
Hexahydrite	$MgSO_4 \cdot 6\,H_2O$
Kieserite	$MgSO_4 \cdot H_2O$
Carnallite	$KMgCl_3 \cdot 6\,H_2O$
Bischofite	$MgCl_2 \cdot 6\,H_2O$
Langbeinite	$K_2Mg_2(SO_4)_3$
Glauberite	$CaSO_4 \cdot Na_2SO_4$

TABLE 7-19 Theoretical Successions of Precipitation (from Bottom to Top) of Salts from Seawater at 0, 25, and 55°C

0°C	25°C	55°C
Potassium and magnesium salts plus halite	Potassium and magnesium salts plus halite	Potassium and magnesium salts plus halite
Polyhalite plus halite	Polyhalite plus halite	Polyhalite plus halite
	Anhydrite plus halite	Anhydrite plus halite
Gypsum plus halite	Anhydrite	
Gypsum	Gypsum	Anhydrite

Source: Data from Stewart (1963).

Thermodynamic calculations (Eugster et al. 1980) indicate that the equilibrium sequence of precipitation from evaporating seawater at 25°C should be gypsum (starting at about 28 percent of original water volume), anhydrite (at about 10 percent), halite (at about 9 percent), and various potassium and magnesium salts (starting at about 8 percent). This sequence is for equilibrium crystallization, that is, there is no occurrence of fractionation. Fractional crystallization, which mainly affects late-forming minerals, occurs when early-formed solid phases are removed from contact with the liquid brine. This theoretical sequence varies with temperature because of the equilibrium relationships between gypsum and anhydrite (Table 7-19 and Figure 7-19). The stable mineral is determined by the activity of H_2O in the brine. The activity of water in a brine decreases with increasing concentration of dissolved species. Decreasing water activity and increasing temperature both tend to make anhydrite the stable phase.

In contrast to these theoretical predictions, anhydrite practically never forms as a primary mineral from modern evaporating brines. Instead, gypsum forms as a metastable mineral for those conditions when anhydrite should form. It is generally believed that the abundant anhydrite found in buried evaporite deposits is a secondary mineral that formed from primary gypsum. Figure 7-19 shows that anhydrite is favored by high temperatures. It is also favored by increasing

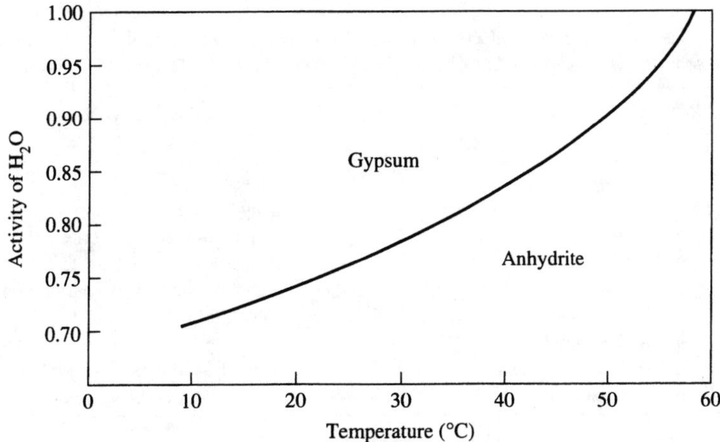

Figure 7-19 Stability fields of gypsum and anhydrite at one atmosphere total pressure. After Hardie (1967).

pressure when the pressure on solids is lithostatic and the pressure on associated pore liquids is hydrostatic. Thus anhydrite forms by replacement of gypsum during diagenesis or after lithification.

Evaporation of 1,000 meters of seawater would produce about 0.75 meter of gypsum and about 13.7 meters of halite. Some of the major evaporite deposits contain as much as 1,000 meters of gypsum or anhydrite and up to 350 meters of halite. It is common to find repetitions of the salt sequence shown in Table 7-19. Thus formation of these deposits must have occurred in a restricted basin where seawater could be continually or periodically added as evaporation proceeds. Some evaporite minerals are forming today, but there are no modern analogs to the extensive evaporite basins that must have existed in the past. Another feature of large evaporite deposits is a rapid lateral variation in thickness and mineralogy of sedimentary layers. Some of these changes can be related to the depositional environment; others are due to secondary changes caused by increases in temperature and pressure during burial. Evidence of recrystallization of primary minerals and of replacement of these minerals by other minerals is commonly found. Replacements may involve water released when gypsum is changed to anhydrite or may be due to the action of percolating groundwater.

Only a few evaporite deposits have significant amounts of the potassium and magnesium salts that precipitate as the last minerals from an extremely concentrated brine. The most abundant of these minerals, polyhalite, is often a secondary rather than a primary mineral. As with the major evaporite minerals, a theoretical sequence can be worked out for the formation of these late-stage minerals (Figure 7-20). Table 7-20 compares the observed mineral zones of the Upper Permian Zechstein II salt formation in Germany to the predicted mineral zones using a thermodynamic model of seawater developed by Harvie and Weare (1980).

Unlike the situation for gypsum, anhydrite, and halite, the order of formation and abundance of these minerals depends on whether or not early-formed minerals can react with the residual liquid. If early-formed minerals remain in contact with the brine from which they have formed, they can become unstable and be changed to new minerals as further crystallization changes the composition of the brine (equilibrium crystallization). Tables 7-20 and 7-21 both are based on the assumption of equilibrium crystallization and Table 7-21 gives more detail on the theoretical brine from which the sequence of minerals shown in Table 7-20 formed.

TABLE 7-20 The Observed Mineral Zones in the Stassfurt Zechstein II Salt Formation Versus the Predicted Mineral Zones Assuming Equilibrium Evaporation of a Brine Chemically Similar to Seawater

Zechstein II zones	Predicted zones
	gypsum
anhydrite	anhydrite
anhydrite-halite	anhydrite-halite
anhydrite-halite-glauberite	anhydrite-halite-glauberite
anhydrite-halite-polyhalite	anhydrite-halite-polyhalite
	anhydrite-halite-polyhalite-epsomite
	anhydrite-halite-polyhalite-hexahydrite
anhydrite-halite-polyhalite-kieserite	anhydrite-halite-polyhalite-kieserite
anhydrite-halite-polyhalite-kieserite-carnallite	anhydrite-halite-polyhalite-kieserite-carnallite
anhydrite-halite-kieserite-carnallite	anhydrite-halite-kieserite-carnallite

Source: Reprinted from *Geochimica et Cosmochimica Acta*, v. 44, Harvie and Weare, The prediction of mineral solubilities in natural waters: The Na-K-Mg-Ca-Cl-SO_4-H_2O system from zero to high concentrations at 25°C, pp. 981–997. Copyright 1980, with kind permission from Elsevier Science Ltd, The Boulevard, Langford Lane, Kidlington OX5 1GB, UK.

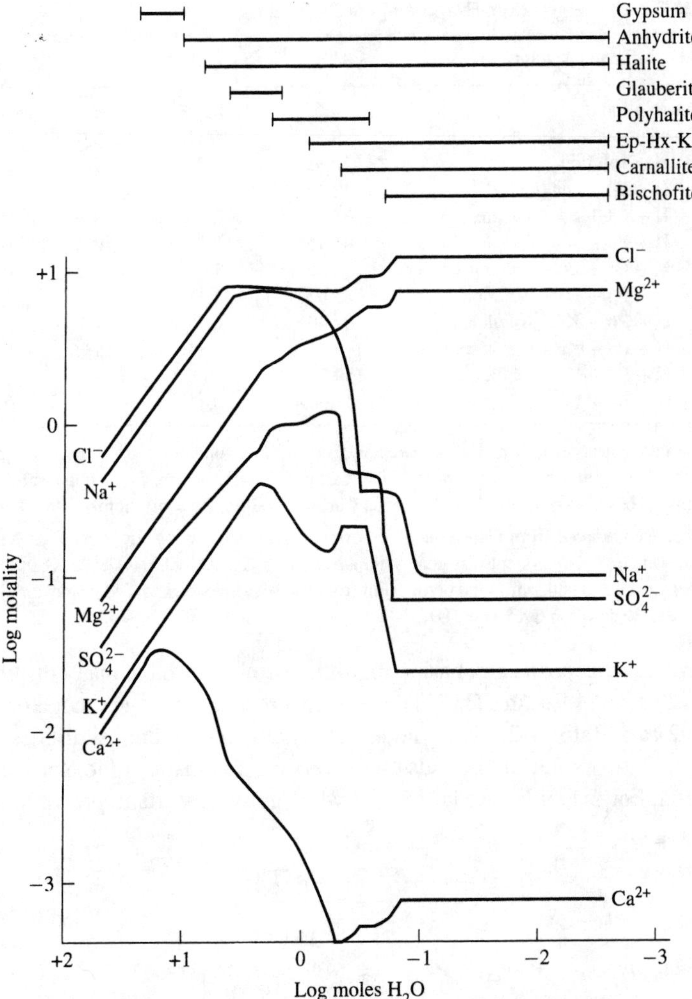

Figure 7-20 Evaporative concentration of modern seawater simulated using the computer program of Harvie and Weare (1980). Shown are the progressive changes in major-ion concentrations in the brine as evaporation and mineral precipitation and resorption proceed. The beginning composition (modern seawater) is on the left and the final composition is on the right. Resorption of a previously formed mineral occurs when it becomes unstable due to changes in the chemical composition of the brine. The order of formation of the minerals is shown at the top of the figure. Ep = epsomite, Hx = hexahydrite, Ks = kieserite. [After Hardie (1991). Reproduced, with permission, from the *Annual Review of Earth and Planetary Sciences*, vol. 19, copyright 1991, by Annual Reviews Inc.]

As with the major evaporite minerals, the succession of late-stage minerals found in evaporite deposits is similar to the predicted sequence, but there are numerous departures from this sequence due to depositional conditions, formation of metastable minerals, diagenetic changes, and changes occurring after lithification and burial. A good summary of the crystallization relationships and natural occurrences of evaporite minerals is given by Sonnenfeld (1984).

TABLE 7-21 Equilibrium Evaporation of Seawater

First appearance of	C.F.	% H_2O left	I	a_{H_2O}
G + solution	3.62	27.63	2.6	0.929
A + solution	9.82	10.18	6.6	0.772
A + H + solution	10.82	9.24	7.2	0.744
A + H + Gl + solution	13.15	7.60	7.5	0.738
A + H + Gl + Po + solution	38.50	2.60	10.1	0.697
A + H + Po + solution	44.76	2.23	10.7	0.685
A + H + Po + Ep + solution	73.56	1.36	13.0	0.590
A + H + Po + Hx + solution	85.05	1.18	13.8	0.567
A + H + Po + Ki + solution	102.40	0.98	14.9	0.498
A + H + Po + Ki + Car + solution	117.11	0.85	15.15	0.463
A + H + Ki + Car + solution	159.74	0.63	15.33	0.457
A + H + Ki + Car + Bi + solution	246.00	0.41	17.40	0.338

C.F. = concentration factor. For seawater, C.F. = 1. I = ionic strength. a_{H_2O} = activity of H_2O. G = gypsum, A = anhydrite, H = halite, Gl = glauberite, Po = polyhalite, Ep = epsomite, Hx = hexahydrite, Ki = kieserite, Car = carnallite, Bi = bischofite.

Source: Reproduced from *Geochimica et Cosmochimica Acta*, v. 44, Eugster et al., Mineral equilibria in a six-component seawater system, Na-K-Mg-Ca-SO_4-Cl-H_2O, at 25°C, pp. 1335–1347. Copyright 1980, with kind permission from Elsevier Science Ltd, The Boulevard, Langford Lane, Kidlington OX5 1GB, UK.

Not all evaporite deposits are marine in origin. Brines can have many different origins and compositions. Hardie and Eugster (1970) developed a model for a wide range of brine compositions (Figure 7-21 and Table 7-22). The model has a number of chemical divides, where the path taken by the evaporating solution depends on the relative abundance of ions involved with a precipitating mineral. For example, in Figure 7-21 when gypsum starts to precipitate, the composi-

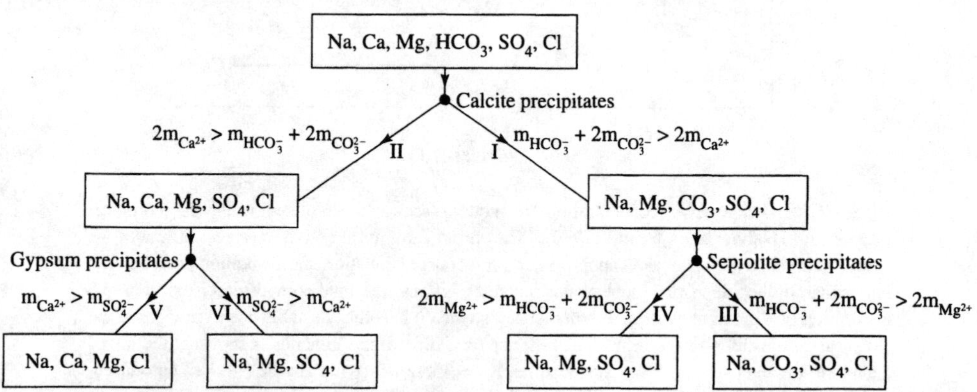

Figure 7-21 Some possible paths for the model evaporation of natural waters [modified from Hardie and Eugster (1970).] The first mineral to precipitate with most natural waters is calcite and this marks the first chemical divide for the evaporating solution. The other two chemical divides develop when gypsum and sepiolite precipitate. The formula for sepiolite is $MgSi_3O_6(OH)_2$. Three different possible final brine compositions are shown. Table 7-22 gives three analyses of brines from saline lakes that are examples of these three types of brine. Marine brines belong to the Na, Mg, SO_4, Cl type. (After Drever, J. I., *The Geochemistry of Natural Waters,* 2d ed. Copyright © 1988 by Prentice Hall, Inc. Adapted by permission of Prentice Hall, Inc., Upper Saddle River, NJ.)

TABLE 7-22 Analyses (mg/kg) of Brines from Saline Lakes in Western North America

	Bristol Dry Lake, California	Great Salt Lake, Utah	Mono Lake, California
SiO_2	—	48	14
Ca	43,296	241	4.5
Mg	1,061	7,200	34
Na	57,365	83,600	21,500
K	3,294	4,070	1,170
HCO_3	—	251	5,410
CO_3	—	—	10,300
SO_4	223	16,400	7,380
Cl	172,933	140,000	13,500
Total	279,150	254,000	56,600

The compositions are related to Figure 7-21 as follows. Bristol Dry Lake is representative of Path V, Great Salt Lake of Path IV, and Mono Lake of Path III.

Sources: Data from Eugster and Hardie (1978). After Drever, J. I., *The Geochemistry of Natural Waters*, 2d ed. Copyright © 1988 by Prentice Hall, Inc., Upper Saddle River, NJ.

tion of the resulting brine depends on whether the ratio of calcium ion to sulfate ion is greater than or less than the one-to-one ratio found in gypsum. If calcium ion is more abundant, then the final brine will have a high concentration of Na, Ca, Mg, and Cl (path V in Figure 7-21). If sulfate ion is more abundant, then path IV is followed and the abundant species are Na, Mg, SO_4, and Cl.

More sophisticated models have been developed since the original work by Hardie and Eugster (1970). However, the basis of their model is still valid: the composition of the final brine is determined by the composition of the original water from which the brine was derived (Figure 7-22). Because the original water chemistry of nonmarine waters is highly variable, the minerals found in nonmarine evaporite deposits are quite diverse. It should be pointed out that nonmarine deposits can have the same minerals as marine deposits and it is not always possible to tell whether a given evaporite mass is marine or nonmarine in origin.

A definite relationship is found in various areas of the world between organic-rich rocks (oil shales, black shales), evaporite deposits, and ore deposits in sedimentary rocks. A syngenetic model for ore formation in the depositional environment of these rocks was proposed by Renfro (1974). Ten years later Eugster (1985) pointed out that additional research indicated that none of these deposits is truly syngenetic and thus the Renfro model needed to be modified. According to Eugster, the connections between organic-rich rocks, evaporites, and ore deposits can be of two kinds: stratigraphic coherence (facies relations in space and time) and geochemical coherence (relationships through chemical processes). Stratigraphic relationships that are generally found consist of a period of red-bed deposition, followed by organic-rich muds (shales), and finally evaporite deposits (transgressive-regressive sequence).

An example of stratigraphic coherence is found at the famous Kupferschiefer deposits of Germany (Figure 7-23). Eugster (1985) states that, in order to form ore bodies, stratigraphic coherence must be combined with geochemical coherence. The latter is based on a particular set of geochemical processes. In the case of the Kupferschiefer deposits, Eugster suggests oxidation-reduction

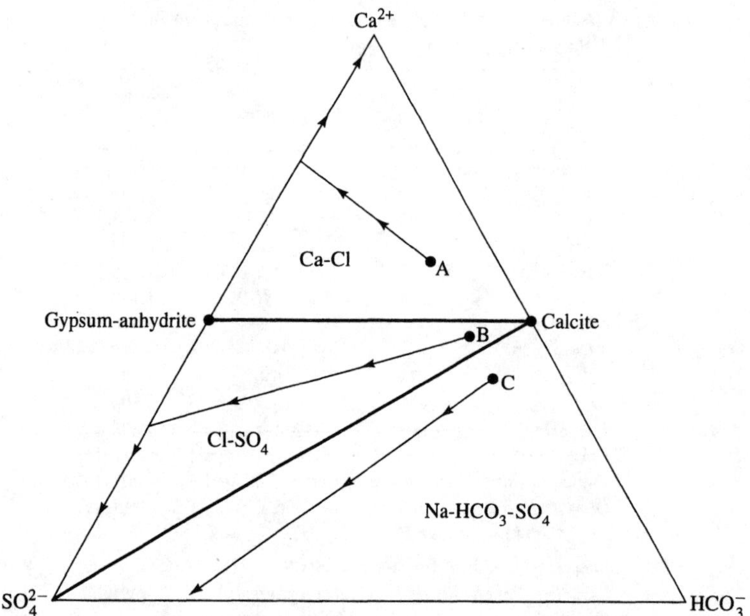

Figure 7-22 Ternary Ca^{2+}-SO_4^{2-}-HCO_3^- phase diagram (in equivalents). Two chemical divides, from calcite to SO_4^{2-} and from calcite to gypsum-anhydrite, separate the Na^+-HCO_3^--SO_4^{2-}, Cl^--SO_4^{2-}, and Ca^{2+}-Cl^- brine fields. The body of the diagram is the primary crystallization field for calcite. Typical evaporation paths for waters of compositions A, B, and C (mixtures of river water and Karst spring water) are shown. During calcite precipitation the fluid composition moves directly away from the calcite compositional point and, as gypsum or anhydrite precipitates, the fluid composition moves away from the $CaSO_4$ compositional point. Therefore evaporating waters are divided into three fields by the two chemical divides. Average world river water has a Na-HCO_3-SO_4 composition, modern seawater is a Cl-SO_4 water, and Karst spring water has a Ca-Cl composition. Average world river water would plot near point C and Karst spring water would plot on the Ca^{2+}-SO_4^{2-} line at the upper end of a line through points A, B, and C. After Spencer et al. (1990).

and acid-base equilibria between underlying red beds, the organic-rich Kupferschiefer, and Cl-SO_4 solutions, with the solutions (brines) provided by the overlying evaporites. He postulates that oxygenated meteoric fluids circulated through the evaporites and then entered the red beds where they became acidic due to mineral precipitation. Following this change, the fluids picked up metals from the oxide-hydroxide coatings on the red-bed clastic particles. Precipitation of sulfides occurred, due to reduction reactions, where the metal-rich solutions later came in contact with organic matter in the Kupferschiefer. This general model of ore formation is outlined in Table 7-23. Eugster describes a number of other sedimentary ore deposits associated with evaporites and/or believed to be deposited by brine solutions. All of the deposits are epigenetic (with or without a diagenetic component), and all can be explained by Eugster's modification of Renfro's syngenetic model.

The first discovery of modern brines containing unusual amounts of metals was made in 1964 in the center of the Red Sea. On the bottom of the basin, three separate depressions were found, overlain by waters of high salinity and temperature. The sediments underlying the deeps are enriched in heavy metals, particularly iron, zinc, copper, and manganese. The depressions are enclosed by the 1,900-meter contour. The largest depression, the Atlantis II Deep, covers about 50 square kilometers

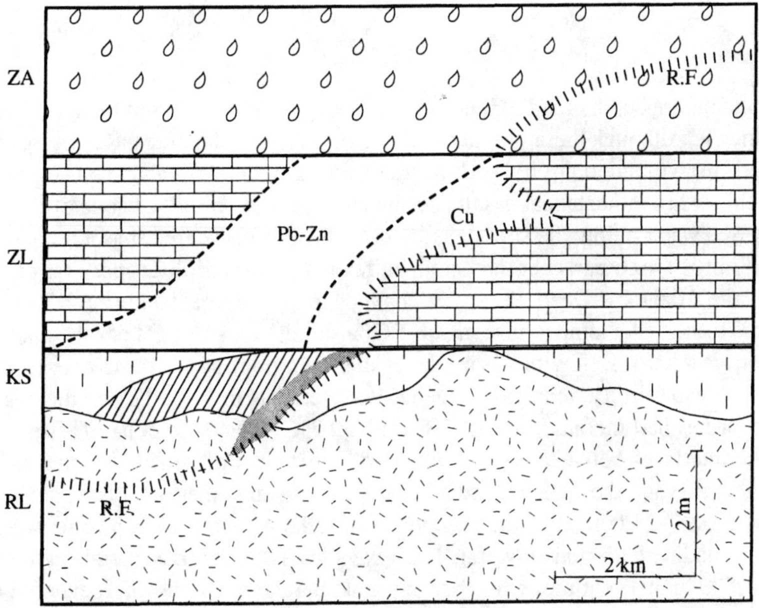

Figure 7-23 Schematic cross-section through the Permian Kupferschiefer ore deposits of Germany. RL = terrestrial red beds. KS = Kupferschiefer (varies from oil shale to marly shale to siliclastic shale). ZL = limestone. ZA = anhydrite. ZL and ZA are part of the massive Zechstein evaporite deposits. To the right of the boundary labeled R.F. the rocks are oxidized. R.F. stands for Rote Fäule facies (red-colored rocks). Rentsch (1974, 409) points out that there is "a clear ore-controlling lateral and vertical zonation around the red-colored rocks . . .". The areas labeled Cu and Pb-Zn have disseminated copper mineralization and disseminated lead-zinc ores, respectively. The black area represents massive copper ores and the ruled area represents massive lead-zinc ores. See Table 7-23 for an explanation of ore formation. After Rentsch (1974). (Reproduced from *Geochimica et Cosmochimica Acta*, v. 49, H. P. Eugster, Oil shales, evaporites, and ore deposits, pp. 619–635. Copyright 1985, with kind permission from Elsevier Science Ltd, The Boulevard, Langford Lane, Kidlington OX5 1GB, UK.)

TABLE 7-23 Origin of Sedimentary Ore Deposits Associated with Evaporites*

		Geochemical coherence				
		Metals	Oxygen	Reducing capacity	Chloride	Sulfur
Stratigraphic coherence						
	Evaporites	—	—	—	X	X
	Black shales	X	—	X	—	X
	Redbeds	X	X	—	—	—

*To form an ore deposit the three rock types listed must occur together *and* appropriate geochemical processes must occur at various times after rock formation. The contribution of each rock type to ore deposit formation is shown. The metals come mainly from leaching of redbeds by oxygenated brines. However, in some cases they can come from areas of black shales where oxidation of sulfides occurs, leading to further acidity and thus mobilization of metals. Precipitation of sulfides occurs where abundant organic matter in the black shales causes reducing conditions to develop in the circulating fluids. See Figure 7-23.

Source: Reproduced from *Geochimica et Cosmochimica Acta*, v. 49, H. P. Eugster, Oil shales, evaporites, and ore deposits, pp. 619–635. Copyright 1985, with kind permission from Elsevier Science Ltd, The Boulevard, Langford Lane, Kidlington OX5 1GB, UK.

and has a maximum depth of 2,170 meters. The average thickness of sediment in the depressions is about 25 meters. Although these sediments are associated with tholeiite basalts, most of the Red Sea is believed to be underlain by sedimentary rocks, including abundant evaporite deposits. Geophysical studies indicate that basalts occur mainly in the central rift zone of the Red Sea.

Red Sea water below the surface and above the hot-brine areas has a temperature of about 22°C, a salinity of 40.6 parts per thousand, and contains 2.0 milliliters per liter of oxygen. In the vicinity of the Atlantis II Deep, the water changes to an intermediate brine below 2,000 meters. In 1966 this brine had a temperature of 44.2°C, a salinity of 123 parts per thousand, and an oxygen content near zero. Below this brine layer the main pool of hot brine averaged 56.5°C, 257 parts per thousand salinity, and zero oxygen. Measurements in 1972 showed that the temperatures of the two brines had increased to 49.7°C and 59.2°C. Concentrations of suspended sediment occur at the interfaces between normal water and the denser intermediate brine and between the intermediate brine and the denser bottom brine. Bacteria and other organisms are absent from the hot brines. Table 7-24 compares the composition of the Atlantis II Deep brine with the water of the Dead Sea and with normal seawater. The table shows that the trace metals in the brine are concentrated 1,500 to 30,000 times over their normal abundance in seawater, even though the brine has only eight times as much chlorine as seawater. Magnesium and sulfate ions are less abundant in the brine than in seawater. The Dead Sea water has formed by evaporation of seawater and clearly has a different composition than the brine. It seems beyond question that the Red Sea brine did not form by evaporation of surface water.

The typical Red Sea sediment outside the hot brine basins consists of calcareous shells and tests of pelagic organisms, plus detrital quartz, feldspar, and clay minerals. This type of sediment

TABLE 7-24 Compositions of Red Sea Brine, Dead Sea Water, and Ordinary Ocean Water

Component (ppt)	Atlantis II Deep	Dead Sea	Ocean water
Na^+	92.60	32.30	10.76
K^+	1.87	6.17	0.39
Ca^{2+}	5.15	13.98	0.41
Mg^{2+}	0.76	34.45	1.29
Sr^{2+}	0.04	0.20	0.008
Cl^-	156.03	178.20	19.35
Br^-	0.13	4.28	0.066
SO_4^{2-}	0.84	0.34	2.71
HCO_3^-	0.14	0.18	0.72
Si	0.03	<0.01	0.004
Fe	0.08	<0.002	0.00002
Mn	0.08	0.004	0.00001
Zn	0.005	<0.02	0.000005
Cu	0.0003	<0.002	0.00001
Co	0.0002	—	—
Pb	0.0006	<0.002	0.000004
Ni	—	<0.002	0.0000001
Salinity	257.76	270.11	35.71
Temperature	56.5°C	22.0°	—
Density	1.178	1.233	1.03

Source: Data from Degens and Ross (1969).

is also found in the Atlantis II Deep, but the sediments of the deep itself are well-bedded silts and clays exhibiting a variety of colors. Four major sediment facies have been recognized within the deep (Table 7-25). At the sediment-water interface of the hot brine basins is a black sediment consisting of clay with some iron oxides and sphalerite (iron-montmorillonite facies). Beneath this is a one-meter layer of goethite and then a layer of black sediment containing iron monosulfide, sphalerite, chalcopyrite, and pyrite (sulfide facies). The sediment of the manganite facies is similar in appearance to that of the sulfide facies, but is more limited in its occurrence. It is characterized by the presence of manganite and several other manganese minerals. The variety of facies is probably due to variations with time in the composition of the brines. The chemical compositions of these facies and of typical Red Sea sediment is given in Table 7-25. In addition to the metals listed in Table 7-25, the hot-brine sediments contain trace quantities of lead, cobalt, nickel, cadmium, bismuth, arsenic, tin, germanium, indium, mercury, silver, and gold.

One approach to ascertaining the origin of the brines and of the metal-rich sediments is the use of isotopic analyses. Zierenberg and Shanks (1986) measured sulfur and oxygen isotope ratios of dissolved sulfate and strontium isotope ratios of dissolved strontium in the Red Sea brines. Analyses were also done on evaporitic rocks considered to be source rocks for the brines. These data were combined with earlier studies to determine the relative importance of the three possible major source reservoirs involved in the formation of the brines: seawater, evaporite rocks that underlie the Red Sea, and rift-zone basalts.

Zierenberg and Shanks (1986) point out that the combined use of several isotopic systems provides increased constraints on the hypotheses for brine formation. They conclude that the three brine pools studied formed by dissolution of evaporite rocks by seawater. Water circulation was probably due to heat from volcanic activity associated with the rifting of the Red Sea. In the case of the Atlantis II Deep, the resulting brine reacted with hot rift-zone basalts (leaching metals and other elements) and then discharged onto the seafloor, forming the metalliferous sediments and the hot lower brine pool. The composition of the sediments is consistent with a basaltic source for the metals. The amount of basaltic strontium added to the main brine was calculated by isotope mass balance. This calculation indicated that approximately 30 percent of the strontium in the

TABLE 7-25 Compositions of Red Sea Sediments

		Hot brine basin sediments			
Component	Typical Red Sea sediment	Iron montmorillonite facies	Goethite facies	Sulfide facies	Manganite facies
SiO_2	27.3*	24.4	8.7	24.7	7.5
Al_2O_3	8.4	1.7	1.1	1.5	0.7
Fe_2O_3	6.5	37.1	64.2	24.3	30.5
FeO	1.4	11.7	2.7	13.4	0.4
Mn_3O_4	0.6	2.1	1.1	1.1	35.5
CaO	23.6	4.8	3.4	2.5	2.9
ZnO	0.1	3.2	0.7	12.2	1.4
CuO	0.0	0.8	0.3	4.5	0.1
CO_2	23.1	8.6	3.6	5.7	2.2
S	0.3	3.9	0.6	16.8	0.6

*All compositions given in weight percent.
Source: Data from Degens and Ross (1969).

brine was derived from basalt during hydrothermal interaction. The intermediate brine appears to have formed by mixing of the hotter lower brine with seawater. The lower brine was as hot as 255°C, as indicated by sulfate-water oxygen isotope fractionation.

SUMMARY

The minerals that are common and abundant in sedimentary rocks are silica minerals (quartz and opal), feldspars (K-feldspar and plagioclase), clay minerals (kaolinite, smectite, and illite), and carbonates (calcite, aragonite, and dolomite). These minerals can be detrital, secondary, precipitated, or authigenic in origin. Some sedimentary rocks are made up of only one or two of these minerals (e.g., limestones) while others are a combination of several minerals with different origins (e.g., shales). The general trend in the formation of sedimentary rocks is to segregate and concentrate elements brought from weathered rocks.

Chemical weathering is a complex mixture of many processes. Four important processes are dissolution, hydrolysis, adsorption, and oxidation-reduction. These reactions can best be understood by looking at the changes that occur at the surfaces of minerals in contact with water. Chemical weathering rates are dependent on the structure and bond strengths of minerals and on the nature of the weathering solution (particularly pH). Microorganisms play an important role in weathering by producing attacking acids and by acting as catalysts for oxidation-reduction and other reactions.

Field study allows us to determine which elements are removed most quickly from a weathering site (usually sodium, calcium, and magnesium) and which elements are least mobile (iron and aluminum). An annual weathering budget for a region can be prepared. Elements lost by weathering of bedrock and introduced by precipitation are either taken up by biomass or removed in runoff (the transfer of elements to soil minerals can be ignored in studying change over one year). If changes in an area are evaluated over time periods much greater than one year, soil chemistry and mass and the rate of physical removal of solid weathered products have to be considered.

Clay minerals are the major products of chemical weathering. They are subdivided into several groups on the basis of the number of sheets making up their basic structure and further subdivided on the basis of the composition of the octahedral sheet. Their overall composition is dependent on (1) the type and number of sheets; (2) the amount of intersheet water; and (3) the amount of substitution of one ion for another. Most of the clay minerals have the ability to adsorb ions on their surfaces and to exchange these surface ions when their environment changes. The clay minerals that occur abundantly in sediments and sedimentary rocks are kaolinite, smectite, illite, chlorite, and mixed-layer clays.

The other major group of minerals formed by chemical weathering, the carbonates, occur as precipitated sediments. The two most important minerals are the polymorphs calcite and aragonite. Two forms of calcite are found: low-magnesium calcite and high-magnesium calcite. A number of factors influence the formation and mineral composition of carbonate sediments. In addition to pressure (total and CO_2) and temperature, kinetic effects and the effect of other ions in solution have to be taken into account. Most shallow-water carbonates form as a result of biological (metabolic) processes and are made up of aragonite and high-magnesium calcite. Deep sea sediments mainly contain low-magnesium calcite.

Diagenesis is a general term for all the processes that take place after deposition of a sediment. Five chemical diagenetic processes that commonly occur are oxidation-reduction, authigenesis, dissolution, diffusion, and chemical compaction. Interstitial water is intimately involved in many changes, as is any organic matter present. Many sediments undergo changes mainly near and at the sediment-water interface. Often the mineralogy of a sediment and the chemistry of the interstitial water changes with depth. A large number of new minerals, such as illite, form during diagenesis. Dolomite, which does not form as a primary mineral under conditions normally found in field environments, can form during diagenesis and also after sediment has become rock. Characteristic features of sedimentary rocks such as concretions, nodules, and styolites can be explained as a result of diagenetic processes.

Equilibrium evaporation of seawater should produce the following sequence of minerals: gypsum, anhydrite, halite, and lastly a variety of potassium and magnesium salts. Anhydrite, however, does not form in modern evaporating brines and must form as a secondary mineral in evaporite deposits. Which late-stage potassium and magnesium minerals form, and in what order, depends on whether or not fractional crystallization occurs. Nonmarine brines of many different compositions can also form by evaporation and a wide variety of nonmarine evaporite deposits is found. Organic-rich rocks, evaporite deposits, and sedimentary ore deposits are often found together. Modern brines are found associated with metal-containing sediments on the bottom of the Red Sea and elsewhere.

QUESTIONS

1. (a) What would you expect to be the major chemical difference between an average clay-rich residual soil and an average shale?

(b) It is unusual for shales to have a higher content of Na_2O than of K_2O. An example of such a shale is the siliceous shale whose analysis is given in Table 7-2. Suggest two different reasons for a shale to have a high Na_2O content.

(c) The carbonaceous shale of Table 7-2 is listed as having 8.06 percent Fe_2O_3 and no FeO. Would you expect most of the iron in this rock to be in the ferric or ferrous state?

2. Suggest a possible mode of occurrence for each of the following trace elements found in sedimentary rocks: (a) manganese; (b) barium; (c) boron; (d) chromium; (e) fluorine; (f) vanadium.

3. (a) Where would a line representing saturation with quartz plot on the diagrams of Figure 7-3? Assume that the temperature is 25°C and the pH is less than 7.

(b) Would you expect most natural waters (other than seawater) to plot to the right or left of the quartz saturation line?

4. The following table gives the chemical composition of the Morton gneiss (shown in Figures 7-8 and 7-9) and of the residual soil formed from it.

Component	Fresh rock	Residual soil
SiO_2	71.54	55.07
TiO_2	0.26	1.03
Al_2O_3	14.62	26.14
Fe_2O_3	0.69	3.72
FeO	1.64	2.53
MgO	0.77	0.33
CaO	2.08	0.16
Na_2O	3.84	0.05
K_2O	3.92	0.14
H_2O^+	0.30	9.75
H_2O^-	0.02	0.64
CO_2	0.14	0.36
	99.82	99.92

(a) Using the fresh rock analysis, we can say that 100 grams of fresh rock contained about 14.62 grams of Al_2O_3. Assuming that no alumina is lost during weathering, what is the weight of the residual soil formed from 100 grams of fresh rock?

(b) Using the result from part (a), how many grams of silica (SiO_2) will there be in the residual soil?

(c) It is clear that the number of grams of silica in the residual soil is less than the number of grams in the original 100 grams of fresh rock. Which oxide, other than H_2O, shows the greatest increase (in grams) in going from the fresh rock to the soil?

5. (a) This question uses data in Table 7-8.Assume that the total amount of iron transported in runoff (as dissolved iron) and incorporated in biomass at Hubbard Brook is 0.12 T/km^2. Also assume that the abundance of iron in the bedrock is 4.4 weight percent. Calculate the relative rate of release of iron in the same manner used to get the figures in the last column of Table 7-8.

(b) Iron in river water is transported as both particulate (suspended) material and dissolved material. For iron, which form would you expect to be the dominant form for most rivers?

6. (a) You are studying the weathering of a given mineral AB and you want to determine if the dissolution of the mineral is diffusion controlled or surface-reaction controlled. You expect the dissolution to be surface-reaction controlled. If so, then the dissolution will follow this kinetic rate law:

$$\text{rate of reaction} = \frac{dC}{dt} = ka$$

where C = concentration of species A^+ or B^- in the solution adjacent to the mineral, t = time, k = reaction rate constant, and a = surface area of the mineral. Assuming the dissolution is surface-reaction controlled, plot a line on the following diagram showing the experimental results you should get.

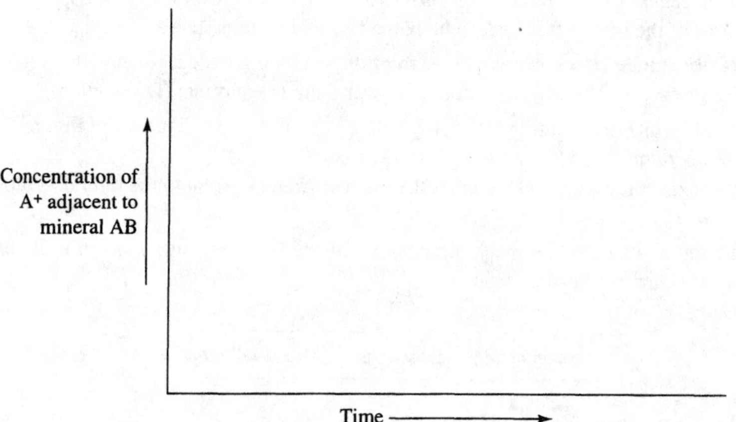

(b) What is the order of the reaction represented by the rate equation in part (a)?

7. (a) Laboratory study of an unknown clay mineral has determined that it has the following formula:

$$KAl_{3.5}Mg_{0.5}[Si_{7.5}Al_{0.5}]O_{20}(OH)_4$$

To which one of the groups in Table 7-9 does this mineral belong?

(b) This clay mineral comes from an altered volcanic ash that is used to stop leakage in soils and rocks. What is the geologic name for this type of altered ash and what property of the clay mineral makes it useful for plugging leaks?

8. Given is the formula for a vermiculite:

$$Ca_{0.4}Mg_{5.2}Fe^{3+}_{0.8}[Si_{6.4}Al_{1.6}]O_{20}(OH)_4$$

(a) Calculate the excess or deficiency of positive charge assuming an ideal formula for vermiculite and assuming that calcium is adsorbed and not substituting in the structure.
(b) How is the excess or deficiency of positive charge neutralized?

9. Analysis 6 in Table 7-10 represents a "mixed-layer illite-smectite." In this material, individual masses are composed of interstratified layers of these two minerals. Since illite and smectite have different compositions, the chemistry of a given sample is determined by the proportions of the two minerals. Would you expect the material represented by analysis six in Table 7-10 to have more illite than smectite or the reverse?

10. **(a)** The mineral glauconite is considered by some geologists to be a member of the mica group of minerals and by other geologists to be a member of the clay mineral group of minerals. Explain why glauconite can be considered to be a member of either the mica group or the clay mineral group.
(b) What does the occurrence of glauconite in a sedimentary rock indicate about the environment in which the rock formed?

11. **(a)** The amount of Sr^{2+} that enters precipitating calcite or aragonite depends on the distribution coefficient

$$K = \frac{N_{Sr}^{mineral}}{N_{Sr}^{seawater}} \quad \text{or} \quad \frac{(\text{ppmSr})_{mineral}}{(\text{ppmSr})_{seawater}}$$

Which would you expect to have the higher value, $(K_{Sr})_{aragonite}$ or $(K_{Sr})_{calcite}$ at 25°C? Explain your answer.
(b) How could knowledge of $(K_{Sr})_{aragonite}$ and of the strontium content of present-day aragonite particles perhaps be used to tell whether the particles are of inorganic or organic origin?
(c) Assume that you have measured $(K_{Sr})_{aragonite}$ in the laboratory at 25°C and gotten values that ranged from 980 to 1,060, where K_{Sr} is expressed in terms of parts per million. If the average concentration of strontium in seawater is 8.1 parts per million, what range of strontium concentrations is possible for inorganically precipitated aragonite? How does this range compare with the strontium content of the madreporarian corals of Table 7-16?

12. The values for CO_3^{2-} in Table 7-15 are given in units of μmol/kg (i.e., in molal units). Because of the density of seawater, molal values for species are not the same as molar values. If we wanted to change the values of CO_3^{2-} in Table 7-15 to molar units, how much would the CO_3^{2-} value for case one be changed?

13. **(a)** For case two of Table 7-15 the water is not in equilibrium with the atmosphere (total CO_2 is kept constant in going from case one to case two). Assume for case two that the temperature is kept at 2.2°C, but the water is exposed to the atmosphere and reaches equilibrium. How would the parameters ΣCO_2, CO_2 (aq), HCO_3^-, CO_3^{2-}, and pH change?
(b) For the situation described in part (a), would the water be more saturated or less saturated with respect to calcite as compared to case two?

14. **(a)** Given aragonite in equilibrium with water that is in contact with the atmosphere, calculate the values of pH, $[Ca^{2+}]$, and $[CO_3^{2-}]$.
(b) Calculate the following ratio: the calcium value from Table 1-11 divided by the calcium value from part (a). What does this number tell us about the saturation of seawater with respect to aragonite?

15. Figure 7-16b points out that non-steady state diagenesis can produce iron-rich layers in sediments. Why couldn't steady-state diagenesis produce an iron-rich layer?

16. (a) Write an equation for the breakdown of organic matter in sediment by bacterial reduction of sulfate ion in interstitial water. Use CH_2O to represent organic matter. Assume H_2S is one of the products.
(b) Fermentation of organic matter can occur under approximately the same reducing conditions as sulfate ion reduction. The most important result of fermentation in marine sediments is the production of methane. As pointed out by Berner (1980), high concentrations of methane in sediments generally occur only below the zone where sulfate ion is present and being reduced. See the following diagram from Berner (1980, 175).

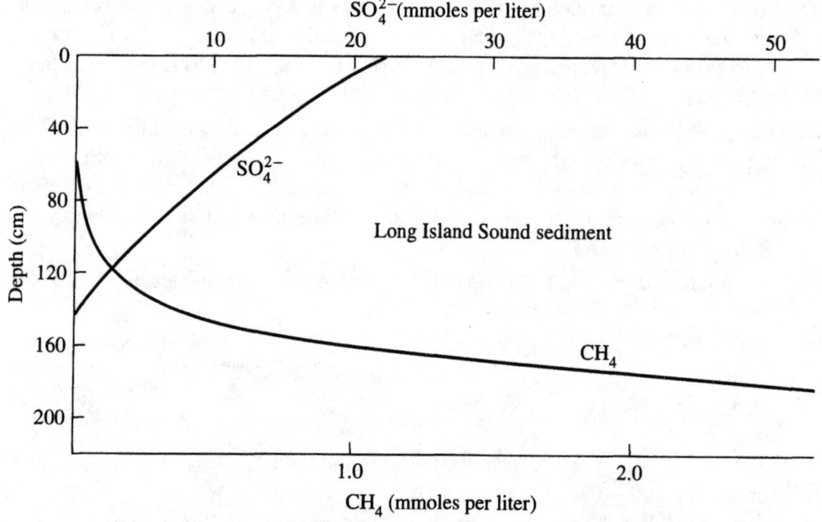

You would expect that some methane might be produced in the zone where sulfate ion is being reduced and also you would expect that some methane would diffuse upward into the sulfate zone after being produced below it. Suggest an explanation for the low levels or complete lack of methane in the sulfate reduction zones of most marine sediments.
(c) The same processes of sulfate ion reduction and methane production also occur in anoxic freshwater sediments. However, methane concentrations occur much closer to the sediment-water interface in freshwater sediments as compared to marine sediments. Suggest an explanation for this.

17. (a) Given is groundwater that is a chloride brine and that contains the following amounts of calcium and magnesium: 35 mg/kg calcium and 33 mg/kg magnesium. The depth at which the groundwater was sampled was two km. If this groundwater were to enter some limestone, would it tend to change the rock to a dolomite rock? Use Figure 7-15 to answer the question.
(b) According to Figure 7-15, could seawater circulating through calcium carbonate sediment cause dolomitization?

18. Figure 7-17 is based on isotopic study of diagenetic minerals. Longstaffe (1987) points out that, at temperatures normal for sedimentary environments, most detrital minerals do not experience significant isotopic exchange with surrounding waters. Diagenetic minerals, however, can form in equilibrium with pore waters and acquire isotopic signatures characteristic of their temperature of formation and related to pore water isotopic composition. They can also be affected by isotopic exchange after their formation.
(a) What mineral properties would you expect to affect the isotopic exchange between authigenic minerals and associated pore waters?
(b) Of the common diagenetic minerals quartz, calcite, and smectite, which one would you expect, subsequent to crystallization, to show oxygen isotopic exchange with pore water at the lowest temperatures?

(c) Which would you expect to show oxygen isotopic exchange at shallow depths and low temperatures, a fine-grained clay fraction or a coarse-grained clay fraction?

19. Answer the following questions about Figure 7-20.

(a) Why does the abundance of Ca^{2+} decrease with continued evaporation after reaching a peak value between +1 and +2 log moles H_2O?

(b) Why does the abundance of K^+ decrease with continued evaporation after reaching a peak value between zero and +1 log moles H_2O?

(c) After hexahydrite has precipitated for a short period of time, it reacts with the remaining brine and is replaced by what mineral?

20. (a) Use Figure 7-19 to explain why anhydrite should form after gypsum in evaporating seawater at 25°C as shown in Table 7-19.

(b) Limestone beds at the base of evaporite deposits are often found to have undergone dolomitization. Suggest an explanation for this.

21. If fractional crystallization occurs during the evaporative concentration of seawater, phases that have crystallized are removed as they form and cannot interact further with the evolving brine. Table 7-21 describes the changes occurring during equilibrium evaporation (no fractional crystallization) of seawater. If fractional crystallization did occur, at what point in Table 7-21 would you expect the equilibrium and fractionation paths to diverge?

22. (a) Given is evaporating water that has a high alkalinity value and low amounts of magnesium and calcium. What paths would the water composition follow in Figure 7-21 as evaporation continued?

(b) In preparing the model shown in Figure 7-21, a different magnesium-containing phase instead of sepiolite could have been used. Name two common magnesium-containing phases that could be used in place of sepiolite.

(c) Given is a brine in which strongly anoxic conditions have developed. How might this situation change the final brine composition from that shown in Figure 7-21?

23. (a) The hot-brine depressions of the Red Sea are similar to the bottom of the Black Sea in being anaerobic. However, the brines and sediments of the Atlantis II Deep have no sulfate-reducing bacteria, in contrast to their abundance in the deep parts of the Black Sea. Suggest two possible reasons why sulfate-reducing bacteria are not found in the Atlantis II Deep.

(b) If the abundant sulfur found in the brine sediments as sulfide minerals does not come from the reduction of sulfate ion in the brine water and sediment, what other sources might provide sulfur?

REFERENCES

ALLAN, J. R., and R. K. MATTHEWS. 1982. Isotope signatures associated with early meteoric diagenesis. *Sedimentology,* v. 29:797–817.

ARTHUR, M. A., T. F. ANDERSON, I. R. KAPLAN, J. VEIZER, and L. S. LAND. 1983. Stable isotopes in sedimentary geology. SEPM Short Course No. 10 Notes, unpaginated.

BERNER, R. A. 1971. *Principles of Chemical Sedimentology.* McGraw-Hill, New York.

BERNER, R. A. 1975. Diagenetic models of dissolved species in the interstitial waters of compacting sediments. *Am. J. Sci.,* v. 275:88–96.

BERNER, R. A. 1980. *Early Diagenesis.* Princeton University Press, Princeton, NJ.

BERNER, E. K., and R. A. BERNER. 1987. *The Global Water Cycle.* Prentice-Hall, Englewood Cliffs, NJ.

BERTHELIN, J. 1988. Microbial weathering processes in natural environments. In *Physical and Chemical Weathering in Geochemical Cycles,* A. Lerman and M. Meybeck, eds. Kluwer Academic Publishers, Dordrecht, The Netherlands. 33–59.

BLATT, H. 1992. *Sedimentary Petrology.* W. H. Freeman and Company, New York.

Boggs, S., Jr. 1992. *Petrology of Sedimentary Rocks.* Macmillan Publishing Company, New York.

Broecker, W. S., and T.-H. Peng. 1982. *Tracers in the Sea.* Eldigio Press, Palisades, New York.

Busenberg, E., and C. V. Clemency. 1976. The dissolution kinetics of feldspars at 25°C and 1 atm CO_2 partial pressure. *Geochim. Cosmochim. Acta,* v. 40:41–49.

Casey, W. H., H. R. Westrich, and G. W. Arnold. 1988. Surface chemistry of labradorite feldspar reacted with aqueous solutions at pH = 2, 3, and 12. *Geochim. Cosmochim. Acta,* v. 52:2795–2807.

Chou, L., R. M. Garrels, and R. Wollast. 1989. Comparative study of the dissolution kinetics and mechanisms of carbonates in aqueous solutions. *Chem. Geol.,* v. 78:269–282.

Colman, S., and D. Dethier, eds. 1986. *Rates of Chemical Weathering of Rocks and Minerals.* Academic Press, New York.

Connor, J. J., and H. T. Shacklette. 1975. Background Geochemistry of Some Rocks, Soils, Plants, and Vegetables in the Conterminous United States. U.S. Geol. Survey Profess. Paper 574–F.

Davison, W. 1985. Conceptual models for transport at a redox boundary. In *Chemical Processes in Lakes,* W. Stumm, ed. Wiley-Interscience, New York. 31–53.

Degens, E. T., and D. A. Ross, eds. 1969. *Hot Brines and Recent Heavy Metal Deposits in the Red Sea.* Springer-Verlag, Berlin.

Dove, P. M., and D. A. Crerar. 1990. Kinetics of quartz dissolution in electrolyte solutions using a hydrothermal mixed flow reactor. *Geochim. Cosmochim. Acta,* v. 54:955–969.

Drever, J. I. 1988. *The Geochemistry of Natural Waters,* 2d ed. Prentice-Hall, Englewood Cliffs, NJ.

Eckhardt, F. E. W. 1985. Solubilization, transport, and deposition of mineral cations by microorganism-efficient rock weathering agents. In *The Chemistry of Weathering,* J. I. Drever, ed. D. Reidel Publishing Co., Dordrecht, The Netherlands. 161–173.

Eugster, H. P. 1985. Oil shales, evaporites, and ore deposits. *Geochim. Cosmochim. Acta,* v. 49:619–635.

Eugster, H. P., and L. A. Hardie. 1978. Saline lakes. In *Lakes—Chemistry, Geology, Physics,* A. Lerman, ed. Springer-Verlag, New York. 237–293.

Eugster, H. P., C. E. Harvie, and J. H. Weare. 1980. Mineral equilibria in a six-component seawater system, Na-K-Mg-Ca-SO_4-Cl-H_2O, at 25°C. *Geochim. Cosmochim. Acta,* v. 44:1335–1347.

Faure, G. 1986. *Principles of Isotope Geology,* 2d ed. John Wiley & Sons, Inc., New York.

Garrels, R. M., and F. T. Mackenzie. 1971. *Evolution of Sedimentary Rocks.* W. W. Norton & Company, Inc., New York.

Garrels, R. M., and M. E. Thompson. 1962. A chemical model for seawater. *Am. J. Sci.,* v. 260:57–66.

Gieskes, J. M. 1975. Chemistry of interstitial waters of marine sediments. In *Ann. Rev. of Earth and Planetary Sciences,* v. 3, F. A. Donath, ed. Annual Reviews Inc., Palo Alto, CA. 433–453.

Goldich, S. S. 1938. A study in rock weathering. *J. Geol.,* v. 46:17–58.

Hardie, L. A. 1967. The gypsum-anhydrite equilibrium at one atmosphere pressure. *Am. Mineralogist,* v. 52:171–200.

Hardie, L. A. 1987. Dolomitization: A critical view of some current views. *J. Sediment. Petrol.,* v. 57:166–183.

Hardie, L. A. 1991. On the significance of evaporites. In *Ann. Rev. of Earth and Planetary Sciences,* v. 19, G. W. Wetherill, ed. Annual Reviews Inc., Palo Alto, CA. 131–168.

Hardie, L. A., and H. P. Eugster. 1970. The evolution of closed-basin brines. *Mineralogical Soc. Am. Special Publication 3*:273–290.

Harvie, C. E., and J. H. Weare. 1980. The prediction of mineral solubilities in natural waters: The Na-K-Mg-Ca-Cl-SO_4-H_2O system from zero to high concentrations at 25°C. *Geochim. Cosmochim. Acta,* v. 44:981–997.

Hellman, R., C. M. Eggleston, M. F. Hochella, Jr., and D. A. Crerar. 1990. The formation of leached layers on albite surfaces during dissolution under hydrothermal conditions. *Geochim. Cosmochim. Acta,* v. 54:1267–1281.

Hering, J. G., and W. Stumm. 1990. Oxidative and reductive dissolution of minerals. In *Mineral-Water Interface Geochemistry,* Rev. in Mineralogy, v. 23, M. F. Hochella, Jr., and A. F. White, eds. Mineralogical Soc. Am., Washington, D.C. 427–465.

Hiebert, F., and P. C. Bennett. 1992. Microbial control of silicate weathering in organic-rich ground water. *Science,* v. 258:278–281.

Klein, C., and C. S. Hurlbut, Jr. 1993. *Manual of Mineralogy,* 21st ed. John Wiley & Sons, Inc., New York.

Krauskopf, K. 1967. *Introduction to Geochemistry.* McGraw-Hill, Inc., New York.

Lasaga, A. C. 1990. Atomic treatment of mineral-water surface reactions. In *Mineral-Water Interface Geochemistry,* Rev. in Mineralogy, v. 23, M. F. Hochella, Jr., and A. F. White, eds. Mineralogical Soc. Am., Washington, D.C. 17–85.

Li, Y. H., and S. Gregory. 1974. Diffusion of ions in sea water and in deep-sea sediments. *Geochim. Cosmochim. Acta,* v. 38:703–714.

Likens, G. E., F. H. Bormann, R. S. Pierce, J. S. Eaton, and N. M. Johnson. 1977. *Biogeochemistry of a Forested Ecosystem.* Springer-Verlag, New York.

Longstaffe, F. J. 1987. Stable isotope studies of diagenetic processes. In *Stable Isotope Geochemistry of Low Temperature Fluids,* Short Course Handbook v. 13, T. K. Kyser, ed. Mineral. Assoc. Canada, Saskatoon, Saskatchewan. 187–257.

Machel, H.-G., and E. W. Mountjoy. 1986. Chemistry and environments of dolomitization—a reappraisal. *Earth Sci. Rev.,* v. 23:175–222.

Machel, H.-G., and E. W. Mountjoy. 1987. General constraints on extensive pervasive dolomitization—and their application to the Devonian carbonates of western Canada. *Bull. Can. Petroleum Geol.,* v. 35:143–158.

Mahaney, W. C., and D. L. Halvorson. 1986. Rates of mineral weathering in the Wind River Mountains, western Wyoming. In *Rates of Chemical Weathering of Rocks and Minerals,* S. M. Colman and D. P. Dethier, eds. Academic Press, New York. 147–167.

Marchand, D. E. 1974. Chemical Weathering, Soil Development, and Geochemical Fractionation in a Part of the White Mountains, Mono and Inyo Counties, California. U.S. Geol. Survey Profess. Paper 352–J.

Mogk, D. W. 1990. Application of Auger electron spectroscopy to studies of chemical weathering. *Rev. of Geophysics,* v. 28:337–356.

Mogk, D. W., and W. W. Locke. 988. Application of Auger electron spectroscopy (AES) to naturally weathered hornblende. *Geochim. Cosmochim. Acta,* v. 52:2537–2542.

Morse, J. W., and F. T. Mackenzie. 1990. *Geochemistry of Sedimentary Carbonates.* Elsevier, Amsterdam.

Mucci, A., R. Canuel, and S. Zhong. 1989. The solubility of calcite and aragonite in sulfate-free seawater and the seeded growth kinetics and composition of precipitates at 25°C. *Chem. Geol.,* v. 74:309–329.

Mucci, A., and J. W. Morse. 1983. The incorporation of Mg^{2+} and Sr^{2+} into calcite overgrowths: influences of growth rate and solution composition. *Geochim. Cosmochim. Acta,* v. 47:217–233.

Nickel, E. 1973. Experimental dissolution of light and heavy minerals in comparison with weathering and intrastratal solution. *Contrib. Sediment.,* v. 1:1–68.

Nissenbaum, A., B. J. Presley, and I. R. Kaplan. 1972. Early diagenesis in a reducing fjord, Saanick Inlet, British Columbia. I. Chemical and isotopic changes in major components of interstitial water. *Geochim. Cosmochim. Acta,* v. 36:1007–1027.

Railsback, L. B. 1993. A geochemical view of weathering and the origin of sedimentary rocks and natural waters. *J. Geol. Ed.,* v. 41:404–411.

REICHE, P. 1950. *A Survey of Weathering Processes and Products.* University of New Mexico Press, Albuquerque, NM.

RENFRO, A. R. 1974. Genesis of evaporite-associated stratiform metalliferous deposits—a sabkha process. *Econ. Geol.,* v. 6:933–45.

RENTSCH, J. 1974. The Kupferschiefer in comparison with the deposits of the Zambian copperbelt. In *Gisements stratiformes et province cuprifères,* P. Bartholomé, ed. Soc. géol. Belgique, Liège. 403–426.

RICHARDSON, S. M., and H. Y. MCSWEEN, JR. 1989. *Geochemistry—Pathways and Processes.* Prentice-Hall, Englewood Cliffs, NJ.

SAYLES, F. L., and P. C. MANGELSDORF. 1977. The equilibration of clay minerals with seawater: exchange reactions. *Geochim. Cosmochim. Acta,* v. 41:951–960.

SCHOTT, J., S. BRANTLEY, D. CRERAR, C. GUY, M. BORCSIK, and C. WILLAIME. 1989. Dissolution kinetics of strained calcite. *Geochim. Cosmochim. Acta,* v. 53:373–382.

SHACKLETTE, H. T., and J. G. BOERNGEN. 1984. Element Concentrations in Soils and Other Surficial Materials of the Conterminous United States: An Account of the Concentrations of 50 Chemical Elements in Samples of Soils and Other Regoliths. U.S. Geol. Survey Profess. Paper 1270.

SHAW, T. J., J. M. GIESKES, and R. A. JAHNKE. 1990. Early diagenesis in differing depositional environments: The response of transition metals in pore water. *Geochim. Cosmochim. Acta,* v. 54:1233–1246.

SHERMAN, G. D. 1952. The genesis and morphology of the alumina-rich laterite clays. In *Problems in Clay and Laterite Genesis.* Am. Inst. of Mining, Metallurgical, and Petroleum Engineers, New York. 154–161.

SHOLKOVITZ, E. 1973. Interstitial water chemistry of the Santa Barbara Basin sediments. *Geochim. Cosmochim. Acta,* v. 37:2043–2073.

SIEVER, R. 1962. Silica solubility, 0°–200°C, and the diagenesis of siliceous sediments. *J. Geol.,* v. 70:127–150.

SONNENFELD, P. 1984. *Brines and Evaporites.* Academic Press, New York.

SPENCER, R. J., T. K. LOWENSTEIN, E. CASAS, and Z. PENGXI. 1990. Origin of potash salts and brines in the Qaidam Basin, China. In *Fluid-Mineral Interactions: A Tribute to H. P. Eugster,* Special Publication No. 2, R. J. Spencer and I.-M. Chou, eds. The Geochemical Society, San Antonio, TX. 395–408.

SPOSITO, G. 1989. *The Chemistry of Soils.* Oxford University Press, New York.

STALLARD, R. F. 1985. River chemistry, geology, geomorphology, and soils in the Amazon and Orinoco Basins. In *The Chemistry of Weathering,* J. I. Drever, ed. NATO ASI Series, D. Reidel Publishing Co., Dordrecht, The Netherlands. 293–316.

STEWART, F. H. 1963. Marine Evaporites, Chapter Y. In *Data of Geochemistry,* 6th ed., M. Fleischer, ed. U.S. Geol. Survey Profess. Paper 440.

STUMM, W., and J. J. MORGAN. 1981. *Aquatic Chemistry,* 2d ed. John Wiley & Sons, New York.

STUMM, W., and R. WOLLAST. 1990. Coordination chemistry of weathering: kinetics of the surface-controlled dissolution of oxide minerals. *Rev. of Geophysics,* v. 28:53–69.

SUESS, E. 1973. Interaction of organic compounds with calcium carbonate. II. Organo-carbonate association in Recent sediments. *Geochim. Cosmochim. Acta,* v. 37:2435–2447.

TREFRY, J. H., and B. J. PRESLEY. 1982. Manganese fluxes from Mississippi Delta sediments. *Geochim. Cosmochim. Acta,* v. 46:1715–1726.

VELBEL, M. A. 1985. Geochemical mass balances and weathering rates in forested watersheds of the southern Blue Ridge. *Am. J. Sci.,* v. 285:904–930.

VELBEL, M. A. 1986. The mathematical basis for determining rates of geochemical and geomorphic processes in small forested watersheds by mass balance: examples and implications. In *Rates of Chemical*

Weathering of Rocks and Minerals, S. Colman and D. P. Dethier, eds. Academic Press, New York. 439–451.

WALTER, L. M., and E. A. BURTON. 1986. The effect of orthophosphate on carbonate mineral dissolution rates in seawater. *Chem. Geol.,* v. 56:313–323.

WHITE, A. F. 1990. Heterogeneous electrochemical reactions associated with oxidation of ferrous oxide and silicate surfaces. In *Mineral-Water Interface Geochemistry,* Rev. in Mineralogy, v. 23, M. F. Hochella, Jr., and A. F. White, eds. Mineraological Soc. Am., Washington, D.C. 467–509.

WHITE, A. F., and A. YEE. 1985. Aqueous oxidation-reduction kinetics associated with coupled electron-cation transfer from iron-containing silicates. *Geochim. Cosmochim. Acta,* v. 49:1263–1275.

ZIERENBERG, R. A., and W. C. SHANKS, III. 1986. Isotopic constraints on the origin of the Atlantis II, Suakin, and Valdivia brines, Red. Sea. *Geochim. Cosmochim. Acta,* v. 50:2205–2214.

ZULLIG, J. J., and J. W. MORSE. 1988. Interaction of organic acids with carbonate mineral surfaces in sea-water and related solutions. I. Fatty acid adsorption. *Geochim. Cosmochim. Acta,* v. 52:1667–1678.

Igneous Rocks

Igneous rocks are extremely varied in occurrence, composition, mineralogy, texture, and origin. We shall be concerned here only with certain chemical aspects of igneous rocks. This chapter begins with a review of the occurrence, chemical composition, and classification of igneous rocks, followed by an explanation of the most important geochemical terminology. Trace elements and isotopes are used a great deal in studying igneous rocks and their use is discussed in some detail. The following section reviews possible origins of magmas, focusing on the nature of the crust and upper mantle where parent and secondary magmas form and evolve. The crystallization of basaltic and granitic magmas is then discussed, using the results of laboratory research on silicate melts.

Knowledge of chemical properties for a specific suite of rocks allows igneous petrologists to understand better the origin and history of the rocks. This improved understanding comes from applying the techniques and laboratory results discussed in this chapter. Four examples of research on specific rock suites are given at the end of the chapter.

OCCURRENCE AND COMPOSITION

Most types of igneous rocks tend to occur abundantly in only a few geologic environments. Some occur mainly on the continents, others mainly in the oceans. Some are found only in areas of past or present tectonic instability. They may be associated with plate boundaries or with intraplate areas. Thus there is a general relationship between composition and environment of formation. In this section we will briefly review the occurrence and composition of the major types of igneous rock and also consider a few rare rocks with unusual compositions. Later in the chapter we will discuss the relationship between the origin of igneous rocks and plate tectonics.

The most abundant of all igneous rocks is basalt. It has formed both on the continents and in the oceans throughout the history of the Earth. Basalts are formed at both diverging and converging plate boundaries and also occur as a result of intraplate volcanism. Petrologists generally subdivide the basalt family into two broad types, *tholeiite basalt* and *alkali olivine basalt*. Tholeiite basalt contains two pyroxenes (augite and a calcium-poor pyroxene) and plagioclase feldspar. Olivine may be present. Alkali olivine basalt differs in having only one pyroxene (augite) along with plagioclase feldspar and olivine. This difference, along with other minor differences in mineralogy, leads to a difference in composition. For example, the total of K_2O plus Na_2O for a given amount of SiO_2 is usually greater for alkali olivine basalts than for tholeiite basalts (Figure 8-1). A distinct difference in trace element abundances is often found for the two rock types, most notably a lower abundance of incompatible elements such as niobium and uranium in tholeiites. However, chemical analysis of the two types overlap each other, and it is not possible to list separate and universal chemical characteristics for each type. On the other hand, the basalts of a specific region of the Earth may have distinctive chemical traits. In some cases, it is even possible to subdivide the basalts of an area on the basis of composition.

Alkali olivine basalts are particularly characteristic of oceanic islands and are also commonly found in zones of continental rifting. A typical analysis is given in Table 8-1, column 1. Tholeiite basalts are mainly found in deep ocean, along oceanic ridges, and as flood basalts on the continents (Table 8-1, columns 2 and 3). Often associated with the flood basalts are intrusive sills and dikes of similar composition. Continental tholeiites tend to have slightly higher K_2O and SiO_2 as compared to oceanic tholeiites.

One extreme in the chemical composition of common igneous rocks is represented by dunites and peridotites. These olivine-rich rocks mainly occur in orogenic areas of the past (fold-mountain

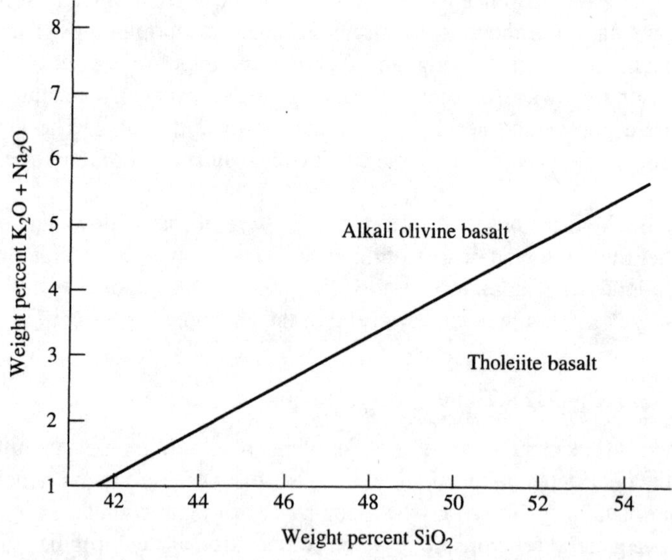

Figure 8-1 Alkali-silica plot showing the fields of the two major types of basalt. Both types occur abundantly in Hawaii, and the dividing line shown here is based on numerous analyses of Hawaiian rocks. Most basalts in Hawaii and elsewhere range in composition from 45 to 52 percent SiO_2 and from 2 to 6 percent K_2O C Na_2O. After MacDonald and Katsura (1964).

TABLE 8-1 Chemical Compositions of Some Common Igneous Rocks

	1 Oceanic alkali olivine basalt	2 Oceanic tholeiite basalt	3 Continental tholeiite basalt	4 Island arc dunite	5 Island arc andesite
SiO_2	50.48	49.20	52.05	39.53	58.60
TiO_2	2.25	2.03	1.70	0.01	0.89
Al_2O_3	18.31	16.09	12.43	0.93	15.38
Fe_2O_3	3.21	2.72	5.18	0.65	2.22
FeO	6.03	7.77	10.08	7.62	6.71
MnO	0.21	0.18	0.24	0.12	0.18
MgO	4.21	6.44	3.95	48.83	3.22
CaO	7.21	10.46	7.33	Trace	7.02
Na_2O	4.80	3.01	2.76	Trace	3.84
K_2O	1.93	0.14	2.07	—	1.46
P_2O_5	0.74	0.23	0.28	—	0.25
H_2O^+	0.46	0.70	1.90	0.89	0.30
H_2O^-	0.38	0.95	0.36	0.16	0.07
Other				1.41	
Total:	100.22	99.92	100.33	100.15	100.14
K/Rb	267	950	498	215	807
Rb/Sr	0.09	0.02	0.04	0.025	0.038
$^{87}Sr/^{86}Sr$	0.7033	0.7034	0.7064	0.7091	0.7036

Sources:

1: Cenozoic alkali olivine basalt, Guadalupe Island; data from A. E. J. Engel et al., 1965, *Bull. Geol. Soc. Am.*, v. 76, p. 725, and Z. E. Peterman and C. E. Hedge, 1971, *Bull. Geol. Soc. Am.*, v. 82, p. 495.

2: Cenozoic tholeiite basalt, Atlantic Ocean, depth 2,910 m, 20°40′ S, 13°16′ W; data from A. E. J. Engel and C. G. Engel, 1964, *Science*, v. 144, p. 1332, and P. W. Gast, 1967, *Basalts (Poldervaart Treatise)*, Wiley (Interscience Division), v. 1, p. 325.

3: Jurassic tholeiite basalt, Nuanetsi region, Karroo Basin, South Africa; data from K. G. Cox et al., 1965, *Phil. Trans. Roy. Soc. Lond., ser. A*, v. 257, p. 71, and W. I. Manton, 1968, *J. Petrol.*, v. 9, p. 30.

4: Permian dunite, Dun Mt., New Zealand; data from J. J. Reed, 1959, *New Zealand J. Geol. Geophys.*, v. 2, p. 16, and A. M. Stueber and V. R. Murthy, 1966, *Geochim. Cosmochim. Acta*, v. 30, pp. 1247—1248.

5: Quaternary pyroxene andesite, Talasea, New Britain; data from G. G. Lowder and I. S. E. Carmichael, 1970, *Bull. Geol. Soc. Am.*, v. 81, p. 27, and Z. E. Peterman et al., 1970, *Bull. Geol. Soc. Am.*, v. 81, p. 314.

belts) and the present (island arcs). They are the most common rock type found as xenoliths of mantle material in volcanic rocks of deep-seated origin. In terms of plate tectonics, these rocks can be described as occurring in ophiolite complexes along converging or formerly converging plate boundaries. They also occur in the lower parts of gabbroic layered intrusions. Other minerals occurring with olivine include enstatite, diopside, and spinel. Often the rocks are partially or wholly altered to serpentinite. The most important chemical traits of these rocks are low SiO_2 and high MgO. Commonly, significant amounts of Cr_2O_3 are present. The dunite analysis in Table 8-1, column 4, includes 1.01 percent Cr_2O_3. Chromium may occur in diopside or as chromite.

Volcanic rocks commonly found associated with orogenic belts include andesite (plagioclase, pyroxene, hornblende) and rhyolite (quartz, K-feldspar, plagioclase). Examples are given in Table 8-1, columns 5 and 6. Andesite has more SiO_2 than basalt and its plagioclase has a larger amount of sodium. These abundant rocks are found, with related volcanic rocks (basalt, dacite, rhyolite), in subduction zones of convergent plate boundaries. In the subduction zones, andesites occur both in island arcs and on continental margins.

TABLE 8-1 *(cont.)*

	6 Island arc rhyolite	7 Continental rift leucite nephelinite	8 Gabbro of continental layered mafic pluton	9 Continental anorthosite	10 Continental batholith quartz monzonite
SiO_2	74.22	39.77	48.08	53.40	65.49
TiO_2	0.28	3.82	1.17	0.77	0.65
Al_2O_3	13.27	12.53	17.22	23.96	14.49
Fe_2O_3	0.88	6.02	1.32	0.91	2.11
FeO	0.92	8.62	8.44	3.02	2.90
MnO	0.05	0.27	0.16	—	0.10
MgO	0.28	4.45	8.62	1.88	2.45
CaO	1.59	11.88	11.38	9.85	4.29
Na_2O	4.24	4.86	2.37	4.17	2.80
K_2O	3.18	5.35	0.25	0.80	3.66
P_2O_5	0.05	1.35	0.10	0.18	0.21
H_2O^+	0.80	0.60	1.01	0.69 }	0.56
H_2O^-	0.23	0.32	0.05		0.05
Other				0.43	0.15
Total:	99.99	99.84	100.17	100.06	99.91
K/Rb	251	352	550	>1,000	232
Rb/Sr	1.0	0.058	0.022	<0.01	0.311
$^{87}Sr/^{86}Sr$	0.7054	0.7051	0.7076	0.7053	0.7086

Sources:

6: Average Quaternary rhyolite, Taupo volcanic zone, New Zealand; data from A. Ewart and J. J. Stipp, 1968, *Geochim. Cosmochim. Acta*, v. 32, pp. 704, 712.

7: Quaternary leucite nephelinite, War Cemetery Flow, Nyiragongo volcano, Congo; data from K. Bell and J. L. Powell, 1969, *J. Petrol.*, v. 10, pp. 546, 549.

8: Chilled border gabbro, Tertiary Skaegaard intrusion, Greenland; data from L. R. Wager and G. M. Brown, 1967, *Layered Igneous Rocks*, Freeman, p. 158, pp. 193–194.

9: Gabbroic anorthosite, Precambrian Adirondack massif, New York; data from A. F. Buddington, 1939, *Geol. Soc. Am. Memoir 7*, p. 30, and S. A. Heath and H. W. Fairbairn, 1969, *N. Y. State Memoir 18*, p. 99.

10: Cretaceous Butte Quartz Monzonite, Kain quarry, Boulder batholith, Mont.; data from A. Knopf, 1957, *Am. J. Sci.*, v. 255, p. 89, and B. R. Doe et al., 1968, *Econ. Geol.*, v. 63, p. 898.

Rhyolite, with its high SiO_2 and low MgO, represents the other end of the scale from dunite. As indicated earlier, rhyolite occurs with andesite and other volcanic rocks at convergent plate boundaries. This group of rocks is sometimes referred to as the calc-alkaline series. The group, in any one area, usually shows a continuous range of composition from basalt to rhyolite. An increase in SiO_2 and K_2O from basalt to rhyolite is paralleled by decreases in Al_2O_3, CaO, FeO, MgO, andTiO_2. As a result, rhyolite mineralogy is dominated by quartz and alkali feldspar.

Tectonic activity in rift zones on the continents is also associated with other types of volcanic rocks, particularly rocks with high amounts of K_2O and Na_2O and low amounts of SiO_2 (Table 8-1, column 7). Some are strongly enriched in K_2O, others in Na_2O, and some in both. These alkaline rocks often contain large amounts of the silica-deficient minerals leucite and nepheline and also have other potassic and sodic minerals, such as pseudoleucite and sodalite. Associated volcanic rocks are alkaline basalts, trachytes, and rhyolites. Unusual associated rocks include carbonatites and lamprophyres (see later discussion).

Large masses of plutonic igneous rock found on the continents occur as layered plutons and as batholiths. The layered intrusions usually have average compositions similar to that of tholeiite basalt, and the major rock type is gabbro (pyroxene, plagioclase). A typical gabbro analysis is given in Table 8-1, column 8. Layered intrusions change from peridotite at their base to gabbro and then ferrogabbro at the top. A special type of gabbro, anorthosite, is made up almost entirely of plagioclase feldspar. This rock occurs abundantly only in batholiths of Precambrian shields. It has the highest Al_2O_3 content of common igneous rocks (Table 8-1, column 9). Most other batholiths are "granitic," meaning that their overall composition is about that of quartz monzonite (Table 8-1, column 10). The major minerals of these batholiths are quartz, K-feldspar, and plagioclase feldspar. A fairly wide variety of rock types makes up the individual plutons of granitic batholiths. Some are true granites, which have a composition similar to the rhyolite of Table 8-1. A true granite is mainly made up of quartz and K-feldspar; a quartz monzonite has quartz plus roughly equal amounts of K-feldspar and plagioclase.

Many plutonic rocks, such as those found along continental margins, have the same mineralogies and chemical compositions as the associated volcanic rocks. Thus the basalt-andesite-dacite-rhyolite series mentioned above is matched by a gabbro-diorite-granodiorite-granite series in associated batholiths. While the granites and other plutonic rocks along continental margins are clearly related to zones of plate convergence, other masses of granitic rocks (anorogenic intrusions) appear to have formed during regional metamorphism but away from areas of subduction, rifting, or other plate tectonic activity. The tectonic environment, if any, for formation of anorthosites and layered intrusions is not well understood. Some layered intrusions may be associated with crustal rifting and others are part of ophiolite complexes.

Table 8-1 shows that the common igneous rocks are composed of approximately 40 to 75 percent SiO_2, 12 to 24 percent Al_2O_3, 0.20 to 49 percent MgO, 1 to 10 percent FeO, 0 to 5 percent K_2O, 0 to 7 percent Na_2O, and small amounts of other oxides. Even more extreme compositions than these are shown by certain rock types, which are of very rare and limited occurrence. These rocks tend to be associated with continental rift zones. Three examples are given in Table 8-2. A few volcanic rocks of continental areas are extremely rich in potassium (Table 8-2, column 1). In addition to the usual major elements, small amounts of zirconium, chromium, strontium, barium, sulfur, and fluorine are found. These rocks have very high K/Na ratios. Rare potassium minerals such as wadeite ($Zr_2K_4Si_8O_{16}$) occur in the rocks. Certain dike rocks known as lamprophyres also have unusual compositions (Table 8-2, column 2). In particular, they tend to have high amounts of H_2O, P_2O_5, and CO_2 and a high total iron content ($FeO + Fe_2O_3$). Trace elements found include sulfur, fluorine, and chlorine. Ferromagnesian minerals such as hornblende, biotite, and augite are very abundant in lamprophyres. Alkaline lamprophyres are closely related chemically and mineralogically to kimberlites (famous as the major source of diamonds) and the two rock types occur together in South Africa and elsewhere.

The most extreme compositions are shown by carbonatites. The main minerals of these exotic rocks are carbonates rather than silicates. They generally occur as small intrusive bodies and as volcanic flows in areas of present or past continental rifting. The analysis in Table 8-2, column 3, is of the rock formed by a 1960 eruption in the East Africa rift zone. This particular carbonatite is unusual in having only a trace of SiO_2 and having a very high Na_2O content. Carbonatites show a wide range of composition, with SiO_2 averaging about 10 percent and Na_2O averaging about 0.5 percent. The CaO and CO_2 values are always high, and important concentrations of certain minor elements are usually found (particularly niobium, titanium, barium, strontium,

TABLE 8-2 Igneous Rocks of Extreme Compositions

	1 Potash-rich volcanic rock		2 Lamprophyre		3 Carbonatite
SiO_2	55.43	SiO_2	40.70	SiO_2	Trace
TiO_2	2.64	TiO_2	5.81	TiO_2	0.10
Al_2O_3	9.73	Al_2O_3	8.99	Al_2O_3	0.08
Fe_2O_3	2.12	Fe_2O_3	4.51	Fe_2O_3	0.26
FeO	1.48	FeO	8.37	FeO	—
MnO	0.08	MnO	0.17	MnO	0.04
MgO	6.11	MgO	9.69	MgO	0.49
CaO	2.69	CaO	9.94	CaO	12.74
Na_2O	0.94	Na_2O	1.58	Na_2O	29.53
K_2O	12.66	K_2O	1.97	K_2O	7.58
P_2O_5	1.52	P_2O_5	1.65	P_2O_5	0.83
H_2O^+	2.07	H_2O^+	4.16	H_2O	8.59
H_2O^-	0.61	H^2O^-	1.42	CO_2	31.75
SO_3	0.46	CO_2	0.64	SO_2	2.00
ZrO_2	0.28	S	0.09	Cl	3.86
Cr_2O_3	0.02	SrO	0.36	F	2.69
SrO	0.27	F	0.15	SrO	1.24
BaO	0.64	Cl	0.02	BaO	0.95
Total:	99.75	NiO, Cr_2O_3, V_2O_3	0.09	Total:	100.73 (corrected for F and Cl)
		Total:	100.31		

Sources:

1: Wyomingite (volcanic rock containing phlogopite, leucite, diopside, apatite), Leucite Hills, Wyo.; data from I. S. E. Carmichael, 1967, *Contrib. Mineral. Petrol.*, v. 15, p. 50.

2: Limburgite dike, Otago, New Zealand; data from C. O. Hutton, 1943; *Roy. Soc. New Zealand Trans.*, v. 73, pt. 1, p. 63.

3: Carbonatite lava, 1960 eruption, Oldoinyo Lengai volcano, Tanzania; data from J. B. Dawson, 1964, *Proc. Geol. Assoc. Can.*, v. 15, p. 106.

fluorine, sulfur, and the rare earths). Calcite is the most abundant carbonate mineral and other major minerals include dolomite, clinopyroxene, and phlogopite.

Chemical analyses of igneous rocks give us fair estimates of the range of compositions of magmas and lavas from which they crystallized. Since material may be gained or lost during or before crystallization, the range of compositions may not represent the true range of compositions of natural silicate liquids. We know that material is lost as a gas phase during the crystallization of lavas, and the components of volcanic gases (volatiles) may also be partially lost from a magma during crystallization. Assimilation, the incorporation of matter from wall rocks, is an important modifier of magma composition. Thus in reviewing the overall composition of igneous rocks we should consider information from analyses of volcanic gases and from studies of contaminated igneous rocks.

The first comprehensive summary of volcanic gases was published by White and Waring (1963). They list about 250 chemical analyses of volcanic gases. White and Waring report that steam (H_2O) is the greatly dominant gas, usually exceeding 90 percent of total gases. Some of the water of volcanic gases is undoubtedly meteoric rather than magmatic (juvenile) water. Probable magmatic components that are often present in gas samples in significant amounts are the compounds CO_2, CO, HCl, HF, H_2S, SO_2, and H_2. Analyses show a great range of concentrations for all these components. Other gases found, such as argon and nitrogen, may be due to contamination of samples by air. All that can be concluded from the 1963 tabulation by White and Waring, and from subsequent gas analyses, is that the gas phase of magmas probably consists mainly of water, with lesser amounts of carbon and sulfur gases. In some cases other gases, such as H_2 or HCl, may be abundant. Some of the elements found in volcanic gases will tend to be concentrated in a magma; others will occur mainly in a coexisting gas phase. Those that tend to stay in the magma (an example is fluorine) will end up in minerals such as biotite, apatite, and hornblende. Those that are generally partitioned to the gas phase (such as chlorine) will occur in rocks in amounts that are not representative of their original magmatic abundance.

A summary of recent research of all types on volatiles in magmas can be found in Carroll and Holloway (1994). The editors define volatile components as "those magma constituents which typically prefer to occur in the gaseous or super-critical fluid state" and indicate that these constituents "may influence virtually every aspect of igneous petrology" (Carroll and Holloway, 1994, p. iii). Individual chapters in the book deal with such topics as methods and results of volcanic-gas analyses, experimental studies of various gases in artificial silicate melts, physical aspects of magmatic degassing, and the effects of dissolved gases on magma density and viscosity. One interesting topic of study (Johnson et al. 1994) deals with differences in the volatile content of basalts from different plate tectonic settings (mid-ocean ridge, island arc, back-arc basin, etc.).

CLASSIFICATION AND GEOCHEMICAL TERMINOLOGY

We shall limit ourselves here to classifications based on chemical properties. At the same time, we shall briefly review some of the commonly used terms that are related to the chemical composition of igneous rocks. Many igneous rock classifications are based on mineral content. This type of classification is not practical for a volcanic rock containing very small mineral grains and glassy material. In this case, a chemical analysis can be used to calculate a theoretical mineral composition based on a standard set of minerals. Artificial minerals with fixed compositions are used for such calculations. This procedure was first suggested by Cross et al. (1902) and has since been modified, so that there is now more than one way to calculate the standard minerals (the *norm*) of a rock. The most commonly used type of norm calculation is the CIPW norm (named after petrologists Cross, Iddings, Pirsson, and Washington), which gives a weight percentage of standard minerals. For a discussion of calculation procedures for the CIPW norm, see Cox et al. (1979, Appendix 3), and Best (1982, Appendix E).

At the present time, norm calculations are not used mainly to classify rocks, but instead to compare them. As long as the norms of a series of rocks are all calculated in the same way, comparisons can be made that are more useful than comparisons of chemical analyses as such. A small difference in the chemical composition of two magmas can sometimes result in a larger difference in actual mineral composition (the *mode*). In many cases the mode of a rock corresponds fairly closely with its CIPW norm (Table 8-3). For some rocks there is poor agreement between actual mineralogy and calculated mineralogy. In part, this is due to the fact that the standard set

TABLE 8-3 Comparison of Normative and Modal Mineral Composition of a Diabase from the Palisades Sill, New Jersey

Chemical composition		CIPW norm		Mode (weight percent)	
SiO_2	47.41	Orthoclase	2.39	Plagioclase	25.2
TiO_2	0.89	Albite	11.53	Augite	16.9
Al_2O_3	8.66	Anorthite	16.21	Bronzite	23.6
Fe_2O_3	2.81	Diopside	13.14	Olivine	24.0
FeO	11.15	Hypersthene	25.91	Opaques	3.0
MnO	0.20	Olivine	23.77	Biotite	3.0
MgO	19.29	Magnetite	4.18	Sphene	Trace
CaO	6.76	Ilmenite	1.65	Apatite	Trace
Na_2O	1.35	Apatite	0.34	Alteration	
K_2O	0.43	H_2O	1.56	products	5.0
P_2O_5	0.10	Total:	100.68	Total:	100.7
H_2O^+	1.45				
H_2O^-	0.11				
Total:	100.61				

Source: Data from Walker (1969, 32, 33, 76).

of minerals used for norm calculations does not include any hydrous minerals, and contains only end-member minerals for groups such as the pyroxenes. Thus some common minerals such as augite are not included in the list of standard minerals. Also, all the standard minerals are "pure" minerals without trace or minor constituents.

A widely used classification for volcanic rocks was proposed by Irving and Baragar in 1971. The rocks are divided into three groups based on their alkali content. The peralkaline (alkali-rich) group is made up of only a few rare rock types with most volcanics belonging to the other two categories, alkaline and subalkaline. In peralkaline rocks, molecular $Na_2O + K_2O$ is greater than molecular Al_2O_3. All other rocks are classified as alkaline or subalkaline using a plot similar to that shown in Figure 8-1; the alkali olivine basalt part of the figure represents the alkaline group and the tholeiite basalt area represents the subalkaline group. These two groups are further subdivided on the basis of chemical composition and normative composition. For example, the subalkaline rocks are divided into the calc-alkaline series and the tholeiite series using the AFM plot shown in Figure 8-2. Further details on the Irving-Baragar classification can be found in their original 1971 paper and in McBirney (1993) and Philpotts (1990).

A plot of total alkalis versus total silica (Figure 8-1) can be used as the basis of a complete classification of volcanic rocks. Le Bas et al. (1986) have subdivided such a plot into fifteen compositional fields. This classification has been recommended by the International Union of Geological Sciences (IUGS). The IUGS suggests that volcanic rocks be classified on the basis of mode if possible; when not possible, a plot of total alkalis versus total silica is a suggested alternative. The IUGS classification of plutonic rocks is based on actual mineralogy (mode in volume percent) and no chemical alternative is needed (Streckeisen 1976).

One important use of norm calculations for igneous rocks is in determining whether or not a rock is saturated. The concept of *rock saturation* was introduced by S. J. Shand and used by him to classify igneous rocks (Shand 1951). Shand divided igneous minerals into two groups, those

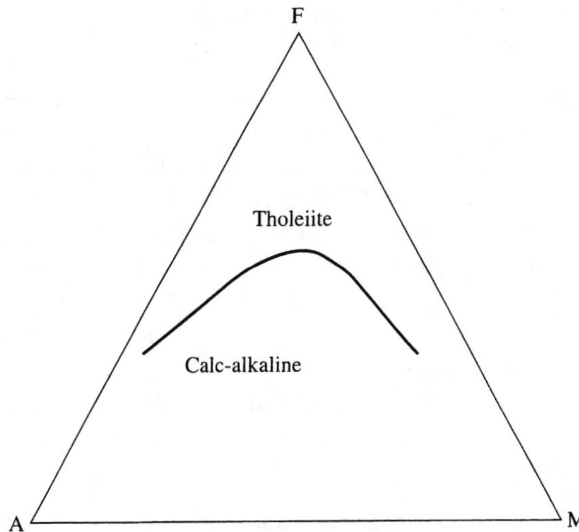

Figure 8-2 AFM plot showing fields of tholeiite series and calc-alkaline series rocks. $A = Na_2 + K_2O$, $F = FeO + 0.8998 \times Fe_2O_3$, and $M = MgO$ (all in weight percent). Data from Irving and Baragar (1971). (After Philpotts, A. R., *Principles of Igneous and Metamorphic Petrology*. Copyright © 1990. Adapted by permission of Prentice Hall, Inc., Upper Saddle River, NJ.)

that can coexist in equilibrium with quartz, such as feldspars and micas, and those that cannot, such as leucite and magnesian olivine. Because of their low SiO_2 content relative to other possible minerals, the latter two minerals are unsaturated with respect to silica. For example, leucite ($KAlSi_2O_6$) reacts with SiO_2 to form K-feldspar ($KAlSi_3O_8$). On this basis the mineralogy of a given rock can be used to classify the rock into one of three categories:

1. Oversaturated (contains quartz).
2. Saturated (contains neither quartz nor an unsaturated mineral).
3. Unsaturated (contains one or more unsaturated minerals).

It should be noted that whether or not a rock is saturated depends not only on its SiO_2 content, but also on the other oxides present, since the overall composition determines the final mineralogy. The CIPW system of calculations for normative minerals was set up so that minerals incompatible with quartz would not appear in the norm of a rock containing normative quartz. Thus the norm of a rock indicates whether or not it is saturated in terms of silica content.

As indicated earlier, the distinction between tholeiite basalt and alkali olivine basalt can be based on actual mineral content (the mode). However, it is also true that, in terms of the CIPW norm, most tholeiite basalts are saturated and most alkali olivine basalts are unsaturated. This makes it possible to classify a basalt whose mineralogy is not easily identified. Norm calculations for the oceanic tholeiite basalt of Table 8-1 show 0.3 normative quartz and no normative olivine; the calculated norm for the alkali olivine basalt in Table 8-1 contains no normative quartz and 2.35 normative olivine. Thus the tholeiite basalt is slightly oversaturated, and the alkali olivine basalt is unsaturated. Note that the alkali olivine basalt actually has a higher SiO_2 content than the oceanic tholeiite basalt.

Another type of saturation proposed by Shand involves Al_2O_3 and its abundance relative to K_2O and Na_2O. Potash feldspar ($KAlSi_3O_8$) has one atom of potassium for each atom of aluminum. Other minerals show a molecular alkali/aluminum ratio that is greater or less than the one-to-one relationship of potash feldspar. The mineralogy of an igneous rock is thus dependent on its overall alkali/aluminum ratio. Shand proposed four groups of rocks in terms of alumina (Al_2O_3) saturation:

1. Peraluminous (molecular Al_2O_3 greater than $Na_2O + K_2O + CaO$); these rocks contain minerals such as muscovite [$KAl_3Si_3O_{10}(OH)_2$] and andalusite (Al_2SiO_5); the norm contains corundum (Al_2O_3).
2. Metaluminous (molecular Al_2O_3 greater than $Na_2O + K_2O$ but less than $CaO + Na_2O + K_2O$); typical minerals are biotite [$K(Mg, Fe)_3AlSi_3O_{10}(OH)_2$] and hornblende [$NaCa_2(Mg, Fe, Al)_5(Si, Al)_8O_{22}(OH)_2$]; the norm contains anorthite ($CaAl_2Si_2O_8$).
3. Subaluminous (molecular Al_2O_3 approximately equal to $Na_2O + K_2O$); these rocks contain nonaluminous minerals such as olivine [$(Mg, Fe)_2SiO_4$] and hypersthene [$(Mg, Fe)SiO_3$]; the norm is also low in aluminum-bearing minerals.
4. Peralkaline (molecular Al_2O_3 less than $Na_2O + K_2O$); alkali minerals such as aegirine ($NaFeSi_2O_6$) and riebeckite [$Na_2Fe_3^{2+}Fe_2^{3+}Si_8O_{22}(OH)_2$] occur in these rocks; the norm has minerals such as acmite ($NaFeSi_2O_6$).

The concept of alumina saturation is used mainly to characterize an individual rock or group of rocks as unusually rich in alumina relative to the alkalis (peraluminous), or as unusually rich in alkalis relative to alumina (peralkaline). Examples of these rock types in Table 8-1 are rhyolite, which is peraluminous, and nephelinite, which is peralkaline.

A common way to classify a group of genetically related igneous rocks is in terms of the alkali-lime index proposed by Peacock (1931). Chemical analyses of a suite of rocks are used to plot the change of ($Na_2O + K_2O$) and CaO in terms of SiO_2. It is typical for magmas or lavas forming from a common primary magma to show an increase in SiO_2, a decrease in CaO, and an increase in $Na_2O + K_2O$ with time. Thus a plot of these will show crossed lines (Figure 8-3). The weight percent value of SiO_2 where the two lines cross characterizes the rock group as belonging to one of four subdivisions: calcic (SiO_2 greater than 61), calc-alkalic (SiO_2 of 56 to 61), alkali-calcic (SiO_2 of 51 to 56), or alkalic (SiO_2 less than 51).

A number of different plots for related rocks have been proposed to show changes in composition with the supposed order of formation from a source magma. Any plot of analytical data for an associated group of rocks, whether they are genetically related or not, is called a *variation diagram.* Figures 8-3 and 8-4 are examples of variation diagrams. Some diagrams are produced by using normative minerals rather than oxides as variables. In the case of rocks that are believed to have all formed from one parent magma, a variation diagram is used in an attempt to show the changing composition of the magma as crystallization occurs. The change in composition is referred to as the liquid line of descent (Wilcox 1979). The details of such a line are obtained by analyses of fine-grained and nonporphyritic rocks. Only the groundmass of a porphyritic rock would be representative of the liquid at the time of formation. For such rocks the chemical analysis can be adjusted by subtracting the oxide percentages represented by the phenocrysts.

Once a liquid line of descent is obtained, various physical models of evolution can be proposed that would produce such a change in composition with time (Figure 8-4). For fractional crystallization, a model would specify the amount and composition of early formed crystals that had been separated from the magma at various times. Other models might involve contamination of a magma by country rock, mixing of two magmas, separation of a magma into two immiscible liquids, crystal accumulation, and so on. In evaluating studies of this type, keep in mind two points. First, there are usually assumptions and uncertainties involved in obtaining the liquid line

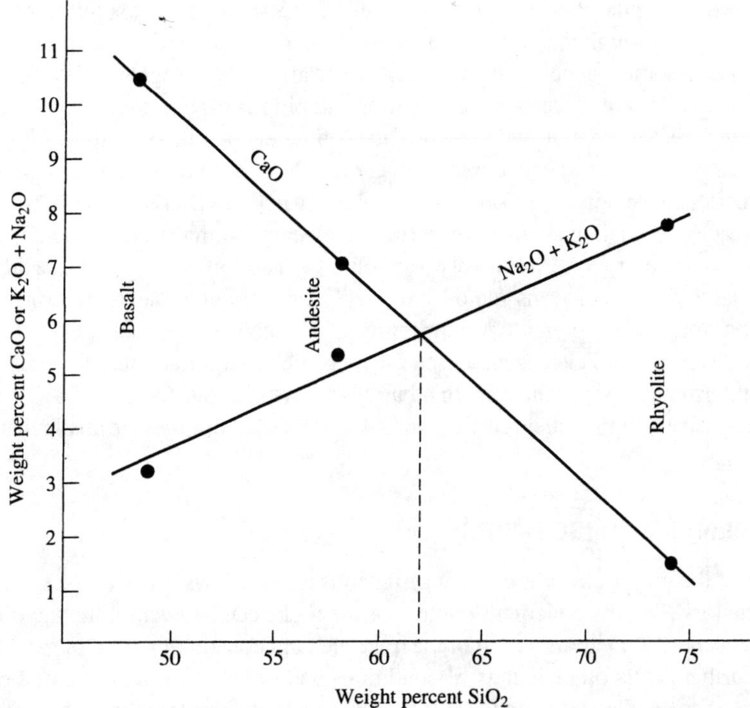

Figure 8-3 Plot of CaO and of $K_2O + Na_2O$ in terms of SiO_2 for the oceanic tholeiite basalt, andesite, and rhyolite of Table 8-1. The intersection of the two lines occurs at a SiO_2 value of 62 weight percent. If these three rocks formed by magmatic differentiation from a single magma, the rock suite would be characterized as calcic according to the alkali-lime index of Peacock (1931). Note that one rock type by itself cannot be classified in terms of the index.

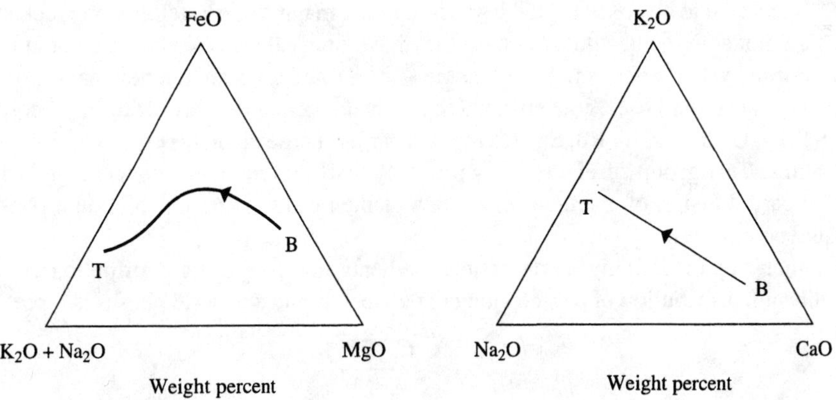

Figure 8-4 Triangular (three-component) variation diagrams for a volcanic rock suite of Gough Island. The variation in composition of whole rocks shown here is consistent with a fractional-crystallization model of rock formation. The rocks are believed to have formed from a parent magma with the composition of alkali olivine basalt. Early formed rocks are picrite basalt and alkali olivine basalt (B), and late-stage rocks are trachytes (T). See text for further discussion of Gough Island rocks. After Le Maitre (1962).

of descent. Second, a number of equally plausible models can often explain a given line of descent. Finally, even though the points on a variation diagram show a regular pattern, this does not prove that the rocks are genetically related. Cox et al. (1979), Ragland (1989), and Rollinson (1993) provide extensive discussions of the uses and pitfalls of variation diagrams.

We conclude this section with a word about the general classification of igneous rocks in terms of silica content. For many years, igneous rocks have been referred to as acid (SiO_2 greater than 66 percent), intermediate (SiO_2 of 52 to 66 percent), basic (SiO_2 of 45 to 52 percent), and ultrabasic (SiO_2 less than 45 percent). This terminology came from the fact that SiO_2 forms an acid when dissolved in water (SiO_2 has a very low solubility and forms a very weak acid). Many geologists feel that the chemical terms "acid" and "base" should not be used to describe the SiO_2 content of igneous rocks. They prefer to use the terms felsic, intermediate, mafic, and ultramafic for the four groups given above. Felsic is also used for rocks rich in light-colored minerals and mafic for those rich in ferromagnesian minerals. In addition to having a low content of SiO_2, mafic and ultramafic rocks have a high content of FeO and MgO. We shall use these terms in this chapter.

TRACE ELEMENTS AND ISOTOPES

Knowledge of the major element content of igneous rocks allows plausible speculation on their origin and history. The trace element content of the rocks can be even more significant than the major element content. Determination of the trace element content of a suite of genetically related rocks puts further limits on estimates of conditions and order of formation based on major element content. A trace element is one that is present in a rock or magma in a concentration of less than 0.1 weight percent.

The trace elements in a magma can be conveniently divided into two groups. In one group are elements (compatible elements) that tend to be taken into minerals crystallizing from the magma. An example is vanadium, which enters readily into the crystal structure of magnetite. Elements of the other group (incompatible elements) do not easily substitute for major elements of minerals and tend to be concentrated in the remaining magma as crystallization proceeds. These elements are not able to substitute because they have unusual sizes (ionic radii) or charges. For example, boron ion has a very small size (about 0.20 Å), and tungsten ion may have a +6 charge. Other elements that tend to be concentrated in residual magmatic fluids for these reasons are Be, Nb, Ta, Sn, Th, U, Pb, Cs, Li, Rb, Sr, and the rare earths. These elements commonly are abundant in pegmatites. This group of elements is probably also preferentially concentrated in magma formed by partial fusion of a source rock. Such elements always prefer the liquid phase in any solid-liquid equilibrium.

Another way of defining incompatible elements involves using distribution coefficients. The equilibrium distribution of a trace element between liquid and solid phases is represented by

$$K_D = \frac{C_i^{\text{mineral}}}{C_i^{\text{melt}}} \qquad (8\text{-}1)$$

where K_D is the Nernst distribution coefficient, C_i^{mineral} is the concentration of element i in a crystallizing mineral, and C_i^{melt} is the concentration of element i in the melt from which the mineral is crystallizing. The concentration values can be in parts per million or in weight percent. Distribution coefficients (also known as partition coefficients) can be measured in two very different ways. One technique involves analysis of phenocrysts and their glassy matrix in volcanic rocks. The other approach consists of laboratory measurement of prepared synthetic materials containing the elements of interest. Depending on the method or sample used, very different values for the same element and mineral are often obtained. Table 8-4 contains some examples of typical values. An extensive compilation of trace element distribution coefficients is given by Rollinson (1993).

In discussing magmatic crystallization, it is useful to calculate, for a given rock, a bulk distribution coefficient D:

$$D = \sum_{i=1}^{n} w_i K_{Di} \tag{8-2}$$

where w_i is the weight proportion of a mineral containing element i in the rock and K_{Di} is the distribution coefficient of element i between the mineral and a coexisting melt. In a rock containing 50 percent olivine, 30 percent orthopyroxene, and 20 percent clinopyroxene, the D value for europium would be

$$D = 0.5 \times 0.01 + 0.3 \times 0.05 + 0.2 \times 0.9 = 0.2$$

The individual K_D values are from Table 8-4.

We can define elements with D values less than one as incompatible elements; they tend to be preferentially concentrated in the liquid phase during crystallization. Similarly, elements with D values greater than one are termed compatible elements and tend to be preferentially taken into the

TABLE 8-4 Mineral-Melt Distribution Coefficients

	Rb	Sr	Ce	Eu	Yb
Amphibole					
mafic rocks	0.3	0.5	0.3	1.0	1.0
felsic rocks	0.01	0.02	1.0	4.0	7.0
Biotite					
mafic rocks	3.0	0.08	0.03	0.03	0.03
felsic rocks	3.0	0.2	0.3	0.3	0.3
Clinopyroxene					
mafic rocks	0.01	0.1	0.3	0.9	1.0
felsic rocks	0.05	0.5	0.9	2.0	2.0
Garnet					
mafic rocks	0.001	0.001	0.05	0.9	30.0
felsic rocks	0.01	0.02	0.6	0.7	40.0
K-feldspar					
felsic rocks	0.4	6.0	0.04	1.1	0.01
Olivine					
mafic rocks	0.004	0.005	0.01	0.01	0.01
Orthopyroxene					
mafic rocks	0.01	0.01	0.02	0.05	0.3
Plagioclase					
mafic rocks	0.1	2.0	0.14	0.3	0.07
felsic rocks	0.06	5.0	0.3	2.0	0.05
Spinel					
mafic rocks	0.01	0.01	0.08	0.03	0.02

Source: From *Basic Analytical Petrology*, by Paul C. Ragland. Copyright ©1989 by Oxford University Press, Inc. Reprinted by permission.

crystallizing minerals of a cooling magma. Two points should be noted about these definitions. First, the minerals forming from a magma will usually change with time and this means that some elements may change from being compatible to being incompatible and vice versa. Second, several other factors have to be considered. Note in Table 8-4 that separate values for a given element are listed for mafic rocks and felsic rocks. This is because distribution coefficients vary in value with melt composition (Figure 8-5). They also vary with temperature and pressure (Figure 8-6).

The above definitions can also be used for the process of partial melting of solid rock (in the mantle, for instance) to form a new magma. Some elements will be preferentially taken into the liquid phase and some will be preferentially left behind in the residual solid mass. This means it is possible to obtain useful information for a rock or group of rocks about both the formation and the crystallization of the parent magma. This type of research is known as trace element modeling (Hart and Allegre 1980; Rollinson 1993, Chap. Four).

Rollinson (1993) reviews the mathematical equations used to model trace element distribution. These can be divided into equations for partial melting processes and equations for magmatic crystallization (Table 8-5). Batch melting refers to the formation of a partial melt in which the liquid remains in equilibrium with the solid residue until it leaves the area as a "batch" of primary magma. Fractional melting is the process by which each infinitesimally small amount of liquid that forms is immediately removed from the source rock. This is the opposite of batch

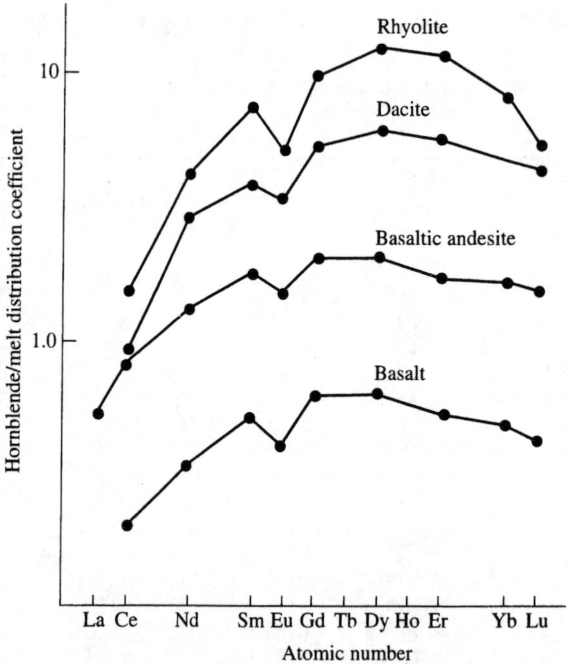

Figure 8-5 A plot of the distribution coefficients for the rare-earth elements between hornblende and melt (log scale) vs. atomic number (normal scale) in basalt, basaltic andesite, dacite, and rhyolite. There is a clear increase in distribution coefficient with increasing silica content of the melt, amounting to an order of magnitude difference between basaltic and rhyolitic melts. (After H. R. Rollinson, *Using Geochemical Data: Evaluation, Presentation, Interpretation*. Copyright © 1993. Used by permission of the Longman Group UK Ltd.)

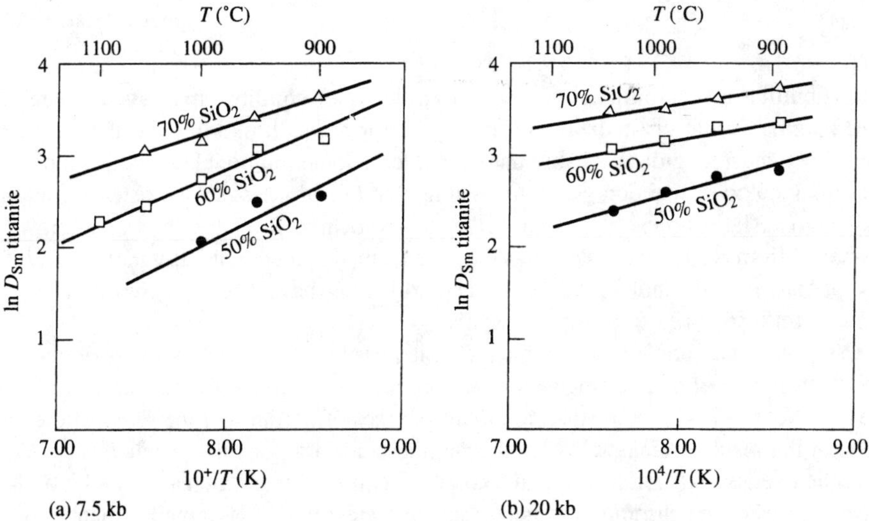

Figure 8-6 The distribution coefficient for samarium in titanite as a function of temperature for liquids with 50, 60, and 70 weight percent SiO_2 at (a) 7.5 kb and (b) 20 kb pressure. (Reprinted from *Chemical Geology*, v. 55, pp. 105–109, T. H. Green and N. J. Pearson, Rare-earth element partitioning between sphene and coexisting silicate liquid at high pressure and temperature. Copyright 1986. Reprinted by permission of Elsevier Science Ltd, Amsterdam, The Netherlands.) After Rollinson (1993).

TABLE 8-5 Equations Governing Trace Element Behavior During Melting and Crystallization

PARTIAL MELTING	CRYSTALLIZATION
Batch Melting	*Equilibrium Crystallization*
1) $C_L/C_O = 1/[D_{RS} + F(1 - D_{RS})]$	3) $C_L/C_O = 1/[D + F(1 - D)]$
2) $C_S/C_O = D_{RS}/[D_{RS} + F(1 - D_{RS})]$	
Fractional Melting	*Rayleigh Fractionation*
4) $C_L/C_O = \frac{1}{D_O}(1 - PF/D_O)^{(1/P-1)}$	6) $C_L/C_O = F^{(D-1)}$
5) $C_S/C_O = \frac{1}{(1-F)}(1 - PF/D_O)^{1/P}$	7) $C_R/C_O = DF^{(D-1)}$

C_L = Weight concentration of a trace element in the liquid.

C_O = In partial melting, the weight concentration of a trace element in the original unmelted solid; in fractional crystallization, the weight concentration in the parental liquid.

C_S = Weight concentration of a trace element in the residual solid after melt extraction.

D_{RS} = Bulk distribution coefficient of the residual solids.

F = Weight fraction of melt produced in partial melting; in fractional crystallization, the fraction of melt remaining.

D = Bulk distribution coefficient of the fractionating assemblage during crystal fractionation.

D_O = Bulk distribution coefficient for the original solid phases.

P = Bulk distribution coefficient of minerals that make up a melt.

C_R = Weight concentration of a trace element in the residual solid during crystal fractionation.

Source: After Rollinson (1993).

melting. Equilibrium crystallization is the process in which equilibrium exists between all solids and the magma during crystallization. This is similar to batch melting and the same equation applies. Since most magmas probably undergo at least some fractional crystallization, equations used for magma crystallization generally assume that this process occurs instead of equilibrium crystallization. The simplest type of model used is known as Rayleigh fractionation, with crystals removed from the magma as soon as they are formed. Other, generally more complex, models for partial melting and for magmatic crystallization have been suggested. The reader is referred to Rollinson (1993) for further details.

Using the equations given in Table 8-5, it is possible to calculate a series of curves representing the behavior of trace elements for each process. An example for batch melting is given in Figure 8-7. Note in Figure 8-7b that, for small degrees of partial melting, compatible elements ($D > 1$) in the residue remain close to their initial concentrations in unmelted rock. Cox et al. (1979) concluded from this that ultramafic nodules in kimberlites, which are samples of the upper mantle, have nickel and chromium contents that are representative of unmelted mantle rock. Even if the nodules came from a mass of rock that had previously undergone one or more partial melting events, their content of the compatible elements nickel and chromium should still be similar to the values of unmelted mantle rock. Knowing these initial values, and assuming a reasonable mantle mineralogy to get bulk distribution coefficients, the curves in Figure 8-7a were used by

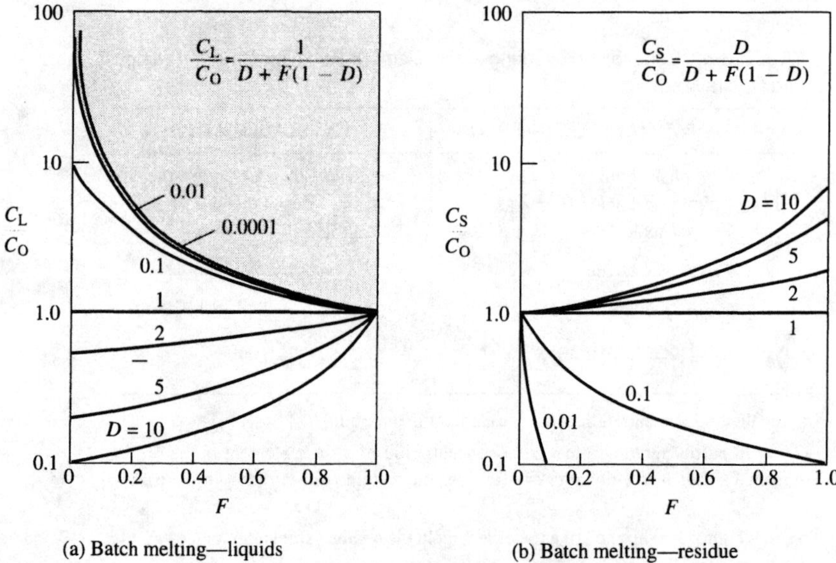

Figure 8-7 (a) The enrichment of a trace element in a partial melt relative to its concentration in the source (C_L/C_O) during batch partial melting with changing degrees of melting (F). The numbered curves are for different values of the bulk distribution coefficient D. At small degrees of melting, compatible elements are greatly depleted relative to the source, whereas incompatible elements are greatly enriched to a maximum of $1/F$. The shaded region is the area in which enrichment is impossible. (b) Enrichment and depletion of a trace element in the residue relative to the original source (C_S/C_O) with changing degrees of melting (F) for different values of bulk distribution coefficient (D). (After H. R. Rollinson, *Using Geochemical Data: Evaluation, Presentation, Interpretation*. Copyright © 1993. Used with permission of the Longman Group UK Ltd.)

Cox et al. to estimate the nickel and chromium contents of primary magmas derived by partial melting of rock in the upper mantle.

It is also possible to calculate specific values for various elements as a result of fractional crystallization of a magma. Table 8-6 lists the concentrations of rubidium and strontium for fractional crystallization of basaltic magma as a function of the percentage of residual liquid remaining. This table is based on equation six in Table 8-5 and bulk distribution coefficients calculated from the data in Table 8-4. Table 8-6 shows that rubidium is more incompatible than strontium. The ratio Rb/Sr increases significantly as crystallization reaches its final stages (F is a small number). Note that the highest values for Rb/Sr in Table 8-1 are shown for rhyolite and quartz monzonite, rocks that are believed to form in the late stages of fractional crystallization.

The rare-earth elements (REE) have been used extensively to study both the formation of magmas by melting of solid rock and the crystallization characteristics of these magmas (Hanson 1989). These elements have atomic numbers from 57 to 71 and exhibit a gradual decrease in radius with increasing atomic number (the lanthanide contraction; see Chapter One). The radii and other data for the REE are given in Table 8-7. It is useful to refer to the elements from $Z = 57$ to $Z = 62$ as the light rare earths (LREE) and those with atomic number higher than europium ($Z = 63$) as the heavy rare earths (HREE).

It was pointed out in Chapter One that elements with an even atomic number are more stable and thus more abundant than elements with an odd atomic number. The REE show this phenomenon. In studying REE distribution, it is convenient to eliminate the zigzag pattern that results from plotting abundance versus atomic number. This can be done by producing a chondrite-normalized

TABLE 8-6 Calculated Enrichments (C_L/C_O) of Rubidium and Strontium Obtained in Basalt by Fractional Crystallization

F	Rb	Sr	Rb/Sr
0.05	18.5	1.8	10.2
0.10	9.3	1.6	5.8
0.20	4.8	1.4	3.4
0.30	3.2	1.3	2.5
0.40	2.4	1.2	2.0
0.50	2.0	1.1	1.8
0.60	1.6	1.1	1.5
0.70	1.4	1.1	1.3
0.80	1.2	1.1	1.1
0.90	1.1	1.0	1.1

F = fraction of melt remaining. C_L = content of element in liquid. C_O = initial content of element in liquid. The bulk distribution coefficients used in the calculations are for a mineral assemblage composed of 50 percent plagioclase, 40 percent calcic clinopyroxene, and 10 percent olivine. Bulk distribution coefficients are $Rb = 0.026$ and $Sr = 0.80$.

Source: Reprinted by permission of the publishers from ORIGINS OF IGNEOUS ROCKS by Paul C. Hess, Cambridge, Mass: Harvard University Press, Copyright © 1989 by the President and Fellows of Harvard College.

TABLE 8-7 Rare-Earth Element Data

Z	REE	r	C	C_M	C/C_M	K_D	40 PL
57	La	1.26	24.2	0.367	65.9	0.14	102.0
58	Ce	1.22	53.7	0.957	56.1	0.14	87.0
59	Pr	1.22	6.5	0.137	47.4	—	—
60	Nd	1.20	28.5	0.711	40.1	0.08	64.1
61	Pm	—	—	—	—	—	—
62	Sm	1.17	6.70	0.231	29.0	0.08	46.4
63	Eu	1.15	1.95	0.087	22.4	0.32	31.7
64	Gd	1.14	6.55	0.306	21.4	0.10	33.9
65	Tb	1.12	1.08	0.058	18.6	—	—
66	Dy	1.11	6.39	0.381	16.8	0.09	26.7
67	Ho	1.10	1.33	0.0851	15.6	—	—
68	Er	1.08	3.70	0.249	14.9	0.08	23.8
69	Tm	1.07	0.51	0.0356	14.3	—	—
70	Yb	1.06	3.48	0.248	14.0	0.07	22.5
71	Lu	1.05	0.55	0.0381	14.4	0.08	23.0

r = ionic radius in angstroms for cubic coordinated, trivalent REE (Whittaker and Muntus 1970)

C = REE concentration in ppm in USGS basaltic rock standard BCR-1 (Taylor and McLennan 1985)

C_M = average REE concentration in ppm in Type 1 carbonaceous chondrites (Evensen et al. 1978); 1.5× original data

K_D = average distribution coefficient for plagioclase in mafic rocks (Henderson 1982)

40 PL = C/C_M in residual melt after 40 percent extraction of plagioclase from original melt composition taken as BCR-1 (40 percent fractionation of plagioclase)

Source: From *Basic Analytical Petrology*, by Paul C. Ragland. Copyright © 1989 by Oxford University Press, Inc. Reprinted by permission.

ratio in which the abundances of the REE in a sample are divided by selected chondrite values believed to be representative of unfractionated original solar system matter (see Chapter One). The result is a smooth pattern rather than a zigzag pattern (Figures 8-8 and 8-9). If any change in the relative abundances of the REE has occurred in the history of a sample, the pattern will be a line or curve that is not horizontal. If no change other than an enrichment or depletion of all of the REE has occurred, then the pattern will be a horizontal line parallel to the chondrite line (Figure 8-8). One point to keep in mind is that there is no generally accepted set of chondrite values to be used for normalization; various sets are used. A table of these values is given by Rollinson (1993, 134).

Because of the changes in ionic radii shown in Table 8-7, some of the properties of the LREE are slightly different from those of the HREE. Generally, the LREE are more incompatible than the HREE. For example, garnet-liquid distribution coefficients are different for the HREE than for the LREE. Partial melting of a garnet peridotite strongly fractionates the REE, with the HREE staying in the solid material and the LREE partitioned to the new melt (Figure 8-8). On the other hand, no sharp separation (fractionation) of HREE from LREE occurs in the melting of a peridotite without garnet. Instead, because of the nature of the mineral-liquid distribution coefficients for minerals such as orthopyroxene and clinopyroxene, the REE pattern produced in the liquid is a smooth but nonhorizontal curve (Figure 8-8). The curve is nonhorizontal because the LREE are partitioned into the melt somewhat more efficiently than the HREE. Thus the fractionation is not as extreme as in the case of garnet peridotite.

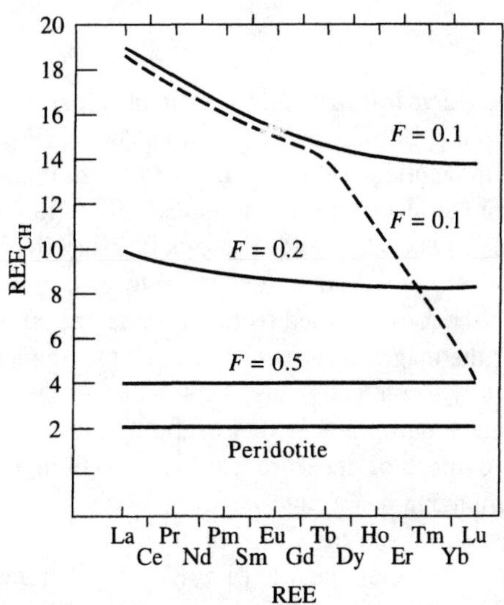

Figure 8-8 Chondrite-normalized REE contents of melts produced by equilibrium melting of peridotite and garnet peridotite. REE$_{CH}$ is obtained by dividing sample concentration by chondrite value for a given element. The original REE content of peridotite is assumed to be a flat pattern at 2× chondrite abundances. The mineral assemblage in peridotite is 55 percent olivine, 25 percent orthopyroxene, and 20 percent calcic clinopyroxene. The mineral assemblage in garnet peridotite is 55 percent olivine, 20 percent orthopyroxene, 20 percent calcic clinopyroxene, and 5 percent garnet. F is the fraction of melt in the rock. Compare the strongly fractionated (HREE)$_{CH}$ pattern produced by melting garnet peridotite (dashed line) with the relatively unfractionated pattern in the melts derived from peridotite (solid lines denoted with F). (Reprinted by permission of the publishers from ORIGINS OF IGNEOUS ROCKS by Paul C. Hess, Cambridge, Mass: Harvard University Press, Copyright © 1989 by the President and Fellows of Harvard College.)

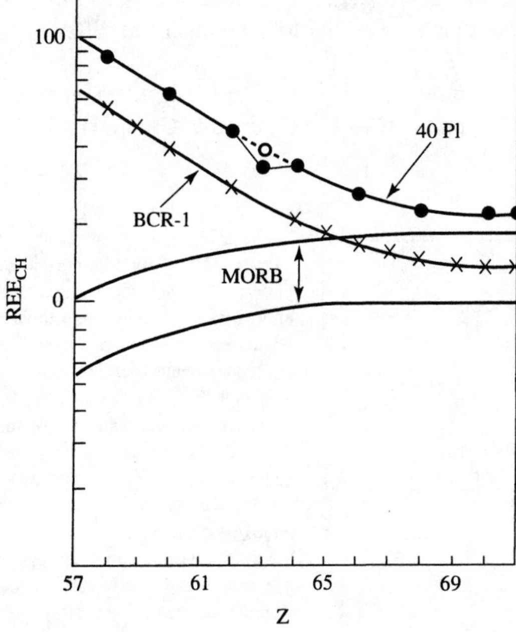

Figure 8-9 Chondrite-normalized REE concentrations for MORB basalts, basalt rock standard BCR-1, and daughter rock (40 Pl) calculated for plagioclase extraction from melt with the composition of BCR-1. The daughter rock has a small negative europium anomaly. (From *Basic Analytical Petrology*, by Paul C. Ragland. Copyright © 1989 by Oxford University Press, Inc. Reprinted by permission.)

How can we use the above information to obtain new insights about igneous rocks? Assume the upper mantle consists of peridotite and has a chondritic REE pattern. It is generally believed that garnet is found only in mantle peridotite at depths greater than 60 km. If this is true, magmas formed by partial melting below 60 km should have REE patterns showing a strongly fractionated pattern for the LREE versus the HREE (depletion in HREE) as in Figure 8-8. Magmas formed by melting above 60 km would show a relatively unfractionated pattern. By extension, this should also be true of the rocks formed from these magmas if no major changes affecting the REE have occurred after the magmas formed. For example, their REE patterns indicate that alkali olivine basalts generally formed at greater depths than tholeiite basalts.

One rare-earth element, europium, is very useful in interpreting the history of magma crystallization. In contrast to most of the rare earths, europium occurs in two valence states, Eu^{+2} and Eu^{+3}. When a magma has a low oxidation state, Eu^{+2} is present and, because of its ionic size and charge, can substitute for Ca^{++} in plagioclase feldspar. Its behavior under these conditions is different from that of all the other REE. As a result, many REE patterns show either a positive (enrichment relative to adjacent REE) or a negative (depletion relative to adjacent REE) anomaly (Figure 8-9). The negative europium anomaly in Figure 8-9 results from formation and removal of plagioclase from a crystallizing melt. A rock consisting of accumulated plagioclase would have a positive europium anomaly (Figure 8-10). Similarly, partial melting of a source rock in which plagioclase stayed in the residual material would produce a magma (and eventually one or more rock masses) with a negative europium anomaly. Other minerals can also cause europium anomalies and this has to be considered in any specific case.

Distinctive REE patterns and europium anomalies are found for various types of igneous rocks. A rock's pattern depends on the origin and history of the magma from which it formed. REE patterns can sometimes be related to tectonic environments. An example is the pattern shown for mid-oceanic ridge basalts (MORB) in Figure 8-9. Ragland (1989, 309) points out that a typical MORB pattern is slightly concave down with a low positive slope, while many other basalts have a pattern that is slightly concave up with a negative slope (see BCR-1 in Figure 8-9). MORB rocks tend to have REE patterns that are parallel to each other but differ in the abundances of total REE.

Ragland (1989) also points out that granitic rocks have steeper negative slopes and higher overall REE contents than basaltic rocks such as BCR-1 in Figure 8-9. Granitic rocks can repre-

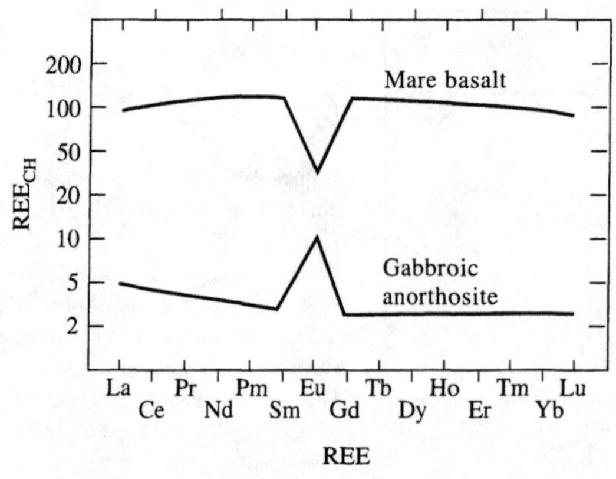

Figure 8-10 Chondrite-normalized REE patterns for basalt with a negative europium anomaly and a gabbroic anorthosite with a positive europium anomaly. Both samples are lunar igneous rocks characterized by very high Eu^{2+}/Eu^{3+} ratios (representative of formation under reducing conditions). The basalt is from the mare regions, and the gabbroic anorthosite is from the lunar highlands of the Moon. The mare basalt is relatively depleted in Eu as a result of the fractional crystallization of plagioclase, whereas the anorthosite has accumulated plagioclase. After Hess (1989).

sent the last liquid resulting from extended fractional crystallization of a parent magma or the first liquid formed by partial melting of a parent rock. In either case, strong fractionation of the REE occurs and a steep slope on a diagram such as Figure 8-9 results.

Let us now summarize our discussion of the REE. One or more of the following changes from chondritic REE values may be seen in a rock:

1. enrichment or depletion of all the REE;
2. enrichment or depletion of the LREE relative to the HREE; and
3. a positive or negative europium anomaly.

Each of these changes, or no change at all, found in a particular rock can tell us something about the formation and history of the magma that produced that rock. Further information on the occurrence of the REE in igneous rocks can be found in Henderson (1984) and Rollinson (1993).

Isotopes, like trace elements, are tools that provide unique data from igneous rocks. Radioactive isotopes are used in two important ways to study these rocks. They were first used to obtain absolute dates for the time of origin of individual rock bodies (see discussion of geochronology in Chapter Two). This led to the dating of geologic events and to an isotopic (absolute) time scale for the Earth. The second, and more recent, use of these isotopes and their daughter elements has been in determining the origin and history of magmas and, by extension, of the crust and mantle. Geochronology is discussed in detail in Chapter Two and the second application of radiogenic isotopes to igneous rocks is briefly reviewed there. Also discussed in Chapter Two is the use of stable isotopes, especially those of oxygen, to determine the interaction of meteoric water with igneous rocks. In this section we will limit ourselves to reviewing the use of isotope geochemistry to study the evolution of the crust and mantle.

Isotopic ratios in igneous rocks are believed to be characteristic of the source region from which the parent magma originated. Whatever the crystallization history of the magma, assuming no metamorphism or outside contamination, the isotopic ratios are not changed. Thus we can learn about the source regions (reservoirs) of igneous rocks and therefore learn about the nature of the crust and upper mantle, since magmas form in both these regions. Even when metamorphism or contamination of a magma occurs, such as by the addition of crustal material to a mantle-derived magma, the process can sometimes be identified by use of isotopic geochemistry. This research has as a primary goal the determination of the number and isotopic character of crust and mantle reservoirs (Table 8-8).

It is important to note that different isotopic parent-daughter pairs in an igneous rock are affected differently by any given geologic process, such as metamorphism or hydrothermal alteration. This is also true in the initial formation of a magma by partial melting of source rock. The elements of some isotopic pairs, such as samarium-neodymium, have very similar geochemical properties and are affected in the same manner by all processes. Thus this pair preserves the parent-daughter isotopic ratio $^{147}Sm/^{143}Nd$ throughout the history of an igneous rock. (The ratio changes, of course, as a result of the regular decay of the parent.) In contrast, rubidium and strontium have different geochemical properties. Rubidium is much more of an incompatible element than strontium and, as a result, enters a magma formed by partial melting of a source rock to a greater degree than does strontium. Rubidium is also more affected by metamorphism and hydrothermal alteration. The pairs uranium-lead and thorium-lead exhibit geochemical behavior that falls between that of samarium-neodymium and rubidium-strontium. Rollinson (1993) gives the following order of incompatibility for these elements: Rb > Th > U > Pb > Nd > (Sr, Sm).

TABLE 8-8 The Isotopic Character of Crust and Mantle Reservoirs. Present-day Isotope Ratios Shown in Parentheses

	^{87}Rb-^{86}Sr	^{147}Sm-^{143}Nd	^{238}U-^{206}Pb	^{235}U-^{207}Pb	^{232}Th-^{208}Pb
Upper crust	High Rb/Sr; high ^{87}Sr/^{86}Sr	Low Sm/Nd; low ^{143}Nd/^{144}Nd (negative epsilon)	High U/Pb; high ^{206}Pb/^{204}Pb	High U/Pb; high ^{207}Pb/^{204}204	High Th/Pb; high ^{208}Pb/^{204}Pb
Lower crust	Rb depletion; Rb/Sr less than about 0.04; low ^{87}Sr/^{86}Sr (0.702–0.705)	Retarded Nd evolution in the crust relative to chondritic source	Severe U depletion; very low ^{206}Pb/^{204}Pb (about 14.0)	Severe U depletion; very low ^{207}Pb/^{204}Pb (about 14.7)	Severe Th depletion; very low ^{208}Pb/^{204}Pb
Depleted mantle (Zindler and Hart 1986)	Low Rb/Sr; low ^{87}Sr/^{86}Sr	High Sm/Nd; ^{143}Nd/^{144}Nd (positive epsilon)	Low U/Pb; low ^{206}Pb/^{204}Pb (about 17.2–17.7)	Low U/Pb; low ^{207}Pb/^{204}Pb (about 15.4)	Th/U = 2.4 ±0.4; low ^{208}Pb/^{204}Pb (about 37.2–37.4)
Enriched mantle (Zindler and Hart 1986)	High Rb/Sr; ^{87}Sr/^{86}Sr > 0.722	Low Sm/Nd; ^{143}Nd/^{144}Nd = (0.511–0.5121)	High ^{207}Pb/^{204}Pb and ^{208}Pb/^{204}Pb at a given ^{206}Pb/^{204}Pb		
Bulk silicate earth	^{87}Sr/^{87}Sr = 0.7052	^{143}Nd/^{144}Nd = 0.51264 (= chondrite)	^{206}Pb/^{204}Pb 18.4 ± 0.3	^{207}Pb/^{204}Pb = 15.58 ±.08	Th/U = 4.2; ^{208}Pb/^{204}Pb = 38.9 ±0.3

^{143}Nd/^{144}Nd values are normalized to ^{146}Nd/^{144}Nd = 0.7219.

Source: After H. R. Rollinson, *Using Geochemical Data: Evaluation, Presentation, Interpretation.* Copyright © 1993. Used by permission of the Longman Group UK Ltd.

It appears that there are at least two different magma reservoirs in the mantle, which are referred to as "depleted" mantle and "enriched" mantle. Depleted mantle is believed to be the source of most mid-oceanic ridge basalts (MORB) and has higher Sm/Nd and lower Rb/Sr ratios than those characteristic of chondritic meteorites (using a reference reservoir referred to as CHUR, for chondritic uniform reservoir—see Figure 2-7). The isotopic compositions of some igneous rocks indicate they came from a reservoir with higher Rb/Sr ratios and lower Sm/Nd ratios than those of chondritic meteorites. Depleted mantle is depleted in incompatible elements and may represent the current upper mantle with depletion due to extraction of material to form the continental crust. In other words, it seems likely that the original upper mantle had a chondritic composition and it was subsequently involved in the formation of the continental crust by partial melting events and removal of incompatible elements to the surface. If this is true, then the mantle below a certain depth (600 km?) is "pristine" or "primitive" mantle, representative of the unchanged, original mantle with a chondritic composition (CHUR). Enriched mantle is postulated as the source of rocks that have high concentrations of incompatible elements. The existence of enriched mantle requires addition of crustal material to areas of the mantle, perhaps by subduction or by interaction of crust and mantle at the Mohorovicic discontinuity.

Figure 8-11 shows the relative locations of crust and mantle reservoirs on a ϵ_{Sr}-ϵ_{Nd} diagram. Additional data on the isotopic character of these reservoirs is given in Table 8-8. An example of the use of three different isotopic systems to study a group of igneous rocks is given in Chapter Two. The ϵ_{Sr}-ϵ_{Nd} plot of the Oslo, Norway, rift rocks (Figure 2-11a) suggests that basaltic magmas came from a slightly depleted mantle reservoir with some magmas later contaminated by crustal rocks. Samples from rocks representing contaminated and uncontaminated magmas can also be separated on a $^{207}Pb/^{204}Pb$ versus $^{206}Pb/^{204}Pb$ diagram (Figure 2-11b).

In order to study igneous rocks of different ages we need to know how the isotopic compositions of magma reservoirs have changed with time. This is not easy to determine, since estimates and assumptions are involved. An example is provided by Figure 2-3, which gives possible evolution curves for $^{87}Sr/^{86}Sr$. These curves can be used, along with absolute ages of rocks, to identify the probable source regions. For example, a rock of a given age will have a very different initial $^{87}Sr/^{86}Sr$ depending on whether it has a source in the crust or in the mantle. Another example of isotopic evolution is shown in Figures 2-5a and 2-5b. Figure 2-5a illustrates the production by partial melting of sources depleted and enriched in neodymium. Calculation of initial $^{143}Nd/^{144}Nd$ ratios for rocks and use of this type of diagram allows estimation of their sources and histories. A related diagram is shown in Figure 2-7. Figure 2-5b is an example of the use of the Sm-Nd system to study the history and present state of the upper mantle. Evolution curves for lead isotopes are shown in Figures 2-8, 2-9, and 2-10. Note that these curves do not plot an isotopic ratio versus time, but instead use isotopic ratios for the two axes. All of these isotopic evolution curves are discussed in more detail in Chapter Two.

ORIGINS OF MAGMAS

Nature Of The Crust And Mantle

Trace element and isotope data both provide information about the nature of the crust and upper mantle—the source regions of magmas. Geophysical data, particularly seismic data, give us additional details about the crust and upper mantle. Before discussing various ideas about magma origins, we need to briefly review our knowledge about the outer region of the Earth.

The manner of formation and later history of magmas is crucially dependent upon the chemical and mineral compositions of the crust and upper mantle (Table 8-9 and Figure 8-12). Educated guesses about the overall composition of the crust and mantle were reviewed in Chapter One. It seems likely that the average composition of the crust beneath the continents is that of the average composition of igneous rocks (Table 1-2). The upper part of the continental crust is felsic rock with an average composition similar to that of granodiorite. The lower part of the continental crust and the thin oceanic crust probably have an intermediate to mafic (basaltic) composition. The felsic to mafic trend with depth for the continental crust is suggested by a general increase in P-wave velocities. Seismic velocities also indicate that the crust is laterally heterogeneous.

Compressional and S-wave velocities in the upper mantle, and estimates from these and from gravity data of the density distribution there, limit its probable composition to that of peridotite, eclogite, or some chemical and mineral composition intermediate between the two. Peridotite is an ultramafic rock composed of olivine and one or more pyroxenes. Eclogite has a composition similar to that of basalt and is composed of garnet and pyroxene. The Mohorovicic discontinuity, which divides the crust and mantle, thus represents either a chemical change from

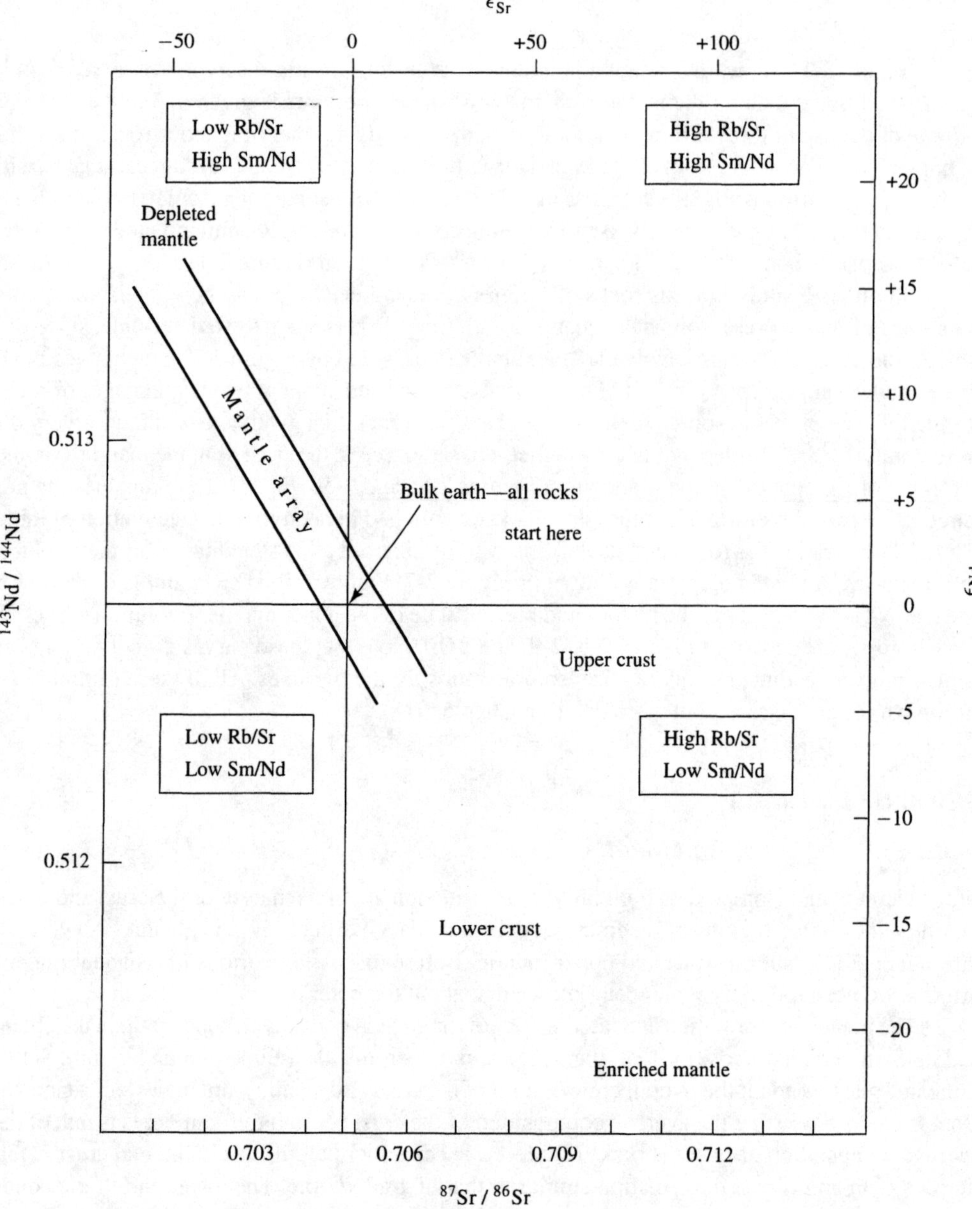

Figure 8-11 ϵ_{Nd}-ϵ_{Sr} isotope correlation diagram showing the relative positions of crust and mantle magma reservoirs. ϵ_{Nd} represents, in units of parts in 10^4, the deviation of the $^{143}Nd/^{144}Nd$ value of a sample from that of CHUR (CHondritic Uniform Reservoir). The line $\epsilon_{Nd} = 0$ represents the present-day isotopic composition of undifferentiated chondritic material. ϵ_{Sr} is defined in the same manner. The intersection of the two zero lines is believed to represent the isotopic composition of primitive mantle. See Chapter Two for further discussion of this type of diagram. In this figure, oceanic basalts mainly plot in the zone labeled "mantle array." This range of isotopic values represents mantle isotopic ratios. Most non-enriched mantle reservoirs plot in the upper left "depleted" quadrant, whereas most crustal rocks plot in the lower right "enriched" quadrant. (Reprinted from *Geochimica et Cosmochimica Acta*, v. 55, D. J. DePaolo and G. J. Wasserburg, Petrogenetic mixing models and Nd-Sr isotopic patterns, pp. 615–627. Copyright 1979, with kind permission from Elsevier Science Ltd, The Boulevard, Langford Lane, Kidlington OX5 1GB, UK.)

TABLE 8-9 Physical and Chemical Properties of the Crust and Upper Mantle

	Crust	Upper mantle below crust	Upper mantle transition zone
Depth (km)	0–70; average 35 km beneath the continents and 7 km beneath the oceans	Mohorovicic discontinuity to 400 km	400 km to 700 km
P-wave velocities (km/s)	1.5 (sediments) to 6.8 (rock)	8.1	9.5
S-wave velocities (km/s)	3.7 (rock)	4.5	5.5
Density (g/cm^3)	1.5 (sediments) to 2.9 (rock)	3.4	3.8
Pressure (kbar)	0–16	2–150	150–350
Temperature (°C)	0–1,000	200–1,700	1,700–2,000
Chemical composition	Varies from felsic to mafic under continents; mafic under oceans	Similar to that of peridotite, eclogite, or a combination of the two; varies vertically and laterally	Same as mantle under crust, but different mineralogy

the composition of basalt to that of peridotite or an isochemical phase change from basalt or gabbro (plagioclase and pyroxene) to eclogite (pyroxene and garnet). Phase studies of eclogite stability and detailed density estimates for the upper mantle indicate rather strongly that it is of peridotitic composition. Direct evidence of mantle material (nodules and xenoliths in mantle-derived rocks) and volcanic rock chemistry also point to a peridotite upper mantle. As with the crust, lateral inhomogeneity seems to exist.

The most common inclusions (nodules and xenoliths) brought in magmas from the mantle are all mafic and ultramafic rocks: mainly spinel and garnet lherzolites plus lesser amounts of pyroxenites and eclogites. The two lherzolites are both a type of peridotite and both are composed of abundant olivine and lesser orthopyroxene and clinopyroxene. As their names indicate, they differ in having either spinel or garnet as an accessory mineral. Pyroxenites have more of the pyroxenes as compared to olivine. Eclogites are composed of pyrope garnet and the pyroxene omphacite.

The Mohorovicic discontinuity lies at a depth of about 7 kilometers beneath the oceans and varies in depth from 30 to 70 kilometers (average 35 kilometers) beneath the continents. Knowledge of the location of the discontinuity comes from seismic data, which show it as a sharp increase with depth in the velocity of compressional and shear waves. Just below the Mohorovicic discontinuity, starting at a depth of about 70 km, there is a low-velocity zone characterized by a decrease and then an increase in S-wave velocity with depth (Figure 8-12). The low-velocity zone may be partially molten and is probably more mobile than the overlying material. The boundary between the lithosphere and the asthenosphere is in this zone. Several types of evidence indicate that the low-velocity zone is the source of most magmas. For example, earthquake foci tend to cluster in this zone. Also, the calculated pressure at these depths generally agree with those suggested by the mineralogy of probable mantle rocks now found at the surface of the Earth.

Two major seismic and density discontinuities occur in the upper mantle at about 410 km and 660 km (Figure 8-12). These discontinuities are believed to be caused by phase transformations.

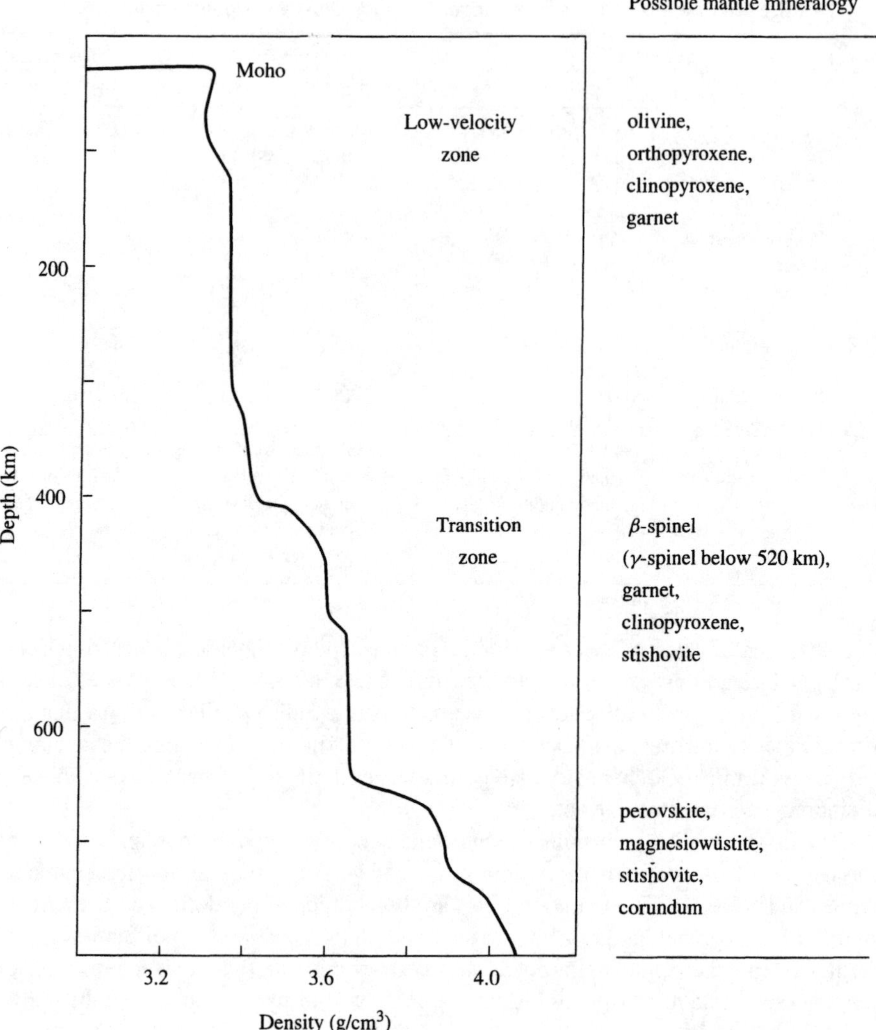

Figure 8-12 Upper mantle density and possible mineralogy. The densities shown by the solid line are zero-pressure densities (corrected for depth), assuming the mantle is composed of pyrolite (three parts peridotite plus one part tholeiite basalt). Density discontinuities correspond to phase transformations inferred from seismic studies. Major seismic velocity discontinuities occur at about 410 km and about 660 km. The Mohorovicic discontinuity (Moho) varies in depth from 5 to 70 km. The bottom of the lithosphere and top of the asthenosphere is at a depth of about 70–120 km. The low-velocity zone is a region of inferred partial melting in which seismic velocity (especially S-wave velocity) decreases. The transition zone is believed to be a region of high-pressure phase changes. Possible new minerals formed in the transition zone are listed on the right. Depending upon whether the mantle is assumed to have the composition of garnet lherzolite or pyrolite, different combinations of minerals are possible in the transition zone. (After S. M. Richardson and H. Y. McSween, Jr., *Geochemistry—Pathways and Processes*. Copyright © 1989. Adapted by permission of Prentice Hall, Inc., Upper Saddle River, NJ.)

Both of them occur in what is known as the transition zone, which is considered to be the bottom section of the upper mantle. An example of a phase change believed to occur at the 410 km discontinuity is the change of orthorhombic olivine to a cubic spinel structure:

$$(Mg, Fe)_2SiO_4 \text{ (olivine)} \rightarrow (Mg, Fe)_2SiO_4 \text{ (β-spinel)} \qquad (8\text{-}3)$$

The density increase is from 3.37 g/cm^3 to 3.63 g/cm^3. At a depth of about 520 km β-spinel changes to a different, more dense structure known as γ-spinel. The major change at the 660-km discontinuity is the breakdown of γ-spinel:

$$(Mg, Fe)_2SiO_4 \text{ (γ-spinel)} \rightarrow (Mg, Fe)SiO_3 \text{ (perovskite)}$$
$$+ (Mg, Fe)O \text{ (magnesiowüstite)} \qquad (8\text{-}4)$$

Other important changes with depth (and thus pressure) include the formation of garnet and perovskite from pyroxenes and the development of the high-density minerals stishovite and corundum. Figure 8-12 summarizes the probable mantle mineralogy at various depths in the upper mantle. Laboratory study of phase relations suggest that the mantle just below the Mohorovicic discontinuity is composed of spinel lherzolite, with garnet lherzolite occurring at greater depths. Locally, a third type of lherzolite containing plagioclase may occur above spinel lherzolite.* Note that spinel lherzolite contains the mineral spinel with a composition of $MgAl_2O_4$; it is not the same mineral as either β-spinel or γ-spinel, mentioned previously as forming from, and having the composition of, olivine.

Magma Formation

Given our present knowledge of the upper mantle, how can magmas form there? Hess (1989, 101) lists three ways by which partial melting in the mantle occurs: (1) the decompression of ascending mantle material; (2) the heating of subducted rock; and (3) the lowering of the melting point of a rock mass through the local accumulation of volatile components. Figure 8-13 describes the first process. The second process involves the intrusion of cold lithosphere into hot mantle. The third process could also be related to subduction, since large quantities of volatiles (H_2O, CO_2, etc.) could be released from subducted minerals such as clay minerals, carbonates, amphiboles, and micas. The last two processes can be used to explain magma formation at plate boundaries while the first process may explain hot spots that cause intraplate igneous activity. The lateral seismic and compositional heterogeneities in the crust and upper mantle are probably related to surface tectonic processes. Unmelted, subducting slabs may go as deep as 700 km (mantle convection) and their seismic velocities would be higher than those of hot, upwelling material elsewhere

*Some research workers divide lherzolites into two types: "fertile" rock capable of producing basaltic magma by partial melting, and "depleted" or "sterile" rock incapable of producing such a magma. These two rocks have slightly different mineral and chemical compositions. For example, "depleted" lherzolite has more olivine and is richer in MgO. Spinel and garnet are no longer present. Partial melting of "fertile" lherzolite would produce a basaltic magma and leave a residue richer in olivine than the original rock. Lherzolite nodules in basaltic rock are believed to represent residual material resulting from partial melting of "fertile" mantle to form basaltic magma. The residual material would be "depleted" lherzolite. The composition of "fertile," primitive mantle rock can be estimated by combining the composition of basalt and lherzolite in the proper proportions. Ringwood (1962) proposed that this hypothetical material be called *pyrolite* (pyroxene-olivine rock). It is generally defined as having a composition made up of three parts harzburgite (olivine-orthopyroxene peridotite) plus one part tholeiite basalt. This composition is similar to that of carbonaceous chondrite meteorites (see Chapter One for a discussion of mantle composition using chondrites as a model).

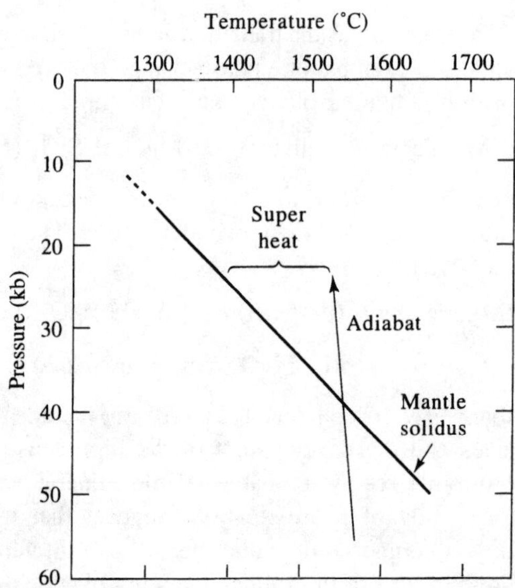

Figure 8-13 Melting of mantle material as it rises along an adiabatic gradient (adiabat). Because the mass of material moves upward along an adiabat, no heat enters or leaves the material. This results in a small decrease in temperature for a given decrease in pressure. Because the melting-point gradient of the general mantle (solidus) is much different from that of an adiabat, the temperature of a mass of mantle material rising along an adiabat will decrease less than general mantle melting temperatures are depressed by lowering of pressure. As the moving mass gets to a lower pressure zone and crosses the mantle solidus, partial melting can occur due to the "superheat" available. Superheat is the heat that can be extracted from the mantle due to the difference between the temperature of the upward moving mass and the temperature represented by the solidus. This type of "decompression melting" can occur under spreading ocean ridges, since a convection cell is likely to have an adiabatic profile. (Reprinted by permission of the publishers from ORIGINS OF IGNEOUS ROCKS by Paul C. Hess, Cambridge, Mass: Harvard University Press, Copyright © 1989 by the President and Fellows of Harvard College.)

(Kirby et al. 1991). The existence of hot solids and partial melts probably explains the lower seismic velocities in the low-velocity zone (Figure 8-12).

We know that several different kinds of magmas can form in the upper mantle. These are often referred to as primary magmas. A primary magma forms by partial melting of preexisting rock; a secondary magma forms as a result of fractional crystallization or some other process affecting a primary magma. The most important primary magmas are those that have developed abundantly throughout geologic time and that form rocks with a wide geographic distribution. At least three types of rock probably are representative of such magmas: tholeiite basalt, alkali olivine basalt, and the granitic rocks of continental batholiths. Other, less important kinds of primary magmas probably have formed in the mantle at certain times or occur in restricted areas of the Earth.

As suggested above, the variety of primary magmas is a result of mantle variations in chemical composition, mineral composition, pressure, and temperature. In addition, different degrees of partial melting of two similar rocks at the same temperature and pressure can produce two different primary magmas. Figure 8-14 is an example of the formation of several magma compositions from different source rocks at different pressures and temperatures. Experimental studies show that tholeiite magmas are produced at shallow depths beneath the Mohorovicic discontinu-

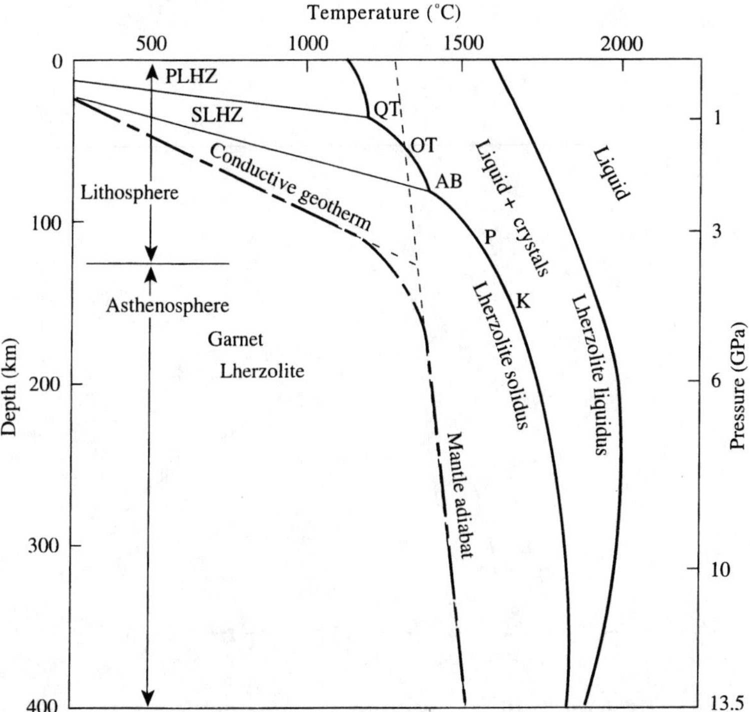

Figure 8-14 Lherzolite compositions, melting relations and products, and temperature conditions beneath a normal (c. 60 Ma old) oceanic area. Heavy solid lines indicate dry solidus and liquidus curves for lherzolite in the absence of volatiles. Compositions of magmas resulting from partial melting at the solidus are QT = quartz tholeiite, OT = olivine tholeiite, AB = alkali basalt, P = picrite, K = komatiite. Faint solid lines indicate boundaries between different forms of lherzolite in the solid state; PLHZ = plagioclase lherzolite, SLHZ = spinel lherzolite. The heavy dot-dash curve indicates change of temperature with depth (geotherm) and is characterized by an upper part where heat transfer is by conduction, leading to a rapid variation of temperature with depth, and a lower part that follows the convective mantle adiabat. The 125 km thickness of the lithosphere beneath this 60 Ma oceanic area is defined by the intersection between the projected conductive and convective portions of the geotherm. The adiabatic temperature gradient (adiabat) represents a situation where no heat enters or leaves a mass of rock that moves from one depth (pressure) to another. See Figure 8-13. (After G. C. Brown and A. E. Mussett, *The Inaccessible Earth*, 2d ed. Copyright © 1993 by Chapman & Hall. Used by permission of Chapman & Hall, London.)

ity, while alkali basalt magmas are produced at greater depths (35–70 km). Magmas representative of the rarest rock types (for example, picrite and komatiite) are produced at depths exceeding 100 km. Thus there is a direct relationship between the abundance and composition of volcanic rocks and the depth of formation of their magmas.

It is possible that partial melting can produce, at a particular depth, two magmas of very different composition from the same parent rock. Yoder (1973) has suggested a mechanism involving fractional fusion of mantle rock. He uses the diopside-forsterite-silica system at $P_{H_2O} = 20$ kilobars as an example (Figure 8-15). For an assumed parent rock of composition X in Figure 8-15, melting begins at 960°C. The initial liquid has the composition A, and magma of

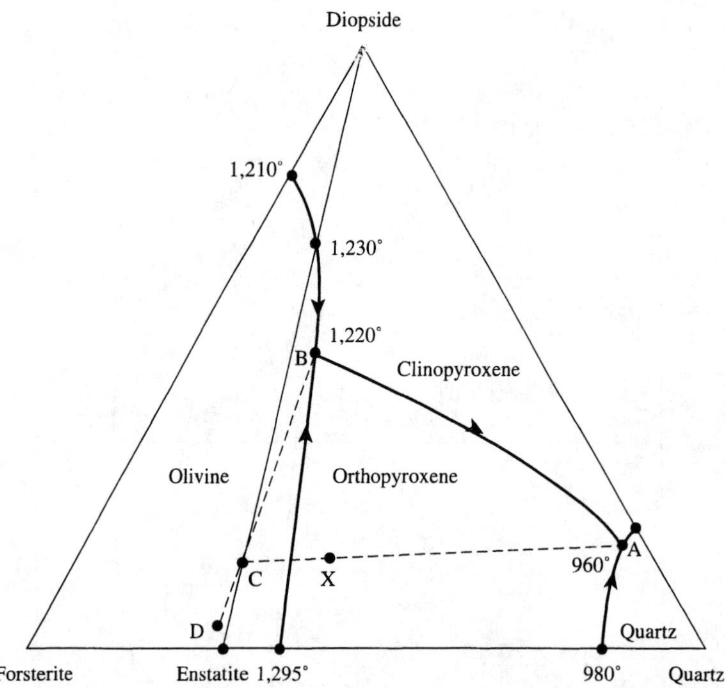

Figure 8-15 The diopside-forsterite-silica system at $P_{H_2O} = 20$ kbar. Fractional fusion of a parent rock of composition X can produce magmas of compositions A and B. Residual rock after removal of liquid A at 960°C will have a composition of C. Partial melting of C at 1,200°C will produce liquid B and residual rock D. Points A, X, and C lie on a straight line, and points B, C, and D lie on a straight line. See further discussion in text. Note the major differences between this system at 1 bar (Figure 8-18) and at $P_{H_2O} = 20$ kbar. After Kushiro (1969) and Yoder (1973).

this composition can be produced by removal of up to 20 percent of the original rock. The remaining rock now has a composition C. Melting ceases at 960°C when all the quartz in the parent rock is exhausted. When the temperature is raised from 960 to 1,220°C, a new liquid of composition B is formed. About 22 percent of rock of composition C can be removed as a magma of composition B. The residual parental material has the composition D. It is thus possible to produce two homogeneous, but very different, liquids without producing liquids of intermediate composition. Liquid A would have a rhyolitic composition and liquid B would have an andesitic composition. Depending upon the abundance and behavior of hydrous minerals in the parent rock, basaltic magma could be formed instead of andesitic magma. Yoder suggests that this type of process may explain the common interbedding of contrasting rocks, such as basalt and rhyolite, found in many parts of the world.

Partial melting can also occur in the crust (Figure 8-16). The causes may be similar to those acting in the mantle, but the parent rocks are different in mineralogy and chemistry. Additional causes for magma formation in the crust are the rise in temperature produced at areas of continent-continent collision and the concentration of heat-producing radioactive elements in the crust.

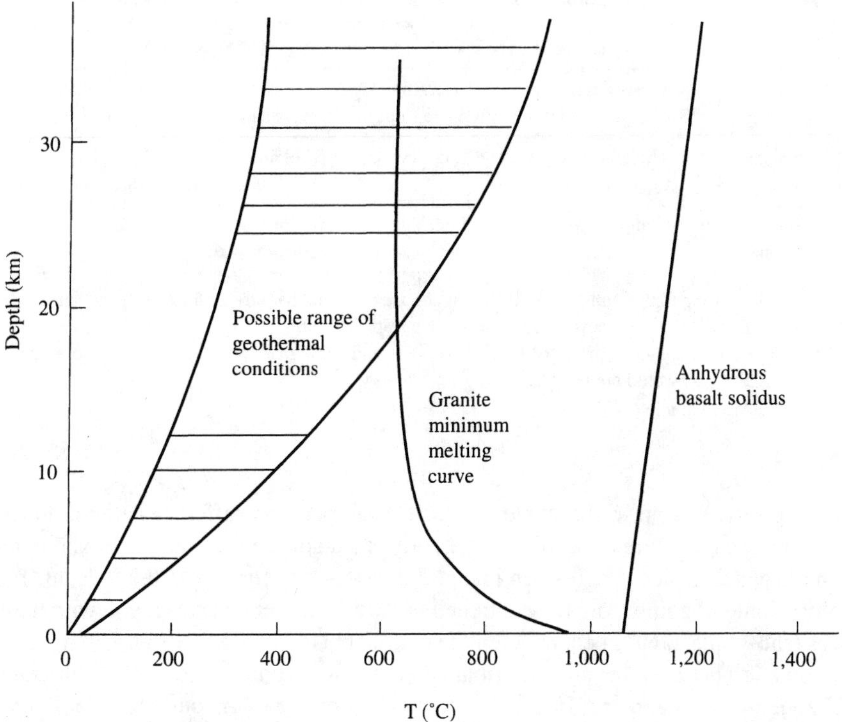

Figure 8-16 Conditions for melting in the crust and upper mantle. The probable range of geothermal conditions is based on calculations by a wide variety of investigators [see, for example, Sclater et al. (1980).] The minimum melting curve of granite is from Figure 8-24 and assumes equilibrium with H_2O. The anhydrous basalt solidus represents the maximum temperatures at which basalts will begin to melt. If H_2O is abundant, melting begins at temperatures as low as 750°C over a wide range of pressures (and thus depths). Data for the basalt curve from Cohen et al. (1967).

Since partial melting is essentially the opposite of crystallization, the first melts to form have the composition of the last liquids resulting from crystallization. Thus melting of intermediate to mafic crustal rocks tends to form felsic (granitic) magmas. For example, migmatites (felsic igneous rock mixed with metamorphic rock composed of mafic minerals) probably form by partial melting of a parent rock in place. Partial melting under high-grade metamorphic conditions is referred to as *anatexis*. Such melting in the crust could explain the formation of felsic rocks such as granite and rhyolite; however, these rocks could also form by a differentiation process acting on a primary basaltic magma formed in the mantle.

As indicated earlier in this chapter, there is a general relationship between the various rock series (related groups of rocks) and the major plate tectonic settings. The three most important rock (magma) series are the tholeiite, alkali olivine, and calc-alkaline series. The most abundant rock types for the first two are tholeiite basalt and alkali olivine basalt. For the calc-alkaline series the dominant volcanic rock is andesite and the dominant plutonic rock is granodiorite. Table 8-10 relates these magma series to plate tectonic settings.

TABLE 8-10 Plate Tectonic Classification of the Three Major Magma Series

	PLATE MARGIN		INTRAPLATE	
Plate setting	Converging (Subduction zone)	Diverging (Oceanic ridge)	Oceanic islands	Continental rifts
Magma series	Tholeiite Calc-alkaline	Tholeiite	Tholeiite Alkali olivine	Tholeiite Alkali olivine
Stress regime	Compressive	Extension	Minor compressive	Extension

Note: According to this simple classification, continental flood basalts (Columbia River Plateau, Deccan Traps, etc.) are considered to form at intraplate continental rifts. These large fissure eruptions are found in a variety of tectonic settings and some of them may be related to plate margins and thus would not be in the intraplate category.
Source: After Condie (1983).

The two series found in subduction zones tend to occur in different parts of an island arc complex. Tholeiite rocks are generally on the convex side of the arc and calc-alkaline rocks are mainly on the concave side. There is an increase in potassium content of the volcanic rocks, for a given silica content, going from the arc trench toward an associated continent, with high-potassium rocks (shoshonites, etc.) occurring on the continent side of the arc. In other words, there is a linear correlation between the silica-normalized potassium content and depth to the subduction (Benioff) zone of the island arc. In contrast to subduction zone igneous rocks, only one major magma series, tholeiite, is associated with oceanic ridges. The volcanic rocks of the tholeiite series at these divergent plate boundaries have a low content of potassium and other incompatible elements, such as the rare earths.

Two magma series, tholeiite and alkali olivine, are found at intraplate igneous rock sites. However, only one series may occur at a particular site. For example, the Hawaiian Islands are composed almost entirely of rocks of the tholeiite series. Some other oceanic islands are made up exclusively of alkali olivine series rocks. Both oceanic islands and continental rifts have minor amounts of igneous rocks when compared to oceanic ridges and to subduction zones. Intraplate rocks tend to be enriched in incompatible elements. Unusual and rare rock types, such as carbonatites and peralkaline granites, are found in continental rift areas. The tholeiite series rocks occurring at intraplate sites have chemical differences from the tholeiite series rocks found at plate margins. (Note in Table 8-10 that the tholeiite series is the only one to occur in all four of the listed plate settings.) For example, intraplate tholeiites are enriched in the light rare-earth elements relative to plate-margin tholeiites and have strontium and neodymium isotope ratios that generally differ from those of the plate-margin rocks (Figure 8-17). These and other differences probably reflect differences in the sources and processes of magma formation. Crustal contamination is also a possible factor.

CRYSTALLIZATION OF SILICATE MELTS

A major method of studying igneous rock formation is to use the results of laboratory studies on the crystallization of artificial melts. This type of approach was pioneered by N. L. Bowen, and his classic book (Bowen 1928) is a useful summary of the significance and application of such

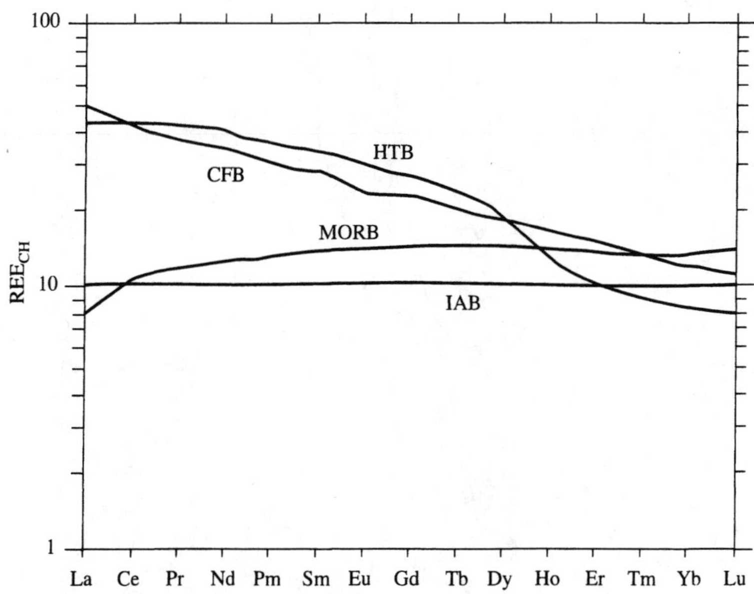

Figure 8-17 Chondrite-normalized plot of rare-earth abundances in typical island arc basalts (IAB), tholeiitic continental flood basalts (CFB), mid-ocean ridge basalts (MORB), and Hawaiian tholeiite basalts (HTB). (After Philpotts, A. R., *Principles of Igneous and Metamorphic Petrology*. Copyright © 1990. Adapted by permission of Prentice Hall, Inc., Upper Saddle River, NJ.)

laboratory experiments. Certain limitations in the use of laboratory results should be kept in mind. Laboratory study by necessity is limited to a few variables and to melts of relatively simple composition. Compositions in nature are complex and affected by many variables. Many laboratory studies have been made using melts that are under low pressures and saturated with water. Most magmas crystallize under high pressures (up to at least 10,000 bars) and are probably not saturated with water. There is always the question of whether or not equilibrium has been obtained, both in the laboratory and in nature. Despite such limitations in carrying out and applying laboratory studies, research along these lines has been very useful and has continued to the present. Only a few examples of such studies will be given here. Detailed discussion of laboratory results can be found in igneous petrology textbooks.

Basaltic Magma

We start with the system Mg_2SiO_4-$CaMgSi_2O_6$-SiO_2 (Figure 8-18), which is pertinent to the crystallization of basaltic magma. Among the first minerals to crystallize from such a magma are olivine and one or more pyroxenes. A crucial feature of magmatic crystallization is the relationship of olivine to pyroxene. In some cases, early formed olivine later changes into pyroxene by reaction with liquid. In other cases, olivine and one or more pyroxenes crystallize together.

Figure 8-18 illustrates the changes in mineral formation with falling temperature. This type of phase diagram is divided into areas on the basis of which mineral forms first from a cooling magma. Along a phase boundary, such as AD of Figure 8-18, two minerals coexist with liquid. Figure 8-18 shows that the first mineral to form from a given liquid will be forsterite (magnesian olivine), pyroxene, or a polymorph of silica, depending upon the initial composition of the liquid.

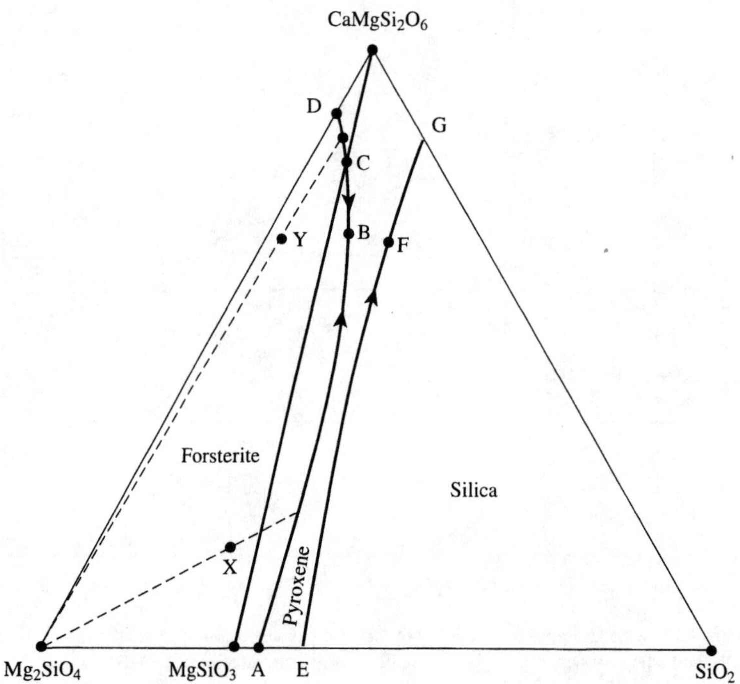

Figure 8-18 The system Mg_2SiO_4-$CaMgSi_2O_6$-SiO_2 at 1 atm pressure. See text for explanation. After Kushiro (1972).

Forsterite would form first from a liquid such as X in Figure 8-18. As crystallization continues by separation of forsterite, the composition of the remaining liquid would move directly away from the forsterite composition point (Mg_2SiO_4) toward the phase boundary AB. At the boundary AB, pyroxene would start forming, while forsterite would begin to react with the liquid and be resorbed. (Since forsterite becomes unstable when pyroxene starts to form, it is said to have a reaction relationship to pyroxene.) These processes would continue as the composition of the liquid changed along the phase boundary AB toward B. A calcium-poor pyroxene would form at first (its composition would plot close to the enstatite point, $MgSiO_3$); as temperature dropped, the pyroxene would become progressively richer in calcium. However, there is a limit to how much calcium can be added to such a pyroxene. Thus at some point two pyroxenes, one calcium rich (its composition would plot close to the diopside point, $CaMgSi_2O_6$) and one calcium poor, will start forming together. Forsterite will continue to be used up, but some will still be present when the last liquid disappears. When equilibrium crystallization is involved, the composition of the final liquid will be somewhere along the line AB, and the final solid will consist of forsterite and two pyroxenes.

Initial liquids whose compositions fall in the area Mg_2SiO_4-B-A will behave as described above. Liquids with compositions in the area Mg_2SiO_4-C-$CaMgSi_2O_6$, such as Y, will have a different history. (Note that C is a maximum or high-temperature dividing point on the line AD.) Again the first mineral to form will be forsterite. The liquid will change in composition directly away from the forsterite point toward line CD as forsterite continues to form. When line CD is reached, a calcium-rich pyroxene begins to form, and the liquid then changes along line CD toward point D, with forsterite and pyroxene both continuing to precipitate. In this case, forsterite does not become unstable when pyroxene begins to form. The final product is a solid consisting of forsterite and a calcium-rich pyroxene.

In a general way, liquids such as X of Figure 8-18 are representative of tholeiite magma, whereas liquids of compositions such as Y are representative of alkali olivine basalt magma. As mentioned earlier, tholeiite basalts have two pyroxenes, whereas alkali olivine basalts have one. Olivine grains in tholeiite basalts tend to have irregular edges, suggesting resorption by surrounding liquid. Pyroxene appears to have formed later than the olivine. In alkali olivine basalts, pyroxenes and olivines often occur as large crystals and appear to have formed together without a reaction relationship.

Fractional crystallization refers to any process that separates the early formed minerals of a magma from the remaining liquid. Let us consider what would happen to a melt such as X of Figure 8-18 if such separation occurred. If the early formed forsterite crystals were separated from the melt (for example, by sinking to the floor of the magma chamber), they could not be resorbed when the liquid composition reached the boundary AB. In effect, the result is a new magma with an overall composition that is closer to the $CaMgSi_2O_6$-SiO_2 side of Figure 8-18. The history of a fractionating melt of initial composition X would be as follows. First, forsterite would form and be separated. The liquid composition would change toward line AB. When it reached the line, a calcium-poor pyroxene would start forming. The liquid composition would change along AB toward B. As is the case for a nonfractionating magma, at some point two different pyroxenes would start to form. Because of the removal of the forsterite crystals, the liquid composition would change all the way to B, the lowest temperature point on line AB. At point B the liquid composition would leave the line AB and move across the pyroxene field toward the line dividing the pyroxene and silica fields (EG) to a point such as F. Two pyroxenes would continue to form as it did this. At F a silica mineral (tridymite) would begin to form, and the liquid composition would change along EG toward G. The final liquid would have a composition between F and G, and the final product would be two pyroxenes and tridymite.

The above discussion shows that a liquid of composition X may produce (1) a rock consisting of forsterite and two pyroxenes by simple equilibrium crystallization of the entire melt, or (2) several different rock types as a result of fractional crystallization of the melt. Possible rock types from fractional crystallization (assuming several episodes of segregation) are peridotite (from segregation of early formed forsterite), tholeiite basalt with olivine (from crystallization and segregation along AB and assuming not all forsterite was previously segregated), tholeiite basalt without olivine (from crystallization and segregation along BF), and tholeiite basalt with a silica mineral (from crystallization along FG). Minerals formed from crystallization along BF and FG could be representative of andesites as well as of tholeiite basalts.

The properties of associated igneous rocks in many parts of the world indicate that fractional crystallization is an important type of *magmatic differentiation* (any process by which two or more different igneous rocks form from an initially homogeneous magma). An example of naturally occurring fractional crystallization is suggested by the variation in rock compositions plotted in Figure 8-4. Most fractional crystallization of basaltic magmas occurs at depths where the pressures are much higher than the 1 atm of Figure 8-18. Laboratory studies indicate that the results of fractional crystallization of magmas are different at high pressures from those occurring at low pressures. The wide compositional diversity of magmas and of igneous rocks seems to be due to two different processes that occur at a variety of depths and pressures: (1) partial melting of various parent rocks and (2) fractional crystallization of primary and derivative magmas.

Kushiro (1979, 200) states that "The compositional range of magma that would be produced by partial melting is probably as wide as that produced by fractional crystallization."

Another variable, in addition to pressure, that has been found to be important is the presence or absence of volatile components such as H_2O and CO_2. Examples of the effects of pressure and volatiles on fractional crystallization of basaltic magma are shown in Figure 8-19. This figure is for the same system as Figure 8-18. Let us compare Figure 8-19a with Figure 8-18. These figures show the liquidus minerals at 1 atm and 20 kb. Note the following major changes caused by an increase in pressure. The field in which a pyroxene is the first mineral to form from a cooling liquid has increased greatly while the field for a silica mineral has decreased greatly. Also the forsterite field is somewhat smaller. These changes imply that two magmas of the same composition will produce very different products if one crystallizes under near-surface conditions and the other crystallizes at depth under a pressure of 20 kb. For example, a liquid whose composition is near that of point Y in Figure 8-18 would first crystallize a pyroxene at 20 kb, but would produce olivine as the first-formed mineral at 1 atm.

Point B of Figure 8-18 represents the composition of liquid coexisting with forsterite, Ca-poor pyroxene, and Ca-rich pyroxene at 1 atm, and this point is also labeled B in Figure 8-19b. The other three points show how the location of the 1-atm point changes for 20-kb pressure under anhydrous, H_2O-saturated, and CO_2-saturated conditions. All four points could represent magma

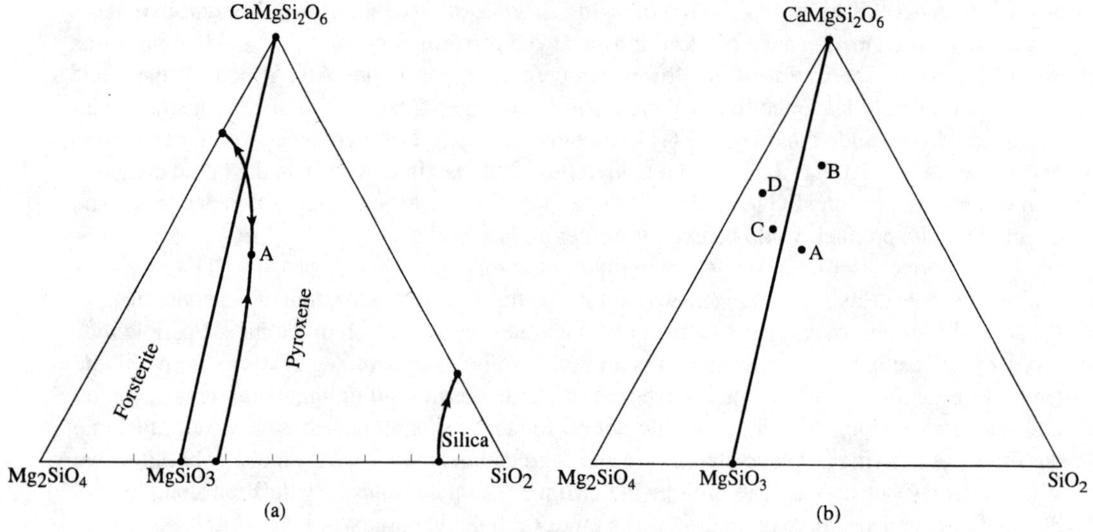

Figure 8-19 (a) System Mg_2SiO_4-$CaMgSi_2O_6$-SiO_2. Projected liquidus features at $P_{H_2O} = 20$ kb. Point A is the composition of melt coexisting with forsterite, low-Ca pyroxene, and high-Ca pyroxene. (After I. Kushiro, *American Journal of Science*. Reprinted by permission of the American Journal of Science.) (b) System Mg_2SiO_4-$CaMgSi_2O_6$-SiO_2. Composition of melt coexisting with forsterite, low-Ca pyroxene, and high-Ca pyroxene at (A) 20 kb, H_2O-saturated; (B) 1 atm; (C) 20 kb, anhydrous; and (D) 20 kb, CO_2-saturated. Liquids A and B are silica-saturated, whereas liquids C and D are silica-undersaturated. Data from Kushiro (1969) and Eggler (1978). (Reprinted by permission of the publishers from ORIGINS OF IGNEOUS ROCKS by Paul C. Hess, Cambridge, Mass: Harvard University Press, Copyright © 1989 by the President and Fellows of Harvard College.)

formed by partial melting of lherzolite mantle composed of olivine and two pyroxenes. Note that raising the pressure from 1 atm (point B) to a dry 20 kb (point C) moves the liquid composition from silica-saturation to silica-undersaturation (the line joining the two pyroxene compositions is the silica saturation line). Also, if the liquid is saturated with H_2O (point A), it is also silica-saturated, while saturation with CO_2 (point D) makes it silica-undersaturated. Each of the four points is a parent magma of a specific composition and fractional crystallization of each of these liquids could produce a variety of new liquids (derivative magmas) and rocks.

Hess (1989, 141) lists the following conclusions related to mantle melting from the results shown in Figure 8-19 and from related laboratory studies:

1. SiO_2-undersaturated magmas (for example, alkali olivine basalt) are produced by small amounts of melting of lherzolite at pressures greater than 15 kb with anhydrous conditions or at similar pressures with a CO_2-rich fluid present.
2. Tholeiite basaltic magmas are products of anhydrous melting of lherzolite at pressures under 15 kb. Quartz tholeiite basalt magma is produced when pressure is less than 5 kb, or at greater depths when a H_2O-rich fluid is present.

We can summarize our brief review of basaltic magma formation and crystallization as follows. The diversity of basaltic magma compositions, and of igneous rocks derived from them, is due to (1) the varying mineralogy of mantle parent rocks at different depths; (2) variation in the extent and conditions of partial melting of those rocks; (3) the degree of fractional crystallization of primary and derivative magmas; and (4) the effect of pressure and volatiles on fractional crystallization. Further information on basaltic magmas can be found in Morse (1980) and BVSP (1981).

Oxygen Fugacity

One of the volatiles that is associated with magmas is much more important than its abundance would indicate. The partial pressure of oxygen coexisting with magmas is on the order of 10^{-12} bar. In the ideal case, at equilibrium, a magma has an oxygen fugacity equal to the partial pressure. (Fugacity can be thought of as a corrected pressure that takes into account the non-ideality of gases. See Chapter Three for a discussion of fugacity.) The oxygen fugacity strongly affects the Fe^{2+}/Fe^{3+} ratio in magmas and, as a result, has a strong effect on the minerals and mineral assemblages that form from magmas. For example, a high oxygen fugacity causes a low Fe^{2+}/Fe^{3+} ratio in a magma and thus the FeO/MgO ratio in a crystallizing mineral such as olivine will also be low.

Experimental studies of the role of oxygen fugacities in magma crystallization use the "oxygen buffers" shown in Figure 3-2. These buffering reactions can be applied to rocks to estimate fugacities that existed when the rocks formed. Figure 8-20 gives values of the relative oxygen fugacities of mafic and felsic magmas. These values were obtained by measuring the concentration of FeO and Fe_2O_3 in glassy volcanic rocks and by measuring the composition of coexisting Fe-Ti oxide microphenocrysts (Carmichael and Ghiorso 1990).

The values in Figure 8-20 are related to the Ni-NiO oxygen buffer:

$$\mathrm{Ni} + 0.5\ \mathrm{O_2} \rightleftharpoons \mathrm{NiO}$$

ΔNNO is defined as

$$\Delta\mathrm{NNO} = \log_{10}\ \mathrm{fugacity}_{\mathrm{O_2}}(\mathrm{sample})\ -\ \log_{10}\ \mathrm{fugacity}_{\mathrm{O_2}}(\mathrm{NNO})$$

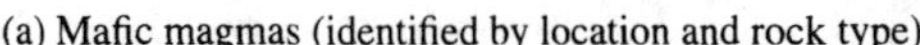

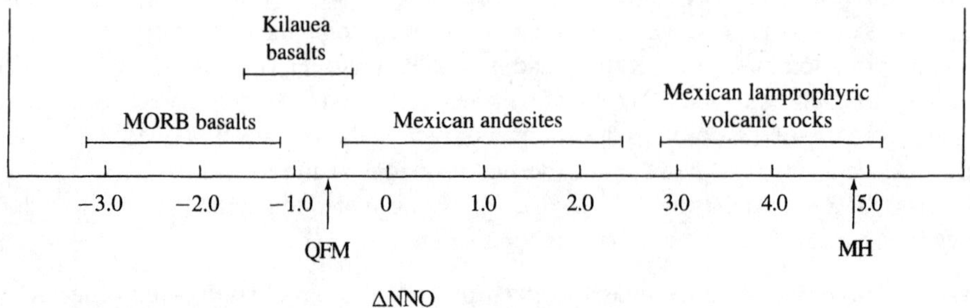

(b) Felsic magmas (identified by phenocryst mineral assemblage)

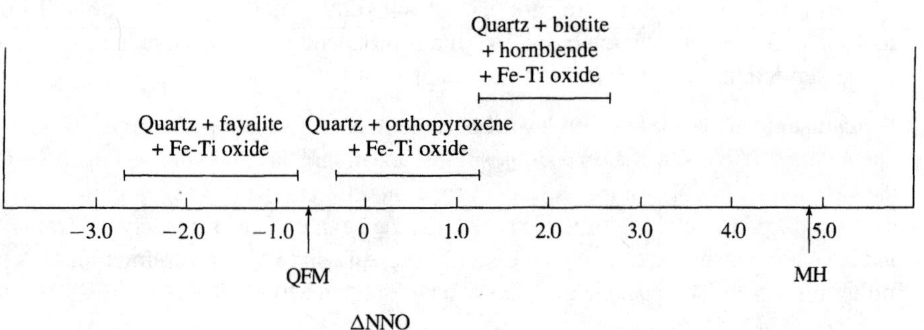

Figure 8-20 Values of the relative oxygen fugacity (ΔNNO) of mafic and felsic magmas. See text for definition of ΔNNO. The locations of the quartz-fayalite-magnetite (QFM) and magnetite-hematite (MH) buffers are shown on the horizontal scale. Data from Carmichael and Ghiorso (1990) and Ghiorso and Sack (1991). (a) Mafic magmas (identified by location and rock type). (b) Felsic magmas (identified by phenocryst mineral assemblage).

at an arbitrary temperature of 1200°C. The Ni-NiO buffer is not shown in Figure 3-2. Its line falls close to, and parallels, the QFM line. As Figure 8-20 shows, most igneous rocks form under fugacities ranging from near the QFM buffer to near the M-H buffer. Note that the range of oxygen fugacities of felsic magmas is smaller than that of mafic magmas.

In addition to calculating oxygen fugacities from phenocryst assemblages (Figure 8-20), it is also possible to determine temperatures. The Fe-Ti oxide geothermometer-oxygen barometer was first outlined by Buddington and Lindsley in 1964. A new and improved formulation of the technique has been prepared by Ghiorso and Sack (1991). Oxygen fugacity and temperature values are obtained by determining the composition of coexisting ilmenite-hematite and magnetite-ulvöspinel solid solutions (phenocrysts). These are related through oxidation reactions such as

$$3\ Fe_2TiO_4\ \text{(ulvöspinel)} + 0.5\ O_2 = 3\ FeTiO_3\ \text{(ilmenite)} + Fe_3O_4\ \text{(magnetite)}$$

The data for felsic magmas in Figure 8-20 was obtained using the new formulation of Ghiorso and Sack, who found that the calculated $\log_{10}$ fugacity$_{O_2}$-temperature relations for felsic volcanic rocks are reflected in the coexisting mineral assemblages. Volcanics with the lowest relative oxygen fugacity are characterized by a phenocryst assemblage of fayalite-quartz and, as the fugacity

TABLE 8-11 Compositions of Natural Volcanic Liquids and Associated Olivines When the Liquids Are Equilibrated at the Oxygen Fugacity of Quartz-Fayalite-Magnetite (A) and in Air (B).

	Liquid A	Liquid B	Olivine A	Olivine B
SiO_2	42.14	42.63	40.08	41.78
FeO	10.00	1.31	14.51	1.95
Fe_2O_3	2.66	11.73	—	—
MgO	7.92	9.75	45.02	55.66
Fo (mol %)	—	—	83.4	97.3
Fa (mol %)	—	—	15.1	1.9

Note: Only a partial list of constituents is given for the liquids. All of the major constituents of olivine are listed. Small amounts of MnO and CaO are reported for the olivines.
Source: After Carmichael and Ghiorso (1990).

of O_2 increases, different assemblages occur in the rocks (Figure 8-20). The values obtained can be plotted on a diagram such as Figure 3-2 and show that most felsic volcanic rocks plot in the area between 650 and 1050°C and between 10^{-18} and 10^{-9} for log fugacity$_{O_2}$.

A specific example of the effect of oxygen fugacity on mineral composition is given by Carmichael and Ghiorso (1990). A natural silicate liquid was crystallized under two different oxygen fugacities. The compositions of newly formed olivines and of the residual liquids were determined and showed that the mol percent forsterite in olivine increased by 14 percent when the oxygen fugacity was increased from that of the QFM buffer to that of air (Table 8-11).

Granitic Magma

As indicated above, iron- and magnesium-bearing minerals are very important in the crystallization of basaltic and other mafic magmas. The major minerals of granitic rocks are potassium feldspar ($KAlSi_3O_8$) and members of the plagioclase solid solution series albite ($NaAlSi_3O_8$)–anorthite ($CaAl_2Si_2O_8$). Laboratory studies over the years have shown the feldspars to be complex minerals with complicated phase relationships. In the laboratory, and undoubtedly in nature, feldspar relationships are strongly affected by variations in the pressure of a water-vapor phase in equilibrium with a silicate melt. An example of this effect on the $KAlSi_3O_8$-$NaAlSi_3O_8$ system is shown in Figure 8-21. Figure 8-21a represents crystallization under volcanic (near-surface) conditions, and Figure 8-21b represents crystallization under possible plutonic (deep-seated) conditions. When this system is under a water pressure of 1 bar (equivalent to the total pressure on the system), the first solid will form from a cooling liquid at a temperature above 1,078°C (point Z of Figure 8-21a). If the liquid is potassium rich, leucite will form first.* As the temperature drops, any leucite that has formed eventually reacts with liquid, and a potassium-sodium feldspar coexists with liquid below about 1,100°C. (At WY, leucite has a reaction relationship to K-feldspar similar to that of olivine-pyroxene mentioned previously. See Figure 3-12). At temperatures below 1,078°C, a homogeneous single feldspar exists with the composition of the initial liquid.

*Leucite does not occur in the true binary system and thus Figure 8-21a is a pseudobinary system.

Thus the cooling history of a liquid of composition X (Figure 8-21a) would be as follows. Leucite would start to form when the liquid reached the solid line (*liquidus*) of Figure 8-21a at a temperature of about 1,200°C. The liquid would then change in composition along the line toward point Y. At the temperature of Y, feldspar would begin to form, and leucite would be completely resorbed by the melt. A potassium-rich feldspar of composition W would be left as the solid phase. As the temperature continued to drop, the liquid and feldspar would both change in composition along the liquidus YZ and *solidus* WZ toward point Z. (The temperature at point Z is 1,078°C, the lowest temperature at which liquid can exist in the system.) Before reaching point Z, all the liquid would be used up, and the solid would consist of homogeneous feldspar of the same composition as X (about $Or_{60}Ab_{40}$).

This is not the end of the story, because feldspar solid solutions of this system become unstable at lower temperatures. The lower curve of Figure 8-21a (the *solvus*) divides the region of stability of one feldspar from that where two feldspars become stable. The high point of the solvus is at about 660°C. At any temperature below 660°C, a horizontal line for the temperature will intersect the solvus at two points that represent the compositions of the two feldspars stable at that temperature. For composition X of Figure 8-21a and a temperature of 500°C, the two coexisting feldspars would have compositions of about $Or_{75}Ab_{25}$ and $Or_{10}Ab_{90}$. The $Or_{75}Ab_{25}$ phase would be more abundant. The separation of these two phases from an initially homogeneous phase (*exsolution*) results in an intergrowth known as perthite. Note that as temperature decreases below 660°C the area of solid solution becomes more and more limited.

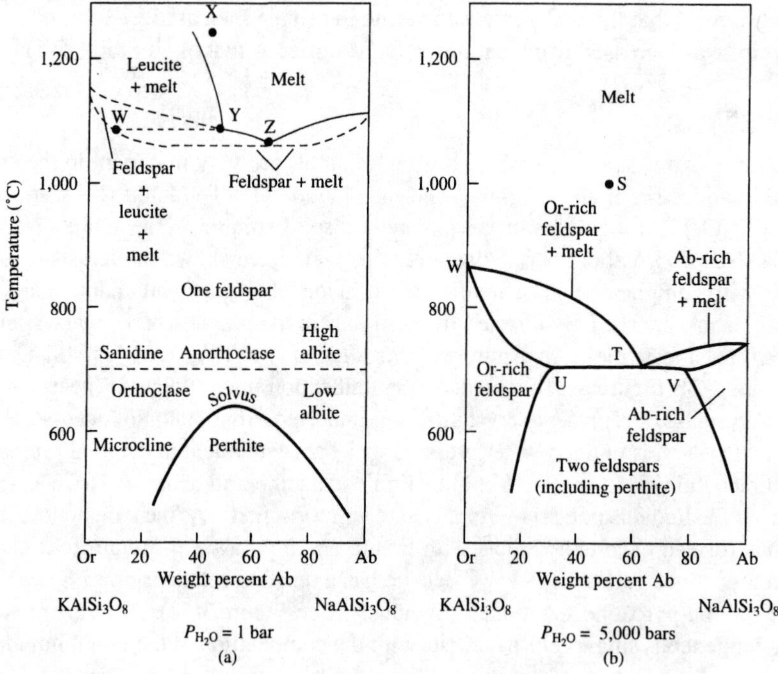

Figure 8-21 The system $KAlSi_3O_8$-$NaAlSi_3O_8$ at water pressures of 1 and 5,000 bars. See text for explanation. (a) After Bowen and Tuttle. (b) After Morse (1970).

Order-disorder relationships of the K-feldspars have been reviewed in Chapter Five. In terms of Figure 8-21a, a potassium-rich feldspar above about 700°C would be sanidine (with complete disorder of aluminum and silicon atoms). Below 700°C, orthoclase would be the stable form, and at the lowest temperatures microcline (with ordered aluminum and silicon atoms) would be stable. Alkali feldspars in the middle of the field of Figure 8-21a are known as anorthoclase [$(K, Na)AlSi_3O_8$]. Sanidine is monoclinic and anorthoclase is triclinic. Sodium-rich feldspars at high temperatures also differ crystallographically from minerals of similar composition at low temperatures. They are referred to as high albite and low albite, respectively. Thus perthite is a mixture of either monoclinic orthoclase or triclinic microcline with low albite. A small amount of sodium substitutes in the K-feldspar and a small amount of potassium in the low albite.

Increasing the water pressure on the $KAlSi_3O_8$-$NaAlSi_3O_8$ system, that is, adding H_2O to the melt, has two major effects. First, and most important, the temperatures of initial crystallization (which determine the liquidus line of Figure 8-21a) are significantly lowered. At 1,000 bars water pressure, point Z of Figure 8-21a occurs at about 850°C and at 2,000 bars it is about 750°C. The other effect of increasing pressure is to increase slightly the temperatures of the solvus line. This lowering of the liquidus line (and of the corresponding solidus line) and rising of the solvus probably results in their intersection at water pressures such as 5,000 bars (Figure 8-21b). This pressure is equivalent to magmatic crystallization at a depth of roughly 20 kilometers.

A liquidus such as S of Figure 8-21b would not start crystallizing a solid until the temperature had dropped to about 800°C and the liquidus line WT was reached. At that temperature a potassium-rich feldspar would start to form. Its composition would be about $Or_{95}Ab_5$ on the solidus line WUT. The liquid composition would then change along WT toward T, and the feldspar composition would change along WUT toward T. When the liquid composition reached T (at a temperature of 703°C), two different feldspars would be forming and coexisting. One would have the composition of point U and the other of point V. When the liquid was all used up, the two feldspars would cool down, and exsolution would occur for each feldspar, giving two perthitic intergrowths. Intergrowths of albite in host K-feldspar crystals (formed from composition U) are referred to as *perthite,* and intergrowths of K-feldspar in host albite crystals (formed from composition V) are known as *antiperthite.* Perthite is much more common in nature.

One way to subdivide granitic rocks involves separating those that crystallize under a low water pressure (indicated by the presence of a single alkali feldspar as in Figure 8-21a) from those that crystallize under a high water pressure (rocks that contain two alkali feldspars as in Figure 8-21b). Those in the first group are termed hypersolvus granites and those in the second group are termed subsolvus granites (Tuttle and Bowen 1958). The feldspars in either group may show exsolution. The two groups can also be described as crystallizing above the feldspar solvus (hypersolvus group) and below the feldspar solvus (subsolvus group). Subsolvus granites are more common because most magmas have an anorthite component (not involved in Figure 8-21) and this favors formation of two feldspars. A more detailed look at the subsolidus portion of the alkali feldspar system is given in the discussion of Figure 5-23 in Chapter Five.

We need to extend the results of Figure 8-21 to the third feldspar component, anorthite ($CaAl_2Si_2O_8$). Figure 8-22 shows relationships for the three-component system. There is complete solid solution, at all temperatures, between albite and anorthite (actually there is a small amount of exsolution at very low temperatures). Very little calcium can substitute into K-feldspar structures, and very little potassium can substitute into the anorthite structure. Thus the phase diagram for 1 bar pressure (Figure 8-22a) shows two feldspars as the final product of crystallization

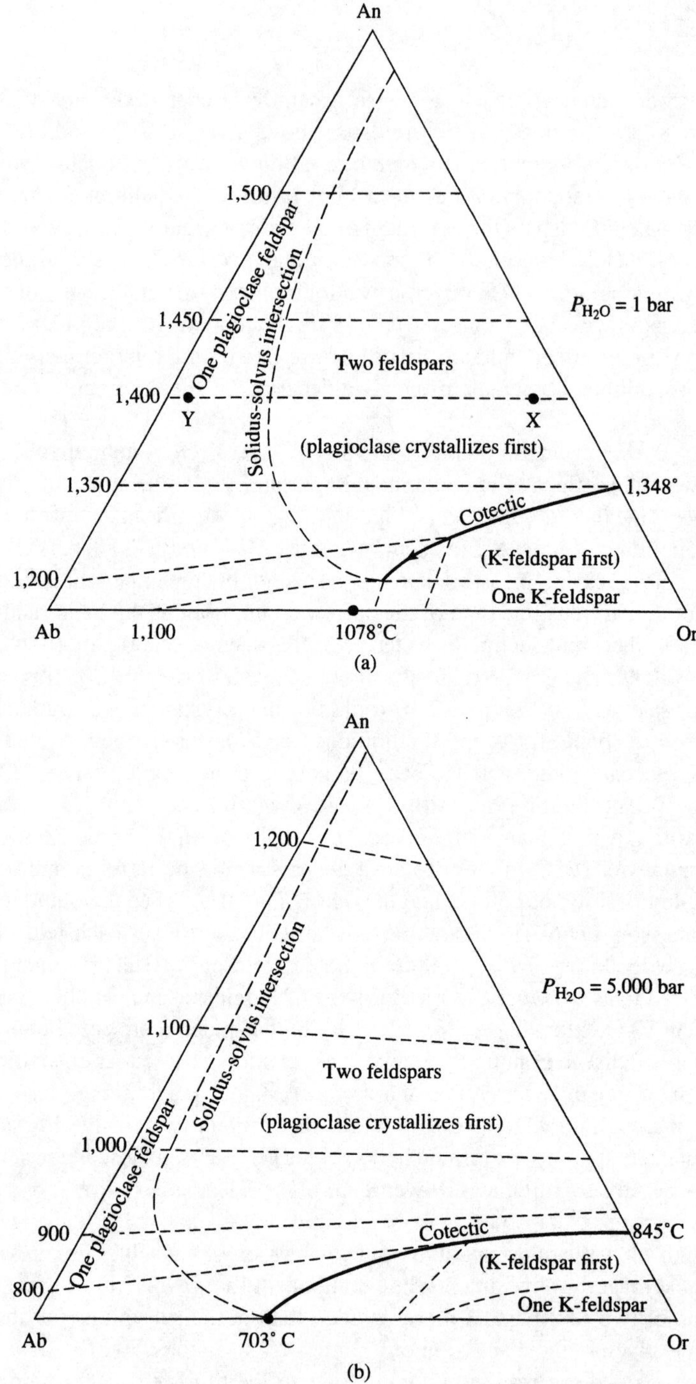

Figure 8-22 Crystallization of the feldspars at (a) 1 and (b) 5,000 bars water pressure. The solidus- solvus intersection for 1 bar in the K-feldspar corner has been extended by assuming the presence of excess SiO_2, thus eliminating the leucite field (see Figure 3-12). Temperature contours are drawn on the liquidus surfaces. (After *Petrology of Igneous and Metamorphic Rocks* by D. W. Hyndman, pp. 1–30. Copyright 1972 by McGraw-Hill Book Company. Used with permission of McGraw-Hill Book Company.)

for most of the area between An and Or. One feldspar is the final product of crystallization in the areas between An and Ab and between Ab and Or. (Note that Figure 8-22 shows only the final products formed from a melt and does not include the possibility of exsolution at lower temperatures.) The division between the one-feldspar field and the two-feldspar field is a line representing solidus-solvus intersection. Although no such intersection occurs in Figure 8-21a at 1 bar pressure, such an intersection does occur at this pressure when calcium is added to the system. Depending on composition, either K-feldspar or plagioclase will form first in the two-feldspar field. The solid line of Figure 8-22a divides these two areas.

A liquid such as X in Figure 8-22a would crystallize plagioclase first at 1,400°C. The liquid composition would change toward the solid (*cotectic*) line. When it reached the line, plagioclase and K-feldspar would form together as the liquid changed in composition toward the 1,078°C point on the Ab-Or line (point Z of Figure 8-21a). The liquid would be used up before reaching this point. The final product would be two feldspars. For a liquid such as Y, plagioclase feldspar would also form at 1,400°C, but the liquid would stay in the one-feldspar field; the final product would be plagioclase feldspar with a small amount of potassium in solid solution. The effect of high water pressure on the three-component system is to lower the liquidus temperatures and to extend the two-feldspar field (Figure 8-22b). The two-feldspar zone in the middle of the Ab-Or line of Figure 8-22b corresponds to the middle of line UTV in Figure 8-21b.

The three-component system that is most representative of granitic magma is shown in Figure 8-23. The SiO_2-$NaAlSi_3O_8$-$KAlSi_3O_8$ system is also shown at 1 bar water pressure (for volcanic crystallization) and 5,000 bars water pressure (for plutonic crystallization). The first mineral to form from a liquid such as X in Figure 8-23a would be a sodium-rich alkali feldspar (see the right side of Figure 8-21a). As the temperature dropped, the composition of the liquid would move toward the solid cotectic line, which defines the low-temperature trough of the system. Feldspar would continue to form and would become richer in potassium (see Figure 8-21a). When the liquid composition reaches the cotectic line, tridymite would begin to form along with alkali feldspar, and then the liquid would change in composition along the line toward m. Point m is the minimum temperature at which liquid can exist in the system. For composition X and equilibrium crystallization, the liquid would be used up before it reached m. The final products would be tridymite and alkali feldspar with a composition of about $Ab_{85}Or_{15}$. Liquids of any composition in this system would change, as crystallization proceeded, toward the cotectic line and then toward m.

At high water pressures (Figure 8-23b), several aspects of the system are different. The liquidus surface is marked by lower temperatures and quartz is now the stable form of silica. The leucite field is eliminated. The most important change is that two different feldspars form from liquids under high water pressure (see Figure 8-21b). A liquid such as X in Figure 8-23b would crystallize a potassium-rich feldspar first. The liquid would then change in composition toward the cotectic line AE. At AE a sodic feldspar would begin forming along with the potassic feldspar. The liquid would then change in composition along AE toward E, the minimum point of the system. At E, quartz would begin to form along with the two feldspars, all three minerals forming until the liquid is used up.* Thus, in cases of both low and high water pressures, the last liquids have a composition somewhere near the center of the SiO_2-$NaAlSi_3O_8$-$KAlSi_3O_8$ triangle.

*E is known as a *eutectic* point because it is an invariant point that represents the composition and temperature of a liquid in a system of three components that is in equilibrium with three solid phases. Applying the phase rule discussed in Chapter Three, $f = c + 1 - p = 3 + 1 - 4 = 0$. Point m of Figure 8-23a is not an invariant point and therefore not a eutectic point, since $f = 3 + 1 - 3 = 1$.

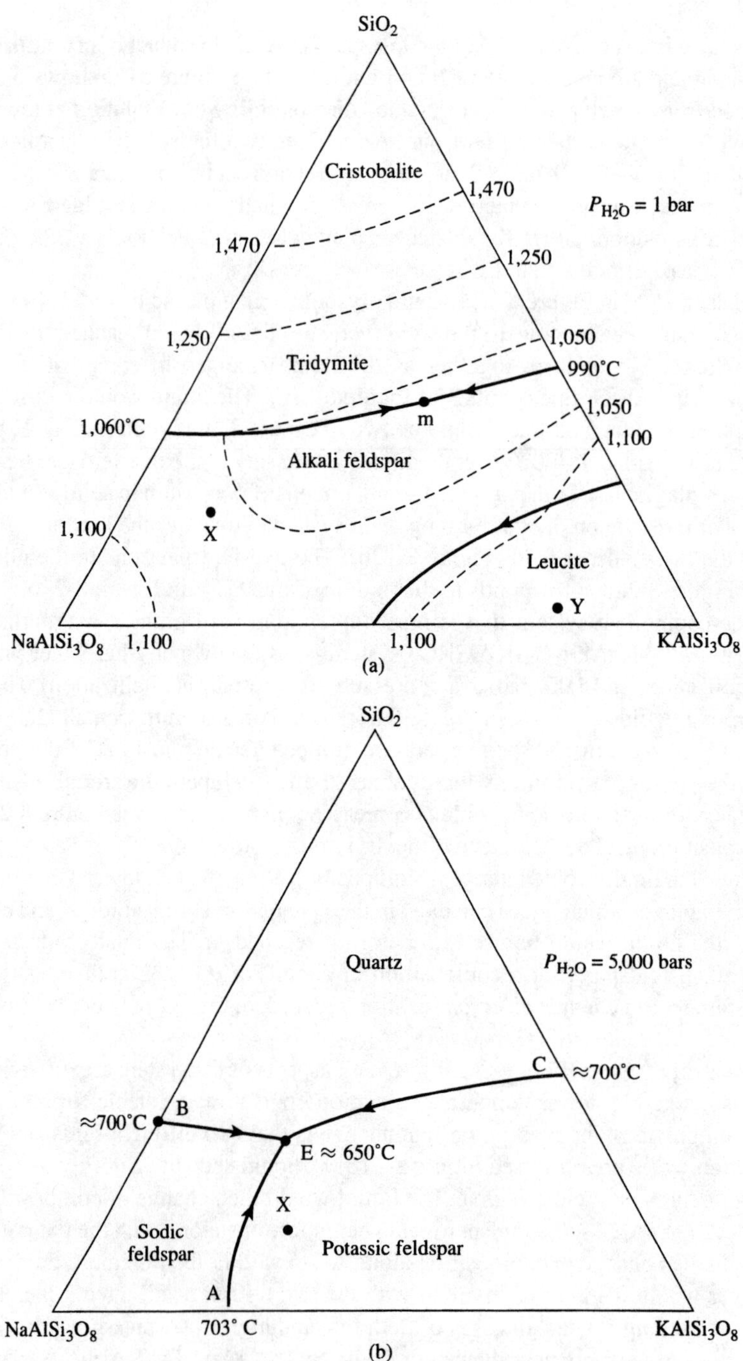

Figure 8-23 The system SiO_2-$NaAlSi_3O_8$-$KAlSi_3O_8$ at (a) 1 and (b) 5,000 bars water pressure. Temperature contours are drawn on the liquidus surface in (a). See further discussion in text. Data from Schairer (1950, 512–517) and Tuttle and Bowen (1958).

Tuttle and Bowen (1958) compared this experimental work with chemical analyses of rhyolites and granites. When only the three components of Figure 8-23 are used (Q, or, and ab of the CIPW classification), most rhyolites and granites have compositions falling between the minimum points of Figures 8-23a and 8-23b. (At water pressures between 1 and 5,000 bars, the minimum point of the system falls between m and E of Figures 8-23a and 8-23b.) Thus it appears that many granitic rocks formed from residual liquids. If this were not so, plots of chemical analyses of granitic rocks would show a large scatter rather than concentrating in the center of the SiO_2-$NaAlSi_3O_8$-$KAlSi_3O_8$ system. Tuttle and Bowen have plotted the minimum point of the system as a function of pressure (Figure 8-24), and this is often referred to as the minimum melting curve for "granite."

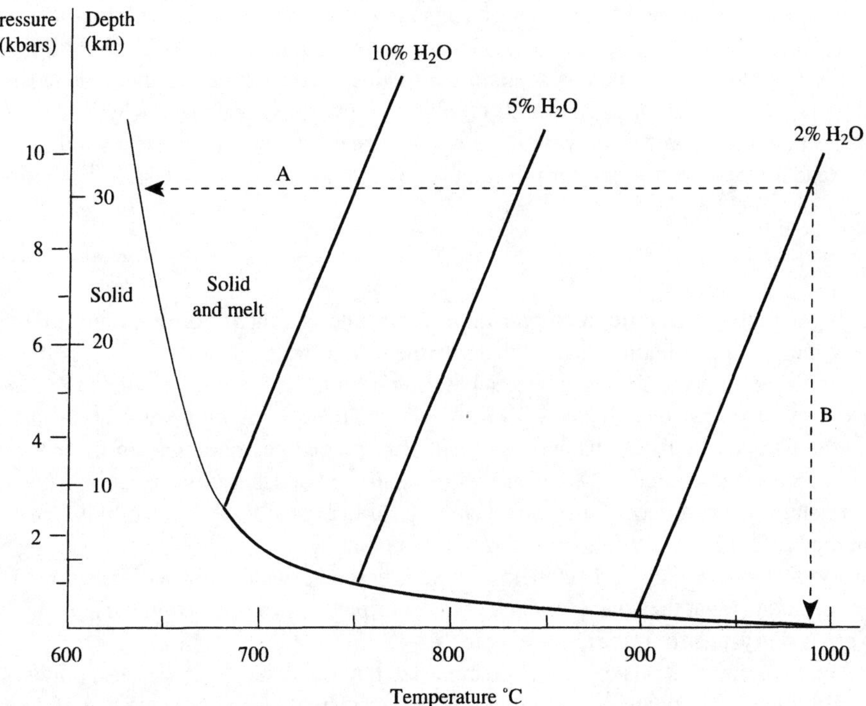

Figure 8-24 Minimum-melt (solidus) temperatures for granitic magma with excess water. The curved line is the minimum temperature, for given pressure, at which melt can exist in the system SiO_2-$KAlSi_3O_8$-$NaAlSi_3O_8$. The curve can be considered to represent temperatures of the last liquids of a crystallization magma or the temperatures of the first liquids to form by partial melting of rocks of granitic composition. The straight lines represent the temperatures and pressures at which crystallization starts for magmas containing 2 weight percent, 5 weight percent, and 10 weight percent H_2O. For example, a granitic magma at 30 km depth (9 kbars pressure) and containing 2 weight percent water would begin to crystallize at about 1000°C and, if the pressure remained constant, it would finish crystallizing at about 650°C (path A). If the temperature stayed at 1000°C, the magma could also rise to the Earth's surface without crystallizing (path B). Solidus curve from Tuttle and Bowen (1958). (After D. W. Hyndman, *Petrology of Igneous and Metamorphic Rocks*, 2d ed. Copyright 1985 by McGraw-Hill, Inc., Used with permission of McGraw-Hill, Inc.)

As pointed out above, the low points of the liquidus surfaces in Figures 8-23a and 8-23b are different. One is a eutectic and one is not. If a rock of the appropriate composition and under 5-kb water pressure is heated enough to start melting, the first liquid formed will have the composition of the eutectic point in Figure 8-23b. Under low water pressure, as in Figure 8-23a, the first liquid formed may *not* have the composition of point m because it is not a eutectic point. In this case the overall composition of the melting solid and other factors will determine what liquid forms first. The existence of a eutectic is more common because adding an anorthite component to the system of Figure 8-23 favors formation of a eutectic.

Whitney (1988) points out that a critical factor in both the formation and crystallization of granitic magmas is the presence of water (Figure 8-24). Two sources of the required water for the anatectic melting of crustal rock are (1) dehydration of hydrous silicates and (2) water in subducted, hydrothermally altered volcanic rock. Experimental work reviewed by Whitney suggests that there are significant differences between vapor-undersaturated and vapor-saturated crystallization. In general, crystallization with low water content leads to long temperature ranges of crystallization (for example, 1000 to 650°C) and coarse grain size. Under water-saturated conditions, a shorter temperature range (for example, 720 to 680°C) and finer grain size are found (Figure 8-24). Whitney estimates that water contents of two to four weight percent are present in most granitic magmas.

Readers interested in further information on granitic rocks are referred to Whitney (1988) and Clarke (1992).

Pegmatites

Most pegmatites are granitic in composition. Jahns and Burnham (1969) proposed a model for the origin of such pegmatites that emphasizes the role of water. Water occurs dissolved in magmas and may also occur as a separate, supercritical fluid. At low temperatures the supercritical fluid will become a water-rich gas and a water-rich liquid (see Figure 4-9 and accompanying discussion in Chapter Four). Jahns and Burnham suggest that pegmatites form from magmas that have a high enough content of water to become saturated under certain conditions (such as lowered pressures due to ascent of magma toward the surface). A silicate liquid of this type could represent the last fluid of a crystallizing magma or the first fluid formed by partial melting of preexisting rocks. Once a magma becomes saturated, a supercritical fluid will separate. Jahns and Burnham suggest that the coarse-grained crystals of pegmatites form from the aqueous fluid and finer-grained crystals from the silicate melt.

More recent work on pegmatites, summarized by London (1992), has cast doubt on the validity of the Jahns-Burnham pegmatite model. Experimental results indicate that H_2O-undersaturated fractional crystallization of granitic systems produces many of the features of pegmatites and that H_2O-saturated experiments do not produce these features. London (1992, 499) states that pegmatites are

> a product of disequilibrium fractional crystallization through liquidus undercooling. The degree of liquidus undercooling, and the concentration of quartz-feldspar incompatible components in the melt (particularly H_2O, B, P, and F), govern the textural development of granitic magmas by controlling the rate and number of stable nuclei formed.

"Liquidus undercooling" refers to the metastable persistence of melt below the temperatures at which the first formation of solids should occur. In other words, crystal nucleation and growth is retarded. "Disequilibrium fractional crystallization" refers to incomplete reaction between earlier-formed crystals and the remaining liquid. Thus the change from a granitic texture to a pegmatitic texture, often seen in granitic bodies, appears to be due to decreases in the nucleation density of the magma (resulting in fewer and bigger crystals) caused by (1) changes in the amount of liquidus undercooling and (2) increasing concentrations in the magma of crystal-incompatible components.

MINERAL DEPOSITS AND MAGMAS

Some pegmatites contain economic concentrations of one or more rare elements (such as lithium and beryllium), and these deposits are probably igneous in origin. The rare elements were concentrated in the residual fluid of a primary magma or were concentrated by some process involved in local magma formation. Other mineral deposits are clearly of direct magmatic origin. Concentrations of chromite [$(Mg, Fe)Cr_2O_4$] and of platinoid metals (Pt, Pd, Ir, Os) occur as direct magmatic segregations in layered mafic intrusions. Other deposits have formed as the result of magma intrusion into carbonate and other reactive rocks. These contact metasomatic deposits were formed by the transfer of ore elements such as iron and copper from magmas to the contact with the invaded rocks. Often non-ore elements (Si, Al, etc.) were also added to limestones, producing ore host rocks known as skarns.

The majority of mineral deposits have formed as a result of the action of hydrothermal solutions. The source of these solutions in some cases was magma; in other cases the source was groundwater, seawater, or solutions resulting from metamorphism. In many cases it appears that *both* high-temperature magmatic (or metamorphic) and low-temperature meteoric waters were involved in the transport and deposition of ore elements. Often ore deposition starts with mainly magmatic waters and finishes with predominantly meteoric waters (Figure 8-25). It is also possible that, for some deposits, the ore elements came from the surrounding country rock or from a crystallized intrusion and not from an invading magma. In some cases, magma provided the heat to initiate circulation of the meteoric water, but nothing else.

A technique now exists that can sometimes identify the source or sources of the ore elements and of the hydrothermal solutions that formed a particular ore deposit. Because different types of water have different isotopic properties, analyses of the oxygen and hydrogen isotopes of minerals associated with ore deposits give us information on the nature of the fluids that deposited the minerals or that interacted with pre-existing minerals of an intrusion and the surrounding country rock. Figure 2-17 shows the isotopic fields for various waters and Figures 8-25, 2-20, 2-21, and 2-22 are examples of the use of oxygen and hydrogen isotopes to study ore deposits. Other isotopes can also be used (Figure 2-26).

A major problem in explaining the transport of metals in hydrothermal solutions is the general insolubility of sulfide minerals in water (see the discussion of solubility products in Chapter Four). Hydrothermal solutions probably have compositions similar to those of natural brines (Table 8-12). Even though the concentrations of metals such as lead and copper in these solutions are low, precipitation from such fluids over long periods of time can apparently form rich ore deposits. It is now believed that the metals are transported in solution as complex ions, such as $PbCl_3^-$ (Eugster 1986). Sulfur may have traveled in the same solutions with the metals (the probable case for an igneous source), or the sulfur may have come from a separate source. Precipitation of sulfides and other ore minerals may be due to temperature and pressure changes, to reaction with wall rocks, or to mixing of different solutions. These processes can cause chemical changes, such as changes in pH or oxygen fugacity, that result in precipitation of ore minerals.

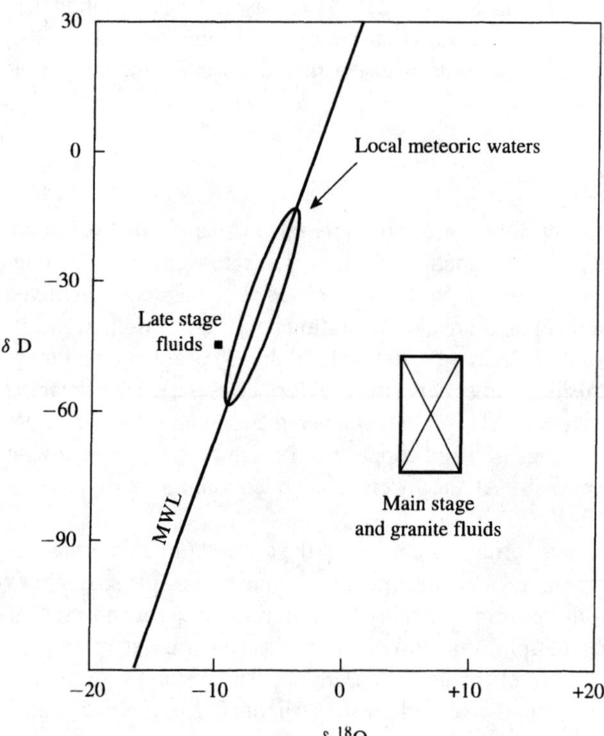

Figure 8-25 Plot of $\delta\,^{18}O$ and δD values of aqueous fluids that would have been in equilibration with granites and ore veins at Dajishan Mine, Southeast China. The tungsten ore occurs in wolframite-quartz veins at the contact between Jurassic granite and Cambrian metasedimentary rocks. The $\delta\,^{18}O$ values of the fluids were calculated from the $\delta\,^{18}O$ of minerals and fluid inclusion temperatures (for granites) or oxygen isotope temperatures (for ore veins). The δD values were obtained from analysis of fluid inclusions or by calculation using isotopic analysis of mineral samples. The main stage hydrothermal vein fluids plot in the same region as the granite fluids; this is the area of magmatic waters shown in Figure 2-17. Thus the data suggest that the ore-forming fluid was derived from the granite magma. Calcite and other minerals formed in the late stage of mineralization plot to the left of the meteoric water line (MWL), indicating that an influx of large amounts of meteoric waters into the hydrothermal system occurred during the final period of hydrothermal activity. Meteoric waters in the mine area plot in the encircled area along the meteoric water line. After Shieh and Zhang (1991).

A detailed review of ore fluids and their relation to magmas can be found in Brimhall and Crerar (1987). Other useful references on hydrothermal ore deposits include Barnes (1979), Robertson (1984), and Berger and Bethke (1985).

CHEMICAL STUDIES OF IGNEOUS ROCKS

Gough Island

Gough Island is one of the southerly islands of the Mid-Atlantic Ridge. The petrology and chemistry of its volcanic rocks and minerals have been studied in detail by Le Maitre (1962). Most of the rocks of the island are varieties of either basalt or trachyte. Almost all the rocks are alkali rich and would plot above the line of Figure 8-1. Le Maitre has classified the rocks into six types on the basis of their alkali contents. Average compositions and norms for these six types are given in Table 8-13. The rocks contain olivine and one pyroxene and probably had a history similar to that given earlier for the liquid of point Y in Figure 8-18. They make up a gradational rock series, and the data assembled by Le Maitre strongly indicates formation of the rocks by fractional crystallization of alkali olivine basalt. Some of the chemical evidence for this conclusion will be briefly reviewed here.

TABLE 8-12 Compositions of Natural Brines Believed to Be Actual or Potential Ore-Forming Solutions

	Salton Sea geothermal brine (1)	Cheleken geothermal brine (2)	Fluid inclusion, fluorite (3)	Fluid inclusion, sphalerite (4)	Fluid inclusion, quartz (5)
Cl	155,000*	157,000	87,000	46,500	295,000
Na	50,400	76,140	40,400	19,700	152,000
Ca	28,000	19,708	8,600	7,500	4,400
K	17,500	490	3,500	3,700	67,000
Sr	400	636	—	—	—
Ba	235	—	—	—	—
Li	215	7.9	—	—	—
Rb	135	1.0	—	—	—
Cs	14	0	—	—	—
Mg	54	3,080	5,600	570	—
B	390	—	< 100	185	—
Br	120	526.5	—	—	—
I	18	31.7	—	—	—
F	15	—	—	—	—
NH_4	409	—	—	—	—
HCO_3^-	> 150	31.9	—	—	—
H_2S	16	0	—	—	—
SO_4^{2-}	5	309	1,200	1,600	11,000
Fe	2,290	14.0	—	—	8,000
Mn	1,400	46.5	450	690	—
Zn	540	3.0	10,900	1,330	—
Pb	102	9.2	—	—	—
Cu	8	1.4	9,100	140	—
As	12	0.03	—	—	—
Ag	1	—	—	—	—

* All compositions are given as concentrations in parts per million.

Note: Dashes indicate that values were not reported. They should not be read as zero.

1: I.I.D. No. 1 well, Salton sea geothermal field, Calif. Brine temperature 300 to 320°C, density 1.21 g/cm^3, pH (25°C) 5.2. Data from L. J. P. Muffler and D. E. White, 1969, *Bull. Geol. Soc. Am.*, v. 80, p. 157.

2: Well in aquifer No. 9, Cheleken Peninsula, USSR. Temperature approximately 80°C at well collar, pH (25°C) 5.5. Data from L. M. Lebedev and I. B. Nikitina, 1968, *Dokl. Akad. Nauk. SSSR,* Eng. transl. earth science section 183, p. 180.

3: Fluid inclusion from a fluorite crystal, Hill Mine, Cave-in-Rock district, Ill. Data from E. Roedder et al., 1963, *Econ. Geol.*, v. 58, p. 353. The metal data are from a different sample and are taken from G. K. Czamanske et al., 1963, *Science*, v. 140, p. 401.

4: Average composition of fluid inclusions in late-stage sphalerite from the OH vein, Creede, Colo. Data from E. Roedder, 1965, *Symp. Probl. Postmagmatic Ore Deposition*, Prague, vol. II. the metal data are from a quartz crystal from the same vein as reported by G. K. Czamanske et al., *op. cit.*

5: Average high-density, multiphase fluid inclusions in quartz from the core zone of the porphyry copper orebody at Bingham Canyon, Utah. Data from E. Roedder, 1971, *Econ. Geol.*, v. 66, p. 98.

Source: After Skinner and Barton (1973).

TABLE 8-13 Average Compositions and Norms of Typical Rock Types, Gough Island

Weight percent	Picrite basalt	Olivine basalt	Trachy-basalt	Trachy-andesite	Trachyte	Aegirine-augite trachyte
SiO_2	46.8	47.7	51.1	56.3	59.5	61.5
TiO_2	1.9	3.2	2.8	1.8	0.9	0.3
Al_2O_3	8.2	15.2	17.6	17.8	19.4	18.3
Fe_2O_3	1.2	2.3	2.8	2.9	1.7	2.6
FeO	9.8	8.7	6.8	4.7	3.6	2.8
MgO	19.8	9.7	4.8	2.3	1.0	0.2
CaO	9.5	8.9	6.9	4.7	2.0	1.5
Na_2O	1.6	2.7	4.0	4.8	5.2	7.0
K_2O	1.2	1.6	3.2	4.7	6.7	5.8
Norms						
Orthoclase	7.2	9.5	18.9	27.8	39.5	34.5
Albite	9.8	21.9	28.3	38.4	39.1	48.9
Anorthite	11.4	24.5	20.5	13.3	9.7	1.1
Nepheline	2.0	0.6	3.1	1.1	2.6	5.6
Diopside	28.6	15.8	11.0	8.1	0.3	5.5
Olivine	35.5	18.4	8.7	3.6	4.5	0.1
Ilmenite	3.6	6.1	5.3	3.5	1.7	0.6
Magnetite	1.9	3.2	4.2	4.2	2.6	3.7
Trace elements (ppm)						
Ni	465	210	52	5	3	3
Co	100	34	29	7	—	—
Cr	1,250	245	51	2	—	—
Zr	100	123	196	323	335	1,000
Rb	30	65	80	200	100	307
Y	10	16	21	28	50	43
La	—	20	42	62	85	150
Ba	340	850	1,020	350	625	—
Sr	450	725	1,010	933	275	15
Ga	5	18	20	18	20	27
Li	3	16	5	13	8	33
V	100	160	128	70	—	—

Source: Data from Le Maitre (1962, 1309–1340).

The chemical analyses in Table 8-13 show gradational changes in composition from the picrite basalt (believed to form first from crystallizing magma) to the aegirine-augite trachyte (formed last). Such changes in composition can be plotted on variation diagrams (see Figure 8-4) to show the continuous nature of the variation. The normative mineral percentages of Table 8-13 also show regular changes throughout the rock series. Trace element analyses of the rocks show several different types of change in going from early to late-formed rocks (Table 8-13). Some elements, such as nickel and cobalt, decrease steadily in concentration. Others, such as lithium and rubidium, show a regular increase. Barium and strontium increase to a maximum in the intermediate rocks and then decrease in amount in the residual rocks. Gallium and yttrium are almost constant throughout the series. As pointed out in Chapter Five, the distribution of most trace elements is controlled by substitution for major elements in crystallizing minerals. For example, nickel and cobalt mainly substitute for the iron and magnesium of olivine. Since olivine is most abundant in the early rocks, so are nickel and cobalt.

The regular and continuous changes of major and trace elements in the rock series indicate origin of the rocks by separation of early formed minerals from a parent magma in a series of steps such that magmas of different compositions are produced. Other evidence for this process is provided by the occurrence and composition of the major minerals of the rocks. Olivine occurs in all the rocks and shows a regular change in composition from 17 mole percent fayalite (Fe_2SiO_4) in the picrite basalt to 70 mole percent fayalite in the aegirine-augite trachyte. Pyroxene compositions show a similar increase in FeO and decrease in MgO toward the felsic end of the series. The Na_2O and K_2O content of the pyroxenes increases in going from basalts to trachytes. Both plagioclase and alkali feldspar occur in the rocks, and both increase in abundance in the late-formed rocks. Plagioclase is zoned from labradorite in the core to andesine or oligoclase at the edges. Plagioclase occurs as phenocrysts in the basalts; both plagioclase and alkali feldspar occur in the groundmass of the rocks. Thus plagioclase formed first, with later formation of both feldspars. This would be similar to the history, given earlier, of the liquid of point X in Figure 8-22a. Phenocrysts in the trachytes consist of interlocking plagioclase and alkali feldspar, representing crystallization of the type that occurs along the cotectic line of Figure 8-22a. Some of the alkali feldspars of the trachytes exhibit exsolution (see Figure 8-21a); others do not. Those that show exsolution are from relatively coarse grained rocks (intrusive plugs) that cooled more slowly than the rocks formed from lava flows. Thus the rate of cooling controlled the amount of exsolution.

Another approach that indicates the occurrence of fractional crystallization involves plotting of the overall composition of various rock types on a diagram such as that of Figure 8-23. The rocks that formed latest in a series should plot in the lower-temperature parts of such a diagram. For the alkali-rich rocks of Gough Island, it is necessary to use the system $NaAlSiO_4$-$KAlSiO_4$-SiO_2 (Figure 8-23 is a part of this system). Le Maitre (1962) plotted chemical analyses for the trachytes of Gough Island and showed that there is a direct relationship between temperature of crystallization and stage of formation. Early formed trachytes plot at higher temperatures and there is a definite trend, with the youngest rock type plotting near a eutectic point (minimum point) in the system.

Zielinski and Frey (1970) have provided further evidence in support of a fractional crystallization model for the Gough Island rocks. They used a computer model of fractional crystallization to calculate the varying proportions of crystallizing minerals to generate the residual liquids represented by each rock type on the island. This allowed prediction of the trace element abundances that should occur in these rocks if the model is correct. Actual abundances of the elements that they studied (rare earths, barium, strontium, nickel, and chromium) show good agreement with the predicted abundances of the model. On the other hand, the model does not explain the isotopic data (lead and strontium isotope ratios) obtained from rocks of Gough Island by Gast et al. (1964). These authors conclude that the rocks could not all be derived from an isotopically homogeneous magma. Possible explanations for the isotope data include wall-rock assimilation and mixing of magmas.

Deccan Traps

Research on continental flood basalts indicates that these rocks form from magma plumes that originated at the base of the continental lithosphere. It is generally believed that these mantle-derived magmas assimilated various amounts of crustal material on their way to the surface. The Deccan traps of India and Pakistan are a well-known example of continental flood basalts (Mahoney 1988). This thick succession of subaerially erupted, horizontal basaltic flows covers an area of about half a million square kilometers and has a maximum thickness of 3500 meters. Detailed mapping of the western portion of the Deccan has resulted in division of the rocks into eleven formations. ^{40}Ar-^{39}Ar dating gives an age of 66 Ma, with all the rocks forming in a very short period of time (< 2 Ma).

The Deccan rocks are primarily tholeiite basalts. Although they have a limited range of major element compositions, they exhibit a large variation in trace element and isotope content. Peng at al. (1994) have reported on a study of isotopes and incompatible elements in the lower six formations. A number of earlier chemical studies reported on the upper five formations. Despite the diversity in isotopic and trace element content, Peng et al. found that the chemical results all converged on a limited range of compositions, which they called the "common signature." For example, it appears that most of the magmas have ϵ_{Nd} values close to the bulk-earth value of zero. The Deccan Nd-Sr-Pb isotopic ratios are different from those of oceanic volcanic rocks, suggesting a different origin and history for the Deccan magmas. Peng et al. develop a model for the formation of the lower six formations. The question they attempt to answer is this: What combinations of mantle-derived magma and crustal rocks (contamination) could produce a group of magmas with the chemical and isotopic properties represented by the common signature values?

One of the upper formations, the Ambenali, has been identified as the least contaminated with crustal material. The Ambenali rocks have high initial ϵ_{Nd} values and low initial $^{87}Sr/^{86}Sr$ values (calculated for 66 Ma), indicating a depleted mantle source in the asthenosphere. The Deccan rocks represent an eruption on the Earth's surface of the Réunion hot spot, named for Réunion Island in the Indian Ocean. After formation of the Deccan basalts, northward movement of the Indian plate over the hot spot produced a chain of submarine volcanoes. The hot spot is currently underneath an active volcano on Réunion Island.* Thus two likely candidates for the mantle-derived magma that formed the Deccan basalts are an Ambenali-type magma and a Réunion-type magma. Analyses of incompatible elements in rocks from the two locations suggest that the magmas differed somewhat in their chemistry, perhaps due to a lower extent of melting for Réunion magma as compared to Ambenali magma (Figure 8-26a).

The composition of the crustal rock to mix with primitive magma in a model was chosen by Peng et al. (1994) as that of Indian Archean amphibolite (which outcrops north and south of the Deccan). These amphibolites have incompatible element patterns that make it possible to mix their average composition with that of the Ambenali or Réunion magmas and produce the pattern of the common signature for the lower six Deccan formations. It was found that mixing 20 percent amphibolite melt and 80 percent parental magma produced a pattern that closely matched the common signature pattern (Figure 8-26b). Table 8-14 gives some isotopic values for the amphibolites, Réunion magma, Ambenali magma, average common signature, and a mixture of 20 percent amphibolite melt, 40 percent Réunion magma, and 40 percent Ambenali magma. This mixture is shown in Figure 8-26b.

Peng et al. (1994) conclude that assimilation of mafic lower crust (represented by the amphibolites) can explain the incompatible element and isotopic characteristics of the Deccan magmas that formed the lower six formations. The divergence of trace element and isotopic values

*Coffin and Eldholm (1994) discuss large igneous provinces such as the Deccan traps and their relationship to hot spots (mantle plumes). These igneous provinces are important because they give us information about regions of the mantle that do not generate normal mid-ocean ridge basalt. It is not clear whether hot spots and plate tectonics operate independently or are linked.

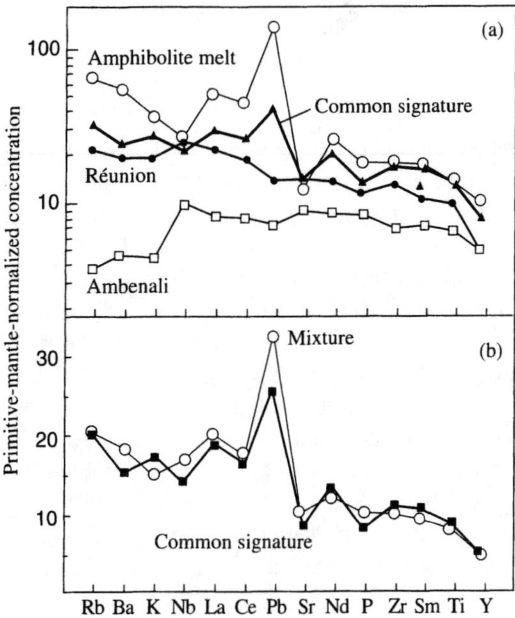

Figure 8-26 (a) Primitive-mantle-normalized incompatible element patterns for the average common signature lava (heavy line), Réunion-type primitive magma, and Ambenali-type primitive magma. Also shown is a calculated pattern for a 40 percent partial melt of average basic Indian amphibolite. Isotopic data for the four materials are in Table 8-14. See discussion in text. (b) Patterns for average of common signature samples and a mixture of 20 percent basic amphibolite melt plus 80 percent of 1 : 1 mixture of Réunion-type magma and Ambenali-type magma. (Reprinted from *Geochimica et Cosmochimica Acta*, v. 58, Peng et al., A role for lower continental crust in flood basalt genesis? Isotopic and incompatible element study of the lower six formations of the western Deccan Traps, pp. 267–288. Copyright 1994, with kind permission from Elsevier Science Ltd, The Boulevard, Langford Lane, Kidlington OX5 1GB, UK.)

TABLE 8-14 Nd-Sr Isotopic Features of Three Hypothetical Crustal End-Members, Average of Common-Signature Basalts, and a Calculated Mixture

	$^{143}Nd/^{144}Nd$	Nd (ppm)	ϵ_{Nd}	$^{87}Sr/^{86}Sr$	Sr (ppm)
Indian shield amphibolite	0.51203	16	−12	0.7161	220
Réunion magma	0.51286	23	+4.4	0.7042	332
Ambenali magma	0.51300	8.8	+7.1	0.7042	202
Average common-signature basalts	0.51249	24	−2.9	0.7075	269
Calculated mixture	0.51259	20	−1.0	0.7065	264

Notes: Réunion magma and Ambenali magma are estimated primitive Réunion and Ambenali magmas. The calculated mixture, which resembles the average for common-signature basalts, consists of 20 percent amphibolite melt, 40 percent Réunion magma, and 40 percent Ambenali magna. Amphibolite melt equals 40 percent partial melt of average basic Indian amphibolite.

Source: Reprinted from *Geochimica et Cosmochimica Acta*, v. 58, Peng et al.; A role for lower continental crust in flood basalt genesis? Isotopic and incompatible element study of the lower six formations of the western Decca Traps, pp. 267–288. Copyright 1994, with kind permission from Elsevier Science Ltd, The Boulevard, Langford Lane, Kidlington OX5 1GB, UK.

for individual formations from their common signature must reflect secondary contamination episodes.

Boulder Batholith

We now turn to an example of the use of isotopes to date igneous rocks. The Boulder batholith of Montana is a composite intrusive mass, made up of at least 12 different plutons ranging in composition from gabbro to alaskite. Most of the batholith consists of granodiorite and quartz monzonite. It outcrops over an area of approximately 5,700 square kilometers. The Butte district, famous for its rich copper ores, is located in the southern part of the batholith.

The ages of the various plutons have been determined using the potassium-argon system (Tilling et al. 1968). Doe et al. (1968) have used the uranium-lead system and the rubidium-strontium system to study the origin and history of the batholith. (These three isotope systems are discussed in detail in Chapter Two.) Tilling (1973) has combined the isotopic data with other chemical data to propose that the batholith formed from two contemporaneous but chemically distinct magma series. Tilling et al. (1968) found that K-Ar ages of biotites and hornblendes from the batholith ranged from 78 to 68 million years (Figure 8-27). There is a direct relationship between the age of a pluton and its silica content, with the more mafic rocks having the older ages. Field relations also indicate that the mafic rocks were first in the intrusive sequence.

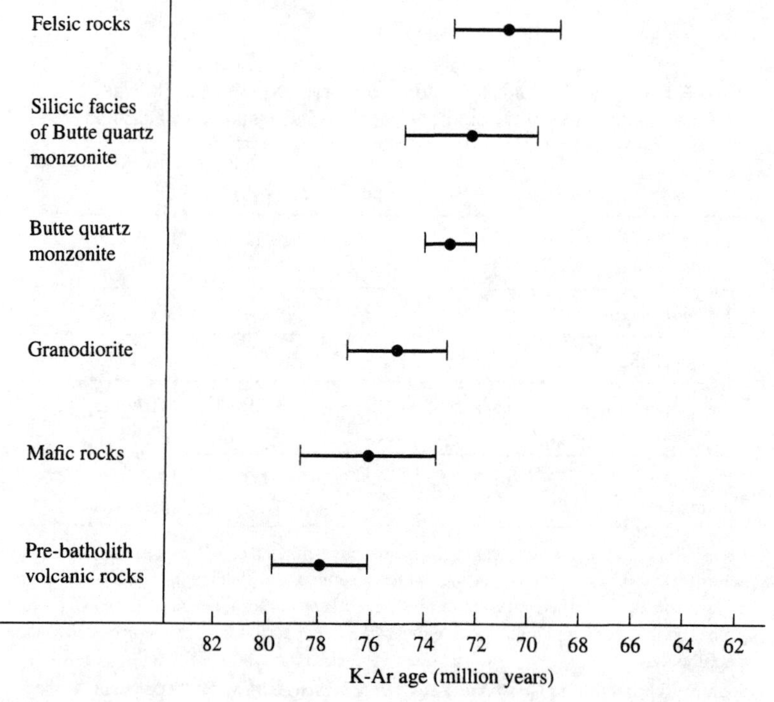

Figure 8-27 Potassium-argon ages of rocks of the Boulder batholith and of associated prebatholith volcanic rocks. The mean ages for each group of rocks is shown by a dot; bars give error of the mean at the 95 percent confidence level. Field mapping also indicates that the rocks were intruded in a sequence from mafic to felsic rocks. After Tilling et al. (1968).

The isotopic composition of lead in K-feldspars varies somewhat in each pluton and varies widely from pluton to pluton, indicating that the plutons came from a source that was not isotopically homogeneous. Variations within plutons may also be due to magma assimilation of country rock. The lead composition of the ore minerals at Butte is very similar to the lead composition of the K-feldspar of the Butte Quartz Monzonite in which the ore deposits occur. This suggests a genetic relationship between host rock and ore. Model lead ages calculated for plutons of the batholith give values as old as 910 million years. There is no systematic relationship between lead isotope ratios and the ages of the rocks obtained from potassium-argon measurements. The younger rocks do tend to have more radiogenic lead. Possible explanations for the lead ages, which are much too old, are (1) loss of uranium sometime after magma formation or (2) gain of lead sometime after magma formation.

The rubidium-strontium data also show a general relationship between the abundance of radiogenic ^{87}Sr and the ages of the rocks as obtained by the K-Ar method. Doe et al. (1968) did not determine Rb-Sr ages for the rocks, but did calculate values of initial $^{87}Sr/^{86}Sr$ in order to estimate the source region of the batholith magma. The initial $^{87}Sr/^{86}Sr$ values for the plutons vary from 0.7055 to 0.7092, suggesting formation from a source in the mantle. The lead and rubidium-strontium data have been explained by Doe et al. as indicating magma formation by partial melting in the lower crust or upper mantle, with assimilation (contamination) of upper crustal material occurring during intrusion. Tilling (1973) presents additional data indicating that two chemically distinct groups of rocks occur in the batholith, and concludes that two different magmas were involved. He suggests that chemical differences occur vertically in the source region and that one magma formed from a deeper source than the other.

Peninsular Ranges Batholith

The Peninsular Ranges batholith is one of a chain of late Mesozoic batholiths occurring on continental margins around the Pacific basin (Figure 8-28). The formation of the batholith is intimately related to plate tectonic processes at a continent-ocean boundary marked by a volcanic island arc (Todd and Shaw 1979). The batholith is a composite intrusive complex made up of hundreds of individual plutons. A variety of rock types is found, with tonalite the most common lithology. (Tonalite is similar to granodiorite, but has more plagioclase and less K-feldspar than granodiorite.) The batholith is about 1600 km long and averages 80 km wide.

Todd and Shaw (1979) have recognized two plutonic series: an older, syntectonic series to the west, and a younger, late syntectonic or post-tectonic series to the east. These two zones, which run northwest-southeast parallel to the long axis of the batholith, have a number of contrasting features. Table 8-15 lists differences between the two zones in lithologies, plutonic structures, emplacement ages, and geochemical properties. We are particularly interested in the geochemical differences summarized in Table 8-15. As indicated earlier in this chapter, isotopic and chemical data can place strong constraints on hypotheses for the origin and history of an igneous body.

The initial $^{87}Sr/^{86}Sr$ ratios in the batholith vary from 0.7030 in the west to 0.7070 in the east (Silver et al. 1979). The regular increase to the east parallels the long axis of the batholith and is independent of rock type. Initial ϵ_{Nd} values change from 8.0 to -6.4 from west to east (DePaolo 1981). Values for $\delta^{18}O$ increase from 6.0 in the west to 12.0 in the east (Silver et al. 1979). The concentrations of incompatible elements such as barium and lead increase from west to east and the rare-earth patterns of the two zones are different (Gromet and Silver 1987). The eastern zone exhibits light REE enrichment and heavy REE depletion compared to the western zone.

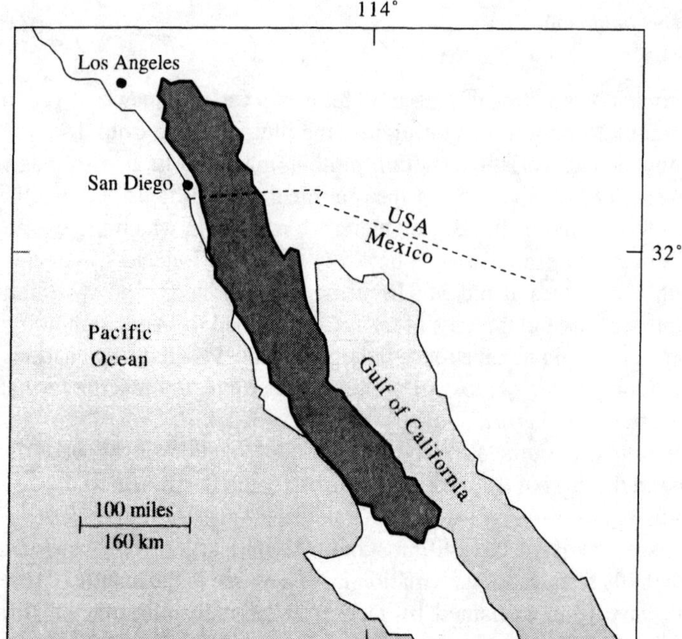

Figure 8-28 Peninsular Ranges batholith in southern California and Baja California. Todd and Shaw (1979) have mapped two contrasting zones in the northern portion of the batholith, which also differ geochemically (see Table 8-15). The zones run northwest-southeast parallel to the long axis of the batholith. After Todd and Shaw (1979).

TABLE 8-15 Summary Comparison of West and East Sides of Peninsular Ranges Batholith near San Diego, California

Features	West Side	East Side
Plutonic structure	Smaller epizonal plutons commonly isolated within prebatholithic rocks	Larger mesozonal plutons mutually contiguous, with prebatholithic rocks minor and confined to isolated pendants and screens
Timing	Static magmatic arc; emplacement ages 105 to more than 120 my	Migrating magmatic arc; emplacement ages near 105 my on west side, becoming progressively younger toward extreme east side, 80 to 90 my
Lithologies	Diverse lithologies ranging from gabbros and quartz gabbros to abundant tonalites with fairly abundant silica-rich leucogranodiorites	Limited range of lithologies dominated by titanite-hornblende-biotite–bearing tonalities and low-K_2O granodiorites
Initial $^{87}Sr/^{86}Sr$	0.7030	0.7070
Initial ϵ_{Nd} (DePaolo 1981)	+8.0	−6.4
$\delta\ ^{18}O$	+6.0	+12.0
Incompatible elements	Low	High
REE	Moderate LREE enrichment (about 35 × chondritic amount) and low HREE enrichment (about 15 × chondritic amount)	Strong LREE enrichment (about 80 × chondritic amount) and very low HREE enrichment (about 4 × chrondritic amount)

Source: After Silver et al. (1979).

What hypotheses for the batholith's formation can be developed from the geochemical and other data in Table 8-15? Hess (1989, Chap. 11) discusses the significance of these data in detail. He points out that the west to east changes in isotopic and incompatible element chemistry of the batholith could be explained by extensive contamination in the east of a mantle-derived, basaltic magma by continental crust. (The calcic character of the batholith rocks requires that the source magma be of basaltic composition.) This hypothesis has the western rocks formed from magma undergoing much less contamination. The original magma would have low $^{87}Sr/^{86}Sr$ ratios, low $\delta\,^{18}O$ values, and high $^{143}Nd/^{144}Nd$ ratios. Relative to mantle material, continental crust has higher $^{87}Sr/^{86}Sr$, higher $\delta\,^{18}O$, and lower $^{143}Nd/^{144}Nd$. Using geochemical reasoning, DePaolo (1981) concludes that mixing 20–30 percent of crust into a primitive basaltic magma could result in crystallization of tonalite. One of the problems with this contamination hypothesis is that it does not explain the constant major-element composition of tonalite across the entire batholith. Another hypothesis discussed by Hess (1989) suggests deriving the western part of the batholith from a shallow, plagioclase-bearing source in the mantle and the eastern part from a greater depth where garnet occurs. This would explain most of the isotopic and incompatible element data and also allows the bulk chemistries of the two source regions to be similar. For example, the negative europium anomalies found in western rocks (and not found in eastern rocks) indicate plagioclase was involved in their formation. Some of the geochemical data, such as the high $\delta\,^{18}O$ values in the eastern zone (which require a near-surface source for part of the eastern magma), are not explained without some modification of the hypothesis. A possible explanation for the oxygen isotope data would be that subducted sediment was transferred to the mantle and added to the parent magma of the eastern zone. Thus, in this hypothesis, there are two parent magmas instead of one, and contamination of a parent magma, if it occurred, is by subducted sediment rather than by continental crust. (Note, however, that past subduction could have added sediment to the lower crust and upper mantle.)

To fully evaluate these two hypotheses, along with others, we need more information on the composition of the lower crust and subcontinental mantle. Such evaluations will involve the combined use of isotopic, chemical, petrologic, and geologic features of the Peninsular Ranges batholith.

SUMMARY

Basalt is the most abundant of all igneous rocks occurring at or near the surface of the Earth. Other common rocks include andesite, rhyolite, peridotite, gabbro, and granitic rocks. These rocks are composed of 40 to 75 percent SiO_2, 12 to 24 percent Al_2O_3, 0.20 to 49 percent MgO, 1 to 10 percent FeO, 0 to 5 percent K_2O, and 0 to 7 percent Na_2O. A continuous range of composition exists for both volcanic and plutonic rocks. For example, an increase in SiO_2 and K_2O from basalt to rhyolite is paralleled by decreases in Al_2O_3, CaO, FeO, and MgO. Often a series of chemically related rocks occur together. The compositions of some rocks reflect their tectonic environment of formation. For example, some are associated with convergent plate boundaries while others are found in continental rift zones.

Igneous rocks can be classified on the basis of mineralogy or chemical composition. Since volcanic rocks contain small mineral grains and glassy material, they are generally classified using chemical composition. Commonly used classifications are based on alkali content or on a plot of total alkalis versus total silica. Both volcanic and plutonic rocks can be described using the concept of rock saturation, with rocks described as saturated or unsaturated with respect to their content of SiO_2 and Al_2O_3. Suites of related rocks can be chemically described using variation diagrams to plot changes in composition from one rock type to another.

A great deal of information about the history of an igneous rock can be obtained from trace element data. Trace elements are described as compatible (taken into minerals crystallizing from a magma) or incompatible (concentrated in remaining magma as crystallization proceeds). Incompatible elements are also preferentially concentrated in a magma formed by partial fusion of a source rock. The equilibrium distribution of a trace element between liquid and solid phases can be represented by a distribution coefficient. These coefficients vary in value with melt composition and with temperature and pressure. Trace element distribution can be modeled using equations for various types of magmatic crystallization and of partial melting processes. The rare-earth elements have been used extensively to study magma formation and crystallization.

Isotopes, like trace elements, provide very useful data for igneous rock research. Radioactive isotopes and their daughter elements are used in two ways. They provide ages for individual rocks and they provide information on the evolution of the crust and mantle. In particular, isotopic geochemistry can be used to identify the number and nature of magma reservoirs. Some reservoirs in the mantle are depleted in incompatible elements while others are enriched in these elements relative to a reference reservoir (CHUR) representative of chondritic meteorites. Since different parent-daughter pairs are affected differently by any given geologic process, each pair provides unique information. In addition, the elements of some pairs are affected in the same manner by all processes (for example, samarium-neodymium) while the elements of other pairs have different geochemical properties and do not react in the same way to a given process (for example, rubidium-strontium).

The overall composition of the crust is similar to the average composition of igneous rocks. The upper part of the continental crust has a granitic composition, while the lower part and the oceanic crust have a basaltic composition. The upper mantle has a composition similar to that of peridotite. Most magmas form in the zone beneath the Moho, which is characterized by a decrease and then an increase in S-wave velocity with depth. The boundary between the lithosphere and the asthenosphere is in this low-velocity zone. Two major seismic and density discontinuities occur in the mantle at depths of about 410 km and 660 km.

Magmas can form in the mantle by the decompression of ascending mantle material, as a result of the heating of subducted rock, and by the lowering of the melting point of a rock mass due to accumulation of volatiles. Several kinds of primary magma form due to mantle variations in chemical composition, mineral composition, pressure, and temperature. A given primary magma may be affected after formation by a process such as fractional crystallization, resulting in one or more secondary magmas with different compositions. Magmas formed in the crust can have compositions that differ from those of the primary and secondary magmas formed in the mantle.

Laboratory studies of the crystallization of artificial melts have provided much information on igneous rock formation. These studies have shown that fractional crystallization is an important type of magmatic differentiation. This process and the process of partial melting of various parent rocks are responsible for the wide compositional diversity of igneous rocks. Volatile components, particularly H_2O, CO_2, and O_2, strongly affect the crystallization of magmas. Iron- and magnesium-bearing minerals are important in the crystallization of basaltic magmas, while the feldspars are the major minerals formed from granitic melts. Crystallization of the feldspars is complicated by the occurrence of solid solutions that become unstable at lower temperatures and by complex order-disorder relationships. The origins of pegmatites and of hydrothermal ore deposits associated with igneous rocks are not well understood.

The volcanic rocks of Gough Island provide strong evidence for formation by fractional crystallization of a parent basaltic magma. There is a regular and continuous change with rock age in both major and trace element content. In contrast to the Gough Island rocks, the Deccan basalts have a limited range of major element composition. However, they exhibit a large variation in trace element and isotope content. The chemical properties of the rocks can be explained by assimilation of large amounts of mafic lower crust material into the parent magma, along with later contamination episodes that produced further variations in trace element and isotope composition among individual basalt flows.

Research on the Boulder batholith provides an example of the use of radioactive isotopes to date and evaluate the plutons of a composite intrusive mass. The data suggest that two different parent magmas were involved and that assimilation of upper crustal material occurred during formation of the batholith. The Peninsular Ranges batholith is another example of a composite intrusive complex. Two plutonic series that differ in lithologies, structures, ages, and geochemical properties have been mapped. Differences in isotopic and incompatible element chemistry of the two series can be explained as follows. One series formed as a result of extensive crustal contamination of a parent magma. The other series is made up of slightly contaminated parent magma. An alternative hypothesis postulates two different parent magmas.

QUESTIONS

1. Given is a rock with the following composition and mineralogy:

(a) Which one of the rock names listed in Table 8-1 would best fit this rock?

		Mode (wt. percent)		Norm	
SiO_2	43.2				
TiO_2	2.7	Olivine	29.9	Orthoclase	9.6
Al_2O_3	13.4	Augite	24.3	Albite	9.7
Fe_2O_3	4.5	Phlogopite	3.1	Anorthite	16.3
FeO	8.2	Nepheline	2.2	Nepheline	10.6
MnO	0.1	Plagioclase	40.5	Diopside	22.2
MgO	10.8			Olivine	16.8
CaO	9.8	Total	100.0	Magnetite	6.6
Na_2O	3.5			Ilmenite	5.1
K_2O	1.6			Apatite	1.6
P_2O_5	0.7				
H_2O^+	1.2			Total:	98.5
H_2O^-	0.2				
Total:	99.9				

(b) The mode for this rock includes augite and phlogopite but these two minerals are not listed in the norm. Why not?

(c) Classify the rock in terms of silica saturation and in terms of alumina saturation.

2. **(a)** Do rocks #2 and #3 in Table 8-1 plot above or below the line in Figure 8-1?

(b) Do rocks #2 and #3 in Table 8-1 plot above or below the line in Figure 8-2?
(c) Note that rock #3 has a higher content of SiO_2 and a higher value of $^{87}Sr/^{86}Sr$ when compared to rock #2. Suggest a reason for the higher values in rock #3.

3. Given are two rocks with the following amounts of FeO, MgO, and ($K_2O + Na_2O$) in weight percent.

	FeO	MgO	($K_2O + Na_2O$)
Rock #1	6%	8%	4%
Rock #2	3%	4%	2%

(a) What numbers would you use to plot each of these rocks on a triangular variation diagram like those shown in Figure 8-4? To get the numbers you have to recalculate the weight percent values to 100 percent.
(b) Why are the numbers the same for both rocks even though their compositions are different?

4. (a) What would be the value of the bulk distribution coefficient for a perfectly incompatible element?
(b) If the bulk distribution coefficient is 1.0 in a case of batch melting, what would be the ratio of C_L to C_O (concentration of a trace element in melt to concentration of the element in the original solid)?

5. Assume fractional crystallization of plagioclase from a basaltic magma occurs until 20 percent of the liquid has crystallized.
(a) Calculate the ratio C_L/C_O for Ce, Sm, Eu, Gd, and Yb in the residual melt. Use equation #3 in Table 8-5. The distribution coefficients needed are in Table 8-7.
(b) Plot C_L/C_O versus atomic number for these elements. Does Eu show an anomaly in the melt relative to the other rare earths? Why or why not?
(c) Repeat the calculation of part (a), assuming 40 percent of the magma has crystallized. Are the rare earths enriched or depleted in the liquid as the amount of residual melt decreases?

6. Note the large difference in distribution coefficients shown for the rare earths in garnet in Table 8-4. Compare these numbers with those for the rare earths in spinel. How could these numbers be used to obtain information about the source of a magma that later produced basalt at the surface of the Earth?

7. (a) Figure 8-17 shows that MORB basalts are enriched in the heavy REE compared to the light REE. What does this suggest about the source areas of these rocks in the mantle?
(b) Using the data in Figure 8-17, would you expect Hawaiian tholeiites to be enriched or depleted in the oxides K_2O, TiO_2, and P_2O_5 relative to MORB basalts?

8. Given is a basalt whose chondrite-normalized REE pattern shows strong enrichment in the light rare-earth elements (LREE) and a negative europium anomaly. What does this suggest about the mineralogy of the mantle source region of this rock?

9. Assume the existence of two groups of igneous rocks:
(A) A group of geographically related rocks that gives the following plot for rubidium-strontium data:

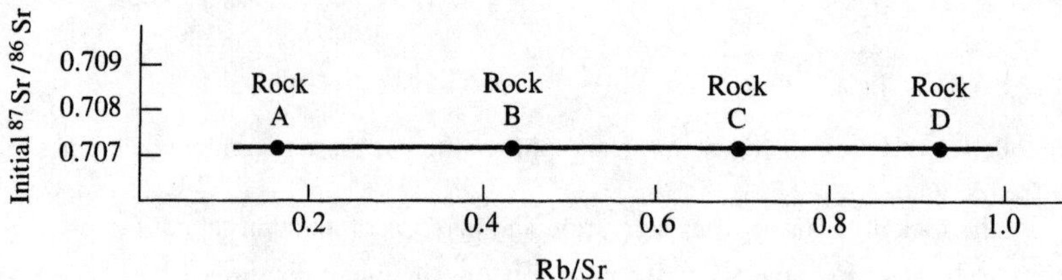

(B) A group of geographically related rocks that gives the following plot for rubidium-strontium data:

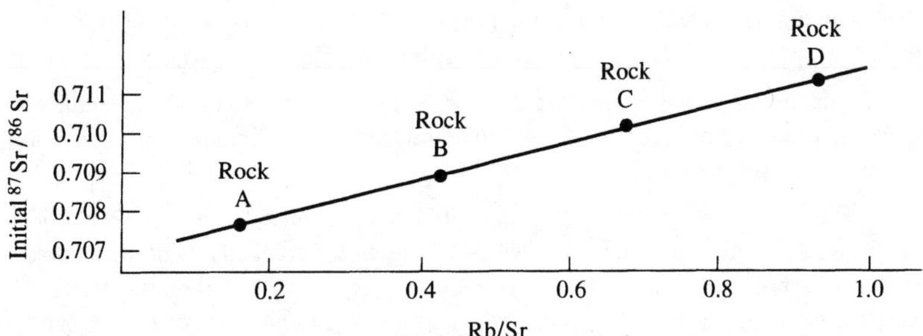

(a) If one of these groups formed as a result of fractional crystallization of a magma, which one is it? Explain your answer.

(b) Suggest an explanation for the origin of the group that did not form by fractional crystallization of a magma.

10. The ratio Sm/Nd in igneous rocks varies from about 0.1 to about 0.4. The ratio Rb/Sr in igneous rocks varies from about 0.06 to about 1.7. Explain why there is a much larger range of values for Rb/Sr.

11. Given is an igneous rock with a positive value of ϵ_{Nd}.

(a) What can you say about the ratio Sm/Nd in the source reservoir of the rock?

(b) Did the rock come from a depleted mantle source reservoir or an enriched mantle source reservoir?

12. **(a)** Using data from Table 8-1, predict whether the oceanic tholeiite basalt came from a depleted or an enriched source. Do the same for the continental tholeiite basalt in Table 8-1.

(b) Using the answers to part (a), which one of the two basalts should have the highest value of Rb/Sr according to Figure 8-11? Which one has the highest according to Table 8-1?

13. Most continental flood basalts have isotopic compositions that plot near the bulk-silicate-earth point in Figure 8-11. Suggest two different interpretations for the source of these rocks using these data.

14. Figure 8-11 shows a negative correlation between Nd isotopes and Sr isotopes in the mantle array. Explain why this occurs.

15. **(a)** In Table 8-8 the lower crust is described as having "retarded Nd evolution" relative to a chondritic source. What does this mean? To answer this question you may want to review the discussion of the Sm-Nd system in Chapter Two.

(b) At the bottom of Table 8-8 is the statement that $^{143}Nd/^{144}Nd$ values are normalized to a certain value of $^{146}Nd/^{144}Nd$. Suggest a reason for this adjustment of analytical values for $^{143}Nd/^{144}Nd$.

16. **(a)** What does Figure 8-14 tell us about the relationship between depth and the chemical composition of magmas produced in the mantle?

(b) Use Figure 8-14 to explain why picrites and komatiites are rare rocks.

17. **(a)** If peridotite rock (olivine plus two pyroxenes) in the mantle with a composition of point X in Figure 8-18 were to undergo partial melting to produce a water-saturated magma at 20 kb, what would be the composition of the first melt produced?

(b) If the liquid formed by partial melting were to be separated from the remaining solid rock, and then crystallizes elsewhere, what minerals would be formed?

18. Use Figure 8-23a to discuss the cooling history of a melt that has a composition (point Y in Figure 8-23a) such that leucite is the first mineral to form. Note that leucite has a reaction relationship to alkali feldspar (see Figure 8-21a) similar to that between olivine and pyroxene in Figure 8-18. Discuss two different situations: (a) fractional crystallization does not occur, and (b) fractional crystallization does occur as a result of separation of early formed leucite.

19. **(a)** What point on Figure 8-22a is equivalent to point Z on Figure 8-21a?
(b) What point on Figure 8-22b is equivalent to point T on Figure 8-21b?

20. Perthites of volcanic rocks tend to be cryptoperthites (optically homogeneous, identified only by X-ray techniques); perthites of plutonic rocks tend to be microperthites (optically recognizable). Suggest an explanation for this difference.

21. **(a)** If a magma were made up in weight percent of 32 percent SiO_2, 32 percent $NaAlSi_3O_8$, 32 percent $KAlSi_3O_8$, and 4 percent H_2O, what would be the mole percent H_2O in the magma?
(b) The chemical potential of a component such as water in a magma is directly related to its mole fraction in the magma. Based on your answer to part (a), would you expect H_2O to have a high chemical potential in most magmas?
(c) A rapid decrease in the pressure acting on a magma can cause rapid crystallization of silicate minerals and thus a fine-grained texture. Explain this in terms of the water content of the magma.

22. The Salton Sea geothermal brine of Table 8-12 may be an example of a hydrothermal solution that could form an ore deposit. It contains 540 parts per million zinc.
(a) How much of this water would be required to deposit 100 metric tons of zinc? Assume that all the zinc in the water is precipitated.
(b) If the water passed through a vein at a rate of 10 liters per second, how long would it take to deposit 100 metric tons of zinc?

23. **(a)** Which one of the following igneous rocks would you expect to have a higher ratio of deuterium to hydrogen: (1) a rock that during its formation interacted only with magmatic water, or (2) a rock that during its formation interacted extensively with meteoric water?
(b) Which of these two rocks would you expect to have a higher ratio of ^{18}O to ^{16}O?

24. Table 8-13 gives the compositions of a group of rocks that are believed to have formed by fractional crystallization of basaltic magma.
(a) Why does the Al_2O_3 content of the rocks increase from early to late-formed rocks?
(b) Why does the rubidium content of the rocks show a general increase from early to late-formed rocks?
(c) Are any of the rocks of Table 8-13 saturated or oversaturated with respect to SiO_2 content?

25. It has been suggested that the chemical differences between the western and eastern zones of the Peninsular Ranges batholith are due to contamination of a primary basaltic magma by crustal material with uncontaminated magma forming the western zone and contaminated magma forming the eastern zone. Do the rare-earth data in Table 8-15 support this hypothesis?

REFERENCES

BARNES, H. L., ed. 1979. *Geochemistry of Hydrothermal Ore Deposits,* 2d ed. John Wiley & Sons, Inc., New York.

BERGER, B. R., and P. M. BETHKE, eds. 1985. *Geology and Geochemistry of Epithermal Systems.* Rev. in Economic Geology, v. 2, Society of Economic Geologists, El Paso, TX.

BEST, M. G. 1982. *Igneous and Metamorphic Petrology.* W. H. Freeman and Company, New York.

BOWEN, N. L. 1928. *The Evolution of the Igneous Rocks.* Princeton University Press, Princeton, NJ.

BOWEN, N. L., and O. F. TUTTLE. 1950. The system $NaAlSi_3O_8$-$KAlSi_3O_8$-H_2O. *J. Geol.,* v. 58: 489–511.

BRIMHALL, G. H., and D. A. CRERAR. 1987. Ore fluids: magmatic to supergene. In *Thermodynamic Modeling of Geological Materials: Minerals, Fluids, and Melts,* Rev. in Mineralogy, v. 17, I. S. E. Carmichael and H. P. Eugster, eds. Mineralogical Soc. Am., Washington, D.C. 235–321.

Brown, G. C., and A. E. Mussett. 1993. *The Inaccessible Earth,* 2d ed. Chapman & Hall, London.

Buddington, A. F., and D. H. Lindsley. 1964. Iron-titanium oxide minerals and synthetic equivalents. *J. Petrol.,* v. 5:310–357.

BVSP (Basaltic Volcanism Study Project). 1981. *Basaltic Volcanism on the Terrestrial Planets.* Pergamon Press, Inc., New York.

Carmichael, I. S. E., and M. S. Ghiorso. 1990. The effect of oxygen fugacity on the redox state of natural liquids and their crystallizing phases. In *Modern Methods of Igneous Petrology: Understanding Magmatic Processes,* Rev. in Mineralogy, v. 24, J. Nicholls and J. K. Russell, eds. Mineralogical Soc. Am., Washington, D.C. 191–212.

Carroll, M. R., and J. R. Holloway, eds. 1994. *Volatiles in Magmas.* Rev. in Mineralogy, v. 30. Mineralogical Soc. Am., Washington, D.C.

Clarke, D. B. 1992. *Granitoid Rocks.* Chapman & Hall, New York.

Coffin, M. F., and O. Eldholm. 1994. Large igneous provinces: crustal structure, dimensions, and external consequences. *Rev. Geophys.,* v. 32:1–36.

Cohen, L. H., K. Ito, and G. C. Kennedy. 1967. Melting and phase relations in an anhydrous basalt to 40 kilobars. *Am. J. Sci.,* v. 265:475–518.

Condie, K. C. 1982. *Plate Tectonics & Crustal Evolution,* 2d ed. Pergamon Press, Inc., New York.

Cox, K. G., J. D. Bell, and R. J. Pankhurst. 1979. *The Interpretation of Igneous Rocks.* George Allen & Unwin, London.

Cross, C. W., J. P. Iddings, L. V. Pirsson, and H. S. Washington. 1902. A quantitative chemico-mineralogical classification and nomenclature of igneous rocks. *J. Geol.,* v. 10:555–690.

DePaolo, D. J. 1981. A neodymium and strontium isotopic study of the Mesozoic calc-alkaline granitic batholiths of the Sierra Nevada and Peninsular Ranges, California. *J. Geophys. Res.,* v. 86:10,470–10,488.

DePaolo, D. J., and G. J. Wasserburg. 1979. Petrogenetic mixing models and Nd-Sr isotopic patterns. *Geochim. Cosmochim. Acta,* v. 43:615–627.

Doe, B. R., R. I. Tilling, C. E. Hedge, and M. R. Klepper. 1968. Lead and strontium isotope studies of the Boulder batholith, southwest Montana. *Econ. Geol.,* v. 63:884–906.

Eggler, D. H. 1978. the effect of CO_2 upon partial melting of peridotite in the system Na_2O-CaO-Al_2O_3-MgO-SiO_2-CO_2 to 35 kb, with an analysis of melting in a peridotite-H_2O-CO_2 system. *Am. J. Sci.,* v. 278:305–343.

Eugster, H. P. 1986. Minerals in hot water. *Am. Min.,* v. 71:655–673.

Evensen, N. M., P. J. Hamilton, and R. K. O'Nions. 1978. Rare earth abundances in chondritic meteorites. *Geochim. Cosmochim. Acta,* v. 42:1199–1212.

Gast, P. W., G. R. Tilton, and C. E. Hedge. 1964. Isotopic composition of lead and strontium from Ascension and Gough Islands. *Science,* v. 145:1181–1185.

Ghiorso, M. S., and R. O. Sack. 1991. Fe-Ti oxide geothermometry: thermodynamic formulation and the estimation of intensive variables in silicic magmas. *Contrib. Min. Petrol.,* v. 108:485–510.

Green, T. H., and N. J. Pearson. 1986. Rare-earth element partitioning between sphene and coexisting silicate liquid at high pressure and temperature. *Chem. Geol.,* v. 55:105–119.

Gromet, L. P., and L. T. Silver. 1987. REE variations across the Peninsular Ranges batholith: implications for batholithic petrogenesis and crustal growth in magma arcs. *J. Petrol.,* v. 28:75–125.

Hanson, G. N. 1989. An approach to trace element modeling using a simple igneous system as an example. In *Geochemistry and Mineralogy of Rare Earth Elements,* Rev. in Mineralogy, v. 21, B. R. Lipin and G. A. McKay, eds. Mineralogical Soc. Am., Washington, D.C. 79–97.

HART, S. R., and C. J. ALLEGRE. 1980. Trace-element constraints on magma genesis. In *Physics of Magmatic Processes,* R. B. Hargraves, ed. Princeton University Press, Princeton, NJ. 121–159.

HENDERSON, P. 1982. *Inorganic Geochemistry.* Pergamon Press, Inc., New York.

HENDERSON, P., ed. 1984. *Rare Earth Element Geochemistry.* Elsevier, Amsterdam.

HESS, P. C. 1989. *Origins of Igneous Rocks.* Harvard University Press, Cambridge, MA.

HYNDMAN, D. W. 1972. *Petrology of Igneous and Metamorphic Rocks.* McGraw-Hill Book Company, New York.

HYNDMAN, D. W. 1985. *Petrology of Igneous and Metamorphic Rocks,* 2d ed. McGraw-Hill Book Company, New York.

IRVING, T. N., and W. R. A. BARAGAR. 1971. A guide to the chemical classification of the common volcanic rocks. *Canad. J. Earth Sci.,* v. 8:523–548.

JAHNS, R. H., and C. W. BURNHAM. 1969. Experimental studies of pegmatite genesis: pt. 1, a model for the derivation and crystallization of granitic pegmatites. *Econ. Geol.,* v. 64:843–864.

JOHNSON, M. C., A. T. ANDERSON, JR., and M. J. RUTHERFORD. 1994. Pre-eruptive volatile contents of magmas. In *Volatiles in Magmas,* Rev. in Mineralogy, v. 30, M. R. Carroll and J. R. Holdaway, eds. Mineralogical Soc. Am., Washington, D.C. 281–330.

KIRBY, S. H., W. B. DURHAM, and L. A. STERN. 1991. Mantle phase changes and deep earthquake faulting in subducting lithosphere. *Science,* v. 252:216–225.

KUSHIRO, I. 1969. The system forsterite-diopside-silica, with and without water at high pressures. *Am. J. Sci.,* v. 267A (Schairer volume):269–294.

KUSHIRO, I. 1972. Determination of liquidus relations in synthetic silicate systems with electron probe analysis: the system forsterite-diopside-silica at 1 atmosphere. *Am. Min.,* v. 57:1260–1271.

KUSHIRO, I. 1979. Fractional crystallization of basaltic magma. In *The Evolution of the Igneous Rocks—Fiftieth Anniversary Perspectives,* H. S. Yoder, Jr., ed. Princeton University Press, Princeton, NJ. 171–203.

LE BAS, M. J., R. W. LE MAITRE, A. STRECKEISEN, and B. ZANETTIN. 1986. A chemical classification of volcanic rocks based on the total alkali-silica diagram. *J. Petrol.,* v. 27:745–750.

LE MAITRE, R. W. 1962. Petrology of volcanic rocks, Gough Island, South Atlantic. *Bull. Geol. Soc. Am.,* v. 73:1309–1340.

LONDON, D. 1992. The application of experimental petrology to the genesis and crystallization of granite pegmatites. *Can. Min.,* v. 30:499–540.

MACDONALD, G. A., and T. KATSURA. 1964. Chemical composition of Hawaiian lavas. *J. Petrol.,* v. 5:82–133.

MAHONEY, J. J. 1988. Deccan Traps. In *Continental Flood Basalts,* J. D. MacDougall, ed. Kluwer Academic Publishers, Norwell, MA. 151–194.

MCBIRNEY, A. R. 1993. *Igneous Petrology,* 2d ed. Jones and Bartlett Publishers, Boston.

MORSE, S. A. 1970. Alkali feldspars with water at 5 kb pressure. *J. Petrol.,* v. 11:221–251.

MORSE, S. A. 1980. *Basalts and Phase Diagrams.* Springer-Verlag, New York.

PEACOCK, M. A. 1931. Classification of igneous rock series. *J. Geol.,* v. 39:54–67.

PENG, Z. X., J. MAHONEY, P. HOOPER, C. HARRIS, and J. BEANE. 1994. A role for lower continental crust in flood basalt genesis? Isotopic and incompatible element study of the lower six formations of the western Deccan Traps. *Geochim. Cosmochim. Acta,* v. 58:267–288.

PHILPOTTS, A. R. 1990. *Principles of Igneous and Metamorphic Petrology.* Prentice Hall, Englewood Cliffs, NJ.

RAGLAND, P. C. 1989. *Basic Analytical Petrology.* Oxford University Press, New York.

RICHARDSON, S. M., and H. Y. MCSWEEN, JR. 1989. *Geochemistry—Pathways and Processes.* Prentice Hall, Englewood Cliffs, NJ.

RINGWOOD, A. E. 1962. A model for the upper mantle. *J. Geophys. Res.,* v. 67:857–867.

ROBERTSON, J. M., ed. 1984. *Fluid-Mineral Equilibria in Hydrothermal Systems.* Rev. in Econ. Geol., v. 1. Society of Economic Geologists, El Paso, Texas.

ROLLINSON, H. R. 1993. *Using Geochemical Data: Evaluation, Presentation, Interpretation.* John Wiley & Sons, Inc., New York.

SCHAIRER, J. F. 1950. The alkali-feldspar join in the system $NaAlSi_3O_8$-$KAlSi_3O_8$-SiO_2. *J. Geol.,* v. 58: 512–517.

SCLATER, J. G., C. JAUPART, and D. GALSON. 1980. The heat flow through oceanic and continental crust and the heat loss of the earth. *Rev. Geophys. Space Phys.,* v. 18:269–311.

SHAND, S. J. 1951. *Eruptive Rocks,* 4th ed. John Wiley & Sons, Inc., New York.

SHIEH, Y., and G. ZHANG. 1991. Stable isotope studies of quartz-vein type tungsten deposits in Dajishan Mine, Jiangxi Province, southeast China. In *Stable Isotope Geochemistry: A Tribute to Samuel Epstein,* Spec. Pub. No. 3, H. P. Taylor, Jr., J. R. O'Neil, and I. R. Kaplan, eds. The Geochemical Society, San Antonio, TX.

SILVER, L. T., H. P. TAYLOR, JR., and B. CHAPPELL. 1979. Some petrological, geochemical, and geochronological observations of the Peninsular Ranges batholith near the International border of the U.S.A. and Mexico. In *Mesozoic Crystalline Rocks,* P. L. Abbott and V. R. Todd, eds. San Diego State University, San Diego, CA. 83–110.

SKINNER, B. J., and P. B. BARTON. 1973. Genesis of mineral deposits. In *Ann. Rev. Earth Planetary Sciences,* v. 1, F. A. Donath, ed. Annual Reviews, Palo Alto, CA. 183–212.

STRECKEISEN, A. 1976. To each plutonic rock its proper name. *Earth Sci. Rev.,* v. 12:1–33.

TAYLOR, S. R., and S. M. MCLENNAN. 1985. *The Continental Crust: its Composition and Evolution.* Blackwell Scientific Publications, Oxford.

TILLING, R. I. 1973. Boulder batholith, Montana: A product of two contemporaneous but chemically distinct magma series. *Bull. Geol. Soc. Am.,* v. 84:3879–3900.

TILLING, R. I., M. R. KLEPPER, and J. D. OBRADOVICH. 1968. K-Ar ages and time span of emplacement of the Boulder batholith, Montana. *Am. J. Sci.,* v. 266:671–689.

TODD, V. R., and S. E. SHAW. 1979. Structural, metamorphic, and intrusive framework of the Peninsular Ranges batholith in southern San Diego County, California. In *Mesozoic Crystalline Rocks,* P. L. Abbott, and V. R. Todd, eds. San Diego State University, San Diego, CA. 177–231.

TUTTLE, O. F., and N. L. BOWEN. 1958. Origin of granite in the light of experimental studies in the system $NaAlSi_3O_8$-$KAlSi_3O_8$-SiO_2-H_2O. *Geol. Soc. Am.* Memoir 74.

WALKER, K. R. 1969. The Palisades sill, New Jersey: A reinvestigation. Geol. Soc. Am. Special Paper no. 111.

WASSERBURG, G. J., S. B. JACOBSEN, D. J. DEPAOLO, M. T. MCCULLOCH, and T. WEN. 1981. Precise determination of Sm/Nd ratios, Sm and Nd isotopic abundances in standard solutions. *Geochim. Cosmochim. Acta,* v. 45:2311–2323.

WHITE, D. E., and G. A. WARING. 1963. Volcanic emanations, U.S. Geol. Survey Profess. Paper 440–K.

WHITNEY, J. A. 1988. The origin of granite: The role and source of water in the evolution of granitic magmas. *Bull. Geol. Soc. Am.,* v. 100:1886–1897.

WHITTAKER, E. J. W., and R. MUNTUS. 1970. Ionic radii for use in geochemistry. *Geochim. Cosmochim. Acta,* v. 34:945–956.

WILCOX, R. E. 1979. The liquid line of descent and variation diagrams. In *The Evolution of the Igneous Rocks—Fiftieth Anniversary Perspectives,* H. S. Yoder, Jr., ed. Princeton University Press, Princeton, NJ. 205–232.

Yoder, H. S., Jr. 1973. Contemporaneous basaltic and rhyolitic magmas. *Am. Mineralogist,* v. 58:153–171.

Zielinski, R. A., and F. A. Frey. 1970. Gough Island: evaluation of a fractional crystallization model. *Contrib. Mineral. Petrol.,* v. 29:242–254.

Zindler, A., and S. R. Hart. 1986. Chemical Geodynamics. In *Ann. Rev. of Earth Planetary Sciences,* v. 14, G. W. Wetherill, ed., Annual Reviews, Palo Alto, CA. 493–571.

Metamorphic Rocks

This chapter begins with a brief review of the origin and classification of metamorphic rocks, followed by a discussion of the variables controlling their formation. Next comes a discussion of the nature of metamorphic reactions, with emphasis on the importance of kinetics. Modern studies of metamorphism are often based on the use of geothermometry, geobarometry, and stable isotope techniques. These important tools are described and then applied in the rest of the chapter. The following sections deal with metasomatism, metamorphic facies, and the relationship of metamorphism to plate tectonics. The mineralogy of common metamorphic rocks is then described, including a discussion of the use of diagrams such as AFM graphs and petrogenetic grids. Finally, use of the various geochemical techniques described in the chapter is illustrated for two aureoles of contact metamorphism and for two areas of regional metamorphism.

ORIGIN AND CLASSIFICATION

Metamorphic rocks result from rock transformation without formation of a melt under pressure-temperature conditions that are not typical of the Earth's surface.* *Contact metamorphism,* which is spatially related to igneous bodies, can occur at the surface of the Earth, although most contact metamorphism appears to have taken place at shallow depths in the crust. *Regional metamorphism* involves the production of altered rocks, on a large scale, from near the Earth's surface to within the upper mantle. Changes occurring near the Earth's surface result from the burial of sediments and volcanic deposits at depths as shallow as 1 kilometer. This very low grade *burial*

*The intensity or degree of metamorphism is described in a general way by the use of the term *grade.* Thus high-grade rocks contain minerals that formed under conditions of relatively high temperature and/or pressure.

metamorphism passes into diagenesis at the Earth's surface. At the other extreme are the changes produced by tectonic movements and high temperatures and pressures in the lower crust and upper mantle. Some of the results of this high-grade metamorphism are found at the surface today as a result of deep erosion and tectonic processes.

Metamorphism may occur without change in the chemical composition of the affected rocks (*isochemical metamorphism*), or chemical changes can be produced by migrating fluids (*allochemical metamorphism*). If the only change is a loss of volatiles, the metamorphism is usually considered to be isochemical. Comparisons of the average composition of shales with that of various rocks formed by the metamorphism of shale suggest that, in many cases, regional metamorphism is approximately isochemical. The only significant change that generally occurs is an overall loss of fluids (particularly H_2O and CO_2). Note, however, that fluids and fluid flow may play an important role in contact metamorphism (Ferry 1991). Changes may occur in the isotopic composition of elements such as oxygen during both regional and contact metamorphism (see later discussion). It should also be kept in mind that internal migration of various elements is not ruled out by the overall constancy of composition of a mass of rock undergoing metamorphism.

In contrast with most regional metamorphism, contact metamorphism may involve substantial addition or subtraction of material. The term *metasomatism* refers to such changes in composition. The extent of metasomatism is mainly determined by the composition and permeability of the affected rock. The existence of a large difference in composition between invading magma and country rock, along with tectonic fracturing of the country rock, can result in extensive chemical changes. The greatest changes occur when limestones and marbles are metamorphosed. Because of the nature of contact metamorphism, it is much easier to prove the occurrence of metasomatic changes for contact metamorphic rocks than it is for regional metamorphic rocks.

Any pre-existing igneous, sedimentary, or metamorphic rock can be altered by metamorphism. The changes produced are both mineralogical and textural (and chemical in the case of allochemical metamorphism). Because of the wide variety of parent rocks and of conditions of alteration, together with the many possible changes in mineralogy and texture, this group of rocks is difficult to classify in any systematic way. Classifications and rock names may be based on texture, mineral composition, chemical composition, or assumed history (parent rock and conditions of metamorphism). Texture and mineralogy are most important.

The characteristics of the common metamorphic rocks are summarized in Table 9-1. Slate, schist, gneiss, and hornfels are defined on the basis of their texture; other rocks of Table 9-1 are defined mainly on the basis of their mineralogy. In terms of texture, metamorphic rocks are either foliated (containing some type of planar structure) or nonfoliated (massive). The foliated rocks may have layers or lenses of contrasting mineralogy (*gneissosity*), preferred orientation of platy minerals (*schistosity*), or planar fracture surfaces (*rock cleavage*). Many metamorphic rock types are typically coarse grained. Exceptions are slate, hornfels, and serpentinite.

Common metamorphic rocks are divided into five chemical classes: (1) pelitic (formed from clay-rich sedimentary rocks), (2) quartzo-feldspathic (formed from sandstones and felsic igneous rocks), (3) calcareous (formed from limestones and dolomites), (4) mafic (formed from mafic igneous rocks), and (5) magnesian (formed from ultramafic igneous rocks). The most abundant rocks are schist and gneiss, which can form from practically any type of parent rock. It is interesting to note that eclogite is both an igneous and a metamorphic rock. Experimental work has shown that, under the very high pressure conditions existing in the deeper parts of the upper mantle, eclogite will crystallize from basaltic magma.

TABLE 9-1 Common Metamorphic Rocks

Name	Typical minerals	Texture	Parent rock
Slate	Muscovite, chlorite, quartz	Fine-grained with well-developed rock cleavage	Pelitic
Schist	Micas, amphiboles, chlorite, quartz	Coarse-grained and schistose	Pelitic, mafic, calcareous
Gneiss	Quartz, feldspars, biotite, hornblende	Coarse-grained and banded	Various
Hornfels	Quartz, micas, feldspars, amphiboles, pyroxenes	Fine-grained and nonfoliated	Pelitic, mafic, calcerous
Quartzite	Quartz	Coarse-grained and nonfoliated	Quartzo-feldspathic
Amphibolite	Hornblende, plagioclase	Coarse-grained, banded or schistose	Calcareous, mafic
Serpentinite	Serpentine	Fine-grained and nonfoliated	Magnesian
Granulite	Quartz, feldspars, pyroxenes	Coarse-grained and semibanded	Quartzo-feldspathic, pelitic
Eclogite	Garnet, omphacite	Coarse-grained and nonfoliated	Mafic

VARIABLES OF METAMORPHISM

The final product of the metamorphism of a given parent rock depends on the variables temperature, load pressure, fluid pressure, and time. The most important is temperature. Metamorphic temperatures may be due to (1) the presence of magma, (2) burial by sediments, (3) a locally high heat flow from radioactive decay, (4) tectonic friction, or (5) conduction from, and convective overturn in, the mantle. Assuming no effect due to increased pressure, the lowest temperature at which metamorphic reactions (as distinct from reactions that can occur in a sedimentary environment) could take place is about 150°C. The upper limit is the temperature at which extensive melting occurs. Partial melting produces migmatites (mixed igneous-metamorphic rocks). This upper limit, of course, is different for different rocks and different pressure conditions. A reasonable figure for this limit is 900°C. Most metamorphism probably occurs between 300 and 800°C.

Heat-flow values for the continents and oceans indicate that the temperature at the crust-mantle boundary (Mohorovicic discontinuity) varies from about 200°C under some parts of the oceans to about 1, 000°C under the thickest parts of the continents. Significant lateral variations in temperature probably occur at any given depth in the Earth. For example, temperatures at the Mohorovicic discontinuity under midoceanic ridges (which have a high heat flow) are probably much higher than at the same depth under other parts of the oceans. Because heat-flow values show a large variation from one geological environment to another, we know that the lower temperature ranges of regional metamorphism occur not only at shallow depths in sedimentary piles, but also at much greater depths near the bottom of the oceanic crust. The higher temperature ranges occur in the lower continental crust and in the upper mantle.

Shallow magmatic intrusion can raise temperatures in the adjacent country rock to values as high as 700°C for basaltic magmas and up to 550°C for granitic magmas (Figure 9-1). The

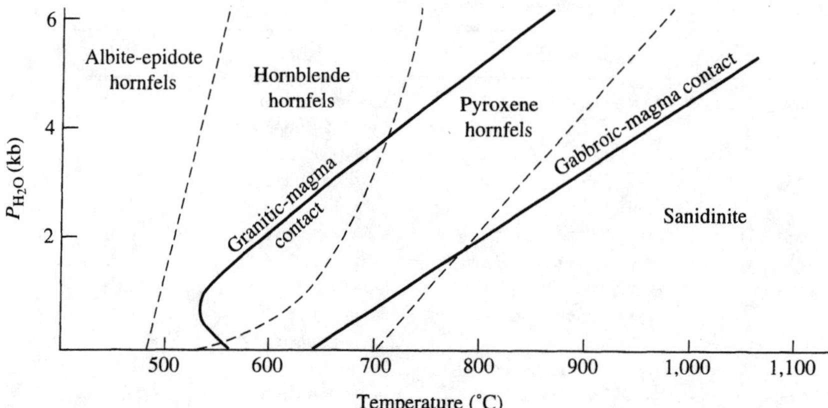

Figure 9-1 Approximate maximum temperatures (solid lines) in wet country rocks (e.g., porous rocks containing water of any origin) immediately adjacent to granitic and gabbroic rock contacts. Assumptions are heat transfer by conduction (heat in moving fluids neglected), magma is 100 percent melt on intrusion, and geothermal gradient is 20°C/km. Estimated boundaries for the four contact metamorphic facies are shown by dashed lines. It is clear that the type of magma and the depth of intrusion will both affect the grade of metamorphism. For example, rocks of the pyroxene hornfels facies would occur only at contacts with granitic rocks where intrusion has occurred at depths greater than about 15 km (4 kbar pressure). After Hyndman (1985, 511).

size of the intrusive is important, with large contact aureoles produced by large intrusives. For small intrusives, temperatures decrease rapidly away from the contact and a small (thin) aureole is formed. At greater depths, the initial temperature of the country rock is higher and wider contact aureoles are formed. Beyond a certain depth, contact metamorphism merges into regional metamorphism.

We have to deal with several types of pressure in studying metamorphic rocks. The most significant pressure is that due to overlying rocks (*load pressure*). This increases at a rate of about 275 bars per kilometer for rocks of average density. The probable load pressure at the Mohorovicic discontinuity thus varies from about 2 kilobars for oceanic regions to about 10 kilobars under the continents. The load pressures of regional metamorphism are estimated to cover about the same range. Contact metamorphism can occur at any load pressure from the Earth's surface to depths where it can no longer be distinguished from regional metamorphism. Load (lithostatic) pressure may be accompanied by a directed pressure. The existence of a shearing stress apparently has little or no effect on the final mineralogy of metamorphic rocks. However, this shearing stress does initiate and accelerate metamorphic reactions by increasing grain contact, rock permeability, and amount of fluid movement. It also plays a major role in the development of foliation and other structural features.

During most metamorphic reactions, a fluid phase is present. The pressure of the fluid phase may be greater than, equal to, or less than the load pressure.* Near the Earth's surface, fluid in fractures open to the surface would have a pressure determined by the height and density of the fluid. In other cases, where fluids are released by metamorphic reactions in rocks without exten-

*When the fluid pressure acts independently from the load pressure, the phase rule (discussed in Chapter Three) becomes $f = c + 3 - p$, since there are two pressure variables.

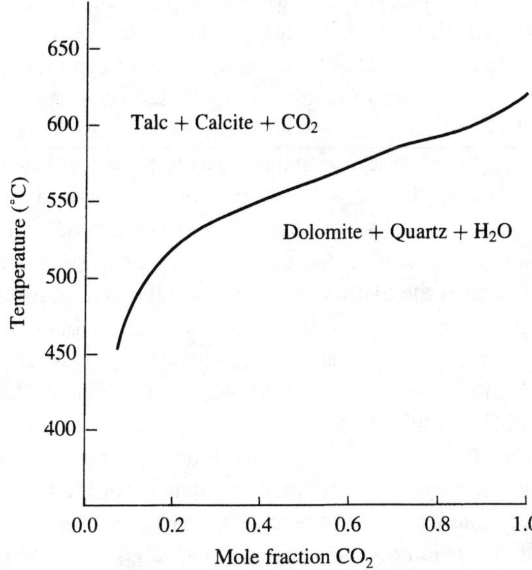

Figure 9-2 Temperature-fluid composition diagram for the equilibrium 3 $CaMg(CO_3)_2$ (dolomite) + 4 SiO_2 (quartz) + $H_2O \rightleftharpoons$ $Mg_3Si_4O_{10}(OH)_2$ (talc) + 3 $CaCO_3$ (calcite) + 3 CO_2. This reaction goes to the right during the metamorphism of siliceous dolomites, with the temperature of the reaction determined by the composition of the fluid phase present. Total fluid pressure ($P_{H_2O} + P_{CO_2}$) equals 5 kbar. After Winkler (1976, 121).

sive fracturing, the fluid pressure may exceed the load pressure. For most situations, fluid pressure probably equals or is less than the load pressure, and the fluid consists essentially of H_2O. In any case, the fugacity of H_2O is a major factor affecting metamorphic reactions [see, for example, Ferry and Baumgartner (1987) and Labotka (1991)]. When metamorphism of carbonate rocks occurs, the fugacity of CO_2 becomes important (see Figure 3-1 for an example). When both H_2O and CO_2 occur in the fluid phase, the temperature of a given reaction depends on the relative abundance of the two components (Figure 9-2).

Another component of the fluid phase that may play a significant role is O_2 (Frost 1991). The fugacity of oxygen is also important for rocks that do not have a fluid phase present (despite extremely low values that are without physical reality). Oxygen fugacity particularly affects the mineralogy of rocks containing large amounts of iron. High oxygen fugacities result in the formation of phases such as magnetite and epidote accompanied by magnesium-rich varieties of the Fe^{2+}-Mg silicates such as biotite. More iron-rich Fe^{2+}-Mg silicates result from low oxygen fugacities. Chemical analyses of iron-bearing minerals can be used to calculate the oxygen fugacities that existed at the time of their formation. A general guide to oxygen fugacity is provided by the iron oxides found in metamorphic rocks (Figure 3-2). The presence of hematite indicates a relatively high f_{O_2}. When magnetite occurs with hematite, f_{O_2} is fixed for any given temperature of metamorphism (load pressure has little effect on the reactions represented in Figure 3-2). Although H_2O and CO_2 are believed to be mobile components (see Chapter Three) in most cases of metamorphism, O_2 appears to be immobile, with f_{O_2} controlled by the composition of the rocks being altered (Rumble 1976).

The presence or absence of carbonaceous material in the rocks being altered is important because of its effect on oxygen and carbon dioxide fugacities. Also, other gas phases, such as CH_4, may form and become part of the fluid phase. A decrease in Fe_2O_3/FeO of metamorphic rocks with increasing grade of metamorphism is probably due to reactions that use up oxygen, such as those that can affect any organic matter present.

The parameter most difficult to treat is time. The duration of a particular metamorphic event, the sequence of formation of rock types, the time relationships of associated igneous rocks, and similar questions are usually difficult to answer. Values of radiometric ages for a set of metamorphic rocks represent different episodes in the overall history of the rocks. Some ages date the time of formation of parent igneous rocks or of high-grade metamorphic rocks affected only by later low-grade metamorphism. Other ages give the time of major metamorphic events. Still others reflect the late stages of metamorphism when temperatures became low enough to "lock in" argon and strontium. Most regional metamorphic rocks have been affected by a number of metamorphic episodes, resulting in a complex history and a confusing pattern of radiometric ages. However, in some cases, it is possible to unravel the history of a metamorphic belt by careful use of several different radiometric techniques, along with stratigraphic and other methods of determining relative ages. Some applications of radiometric techniques to metamorphic rocks can be found in Chapter Two. The $^{40}Ar/^{39}Ar$ method is one that has been used a great deal in the study of metamorphic terranes (McDougall and Harrison 1988).

The pressure-temperature conditions that are imposed in metamorphic terranes vary over time. Metamorphic petrologists describe the history of the pressure-temperature changes as a *pressure-temperature-time path* (Haugerud and Zen 1991; Spear 1993). An example is shown in Figure 9-3. The mineral assemblage seen in a metamorphic rock today (which is metastable at the Earth's surface) probably represents the maximum pressure-temperature conditions to which it was subjected. In some cases, the rock's pressure-temperature-time path can be identified by the

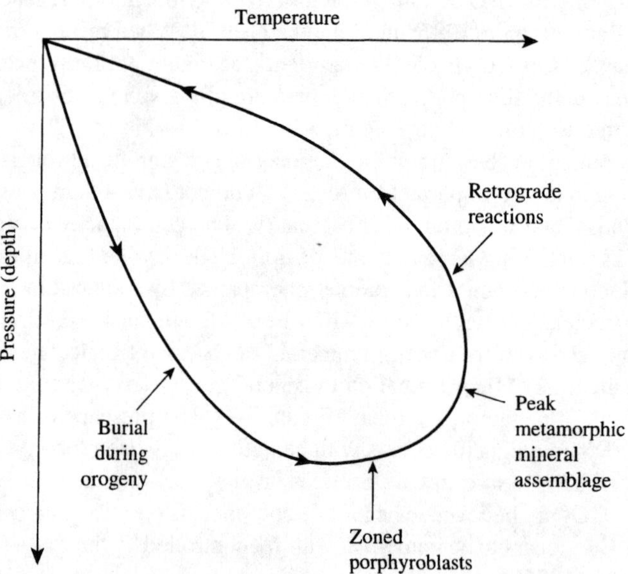

Figure 9-3 Pressure-temperature-time path that a sedimentary rock might follow as a result of burial during orogenesis, followed by metamorphism, and eventually uplift due to erosion. The mineral assemblage that is likely to be preserved is developed near the maximum temperature. Zoned porphyroblasts may preserve a record of conditions just prior to the peak temperatures, and retrograde minerals may develop as rims on the high-temperature minerals during cooling. (After Philpotts, A. R., *Principles of Igneous and Metamorphic Petrology*, p. 413. Copyright © 1990 by Prentice Hall, Inc. Adapted by permission of Prentice Hall, Inc., Upper Saddle River, NJ.)

study of zoned minerals present, identification of incomplete mineral reactions (shown by some minerals forming rims around others), and by interpreting relict textures. Ridley and Thompson (1986) discuss how rock textures may give information about pressure-temperature-time paths of metamorphism. More often, textural and other features are not helpful, and only the mineral assemblage formed at the highest grade of metamorphism is found. This assemblage represents only one point on a rock's pressure-temperature-time path.

NATURE OF METAMORPHIC REACTIONS

Rocks that are undergoing metamorphism change in ways such that they approach chemical and physical equilibrium for the prevailing pressure-temperature conditions. The chemical reactions that occur are governed by the requirements of thermodynamics (Chapter Three). For example, reactions in which water vapor is produced tend to take place, since these reactions usually have negative values for ΔG (because of the large entropy of water vapor). The regularity of metamorphic assemblages and the successful application of thermodynamic principles to many metamorphic rocks suggest that overall chemical equilibrium is closely approached in many cases. However, in detail, the only true equilibrium is often a local equilibrium, measured in centimeters or meters. Electron-probe analyses have shown that even individual mineral grains may have local zones of equilibrium. Thus the thermodynamic system to be studied can range from part of one mineral grain to a hand specimen to a rock mass whose volume is measured in cubic kilometers. Any given system may or may not have been an open system when metamorphism occurred, and equilibrium may or may not have been reached in that system. The problem of determining the extent of chemical equilibrium in rocks has been summarized very well by Zen (1963). A specific study of the spatial extent of chemical equilibrium in some high-grade metamorphic rocks is given by Blackburn (1968).

Lasaga and Rye (1993) suggest that many metamorphic processes involve reactions between mineral surfaces and moving fluids. For a set of reacting minerals and a coexisting, moving fluid, the fluid composition can reach a steady state (i.e., production of various components by all processes equals consumption of these components) that in general will not be the same as the equilibrium composition (Figure 9-4). Thus equilibrium between rocks and moving fluids should not be assumed and the kinetics of pertinent chemical reactions must be considered. This is particularly true for metasomatism, which involves definite changes in composition produced by the flow of fluids through rocks.

The rates of metamorphic reactions are discussed in Fyfe et al. (1979, Chap. 5). The authors note that most mineral reactions proceed by a series of complex consecutive steps, with one of the steps being slower than the others and thus the rate-controlling step. For metamorphic processes, we would like to know the rates of mineral dissolution in aqueous fluids, the various rates of diffusion (of species through aqueous fluids, along grain boundaries, and through mineral lattices), the rates of mineral nucleation and growth, the rates of hydration and dehydration reactions, and the rates of solid-solid reactions. We do not know much about these rates under conditions of high temperature and pressure. Fyfe et al. point out that the presence of an aqueous fluid and of shearing stress have a strong catalytic effect on reaction rates.

A common type of rate-controlling step is that of surface reactions at mineral-mineral and mineral-fluid interfaces. Other possible rate-controlling steps involve lattice diffusion and mineral nucleation. The nucleation and growth of metamorphic minerals are not well understood, but

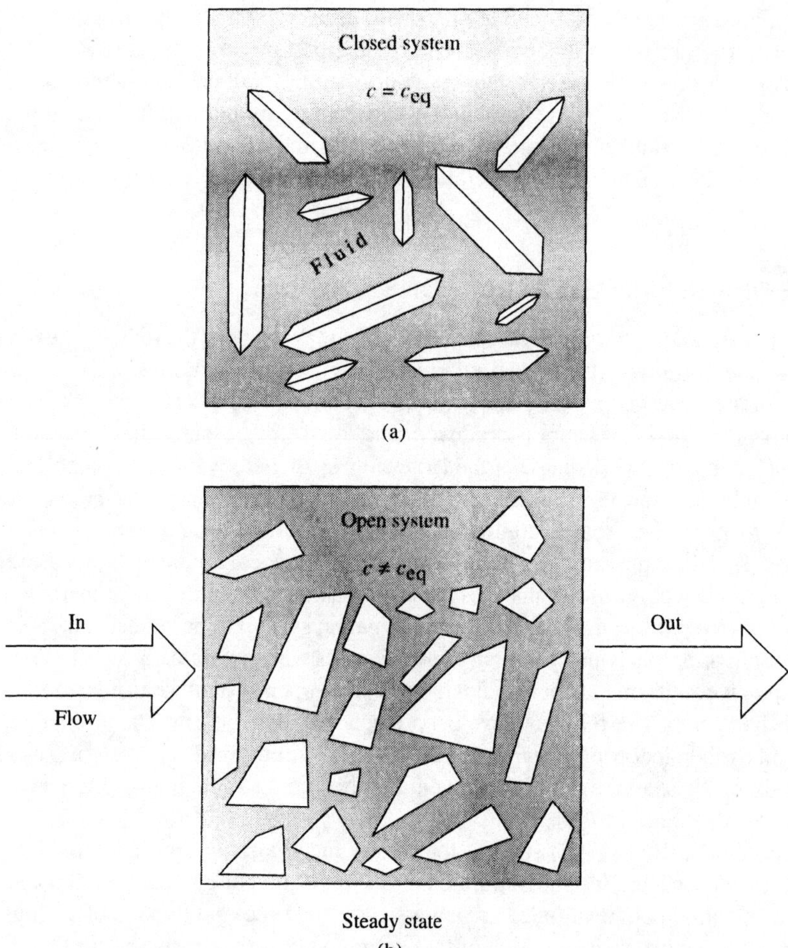

Figure 9-4 (a) Fluid reacting with mineral assemblage in a closed system. The fluid will eventually reach equilibrium with each mineral. $c =$ fluid composition and $c_{eq} =$ fluid composition at equilibrium between minerals and fluid. The equilibrium composition is reached as a special case of steady state, where for each mineral present in the system the input rate to the fluid from the reacting mineral surface is balanced by an equal output rate from the fluid to the mineral surface. (b) Fluid reacting with mineral assemblage in an open system. The fluid will approach a nonequilibrium steady state. A steady state composition for the fluid means that, for each chemical component, the input and output from all processes (mineral surface reactions, advection, dispersion, etc.) are balanced. [After Lasaga and Rye (1993, 363). Reprinted by permission of American Journal of Science.]

it is clear that they are controlled by a number of factors, such as temperature, diffusion parameters, and the chemical potentials of components in pore fluids (Joesten and Fisher 1988). The increasing temperatures and circulating fluids of *prograde* metamorphism speed up reaction rates and grain growth, whereas the decreasing temperature and lack of fluids of *retrograde* metamorphism tend to produce incomplete reactions or no change at all (thus allowing us to find high-grade metamorphic rocks at the surface of the Earth). At any particular temperature, if nucleation and growth of new minerals cannot occur easily, the rate of the metamorphic reactions that should take place at that temperature is slowed.

The other factor causing nonattainment of equilibrium is limited diffusion of atoms to and from a reaction site. Diffusion may occur by movement through a pore fluid, by transfer along grain boundaries, or by passage through crystals. Extensive diffusion may not be possible because of the absence of a pore fluid, and because of the physical and chemical nature of the solid phases. Limitations on diffusion are responsible for the local rock equilibrium mentioned above and for compositional zoning in metamorphic minerals. Some components may move more easily than others owing to differences in chemical potential gradients. The differential movement of material leads to *metamorphic differentiation,* the development of contrasting mineral assemblages (often as layers) from an initially homogeneous parent rock. Since the trend of metamorphic chemical reactions is to eliminate compositional contrasts and to produce equality of element chemical potentials, metamorphic differentiation must be due to the overriding influence of some factor, such as grain size or a high temperature or pressure gradient.

The presence or absence of a fluid phase (usually H_2O-rich) during metamorphism is probably a critical factor in determining the types of reactions that take place, the extent to which they take place, and the degree of equilibrium attained. Most sediments and sedimentary rocks provide interstitial and combined water for their metamorphism, whereas the only water available in igneous rocks comes from the breakdown of their relatively scarce hydrous minerals. Water involved in contact metamorphism generally is provided by the magma, with granitic magmas having the greatest amount of associated fluid phase. Circulating water, with its high solvent power, serves as a means of transportation of elements to and from reaction sites and thus as a catalyst for chemical reactions. At high temperatures, solubility is strongly dependent on the density of the fluid phase (see Chapter Four).

Experiments have shown that nucleation and grain growth are much faster in the presence of water. Breakdown of pre-existing grains is also faster. In addition, water takes part in the formation and breakdown of the large number of hydrous minerals involved in metamorphism. Many of the reactions of prograde metamorphism are dehydration reactions, since increasing temperature favors high-entropy products such as water vapor. One major reason why retrograde metamorphic reactions are limited and incomplete is the absence of an abundant fluid owing to previous expulsion of most water during prograde metamorphism.

The physical adjustment of a rock to imposed metamorphic conditions includes both rearrangement of pre-existing mineral grains and growth of new grains in preferred orientations due to differential stress. Old minerals recrystallize, rotate, and are deformed; new minerals grow in the directions of least resistance. The presence or absence of water affects the behavior of rocks undergoing such changes. The results of these processes are the various types of foliation found in metamorphic rocks. Vernon (1975, Chaps. 6 and 7) provides an extensive discussion of the development of preferred orientation in metamorphic rocks. He points out that chemical reactions affect the mechanical properties of a rock and that deformation can determine the mineral assemblage that is formed. Formation of new minerals and changes in the abundance of the vapor phase result in new mechanical properties. Deformation controls mineralogy to some extent by causing preferential solution in stressed areas and by contributing to the general movement of material within a rock mass.

For further discussion of the kinetics of various metamorphic reactions, see Ridley and Thompson (1986), Walther and Wood (1986), Joesten (1991), and Kerrick et al. (1991). Lasaga and Kirkpatrick (1981) contains a general review of the kinetics of geochemical processes.

GEOTHERMOMETRY AND GEOBAROMETRY

Temperatures and pressures of metamorphism can be estimated in a variety of ways (Essene 1982, 1989). For temperature the two most commonly used methods are (1) measurement of the distribution of elements between coexisting minerals and (2) measurement of the distribution of elements across a solvus for two structurally related minerals. The first method requires use of an exchange reaction such as

$$\begin{aligned} &KFe_3AlSi_3O_{10}(OH)_2 \text{ [annite]} + Mg_3Al_2Si_3O_{12} \text{ [pyrope]} \\ &= KMg_3AlSi_3O_{10}(OH)_2 \text{ [phlogopite]} + Fe_3Al_2Si_3O_{12} \text{ [almandine]} \end{aligned} \tag{9-1}$$

The following discussion explains the procedure when we apply thermodynamics to such an exchange reaction.

The free-energy change of the reaction can be related to the distribution coefficient for iron and magnesium (see equation 3-24) assuming both phases are ideal solutions

$$\Delta G^\circ = -RT \ln K_D \tag{9-2}$$

where

$$K_D = \frac{(\text{Fe/Mg})_{\text{garnet}}}{(\text{Fe/Mg})_{\text{biotite}}} = \frac{N_{\text{Fe}}^{\text{gar}}/N_{\text{Mg}}^{\text{gar}}}{N_{\text{Fe}}^{\text{bio}}/N_{\text{Mg}}^{\text{bio}}}$$

and $N_{\text{Fe}}^{\text{gar}}$ = the mole fraction of iron in garnet, $N_{\text{Mg}}^{\text{gar}}$ = the mole fraction of magnesium in garnet, $N_{\text{Fe}}^{\text{bio}}$ = the mole fraction of iron in biotite, and $N_{\text{Mg}}^{\text{bio}}$ = the mole fraction of magnesium in biotite. By expressing ΔG° in terms of enthalpy and entropy, we get (Philpotts 1990, 360)

$$\ln K_D = -\left(\frac{\Delta H^\circ}{3R}\right)\frac{1}{T} + \frac{\Delta S^\circ}{3R} \tag{9-3}$$

We have ignored the change in volume of reaction (9-1) since it is very small. The numeral 3 in equation (9-3) is for the three iron-magnesium sites in each of the minerals that are involved in the exchange. Thus we have an expression relating the composition of coexisting garnet and biotite to temperature. By analyzing biotite and garnet in metamorphic rock, and by calibrating this geothermometer (Figure 3-4), we can obtain the temperature at which equilibrium was reached. The calibration of a geothermometer involves using either (1) thermodynamic data or (2) independent determinations of temperatures for samples. Figure 3-4 shows three different calibrations of the biotite-garnet thermometer.

Note that distribution coefficients are functions of temperature and pressure, but do not behave exactly like equilibrium constants. The relationship between the two is

$$K_{\text{equil}} = K_D K_\lambda \tag{9-4}$$

where, for reaction (9-1),

$$K_\lambda = \frac{\lambda_{\text{Fe}}^{\text{gar}}/\lambda_{\text{Mg}}^{\text{gar}}}{\lambda_{\text{Fe}}^{\text{bio}}/\lambda_{\text{Mg}}^{\text{bio}}}$$

and $\lambda_{\text{Fe}}^{\text{gar}}$ is the activity coefficient of Fe^{2+} in garnet, and so forth. If both minerals exhibit ideal behavior, then the activity coefficients are all 1.0 and $K_D = K_{\text{equil}}$. K_λ takes into account compositional variation (other than the actual exchange reaction) in the coexisting minerals.

Because the relation (equation 3-39)

$$\frac{dP}{dT} = \frac{\Delta S}{\Delta V} \tag{9-5}$$

describes the slope of the line for a univariant reaction such as that of equation (9-1), we see that, for a reaction to be a good geothermometer, it should have a ΔS value that is much greater than its ΔV value. For many exchange reactions, volume changes are small and entropy changes are relatively large. As a result, the slope of the reaction line on a pressure-temperature diagram is steep and pressure has little effect on K_D values (Figure 5-21). This is true for the garnet-biotite reaction. Examples of its use as a geothermometer are given by Hodges and Spear (1982) and Indares and Martignole (1985). For discussion of another exchange reaction geothermometer (garnet-pyroxene), see Chapter Five.

An example of a solvus geothermometer is the use of the distribution of potassium and sodium between coexisting alkali feldspar and plagioclase (Figure 5-23). Essene (1989) reviews various versions of a feldspar geothermometer. Many versions deal only with albite-anorthite and albite-orthoclase. Fuhrman and Lindsley (1988) provide a thermodynamically based thermometer using feldspars that takes into account ternary solid solution in both plagioclase and alkali feldspar. Their thermometer yields three temperatures, one each for albite, anorthite, and orthoclase equilibria. Other solvus thermometers have been based on the distribution of calcium and magnesium between coexisting clinopyroxene and orthopyroxene (Davidson and Lindsley 1985) and between coexisting calcite and dolomite (Anovitz and Essene 1987).

Experimental study of mineral reactions such as

$$\begin{aligned} &3\ CaAl_2Si_2O_8\ \text{[anorthite]} = Ca_3Al_2Si_3O_{12}\ \text{[grossularite]} \\ &+ 2\ Al_2SiO_5\ \text{[kyanite]} + SiO_2\ \text{[quartz]} \end{aligned} \tag{9-6}$$

has been used to develop geobarometers. These types of solid-solid reactions have large volume changes and thus the equilibrium constant is pressure sensitive. The equilibrium constant for the above reaction can be written (equation 3-24)

$$-RT \ln K_{\text{equil}} = -RT \ln \frac{(a_{\text{gros}}^{\text{gar}})}{(a_{\text{anor}}^{\text{plag}})^3} \tag{9-7}$$

where the activities of quartz and kyanite are set at 1.0 because they are essentially pure phases and where $a_{\text{gros}}^{\text{gar}}$ = the activity of grossularite in garnet and $a_{\text{anor}}^{\text{plag}}$ = the activity of anorthite in plagioclase. An experimental study of this reaction (Koziol and Newton 1988) can be used to develop a pressure-temperature diagram (Figure 9-5) contoured for values of K_{equil} (Spear 1993, 525–527).

To apply this barometer, the compositions of coexisting garnet and plagioclase are measured, mole fractions are found, and the required activities are calculated from the mole fractions as follows:

$$a_{\text{gros}}^{\text{gar}} = (N_{\text{gros}}^{\text{gar}})^3 \tag{9-8}$$

and

$$a_{\text{anor}}^{\text{plag}} = (N_{\text{anor}}^{\text{plag}}) \tag{9-9}$$

Note that the activity of an ideally mixed component i on n equivalent sites in a mineral is $a_i = (N_i)^n$ and this is why the number 3 appears for garnet activity. Because garnet and plagioclase

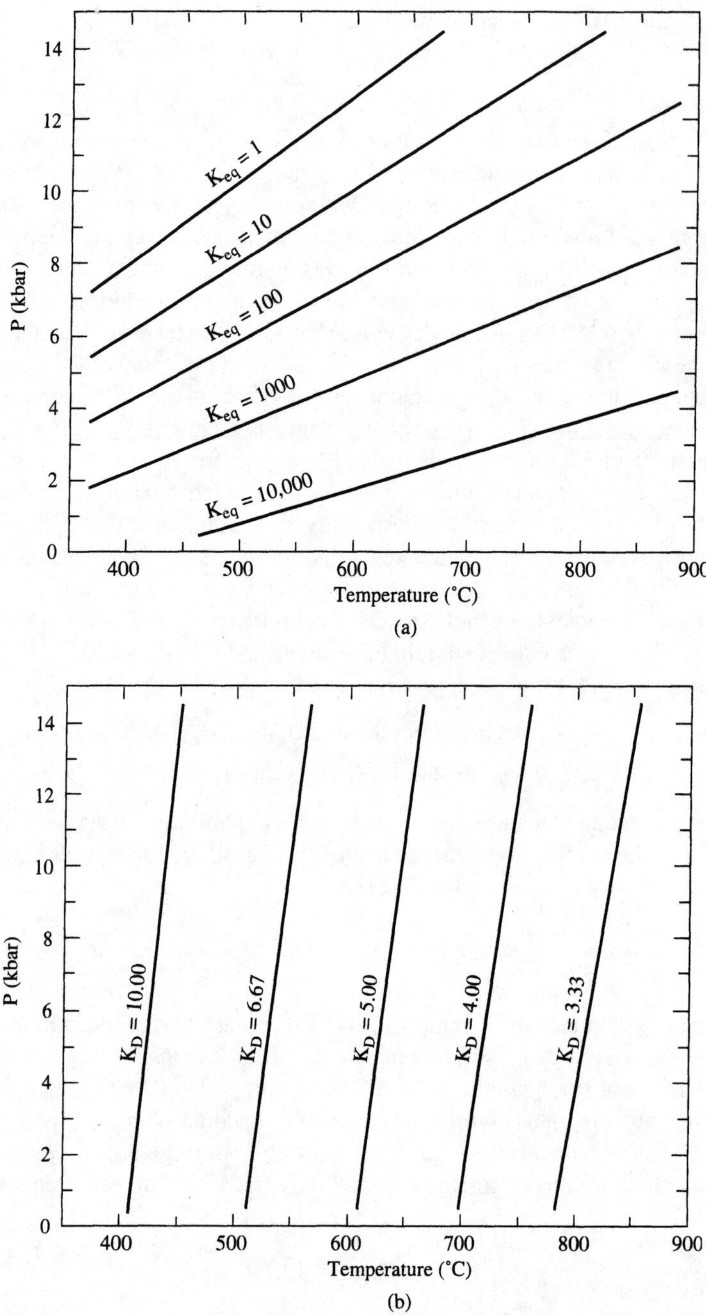

Figure 9-5 (a) Pressure-temperature diagram contoured for values of K_{eq} for the GASP geobarometer reaction. GASP stands for garnet-aluminosilicate-silica-plagioclase. See text for discussion of the reaction. After Spear (1993, 527). (b) Pressure-temperature diagram in which lines of constant K_D for the garnet-biotite Fe-Mg exchange geothermometer have been plotted. See text for discussion of the geothermometer. After Spear (1993, 524).

are not ideal solutions, the above equations should be corrected by assuming a particular model for their behavior (Spear 1993, 526). The calculated activities are used to determine a value for K_{equil} and, using an independent estimate of temperature, Figure 9-5a can be used to estimate pressure for the sample. It is possible to do similar calculations for K_D for the biotite-garnet geothermometer and these are shown in Figure 9-5b.

Different calibrations of both a geothermometer and a geobarometer can be applied to a single set of mineral analyses to show the uncertainties in pressure-temperature determinations. Figure 9-6 gives results for a metamorphic rock in New Hampshire using the garnet-biotite geothermometer and the garnet-plagioclase geobarometer discussed above. The total range of calculated temperatures in Figure 9-6 is about 80°C, and the total range of calculated pressures is about 3.5 kbar. Spear (1993, 528) suggests that the spread of values mainly comes from use of different activity models (different assumptions about the solid solutions and different corrections for nonideality) for garnet, biotite, and plagioclase. Holdaway and Mukhopadhyay (1993, 691) state that, with further research on activity models, it should be possible to approach accuracies of ±25°C and ± 250 bars in low- to medium-pressure rocks.

The most important solid-solid reactions involve the Al_2SiO_5 polymorphs (Figure 3-7). The triple point of this system has been controversial (Kerrick 1990). The two most preferred locations come from Richardson et al. (1969) [5.5 kbar and 620°C] and Holdaway (1971) [3.8 kbar and 501°C]. More recently, Holdaway and Mukhopadhyay (1993) have evaluated all experimental work on aluminum silicates and they place the triple point at 504 (±20)°C and 3.75 (±0.25) kbar. This is now the accepted location. Kerrick (1990) discusses in detail the

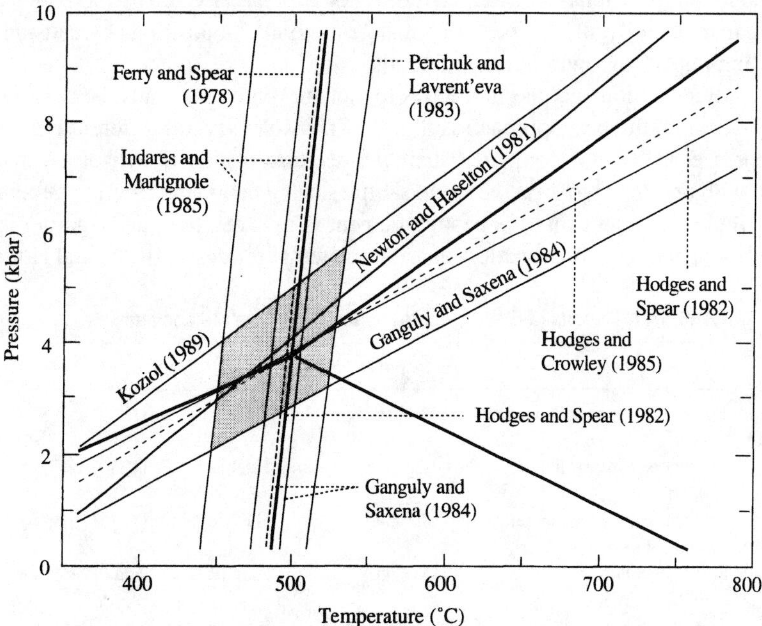

Figure 9-6 Pressure-temperature diagram showing the range of results from garnet-biotite thermometry (steep lines) and GASP barometry (gentle lines). All curves are calculated from a single set of mineral analyses from Mt. Moosilauke, New Hampshire. Each labeled curve represents a different calibration of the geothermometer or geobarometer. After Spear (1993, 530).

experimental difficulties in locating this triple point. Because of the common occurrence of one or more of the Al_2SiO_5 minerals in metamorphic rocks, the system phase diagram (Figure 3-7) is often used, alone or with independent thermometers, to estimate pressures and temperatures.

OXYGEN ISOTOPES AND METAMORPHISM

Stable isotope geochemistry has been used to obtain several kinds of information about metamorphic rocks. Isotope geothermometers using coexisting mineral pairs (discussed in Chapter Two) gives us estimates of metamorphic temperatures. Evidence on the nature and extent of fluid flow through rocks during metamorphism can be obtained from isotopic measurements of mineral pairs and of whole rocks. A third type of information, provided by the stable isotope composition of whole rocks, is the nature of the pre-metamorphic protolith. In this section we will briefly discuss these three uses of stable isotope geochemistry.

An example of oxygen-isotope geothermometry applied to metamorphic rocks is given in Table 2-18. Most of the temperatures using different mineral pairs from the same sample do not show good agreement. There are several possible explanations for the lack of agreement. A major problem, not limited to metamorphic rocks, is correct calibration of the temperature dependence of the partitioning of oxygen isotopes between coexisting minerals (Figure 2-19). Some mineral pairs (e.g., quartz-magnetite) apparently undergo subsolidus reequilibration (adjustment of oxygen isotope ratios). This probably occurs during retrograde changes in temperature. Isotopic disequilibrium could also occur if one of the two minerals in a mineral pair continues to change its $^{18}O/^{16}O$ ratio as temperature decreases, while the partner mineral has reached its "closure temperature" and its composition has become fixed. When a rock is affected by more than one episode of metamorphism, later events can partially change mineral compositions by causing interaction with infiltrating fluids or growth of new minerals.

Thus applying isotope geothermometers to metamorphic rocks must be done with caution. As with any research, the best approach is to use, if possible, several different methods. For example, Huebner et al. (1986) compared their temperature results from isotope measurements (Table 2-18) with previous work on the same samples using cation distribution geothermometers (Table 9-2). Table 9-2 shows that the best agreement is between the quartz-garnet isotope temperatures of Huebner et al. and the cation-distribution temperatures of Harris and Holland (1984).

TABLE 9-2 Results of Geothermometry for Zimbabwe Metapelites

Geothermometer	Temperature (°C)	Source
Oxygen isotope exchange, quartz-feldspar	384–562	Huebner et al. (1986)
Oxygen isotope exchange, quartz-garnet	621–846	Huebner et al. (1986)
Oxygen isotope exchange, quartz-biotite	505–643	Huebner et al. (1986)
Fe-Mg exchange, biotite-garnet	600–950	Harris and Holland (1984)
Fe-Mg exchange, cordierite-garnet	<650	Harris and Holland (1984)
Al distribution, garnet-orthopyroxene	740–820	Harris and Holland (1984)

A possible reason that the quartz-feldspar isotope temperatures in Table 9-2 are lower than the other calculated temperatures is the occurrence of significant isotope exchange by feldspar to lower temperatures as compared to garnet and biotite.

Huebner et al. (1986) suggest that the order of oxygen-isotope closure for these minerals, from higher to lower temperature, is garnet, biotite, quartz, and feldspar. They believe that the isotope ratio of quartz in the Zimbabwe metapelites did not change significantly as the temperature fell below the closure temperature of garnet (even though quartz was above its closure temperature). If so, then the isotopic composition of the fluid present must have been buffered by the rock under conditions of a low fluid/rock ratio. The lower temperatures for the quartz-feldspar and quartz-biotite pairs in Table 9-2 could be due to reequilibration caused by a later, lower-grade metamorphic event. The Archean rocks studied are from an area in Zimbabwe that has undergone at least two major metamorphic events, the first in the granulite facies and the second, a retrograde event, in the upper-amphibolite facies. Freer and Dennis (1982) suggest that quartz-garnet pairs record near-peak metamorphic temperatures, whereas quartz-biotite and quartz-feldspar pairs do not.

The pattern of fluid movement in regional metamorphism can be determined using oxygen-isotope measurements. Chamberlain and Rumble (1988) discuss the role of fluids and how isotopes can be used to map fluid flow. In addition to fluid isotopic composition, the pattern of fluid flow and the fluid/rock ratio are important parameters affecting isotopic values in metamorphic rocks (the fluid is generally considered to be mainly H_2O). Fluid/rock ratios can vary from very high to very low values. A very high ratio involving a fluid with a uniform isotopic composition acting over a long period of time would lead to isotopic equilibrium throughout a group of metamorphosed rocks. On the other hand, a very low fluid/rock ratio acting over a short time period could produce local isotopic equilibrium between adjacent mineral grains or within one rock type, but would not necessarily produce isotopic equilibrium for separated mineral grains or for masses of adjacent, but different, rock types. Valley (1986) points out that sharp gradients for oxygen isotopes occur in some metamorphic areas while other areas exhibit pervasive homogenization. He suggests that the exact nature of fluid migration is highly variable among metamorphic rocks.

Dipple and Ferry (1992) state that fluid flow during metamorphism normally occurs along temperature gradients (i.e., as temperature increases or decreases along the flow path). They believe that the stable isotopic alteration commonly found in metamorphic terranes is due to this type of flow. The alteration is usually one of ^{18}O-depletion. Dipple and Ferry indicate that contact metamorphic zones commonly exhibit $\delta^{18}O$ depletions of 5–8 per mil compared to unmetamorphosed parent rocks, while the $\delta^{18}O$ values of regionally metamorphosed rocks is usually 2–6 per mil less than that of unmetamorphosed or low-grade equivalents. The degree of ^{18}O-depletion often increases with increasing grade of metamorphism. Dipple and Ferry suggest that, for both contact and regional metamorphism, the change in $\delta^{18}O$ values is due to fluid flow from lower to higher temperatures.

Figure 2-17 outlines the isotopic field for metamorphic waters. As discussed in Chapter Two, the nature of the water involved in ore formation can be identified by measuring the oxygen and hydrogen isotopic ratios of minerals assumed to have formed in equilibrium with the fluid (Figure 2-20). In a similar manner, the isotopic nature of fluids involved in metamorphism can be identified and plotted on a δD-$\delta^{18}O$ diagram like Figure 2-17. For example, Huebner et al. (1986) found that the isotopic compositions of waters calculated to be in equilibrium with the Zimbabwe samples discussed earlier plot near the lower left corner of the metamorphic water box in Figure 2-17. Thus it appears that the waters were mainly of metamorphic origin, but they may have had a substantial meteoric water component.

Petrologic study of the Zimbabwe rocks discussed by Huebner et al. (1986) indicates that the rocks are metapelites formed from clastic sediments composed of detrital quartz, feldspars, clay minerals, and rock fragments. The original rocks (protoliths), prior to metamorphism, probably had a $\delta^{18}O$ value of +10 to +15, since mixtures of the isotopic compositions of these minerals plus igneous rock fragments using data such as that in Table 2-17 would produce this range of values. The range of whole-rock $\delta^{18}O$ values found in the metapelites is +8.8 to +9.9. Thus depletion of ^{18}O has occurred during metamorphism.

In a case where a particular metamorphic rock could have formed from two or more different protoliths, measurement of whole-rock $\delta^{18}O$ values could identify the original rock, since different rock types have different $\delta^{18}O$ values (Table 2-17). However, adjustment needs to be made for possible ^{18}O-depletion during metamorphism. Also, fluids passing through a group of rocks undergoing metamorphism could redistribute the oxygen isotopes among various lithologic units. The Zimbabwe metapelites analyzed by Huebner et al. (1986) occur interlayered with calc-silicates ($\delta^{18}O = +6.7$ to $+8.9$) and metabasites ($\delta^{18}O = +7.5$ to $+8.7$). Huebner et al. believe redistribution of the oxygen isotopes has occurred among these rock units and that pervasive metamorphic water was the agent of change.

The majority of the quartz samples from the metapelites has $\delta^{18}O$ values near +11.2. This suggests an approach to isotopic equilibrium among the metapelite units. However, the $\delta^{18}O$ values of all quartz samples from the metapelites, metabasites, and calc-silicates range from +8.5 to +11.6 (Huebner, unpublished). This suggests that isotopic equilibrium was not reached among the lithologic units that were affected by the regional metamorphism. Huebner et al. (1986) believe that there was restricted intercommunication of fluids between the lithologic units, leading to the development of localized mineral equilibrium and distinct $\delta^{18}O$ values within each unit.

METASOMATISM

Metasomatism is defined by Barton et al. (1991) as referring to changes in the bulk composition of rocks in the solid state. As indicated earlier in this chapter, metasomatism clearly occurs in many cases of contact metamorphism. Its occurrence during regional metamorphism is much more difficult to prove. When it does occur, chemical transfer can occur by diffusion in a stationary solution or by infiltration of a moving solution. In addition to overall changes in composition, metasomatic processes can result in metamorphic differentiation: the development of contrasting layers in metamorphic rocks due to segregation of components. A characteristic group of ore deposits known as skarn deposits is produced by contact metasomatism (Einaudi et al. 1981).

Table 9-3 lists some of the common types of metasomatism. For the following discussion, keep in mind that the width of most contact metasomatic zones is small, varying from less than one centimeter (adjacent to veins and thin dikes) to a few tens of meters (adjacent to a large intrusive that has invaded a chemically contrasting rock). Some contact zones may be even wider. In certain areas of regional metamorphism, metasomatism appears to have occurred over very large areas.

Alkali metasomatism occurs in a variety of environments. It develops in seafloor hydrothermal systems (Harper et al. 1988), in schists invaded by granitic plutons (forming K-feldspar porphyroblasts), and in gneisses and other rocks invaded by alkaline igneous rocks such as nepheline syenite and carbonatite (forming fenites). Fenites are rocks characterized by alkali feldspar and aegirine. These rocks are the result of a process, known as fenitization, that forms contact zones that

TABLE 9-3 Types of Metasomatism

Type	Typical chemical change	Typical mineralogical changes
Alkali metasomatism	Na for K, Ca	plagioclase to albite; mafic minerals to aegirine
Hydrogen metasomatism	H for Ca, Na, K	feldspar to sericite; mafic minerals to chlorite
Serpentinization	Addition of silica and water	formation of serpentine
Skarn formation	Addition of silica	formation of calcsilicates

Source: After Barton et al. (1991).

range in width from less than a centimeter to several thousand meters (Kresten 1988). Other elements, such as iron and titanium, are also often added to altered rocks along with sodium.

Hydrogen metasomatism is commonly associated with wallrock alteration produced by ore-forming, hydrothermal vein solutions. The four major types of alteration have been named sericitic, intermediate argillic, advanced argillic, and propylitic (Meyer and Hemley 1967). The characteristic minerals are as follows: sericitic (sericite, quartz, and pyrite); intermediate argillic (clay minerals); advanced argillic (dickite, kaolinite, and pyrophyllite); and propylitic (epidote, albite, and chlorite). Sericitic alteration of orthoclase, forming sericite and quartz, can be represented by the following equation:

$$\begin{aligned} &3\ KAlSi_3O_8\ \text{(orthoclase)} + 2\ H^+ \rightarrow KAl_3Si_3O_{10}(OH)_2\ \text{(sericite)} \\ &\quad + 2\ K^+ + 6\ SiO_2\ \text{(quartz)} \end{aligned} \tag{9-10}$$

Hydrous minerals of the other types could form in a similar manner. It appears that hydrogen and other elements move by diffusion into the walls of cracks carrying hydrothermal solutions. These four types of alteration are commonly found at porphyry copper ore deposits.

Serpentinites are rocks composed mainly of the serpentine minerals antigorite and chrysotile. They are metamorphic rocks that may have formed by addition of water and silica to olivine-rich rocks, as shown by the following equation:

$$3\ Mg_2SiO_4\ \text{(olivine)} + SiO_2 + H_2O \rightarrow 2\ Mg_3Si_2O_5(OH)_4\ \text{(serpentine)} \tag{9-11}$$

The details of the process of serpentinization are not known. Other reactions, such as the following involving the addition of water alone, could also be involved:

$$\begin{aligned} &2\ Mg_2SiO_4\ \text{(olivine)} + 3\ H_2O \rightarrow Mg_3Si_2O_5(OH)_4\ \text{(serpentine)} \\ &\qquad + Mg(OH)_2\ \text{(brucite)} \end{aligned} \tag{9-12}$$

A major question involves the amount of volume change during serpentinization. A large volume increase results from the first reaction above, while the volume change in the second reaction depends on whether or not some of the magnesium is removed or stays as brucite (often found in serpentinites). Another question involves the source of the water; it could be of magmatic, metamorphic, or meteoric origin. Contact metamorphism of serpentinite by felsic igneous rocks produces, as a result of metasomatic diffusion, monomineralic zones (Figure 3-5).

Skarns have been the object of much research because they often exhibit the results of extensive chemical change and because they are commonly associated with valuable ore deposits

(Einaudi et al. 1981). These rocks are composed almost entirely of calcium-bearing silicates formed by reactions between emanations from intrusive magmas and adjacent limestone or dolomite. Although silica is the most important element added to the carbonate rock, many other elements, such as iron and aluminum, are also added. Minerals typically found in skarns include garnet, hedenbergite, diopside, tremolite, wollastonite, epidote, idocrase, hornblende, and epidote. Skarns are coarse-grained and often exhibit zoning (monomineralic or bimineralic) with respect to the contact with the intrusive igneous rock. A typical sequence, from intrusive to unaltered limestone, is andradite garnet–hedenbergite–diopside–wollastonite–calcite. Economic amounts of iron, tungsten, copper, lead, zinc, molybdenum, and tin have been found in skarn deposits.

An interesting example of small-scale metasomatism has been described by Joesten and Fisher (1988). Limestone containing chert nodules occurring in the Christmas Mountains of Texas has undergone contact metamorphism by a composite stock of alkali gabbro. As a result of the metamorphism, the chert nodules reacted with the surrounding limestone to form reaction rims between the two materials. In some areas of the contact aureole a rim of wollastonite ($CaSiO_3$) separates the nodules from the surrounding calcite. In other areas the chert is surrounded by wollastonite, which is in turn separated from the calcite by tilleyite (Figure 9-7). The formula for tilleyite is $Ca_5Si_2O_7(CO_3)_2$.

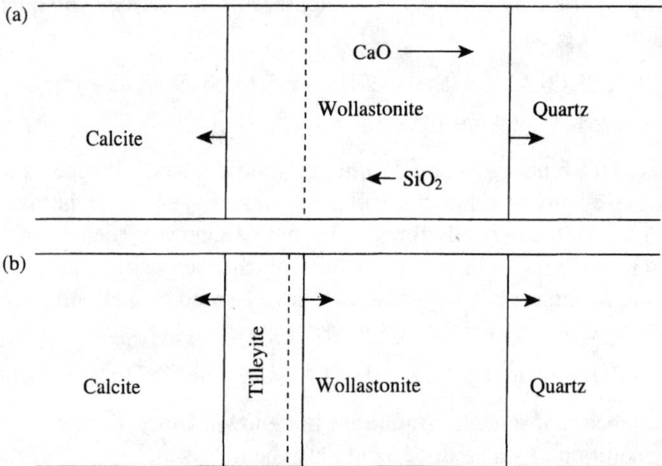

Figure 9-7 (a) Heating of chert nodules in limestone results in the formation of a reaction rim of wollastonite on the nodules. Growth of the wollastonite is caused by the diffusion of CaO and SiO_2 through the reaction rim. Because CaO diffuses more rapidly than SiO_2, wollastonite grows more rapidly on the quartz side of the rim than it does on the calcite side. The relative magnitudes of the diffusion coefficients for CaO and SiO_2 are indicated by the lengths of the two arrows. The original boundary between the chert nodule and the limestone is indicated by the dashed line. (b) Because of higher temperatures nearer an igneous intrusive, a reaction rim of tilleyite formed between the calcite and wollastonite of (a). The dashed line represents the original limestone-chert boundary. The wollastonite layers are up to about 100 mm in thickness and the tilleyite layers are up to about 25 mm in thickness. [(a) and (b) after Philpotts, A. R., *Principles of Igneous and Metamorphic Petrology*, pp. 384, 386. Copyright © 1990 by Prentice Hall, Inc. Adapted by permission of Prentice Hall, Inc., Upper Saddle River, NJ.] After Joesten and Fisher (1988).

Diffusion of CaO and SiO_2 through the reaction layers in an intergranular fluid is caused by chemical potential gradients due to the compositional differences between the chert and calcite. The relative stability of the minerals can be shown in terms of the chemical potentials of CaO and SiO_2. Figure 9-8 shows plots of the chemical potentials of CaO and SiO_2 in fluids in equilibrium with calcite, wollastonite, tilleyite, and quartz (solid lines). Thus calcite is stable from point A to point B in Figure 9-8a as the chemical potential of SiO_2 is increased. Beyond point B the chemical potential of SiO_2 in the fluid becomes high enough to cause the reaction

$$CaO + SiO_2 \rightarrow CaSiO_3 \text{ (wollastonite)} \tag{9-13}$$

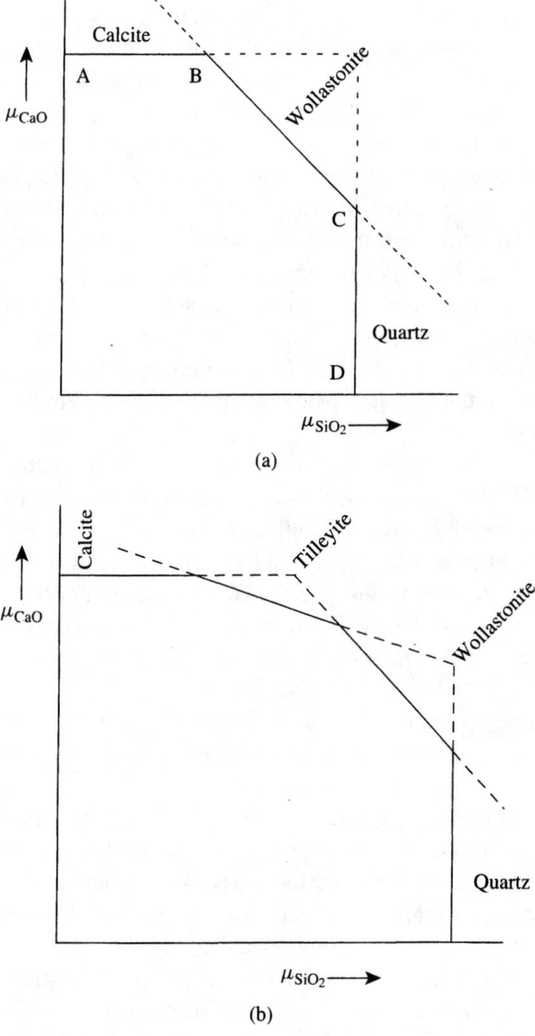

Figure 9-8 (a) Plot of the chemical potentials of CaO and SiO_2 in fluids in equilibrium with calcite, wollastonite, and quartz (solid lines). The chemical potentials of CaO and SiO_2 are uniquely defined where calcite and wollastonite are in contact and where wollastonite and quartz are in contact. Differences in chemical potential between these two boundaries drive the diffusion through the wollastonite layer. The exact location of line BC depends on temperature. Below 600°C ($P = 325$ bars) it does not intersect lines AB and CD and wollastonite is unstable with respect to calcite and quartz. (b) Same as (a), but at a higher temperature (above 941°C at $P = 325$ bars). Tilleyite becomes stable under certain conditions of μ_{CaO} and μ_{SiO_2} and forms between calcite and wollastonite. The formula for tilleyite is $Ca_5Si_2O_7(CO_3)_2$. [(a) and (b) after Philpotts, A. R., *Principles of Igneous and Metamorphic Petrology*, pp. 385, 386. Copyright © 1990 by Prentice Hall, Inc. Adapted by permission of Prentice Hall, Inc., Upper Saddle River, NJ.] After Joesten and Fisher (1988).

A similar argument holds as the chemical potential of CaO is increased from point D to point C. As shown in Figure 9-8b, a surface representing tilleyite stability develops under higher temperature conditions and thus two reaction zones separate calcite and quartz, as shown in Figure 9-7b.

METAMORPHIC ZONES AND METAMORPHIC FACIES

The first development in the modern study of metamorphic rocks was the recognition of mineralogical zones within larger masses of these rocks. Such zones were found in both contact and regional metamorphic rocks. Barrow (1893) mapped regional zones in describing progressive metamorphism of pelitic rocks in the Scottish Highlands. The beginning of each zone, starting with unmetamorphosed rocks, was marked by the appearance of distinctive mineral (*index mineral*). These zones were soon found to occur in similar rocks in other parts of the world, and are now referred to as the chlorite, biotite, garnet, staurolite, kyanite, and sillimanite zones.

The first detailed study of contact metamorphic zones was by Goldschmidt (1911), who demonstrated a consistent relationship among metamorphic zoning, rock composition, and rock mineralogy for hornfelses of the Oslo region of Norway. In most of the contact aureoles he studied, there are two zones, an inner zone next to the intrusive rock and a lower-grade zone farther away. The hornfelses formed from pelitic, quartzo-feldspathic, and calcareous sedimentary rocks. Goldschmidt concentrated on the inner zones, for which he compared chemical composition with mineralogy. A consistent correlation was found, such that all the hornfelses of the inner zones could be classified on the basis of their mineral assemblages (Figure 3-11). Later studies of other contact aureoles showed this approach to be of general use, with the same direct relationship between chemical and mineralogical composition holding true. The simplicity and uniformity of his mineral assemblages led Goldschmidt to assume attainment of equilibrium in the rocks. To test this, he successfully applied the phase rule in the first application of thermodynamics to the study of metamorphic rocks (see Chapter Three).

Goldschmidt's results provided the groundwork for the development of the concept of metamorphic facies by Eskola (1915, 1920). Eskola studied contact metamorphic rocks in Finland and, while finding a similar regular relationship between chemical composition and mineralogy, found some differences between the mineralogy of his rocks and those of Goldschmidt. For example, muscovite and biotite occurred in Finland in the place of K-feldspar and cordierite in Norway. Eskola concluded that, since the rock compositions were similar, his contact rocks formed under different conditions of metamorphism. He defined a metamorphic facies as including all rocks that have been metamorphosed under identical conditions (Eskola 1915). As a result of formation under specific temperature and pressure conditions, the mineral composition of each rock in a facies is dependent only on chemical bulk composition, with a given composition always producing the same set of minerals.

Following the original formulation of metamorphic facies in 1915, Eskola and others have given various definitions for the concept. Some confusion has developed as a result of not making a distinction between the assignment of rocks to a particular facies and the interpretation of the physical conditions represented by that facies. A modern definition for *metamorphic facies* is that a facies is a set of metamorphic mineral assemblages, repeatedly associated in space and time, such that there is a constant and therefore predictable relation between mineral composition and chemical composition (Fyfe and Turner 1966). The assemblages that belong together in one facies can be determined by field mapping of associated rocks of contrasting compositions. Thus all the dif-

TABLE 9-4 Mineral Assemblages for Different Grades of Metamorphism of Three Common Rocks

Unmetamorphosed rock	Facies A	Facies B	Facies C	Facies D	Facies E
Shale	Chlorite	Biotite	Garnet	Staurolite	Sillimanite
		Chlorite	Biotite	Garnet	Garnet
	Muscovite	Muscovite	Muscovite	Biotite	Biotite
	Albite	Albite	Albite	Plagioclase	Plagioclase
	Quartz	Quartz	Quartz	Quartz	Quartz
	Slate or phyllite		Phyllite or schist		Schist or gneiss
Andesitic volcanic tuff	Actinolite	Actinolite	Hornblende	Hornblende	Hornblende
	Albite	Albite	Albite	Plagioclase	Plagioclase
	Epidote	Epidote	Epidote	Quartz	Quartz
	Chlorite	Chlorite	Quartz		
	Quartz	Quartz			
	Chlorite or actinolite schist			Amphibolite	
Sandy limestone or siliceous dolomite	Dolomite	Tremolite	Tremolite	Diopside	Diopside
	Calcite	Calcite	Calcite	Calcite	Calcite
	Quartz	Quartz	Quartz	Quartz	Quartz
	Marble	Tremolite marble		Diopside marble	

→ Increasing grade

Source: After Hyndman (1985).

ferent rocks found in the inner zones of Goldschmidt's contact aureoles belong to the same facies. Table 9-4 gives examples of different mineral assemblages that belong to the same facies.

Turner (1968) and others have emphasized the following characteristics of a given metamorphic facies:

1. It is not defined on the basis of assumed physical conditions, but is purely descriptive, with interpretation of its physical conditions subject to change.
2. Equilibrium may or may not have been reached in the formation of individual rocks of the facies (however, no evidence of replacement of one mineral by another should be present).
3. The mineral composition of each rock in the facies is a function of the present bulk composition of the rock (which may have been changed from its original premetamorphic composition by metasomatism).
4. It cannot be defined in terms of a single rock type (even though each facies is usually named after a characteristic rock).
5. Some or all of the mineral assemblages have formed in many different regions and in rocks of varied ages.
6. Some of the mineral assemblages in the facies may occur in other facies (emphasizing the fact that it is the entire group of assemblages that defines the facies).

Since most metamorphic rocks satisfy the above definition and characteristics, two interpretations of metamorphic facies are generally accepted: (1) rocks belonging to any particular facies formed under the same range of physical conditions, and (2) equilibrium was reached or closely

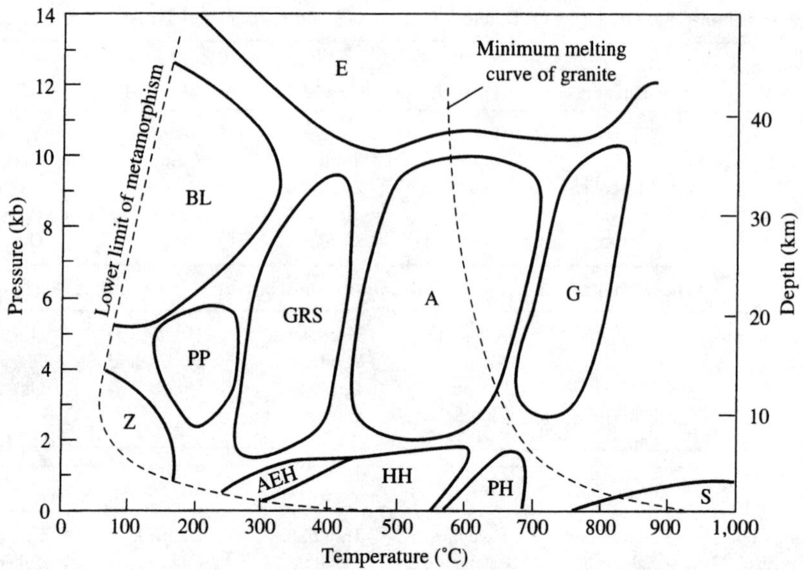

Figure 9-9 Metamorphic facies in relation to temperature and total pressure ($P_{total} = P_{fluid}$). When P_{fluid} is less than P_{total}, the temperature of the facies boundaries is lower. All boundaries are gradational. The minimum melting curve of granitic rocks is discussed in Chapter Eight and shown in Figure 8-24. Abbreviations for the facies are Z, zeolite; PP, prehnite-pumpellyite; BL, blueschist; E, eclogite; GRS, greenschist; A, amphibolite; G, granulite; AEH, albite-epidote hornfels; HH, hornblende hornfels; PH, pyroxene hornfels; S, sanidinite. After Turner (1968).

approached in the formation of the individual mineral assemblages. Keep in mind that neither of these interpretations is a part of the definition of metamorphic facies given above.

Eskola in 1920 proposed five metamorphic facies (greenschist, amphibolite, hornfels, sanidinite, and eclogite) and later added three more (epidote-amphibolite, blueschist, and granulite). His scheme was expanded by Turner and others (see Fyfe et al. 1958) to 10 facies by adding the zeolite facies and by specifying three types of hornfels facies (albite-epidote hornfels, hornblende hornfels, and pyroxene hornfels). By 1968, 11 facies were generally recognized, the newest one being the prehnite-pumpellyite facies (Turner 1968). The generally accepted relationship of these 11 facies to pressure-temperature conditions of formation is shown in Figure 9-9. Figure 9-1 gives the relationship of the contact metamorphic facies to intrusive magmas. Typical minerals of the major metamorphic facies are listed in Table 9-5.

As more fieldwork led to more facies, it became clear that no sharp division could be made between the facies of regional metamorphism and those of contact metamorphism. Furthermore, sequences of regional metamorphic rocks were found that differed from those described by Barrow in the Scottish Highlands. This led Miyashiro (1961) to propose the concept of *facies series,* which classifies metamorphism into three pressure types (Figure 9-10). Table 9-6 relates the three facies series of Figure 9-10 to the location and age of various metamorphic belts. Spear (1993, 24) points out that a metamorphic facies series only represents a sequence of peak metamorphic conditions. It does not necessarily represent the actual pressure-temperature path of any of the metamorphic rocks making up the various metamorphic facies.

TABLE 9-5 Typical Minerals of the Major Metamorphic Facies

Facies	Protolith (precursor) rock type		
	Mafic igneous	Pelite	Calcareous
Blueschist	Glaucophane, lawsonite pumpellyite, jadeite, chlorite	Glaucophane, lawsonite, chlorite, muscovite, quartz	Tremolite, aragonite, muscovite, glaucophane
Greenschist	Chlorite, actinolite, epidote, albite	Chlorite, muscovite, albite, quartz	Calcite, dolomite, tremolite, phlogopite, epidote, quartz
Epidote-amphibolite	Hornblende, epidote, albite, almandine garnet, quartz	Almandine garnet, chlorite, muscovite, biotite, quartz	Calcite, dolomite, epidote, plagioclase, tremolite, forsterite or quartz
Amphibolite	Hornblende, andesine, garnet, quartz	Garnet, biotite, muscovite, sillimanite, quartz	Calcite, dolomite, diopside, plagioclase, quartz or forsterite, wollastonite
Granulite	Diopside, hypersthene, garnet, intermediate plagioclase	Garnet, orthoclase intermediate plagioclase, quartz, kyanite or sillimanite	Calcite, plagioclase, diopside, wollastonite, forsterite or quartz
Eclogite	Jadeitic pyroxene, pyropic garnet, ± kyanite	—	—
Hornfels	Diopside, hypersthene, plagioclase	Biotite, orthoclase, quartz, cordierite, andalusite	Calcite, wollastonite, grossularite

Source: From PETROLOGY: IGNEOUS, SEDIMENTARY, AND METAMORPHIC by Ehlers and Blatt. Copyright ©1982 by W. H. Freeman and Company. Used with permission.

Use of the metamorphic facies concept is limited to a general classification of the rocks in a given region. A more useful approach to estimating pressures and temperatures for metamorphic rocks is use of a petrogenetic grid. This type of grid is constructed by plotting univariant reaction curves (from experimental studies or thermodynamic calculations) on a pressure-temperature diagram (Figure 9-11). The mineral compositions of rocks in a region can be compared with the grid curves to estimate the pressure-temperature conditions of metamorphism. Petrogenetic grids are discussed in more detail later in the chapter.

Tilley (1924) used the term *isograd* for a line drawn on a map to enclose all occurrences of the characteristic mineral of a metamorphic zone. The zones are named after these index minerals as explained earlier for the rocks mapped by Barrow (1893). Ideally, an isograd joins points of similar pressure-temperature conditions at the time of metamorphism. It is now recognized that the use of a single mineral to define an isograd is not practical, since the first appearance of a given mineral depends on a number of factors. The study of assemblages of minerals is more useful in locating isograds. Thus, although index minerals can be used to outline metamorphic zones, they cannot be used to estimate the pressure-temperature conditions of metamorphism. Zen and Thompson (1974) use the term *zone marker* in describing the appearance or disappearance of minerals at the boundaries of metamorphic zones. They use isograd to refer to a specified first-

TABLE 9-6 Pressure Types of Regional Metamorphism and Geologic Ages

Metamorphism	Precambrian	Paleozoic	Mesozoic-Cenozoic
Low pressure	Svecofennides Karelides Canada (partly) Australia Northeast China	Hercynides Appalachians (partly) Eastern and south Australia Hida belt (Japan) Pichilemu series (Chile)	Ryoke-Abukma belt (Japan) Hidaka belt (Japan)
Medium pressure	Canada (partly)	Caledonides Appalachians (partly)	North America Cordillera (partly)
High pressure	Anglesey (Wales)	Kiyama (Japan) Sangun belt (Japan) Curepto series (Chile) Penjna Range (Northwest Kamchatka)	Alps Franciscan group (California) Sanbagawa belt (Japan) Kamuikotan belt (Japan) New Caledonia Central Kamchatka

Source: After Miyashiro (1973).

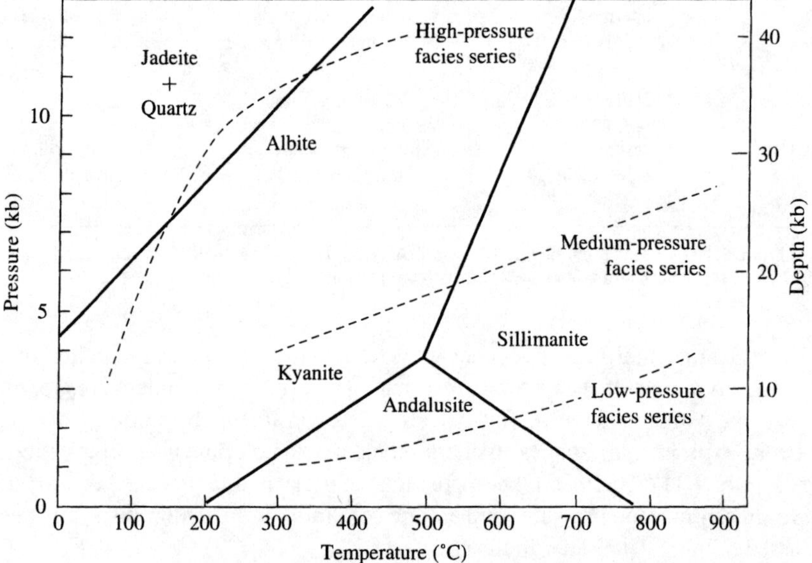

Figure 9-10 Possible pressure-temperature gradients for facies series in relation to the stability fields of Al_2SiO_5 minerals and jadeite. The low-pressure type is characterized by the presence of andalusite, the medium-pressure type by the presence of kyanite and absence of glaucophane, and the high-pressure type by the presence of glaucophane, jadeite, and lawsonite. The sequence of facies would be similar for the two lower-pressure series (greenschist → amphibolite → granulite) and would usually be prehnite-pumpellyite → glaucophane schist for the high-pressure series. The low-pressure sequence of mineral assemblages is known as the Buchan type and the medium-pressure sequence as the Barrovian type for two different areas of the Scottish Highlands. Any given area of regional metamorphic rocks probably has its own unique P-T gradient. After Miyashiro (1973).

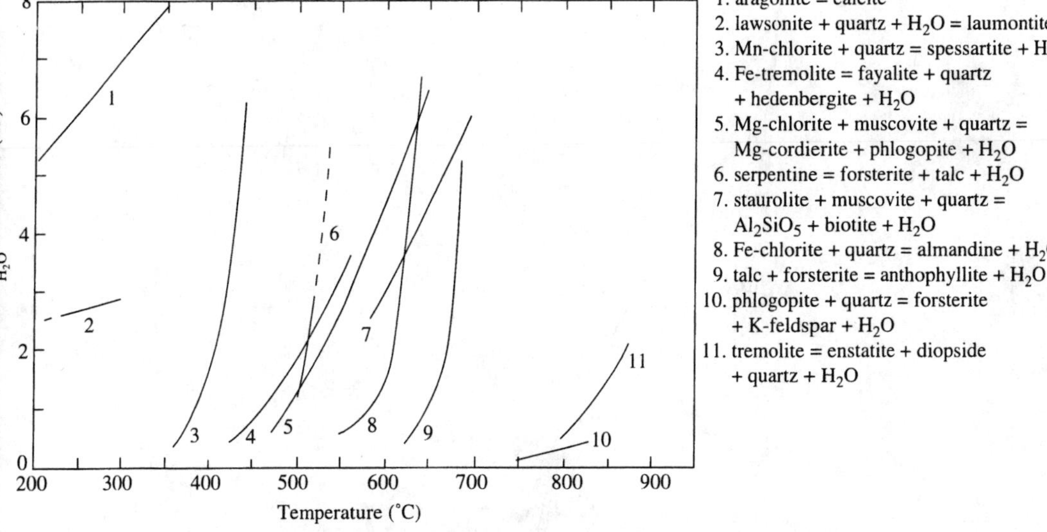

Figure 9-11 Univariant equilibria for some end-member systems in which $P_{H_2O} = P$. Curves such as these can be used to construct petrogenetic grids. See Figure 9-17. Note that, for reactions involving hydrous minerals, the univariant curves are convex to the temperature axis (because their stability depends on P_{H_2O}). (From IGNEOUS AND METAMORPHIC PETROLOGY by Best. Copyright © 1982 by W. H. Freeman and Company. Used with permission.) After Greenwood (1976) and Fyfe et al. (1978).

order heterogeneous reaction leading to an abrupt change in mineral assemblage and to a change in the topology of the pertinent phase diagram, as proposed by Thompson (1957). A change of this type can be considered to be the boundary between two different facies (Figure 9-12).

REGIONAL METAMORPHISM AND PLATE TECTONICS

The plate-tectonic model provides a framework for understanding the formation and spatial distribution of the various regional metamorphic facies. Plate junctures can be divided into three types: (1) divergent (characterized by growth of plate margins by addition of material from below); (2) convergent (two plates collide, forming a mountain belt, or one plate overrides another, which moves downward along a subduction zone); and (3) shear (plates move more or less horizontally past one another). The plates are about 100 kilometers thick and are composed of surficial sediments, crustal rocks, and upper-mantle material (known collectively as the lithosphere). Beneath the lithosphere is the asthenosphere, which is less rigid and allows horizontal and vertical movement of the plates to occur.

Figure 9-13 relates the approximate areas of the facies series shown in Figure 9-10 to three different tectonic environments. Convergent plate boundaries are associated with the high-pressure (subduction zones) and medium-pressure (continental collisions) facies series (Figure 9-14). The low-pressure facies series is found at ocean ridges associated with divergent plate boundaries and in the island arc environment (Figure 9-15). Ancient subduction zones have paired metamorphic

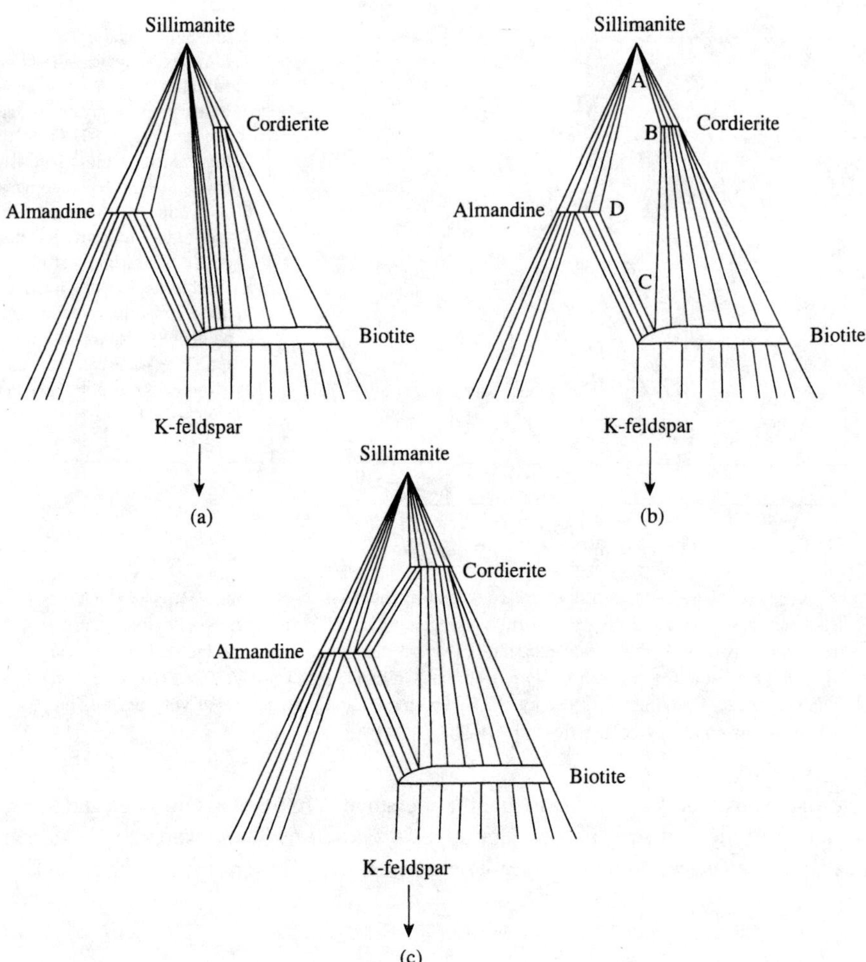

Figure 9-12 Example of a reaction that can be used to define an isograd. The sequence (a) → (b) → (c) is that of rising metamorphic grade, and the reaction at (b) is sillimanite + biotite (+ muscovite + quartz) → almandine + cordierite (+ muscovite + quartz). All compositions in the quadrilateral ABCD would show the discontinuity, which could be used as an isograd. In contrast, a new mineral might appear at a constant metamorphic grade if the bulk composition of a rock changed across the sillimanite-biotite boundary of (a) or the cordierite-almandine boundary of (c). In this case a new zone could be mapped, but its boundary would not be an isograd. The various mineral assemblages of (a) could be used to define one facies, and the set of assemblages of (c) used to define another facies. See text and Figure 9-16 for description of these AFM diagrams. After Thompson (1957).

belts, with high-pressure facies series (subduction) rocks adjacent to low-pressure facies series (island arc) rocks that have been thrust up against the subduction rocks. For example, the high-pressure Sanbagawa belt in Japan is adjacent to the low-pressure Ryoke-Abukuma belt (Table 9-6).

Most regional metamorphism has occurred at convergent plate boundaries. Figure 9-14 is a schematic diagram of the distribution of metamorphic facies in the vicinity of a subduction

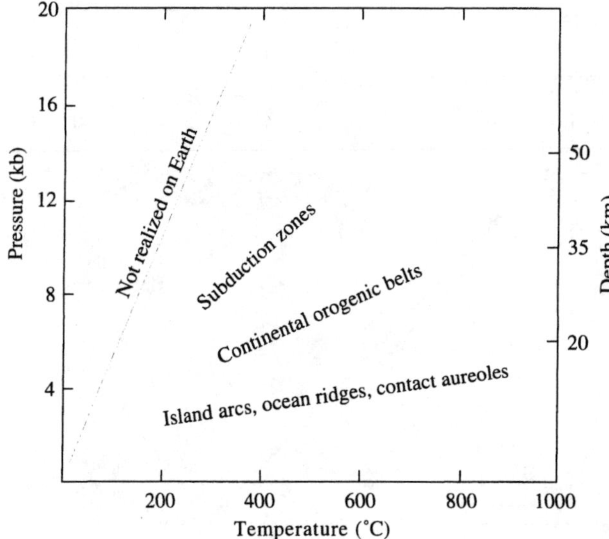

Figure 9-13 Pressure-temperature diagram showing the general regions for metamorphism of the crust. After Spear (1993, 9).

zone. Near the surface, rocks of the zeolite and prehnite-pumpellyite facies occur in sediments as a result of burial metamorphism. Along the edge of the descending plate, high-pressure, low-temperature metamorphism produces rocks of the blueschist facies. Closer to the zone of magmatic activity (due to partial melting of downward-moving lithosphere), higher-temperature metamorphism results in rocks of the greenschist and amphibolite facies. At greater depths, rocks of the eclogite facies develop, with granulite-facies rocks forming in the highest-temperature zones.

Most convergent plate junctions recognized on the Earth today are characterized by deep-sea trenches, volcanic island arcs, and two major zones of metamorphic rocks (Miyashiro 1961). On the oceanic side of the subduction zone is a high-pressure, low-temperature belt defined by rocks of the blueschist facies, and occurring with rocks of the *ophiolite suite* (spilitic pillow lavas, gabbros, serpentinized peridotites, cherts, deep-sea sediments). This mixture of rock types is believed to represent oceanic crust and sediments that have gone down to great depths in a subduction zone and suffered metamorphism and tectonic movement. The metamorphic rocks belong to the high-pressure facies series of Figure 9-10. On the landward side of the subduction zone, the metamorphic rocks have formed under lower-pressure, but higher-temperature, conditions. They belong to the greenschist and amphibolite facies and are indicative of the low-pressure facies series of Figure 9-10. These metamorphic rocks occur with the abundant calc-alkaline intrusive and volcanic igneous rocks of island arcs and continental crust. The occurrence of paired metamorphic belts seems to be limited to Phanerozoic metamorphic belts, with rocks of the blueschist facies being very rare in Precambrian terranes and progressively more abundant in younger regions (Table 9-6).

Metamorphic rocks also form in the vicinity of divergent plates (Figure 9-15). Dredge hauls from the ocean floor have yielded metamorphosed basalts and gabbros whose mineralogy puts them in the zeolite, greenschist, and amphibolite facies. Rocks of the zeolite facies form near the surface, and those of the greenschist facies at greater depth. The amphibolite facies rocks are

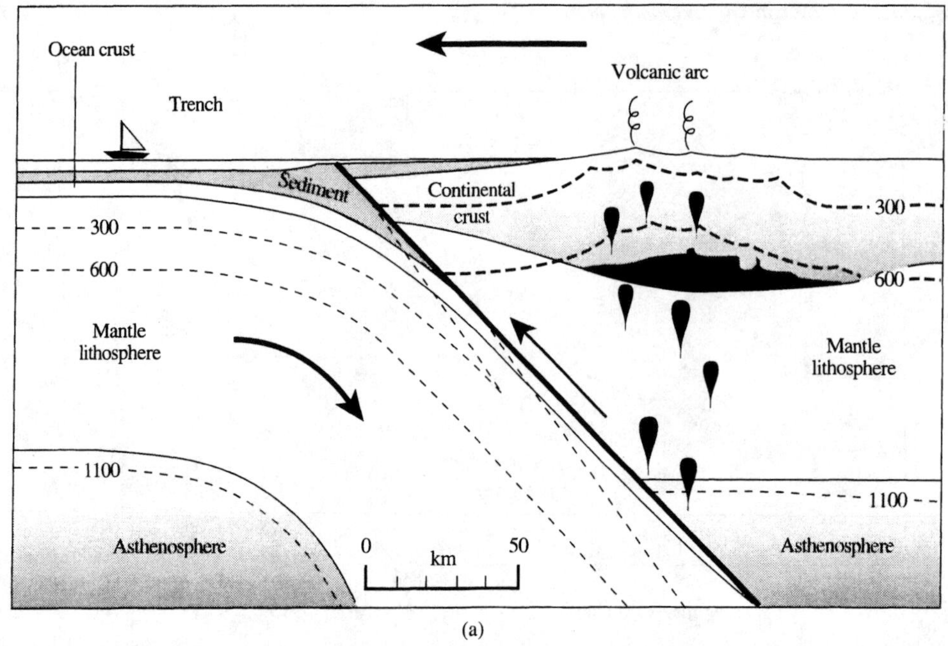

(a)

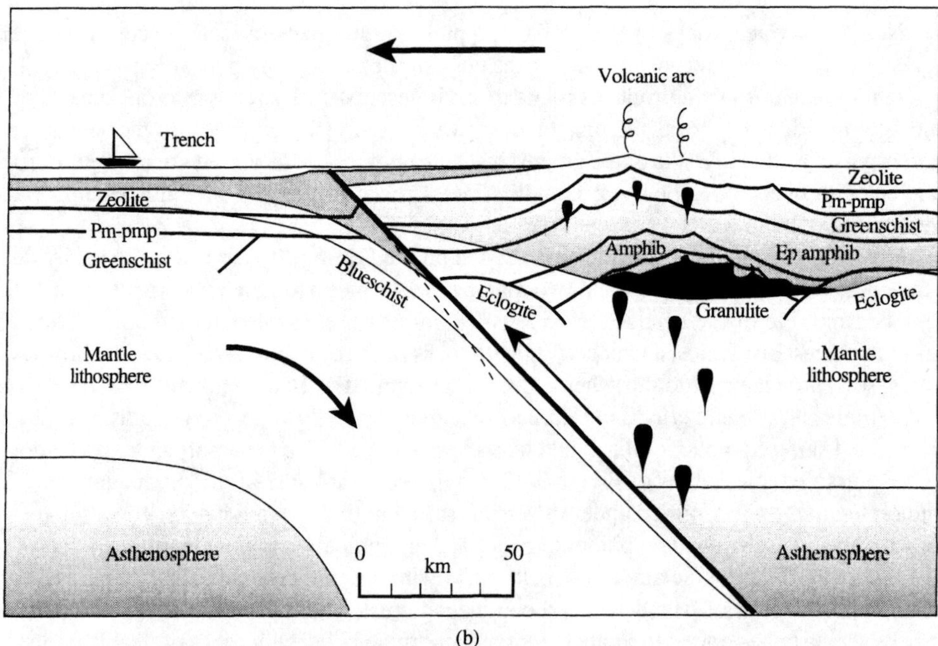

(b)

Figure 9-14 A schematic cross section of an ocean-island arc collision zone. (a) Isotherms are bowed downward in the subduction zone because cool oceanic lithosphere is being subducted. Isotherms in the arc are bowed upward because of the advection of heat by rising magmas. (b) The distribution of metamorphic facies in the subduction zone and island arc. The epidote amphibolite facies is currently not considered to be a separate facies. Mineral abbreviations used above are as follows: Ep amphib = epidote amphibolite; Prn-pmp = prehnite-pumpellyite; Amphib = amphibolite. After Ernst (1976) and Spear (1993, 21).

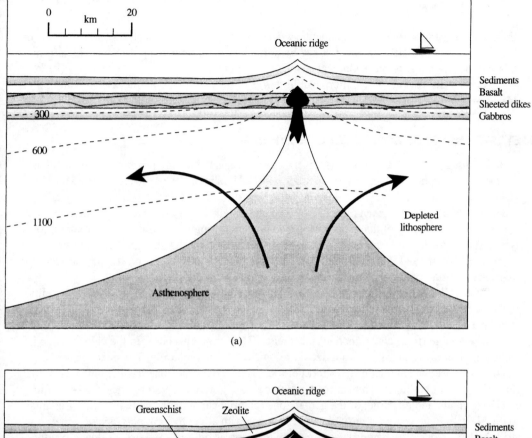

0 20
km
Oceanic ridge
Sediments
Basalt
Sheeted dikes
Gabbros
300
600
1100
Depleted lithosphere
Asthenosphere

(a)

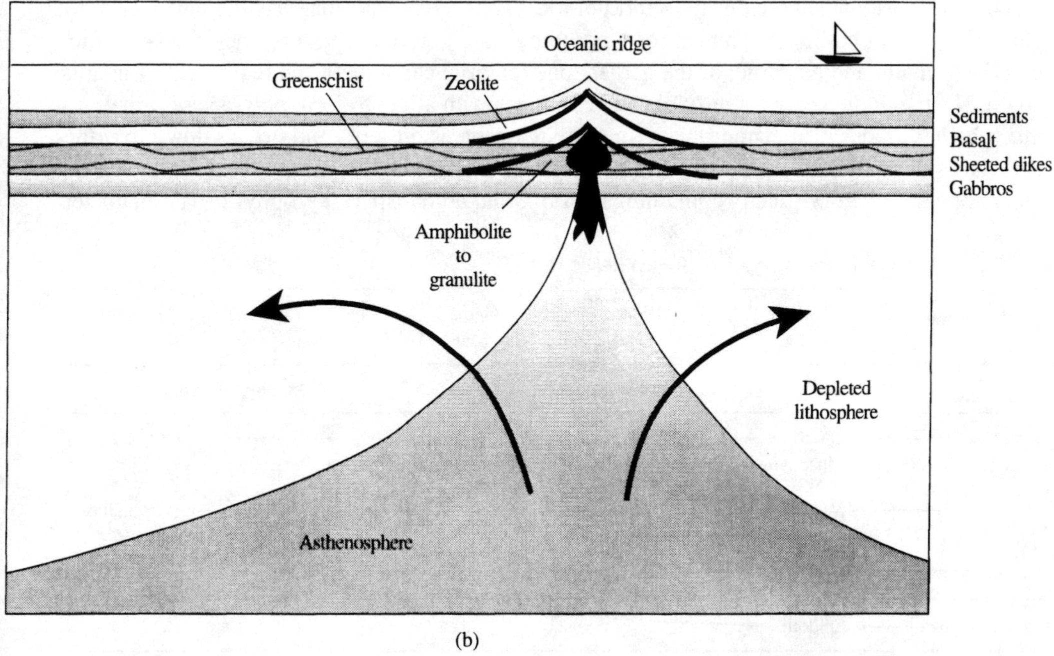

(b)

Figure 9-15 A schematic cross section of an oceanic ridge. (a) Isotherms are bowed upward in the vicinity of the ridge owing to the advection of heat by the intrusion of magmas. (b) The distribution of metamorphic facies in the ocean crust encountered in the vicinity of the ridge. After Ernst (1976) and Spear (1993, 20).

limited to the central zone of rising magma where temperatures are highest. Metamorphism is of least importance at plate junctures marked by horizontal shear movement. Mechanical deformation without chemical or mineralogical change is the main result.

Further discussion of metamorphism in terms of the plate-tectonic model can be found in Miyashiro (1994), Ernst (1975a, 1975b, 1988), Platt (1986), and Daly et al. (1989).

METAMORPHISM OF COMMON ROCK TYPES

Detailed discussions of the numerous mineralogical changes that take place during regional and contact metamorphism can be found in Turner (1981), Hyndman (1985), Spear (1993), Miyashiro (1994), and Bucher and Frey (1994). A good review of mineral reactions during low-grade regional metamorphism has been provided by Zen and Thompson (1974). We shall limit ourselves here to a brief summary of the major changes that occur during metamorphism of common rock types. These changes are summarized in Tables 9-7 through 9-11. Although the tables are gross oversimplifications and indicate little about the actual reactions that take place, they do serve to indicate the nature of the changes during progressive metamorphism.

Table 9-7 summarizes changes for progressive regional metamorphism of K-rich pelitic rocks. The starting material (shales and related rocks) consists of quartz, clay minerals, feldspars, chlorite, and calcite. The first new metamorphic minerals to form are muscovite (from clay minerals) and epidote (from clay minerals and calcite). These minerals occur in the chlorite zone of the greenschist facies along with quartz, albite, microcline, and chlorite. The next change is the reaction of muscovite and chlorite to form biotite. Thus the biotite zone has the minerals of the chlorite zone plus biotite. At a higher metamorphic grade, some chlorite combines with other minerals to form almandine garnet of the garnet zone (chlorite can remain stable into the sillimanite zone). Next, epidote becomes unstable and combines with albite to form plagioclase. Finally, at much higher grades of metamorphism, muscovite becomes unstable and breaks down to orthoclase and sillimanite. Staurolite forms in pelitic rocks when enough iron is present for it to be stable. For pelitic rocks originally containing kaolinite and no feldspars, pyrophyllite is a character-

TABLE 9-7 Regional Metamorphism of K-Rich Pelitic Rock

	Chlorite zone	Biotite zone	Garnet zone	Staurolite-kyanite zones	Sillimanite zone
	Greenschist Facies			Amphibolite Facies	
Quartz	Quartz				Quartz
K-bearing clays	Muscovite	Muscovite			Sillimanite
Detrital K-feldspars	Microcline				Orthoclase
		Biotite			Biotite
Chlorite	Chlorite	Chlorite	Almandine		Almandine
Albite	Albite			Plagioclase	Plagioclase
Calcite	Epidote				

Note: Increasing grade of metamorphism is from left to right.

Source: After *Textbook of Lithology* by Kern C. Jackson. Copyright © 1970 by McGraw-Hill, Inc. Used with permission of McGraw-Hill Book Co.

TABLE 9-8 Contact Metamorphism of Pelitic Rocks

	Albite-epidote facies	Hornblende hornfels facies	Pyroxene hornfels facies	Sanidinite facies
Quartz	Quartz		Quartz	Tridymite
		Microcline	Orthoclase	Sanidine
K-bearing clays	Muscovite	Muscovite		
	Andalusite	Andalusite	Andalusite → Sillimanite	Mullite
Kaolinitic clays				
	Cordierite		Cordierite	Cordierite
		Biotite		
Chlorite	Chlorite			
		Chlorite	Hypersthene	Hypersthene
Albite	Albite	Plagioclase	Plagioclase	Plagioclase
Calcite	Epidote			

Note: Increasing grade of metamorphism is from left to right.

Source: After *Textbook of Lithology* by Kern C. Jackson. Copyright © 1970 by McGraw-Hill, Inc. Used with permission of McGraw-Hill Book Co.

istic mineral formed by low-grade metamorphism. At higher grades, kyanite (or andalusite) forms from pyrophyllite.

Table 9-8 lists the changes that occur during contact metamorphism of pelitic rocks. As in regional metamorphism, muscovite and epidote form at low temperatures. However, for rocks low in potassium and high in iron, magnesium, and aluminum, cordierite and andalusite also form. At higher temperatures, biotite and plagioclase develop. Anthophyllite or almandine may form if enough iron is available. In the inner zone of a contact aureole, at the highest temperatures, orthoclase and hypersthene form and biotite and muscovite break down. If the temperature is high enough, sillimanite replaces andalusite (Figure 3-7). Rocks of the sanidinite facies form when xenoliths are incorporated in high-temperature magmas. Sanidine, tridymite, and mullite may form from orthoclase, quartz, and sillimanite, respectively.

Another rock type that is very susceptible to metamorphism is the impure calcareous rock. When quartz and clay minerals are the impurities and dolomite is not present, the changes shown

TABLE 9-9 Regional Metamorphism of Impure Limestones

	Greenschist facies	Amphibolite facies	Granulite facies
Calcite			Calcite
Quartz		Grossularite	Quartz
	Epidote		
		Plagioclase	Plagioclase
Clay minerals	Tremolite	Diopside	Diopside
	Muscovite	Microcline	Orthoclase

Note: Increasing grade of metamorphism is from left to right.

Source: After *Textbook of Lithology* by Kern C. Jackson. Copyright © 1970 by McGraw-Hill, Inc. Used with permission of McGraw-Hill Book Co.

TABLE 9-10 Contact Metamorphism of Impure Limestones

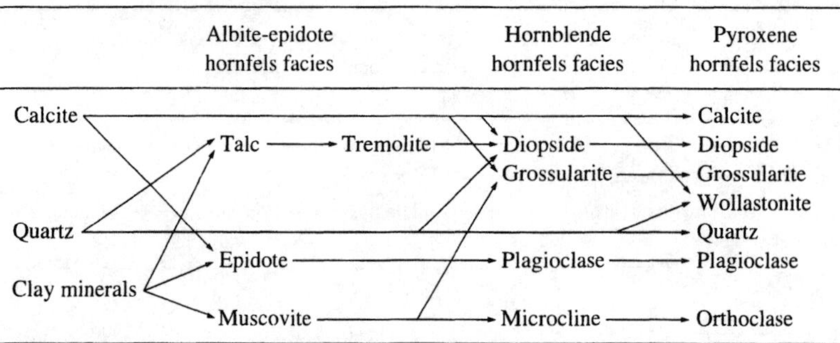

Note: Increasing grade of metamorphism is from left to right.

Source: After *Textbook of Lithology* by Kern C. Jackson. Copyright © 1970 by McGraw-Hill, Inc. Used with permission of McGraw-Hill Book Co.

in Tables 9-9 and 9-10 occur with increasing grade of metamorphism. At low grades of regional metamorphism, tremolite and epidote are the characteristic new minerals. Because of the availability of magmatic water, talc may form in contact metamorphic rocks. At higher grades, diopside, plagioclase, and grossularite develop, and the highest-grade rocks have, in addition, orthoclase and wollastonite. Wollastonite can also form at lower grades, depending on CO_2 pressure. The presence of dolomite in the original rock results in an increased abundance for Mg-bearing phases, and may allow the formation of forsterite, periclase, and brucite in addition to the minerals listed in Tables 9-9 and 9-10. Besides all these minerals, other minerals such as scapolite may be present if there have been metasomatic additions of components (especially Cl_2) from a contact metamorphic magma or from a circulating fluid during regional metamorphism. Carbonate rocks are particularly susceptible to metasomatic additions and subtractions. If metasomatism has occurred, the affected rock is known as a *skarn*. If no change has occurred other than loss of CO_2, the rock is referred to as a *calc-silicate rock*.

Mafic igneous rocks are the third common group of rocks of great importance in the study of metamorphic changes. The major changes for regional metamorphism of these rocks are listed in Table 9-11. The beginning mineralogy is simple, consisting mainly of labradorite and augite (i.e., a basalt). At an early stage of metamorphism, labradorite breaks down to epidote and albite,

TABLE 9-11 Regional Metamorphism of Mafic Igneous Rocks

	Greenschist facies		Amphibolite facies		Granulite facies
	Albite		Plagioclase		Plagioclase
Plagioclase	Epidote				Diopside
Augite	Actinolite		Hornblende		Hypersthene
	Chlorite		Almandine		Almandine-pyrope
		Biotite		Biotite	Orthoclase

Note: Increasing grade of metamorphism is from left to right.

Source: After *Textbook of Lithology* by Kern C. Jackson. Copyright © 1970 by McGraw-Hill, Inc. Used with permission of McGraw-Hill Book Co.

and augite breaks down to chlorite and actinolite. Next biotite forms (if sufficient K_2O is present) and then hornblende. The upper limit of the greenschist facies is marked by the formation of calcic plagioclase from albite and epidote. The rocks of the greenschist facies are characterized by the green minerals epidote, actinolite, and chlorite. Calcite is often present, indicating addition of CO_2. In the amphibolite facies the rocks commonly consist of hornblende and plagioclase. At the highest grades of metamorphism, hydroxyl-bearing minerals such as hornblende and biotite become unstable, and hypersthene, diopside, and orthoclase form. The changes produced by contact metamorphism are generally similar to those listed above for regional metamorphism.

Altered mafic igneous rocks have also been used to describe three less common facies, the prehnite-pumpellyite, blueschist, and eclogite facies. Under the low-temperature, moderate-pressure conditions occurring in thick accumulations of sediments and volcanic flows, the minerals of mafic volcanic rocks are changed to pumpellyite and prehnite (both similar in composition to epidote), which occur with more common metamorphic minerals such as albite and chlorite. At higher pressures, rocks of the blueschist facies develop. In addition to glaucophane [$Na_2(Mg, Fe)_3(Al, Fe)_2Si_8O_{22}(OH)_2$], these rocks contain such minerals as lawsonite [$CaAl_2Si_2O_7(OH)_2 \cdot H_2O$], jadeite [$NaAlSi_2O_6$], and aragonite. They form from the minerals of the greenschist and prehnite-pumpellyite facies in regions of high pressure due to tectonic activity. Under very high pressures and temperatures, the rock eclogite is formed from basalt. It is made up of omphacite pyroxene (jadeite-diopside) and garnet (grossularite-almandine). Eclogites are essentially the only rocks of the eclogite facies, since their manner of occurrence (as isolated blocks in metamorphic rocks of several different facies and as xenoliths in igneous rocks) makes it difficult to define coexisting mineral assemblages.

It is important to note that, in addition to changes in mineralogy with increasing metamorphic grade, changes in the composition and properties of individual minerals also occur. Biotites and hornblendes typically change from green colors at low grades to brown colors at high grades. Plagioclase feldspar increases in calcium content; garnet decreases in calcium and manganese content with increasing grade. Some minerals, such as chlorite and biotite, tend to show regular changes in their MgO/FeO ratios. In any detailed study of a group of metamorphic rocks, it is necessary to determine the compositions of the coexisting minerals in order to estimate the relative importance of rock composition, pressure factors, and temperature in producing the final assemblages.

GRAPHICAL REPRESENTATION OF MINERAL ASSEMBLAGES

Many of the diagrams discussed in Chapter Three, involving various combinations of intensive and extensive variables, may be used in studying either igneous or metamorphic rocks. One type of diagram mentioned there, the *ACF diagram,* is used mainly for metamorphic rocks (Figure 3-11). Coexisting mineral assemblages are plotted in terms of oxides using a triangular diagram. As explained in Chapter Three, Figure 3-11 is a representation of the various four-phase assemblages (all including quartz) for the contact metamorphic rocks described by Goldschmidt in 1911. The lines connecting the composition points of coexisting phases are known as *tie lines.* The A corner of the triangle represents ($Al_2O_3 + Fe_2O_3$) which is not combined with sodium or potassium in feldspars. The second corner, C, is CaO, and the third corner, F, is (Fe, Mg)O. SiO_2 is usually considered to be in excess (indicated by the presence of quartz). Thus andalusite would plot at A, wollastonite at C, and hypersthene at F. Other minerals would plot along the boundaries of the triangle (for example, staurolite) or within the triangle (for example, pumpellyite). Minerals with a variable composition plot as a line or an area. Only abundant minerals are plotted, with accessory minerals such as magnetite not plotted (although corrections are made for minor components such as TiO_2 in the chemical analysis of a rock).

The advantage of the ACF diagram lies in the fact that it provides a two-dimensional representation of the assemblages of many metamorphic rocks, and shows the gross relationships of conditions of metamorphism, rock chemical composition, and mineralogy. Thus the ACF diagrams for rocks of the same composition occurring in two different metamorphic facies would show the contrasting mineral assemblages of the two facies. These diagrams have a number of disadvantages. The treatment of (Mg, Fe)O as a single component is clearly incorrect in most cases. Iron and magnesium are thermodynamically independent (e.g., magnesium does not substitute for iron in the mineral staurolite). Similarly, Al^{3+} and Fe^{3+} do not always substitute for each other. Occurrence of biotite or muscovite in a rock requires corrections to the standard procedure for plotting the diagrams. Because of their simplicity, comparison of ACF diagrams for a related group of rocks often does not show what reactions have taken place in going from one zone to the next. Despite these drawbacks, ACF diagrams have been found useful ever since Eskola first proposed them in 1915. Details of their calculation can be found in Ehlers and Blatt (1982, Chap. 18).

Any three components can be used in constructing a triangular diagram. The other commonly used diagram of this type, also introduced by Eskola, is the *AKF diagram.* The CaO point of the ACF diagram is replaced by K_2O, allowing the plotting of potassium-bearing minerals (and ruling out the plotting of calcium minerals). The A corner of the triangle now represents $(Al_2O_3 + Fe_2O_3)$, minus potassium, sodium, and calcium in feldspars. The general procedure for constructing an AKF diagram is similar to that for the ACF diagram. Corrections have to be made for the presence of calcium-bearing minerals such as grossularite and epidote.

AKF diagrams have been used mainly for pelitic rocks. A better diagram for these rocks is the *AFM projection diagram* of Thompson (1957). This is a method of representing mineral assemblages in the system SiO_2-Al_2O_3-MgO-FeO-K_2O-H_2O. These six components (variables) can be reduced to five by considering SiO_2 to be always present in excess (indicated by the occurrence of quartz in the various assemblages). H_2O can be neglected by assuming its presence as a pure phase at the time each assemblage is formed, or by assuming an open system with the chemical potential of H_2O externally controlled (see discussion of the phase rule in Chapter Three). That leaves us with four components, which could be shown in the form of a tetrahedron (an *AKFM diagram*). Variations within a tetrahedron, as shown on paper in two dimensions, are difficult to visualize; thus Thompson suggested a projection onto an appropriate surface of minerals in equilibrium with muscovite (which is commonly found in metamorphosed pelitic rocks and that shows relatively little variation in composition in terms of AKFM components). The surface of projection is the plane Al_2O_3-MgO-FeO on one side of the AKFM tetrahedron (Figure 9-16).

The value of the AFM projection comes from its ability to show the relative compositions of coexisting minerals of variable composition such as biotite, staurolite, and almandine garnet. In contrast to ACF and AKF diagrams, which mainly relate rock composition to mineralogy, the emphasis in Thompson AFM diagrams is on the relationship of mineral compositions [particularly the ratio MgO/(MgO + FeO)] to metamorphic zones and metamorphic facies (Figure 9-12). The relative abundances of the minerals of an assemblage depend on the bulk composition of the rock, whereas the relative compositions of the coexisting minerals depend on the physical condi-

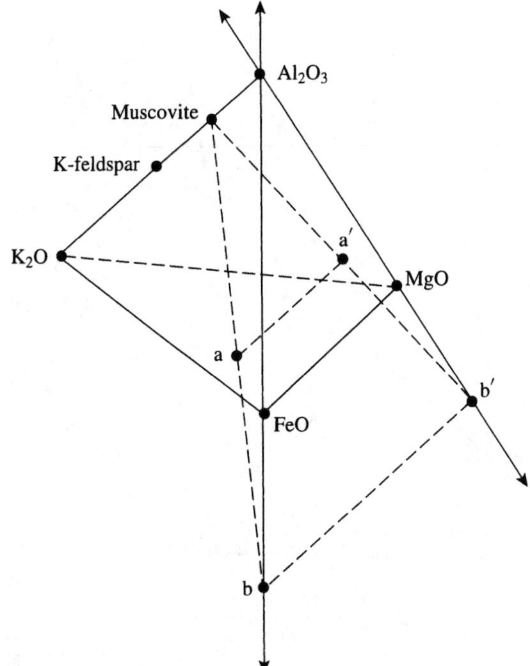

Figure 9-16 The tetrahedron Al_2O_3-K_2O-FeO-MgO showing projection through the muscovite point onto the Al_2O_3-FeO-MgO plane. K-feldspar projects to the Al_2O_3 point and to the indefinite extension of the Al_2O_3-FeO-MgO plane beyond bb′. Biotite compositions project from aa′ to bb′ and to the indefinite extension of the Al_2O_3-FeO-MgO plane beyond the Al_2O_3 point. The result is a diagram such as those shown in Figure 9-12. After Thompson (1957).

tions of metamorphism. Thus mineral composition is more important than rock composition in studying metamorphic rocks. Another advantage of AFM diagrams is that they represent five-phase assemblages (three projected minerals plus muscovite and quartz), whereas ACF and AKF diagrams represent only four-phase assemblages (three plotted minerals plus quartz). However, ACF and AKF diagrams can be used for a variety of rock types, whereas AFM diagrams are limited to pelitic rocks. It should be kept in mind that AFM diagrams, as with other diagrams, are less useful when other components, not treated in the diagram, occur in the minerals being studied. Further details on AFM diagrams can be found in Thompson (1957) and Ehlers and Blatt (1982).

It sometimes happens that the plotting of an ACF or other diagram results in crossing tie lines. There are several possible explanations. The mineral assemblages may be disequilibrium assemblages. Another possibility is that some of the assumptions required by the use of a particular diagram (such as free substitution of Mg and Fe for each other in ACF diagrams) may not hold true. Often the cause of the crossing tie lines is the occurrence of a minor component, such as manganese or titanium, in one or more minerals, resulting in a change in their stability relationships. If minerals have been plotted without the use of detailed mineral analyses, the crossing tie lines could be due to inexact location of the mineral compositions.

As mentioned earlier, a petrogenetic grid consists of intersecting univariant curves on a pressure-temperature diagram (Figure 9-17). Bowen (1940) proposed that such a grid, for a given bulk composition, could be used to determine the conditions of metamorphism for rocks of that composition. The mineral composition of a particular rock, assuming thermodynamic equilibrium and a constant activity of water, should fit into a specific part of the appropriate grid for specific pressures and temperatures of metamorphism.

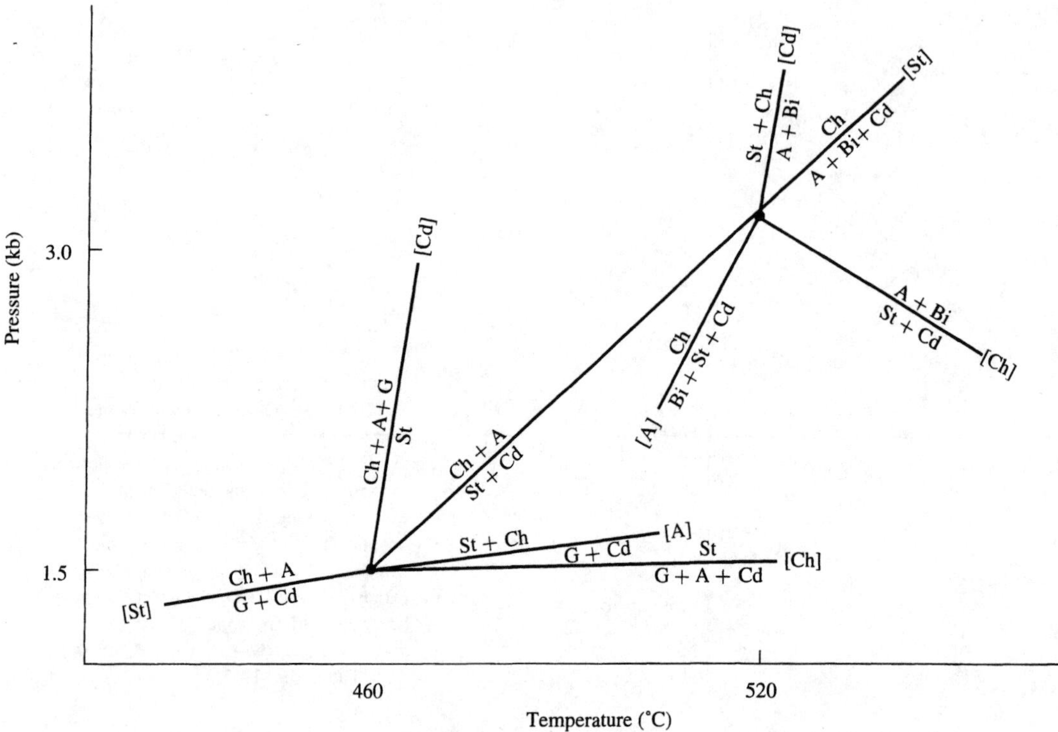

Figure 9-17 A partial petrogenetic grid for the system Al_2O_3-K_2O-FeO-MgO. Two invariant points are shown with five specific minerals in equilibrium at each point. The lower-temperature invariant point is for the minerals garnet, chlorite, staurolite, cordierite, and one of the Al_2SiO_5 minerals. The upper-temperature invariant point is for the minerals cordierite, staurolite, chlorite, biotite, and one of the Al_2SiO_5 minerals. The four minerals (plus/minus muscovite, quartz, and an aqueous fluid) that are in equilibrium at each univariant line are shown with the following labels: [Cd] = cordierite; [St] = staurolite; [Ch] = chlorite; [Bi] = biotite; [G] = garnet; [A] = one of the Al_2SiO_5 polymorphs. Depending on the pressure and temperature, andalusite, kyanite, or sillimanite could be the Al_2SiO_5 phase that occurs with the other minerals. The various univariant lines radiating from the two invariant points could be extended to intersect each other and to intersect added univariant lines to create other invariant points. In this way a petrogenetic grid is developed. (After Philpotts, A. R., *Principles of Igneous and Metamorphic Petrology*. Copyright © 1990 by Prentice Hall, Inc. Adapted by permission of Prentice Hall, Inc., Upper Saddle River, NJ.)

The simplest petrogenetic grid involves a one-component system, such as that shown in Figure 3-7. The triple point in Figure 3-7 is an invariant point and the three curves radiating out from this point are univariant curves representing two-phase equilibria. Only one phase is stable between the curves. Another way of describing this diagram is by the use of the phase rule (see Chapter Three). The number of phases coexisting in the areas between the curves is equal to the number of components in the system, c. The number of phases coexisting along the univariant lines is $c + 1$ and the number of phases coexisting at the invariant point is $c + 2$.

The system shown in Figure 9-17 is for reactions involving common minerals in the AFM diagram of Thompson (Philpotts 1990). The number of components that can vary is three (Al_2O_3, FeO, MgO) and thus the number of minerals coexisting at an invariant point is $c + 2 = 5$. Four minerals ($c + 1$) are in equilibrium along each univariant line of Figure 9-17. As the temperature and pressure vary, reactions among the various minerals occur and the spaces between the univariant lines represent the fields of stability of different three-phase mineral assemblages. An example of a reaction that takes place at one of the univariant lines in Figure 9-17 is

$$\underset{[St]}{\text{staurolite}} + \underset{[Ch]}{\text{chlorite}} = \underset{[Bi]}{\text{biotite}} + \underset{[A]}{\text{kyanite}} + H_2O \qquad (9\text{-}14)$$

This line is labeled "Cd" in Figure 9-17 (upper right corner) because cordierite is the one mineral of the five coexisting at the invariant point that is not involved in the above reaction. The other univariant lines are labeled in a similar manner. The three minerals that coexist in the area between the [Cd] and [St] lines in the upper right corner of the diagram are kyanite (A), biotite (Bi), and chlorite (Ch).

A complete petrogenetic grid is much more complicated than Figure 9-17. A complete grid has to show all the possible reactions among all the possible minerals in a system. A number of different grids have been proposed for different rock compositions. These are reviewed in the books by Spear (1993), Miyashiro (1994), and Bucher and Frey (1994). Certain assumptions and limitations pertain to all petrogenetic grids. For example, it is usually assumed that water has a fixed activity (most commonly $a_{H_2O} = 1$). This may not be a good assumption for some metamorphic rocks. The occurrence of solid solution in many minerals affects the location of univariant lines. Even without the limitation of solid solution, the lines are difficult to locate precisely, whether done by experiment or by thermodynamic calculation. Despite such problems, when used properly, petrogenetic grids provide a useful method for estimating the temperatures and pressures of metamorphism for a particular mineral assemblage.

EXAMPLES OF CONTACT METAMORPHIC ROCKS

Marysville Stock, Montana

The Marysville granodiorite is a small stock associated with the Late Cretaceous Boulder batholith. The stock intruded the Precambrian silica-rich, argillaceous Helena Dolomite, producing a contact aureole up to three kilometers in width (Figure 9-18). Rice (1977) reported the occurrence of divariant and univariant mineral assemblages separated by isograds characterized by invariant assemblages. He concluded that fluid compositions (H_2O and CO_2) were buffered along isobaric univariant reaction lines, with externally derived fluid not involved. This model assumes instantaneous equilibrium between fluids and rocks (infinite chemical reaction rates). The initial rock is assumed to have the composition dolomite–calcite–quartz–K-feldspar.

The mineralogical evolution suggested by Rice (1977) can be depicted on an isobaric temperature-X_{CO_2} diagram (Figure 9-19). This diagram shows the change in temperature and fluid composition with time at a fixed point in the contact aureole. Starting with an initial fluid composition of $X_{CO_2} = 0.1$, addition of heat increases the temperature until the following univariant reaction boundary is reached:

$$3\ \text{dolomite} + \text{K-feldspar} + H_2O = \text{phlogopite} + 3\ \text{calcite} + 3\ CO_2 \qquad (9\text{-}15)$$

As more heat is added, the mole fraction of CO_2 in the fluid increases, the mole fraction of H_2O decreases, and the fluid composition moves along the univariant reaction line toward point C (an invariant point). This assumes the fluid remains in contact with, and in equilibrium with, the local rock mass. According to Rice and Ferry (1982), only a small amount of phlogopite would be produced by the time the invariant point is reached.

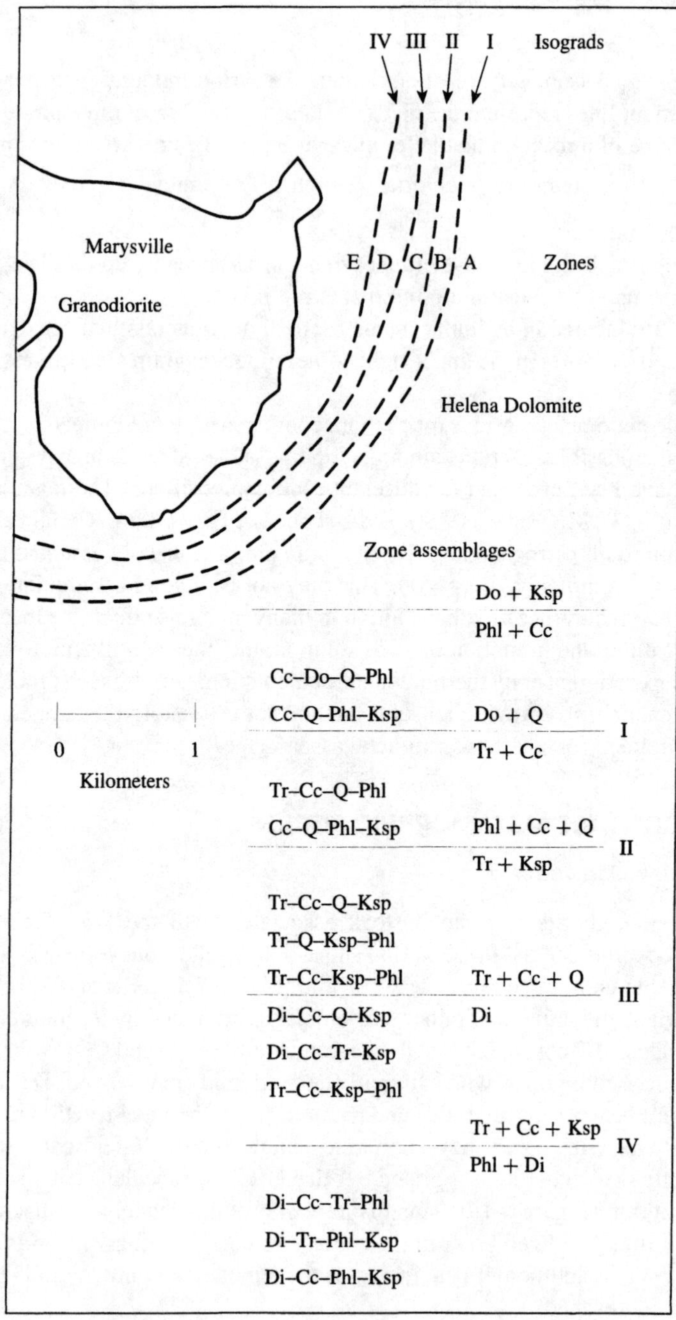

Figure 9-18 Map of the eastern portion of the Marysville stock, Montana, showing mineral zones A–E, diagnostic zone mineral assemblages, and isograds I–IV in the Marysville aureole. Univariant reactions that characterize the isograds are shown with the zone assemblages by listing the pertinent minerals above and below the isograd lines. Mineral abbreviations are as follows: Cc = calcite; Q = quartz; Do = dolomite; Ksp = K-feldspar; Phl = phlogopite; Tr = tremolite; Di = diopside. [After Rice (1977). Reprinted by permission of American Journal of Science.]

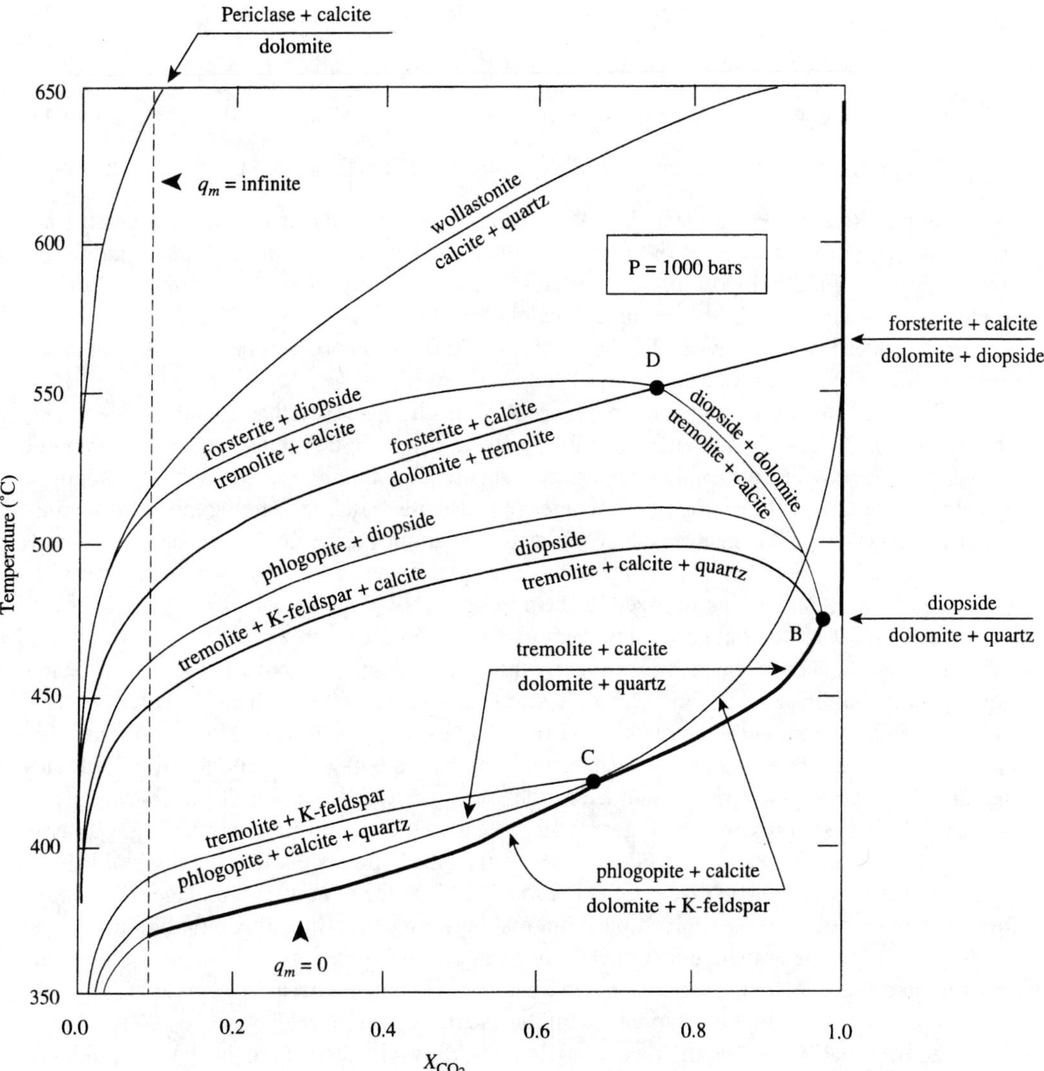

Figure 9-19 Isobaric temperature-X_{CO_2} conditions of some equilibria in the system $KAlO_2$-CaO-MgO-SiO_2-CO_2-H_2O (solid curves). Points B, C, and D are invariant points. Thick shaded curve illustrates the temperature-X_{CO_2} evolution of a hypothetical siliceous dolomitic limestone (such as the Helena Dolomite in Table 9-12) with near-zero porosity and fluid of initial composition $X_{CO_2} = 0.1$ heated from 350°C to 650°C without infiltration of fluid ($q_m = 0$). Dashed vertical line illustrates the temperature-X_{CO_2} evolution of the same rock heated from 350°C to 650°C and infiltrated by an infinite time-integrated flux of CO_2-H_2O fluid with composition $X_{CO_2} = 0.1$. After Ferry (1991).

At the invariant point C, the following reactions would occur simultaneously at constant temperature (Rice and Ferry 1982):

$$3 \text{ dolomite} + \text{K-feldspar} + H_2O = \text{phlogopite} + 3 \text{ calcite} + 3\ CO_2 \qquad (9\text{-}16)$$

$$5 \text{ dolomite} + 8 \text{ quartz} + H_2O = \text{tremolite} + 3 \text{ calcite} + 7\ CO_2 \qquad (9\text{-}17)$$

$$\text{phlogopite} + 2 \text{ dolomite} + 8 \text{ quartz} = \text{tremolite} + \text{K-feldspar} + 4\ CO_2 \qquad (9\text{-}18)$$

$$5 \text{ phlogopite} + 6 \text{ calcite} + 24 \text{ quartz} = 3 \text{ tremolite} + 5 \text{ K-feldspar} + 6\ CO_2 + 2\ H_2O \qquad (9\text{-}19)$$

According to Rice and Ferry, the net result at invariant point C is the production of tremolite and K-feldspar and the consumption of phlogopite, quartz, calcite, and dolomite. Further addition of heat causes the fluid composition to move along the univariant curve of reaction 9-17 above (dolomite + quartz + H_2O $\rightarrow$ tremolite + calcite + CO_2) to invariant point B, where four more reactions proceed simultaneously. Further details of the fluid composition path for Marysville are given in Rice (1977) and Rice and Ferry (1982).

The predicted mineralogy from Figure 9-19 is the same as that found at Marysville (Figure 9-18). Isobarically invariant assemblages (e.g., calcite + dolomite + quartz + tremolite + phlogopite + K-feldspar), univariant assemblages (e.g., dolomite + K-feldspar + phlogopite + calcite + quartz), and divariant assemblages (e.g., calcite + quartz + phlogopite + dolomite) occur at Marysville. The changes occurring at the invariant points can be mapped as isograds (lines of equal intensity of metamorphism). For example, in Figure 9-18, isograd I separates mineral zones A and B and is characterized by the appearance of tremolite and K-feldspar. Thus isograd I in Figure 9-18 corresponds to invariant point C in Figure 9-19.

The significance of the path followed by the fluid as described above is that it is an example of internal buffering of metamorphic fluid composition by reaction with mineral assemblages. In other words, the mineral assemblages control the composition of the coexisting fluid. The other extreme would be a case where an abundance of external fluid of a given composition infiltrates the contact metamorphic aureole and determines the mineralogy that develops. If this were to occur for the rocks represented by Figure 9-19, then the fluid composition would follow a vertical line such as the dashed line in Figure 9-19. Ferry (1991) points out that the mineral assemblages for the two cases would be different. Table 9-12 lists the mineral assemblages that would form in the two extreme cases of complete internal buffering (zero flux of external fluid) and infiltration of an infinite amount of external fluid. Ferry also notes that, except initially at low temperatures and finally at high temperatures, no divariant assemblages occur in the case of zero flux. In the case of infinite flux, no invariant assemblages are likely to develop.

Lattanzi et al. (1980) reported isotopic data for Marysville contact rocks that show a large depletion in both ^{13}C and ^{18}O. In many contact metamorphic aureoles involving carbonate rocks, values of $\delta^{13}C$ and $\delta^{18}O$ decrease as the degree of metamorphism increases (Valley 1986). Lattanzi et al. explained their results for Marysville as due to decarbonation (continuous loss of CO_2) combined with silicate disequilibrium. While the ^{13}C data can be explained by continuous loss of CO_2 that was in local isotopic equilibrium with carbonate minerals, silicate disequilibrium is necessary to explain the ^{18}O data. The amount of locally-derived pore waters is insufficient to explain the large change in ^{18}O values and therefore external fluids (meteoric waters) must have been involved. Thus the petrologic study by Rice (1977) supports the occurrence of local chemical equilibrium in the aureole at Marysville while the isotopic results reported by Lattanzi et al. indicate isotopic disequilibrium during contact metamorphism.

TABLE 9-12 Mineral Assemblages Developed in Model Siliceous Carbonate Rocks between 350° and 650°, $P_{CO_2} + P_{H_2O} = P_{fluid} = 1$ kb.

Siliceous dolomite (example: Notch Peak dolomitic limestone)		K-Al-bearing siliceous dolomitic limestone (example: Helena Dolomite)	
Zero flux	**Infinite flux**	**Zero flux**	**Infinite flux**
Dol-Qtz	Dol-Qtz	Dol-Cal-Qtz-Kfs	Dol-Cal-Qtz-Kfs
Dol-Cal-Qtz-Tlc	Dol-Cal-Qtz-Tlc	Dol-Cal-Qtz-Kfs-Phl	Dol-Cal-Qtz-Kfs-Phl
Dol-Cal-Qtz-Tlc-Tr	Dol-Cal-Tlc	Dol-Cal-Qtz-Kfs-Phl-Tr	Dol-Cal-Qtz-Phl
Dol-Cal-Qtz-Tr	Dol-Cal-Tlc-Tr	Dol-Cal-Qtz-Kfs-Tr	Dol-Cal-Qtz-Phl-Tr
Dol-Cal-Qtz-Tr-Di	Dol-Cal-Tr	Dol-Cal-Qtz-Kfs-Tr-Di	Cal-Qtz-Phl-Tr
Dol-Cal-Qtz-Di	Dol-Cal-Tr-Fo	Dol-Cal-Qtz-Kfs-Di	Cal-Qtz-Phl-Tr-Kfs
Dol-Cal-Di	Dol-Cal-Fo	Cal-Qtz-Kfs-Di	Cal-Qtz-Tr-Kfs
Dol-Cal-Di-Fo	Dol-Cal-Fo-Per		Cal-Qtz-Tr-Kfs-Di
Dol-Cal-Fo	Cal-Fo-Per		Cal-Qtz-Kfs-Di
			Cal-Qtz-Kfs-Di-Wo
			Cal-Kfs-Di-Wo

Mineral assemblages listed in order of increasing temperature. Zero flux corresponds to increase in temperature in rock with small porosity and no infiltration; initial pore fluid composition $X_{CO_2} = 0.1$. Infinite flux corresponds to infiltration of rock by infinite time-integrated fluid flux with composition $X_{CO_2} = 0.1$ (flow direction irrelevant).

Definition of mineral symbols: Tlc–talc; Cal–calcite; Qtz– quartz; Tr–tremolite; Dol–dolomite; Di–diopside; Phl–phlogopite; Kfs–K-feldspar; Fo–forsterite; Wo–wollastonite; Per–periclase.

Source: After Ferry (1991).

A model to explain both the petrologic and isotopic data for Marysville has been proposed by Lasaga and Rye (1993). Lasaga and Rye calculated a number of possible paths for metamorphic fluids assuming an open, nonequilibrium system. They included fluid transport properties and the kinetics of heterogeneous chemical reactions in their general model. Using this model they developed other possible paths on a diagram such as Figure 9-20. The paths crossed or followed metastable reaction lines (not shown in Figure 9-20) and required that several reactions occur simultaneously at many places along the paths. In contrast, the equilibrium model described above suggests that little reaction takes place along univariant lines and that most of the change in mineralogy occurs at invariant points. The model of Lasaga and Rye proposes that extensive reaction can occur far from invariant points and that these points may never be reached. They state that the mineral assemblages exhibited at Marysville could have been produced by either an equilibrium model or by a kinetic, nonequilibrium model. If they are correct, then the identification of isograds using isobaric invariant points could lead to erroneous interpretations.

Notch Peak Pluton, Utah

One of the most thoroughly studied contact aureoles is located in west-central Utah. In this area the Jurassic Notch Peak pluton discordantly intrudes Middle and Upper Cambrian limestones and calcareous argillites (Hover-Granath et al. 1983). Figure 9-21 shows part of the aureole formed in the Big Horse Limestone Member of the Orr Formation. Two lithologies of the Big Horse have been studied in detail in the contact zone, low-silica dolomitic limestones and impure

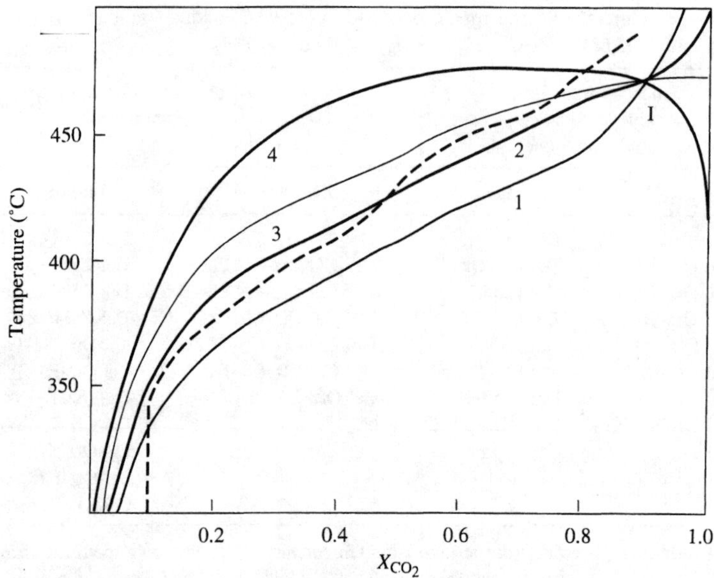

Figure 9-20 Temperature-X_{CO_2} evolution diagram for an H_2O-CO_2 metamorphic fluid 750 meters from the Marysville igneous contact. Each of the four lines represents a reaction that is important in the Marysville contact aureole as follows:

1. $3\ Do + Ksp + H_2O = Phl + 3\ Cc + 3\ CO_2$
2. $5\ Do + 8\ Q + H_2O = Tr + 3\ Cc + 7\ CO_2$
3. $Phl + 2\ Do + 8\ Q = Tr + Ksp + 4\ CO_2$
4. $5\ Phl + 6\ Cc + 24\ Q = 3\ Tr + 5\ Ksp + 6\ CO_2 + 2\ H_2O$

Point I is an invariant point. Possible reaction paths, such as that shown by the dashed line, can be calculated by using assumptions for variables such as fluid flow rates, diffusion coefficients, and kinetic rate constants. The dashed line shows the changes in fluid temperature and X_{CO_2} with time at a fixed point in the aureole. It crosses the univariant lines (parts of which are for metastable reactions) and represents a possible path for a nonequilibrium, open system where reaction kinetics are important. Another possible path for the fluid is as follows. Assume the path of the fluid, starting below line 1, were to intersect line 1, then move along line 1 to the invariant point I, then move along a line running from point I to another invariant point (not shown) and so forth. At some point the path could cross from one line to another, then follow that line. This would be the expected overall path for local equilibrium in a closed system between the fluid and the rocks through which it passed (Rice 1977). This interpretation is shown in Figure 9-19. According to Lasaga and Rye (1993), nonequilibrium reaction paths, such as the one shown here, can produce a situation where *several* metamorphic reactions (including metastable reactions) are operating at the same time. An equilibrium path, such as that described above, would generally follow the reaction lines and thus have only one reaction occurring at a given time. The results, as mineral assemblage zones, could be similar for both situations. Mineral abbreviations for the above equations are as follows: Do = dolomite; Ksp = K-feldspar; Phl = phlogopite; Cc = calcite; Q = quartz; Tr = tremolite. [After Lasaga and Rye (1993). Reprinted by permission of American Journal of Science.]

argillaceous limestones (argillites). In addition to detailed documentation of the petrology and mineralogy of the contact aureole, a large amount of isotopic data is also available (Nabelek et al. 1984).

The dolomitic limestones are representative of rocks that would show the mineralogical evolution listed in Table 9-12. The two groups of mineral assemblages (for no infiltration of fluid

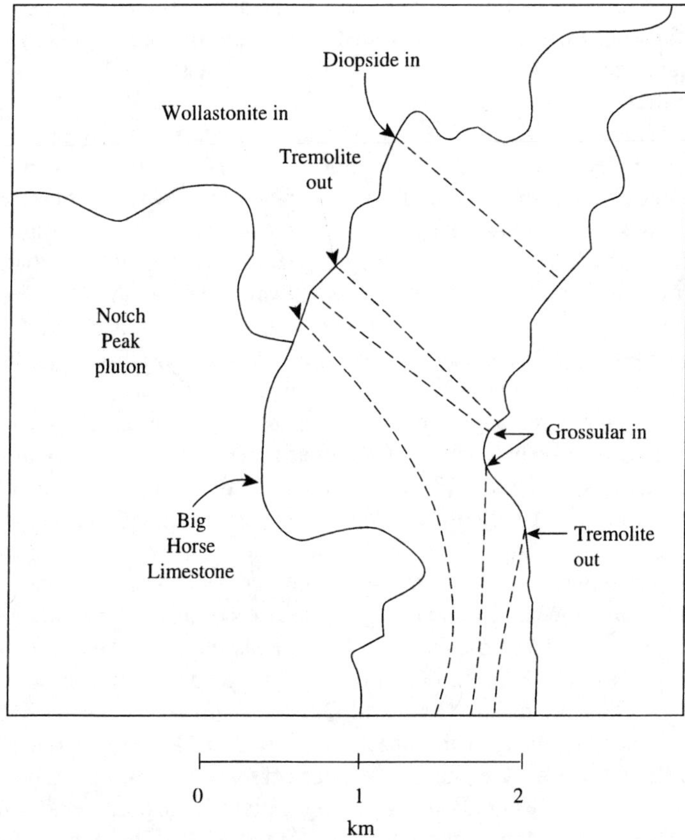

Figure 9-21 Geologic and metamorphic map of the eastern and northeastern portion of the Notch Peak contact metamorphic aureole, Utah. Models described by Ferry and Dipple (1992) attempt to reproduce the observed sequence of four isograds, the spacing between isograds, and the occurrence of $\delta^{18}O$ values that vary from as low as 8 per mil next to the pluton to 19 per mil in unmetamorphosed limestone two or more kilometers from the pluton (see Figure 9-22). The isograds are labeled as follows: "mineral in" indicates the appearance of that mineral with increasing grade of metamorphism and "mineral out" indicates the disappearance of the mineral with increasing grade of metamorphism. After Hover-Granath et al. (1983) and Ferry and Dipple (1992).

and for an infinite flux of fluid over time) can be compared to the two assemblages shown in Table 9-12 for a bulk composition similar to that of Helena Dolomite at Marysville (discussed in the previous section). The Helena Dolomite contains significant amounts of quartz and K-feldspar while the Notch Peak dolomitic limestones have little or no quartz and K-feldspar.

Ferry (1991) uses the results shown in Table 9-12 and other calculated data in conjunction with results of various field studies to conclude the following about contact metamorphism:

1. fluid-rock interaction occurs in all contact metamorphic aureoles, but the amount of fluid involved is finite; and
2. some rocks (for example, calc-silicate hornfelses) are permeable and act as metamorphic aquifers while other rocks (for example, carbonate-rich marbles) are relatively less permeable and act as metamorphic aquitards.

In the Notch Peak aureole the limestones have pristine and unaltered oxygen-isotope compositions, while the argillites, which are interbedded with the limestones, exhibit ^{18}O depletions caused by metamorphic fluid-rock interaction. Thus the limestones acted as aquitards and the argillites as aquifers.

Ferry and Dipple (1992) tested several models of contact metamorphism in an attempt to reproduce the prograde mineralogical and oxygen-isotopic evolution of the argillites in the Notch Peak aureole. They used equations grounded in transport theory to predict the changes that would result from various types of fluid-rock interaction. The three types of models studied were: (1) no fluid flow; (2) flow in the direction of increasing temperature (toward the Notch Peak pluton); and (3) flow in the direction of decreasing temperature (away from the pluton). The specific features they tried to predict with each model were (1) the occurrence and position (in map view) of the isograds shown in Figure 9-21 and (2) irregular depletions of ^{18}O adjacent to the pluton contact and a gradual increase of $\delta\,^{18}O$ with increasing distance from the pluton.

Table 9-13 lists the observed distances from the pluton contact at which various mineralogical changes occurred and also lists the $\delta\,^{18}O$ values near the contact and approximately one kilometer from the contact. Also in Table 9-13 are the predicted values for three of the models calculated by Ferry and Dipple (1992). Their no-flow models, including the one in Table 9-13, failed to predict the formation of either grossular garnet or wollastonite. They also predicted virtually no change in whole-rock oxygen-isotope compositions during metamorphism. Thus the no-flow models do not reproduce the mineralogy and isotopic properties of the Notch Peak contact aureole.

Metamorphism with fluid flow from the Notch Peak pluton in the direction of decreasing temperature was proposed by Labotka et al. (1988). The magmatic fluid was assumed to be pure H_2O with $\delta\,^{18}O = +8.5$ per mil by Ferry and Dipple (1992) in the calculated model listed in Table 9-13. This model reproduced the observed isograds in the proper order and maximum whole-rock ^{18}O depletions near the pluton of the correct magnitude (about 10 per mil). However, the predicted positions of the isograds are closer to the pluton than observed. Further, the continuous range of $\delta\,^{18}O$ values in the contact aureole is not predicted. Instead the model predicts a narrow range of low $\delta\,^{18}O$ values near the pluton contact and high $\delta\,^{18}O$ values further away. Related models for fluid flow away from the pluton also do not predict the observed isograds and isotopic values in a specific fashion. Thus Ferry and Dipple conclude that models involving fluid flow away from the pluton are quantitatively unsatisfactory.

TABLE 9-13 First-order Isotopic and Petrologic Features of the Notch Peak Contact Aureole, Utah. Distances Measured away from the Pluton with Zero at the Pluton-Argillite Contact

Feature	Observed	No-flow model	Up-temperature flow model	Down-temperature flow model
Appearance of diopside	1.4–2.2 km	1.80 km	1.80 km	0.59 km
Disappearance of tremolite	0.7–1.2 km	0.21 km	0.98–1.18 km	0.59 km
Appearance of grossular garnet	0.6–1.1 km	Not predicted	0.86–1.16 km	0.43 km
Appearance of wollastonite	0.4–0.8 km	Not predicted	0.47–0.66 km	0.42
$\delta\,^{18}O$ at or near contact	8–15‰	18.7‰	9–15.4‰	9.3‰
$\delta\,^{18}O$ one km from contact	12–19‰	19.0‰	11.8–15.4‰	19.3‰

The down-temperature flow model was proposed by Labotka et al. (1988).

Source: Data from Hover-Granath et al. (1983), Nabelek et al. (1984), and Labotka et al. (1988). After Ferry and Dipple (1992).

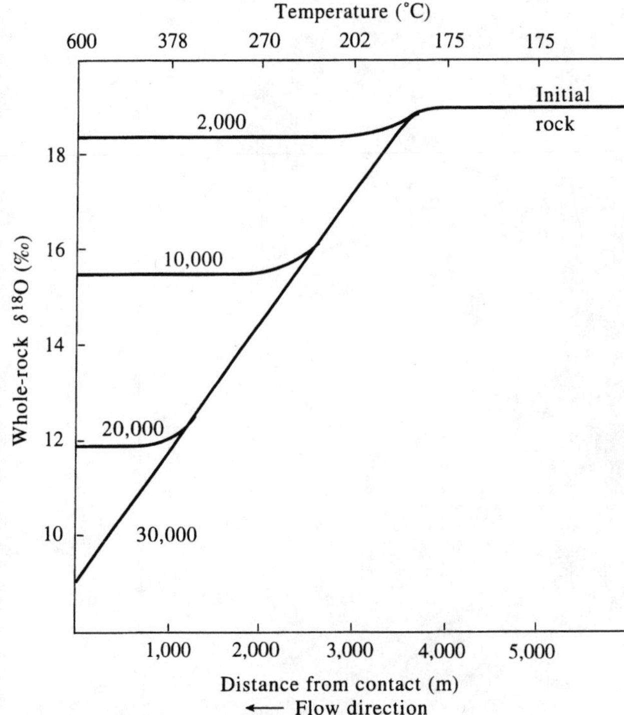

Figure 9-22 Calculated whole-rock $\delta\,^{18}O$ values of calcareous argillites in the Notch Peak contact metamorphic aureole, Utah, as a function of distance from the pluton-argillite contact. The curves are calculated for four different values of time-integrated mole fluid flux and for flow in the direction of increasing temperature (toward the contact). Curves for the four different fluxes superimpose along their inclined portions and at distances greater than 4000 m. Most observed values of $\delta\,^{18}O$ in the aureole can be explained by flow of 15,000–25,000 mol/cm^2 fluid through the contact aureole. After Ferry and Dipple (1992).

The third model in Table 9-13, involving flow of fluid inward toward the pluton (in the direction of increasing temperature), also reproduces all observed isograds in the proper order and maximum whole-rock ^{18}O depletions near the pluton of the correct magnitude. However, in this case, the model also predicts locations for the isograds that match observed locations in the field. For a flux of fluid equal to about 20,000 mol/cm^2, the model reproduces 70 percent of the measured whole-rock isotopic compositions to within about 1 per mil $\delta\,^{18}O$ (Figure 9-22). The other 30 percent of the values can be explained if some of the argillites had a lower permeability compared to the majority of the argillites. Thus this third model explains both qualitatively and quantitatively the first-order isotopic and mineralogical features of the Notch Peak aureole. One possible source of the fluid in the third model is devolatilization reactions in the sedimentary rocks that produced H_2O and CO_2. (For examples, see the reactions listed in the previous section on the Marysville aureole.) An additional source could have been connate water contained in pores in the rocks. Finally, some magmatic fluid could have mixed with the main fluid in the inner portion of the contact aureole.

EXAMPLES OF REGIONAL METAMORPHIC ROCKS

Western California

Two major units of Jurassic-Cretaceous rocks occur in western California (Figure 9-23). The Franciscan assemblage consists of up to 15,000 meters of graywacke (by far the major rock type), shale, mafic volcanic rock, chert, and limestone (Bailey et al. 1964). Some of these rocks have

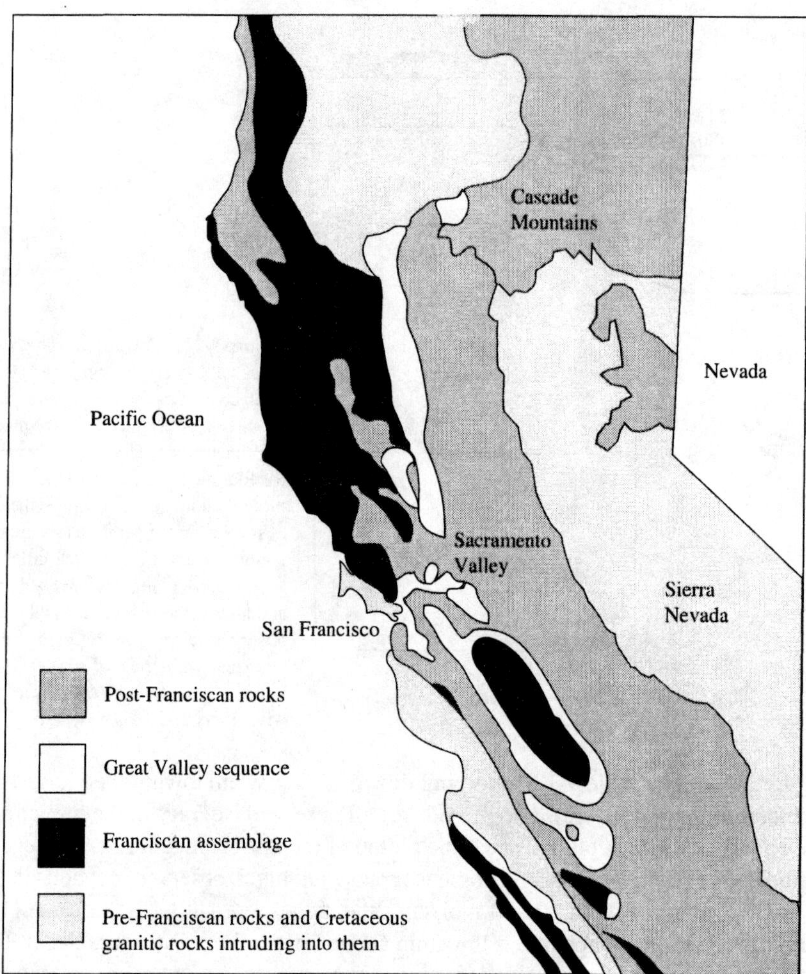

Figure 9-23 Geologic map of California showing the relationship of the Great Valley sequence to the Franciscan assemblage. The two units are everywhere separated by thrust faults, including a regional thrust fault along the eastern boundary of the Franciscan rocks. Metamorphic zones within the Franciscan north of San Francisco consist of the following facies, from east to west: blueschist, prehnite-pumpellyite, zeolite. After Bailey et al. (1970).

been altered to metamorphic rocks of the zeolite, prehnite-pumpellyite, and blueschist facies (Jayko et al. 1986). The ages of metamorphism range from Late Jurassic to Late Cretaceous. The base of the assemblage is not exposed. East of the Franciscan rocks is the Great Valley sequence (also up to 15,000 meters thick) consisting of graywacke, shale, and conglomerate. The sedimentary rocks are essentially unmetamorphosed except for the beginnings of burial metamorphism at the base of the sequence. In areas near the Franciscan rocks, the Great Valley sequence lies depositionally on top of mafic volcanic rocks (900 meters thick), which in turn rest on up to 1,500 meters of serpentinized peridotite (Bailey et al. 1970). A low-angle, regional thrust fault separates the serpentinite at the base of the Great Valley sequence from the adjacent Franciscan rocks. Movement along the thrust has brought the Great Valley rocks over the Franciscan assemblage. Imbricated klippen of the Great Valley rocks lie discordantly above deformed Franciscan rocks along the thrust. Throughout the Franciscan, faults are abundant.

The metamorphic rocks of the Franciscan assemblage are zoned, with the blueschist rocks next to the thrust fault, prehnite-pumpellyite rocks further west, and zeolite-facies rocks along the present coast. The metamorphic zonation is "upside down," since metamorphic grade increases upward in the rock section from zeolite facies to prehnite-pumpellyite facies to blueschist facies, rather than downward as might be expected for burial metamorphism (Blake et al. 1967). The most abundant blueschist rocks consist of masses characterized by jadeite and lawsonite (found in metamorphosed graywackes) and masses characterized by glaucophane (found in metamorphosed mafic volcanics). Metavolcanics and metagraywackes of the prehnite-pumpellyite facies contain abundant pumpellyite. Rocks of the zeolite facies show metamorphism chiefly by the presence of laumontite. In addition to these rocks, eclogites and related rocks occur as isolated blocks (up to 100 meters in diameter) in unmetamorphosed and metamorphosed Franciscan rocks. These blocks were obviously tectonically emplaced and are not now in their place of origin.

In contrast to the Franciscan rocks, metamorphism is directly related to depth of burial in the Great Valley sequence (Dickinson et. al 1969). Postdepositional albitization of plagioclase and chloritization of biotite in detrital sediments increase systematically with age. Widespread alteration is found in rocks that were buried at depths of more than 6,000 meters. Prehnite and pumpellyite have been found in rocks that were buried to depths of 13,000 meters. Thus the Great Valley sequence shows the typical metamorphism resulting from deep burial of volcanic and sedimentary rocks.

A tectonic model for the relationship between the Franciscan assemblage and the Great Valley sequence has been proposed by Bailey et al. (1970) and Ernst (1971). According to this model, these rocks represent material of a Mesozoic arc-trench subduction zone, such as that shown in Figure 9-14. The Franciscan rocks formed on the oceanic side of the trench and consisted of ocean-floor sediments mixed depositionally and tectonically with oceanic crust. The Great Valley sequence represents detritus collected in a sedimentary zone extending from the arc-trench area to the continental margin. Part of the sequence was deposited on oceanic crust, represented today by the chert, basalt, and serpentinite at the base of the sequence and next to the regional thrust fault that separates the Great Valley sequence from the Franciscan rocks. Farther east, deposition was on continental crust. Because of sea-floor spreading, the Franciscan rocks moved toward and beneath the Great Valley sequence. The Franciscan rocks experienced high-pressure, low-temperature metamorphism near the thrust fault zone, producing rocks of the blueschist facies.

Ernst (1988) discusses how the tectonic history of subduction zones can be inferred from retrograde blueschist pressure-temperature paths. He describes a range of conditions defined by two end members: (1) the retrograde path (decreasing pressure with nearly constant temperature) causes blueschist rocks to be overprinted by greenschist and/or epidote amphibolite facies assemblages; and (2) the retrograde path retraces the prograde path in such a way that the prograde mineral assemblages are preserved. The first type of blueschist belt is found in the western Alps, while the second type is represented by the Franciscan blueschists (Figure 9-24). The blueschist rocks, after decoupling from the descending slab in a subduction zone, can ascend toward the surface in several different tectonic modes. For example, ascent can be due to buoyancy as imbricate slices (Ernst 1977) or as a result of forced convection (Cloos 1982).

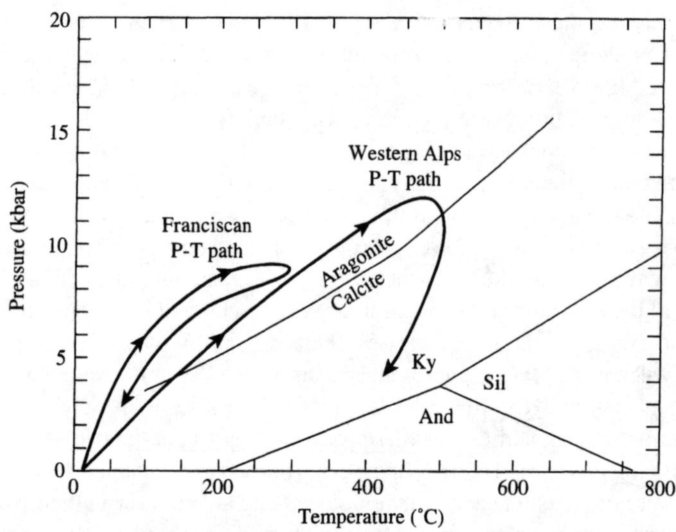

Figure 9-24 Pressure-temperature-time paths for metamorphism in the western Alps and for the Franciscan rocks of California (Ernst 1988). Also shown are the Al_2SiO_5 phase relations (Holdaway 1971) and the calcite-aragonite phase boundary (Johannes and Puhan 1971). The western-Alps-type pressure-temperature-time path crosses into the calcite stability field at sufficiently high temperature that all aragonite is converted to calcite. The Franciscan-type pressure-temperature-time path crosses into the calcite field at a temperature sufficiently low to preserve metamorphic aragonite. After Spear (1993).

Ernst (1988) suggests that the blueschist rocks in the western Alps ascended as a result of deceleration or cessation of plate underflow and buoyant rise of the stranded subduction complex. The change in subduction dynamics could be due to a different type of material (e.g., a continental fragment) entering the subduction zone. For the Franciscan rocks, Ernst (1988) describes a slow migration back up the subduction zone in response to isostatic forces during continued plate descent (Figure 9-25). Thus a correlation seems to exist between the nature of retrograde pressure-temperature paths and long-continued, unchanging subduction flow (Franciscan type) versus irregular and nonuniform plate descent (western Alps type).

Cloos (1985) has modeled heat flow through a wedge-shaped accretionary complex emplaced beneath the overriding plate at a convergent plate boundary and related the results to K-Ar and $^{40}Ar/^{39}Ar$ ages of blueschists in the Franciscan rocks. He points out that material recrystallized at depth where the wedge is narrow (see Figure 9-25) can be rapidly heated to high temperatures and then rapidly cooled to temperatures low enough to prevent argon loss from mineral structures.* At shallower depths, where the wedge is wider, maximum temperatures are lower and the periods of heating and cooling are longer. Because of this difference in thermal history, Cloos suggests that K-Ar and $^{40}Ar/^{39}Ar$ recrystallization ages for blueschists from the upper part of the wedge can be younger by tens of millions of years than recrystallization ages of material at greater depth, even though both areas were metamorphosed at the same time. In the case of the Franciscan

*A cooling age represents the time elapsed since a mineral cooled to a temperature below which it can effectively retain radiogenic daughter isotopes. The ages of many minerals in high grade metamorphic rocks reflect the time of cooling from the last metamorphism. These ages are sometimes referred to as recrystallization ages.

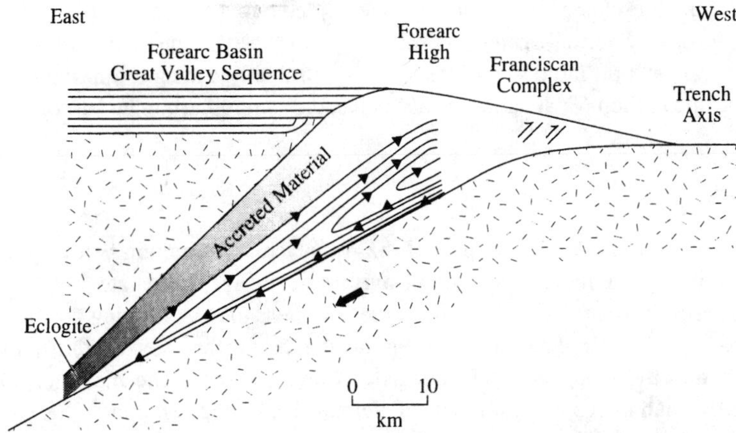

Figure 9-25 Schematic cross section of the subduction zone involved in the formation of the Franciscan assemblage. Subducted oceanic sediments and crust form a wedge in which high-pressure, low-temperature conditions develop, causing blueschist and eclogite formation. As subduction continues, according to the model shown here, forced convection occurs and carries masses of blueschist and eclogite back to the surface. The accreted material has been welded to the plate on the left. After Cloos (1982).

complex, this interpretation implies that eastern rocks with Ar-isotopic ages of 110–125 million years could have formed at the same time as the more strongly metamorphosed western rocks with Ar-isotopic ages of 140–150 million years. An alternative explanation of the 20–30 million year age difference invokes the occurrence of two different subduction events with subduction starting, stopping, and starting again. Cloos suggests that the age patterns found across many blueschist belts are the result of one continuous period of subduction and are controlled by (1) the time since initiation of subduction and (2) the position of individual rock masses within the accretionary complex.

Figure 9-24 shows the calcite-aragonite boundary at high temperatures and pressures. Spear (1993, 738) points out that all blueschist terranes should contain aragonite as the stable high-pressure polymorph. Often calcite is found instead of aragonite. Which mineral occurs in blueschists depends on the pressure-temperature conditions at the point where the pressure-temperature path crosses the phase boundary. For the western Alps the boundary is crossed at high pressure and temperature values. Under these conditions, aragonite is changed to calcite (Figure 9-24). The Franciscan path crosses the phase boundary at much lower values of pressure and temperature; the kinetics of the reaction prevent the phase change from occurring and metastable aragonite is preserved in the rocks (Carlson and Rosenfeld 1981).

Magaritz and Taylor (1976) have studied hydrogen, oxygen, and carbon isotopic variations in Franciscan rocks and minerals. They found that metamorphism produces no change in the $\delta^{18}O$ of graywackes, enrichment in the ^{18}O content of igneous rocks, and depletion of the ^{18}O content of cherts. Mineral assemblages are generally not in oxygen isotopic equilibrium, thus preventing the estimation of metamorphic temperatures using coexisting mineral pairs (Chapter Two). The δD values of Franciscan rocks are uniform and do not show any significant change with degree of metamorphism. The general uniformity of the $\delta^{18}O$ and δD rock values suggests that the H_2O involved in Franciscan metamorphism must have been isotopically quite uniform.

Fluids involved in blueschist metamorphism were depleted in deuterium and enriched in ^{18}O relative to fluids (probably normal ocean water) involved in lower-grade metamorphism. Franciscan carbonate minerals exhibit large variations in $\delta\,^{13}C$, probably owing to mixing of variable proportions of organic carbon ($\delta\,^{13}C$ about -25) and sedimentary carbonate carbon ($\delta\,^{13}C$ approximately zero).

New England

New England metamorphic rocks are part of the Appalachian mountain belt (Figure 9-26). The rocks vary widely in chemical composition and range in age from late Precambrian to late Paleozoic. The Precambrian rocks occur as a series of linear masses running north-south through the western part of New England and as irregular masses in the most eastern parts of Maine, Massachusetts, and the Connecticut–Rhode Island area. Most of the metamorphic rocks are Cambrian to Devonian in age, with some late Paleozoic rocks occurring in scattered basins along the eastern coast. Calc-alkalic plutons occur throughout the region, including rocks of probable Ordovician, Silurian, Devonian, and Jurassic-Cretaceous ages.

At least three major episodes of metamorphism are believed to have affected the New England region: the Taconic orogeny (Ordovician), the Acadian orogeny (Silurian and Devonian), and the Alleghanian orogeny (Permian). The major metamorphic features of most of New England can be related to the Acadian orogeny, with Taconic effects best shown on the western edge and Alleghanian effects clearest in the rocks along the southeastern coast. In detail, relationships are very complex. Multiple deformations and several metamorphisms are known to have occurred during the Acadian orogeny [see, for example, De Yorea et al. (1989)].

The metamorphic rocks of western California, discussed in the previous section, belong to the high-pressure facies series of Miyashiro (Figure 9-10). On the other side of the continent, in New England, are rocks belonging to the low-pressure and medium-pressure facies sequences (Figure 9-26). The isobaric trace of the aluminosilicate triple point in Figure 9-26 divides a low-pressure facies series (andalusite $\rightarrow$ sillimanite) in northeastern New England from a medium-pressure facies series (kyanite $\rightarrow$ sillimanite) in southwestern New England. Thus the rocks in the southwest must have been more deeply buried during metamorphism. If the metamorphic rocks are divided on the basis of grade of metamorphism, three high-grade zones (New York City through the Green Mountains, central Connecticut through western Maine, and Rhode Island through the southern tip of Maine) are separated by areas of lower-grade metamorphism (Morgan 1972). This pattern could be due to varying metamorphic conditions during the major phase of Acadian orogeny, to structural deformation during and after metamorphism, or to formation of the rocks during several different metamorphic events. Thompson et al. (1968) demonstrate that isogradic surfaces in west-central New England have been overturned and deformed after metamorphism. Detailed mapping has shown that, in addition to metamorphism and extensive plutonism, parts of the New England area have been strongly affected by nappe formation and gneiss doming (Robinson et al. 1991).

The full set of isograds commonly used for pelitic rocks can be mapped throughout much of New England. They tend to run north-south or northeast-southwest in the southern part of the region. In northern Vermont and New Hampshire and in west-central Maine, isograd locations are irregular and spatially related to numerous plutonic masses. Farther north in Maine, metamorphic intensity dies out, and locally rocks of the prehnite-pumpellyite facies are found (Richter and Roy

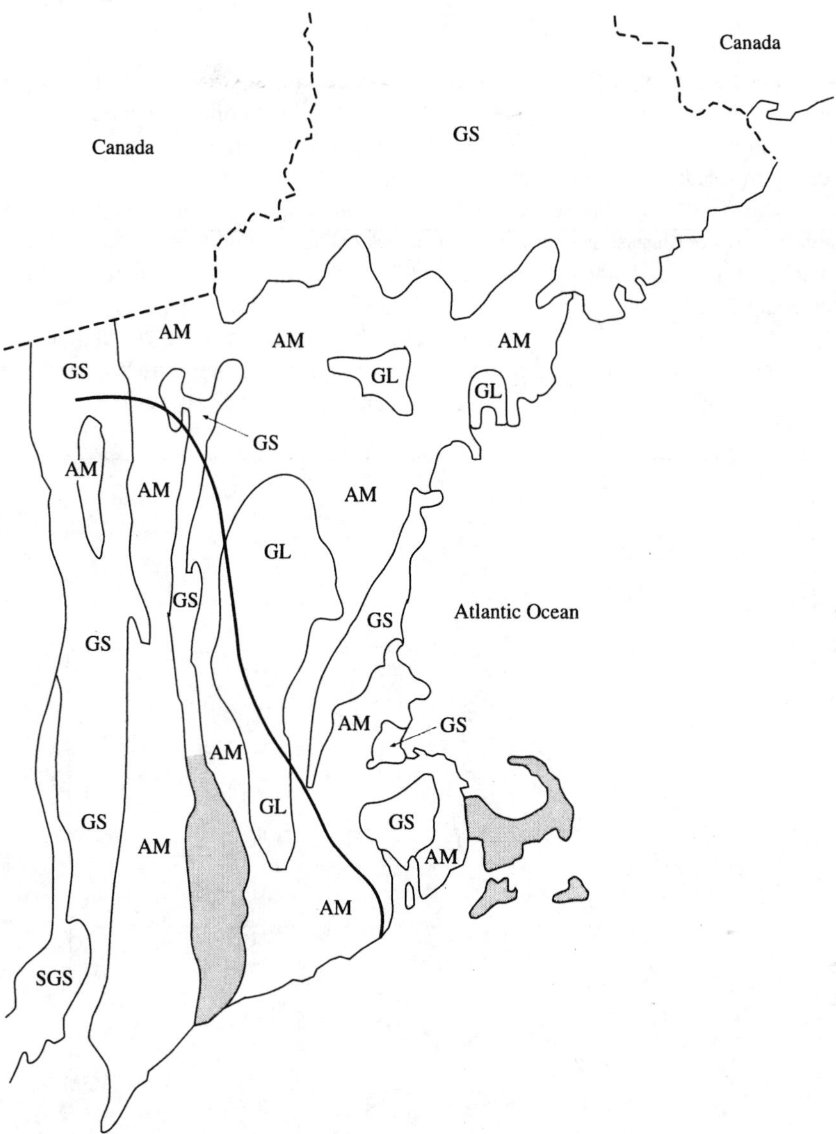

Figure 9-26 Metamorphic map of New England. Paleozoic regional metamorphism has produced the following metamorphic facies: subgreenschist (SGS), greenschist (GS), amphibolite (AM), and granulite (GL). The Triassic rocks of the Connecticut River Valley and the Pleistocene deposits of Cape Cod are shown by shading. The northern part of Maine (beyond the top edge of the map) has a large area of subgreenschist facies rocks. Not shown on the map are the Precambrian rocks of western New England (affected by Precambrian and Paleozoic metamorphism) and the numerous calc-alkalic plutons scattered throughout New England. The heavy line cutting across the map from northwest to southeast is the isobaric trace of the aluminosilicate triple point. Andalusite and sillimanite occur to the northeast of this line, and kyanite and sillimanite to the southwest of the line. The rocks labeled granulite facies are considered by most New England geologists to represent high-grade rocks of the amphibolite facies. Much of the "AM" field in eastern New England is now known to be greenschist facies. The rocks on the eastern edge of New England are part of the Late Proterozoic Avalon terrane, which is a continental fragment that was added to the other New England rocks during a Paleozoic continent-continent collision. After Morgan (1972).

1976). In some parts of New England, plutonic rocks occur in areas with the highest-grade metamorphic rocks; in other parts there is no correlation between plutonic activity and grade of metamorphism. Plutonic rocks of Devonian age show the best correlation and appear to be contemporaneous with Acadian metamorphism.

Carmichael (1978) proposed a method of geobarometry for metamorphic rocks that is based solely on mineral assemblages in pelitic rocks. The method works well in New England. Figure 9-27a outlines six bathozones defined by the intersection of various reaction curves. The bathozones are separated by bathograds, which are lines dividing higher-pressure assemblages from lower-pressure assemblages. For example, the Al_2SiO_5 triple point is used to separate bathozone 3 from bathozone 4. This bathograd is the heavy line shown in Figure 9-26 and it represents a pressure of 3.75 kbar using the Al_2SiO_5 phase diagram of Holdaway (1971).

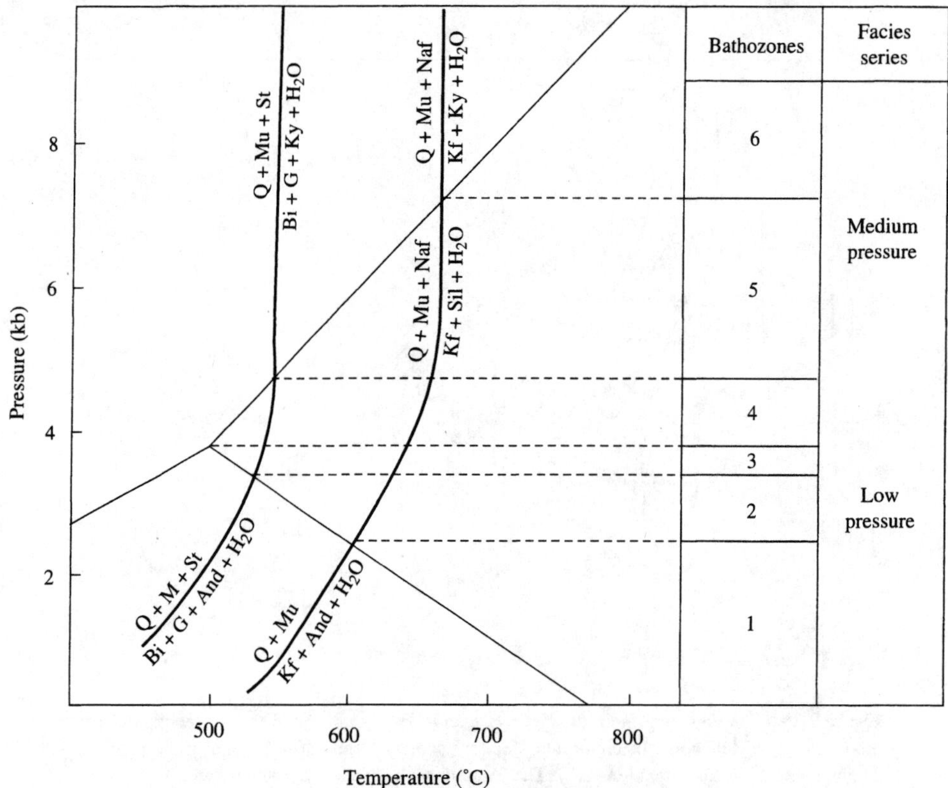

Figure 9-27a Classification of pressure (depth) zones by Carmichael (1978). The boundaries between bathozones are termed bathograds and are defined by the intersection of reaction curves with the Al_2SiO_5 polymorphic transformation lines. One boundary is defined by the Al_2SiO_5 triple point. Also shown is the approximate relationship of bathozones to the low pressure and medium pressure facies series of Miyashiro (1973, 1994). Mineral abbreviations are as follows: Q = quartz; Mu = muscovite; St = staurolite; Bi = biotite; G = garnet; And = andalusite; Kf = K-feldspar; Naf = Na-feldspar; Ky = kyanite; Sil = sillimanite. (After Philpotts, A. R., *Principles of Igneous and Metamorphic Petrology*. Copyright © 1990 by Prentice Hall, Inc. Adapted by permission of Prentice Hall, Inc., Upper Saddle River, NJ.)

Other bathozones are defined by the intersection of certain reaction lines with the Al_2SiO_5 lines. For example, the point where the reaction line

$$\text{quartz} + \text{muscovite} + \text{staurolite} = \text{biotite} + \text{garnet} + Al_2SiO_5 + H_2O \qquad (9\text{-}20)$$

crosses the sillimanite-andalusite line separates bathozone 2 from bathozone 3. Bathozone 2 has mineral assemblages containing andalusite, while similar rocks in bathozone 3 can have the same mineral assemblages but with sillimanite instead of andalusite. Note that it is a group of minerals occurring together that defines the separation line between the two zones. The line between the two zones represents a pressure of about 3.3 kbar. Figure 9-27b shows the distribution of bathograds for New England. They indicate that the highest pressures of metamorphism occurred in southern and western New England. In terms of the facies series of Miyashiro (1973, 1994), the medium-pressure facies series occurs in bathozone 5 (southern and western New England) and the low-pressure facies series occurs in bathozones 1 and 2 (northern New England).

In addition to mapping lines of identical pressure (bathograds), lines of identical temperature have been located for some of the metamorphic rocks of New England. These isotherms tend to mimic the pattern shown by mapped isograds. Thus different grades of metamorphism generally indicate different temperatures of metamorphism, and pressure variations have not been as

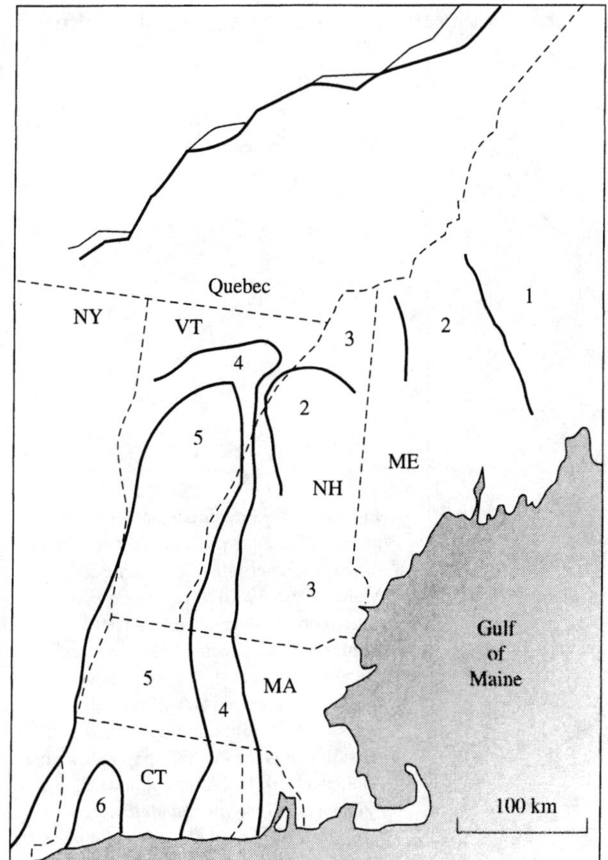

Figure 9-27b Bathozones of New England. Numbers for bathozones same as in (a). Greater depths of metamorphism, and thus greater uplift, have occurred in western and southern New England. (After Philpotts, A. R., *Principles of Igneous and Metamorphic Petrology*. Copyright © 1990 by Prentice Hall, Inc. Adapted by permission of Prentice Hall, Inc., Upper Saddle River, NJ.)

important as temperature variations. However, the area is one where a variety of metamorphic events have to be related to sequences of complex tectonic deformation. Thus it would be helpful to know in detail how pressure and temperature changed everywhere during the various periods of metamorphism.

Philpotts (1990, 364–366) compares two different areas of New England where metamorphic conditions have been studied in detail. In south-central Maine metamorphism followed a pattern typical of the low-pressure facies series of Miyashiro (1973, 1994). Ferry (1980) compared eight different geothermometers/geobarometers in this part of Maine and Figure 9-28 shows the probable pressure-temperature-time trajectory that he found. Apparently, temperature increased at a nearly constant pressure and, after the highest grade of metamorphism was reached, kinetic factors prevented any major retrograde changes as temperature and pressure dropped quickly. This type of pressure-temperature-time path has been assumed for many areas of metamorphic rocks. Recent research is now showing that other, nonlinear pressure-temperature-time trajectories involving both prograde and retrograde metamorphism have occurred.

In central Massachusetts two contrasting, nonlinear pressure-temperature-time trajectories have been identified (Schumacher et al. 1989). The Bronson Hill anticlinorium near the New Hampshire border has a clockwise pressure-temperature-time trajectory, while the associated Merrimack synclinorium near the Connecticut border has a counterclockwise pressure-temperature-time trajectory (Figure 9-28). The contrasting trajectories appear to be the result of different

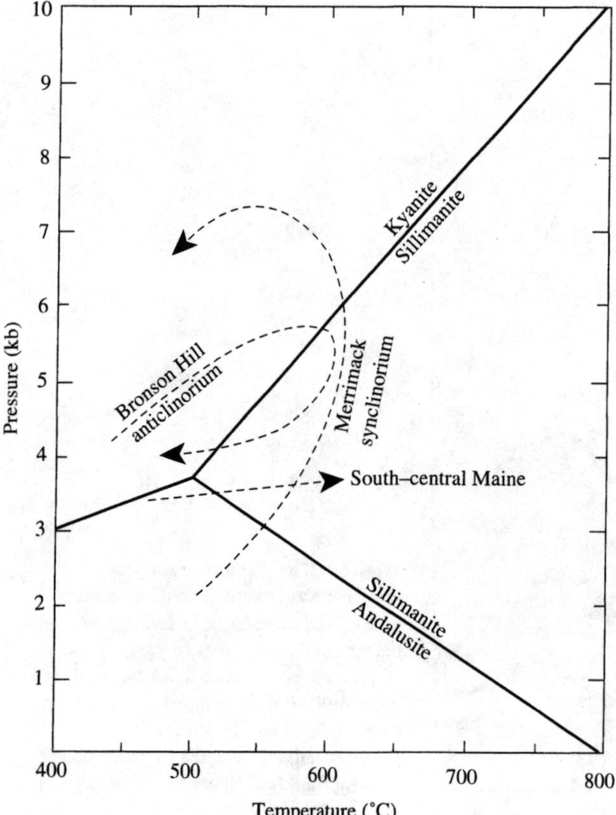

Figure 9-28 Pressure-temperature diagram showing the pressure-temperature-time paths determined for south-central Maine (Ferry 1980), the Bronson Hill anticlinorium in southwestern New Hampshire and in central Massachusetts (Schumacher et al. 1989), and the Merrimack synclinorium in south-central Massachusetts (Schumacher et al. 1989). Also shown is the Al_2SiO_5 phase diagram (Holdaway 1971). (After Philpotts, A. R., *Principles of Igneous and Metamorphic Petrology*. Copyright © 1990 by Prentice Hall, Inc. Adapted by permission of Prentice Hall, Inc., Upper Saddle River, NJ.)

tectonic histories for the two areas. The counterclockwise trajectory in Figure 9-28 could be caused by increasing tectonic pressure and/or rapid burial occurring with little change in temperature. In contrast, the clockwise trajectory could represent slow burial with pressure and temperature increasing together. Schumacher et al. (1989) suggest that nappe formation, gneiss doming, and backfolding of earlier structures occurred at different times and in a different order for the Bronson Hill and Merrimack rocks.

A possible cause of the Acadian orogeny is the collision of the North American continental margin with a continental fragment (the Avalon block) coming from the east. It is generally agreed that the collision occurred sometime during the Paleozoic and resulted in underthrusting of Avalonian rocks beneath rocks to the west. Wintsch et al. (1992) have combined mineral cooling ages with pressure-temperature data from mineral thermobarometry to study this problem in south-central New England. They found that the cooling history of some non-Avalonian rocks is very different from that of underlying Avalonian rocks. For example, peak metamorphic conditions in certain overlying rocks occurred in the Silurian, while in the underlying Avalonian rocks these conditions were reached in the late Paleozoic and the underlying rocks do not show the Silurian-Devonian Acadian effects. It would be difficult for the Avalonian rocks lying below the others to have escaped the Acadian effects if they were in place in Silurian-Devonian time. Thus Wintsch et al. concluded that significant assembly of all the Paleozoic rocks of New England post-dates most Paleozoic metamorphism and that at least the last stages of assembly occurred during the late Paleozoic Alleghanian orogeny.

SUMMARY

Metamorphism can occur with or without change in the chemical composition of the rocks being transformed. Often the only change involves loss of fluids or alteration due to the action of fluids (particularly H_2O and CO_2). The term metasomatism refers to changes in the bulk composition of rocks through which fluid flow has occurred. Common types of metasomatism involve addition of alkalis, hydrogen, silica, and water to rocks. Many ore deposits occur in rocks affected by metasomatism.

The final product of the metamorphism of a given rock depends on the variables temperature, load pressure, fluid pressure, and time. The pressure-temperature conditions of a metamorphic episode may be due to magmatic intrusion (contact metamorphism) or to orogenesis (mountain building and regional metamorphism). The history of the episode is represented by a pressure-temperature-time path. Because of the wide variety of parent rocks, conditions of alteration, and final mineralogies and textures, metamorphic rocks are difficult to classify. They can be classified in terms of mineralogy, texture, or chemical composition.

Often the chemical changes produced by metamorphism involve fluid-solid reactions controlled by kinetics rather than by equilibrium thermodynamics. Reaction rates tend to be faster during prograde metamorphism as compared to retrograde metamorphism. The nature of the changes producing the final rock are probably controlled by the maximum pressure-temperature conditions of metamorphism.

Temperature estimates can be obtained by measurement of the distribution of elements between coexisting minerals and by measuring the distribution across a solvus for two structurally related minerals. Two exchange reaction geothermometers that have been applied to metamorphic rocks use the mineral pairs garnet-biotite and garnet-pyroxene. An example of a solvus geothermometer is the distribution of potassium and sodium between coexisting alkali feldspar and plagioclase. Isotope geothermometers can also provide estimates of metamorphic temperatures. In addition, isotopic data give evidence on the nature and extent of fluid flow and on types of premetamorphic protoliths. Most studies have used $\delta\,^{18}O$ and $\delta\,D$ values in minerals. ^{18}O depletion is often a result of both contact and regional metamorphism.

Geobarometers use mineral reactions that are pressure sensitive. The reaction anorthite = grossularite + kyanite + quartz is an example. One method of geobarometry defines bathozones separated by bathograds, which are lines dividing higher-pressure assemblages from lower-pressure assemblages. The most useful minerals for pressure-temperature estimates are the Al_2SiO_5 polymorphs with their triple point at 3.75 ($\pm$0.25) kbar and 504 ($\pm$20) °C. Future research will allow geobarometers and geothermometers to estimate pressures and temperatures for some rocks to $\pm$250 bars and $\pm$25 °C.

A metamorphic facies is a set of metamorphic mineral assemblages, repeatedly associated in space and time, that displays a constant relation between mineral composition and chemical composition. Rocks belonging to any particular facies formed under the same physical conditions. A general range of pressure-temperature conditions can be assigned to each metamorphic facies. More precise conditions can be specified using a petrogenetic grid, which consists of intersecting univariant curves on a pressure-temperature diagram. Other diagrams commonly used in metamorphic petrology are the ACF diagram and the AFM projection diagram.

The concept of facies series classifies metamorphism into three pressure types. Convergent plate boundaries are associated with the high-pressure (subduction zones) and medium-pressure (continental collisions) facies series. The low-pressure facies series is found at ocean ridges associated with divergent plate boundaries and in the island arc environment. Most regional metamorphism has occurred at convergent plate boundaries. Regional metamorphism of pelitic rocks results in slates, schists, and gneisses whose typical minerals include micas, chlorite, garnet, amphiboles, and the Al_2SiO_5 polymorphs andalusite, kyanite, and sillimanite. For impure limestones, schists and amphibolites form with minerals such as epidote, tremolite, diopside, wollastonite, and grossularite. Regional metamorphism of mafic igneous rocks produces schists and amphibolites containing hornblende, garnet, actinolite, glaucophane, and plagioclase among others. A wide variety of minerals are found in contact metamorphic rocks, particularly in altered impure limestones and dolomites.

Contact metamorphism by the Marysville stock has produced mineral assemblages that can be explained by an equilibrium model that postulates internal buffering of fluids along isobaric univariant reaction lines. In this case the mineral assemblages control the compositions of coexisting fluids, as opposed to the fluids determining the mineralogies. A kinetic, nonequilibrium model could also explain the Marysville rocks. Stable isotope data for these rocks indicate isotopic disequilibrium during metamorphism. The contact aureole of the Notch Peak pluton differs from that at Marysville because the precursor rocks were different. The mineralogy and $\delta\,^{18}O$ data for the Notch Peak aureole indicate that fluid-rock interaction occurred as a result of fluid flow in the direction of increasing temperature (toward the pluton). Research on the Notch Peak and other aureoles suggests that, although fluid-rock interaction occurs in all contact metamorphic aureoles, only a limited amount of fluid is actively involved in the metamorphism.

The Franciscan rocks of California represent part of a Mesozoic arc-trench subduction zone. The rocks experienced high-pressure, low-temperature metamorphism, resulting in assemblages characteristic of the blueschist facies. The tectonic history of the subduction zone can be inferred by interpretation of the retrograde pressure-temperature-time path. Eastern rocks of the Franciscan complex are 20–30 million years younger than the more strongly metamorphosed western rocks. This difference in Ar-isotopic ages can be explained by the occurrence of two different subduction events at two different times, or by differences in thermal history even though metamorphism occurred at the same time. The metamorphic rocks of New England vary widely in chemical composition, age, and grade of metamorphism. At least three major episodes of metamorphism affected the region. In addition to metamorphism and extensive plutonism, parts of the area have been strongly affected by nappe formation and gneiss doming. Different tectonic histories in different parts of New England can be related to contrasting, nonlinear pressure-temperature-time paths.

QUESTIONS

1. (a) Given is a rock consisting of dolomite, quartz, tremolite, and calcite. If a gas phase containing H_2O and CO_2 was present when these minerals formed, how many variables can be considered independent according to the phase rule?

(b) Given is a rock consisting of dolomite, quartz, and diopside. If a gas phase containing H_2O and CO_2 was present when these minerals formed, how many variables can be considered independent according to the phase rule?

2. (a) Write a balanced equation for the reaction forsterite + talc $\rightleftharpoons$ enstatite + water.

(b) Would the slope of the equilibrium curve for this reaction on a pressure-temperature diagram be positive or negative at very low total pressure with $P_{total} = P_{H_2O}$? Would the slope be steep or gentle?

(c) How many degrees of freedom would the system have when P_{H_2O} is less than total pressure? Assume no iron in any of the minerals.

3. (a) How many degrees of freedom are represented by a closed system in which the reaction of Figure 9-12 occurs?

(b) How many degrees of freedom exist for this system if it is open, and H_2O is a mobile component whose chemical potential is not externally controlled (fixed) by the surroundings?

(c) How many degrees of freedom exist if the system is open with the chemical potential of H_2O fixed by the surroundings?

(d) If albite also occurred in the assemblage of part (a), how would the degrees of freedom change?

4. (a) The following table gives the mole fractions of iron and magnesium of coexisting garnet and biotite in a high-grade gneiss. Using the Ferry and Spear (1978) calibration of the garnet-biotite geothermometer (Figure 3-4), determine the temperature recorded by these mineral samples.

	Garnet	Biotite
N_{Fe}	0.34	0.18
N_{Mg}	0.06	0.12

(b) Using the calibration of Goldman and Albee (1977) in Figure 3-4, what would the temperature be using the data in part (a)?

5. (a) The following mole fractions for coexisting garnet and plagioclase are found in a metamorphic rock: $N_{gross}^{gar} = 0.050$ and $N_{anor}^{plag} = 0.023$. An independent estimate of temperature for this rock gives a value of 500°C. Use Figure 9-5a to estimate the pressure recorded by these mineral samples.

(b) Assume the value for N_{gross}^{gar} in part (a) is wrong due to analytical error and the correct value is 0.04. About how much would the pressure value found in (a) be in error?

6. You want to determine if a large mass of pelitic schist has a homogeneous distribution of oxygen isotope ratios over the entire outcrop area of the schist. Can this be done by determining the isotopic ratios of whole rock samples? Why or why not?

7. (a) How could measurements of the isotopic composition of oxygen in regional metamorphic rocks be used to determine if changes had occurred from the original composition of the parent rock?

(b) Given is a metamorphic aureole made up of hornfels that is in contact with quartz monzonite. Oxygen isotope measurements show that the hornfels has the same $\delta\,^{18}O$ values as the average value found for shales. However, the minerals of the quartz monzonite near the contact show enrichment in ^{18}O compared to the rest of the intrusive. How would you interpret this?

8. The Pine Creek Mine in California contains scheelite-bearing skarns at the contact between marble and a quartz monzonite pluton (Newberry 1982). The grade of tungsten ore varies with the distance from the pluton as shown in the following diagram. The skarn zones have the following mineralogy: inner zone (garnet + quartz); main zone (garnet + pyroxene + scheelite); and outer zone (idocrase + wollastonite + scheelite). Suggest a reason why scheelite ($CaWO_4$) is most abundant in the outer zone skarn.

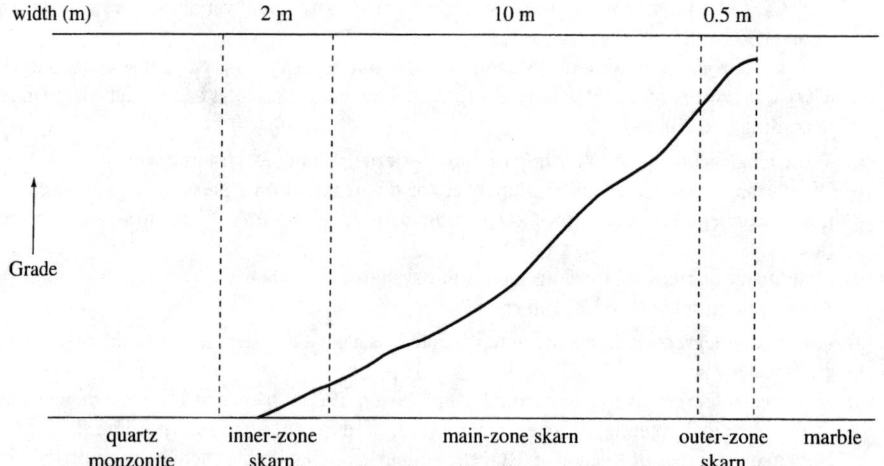

9. Refer to Figure 9-7. Use the phase rule to explain why only one phase is present in each zone shown in this figure.

10. (a) The curves in Figure 9-11 are all dehydration reactions with H_2O as one of the products. They are plotted for systems in which $P_{H_2O} = P_{total}$. How would the location of the curves change if $P_{H_2O} = \frac{1}{2} P_{total}$?

(b) For what type of precursor rock, pelitic, mafic, or calcareous, would the curve involving Mg-chlorite, muscovite, and quartz as reactants in Figure 9-11 be used on a petrogenetic grid?

11. (a) To what metamorphic facies do each of the mineral assemblages of Table 9-4 belong?

(b) To what facies do the rocks described by Goldschmidt (1911) belong?

(c) The rocks described by Eskola (1915) in Finland have similar compositions to those of Goldschmidt in Norway, but are characterized by hydrous minerals such as biotite, muscovite, hornblende, and anthophyllite. In what facies would you place them?

12. (a) Why isn't hornblende usually found in pelitic rocks that belong to the hornblende-hornfels facies?

(b) Given is a regional metamorphic rock composed of antigorite, diopside, and calcite in equal proportions. What minerals would you guess made up the original rock?

13. Why is it incorrect to plot biotite on an ACF diagram?

14. (a) In using the Thompson AFM diagram, it is assumed that SiO_2 is present in excess. This is equivalent to assuming wh a value for the activity of SiO_2?

(b) In using the Thompson AFM diagram, it is often assumed that H_2O can be neglected because the system under consideration was an open system for H_2O. This is equivalent to assuming what about the activity of H_2O?

15. (a) What type of rock did each of the following contact metamorphic rocks probably form from? Assume no metasomatism has occurred.

(1) Andalusite-cordierite-biotite-orthoclase-quartz rock.
(2) Wollastonite-calcite rock.
(3) Diopside-grossularite-anorthite rock.
(4) Corundum-iron spinel rock.
(5) Plagioclase-hornblende-biotite rock.

(b) Given is a contact metamorphic rock containing by weight, 8 percent Al_2O_3, 32 percent CaO, 40 percent (Mg, Fe)O, and 20 percent SiO_2. What would be the equilibrium mineral assemblage of the rock if it formed under conditions corresponding to the ACF diagram of Figure 3-11?

16. (a) This question involves the wollastonite rims that formed on chert nodules as discussed in the text and in Joesten and Fisher (1988). Where in the contact aureole would you expect the wollastonite rims to be the thickest?

(b) Joesten and Fisher (1988) found that rim growth curves (calculated by use of kinetic equations) for nodules in different parts of the aureole had the same general shape but different slopes during early growth. Two typical curves plot as shown below. Suggest a reason why the early-growth slope of curve A is steeper than the early-growth slope of curve B.

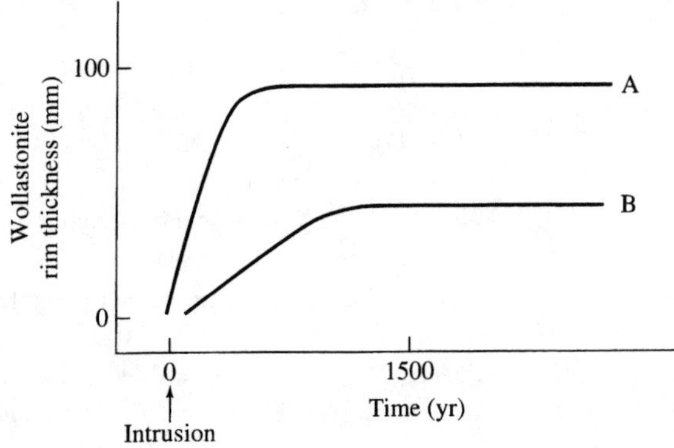

17. (a) How should the line in Figure 9-17 going from the higher-temperature invariant point toward the lower-temperature invariant point be labeled? In other words, what is the absent phase?

(b) Should the line going from the lower-temperature invariant point to the higher-temperature invariant point have the same label as your answer to part (a)?

(c) Using mineral names, write out the reaction representing the univariant line connecting the two invariant points in Figure 9-17.

18. (a) Which mineral assemblage in Table 9-12 represents invariant point B in Figure 9-19 for the case of zero flux?

(b) According to Table 9-12, what mineral is destroyed at invariant point B of Figure 9-19 for the case of zero flux?

19. (a) One of the reactions that occurs during contact metamorphism of siliceous carbonate rocks is as follows:

$$\text{tremolite} + 2\text{ quartz} + 3\text{ calcite} = 5\text{ diopside} + 3\ CO_2 + H_2O$$

Where would you expect the reactants and products of this reaction to occur together in the metamorphic aureole shown in Figure 9-21?

(b) If anorthite, calcite, and quartz in the aureole rocks of Figure 9-21 were to react to form a new metamorphic mineral, what mineral would you expect to form?

20. Given are two different contact metamorphic aureoles at two different locations. In aureole A the sedimentary rocks that have been metamorphosed by a granitic pluton have vertical dips and have strikes that are parallel to the pluton margin. In contrast, the layering of the sedimentary rocks in aureole B is horizontal and perpendicular to the contact with a granitic pluton. If you were to determine the oxygen isotope values for the two aureoles, what differences in the results might be found?

21. (a) One cause of discordant K-Ar ages when dating metamorphic rocks is the presence of excess radiogenic argon in minerals. This excess argon is due to a second period of metamorphism, which liberates previously formed argon for inclusion in newly formed minerals. Although several periods of metamorphism affected some of the Franciscan rocks of California, enrichment in radiogenic argon probably did not occur. Why not?

(b) Suggest two types of evidence from dating of the rocks and minerals of the Franciscan assemblage that would confirm the absence of excess radiogenic argon.

22. How many degrees of freedom are there at the point where reaction curve (9–20) crosses the sillimanite-andalusite boundary line in Figure 9-27a?

REFERENCES

ANOVITZ, L. M., and E. J. ESSENE. 1987. Phase equilibria in the system $CaCO_3$-$MgCO_3$-$FeCO_3$. *J. Petrol.*, v. 28:389–414.

BAILEY, E. H., W. P. IRWIN, and D. L. JONES. 1964. Franciscan and related rocks, and their significance in the geology of western California. *Calif. Div. Mines Geol. Bull.* 183:89–112.

BAILEY, E. H., M. C. BLAKE, JR., and D. L. JONES. 1970. On-land Mesozoic Oceanic Crust in California Coast Ranges. U.S. Geol. Survey Profess. Paper 700-C. 70–81.

BARROW, G. 1893. On an intrusion of muscovite-biotite gneiss in the southeast Highlands of Scotland, and its accompanying metamorphism. *Geol. Soc. London Quart. J.*, v. 49:330–358.

BARTON, M. D., R. P. ILCHIK, and M. A. MARIKOS. 1991. Metasomatism. In *Contact Metamorphism*, v. 26, Rev. in Mineralogy, D. M. Kerrick, ed. Mineralogical Society of America, Washington, D.C. 321–350.

BEST, M. G. 1982. *Igneous and Metamorphic Petrology.* W. H. Freeman and Company, New York.

BLACKBURN, W. H. 1968. The spatial extent of chemical equilibrium in some high-grade metamorphic rocks from the Grenville of southeastern Ontario. *Contrib. Mineral. Petrol.*, v. 19:72–92.

BLAKE, M. C., JR., W. P. IRWIN, and R. G. COLEMAN. 1967. Upside-down Metamorphic Zonation, Blueschist Facies, Along a Regional Thrust in California and Oregon. U.S. Geol. Survey Profess. Paper 575-C. 1–9.

BOWEN, N. L. 1940. Progressive metamorphism of siliceous limestone and dolomite. *J. Geol.*, v. 48:225–274.

BUCHER, K., and M. FREY. 1994. *Petrogenesis of Metamorphic Rocks*, 6th ed. Springer-Verlag, New York.

CARLSON, W. D., and J. L. ROSENFELD. 1981. Optical determination of topotactitic aragonite-calcite growth kinetics: metamorphic implications. *J. Geol.*, v. 89:615–638.

CARMICHAEL, D. M. 1978. Metamorphic bathozones and bathograds: a measure of post-metamorphic uplift and erosion on a regional scale. *Am. J. Sci.*, v. 278:769–797.

CHAMBERLAIN, C. P., and D. RUMBLE, III. 1988. Thermal anomalies in a regional metamorphic terrane: the role of advective heat transport during metamorphism. *J. Petrol,* v. 29:1215–1232.

CLOOS, M. 1982. Flow melanges: numerical modeling and geologic constraints on their origin in the Franciscan subduction complex, California. *Bull. Geol. Soc. Am.,* v. 93:330–345.

CLOOS, M. 1985. Thermal evolution of convergent plate margins: thermal modeling and reevaluation of isotopic Ar-ages for blueschists in the Franciscan Complex, California. *Tectonics,* v. 4:421–433.

DALY, J. S., R. A. CLIFF, and B. W. D. YARDLEY. 1989. *Evolution of Metamorphic Belts.* The Geological Society, London.

DAVIDSON, P. M., and D. H. LINDSLEY. 1985. Thermodynamic analysis of quadrilateral pyroxenes. Part II. Model calibration from experiments and applications to geothermometry. *Contrib. Mineral. Petrol.,* v. 91:390–404.

DE YOREA, J. J., D. R. LUX, C. V. GUIDOTTI, E. R. DECKER, and P. H. OSBERG. 1989. The Acadian thermal history of western Maine. *J. Met. Geol.,* v. 7:169–190.

DICKINSON, W. R., R. W. OJAKANGAS, and R. J. STEWART. 1969. Burial metamorphism of the late Mesozoic Great Valley sequence, Cache Creek, California. *Bull. Geol. Soc. Am.,* v. 80:519–526.

DIPPLE, G. M., and J. M. FERRY. 1992. Fluid flow and stable isotopic alteration in rocks at elevated temperatures with applications to metamorphism. *Geochim. Cosmochim. Acta,* v. 56:3539–3550.

EHLERS, E. G., and H. BLATT. 1982. *Petrology.* W. H. Freeman and Company, San Francisco.

EINAUDI, M. T., L. D. MEINERT, and R. J. NEWBERRY. 1981. Skarn deposits. In *Seventy-Fifth Anniversary Volume,* Econ. Geol., B. J. Skinner, ed. The Economic Geology Publishing Company, El Paso, TX. 317–391.

ERNST, W. G. 1971. Do mineral parageneses reflect unusually high-pressure conditions of Franciscan metamorphism? *Am. J. Sci.,* v. 270:81–108.

ERNST, W. G., ed. 1975a. *Metamorphism and Plate Tectonic Regimes.* Dowden, Hutchinson & Ross, Inc., Stroudsburg, PA.

ERNST, W. G., ed. 1975b. *Subduction Zone Metamorphism.* Dowden, Hutchinson & Ross, Inc., Stroudsburg, PA.

ERNST, W. G. 1976. *Petrologic Phase Equilibria.* W. H. Freeman and Company, San Francisco.

ERNST, W. G. 1977. Tectonics and prograde versus retrograde P-T trajectories of high-pressure metamorphic belts. *Rediconti Soc. Italia Mineral. Petrol.,* v. 33:191–220.

ERNST, W. G. 1988. Tectonic history of subduction zones inferred from retrograde blueschist P-T paths. *Geology,* v. 16:1081–1084.

ESKOLA, P. 1915. On the relation between chemical and mineralogical composition in the metamorphic rocks of the Orijärvi region. *Comm. Geol. Finlande Bull.,* no. 44.

ESKOLA, P. 1920. The mineral facies of rocks. *Nord. Geol. Tidsskr.,* v. 6:143–194.

ESSENE, E. J. 1982. Geologic thermometry and barometry. In *Characterization of Metamorphism through Mineral Equilibria,* v. 10, J. M. Ferry, ed. Rev. in Mineralogy, Mineralogical Society of America, Washington, D.C. 153–206.

ESSENE, E. J. 1989. The current status of thermobarometry in metamorphic rocks. In *Evolution of Metamorphic Belts,* J. S. Daly, R. A. Cliff, and B. W. D. Yardley, eds. The Geological Society, London.

FERRY, J. M. 1980. A comparative study of geothermometers and geobarometers in pelitic schists from south-central Maine. *Am. Min.,* v. 65:720–732.

FERRY, J. M. 1991. Dehydration and decarbonation reactions as a record of fluid infiltration. In *Contact Metamorphism,* v. 26, Rev. in Mineralogy, D. M. Kerrick, ed. Mineralogical Society of America, Washington, D.C. 351–393.

Ferry, J. M., and L. Baumgartner. 1987. Thermodynamic models of molecular fluids at the elevated pressures and temperatures of crustal metamorphism. In *Thermodynamic Modeling of Geological Materials: Minerals, Fluids, and Melts,* v. 17, Rev. in Mineralogy, I. S. E. Carmichael and H. P. Eugster, eds. Mineralogical Society of America, Washington, D.C. 323–365.

Ferry, J. M., and G. M. Dipple. 1992. Models for coupled fluid flow, mineral reaction, and isotopic alteration during contact metamorphism: The Notch Peak aureole, Utah. *Am. Min.,* v. 77:577–591.

Ferry, J. M., and F. S. Spear. 1978. Experimental calibration of the partitioning of Fe and Mg between biotite and garnet. *Contrib. Mineral. Petrol.,* v. 66:113–117.

Freer R., and P. F. Dennis. 1982. Oxygen diffusion studies. I. A preliminary ion microprobe investigation of oxygen diffusion in some rock-forming minerals. *Min. Mag.,* v. 45:179–192.

Frost, B. R. 1991. Magnetic petrology: The factors that control the occurrence of magnetite in crustal rocks. In *Oxide Minerals: Petrologic and Magnetic Significance,* v. 25, Rev. in Mineralogy, D. H. Lindsley, ed. Mineralogical Society of America, Washington, D.C. 489–509.

Fuhrman, M. L., and D. H. Lindsley. 1988. Ternary-feldspar modeling thermometry. *Am. Min.,* v. 73:201–215.

Fyfe, W. S., F. J. Turner, and J. Verhoogen. 1958. *Metamorphic Reactions and Metamorphic Facies.* Memoir 73, Geological Society of America, New York.

Fyfe, W. S., and F. J. Turner. 1966. Reappraisal of the concept of metamorphic facies. *Contrib. Mineral. Petrol.,* v. 12:354–364.

Fyfe, W. S., N. J. Price, and A. B. Thompson. 1978. *Fluids in the Earth's Crust.* Elsevier Scientific Publishing Company, Amsterdam.

Ganguly, J., and S. K. Saxena. 1984. Mixing properties of alumino-silicate garnets: constraints from natural and experimental data, and applications to geothermo-barometry. *Am. Min.,* v. 69:88–97.

Goldman, D. S., and A. L. Albee. 1977. Correlation of Mg/Fe partitioning between garnet and biotite with $^{18}O/^{16}O$ partitioning between quartz and magnetite. *Am. J. Sci.,* v. 277:750–761.

Goldschmidt, V. M. 1911. Die Kontaktmetamorphose im Kristianiagebeit, Oslo Vidensk. Skr., I. *Mat.-Naturv. Kl.,* no. 11.

Greenwood, H. J. 1976. Metamorphism at moderate temperatures and pressures. In *The Evolution of the Crystalline Rocks,* D. K. Bailey and R. Macdonald, eds. Academic Press, New York.

Harper, G. D., J. B. Bowman, and R. Kuhns. 1988. A field, chemical, and stable isotope study of subseafloor metamorphism of the Josephine ophiolite, California-Oregon. *J. Geophys. Res.,* v. 93:4625–4656.

Harris, N. B. W., and T. J. B. Holland. 1984. The significance of cordierite-hypersthene assemblages from the Beitbridge region of the central Limpopo belt; evidence for rapid decompression in the Archean. *Am. Min.,* v. 69:1036–1049.

Haugerud, R. A., and E-an Zen. 1991. An essay on metamorphic path studies or Cassandra in P-T-t space. In *Progress in Metamorphic and Magmatic Petrology,* L. L. Perchuk, ed. Cambridge University Press, Cambridge, MA. 323–348.

Hodges, K. V., and F. S. Spear. 1982. Geothermometry, geobarometry, and the Al_2SiO_5 triple point at Mt. Moosilauke, New Hampshire. *Am. Min.,* v. 67:1118–1134.

Hodges, K. V., and P. Crowley. 1985. Error estimation and empirical geothermobarometry for pelitic systems. *Am. Min.,* v. 70:702–709.

Holdaway, M. J. 1971. Stability of andalusite and the aluminum silicate phase diagram. *Am. J. Sci.,* v. 271:97–131.

Holdaway, M. J., and B. Mukhopadhyay. 1993. Geothermobarometry in pelitic schists: a rapidly evolving field. *Am. Min.,* v. 78:681–693.

Hover-Granath, V. C., J. J. Papike, and T. C. Labotka. 1983. The Notch Peak contact metamorphic aureole, Utah: Petrology of the Big Horse Limestone Member of the Orr Formation. *Bull. Geol. Soc. Am.,* v. 94:889–906.

Huebner, M., T. K. Kyser, and E. G. Nisbet. 1986. Stable-isotope geochemistry of high-grade metapelites from the Central zone of the Limpopo belt. *Am. Min.,* v. 71:1343–1353.

Hyndman, D. W. 1985. *Petrology of Igneous and Metamorphic Rocks,* 2d ed. McGraw-Hill Book Company, New York.

Indares, A., and J. Martignole. 1985. Biotite-garnet geothermometry in the granulite facies: the influence of Ti and Al in biotite. *Am. Min.,* v. 70:272–278.

Jackson, K. C. 1970. *Textbook of Lithology.* McGraw-Hill Book Company, New York.

Jayko, A. S., M. C. Blake, Jr., and R. N. Brothers. 1986. Blueschist metamorphism of the Eastern Franciscan belt, northern California. In *Blueschists and Eclogites,* Memoir 164, B. W. Evans and E. H. Brown, eds. Geological Society of America, Boulder, CO. 107–123.

Joesten, R. L. 1991. Kinetics of coarsening and diffusion-controlled mineral growth. In *Contact Metamorphism,* v. 26, Rev. in Mineralogy, D. M. Kerrick, ed. Mineralogical Society of America, Washington, D.C. 507–582.

Joesten, R. L., and G. Fisher. 1988. Kinetics of diffusion-controlled mineral growth in the Christmas Mountains (Texas) contact aureole. *Bull. Geol. Soc. Am.,* v. 100:714–732.

Johannes, W., and D. Puhan. 1971. The calcite-aragonite transition, reinvestigated. *Contrib. Mineral. Petrol.,* v. 31:28–38.

Kerrick, D. M. 1990. *The Al_2SiO_5 Polymorphs,* v. 22, Rev. in Mineralogy, Mineralogical Society of America, Washington, D.C.

Kerrick, D. M., A. C. Lasaga, and S. P. Raeburn. 1991. Kinetics of heterogeneous reactions. In *Contact Metamorphism,* v. 26, Rev. in Mineralogy, D. M. Kerrick, ed. Mineralogical Society of America, Washington, D.C. 583–671.

Koziol, A. M. 1989. Recalibration of the garnet-plagioclase-Al_2SiO_5-quartz (GASP) geobarometer and application to natural parageneses, EOS. *Trans. Am. Geophys. Union,* v. 70:493.

Koziol, A. M., and R. C. Newton. 1988. Redetermination of the garnet breakdown reaction and improvement of the plagioclase-garnet-Al_2SiO_5-quartz geobarometer. *Am. Min.,* v. 73: 216–223.

Kresten, P. 1988. The chemistry of fenitization, examples from Fen, Norway. *Chem. Geol.,* v. 68:329–349.

Labotka, T. C. 1991. Chemical and physical properties of fluids. In *Contact Metamorphism,* v. 26, Rev. in Mineralogy, D. M. Kerrick, ed. Mineralogical Society of America, Washington, D.C. 43–104.

Labotka, T. C., P. I. Nabelek, and J. J. Papike. 1988. Fluid infiltration through the Big Horse Limestone member in the Notch Peak aureole, Utah. *Am. Min.,* v. 73:1302–1324.

Lasaga, A. C., and R. J. Kirkpatrick, eds. 1981. *Kinetics of Geochemical Processes,* v. 8, Rev. in Mineralogy. Mineralogical Society of America, Washington, D.C.

Lasaga, A. C., and D. M. Rye. 1993. Fluid flow and chemical reaction kinetics in metamorphic systems. *Am. J. Sci.,* v. 293:361–404.

Lattanzi, P., D. M. Rye, and J. M. Rice. 1980. Behavior of ^{13}C and ^{18}O in carbonates during contact metamorphism at Marysville, Montana: Implications for isotope systematics in impure dolomitic limestones. *Am. J. Sci.,* v. 280:890–906.

Magaritz, M., and H. P. Taylor, Jr. 1976. Oxygen, hydrogen, and carbon isotope studies of the Franciscan formation, Coast Ranges, California. *Geochim. Cosmochim. Acta,* v. 40:215–234.

McDougall, I., and T. M. Harrison. 1988. *Geochronology and Thermochronology by the $^{40}Ar/^{39}Ar$ Method,* Oxford University Press, New York.

MEYER, C., and J. J. HEMLEY. 1967. Wall rock alteration. In *Geochemistry of Hydrothermal Ore Deposits,* H. L. Barnes, ed. Holt, Rinehart and Winston, New York. 166–235.

MIYASHIRO, A. 1961. Evolution of metamorphic belts. *J. Petrol.,* v. 2:277–311.

MIYASHIRO, A. 1973. *Metamorphism and Metamorphic Belts.* Halstead Press, New York.

MIYASHIRO, A. 1994. *Metamorphic Petrology.* Oxford University Press, New York.

MORGAN, B. A. 1972. *Metamorphic Map of the Appalachians.* U.S. Geol. Survey Misc. Geol. Invest. Map I-724.

NABELEK, P. I., T. C. LABOTKA, J. R. O'NEIL, and J. J. PAPIKE. 1984. Contrasting fluid/rock interaction between Notch Peak granitic intrusion and argillites and limestones in western Utah: Evidence from stable isotopes and phase assemblages. *Contrib. Mineral. Petrol.,* v. 86:25–34.

NEWBERRY, R. J. 1982. Tungsten-bearing skarns of the Sierra Nevada. I. The Pine Creek Mine, California. *Econ. Geol.,* v. 77:823–844.

NEWTON, R. C., and H. T. HASELTON. 1981. Thermodynamics of the garnet-plagioclase-Al_2SiO_5-quartz geobarometer. In *Thermodynamics of Minerals and Melts,* R. C. Newton, ed. Springer-Verlag, New York. 131–147.

PERCHUK, L. L., and I. V. LAVRENT'EVA. 1983. Experimental investigation of exchange equilibria in the system cordierite-garnet-biotite. In *Kinetics and Equilibrium in Mineral Reactions,* S. K. Saxena, ed. Springer-Verlag, New York. 199–240.

PHILPOTTS, A. R. 1990. *Principles of Igneous and Metamorphic Petrology.* Prentice Hall, Englewood Cliffs, NJ.

PLATT, J. P. 1986. Dynamics of orogenic wedges and the uplift of high-pressure metamorphic rocks. *Bull. Geol. Soc. Am.,* v. 97:1037–1053.

RICE, J. M. 1977. Progressive metamorphism of impure dolomitic limestone in the Marysville aureole, Montana. *Am. J. Sci.,* v. 277:1–24.

RICE, J. M., and J. M. FERRY. 1982. Buffering, infiltration, and the control of intensive variables during metamorphism. In *Characterization of Metamorphism through Mineral Equilibria,* v. 10, Rev. in Mineralogy, J. M. Ferry, ed. Mineralogical Society of America, Washington, D.C. 263–326.

RICHARDSON, S. W., P. M. BELL, and M. C. GILBERT. 1969. Experimental determination of kyanite-andalusite and andalusite-sillimanite equilibria: the aluminum silicate triple point. *Am. J. Sci.,* v. 267:259–272.

RICHTER, D. A., and D. C. ROY. 1976. Prehnite-pumpellyite facies metamorphism in central Aroostook County, Maine. In *Studies in New England Geology,* Memoir 146, P. C. Lyons and A. H. Brownlow, eds. Geol. Soc. Am., Boulder, CO. 239–261.

RIDLEY, J., and A. B. THOMPSON. 1986. The role of mineral kinetics in the development of metamorphic microtextures. In *Fluid-Rock Interactions during Metamorphism,* J. V. Walther and B. J. Wood, eds. Springer-Verlag, New York. 154–193.

ROBINSON, P., P. J. THOMPSON, and D. C. ELBERT. 1991. The nappe theory in the Connecticut Valley region: Thirty-five years since Jim Thompson's first proposal. *Am. Min.,* v. 76:689–712.

RUMBLE, D., III. 1976. Oxide minerals in metamorphic rocks. In *Oxide Minerals,* v. 3, Rev. in Mineralogy, D. Rumble, III, ed. Mineralogical Society of America, Washington, D.C. R1–R24.

SCHUMACHER, J. C., R. SCHUMACHER, and P. ROBINSON. 1989. Acadian metamorphism in central Massachusetts and south-western New Hampshire: evidence for contrasting P-T trajectories. In *Evolution of Metamorphic Belts.* J. S. Daly, R. A. Cliff, and B. W. D. Yardley, eds. The Geological Society, London. 453–460.

SHIEH, Y. N., and H. P. TAYLOR, JR. 1969. Oxygen and carbon isotope studies of contact metamorphism of carbonate rocks. *J. Petrol.,* v. 10:307–331.

SPEAR, F. S. 1993. *Metamorphic Phase Equilibria and Pressure-Temperature-Time Paths.* Mineralogical Society of America, Washington, D.C.

Thompson, J. B., Jr. 1957. The graphical analysis of mineral assemblages in pelitic schists. *Am. Min.,* v. 42:842–858.

Thompson, J. B., Jr., P. Robinson, T. N. Clifford, and N. J. Trask, Jr. 1968. Nappes and gneiss domes in west-central New England. In *Studies of Appalachian Geology: Northern and Maritime,* E-an Zen, W. S. White, J. B. Hadley, and J. B. Thompson, Jr., eds. John Wiley & Sons, Inc., New York. 203–218.

Tilley, C. E. 1924. The facies classification of metamorphic rocks. *Geol. Mag.,* v. 61:167–171.

Turner, F. J. 1968. *Metamorphic Petrology.* McGraw-Hill Book Company, New York.

Turner, F. J. 1981. *Metamorphic Petrology,* 2d ed. McGraw-Hill Book Company, New York.

Valley, J. W. 1986. Stable isotope geochemistry of metamorphic rocks. In *Stable Isotopes in High Temperature Geological Processes,* v. 16, Rev. in Mineralogy, J. W. Valley, H. P. Taylor, Jr., and J. R. O'Neil, eds. Mineralogical Society of America, Washington, D.C. 445–489.

Vernon, R. H. 1975. *Metamorphic Processes.* John Wiley & Sons, Inc., New York.

Walther, J. V., and B. J. Wood, eds. 1986. *Fluid-Rock Interactions during Metamorphism.* Springer-Verlag, New York.

Winkler, H. G. F. 1976. *Petrogenesis of Metamorphic Rocks,* 4th ed. Springer-Verlag, New York.

Wintsch, R. P., J. F. Sutter, M. J. Kunk, J. N. Aleinikoff, and M. J. Dorais. 1992. Contrasting P-T-t paths: thermochronologic evidence for a late Paleozoic final assembly of the Avalonian composite terrane in the New England Appalachians. *Tectonics,* v. 11:672–689.

Zen, E-an. 1963. Components, phases, and criteria of chemical equilibrium in rocks. *Am. J. Sci.,* v. 261:929–942.

Zen, E-an, and A. B. Thompson. 1974. Low-grade regional metamorphism: mineral equilibrium relations. In *Annual Review of Earth and Planetary Sciences,* v. 2, F. A. Donath, ed. Annual Reviews, Inc., Palo Alto, CA. 179–212.

Answers to Questions

CHAPTER ONE

1. a-h, b-e, c-g, d-f
2. The five 3*d* electrons and the two 4*s* electrons have about the same energy level. Thus two or more electrons up to a maximum of seven can be lost (see Figure 1-4).
3. These chemists would define the transition elements as those that use inner *d* orbital electrons in bonding. Since these three elements use only their outermost electrons in bonding, they would not be considered transition elements according to this definition.
4. Element 104 represents the continuation of the transition series begun by actinium and interrupted by the development of the actinide series. Thus it should have properties similar to those of hafnium or zirconium.
5. A "short-period" form of the periodic table was proposed in 1871 independently by Mendelyeev and Meyer. Since the first two regular periods had seven elements in them (the inert gases had not yet been discovered at that time), they proposed to break up the 17 elements of the long periods into short periods of seven and seven plus an eighth group of three elements. This arrangement became generally used, but has been replaced in more recent times by the "long-period" form of Figure 1-1.
6. These elements have eight electrons in their outer orbits (helium has two). This is a more stable arrangement than any other grouping of electrons in atoms. Since these elements were not known, at the time of their discovery, to combine with any other elements, they were said to have a valence of zero and were put in a new group 0.
7. All four elements occur in the same column (IIIB) of the periodic table. The electron structure of the outer three shells of all four elements is similar and thus they exhibit similar geochemical behavior. The ions of these elements all have similar sizes (see Chapter Five).

8. (ppm by weight) $= 1 \times 10^{-2} \times$ atomic weight $\times$ atomic abundance

$$1.268 = 1 \times 10^{-2} \times 195.09 \times \text{atomic abundance}$$

atomic abundance $= 0.65$

9. Pallasites. These stony-iron meteorites are made up of roughly equal amounts of olivine (probably a common mineral in the Earth's mantle and in the mantles of the parent bodies of differentiated meteorites) and nickel-iron (probably the main phase of the Earth's core and of the cores of the meteorite parent bodies).

10. The behavior of the REE is generally controlled by the two electrons in the P shell and one easily lost electron in the O shell. Thus the REE usually have a $+3$ ionic state. However, europium under somewhat reducing conditions tends to lose only the two outer electrons in the P shell and to have a $+2$ ionic state. The other REE are always incompatible elements because their large ionic size prevents them from substituting for common mineral ions of similar charge. Europium in the $+2$ state has a size and charge such that it can substitute for calcium in minerals such as anorthite. This explains why anorthosites from the lunar highlands have a positive europium anomaly since they have abundant plagioclase feldspar. Mare basalts show a negative anomaly (plotted in the diagram for this question), probably indicating that they were derived from the source region from which the anorthosites were earlier subtracted. The formational history of a given meteorite would determine whether its europium anomaly is positive or negative.

11. The partitioning of ^{18}O atoms versus ^{16}O atoms (mass difference of two) should produce twice as great an effect as the partitioning of ^{17}O atoms versus ^{16}O atoms (mass difference of one). Thus we get a slope of 1:2 for a plot of $^{17}O/^{16}O$ versus $^{18}O/^{16}O$. The best explanation for the lower curve is that it represents the mixing, in various proportions, of two different materials, each with its own unique isotopic composition. It has been suggested that the two components were the original dust and gas of the nebula. Evidence of this difference has been retained in the more primitive carbonaceous chondrites but not in the younger and more homogenized Earth and lunar samples.

12. **(a)** Helium
(b) Sodium
(c) Iron, silicon, oxygen, helium, hydrogen

13. Barium is an involatile element and rubidium is a volatile element. The high Ba/Rb ratio indicates that the Moon is depleted in volatile elements relative to the material from which the solar system formed. The constant ratio for both mare and highland samples suggests that the Moon formed by homogeneous accretion or, if not, that homogenization followed accretion.

14. A hot gas of solar composition would have a reducing environment and would form metallic iron as a condensation product (Figure 1-15). Thus the chondrules formed at high temperatures while the matrix material must have formed at lower temperatures in a more oxidizing environment.

15. The oxidation-reduction conditions would determine how much iron would be in the metallic state versus how much would be in the more oxidized forms found in iron silicates, oxides, and sulfide. The densities of these various forms of iron are significantly different. See Table 1-9 for estimates of iron abundance in the inner planets. As shown there, Mars has the smallest core percentage (metallic iron) and the largest amount of iron (probably mostly oxidized) in its crust-mantle, resulting in a low overall density as compared to the other inner planets.

16. The ratios of Ca (mantle) to Ca (Type I) and Al (mantle) to Al (Type I) are similar (about 1.03) and could be used with Type I data for the trace elements to estimate the trace element abundances in the primitive mantle. For example, the primitive mantle abundance for barium would be calculated as 4.48 (Type I abundance) $\times 1.03$, or 4.61 atoms (relative to Si $= 10^6$ atoms). Many of the refractory trace elements are incompatible elements and have been extracted from the primitive mantle and strongly concentrated in the crust of the Earth. Thus their concentration in the present mantle is probably significantly different from their original concentration in the primitive mantle.

17. The local abundance in seawater of elements such as phosphorus and silicon is strongly controlled by biological activity. In general, they are less abundant in near-surface waters (due to incorporation into organisms) and more abundant in deep waters (due to decay of organisms).

18. $$T = \frac{A}{dA/dt} = \frac{0.2 \times 10^{11} \text{ moles} \times 64 \text{ g/mole}}{70 \times 10^{12} \text{ g/yr} \times 2} = 0.009 \text{ yr}$$

$$= 3.29 \text{ days}$$

The residence time in Table 1-12 is given as hours to weeks.

19.

H																	He
Li	Be											B	C	N	O	F	Ne
Na	Mg											Al	Si	P	S	Cl	Ar
K	Ca	Sc	Ti	V	Cr	Mn	Fe	Co	Ni	Cu	Zn	Ga	Ge	As	Se	Br	Kr
Rb	Sr	Y	Zr	Nb	Mo	Tc	Ru	Rh	Pd	Ag	Cd	In	Sn	Sb	Te	I	Xe
Cs	Ba	La	Hf	Ta	W	Re	Os	Ir	Pt	Au	Hg	Tl	Pb	Bi	Po	At	Rn
Fr	Ra	Ac															

lithophile — siderophile — chalcophile — atmophile

The rare-earth elements and uranium and thorium are lithophile. Because some of the elements exhibit more than one type of behavior, slightly different outlines are equally correct.

CHAPTER TWO

1. As a result of an international agreement, ^{12}C atoms are assigned a mass of exactly 12 atomic mass units (amu). The atomic weight (mass) of all other nuclides is determined by comparing them with ^{12}C. For instance, the atomic weight (mass) of ^{1}H is 1.008 amu. In the case of uranium, the nuclidic mass of ^{238}U is about 238.03 amu relative to ^{12}C. Since ^{238}U makes up 99.28 percent of present-day uranium, the atomic weight (mass) of uranium is also about 238.03.

2. 2.77×10^9 years

3. Age is approximately 3.70×10^9 years and initial $^{87}\text{Sr}/^{86}\text{Sr}$ about 0.68.

4. To get a precise age, it is necessary to determine the slope of the isochron line as correctly as possible. Having data points widely spaced on the line allows precise measurement of the slope. Obviously, very precise measurements of $^{143}\text{Nd}/^{144}\text{Nd}$ are also needed. Having a wide range of $^{147}\text{Sm}/^{144}\text{Nd}$ and precise measurements of both ratios allows precise determination of the initial $^{143}\text{Nd}/^{144}\text{Nd}$ value as well as a closely defined age. A typical range of the $^{147}\text{Sm}/^{144}\text{Nd}$ ratio in the minerals of a rock is shown in Figure 2-4.

5. Point A represents the value of $^{143}\text{Nd}/^{144}\text{Nd}$ for the chondritic uniform reservoir (and the whole Earth) at the time the Earth formed (4.55 billion years). It is unlikely that samarium or neodymium would partition to the core of the Earth during nebular condensation and thus all of these two elements was probably concentrated in the original Earth mantle. The time of magma formation by partial melting for the igneous rock sampled is shown at point B. This point is also the value of initial $^{143}\text{Nd}/^{144}\text{Nd}$ calculated from the rock analysis. The fact that point B falls on the CHUR line indicates that the rock's parent magma formed from the primary chondritic uniform reservoir and not from material that has undergone an earlier differentiation at some time after the Earth formed 4.55 billion years ago. Point C is the present, measured value of $^{143}\text{Nd}/^{144}\text{Nd}$, and point D is the present value of $^{143}\text{Nd}/^{144}\text{Nd}$ for the entire Earth.

6. **(a)** The plotted mineral isochron gives an age of 1948×10^6 years and an initial $^{143}\text{Nd}/^{144}\text{Nd}$ value of 0.5098.

(b) $\epsilon = \left[\frac{0.5098}{0.5110} - 1\right] \times 10^4 = -23.5$

(c) The large negative value for ϵ suggests that the magma formed in the crust rather than in the mantle or, if initially formed in the mantle, that the magma was strongly contaminated by crustal rocks during intrusion into the crust.

7. $\frac{^{238}\text{U (atoms)}}{^{204}\text{Pb (atoms)}} = \frac{204}{238} \times \frac{^{238}\text{U (ppm)}}{^{204}\text{Pb (ppm)}}$

8. Accuracy and precision of measurement of isotopes and of isotope ratios in a sample depend on abundance of the isotopes in the sample. In young samples there is very little ^{207}Pb compared to ^{206}Pb (which forms from the much more abundant ^{238}U); thus ^{207}Pb cannot be determined with much reliability. This results in a low reliability for the ^{207}Pb/^{206}Pb method. In older samples there is more ^{207}Pb because of the greater age, and there is more ^{207}Pb relative to ^{206}Pb because the shorter half-life of ^{235}U allowed ^{207}Pb to accumulate more rapidly than ^{206}Pb.

9. The slope of the isochron equals 0.19138 and the age is approximately 2.76×10^9 years. The age can be obtained by using Equation 2-30 and solving by successive approximations or by using Table 2-9.

10. Atmospheric argon contains ^{36}Ar as well as ^{40}Ar. Determination of the amount of ^{36}Ar in a sample, and knowledge of the ^{40}Ar/^{36}Ar ratio in air, allows a correction for the amount of atmospheric ^{40}Ar present in the sample.

11. Approximately 7.87×10^6 years

12. Approximately 51,900 years

13. (a) $T(°\text{C}) = 87.5 - 2.5\delta$ or $\delta = 35 - 0.4T(°\text{C})$

(b) 0.00205503

14. Isotopic theory predicts that the bond strength of oxygen with cations should affect ^{18}O/^{16}O ratios, with ^{18}O concentrated in compounds that have the most strongly bonded oxygen. Silicon-oxygen bonds in silicates are stronger than aluminum-oxygen bonds; thus ^{18}O should be concentrated in quartz as compared to K-feldspar.

15. Post-depositional alteration of sedimentary rocks usually causes a lowering of ^{18}O/^{16}O ratios. The waters causing the alteration generally contain at least some groundwater and this water has less ^{18}O than ocean water.

16. If only equilibrium isotopic fractionation occurred during evaporation of seawater, then the MWL would pass through the SMOW point. However, kinetically controlled evaporation (and fractionation) also occurs and this causes the MWL to be offset from the SMOW point.

17. −41.3 ppt; sedimentary

18. Near the surface some pore-water sulfate in a layer of sediment is changed to dissolved hydrogen sulfide by bacterial action. Fractionation occurs, concentrating ^{32}S in the hydrogen sulfide. As time goes on, the remaining sulfate ion in the layer occurs at a greater depth due to further sedimentation at the sediment-seawater interface. When more of the remaining sulfate ion, now enriched in ^{34}S, is changed to dissolved hydrogen sulfide, the product is hydrogen sulfide enriched in ^{34}S compared to the hydrogen sulfide in sediment pore water above it.

19. (a) 1.00110

(b) 52 degrees too high

20. (a) Photosynthesis concentrates ^{12}C relative to ^{13}C in organic matter formed in surface waters. Thus surface waters are enriched in ^{13}C. Sinking of the organic matter, followed by degradation, transfers ^{12}C to bottom waters.

(b) The lake exhibits strong thermal stratification in the summer. In the fall and winter overturn occurs and all the water is completely mixed. Thus there is no variation in $\delta\,^{13}$C values at that time.

21. $F_m = \left[\frac{\delta_s - \delta_T}{\delta_m - \delta_T}\right] \times 100$

where

$$F_m = \text{percent N in sediment from marine source}$$
$$\delta_s = \delta\ ^{15}\text{N for total N in sediment}$$
$$\delta_T = \delta\ ^{15}\text{N terrestrial component}$$
$$\delta_m = \delta\ ^{15}\text{N marine component}$$

For 6.5 per mil, $F_m = 53.3$ percent.

CHAPTER THREE

1. C (graphite) + $O_2 = CO_2 + 94{,}051$ cal.

   $CO + \frac{1}{2}O_2 = CO_2 + 67{,}635$ cal.

   By subtraction, C (graphite) + $\frac{1}{2}O_2 = CO + 26{,}416$ cal.

   The heat of formation of carbon monoxide is $-26{,}416$ cal.

2. **(a)** In general, the greater the hardness of a mineral, the lower its entropy. Hard minerals like diamond have their atoms (ions) tightly bonded together, and this limits the random thermal motion of the atoms.

   (b) Sillimanite. The stable high-temperature form of a substance is always the form with the largest molar entropy. In this case, sillimanite has the largest entropy at both low and high temperatures. By contrast, the SiO_2 polymorphs have a different entropy ranking at low and high temperatures.

3. **(a)** $\Delta G^0 = \Delta G_f^0$ (calcite) $-$ ΔG_f^0 (aragonite) $= -230$ cal/mol; calcite

   (b) Calcite, because of higher molar entropy

   (c) Aragonite, because of lower molar volume

   (d)

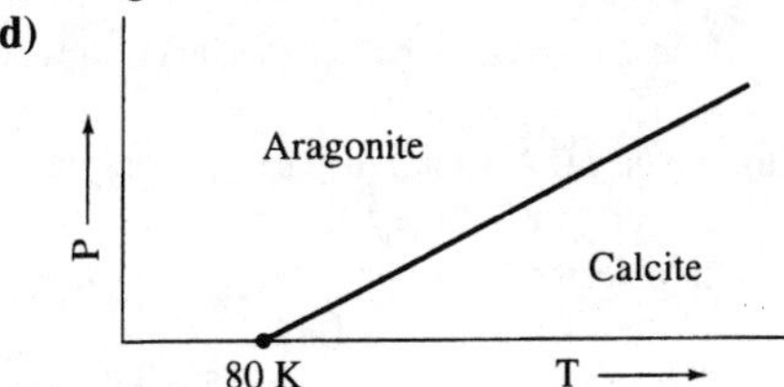

4. True. Because the system is adiabatic (thermally insulated) $dQ = 0$ and the change in internal energy is equal to the work done. U is a "state function" and thus the integral of dU is independent of the way in which a system changes. In this special case, the integral of dW also does not depend on the way the work is done.

5. **(a)** O_2 is the stable form of oxygen under the chosen standard conditions; S_2 is not the stable form of sulfur under these conditions.

   (b) Gases represent a more disordered state than that of solids.

   (c) The values would be much smaller negative numbers. They can be calculated from the data in Table 3-2.

6. 25°C: $\Delta G^0 = \Delta H$ (products minus reactants) $- T\,\Delta S$ (products minus reactants) $= +285$ cal/mol; $\Delta G^0 = \Delta G$ (products minus reactants) $= +286$ cal/mol; gypsum is stable.

   50°C: $\Delta G^0 = -36$ cal/mol; anhydrite is stable. To get this answer, it is necessary to calculate the change in ΔH of the reaction and in ΔS of the reaction with temperature.

7. True, because $dG = \int_{P_1}^{P_2} VdP$ and V is always greater than zero.

8. (a) $-0.0000027 \text{ bar}^{-1} \times 10{,}000 \text{ bar} = -0.027$;

$$-0.027 \times 22.69 \text{ cm}^3 = -0.61 \text{ cm}^3$$

(b) $\Delta G = \int_1^{10{,}000} VdP = 5,422\text{cal/mol}$

9. $\Delta G = 2{,}658$ cal/mol. This result is fairly close to $\Delta G = 0$ and thus to the equilibrium line between kyanite and sillimanite shown in Figure 3-7. Since ΔG is positive, kyanite is stable.

10. (a) $\Delta S = \int_{273.15}^{373.15} \frac{C_P}{T} dT = 5.59$ cal/mol-deg

(b) $\Delta S = \frac{Q}{T} = \frac{540 \times 18}{373.15} = 26$ cal/mol-deg

11. (a) $\frac{dP}{dT} = \frac{\Delta S}{\Delta V}$; $T(\text{equilibrium}) = \frac{\Delta H}{\Delta S}$

(b) Variation of H, V, and S with temperature

12. $K = 3.25 \times 10^{-7} \text{ sec}^{-1}$ at 250°C and 198.76×10^{-7} at 300°C. Thus increasing the temperature by 50 degrees increases the rate of reaction by a factor of about 61.

13. (a) Very small
(b) Reactants
(c) The activity product expression must equal 1. Some activities may be larger and others smaller than 1 as long as their combination in the expression gives a value of 1.

14. (a) $\Delta G^0 = \Delta G_f^0 \text{ (dolomite)} + \Delta G_f^0(Ca^{2+}) - [2\Delta G_f^0 \text{ (calcite)} + \Delta G_f^0(Mg^{2+})]$

$$= -2{,}318 \text{ cal}$$

$$\Delta G = -RT \ln K;$$

$$K = 10^{1.7} = 50$$

(b) Since the value for ΔG^0 is based on unit activities, it does not tell us whether the reaction would go to the right or left for nonunit activities.
(c) $Ca^{2+}/Mg^{2+} = 0.147$; since this value is less than the value of K found in part (a), the reaction should go to the right. However, dolomite is not found forming in ordinary seawater, even though it is more stable than calcite under the conditions that characterize seawater.

15. (a) $10^{-8.3}$ mol/l
(b) The value in Table 1-11 is slightly less than the value of part (a). Thus malachite should not precipitate from seawater.

16. (a) $\Delta G^0 = -RT \ln K$; $K = 10^{-5.08}$
(b) $(Ca^{2+})(SO_4^{2-}) = 4.5 \times 10^{-6} = 10^{-5.35}$; yes

17. (a) $KFe_3AlSi_3O_{10}(OH)_2$ (annite) $+ \frac{1}{2}O_2 \rightarrow KAlSi_3O_8$ (K-feldspar) $+ Fe_3O_4$ (magnetite) $+ H_2O$
(b) Lower part. This represents lower oxygen fugacities. As oxygen is added, annite would break down to phases such as magnetite that contain ferric iron. The mineral assemblage of part (a) would maintain a fixed oxygen fugacity (buffer the system) as more oxygen was added to the system until eventually enough oxygen had entered to use up all of the annite.

18. $a_{Fe^{2+}} = 0.2 \times 0.8 \times 0.6 = 0.096$ mol/kg

19. (a) The mineral combination can exist in equilibrium and has one degree of freedom. A change in temperature would throw the system out of equilibrium, unless there is a specific change in pressure that is the exact amount needed to bring the system back to equilibrium.
(b) This combination can also exist in equilibrium, but with zero degrees of freedom. Changing the temperature would throw the system out of equilibrium.

20.

One	point	three
Two	point	four
Three	point	five
One	line	two
Two	line	three
Three	line	four
One	region	one
Two	region	two
Three	region	three

21. **(a)** annite + pyrope $\rightarrow$ phlogopite + almandine

(b) $K_D = \dfrac{X_{\text{Fe}}^{\text{gar}}/X_{\text{Mg}}^{\text{gar}}}{X_{\text{Fe}}^{\text{bio}}/X_{\text{Mg}}^{\text{bio}}}$

(c) $K_D = 7.68$. Curve one of Figure 3-4 gives a temperature of about 769 K, curve two a temperature of about 763 K, and curve three a temperature of about 757 K.

(d) Additional components such as calcium and titanium may occur in the solid solutions and affect the calculated temperature. Also the occurrence of both Fe^{2+} and Fe^{3+} in solid solution could affect the final result unless the two are measured separately and only Fe^{2+} is used in calculating temperature.

22. **(a)** 2
(b) No
(c) Leucite; K-feldspar
(d) Leucite + K-feldspar; K-feldspar + tridymite
(e) The composition represented by point R

CHAPTER FOUR

1. **(a)** $10^{-5.25}\,M$; 649×10^{-6} gm/l; 0.649 ppm
(b) $10^{-5.7}\,M$
(c) $10^{-4.8}\,M$

2. **(a)** $10^{-2.25}\,M$; less
(b) $K = [H_2O]^2$
(c) Gypsum is stable.
(d)

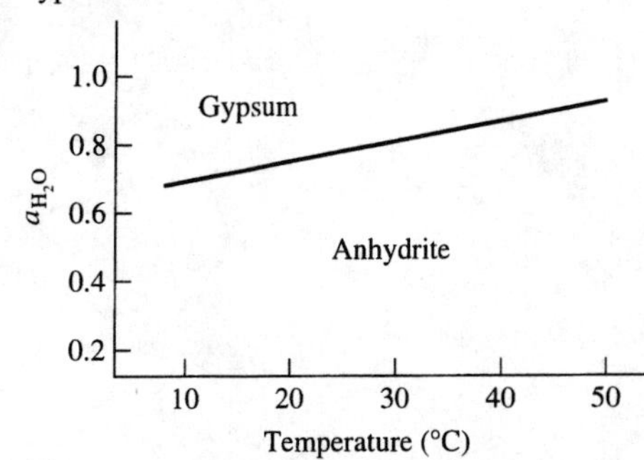

The upper field must be the gypsum field since gypsum is the stable form found using standard-state free energies, and the free energy of formation of water is based on unit activity of water.

3. A solubility product is the product of the activity values for a saturated solution. An activity product is the product of the activities for any solution, whether saturated or not.

4. **(a)** $K = \dfrac{[Ca^{2+}]}{[Mg^{2+}]} = 50$

(b) Yes

5. $K = 0.02$ mol/liter-atm

6. **(a)** 4.7

(b) $10^0 M = 1 M$

7. 0.25 V

8. **(a)** $10^{+0.67} M$

(b) Increase, because $[H^+]$ would be lower and the reaction would tend to go farther to the right.

9. **(a)** Mn^{2+}

(b) $Mn^{2+}/Mn^{3+} = 10^{10}$

10. **(a)** $2\ Fe^{2+} + 3\ H_2O \rightleftharpoons Fe_2O_3\ (s) + 6\ H^+ + 2\ e^-$

(b) +0.7 V

(c)

Eh
+0.4
+0.2
0.0
−0.2
−0.4
1 3 5 7 9 11
Hematite
Fe_2O_3
Fe^{2+}

(d) $[Fe^{2+}] = 10^{-14} M$

11. **(a)** Low; **(b)** High; **(c)** Low; **(d)** High; **(e)** High; **(f)** Low

12. **(a)** A higher pH would be expected in the summer when photosynthesis would have its greatest effect. Dissolved carbon dioxide is used up by photosynthesis, and one result is to reduce the abundance of hydrogen ion [equation (4-48)].

(b) During the day photosynthesis by plants removes carbon dioxide and raises the pH. At night plant and animal respiration produces carbon dioxide and lowers the pH.

13. Figure 4-9b indicates that, at constant pressure, higher temperatures result in lower densities and thus lower solubility of ionic materials. So log K will decrease with increase in temperature. Experimental data [Eugster and Baumgartner (1987)] give the following result.

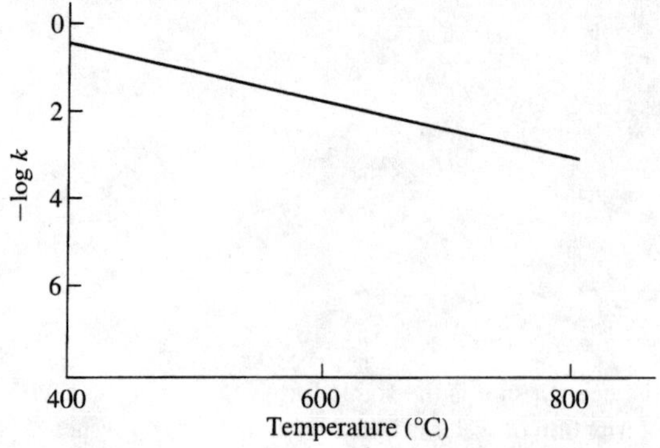

14. **(a)** $t_{1/2} = \frac{\ln 2}{k}$

(b) 693 seconds. The half-life of aqueous reactions ranges from fractions of a second to thousands of years.

(c) 2.73 kcal/mol

15. **(a)** $2\,FeS_2\,(s) + 7\,O_2 + 2\,H_2O \rightarrow 2\,Fe^{2+} + 4\,SO_4^{2-} + 4\,H^+$

$FeS_2\,(s) + 14\,Fe^{3+} + 8\,H_2O \rightarrow 15\,Fe^{2+} + 2\,SO_4^{2-} + 16\,H^+$

(b) $4\,Fe^{2+} + O_2 + 4\,H^+ \rightarrow 4\,Fe^{3+} + 2\,H_2O$

(c) $Fe^{3+} + 3\,H_2O \rightarrow Fe(OH)_3\,(s) + 3\,H^+$; yes

(d) The oxidation of ferrous ion to ferric ion would be the rate-determining step. Stumm and Morgan (1981, 470) report experimental data indicating that, at pH 3, half-lives for the oxidation of Fe^{2+} are on the order of 1,000 days. For the oxidation of pyrite by Fe^{3+}, half-lives of 20 to 1,000 minutes are observed. This explains why the overall rate of dissolution of pyrite is independent of surface area. Microorganisms influence the overall rate of reaction by mediating the oxidation of Fe^{2+}.

16. **(a)** $\frac{(HCO_3^-)}{(H_2CO_3)} = \frac{10^{-6.4}}{(H^+)} = 10^{1.9}$

$\frac{(CO_3^{2-})}{(HCO_3^-)} = \frac{10^{-10.3}}{(H^+)} = 10^{-2}$

There is approximately 100 times as much (HCO_3^-) as (H_2CO_3) and 1/100 times as much (CO_3^{2-}) as (HCO_3^-). Thus (H_2CO_3) and (CO_3^{2-}) each make up about 1 percent of the total dissolved CO_2, and the rest is (HCO_3^-).

(b) $(HCO_3^-) = 0.0008\,M$; $(H_2CO_3) = 0.0002\,M$; (CO_3^{2-}) is negligible.

17. **(a)** 8.34 meq/l

(b) 417 mg/l as $CaCO_3$

18.

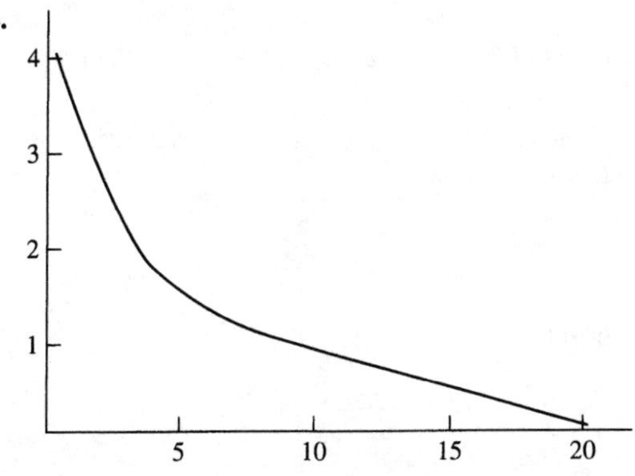

This is an example of the type of curve shown in Figure 4-14a for removal from solution of a component in river water as it enters an estuary. The mixing of river water and ocean water causes flocculation of colloidal iron and precipitation of dissolved iron. One result of the mixing is neutralization of negative colloid charges by seawater cations and another result is an increase of pH. The first produces flocculation of river-borne colloidal iron compounds and the second causes precipitation of iron hydroxide. In most estuaries, more than 50 percent of the iron entering from rivers is added to estuarine bottom sediments.

19. **(a)** Carbon dioxide has escaped from the water. The following reactions go to the left.

$$CO_2(g) + H_2O \leftrightharpoons H_2CO_3$$

$$CaCO_3(s) + H_2CO_3 \rightleftharpoons Ca^{2+} + 2HCO_3^-$$

This results in a decrease of Ca^{2+} and HCO_3^-, and the decrease in HCO_3^- causes a decrease in H^+ (increase in pH) as H^+ reacts with CO_3^{2-} to replace some of the HCO_3^-.

(b) Contamination by seawater has occurred.

20. Model RP-2 has a fixed CO_2 pressure (the system is open to CO_2) and this keeps the pH depressed with the reaction $CO_2 + H_2O \rightarrow H^+ + HCO_3^-$ occurring to replace some of the H^+ used up in water-rock reactions. The reaction causes bicarbonate concentration to increase rapidly with time. In model RP-1 the CO_2 pressure is completely controlled by water-rock reactions (the system is closed to CO_2) and thus the pH rises rapidly as CO_2 and H^+ are used up in these reactions. CO_2 pressure decreases and pH increases as the water moves through the stratified drift. Bicarbonate concentration increases slowly since there is no outside source of CO_2.

CHAPTER FIVE

1. **(a)** Ionic and covalent
(b) Covalent
(c) Ionic
(d) Covalent
(e) Ionic

2. **(a)** Iron has a different valence in the two minerals.
(b) Aluminum has a different coordination number in the two minerals.

3. **(a)** 12
(b) 6
(c) Because you have directed bonds that involve only certain atoms around a given atom.

4. **(a)** Radius ratios (cation/anion) give the following coordination numbers: potassium, 12; rubidium, 12; calcium, 8; sodium, 8; and boron, 3.
(b) Yes. Potassium occurs with coordination numbers of 8 and 12, rubidium 8 and 12, calcium 6 and 8, sodium 6 and 8, and boron 3 and 4.

5. The potassium ion has a larger radius than the sodium ion. This makes it easier to pack a large number of anions around the potassium ion as compared to the sodium ion.

6. (d) +2 charge, coordination number 6, high-spin state

7. **(a)** Se^{2-}
(b) Cu^+
(c) They should have similar sizes because of the lanthanide contraction.
(d) Cd^{2+}
(e) High-spin state
(f) B^{3+}

8. Four formula units

9. **(a)** Inosilicate
(b) Tektosilicate
(c) Tektosilicate
(d) Nesosilicate
(e) There are nine silicon atoms in the formula of vesuvianite. Four of these form two Si_2O_7 groups, and the other five form independent SiO_4 groups. Thus this mineral is both a nesosilicate and a sorosilicate.
(f) Cyclosilicate

10. $Z = 4$; thus there are 12 aluminum atoms in the unit cell.

11. **(a)** $Mg_{0.3}Fe_{1.7}SiO_4$
(b) Fa, 0.85; Fo, 0.15
(c) 45.98 cm^3

12. **(a)** 760°C
(b) 757 to 764°C

13. **(a)** $\gamma_i = \frac{a_i}{N_i}$ and $\gamma_j = \frac{a_j}{N_j}$. In this case, γ_i and γ_j both equal 1.0 and activity and mol fraction are identical throughout the compositional range (Figure 5-20b).
(b) γ_i is a constant (but not 1.0) and $a_i = \gamma_i N_i = h_i N_i$.

14. **(a)** Sanidine, because it is more disordered and thus less compact.
(b) Microcline, because it is more ordered and thus more compact.
(c) Microcline, because it is more compact and thus would have stronger bonding.

15. **(a)** As shown by Figure 5-23c, sample B occurs higher on the solvus curve.
(b) The chemical potential of $NaAlSi_3O_8$ is equal in the two phases if equilibrium is maintained.

16. **(a)** High quartz
(b) High quartz

17. **(a)** The decrease in volume is much greater perpendicular to the layers with most of the decrease involving the more weakly bonded potassium layers as opposed to the tetrahedral and octahedral layers. Change parallel to the layers is limited by the most rigid layer, which is the octahedral layer.
(b) The increase in volume is much greater perpendicular to the layers for the same reasons as in part (a).

18. **(a)** K
(b) Ca
(c) Al
(d) Mg, Fe, Al
(e) Mg
(f) K
(g) Si
(h) Ca
(i) K
(j) Mg
(k) Mg, Fe

19. **(a)** The first problem is deciding what ionic radius to use for Fe^{3+}. For most spinels, Fe^{3+} is probably in the high-spin state and, for octahedral (6) coordination, Table 5-2 gives a radius of 0.73 Å for Fe^{3+}. The radius of Cr^{3+} for octahedral coordination is 0.70 Å. If the difference between 0.70 and 0.73 Å is significant (which probably isn't true), Goldschmidt's second rule indicates that Cr^{3+}, as the smaller ion, should be preferentially concentrated in early formed crystals.
(b) Chromium has a lower electronegativity (Table 5-1) and, according to Ringwood's rule, should be preferentially concentrated in early formed crystals. Analytical work on chromite (a member of the spinel group of minerals) from Hawaiian lavas (Evans and Wright 1972) shows that, with falling temperature, decreasing Cr_2O_3 content is accompanied by increasing Fe_2O_3 content. In this case, predictions based on two different approaches agree with what is found in nature.

CHAPTER SIX

1. **(a)** 1 (for a chain structure)
(b) 0 (for a ring structure)
(c) 2 for oxygen and 3 for nitrogen

2. $CH_4 + SO_4^{2-} + 2\,H^+ \rightarrow H_2S + CO_2 + 2\,H_2O$
3. **(a)** Increasing
   (b) Volcanic gases from the interior of the Earth (juvenile carbon)
4. SO_2, SO_4^{2-}, H_2S
5. **(a)** N
   (b) Na
   (c) atmospheric input + weathering release = biomass accumulation + streamflow loss
   (d) It has to be at steady state, not growing or getting smaller.
6. Spartina detritus = 67 percent, phytoplankton = 33 percent
7. **(a)** 50
   (b) 15×10^9 gm/yr
   (c) Shortest residence time: atmosphere; longest residence time: sediments
8. **(a)** Freshwater
   (b) Nonmarine sediments
   (c) Hydrocarbons
9. **(a)** $C_6H_{12}O_6 + 6\,O_2 \rightarrow 6\,CO_2 + 6\,H_2O$
   (b) $\Delta G^0 = -687{,}952$ cal
   (c) The free-energy change is significantly less for a fermentation reaction; thus there is a slower decomposition of organic matter in anaerobic sediments as compared to aerobic sediments. Because of this, organic matter in aerobic sediments tends to be completely destroyed, whereas some of it in anaerobic sediments is not destroyed and becomes part of the rocks that form from the sediments.
   (d) $C_6H_{12}O_6 \rightarrow 3\,CO_2 + 3\,CH_4$
10. Petrochemical industrial wastes would have no ^{14}C, since petroleum is too old to have any remaining ^{14}C. Carbon from sewage wastes and carbon in plants and animals has a ^{14}C content reflecting equilibrium with the present atmosphere. Thus measurement of the amount of ^{14}C would indicate the relative importance of these two sources of organic carbon.
11. **(a)** Pyrite
    (b) Native copper
    (c) Uraninite (UO_2)
    (d) Native sulfur. There are other possible answers.
12. Lower, because metamorphism increases the carbon content of the remaining organic material.
13. **(a)** 0.82 percent
    (b) 0.73 percent
    (c) Higher
14. **(a)** Naphthenes
    (b) Aromatics
    (c) Aliphatics
    (d) Olefins
    (e) Aliphatics
15. Land plants of all ages have apparently maintained the same isotopic composition. Furthermore, the various processes involved in diagenesis and coalification have caused little isotopic fractionation.
16. Beyond the bituminous coal stage, hydrogen content decreases significantly, as shown by the vitrinite path of Figure 6-14.
17. 162,737 ml CO_2 and 757 ml SO_2
18. According to Hunt (1979), oil has not formed in young, cold basins and it has been destroyed in old, hot basins. Therefore, the places to prospect are in young, hot basins (where oil can form fairly quickly) and

in old, cold basins (where there has been enough time to form hydrocarbons even though the temperatures have never gotten very high).

19. One possible interpretation is as follows. Because of their polar nature, compounds containing N, O, and S in addition to H and C (example: porphyrins) interact more strongly with interstitial water and mineral surfaces during migration. (A polar molecule has an uneven charge distribution even though it is electrically neutral overall.) In this case, these compounds were selectively left behind as the petroleum (which formed in the source rock) migrated through the shale to the sandstone.

20. The diagram represents a kinetic isotope effect due to the fact that bonds containing ^{12}C generally react more readily than bonds containing ^{13}C. Thus in kinetically controlled reactions ^{13}C tends to be depleted in the product relative to ^{12}C. Figure 6-5 shows a depletion of ^{13}C in going from atmospheric CO_2 to living matter and from oceanic HCO_3^- to living matter. Schidlowski et al. (1983) suggest that the main reason for the ^{12}C enrichment observed in Precambrian kerogens lies with biological activity on the Earth beginning about 3,500 million years ago. In other words, the isotopic evidence suggests that carbon fixation by autotrophic life forms gained control of the terrestrial carbon cycle at least as early as 3,500 million years ago.

21. This suggests that free O_2 was present in the atmosphere starting about two billion years ago. If oxygen had been present earlier, uraninite grains would probably have been altered to carnotite or some other oxidized uranium mineral.

CHAPTER SEVEN

1. **(a)** The residual soil would have less potassium, calcium, and magnesium. These elements are mostly removed during weathering and later added to sediment material during transportation, deposition, and diagenesis.

(b) Glacial clays contain large amounts of fine-grained detrital material rather than the large amounts of clay minerals found in most fine-grained sediments. These glacial deposits often have more Na_2O than K_2O and would form shales with this characteristic. The siliceous shale of Table 7-2 has volcanic ash as a component. Shales containing volcanic material often have a high Na_2O content.

(c) Carbonaceous shales form in a reducing environment and thus the iron mainly occurs in the ferrous state. The chemical analysis probably reports total iron content of the rock expressed as Fe_2O_3.

2. Not all the possibilities are listed here.

(a) In manganese oxides such as pyrolusite, as the carbonate mineral magnesite, and as a trace element substituting for calcium and magnesium in carbonate minerals.

(b) As a trace element in carbonate minerals and in the mineral barite.

(c) As adsorbed ions on clay minerals, as a trace element substituting for potassium in clay minerals, and in detrital tourmaline.

(d) As detrital chromite and as a trace element in aluminum hydroxides such as gibbsite.

(e) As adsorbed ions on clay minerals, in the minerals apatite and fluorite, and in the phosphates of bones, teeth, and brachiopod shells.

(f) As a trace element in detrital tourmaline and in precipitated aluminum hydroxides, in minerals such as carnotite, and in residual organic matter.

3. **(a)** The saturation value is 10 ppm (Figure 7-2), which is 10 g/1,000,000 g or 0.010 g/liter. This is about 0.000166 mol/liter or $10^{-3.78}$ mol/liter. Thus quartz saturation would be represented by a vertical line at 3.78 on the diagrams.

(b) Most waters other than seawater are supersaturated with respect to quartz but not with respect to amorphous silica. Thus they would plot between the quartz and amorphous silica lines.

4. **(a)** The residual soil will also contain 14.62 g of alumina. The amount of alumina in the residual soil is 26.14 percent of the soil; thus the weight of the residual soil is 14.62/26.14 times 100 = 55.88 g.

(b) 55.88/100 times 55.07 = 30.83 g

(c) Fe_2O_3 increases by 1.39 g

5. **(a)** 1.8 percent

(b) Usually over 99 percent of the total iron is particulate material.

6. **(a)** The concentration of A^+ adjacent to the solid should be the same as in the bulk solution at any given time and will increase linearly with time.

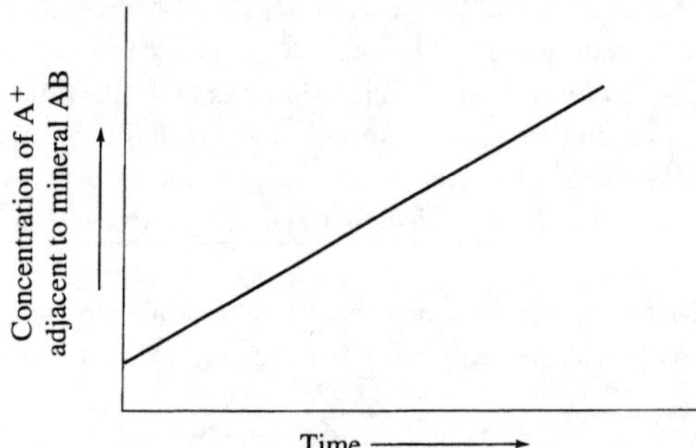

(b) The reaction is a zero-order reaction. The rate of this kind of reaction is a constant.

7. **(a)** The smectite group; it is a montmorillonite.

(b) Bentonite. The montmorillonite in bentonite can expand to several times its original volume when wetted. Water layers are adsorbed between montmorillonite layers to cause the expansion.

8. **(a)** In the tetrahedral layer there is a deficiency of positive charge of 1.6 due to substitution of Al^{3+} for Si^{4+}. In the octahedral layers there is an excess of positive charge of 0.8 due to substitution of Fe^{3+} for Mg^{2+}. Thus overall there is a deficiency of 0.8 positive charge.

(b) The adsorbed Ca^{2+} has a positive charge of 0.8, which neutralizes the deficiency of 0.8 due to substitutions in the structure.

9. The sample contains about 90 percent illite layers and 10 percent smectite layers.

10. **(a)** Glauconite can be considered a variety of illite in which Fe^{3+} substitutes for Al^{3+}. Illite is similar in structure to muscovite, but it has a smaller amount of potassium than muscovite, and its potassium content varies, whereas that of muscovite does not. Thus, from the viewpoint of structure, glauconite can be considered to be a mica, but in terms of composition and ion-exchange capacity it belongs to the clay-mineral group.

(b) Glauconite is an authigenic mineral that usually forms under the following conditions: shallow, marine environment; intermediate Eh conditions (glauconite contains both Fe^{2+} and Fe^{3+}); low rate of sedimentation.

11. **(a)** $K_{aragonite}$, because aragonite is known to contain larger amounts of strontium. Laboratory experiments have shown that $(K_{Sr})_{aragonite}$ is significantly larger than $(K_{Sr})_{calcite}$.

(b) Laboratory measurements of $(K_{Sr})_{aragonite}$ can be combined with knowledge of the amount of Sr^{2+} in seawater to predict the expected concentration of Sr^{2+} in aragonite. If the measured amounts in the particles agree with this prediction, inorganic precipitation is indicated. If not, biological fractionation (i.e., nonequilibrium) has probably occurred, and thus the particles formed by organic precipitation. Important factors that are not taken into account by this approach are variations in temperature and particle growth rate.

(c) 7,938 to 8,586 ppm. This range is slightly higher than the value for the corals (7,847 ppm).

12. It would change from 293 μmol/kg to 290 μmol/l.

13. **(a)** $\sum CO_2$, $CO_2(aq)$, and HCO_3^- would increase, and CO_3^{2-} and pH would decrease.
(b) Less saturated

14. **(a)** pH = 8.3, $[CO_3^{2-}] = 10^{-5.04}\ M$, and $[Ca^{2+}] = 10^{-3.18}\ M$
(b) 6.79. This suggests that seawater is oversaturated by a factor of about seven with respect to aragonite. Notice the similarity of this number to the Ω value for case one in Table 7-15.

15. For a steady-state situation the redox front would stay at a constant depth because the flux of O_2 downward would balance with the flux of Fe^{2+} upward. In general, steady-state diagenesis occurs when all diagenetic processes balance one another so that there is no change with time at a given depth.

16. **(a)** $2\ CH_2O + SO_4^{2-} \rightarrow H_2S + 2\ HCO_3^-$
(b) Berner (1980) states that any methane occurring in the sulfate ion zone is apparently consumed by sulfate-reducing bacteria and associated fermentative bacteria. Calculations of possible methane consumption rates are of the same order of magnitude as the rates of organic matter consumption calculated for bacterial sulfate reduction in the Long Island Sound sediment diagramed in this question.
(c) Berner (1980) points out that in the interstitial water of freshwater sediments sulfate ion concentrations are much lower. Thus complete sulfate reduction under anoxic conditions occurs quickly during diagenesis, enabling a build-up of methane near the sediment-water interface.

17. **(a)** $mCa^{2+}/mMg^{2+} = 0.643$; yes
(b) $mCa^{2+}/mMg^{2+} = 0.185$; yes

18. **(a)** Chemical composition, structure, and grain size of the minerals
(b) Calcite, because of its greater ease of recrystallization
(c) The fine-grained fraction

19. **(a)** Gypsum starts to precipitate and soon after anhydrite replaces it as a calcium-bearing precipitate.
(b) Polyhalite starts to precipitate.
(c) Kieserite

20. **(a)** With continued evaporation, the activity of H_2O in the brine would decrease, and eventually the line dividing the stability fields of gypsum and anhydrite would be crossed.
(b) The brine formed by evaporation, which is magnesium rich, has interacted with the underlying rock.

21. As long as minerals that do not interact are precipitated, equilibrium and fractionation paths coincide. In the equilibrium crystallization of Table 7-21, the first interaction occurs when glauberite forms at the expense of some of the anhydrite:

$$CaSO_4\ (\text{anhy}) + 2\ Na^+ + SO_4^{2-} \rightarrow Na_2Ca(SO_4)_2\ (\text{glaub})$$

If the anhydrite had been removed from contact with the brine, the brine composition would have followed a different path. See Eugster et al. (1980).

22. **(a)** Paths one and three
(b) Possibilities are dolomite, high-magnesium calcite, and magnesium-rich smectite.
(c) Sulfate ion reduction would tend to occur, thus producing an $Na\text{-}CO_3\text{-}Cl$ brine or an Na-Mg-Cl brine, depending on the original composition.

23. **(a)** 1. The brine waters and sediments have a low content of organic matter because the density layering of the brines prevents most organic particles from sinking into the brine zone. Thus there may not be enough organic matter present to support sulfate-reducing bacteria.
2. The hot temperatures, high salinities, and high metal concentrations of the hot brine areas may make it impossible for sulfate-reducing bacteria to exist.
(b) 1. If the brine water is injected into the bottom of the Atlantis II Deep from a deeper, crustal source or from a lateral source in the sediments of the Red Sea Basin, it may bring not only the metals with it but also the sulfur.

2. Significant sulfate reduction may occur in the water immediately above the brine zones with H_2S diffusing into the brine itself, followed by sulfide precipitation. Sulfate-reducing bacteria have been found in the water above the brine zones.

CHAPTER EIGHT

1. (a) Alkali olivine basalt

(b) Hydrous minerals, such as phlogopite, are not included in the standard set of minerals used for norm calculations. Only end-member minerals for the pyroxenes are used and thus augite is not included in the standard set.

(c) Unsaturated and metaluminous

2. (a) #2 plots below the line while #3 plots on the line.

(b) #2 plots just below the line and #3 plots above the line.

(c) The primary magmas for both rocks could have been similar. On the way to the surface, the continental basalt (#3) probably was contaminated with granitic crustal material. This would result in higher values for both SiO_2 and $^{87}Sr/^{86}Sr$. The oceanic basalt (#2) was not contaminated by such material.

3. (a) $FeO = 33.3$, $MgO = 44.4$, $(K_2O + Na_2O) = 22.2$

(b) They would both plot at the same point on the diagram because such diagrams use only ratios of constituents.

4. (a) $D = 0$

(b) $C_L/C_O = 1.0$

5. (a) Ce, 1.208; Sm, 1.217; Eu, 1.163; Gd, 1.225; Yb, 1.229

(b)

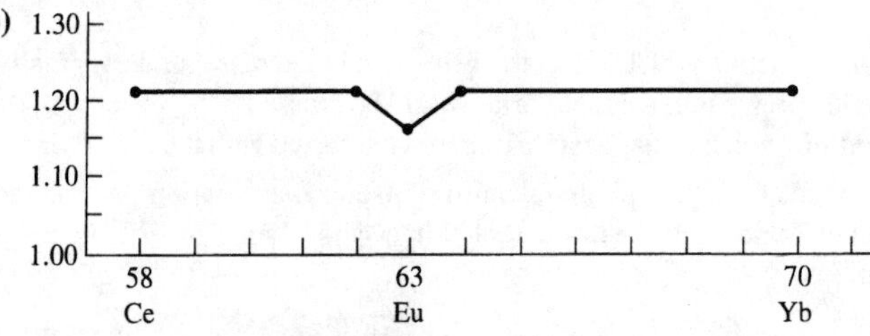

Yes. There is a negative Eu anomaly because crystallization of plagioclase depletes the melt in Eu due to substitution of Eu^{2+} for Ca^{2+} in the plagioclase.

(c) Enriched

6. As shown by Figure 8-8, a melt produced by partial melting of a garnet-bearing peridotite has an REE pattern with a steep slope (particularly for the HREE) because the heavier REE are retained in the source rock relative to the lighter REE. The REE pattern for magma from a spinel-bearing peridotite is similar to the less steep lines in Figure 8-8, because the distribution coefficients for the rare earths in spinel are similar for all the rare earths. Spinel is stable at shallower levels in the mantle than garnet (Figure 8-14). Thus the nature of the REE pattern of a basalt gives information about the depth where the parent magma formed. Steeper slopes indicate greater depths of formation.

7. (a) One possible interpretation is as follows. Because of its mineralogy, partial melting of the source rock causes formation of a magma enriched in the LREE (see question six). Subsequent melts derived from this depleted source (depleted in incompatible elements) will be enriched in the HREE relative to the LREE.

(b) The enrichment of the Hawaiian tholeiites in the LREE suggests that they should also be enriched in the incompatible elements K, Ti, and P. They are found to be strongly enriched in these elements relative to MORB basalts.

8. The enrichment in LREE suggests that garnet is present in the source region (see question six), while the negative europium anomaly suggests that plagioclase has been removed from the region prior to magma formation which led to the basalt (see question five).

9. **(a)** Group A rocks. These all give the same initial $^{87}Sr/^{86}Sr$, suggesting that all the rocks formed from the same magma. The magma developed from an isotopically homogeneous source.

(b) The rocks of group B probably formed from different magmas with different initial $^{87}Sr/^{86}Sr$ values. The magmas could have come from a source region that is not isotopically homogeneous, or they could be from different source regions.

10. The elements Sm and Nd are much more similar in their geochemical properties (they are both REE) than are Rb and Sr. Thus geological processes tend to fractionate (separate) Rb from Sr more than Sm from Nd.

11. **(a)** The source had a larger Sm/Nd than CHUR.

(b) Depleted

12. **(a)** The oceanic basalt $^{87}Sr/^{86}Sr$ is less than 0.7048, thus a depleted source. The continental basalt $^{87}Sr/^{86}Sr$ is more than 0.7048, thus an enriched source.

(b) The continental basalt should have the highest value and it does.

13. The source could be "pristine" mantle with a composition representative of the original, unchanged mantle. Alternatively, the source could be depleted mantle magma contaminated with crustal material, resulting in a bulk-silicate-earth composition. Other interpretations are possible.

14. Assuming the crust consists of material extracted from the upper mantle, the residual mantle material would have a higher Sm/Nd ratio and a lower Rb/Sr ratio. Since ^{147}Sm decays to ^{143}Nd and ^{87}Rb decays to ^{87}Sr, the more depleted mantle material becomes, the greater the value of $^{143}Nd/^{144}Nd$ that will form and the smaller the value of $^{87}Sr/^{86}Sr$ that develops. In other words, there is a negative correlation between the isotope ratios of Nd and Sr.

15. **(a)** Magma formed by partial melting in the mantle has a lower Sm/Nd ratio than CHUR (chondritic uniform reservoir) and thus crustal material formed from such magmas has lower $^{143}Nd/^{144}Nd$ ratios (retarded Nd evolution) than the source rock in the mantle.

(b) It is necessary to correct analytical values for isotope (mass dependent) fractionation that occurs in the analytical process. This is done by normalizing (adjusting) measured ratios to agree with a specific value for a ratio of unradiogenic isotopes (Wasserburg et al. 1981). Unfortunately, different laboratories have used different values for the normalizing factor. Thus $^{143}Nd/^{144}Nd$ values from different sources may not be directly comparable.

16. **(a)** Increasing depth results in decreasing SiO_2 and increasing MgO, that is, more mafic magmas.

(b) Their magmas are formed deep in the mantle and thus do not often reach the surface.

17. **(a)** The first melt would have the composition of point A in Figure 8-19.

(b) Two pyroxenes and a silica material. The $CaMgSi_2O_6$-$MgSiO_3$ line in Figures 8-18 and 8-19 divides the final products of liquid crystallization (and the first products of partial melting) into two groups: mineral assemblages that coexist with olivine and mineral assemblages that coexist with a silica mineral.

18. **(a)** Leucite starts to form between 1,100 and 1,200°C, and the liquid changes in composition toward the line dividing the leucite and alkali feldspar fields. At the line, leucite passes into solution and alkali feldspar crystallizes. The liquid changes in composition along the line until leucite is completely eliminated. It then enters the alkali feldspar field and changes in composition toward the cotectic line dividing the alkali feldspar and tridymite fields. Alkali feldspar continues to form. When the liquid composition reaches the line, tridymite and feldspar form together, and the liquid composition changes along the line until the whole mass is crystalline. The end product consists of alkali feldspar and tridymite.

(b) If the early formed leucite is separated from the liquid, then the liquid composition will change directly across the leucite-alkali feldspar line, reach the tridymite-alkali feldspar cotectic line, and move along the line to eventually reach the minimum point *m*. The final products would be separate masses of leucite and of tridymite-alkali feldspar. More tridymite and less alkali feldspar would be present since SiO_2 was not used in the reaction: $KAlSi_2O_6$ (leucite) + $SiO_2 \rightarrow KAlSi_3O_8$ (K-feldspar).

19. **(a)** The low point of the Ab-Or line at 1,078°C
(b) The low point of the cotectic line at 703°C

20. Plutonic rocks cool more slowly; thus more extensive separation of the two feldspars can occur.

21. **(a)** 22.4 percent. The low molecular weight of H_2O compared to the other components results in a high mole percent for H_2O.
(b) Yes
(c) A decrease in pressure causes loss of water, because the solubility of water in a magma decreases with decreasing pressure. When a small amount of water is dissolved in a magma, it has a high chemical potential and a strong effect on the chemical potentials of other components. For example, the crystallization (melting) temperature of most silicates is lowered by hundreds of degrees. A sudden decrease in water content of a magma at a low temperature thus raises the crystallization temperatures of these silicate minerals, and they then crystallize rapidly.

22. **(a)** 1.85×10^8 liters
(b) 1.85×10^7 seconds, which is approximately seven months

23. **(a)** The rock that interacted only with magmatic water
(b) The rock that interacted only with magmatic water

24. **(a)** The major aluminum-bearing minerals in the rocks are the feldspars, and their abundance increases as you go from early to late formed rocks.
(b) The crystal chemical properties of rubidium are similar to those of potassium. Thus it occurs mainly in K-feldspar, which increases in abundance from early to late formed rocks.
(c) No. All the norms list nepheline, a mineral that is unsaturated with respect to silica.

25. Crystal rocks tend to have REE patterns that are LREE enriched relative to mantle material. Thus the steeper pattern of the eastern zone (shown below) could indicate crustal contamination. However, the rocks of the eastern zone have a very low HREE content. Upper continental crust has an HREE content that is about 17 times that of chondrites. Assimilation of these rocks would *not* result in the low HREE values of the eastern zone. Hess (1989, 176) points out that even crust with *no* HREE could only produce the eastern zone values if 80 percent or more contamination occurred. Thus the REE data require some other explanation for the differences between the two zones.

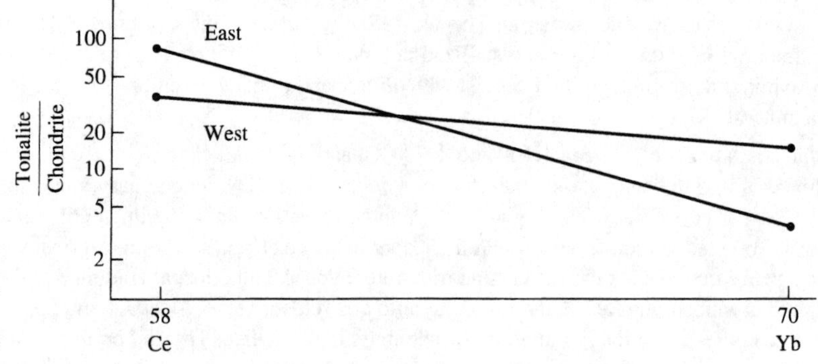

CHAPTER NINE

1. **(a)** $f = c + 2 - p = 5 + 2 - 5 = 2$
   (b) Since dolomite and diopside have the same CaO/MgO ratio, there are only four components. $f = 4 + 2 - 4 = 2$.
2. **(a)** $Mg_2SiO_4 + Mg_3Si_4O_{10}(OH)_2 \rightleftharpoons 5\ MgSiO_3 + H_2O$
   (b) Positive with a gentle slope, since $dP/dT = \Delta S/\Delta V$, and ΔV would be very large.
   (c) $f = c + 3 - p = 3 + 3 - 4 = 2$
3. **(a)** The components are Al_2O_3, SiO_2, MgO, FeO, and $K(OH)_2$. The phases are sillimanite, biotite, muscovite, quartz, almandine, and cordierite. Thus $f = 5 + 2 - 6 = 1$.
   (b) There is now an additional component, since $K(OH)_2$ has to be split into K_2O and H_2O. $f = 6 + 2 - 6 = 2$.
   (c) H_2O is not counted as a component, and $f = 5 + 2 - 6 = 1$.
   (d) There would be no change, since the additional phase has an additional component (Na_2O). $f = 6 + 2 - 7 = 1$.
4. **(a)** About 737°C
   (b) About 616°C
5. **(a)** About 8.2 kb
   (b) P is about 9.2 kb; error about 1 kb
6. Isotopic ratios of whole-rock samples vary according to differences in the abundances of minerals. Thus, even if all the rock samples have equilibrated with a common, homogeneous fluid, analyses of the rocks will *not* demonstrate the occurrence of a homogeneous distribution. This can only be done by analysis of mineral separates.
7. **(a)** If metamorphic reactions caused by circulating water produced changes in the original composition, a metamorphic rock such as a pelitic schist should have $\delta\,^{18}O$ values significantly different (probably lower) from the average $\delta\,^{18}O$ values of shales. If premetamorphic differences in isotopic ratios occurred, and if a homogeneous distribution of ratios is now found, then there may have been a large amount of fluid flow through the rock during metamorphism (Valley 1986). Fluid-absent metamorphism will tend to preserve premetamorphic ratios.
   (b) One possible explanation is that a flow of water moved into the magma or cooling quartz monzonite mass from the surrounding sedimentary rock. This would cause enrichment in ^{18}O. No outward flow of water from the magma took place, and thus no change occurred in the isotopic composition of the sedimentary rocks when they were metamorphosed.
8. Newberry (1982) suggests that the metasomatic solutions were slightly undersaturated with respect to $CaWO_4$ and deposited scheelite where the activity of calcium ion in solution was raised by interaction with calcium-rich rocks. This resulted in continuous dissolution of scheelite close to the pluton (where activity of calcium ion was low) and precipitation at the outward-moving skarn-marble contact (where activity of calcium ion was high).
9. Two components, CaO and SiO_2, are necessary to describe all the phases if we assume a phase of pure CO_2 was present when the minerals formed. Only one of these two components is an independent variable. Thus, using $f = c + 2 - p$, and assuming two degrees of freedom (P and T), $2 = 1 + 2 - p$, and $p = 1$. Therefore only one phase is present in each layer.
10. **(a)** The curves would move to the left (plot at lower temperatures). Hyndman (1985, 529) points out that, if the water pressure is half the load pressure, the temperature of a reaction boundary is approximately that at half the load or total pressure.
    (b) Pelitic rocks

11. **(a)** The assemblages of "facies" A, B, and C belong to the greenschist facies. Those of D and E belong to the amphibolite facies.

(b) Tables 9-8 and 9-10 indicate that Goldschmidt's rocks (represented in Figure 3-11) belong to the pyroxene-hornfels facies.

(c) Hornblende-hornfels facies

12. **(a)** Not enough CaO in the rocks to allow hornblende to form

(b) Dolomite, calcite, and quartz. The antigorite represents alteration of forsterite after its formation during metamorphism.

13. Because biotite and other minerals containing potassium cannot be plotted in an exact manner on the diagram. The potassium of these minerals is corrected for in plotting the molecular ACF ratios of a rock.

14. **(a)** $a_{SiO_2} = 1$

(b) $a_{H_2O} = \text{constant}$

15. **(a)** (1) shale; (2) siliceous limestone; (3) shaley dolomite; (4) bauxite; (5) mafic igneous rock

(b) Anorthite-diopside-hypersthene-quartz

16. **(a)** Next to the intrusive contact

(b) The slope of the curve during the period of rapid growth is a function of the rate of temperature rise as a result of the intrusion. Curve A represents a nodule closer to the intrusive than the nodule of curve B (therefore it has a thicker rim) and thus temperature rose faster for nodule A. Note that the duration of growth increases with distance from the intrusive contact, with growth of nodule B occurring over a longer period of time compared to nodule A. The time span for the transition from rapid to infinitesimally slow growth is a function of cooling rate. Nodule B cooled more slowly than nodule A. Joesten and Fisher (1988) estimate that all growth of wollastonite rims effectively ceased 1,500 years after emplacement of the intrusive.

17. **(a)** [Bi] for biotite

(b) No. It should be labeled [G] for garnet.

(c) chlorite + andalusite = staurolite + cordierite + H_2O

18. **(a)** Dolomite–calcite–quartz–K-feldspar–tremolite–diopside

(b) Tremolite

19. **(a)** The minerals would coexist between the "diopside in" and "tremolite out" isograds.

(b) anorthite + 2 calcite + quartz $\rightarrow$ grossularite + 2 CO_2

20. The $\delta^{18}O$ values would probably show much more variation in aureole B while little change from original values would be found in aureole A. Any fluid flow that occurred, either toward or away from the pluton at aureole B, would tend to be parallel to the lithologic layering and thus more extensive at aureole B than at aureole A. The Notch Peak aureole in Utah, discussed in this chapter, is an example of the situation described for aureole B. The Birch Creek aureole in California, described by Shieh and Taylor (1969), is an example of an aureole A situation. At Birch Creek, the $\delta^{18}O$ values of the metasediments were little affected by the intrusion.

21. **(a)** The Franciscan sediments did not form from old volcanic and plutonic rocks, but rather from young igneous rocks formed shortly before the beginning of sedimentation. Thus the sediment was not enriched in radiogenic argon. In addition, the length of time between metamorphic periods was not long enough for large amounts of radiogenic argon to form before a succeeding metamorphism.

(b) 1. Whole-rock samples give younger dates than mineral samples. If there was excess argon in one or more minerals, the whole-rock ages would tend to be older than some mineral ages.

2. Whole-rock dates of fossiliferous metasediments are all younger than the age of the fossils. Excess argon would cause the radiometric dates to be greater than the paleontological dates.

22. Zero degrees of freedom; it is an invariant point. The other three points where the reaction curves cross the Al_2SiO_5 lines are also invariant points.

Author Index

Subject Index

SI Base Units

Quantity	Name	Symbol
length	meter	m
mass	kilogram	kg
time	second	s
electric current	ampere	A
thermodynamic temperature	kelvin	K
amount of substance	mole	mol

SI Derived Units

Quantity	Name	Symbol	Equivalent
force	newton	N	$kg\text{-}m/s^2$
pressure, stress	pascal	Pa	N/m^2
work, energy, heat	joule	J	$kg\text{-}m^2/s^2$
electric charge	coulomb	C	A-s
electric potential, emf	volt	V	J/C
conductance	siemens	S	A/V
Celsius temperature	degree Celsius	°C	K
radioactivity	becquerel	Bq	s^{-1}

Fundamental Constants

Quantity	Symbol	Value	Units
molar gas constant	R	8.31451	$Jmol^{-1}K^{-1}$
Faraday constant	F	9.64853×10^4	$Cmol^{-1}$
Avogadro constant	N_A	6.02213×10^{23}	mol^{-1}
Planck constant	h	6.62607×10^{-34}	Jsec
molar volume (ideal gas at 273.15 K, 101,325 Pa)	V_m	2.24141×10^4	cm^3mol^{-1}

Conversion Factors

1 lb = 453.6 g
1 in = 2.54 cm
1 mi = 1.609 km
1 km = 0.6215 mi
1 Å = 10^{-8} cm
1 atm = 101,325 Pa
1 Pa = 10^{-5} bar
1 cal = 4.184 J
1 U.S. gal = 3.785 liters (L)
1 curie (Ci) = 3.7×10^{10} dis.s^{-1}
1 electron volt (eV) = 1.6021×10^{-19} J